RECENT ADVANCES IN NANOFABRICATION TECHNIQUES AND APPLICATIONS

Edited by **Bo Cui**

INTECHOPEN.COM

Recent Advances in Nanofabrication Techniques and Applications
Edited by Bo Cui

Published by InTech

Janeza Trdine 9, 1000 Rijeka, Croatia

Edition 2014

Publishing Process Manager Alenka Urbancic

Technical Editor Teodora Smiljanic

Cover Designer InTech Design Team

Image Copyright 2011. Used under license from Shutterstock.com

Additional hard copies can be obtained from orders@intechopen.com

Recent Advances in Nanofabrication Techniques and Applications, Edited by Bo Cui
p. cm.
978-953-307-602-7

INTECH

open science | open minds

Contents

Preface

Nanotechnology has experienced a rapid growth in the past decade, mostly because of the rapid advances in nano-fabrication techniques employed to fabricate the nano-devices. Nano-fabrication can be divided into two categories: the first is the so-called "bottom up" approach, in which nano-structures are created either by chemical synthesis of nano-wires, nanotubes and nano-particles, or by self-assembly of nano-objects or thin layers to form quasi-periodic arrays or phase separation patterns. The second category is the "top down" approach, where nano-structures are created or duplicated from a master mask or mold, using nano-lithography, thin film deposition and etching techniques. Both topics are covered, though with a focus on the second category.

This book aims to provide the fundamentals and the recent advances of nano-fabrication techniques and its device applications. The thirty chapters are a result of contributions from roughly 100 researchers worldwide. Most chapters focus on in-depth studies of a particular research field, and are thus targeted for researchers, though some chapters focus on the basics of lithographic techniques accessible for upper year undergraduate students.

The chapters are divided into five parts. Part I covers lithography based on charged beam, namely electron and focused ion beam lithography, which are the most popular nano-lithographies for R&D. Part II focuses on nano-imprint and soft lithography, which duplicate in parallel a pattern on the mold into a resist layer, and thus have high throughput making them suitable for volume production of nano-devices. Part III contains information on several lithographies using light (UV and X-ray) other than conventional optical lithography. Part IV deals with lithographic techniques relevant only to the next generation integrated circuit fabrication, including extreme UV and deep UV lithography with resolution enhancement techniques. All other lithographic techniques, both bottom up and top down, are grouped in Part V.

In conclusion, I would like to express my sincere gratitude to the contributing authors of each chapter, as well as the publishing process manager, Ms Alenka Urbancic, who spent tremendous time in communicating with me and the authors.

Bo Cui
University of Waterloo
Canada

Part 1

Electron and Ion Beam Lithography

1

Electron Beam Lithography for Fine Dot Arrays with Nanometer-Sized Dot and Pitch

Sumio Hosaka
Gunma University
Japan

1. Introduction

Recently, electron beam (EB) lithography has been applied to mask and reticle pattern draw, for fabricating semiconductor devices, and nanometer-sized pattern direct writing for developing of new concept nano-device. Mainly, the developing of practical EB drawing system has been started since 1960s, and fine pattern formation has also been studied together with the system development [1-3]. Regarding EB-drawn pattern size, at first, micron and submicron-sized pattern has been drawn on mask blank and directly on the device [4]. Today, the pattern size miniaturizes to nanometer-size of less than 20 nm in research [5, 6].
Especially, I have focused the EB lithography into the possibility to form fine dot and fine pitched dot 2-dimensional arrays for patterned media and quantum devices. The research has been done by dependences of resist material and thickness on drawing of fine dot arrays with nanometer-sized pitch in EB drawing, theoretically and experimentally. I have used Monte Carlo simulation and a conventional EB drawing system combined with scanning electron microscope (SEM) and EB drawing controller [7].
In this chapter, I describe key factors such as resist type, resist thickness, proximity effect, etc for a formation of nanometer-pitch dot arrays, a limitation of the EB-drawn size theoretically and experimentally, and demonstrate the applications to dry-etching process and nano-imprinting.

2. Monte Carlo simulation of electron scattering in solid for EB lithography [8]

Electron scattering in the resist and substrate is described based on its scattering angle, mean free path, energy loss, etc. Trajectories of incident electrons and energy deposition distributions (EDDs) in the resist are calculated. From the EDD, EB drawn resist dot profiles are estimated. The formation of nanometer sized pattern for electron energy, resist thickness and resist type can be studied. The EDD in 100 nm-thick resist on Si substrate were calculated for small pattern drawing. The calculations show that 4 nm-wide pattern will be formed when resist thickness is less than 30 nm. Furthermore, a negative resist is more suitable than positive resist by the estimation of a shape of the EDD.

2.1 Calculation model for Monte Carlo simulation of electron scattering

For the treatment of electron elastic scattering, the screened Rutherford scattering model [11] is employed as follow,

$$\frac{d\sigma_i}{d\Omega} = \frac{e^4 Z_i(Z_i+1)}{4E^2(1-\cos\theta+2\beta)} \tag{2.1}$$

where the β is the screening parameter which is given by

$$\beta = \frac{1}{4}\left(\frac{1.12\lambda_0 h}{2\pi p}\right), \ \lambda_0 = Z^{1/3} / 0.885a_0 \tag{2.2}$$

where the e is electronic charge, th the θ is scattering angle, the a_0 is Bohr radius, the h is Planck's constant and the p is the electron momentum. Step length is calculated based on the electron mean free path Λ. The Λ is given by

$$\Lambda = \frac{1}{n\sigma} = \frac{A}{N\rho\sigma} \tag{2.3}$$

where the n is volume density of atoms, the σ is total cross section calculated from differential scattering cross section, the N is Avogadro's number, the A is atomic weight and the ρ is mass density. Scattering angle θ and azimuthal angle ϕ can be obtained using the following equations:

$$\theta = \cos^{-1}\left(1-\frac{2\beta R_1}{1-\beta-R_1}\right), \ \phi = 2\pi R_2 \tag{2.4}$$

where the R_1 and R_2 are independent equidistributed random number between 0 and 1. Since the electron suffers scattering along its trajectory, it continuously loses its kinetic energy along its trajectory. In Monte Carlo simulation, incident electron is slowing down following Bethe's formula, which is a good empirical method of calculating this energy loss in electron solid interaction. The Bethe's approximation is given by

$$-\frac{dE}{ds} = \frac{2\pi e^4}{E}\sum_i n_i Z_i \ln\left(\frac{1.166E}{J_i}\right) \tag{2.5}$$

where the n_i is volume density of atoms, the J_i is mean ionization energy of atom i . The terminal energy of the nth scattering is:

$$E_{n+1} = E_n - |dE / ds|_{E_n} \cdot \Lambda_n \tag{2.6}$$

where the E_n is the energy of the (n-1)th scattering, the Λ_n is step length, and the $|dE / ds|_{E_n}$ is the energy loss rate which can be obtained from Eq. (2.5).

The EDD is an important parameter in consideration of EBL. In order to calculate the EDD in resist, we use cylindrical coordination system. We divide the resist layer along Z-axis into several thin sub-layers. The EDD was calculated in a radius-depth coordination system. This means that the resist layer was devided into many small concentric rings. The simulation was excuted to calculate the total energies $E(r,z)$ in every unit ring for EDD function. The ring volume ΔV is given by following equation,

$$\Delta V = (\pi(r+\Delta r)^2 - \pi r^2) \cdot \Delta Z \tag{2.7}$$

where the ΔZ is the thickness of sub-layer and the Δr is increment in radius direction, From the volume, the EDD function is given by following equation,

$$EDD(r,z) = E(r,z) / (\Delta V \cdot N_0) \tag{2.8}$$

where the N_0 is total number of incident electron.

2.2 Simulation results and discussion

2.2.1 Description and electron trajectories

Monte Carlo simulation has been excuted in energetic electrons impinging in thin film of Si covered with resist material. By using uniform random numbers between 0 and 1, the scattering angles θ and ϕ can be calculated by using Eq. (2.4). Using Eqs. (2.5) and (2.6) ,we can calculate the energy loss ΔE due to scattering of the electrons with atoms in the sample along its trajectory. The trajectory of the electron was traced till its energy slowed down to 50 eV. PMMA $(C_5H_8O_2)$ with a compound of carbon(C), hydrogen(H) and oxygen(O) was used as the typical resist. In the simulation, we use random sampling method to determine the scattering center, the step length and use a new coordinate conversion method [12] for calculating the trajectories of electrons. The initial energies of the incident electrons are taken to be 30 keV and 10 keV. The scattering trajectories of electrons with different incident beam energies in the resist material was used with the same as PMMA resist layer on Si target are shown in Figs. 2. 1- 2. 4. In the simulation, the thickness of the resist layer of 100 nm and the number of incident electrons of 500 was used. With incident energy of 30 keV, the penetration depth was about 3.5μm and lateral range was about 1.5 μm in Si (Fig. 2. 1). In the resist layer, the electron scattering was expanded only to about 20 nm in radius direction (Fig. 2. 2). Although using 10 keV incident electrons can diffuse as deep as 0.5 μm into the sample (Fig. 2. 3) but lateral range was about 50 nm in the resist layer which is larger than that of 30 keV (Fig. 2. 4). It indicated that as the energy decreases, the electrons scattering lateral range is expanded in the thin resist layer at the top.

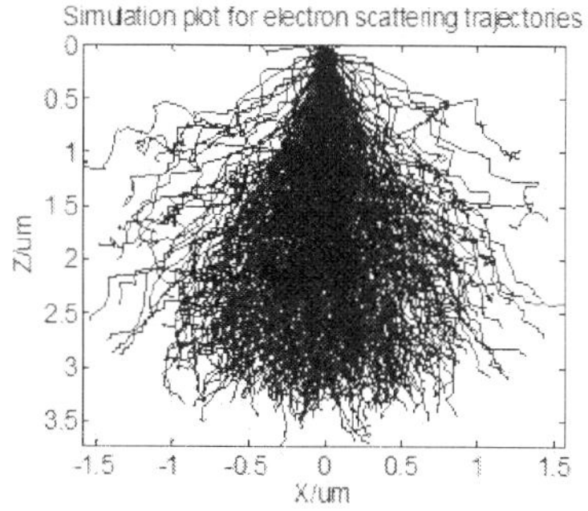

Fig. 2.1. Electron scattering trajectories at incident energy 30 keV

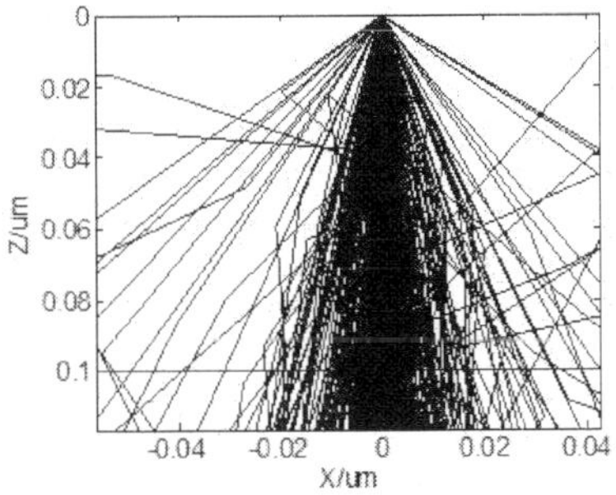

Fig. 2.2. Trajectories in resist (100 nm) at incident energy 30 keV.

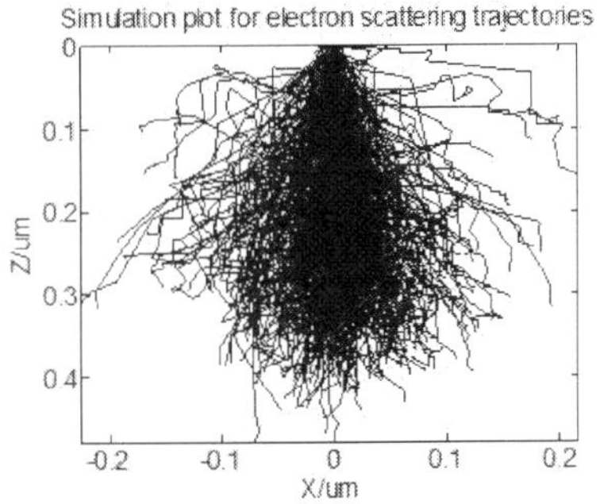

Fig. 2.3. Electron scattering trajectories at incident energy 10 keV.

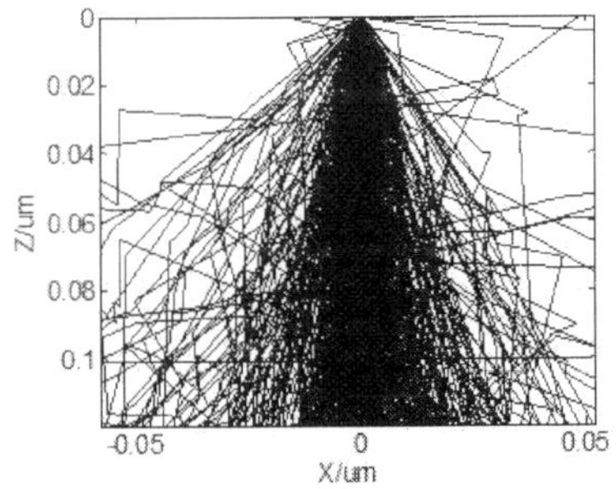

Fig. 2.4. Trajectories in resist (100 nm) at incident energy 10 keV.

2.2.2 Energy deposition distribution (EDD)

The energy deposition density at various positions in depths and radius in the thin resist was calculated. The thickness of the resist was 100 nm, and the incident energy was 30 keV. The ΔZ and Δr were 2nm, and the number of electrons was 30000. Fig. 2. 5 shows the EDD in the resist layer of various depths 10 nm, 50 nm and 100 nm. It can be clearly seen that the shallower the depth from the surface of the resist, the narrower and the shaper the EDD. Fig. 2.6 shows the relationship between resist depth and standard deviation σ of the EDD assuming that the EDD is approximated by Gauss distribution.

$$EDD(r) = \frac{1}{\sqrt{2\pi}} \cdot \frac{1}{\sigma} \exp\left(-\frac{(r - r_0)^2}{2\sigma^2} \right) \tag{2.9}$$

It indicates that small pattern could be produced by using thin resist. It can effectively reduce proximity effects and thus greatly improve resolution.

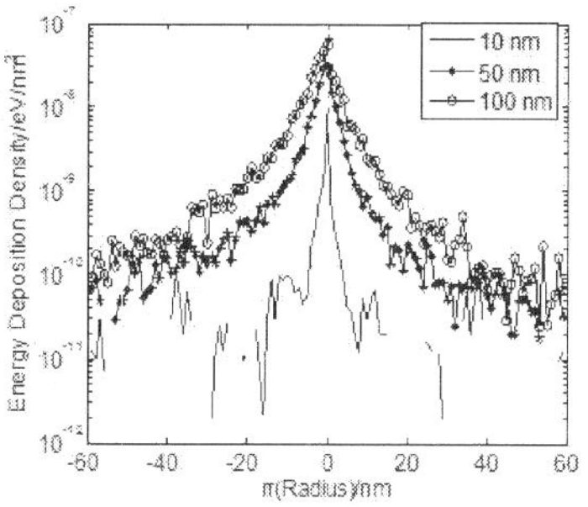

Fig. 2.5. Energy deposition distribution of different depth of resist.

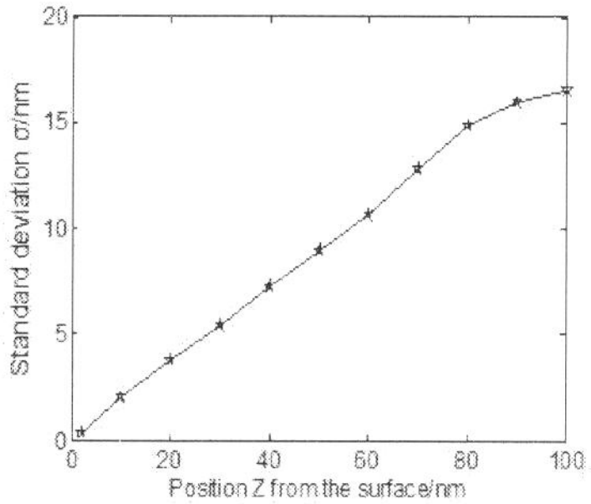

Fig. 2.6. The relationship between resist depth and width of dot.

2.2.3 Consideration for resist development based on EDD

Resist development is assumed as the resist molecule are solved and linked over the critical energy density so that they are soluble and insoluble in positive and negative resists, respectively. Figs. 2. 7(a)-(d) show the area over the critical energy density of 28.125 keV/cm^3 - 0.5 keV/cm^3. It is clear that small pattern formation is possible by selecting large critical energy density, which corresponds to small exposure dosage in experiment. In the positive resist, however, it is very important to solve the top layer at first. The energy densities of the top layer do not reach to the critical energy density in Figs. 2. 7(a)-(c). As the result, no patterning occurs in the energy region as shown in Figs. 2. 7(e)-(g). When the critical energy density is less than $0.5\,keV/cm^3$, the hole pattern appears as Fig. 2. 7(h). The hole diameter increases with the depth in the resist layer. In our experiment, however, the small diameter of about 4 nm disappears. This may be caused by capillary force. The minimum diameter of about 7 nm was obtained in previous experiment using ZEP520 positive resist [7]. On the other hand, Figs. 2. 7(i)-(l) show the developed resist profiles at various critical energy densities. As the linked molecule is remained on the substrate based on negative resist development mechanism, nanometer-sized patterns are formed as shown in Fig. 2.7(i), although the height of the resist pattern is not complete and short. It is clear that the smaller pattern size is obtained by selecting the heigher critical energy density,but the hight of the resist pattern decreases as the critical energy density increases. Therefore, negative resist is very suitable to form nanometer-sized pattern.

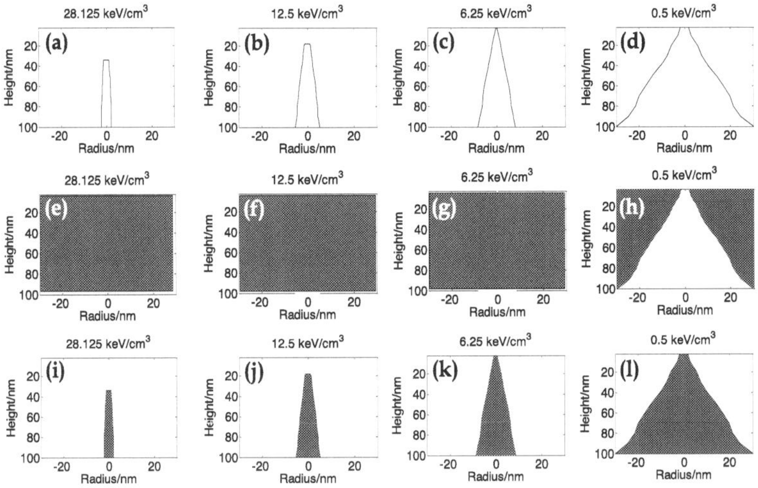

Fig. 2.7. Simulated resist profiles at various critical energies; (a) resist region over 28.125 keV/cm2, (b) 12.5 keV/cm2, (c) 6.25 keV/cm2, (d) 0.5 keV/cm2, (e)-(h) positive resist region remained by development, (i)-(l) remained negative resist regions.

3. Formation of highly packed fine pit and dot arrays using EB drawing with positive and negative resists [7, 13]

Fine bit arrays formation for an ultrahigh density optical and magnetic recordings has experimentally been studied using EB drawing with a high resolution scanning electron microscope (HR-SEM), a drawing controller [7] and positive and negative EB resists. As experimental results, calixarene negative EB resist is very suitable to form an ultrahigh packed resist dot arrays pattern, comparing with ZEP520 positive resist as same as described in previous section. We obtained very fine dot arrays with a diameter of <15 nm, and 2-dimensional pitch of <30 nm.

3.1 EB-drawing system [7]

I have already reported key technologies which enable ultrahigh packing over 1 Tb/in^2: (1) the use of an EB drawing system with a fine probe and large probe current, (2) the thinning of the resist to prevent the spread of incident electron scattering in the resist, and (3) the design of a highly packed pattern with a hexagonal or centered rectangular lattice structure to avoid the proximity effect (Fig. 3. 1). The EB drawing system consists of a HR-SEM (JSM6500F, JEOL, Ltd.) and a writing controller (Tokyo Technology Co., Ltd.) (Fig. 3. 2). The drawing was done on a resist coating on a piece of Si placed on the XY table, which was not moved during the EB writing. This type system is very suitable for checking the limit of the drawing pattern size, because the stage position error can be neglected. The system provides a high probe current of 2 nA at a resolution of 2 nm. We used the system under a probe current of 100 pA and an acceleration voltage of 30 kV. In the drawing, the address resolutions were 10 nm and 2.5 nm for ZEP520 and calixarene [14, 15] resists, respectively. Development was carried out using the commercial developers ZED-N50 (MIBK+IPA) and ZEP-RD (xylene) for 210 sec and 180 sec for ZEP520 and calixarene, respectively. In particular, we adjusted the focus to get the fine probe on the sample surface at a magnification of 250k-200k.

In order to achieve 1-Tb/in^2 storage patterned media, we carried out experiments to confirm whether EB writing can form very fine pit or dot arrays with a pitch of <60 nm or not. At first, we coated the resist on a piece of Si substrate with a thickness of 70 nm and 15 nm in ZEP520 and calixarene, respectively. After pre-baking, we drew ultrahigh-packed pit and dot patterns with a pitch of 20-60 nm. After developing and rinsing the resists, we checked the drawn patterns using the same HR-SEM. Thinning allows not only to prevent electron scattering extension in the resist, but also to avoid electron charging. We determined the minimum thickness at which there is sufficient contrast and no damage to the resist during SEM observation. The thickness of 70 nm in ZEP520 resist was decided by electron damage and contrast signal in SEM. I could not obtain the high magnitude SEM image in a thickness range of less than 70 nm.

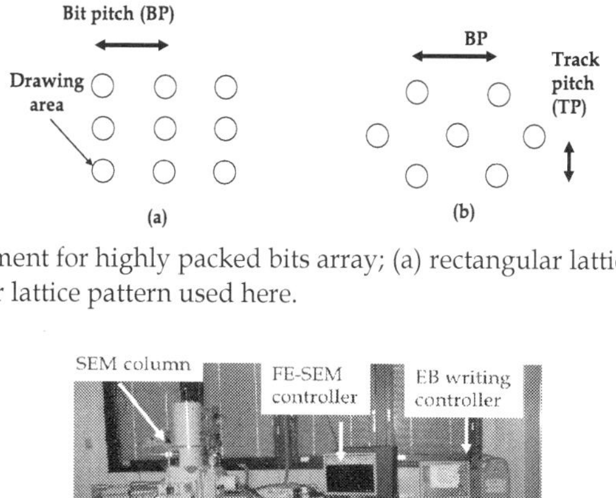

Fig. 3.1. Bit arrangement for highly packed bits array; (a) rectangular lattice pattern and (b) centered rectangular lattice pattern used here.

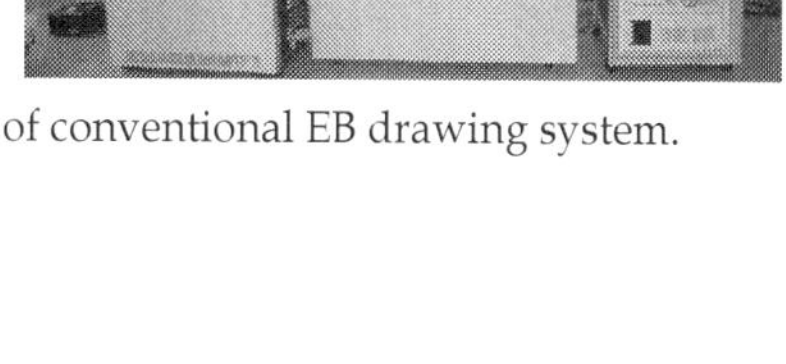

Fig. 3.2. Overview image of conventional EB drawing system.

3.2 EB drawing [7, 13]

3.2.1 Using ZEP520 positive EB resist

Figure 3.3 shows SEM images of ZEP520 pit patterns drawn at an exposure dosage of around 180 μC/cm². The figure shows that the minimum pit arrays were drawn with a minimum pit diameter of <20 nm at a pitch of 40 nm x 60 nm [13]. We could not form higherpacked pit patterns than this. Furthermore, the pit size drastically changed, with a fluctuation of about 18 nm in this pitch arrays pattern and about 11 nm in the case of a pitch of 100 nm x 60 nm (Fig. 3. 4). This indicates that the pit array pattern with a pitch of 40 nm x 60 nm is the limit for ZEP520. This pattern corresponds to about 270 Gb/in². But it is not practically usable because the fluctuations are too large. In addition, the minimum pit diameter is about 7 nm in Fig. 3. 3. Comparing with Monte Carlo Simulation result, the pit diameter is larger than simulated size of 4 nm. This is caused by capillary force between pit and developer.

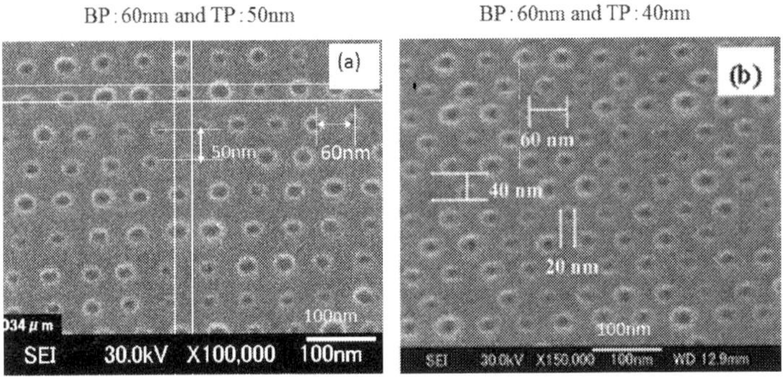

Fig. 3.3. SEM images of ultrahigh packed pit resist pattern using ZEP520 (180 μC/cm², 30 kV), (a) pitch of 60 nm x of 50 nm, (b) 60 nm x 40 nm.

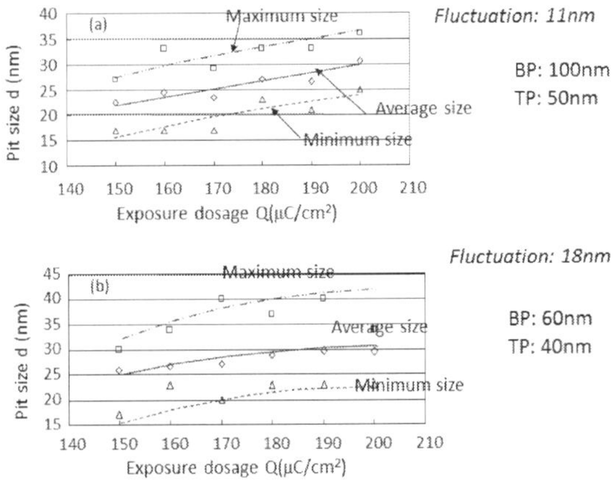

Fig. 3.4. Variations of ZEP520 resist pit width in ultrahigh packed pit arrays (a) with a pitch of 100 nm x 50 nm and (b) 60 nm x 40 nm for exposure dosage at 30 kV.

3.2.2 Using calixarene negative EB resist

Figure 3.5 shows SEM images of ultrahigh-packed dot arrays resist patterns (a) at a pitch of 30 nm x 30 nm, and (b) 30 nm x 25 nm. The exposure dosage was 40 mC/in^2. In these experiments, we succeeded in obtaining high packed dot arrays pattern (Fig. 3. 5(b)). The dot size was 11 to 14 nm in diameter. The size fluctuation was about 2 or 3 nm and almost constant in the range of 30 to 45 mC/cm^2. This resist is very suitable to nano-fabrication (Fig. 3. 6).

Calixarene resist, however, has the drawback that its sensitivity is too low for mass production purposes. It takes much time to draw the dot arrays pattern over large sample. We have to develop a new resist with sensitivity higher by 2 orders of magnitude. The reason for its low sensitivity is in related with its molecule size. A lot of electrons are required to change the calixarene molecule to large molecular (weight: >several 10000s) for insolubility by linking many calixarene molecular (weight: about 600).

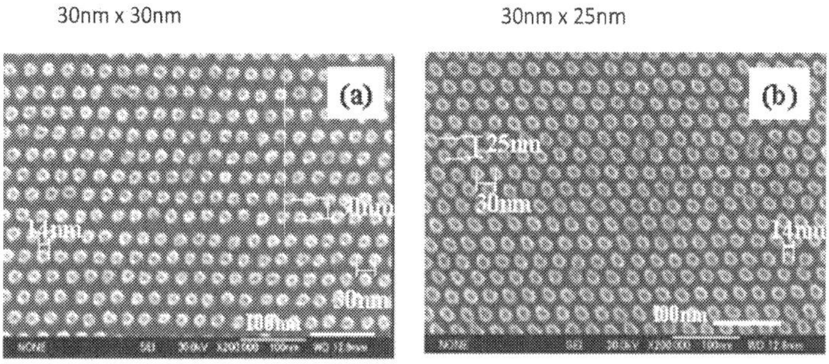

Fig. 3.5. SEM images of ultrahigh packed dot resist pattern using calixarene (14mC/cm^2, 30kV), (a) pitch of 30 nm x 30 nm, (b) 30 nm x 25 nm.

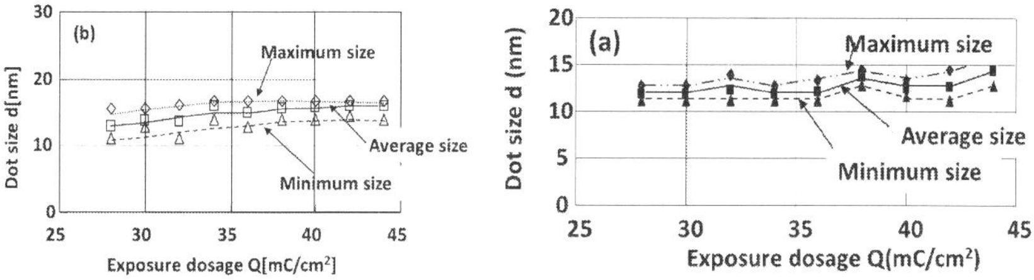

Fig. 3.6. Variations of the calixarene dot size in ultra-high-packed dot arrays; (a) pitch of 30 nm x 30 nm and (b) 25 nm x 25 nm.

3.3 Consideration of the different limitations in ZEP520 and calixarene

The difference between the limitations has been investigated using the exposure intensity distribution (EID) and EDD in Monte Carlo simulation. The EID functions were determined by measuring the widths of one-line patterns drawn under various exposure dosages. We obtained the change of the required exposure dosage for the line-width, and then the EID from the change.

The EID is defined by Eq. (3.1), assuming that the distribution is Gaussian. The 1st and 2nd terms in Eq. (3.1) represent the energy depositions due to electron forward scattering (FS) and backward scattering (BS), respectively.

$$E(r) = C_1 \exp(-\frac{r^2}{\sigma_1^2}) + C_2 \exp(-\frac{r^2}{\sigma_2^2}) \quad (3.1)$$

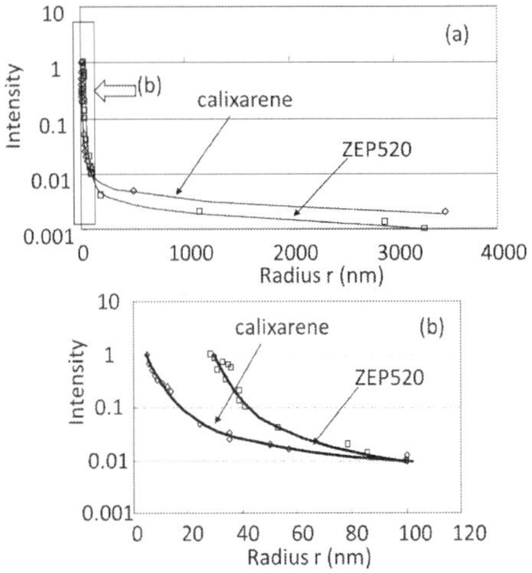

Fig. 3.7. EID functions measured by using ZEP520 and calixarene EB-drawing at 30 keV.

For the miniaturization of the bit size, the σ_1 value in the 1st term is crucial. In the experiments, the exposure dosages were changed from 10 μC/cm² to 5 C/cm². Then, we obtained the EIDs using ZEP520 and calixarene, as shown in Fig. 3. 7. Figure 3. 7 shows the 1st term in detail. When we roughly fit the EID function to Eq. (3.1), the σ_1 values were about 5 nm and 20 nm in calixarene and ZEP520, respectively. In the Monte Carlo simulation, the σ_1 values were about 2 nm and 3 nm at a resist thickness of 15 nm and 70 nm, respectively (Fig. 4. 1). Although the values were calculated under the condition of 30-keV electron incidence on PMMA resist on a Si substrate and are slightly smaller than those using calixarene and ZEP520, we consider that they are roughly the same as those using PMMA because the resist thicknesses are very thin. When the electron probe size of 2 nm and the molecular size of 1 and 3 nm in calixarene and ZEP520, respectively, are considered, the experimental value becomes 13 nm larger than the estimated value in ZEP520, while the values are almost same in calixarene. This may be due to the molecular size, structure of the ZEP520 resist and capillary force in the development. The size of ZEP520 is a few nm assuming to be spherical. Sometimes, ZEP520 may be in a chain structure when the molecule is not solved after EB exposure. This comparison indicates that the smallest pattern in EB writing may be determined by the resist's molecular size, structure and resist type.

4. Challenge of formation of less than 20 nm x 20 nm very fine pitch dot arrays using 30 keV EB drawing with thin calixarene resists [5, 16, 17]

The possibility to achieve very fine dot arrays with a pitch of less than 20 nm x 20 nm using 30 keV EB drawing with calixarene resist has been investigated. In order to form such a pattern, the resist thickness dependence on dot size and packing has been studied. In this section, EB drawing with extremely thin film for very highly packed dot arrays formation is described. As the experimental results, it is demonstrated to form higher packed dot arrays pattern with a pitch of 20 nm x 20 nm and 18 nm x 18 nm in 13-nm-thick resist using 30 keV electron beam.

4.1 Electron scattering in thin resist film for fine dot arrays [16]

In order to check the resist thickness dependence on formation of very fine pitch dot arrays using 30 keV electrons, the EDD was calculated using Monte Carlo simulation. From the simulation, the possibility to achieve a very fine pitch dot arrays is studied. It is very important to study the EDD because the resist dot formation occurs by linking the molecula enhanced by the energy in a case of using negative resist. The EDD function is represented as Eq. (3. 1), assuming that the distribution consists of 2 Gaussian distributions as described in last section. At least, the miniaturization of the dot size, the σ_1 value in the first term is very crucial.
Figure 4. 1 shows the EDDs with PMMA resist thickness of 15 nm, 70 nm and 200 nm on Si substrate with 30 keV electrons using Monte Carlo simulation. From the result, the σ_1 values are about 2 nm, 3 nm and 8 nm at a resist thickness of 15 nm, 70 nm and 200 nm, respectively. Figure 4. 2 shows a variation of the σ_1 value for the resist thicknesses. As the resist thickness decreases, the σ_1 value becomes small. We can estimate the σ_1 value of about 1.5 nm at a resist thickness of 10 nm. This means that it is possible to form finer pitch dot arrays than that in previous section using such a thin film. Therefore, it is very important to use resist film as thin as possible for formation of very fine pitch dot arrays.

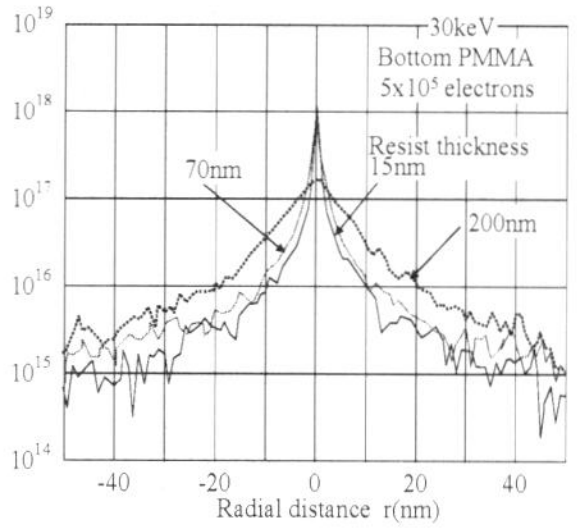

Fig. 4.1. Monte Carlo calculated result of energy deposition distributions (EDDs) for various resist thickness of PMMA on Si substrate.

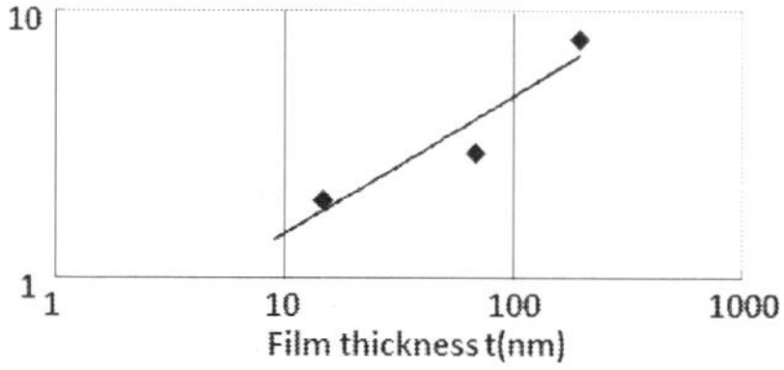

Fig. 4.2. A variation of the σ_1 value in EDD for the resist thicknesses.

4.2 EB drawing and sample preparation

The conventional EB drawing system was used as described in section 3. 1. The drawing was done on a resist layer coating on a piece of Si after we set the sample on the XY table in the system. The EB drawing was done under a probe current of 100 pA at an acceleration voltage of 30 kV. In the other system parameters, the address resolution was 2.5 nm in a drawing field of 25 µm x 25 µm.

The calixarene resist films on Si substrate with a thickness of 11.8 nm to 16.9 nm, which were controlled by the spin coating at a speed of 3000 rpm to 8000 rpm for 190 s was prepared as shown in Fig. 4. 3. The thicknesses were measured by contact mode atomic force microscope (AFM).

The resist process is as follows. After coating the resist on the Si substrate, pre-baking was done at the 110 °C for 3 min in air. Then, the EB drawing was done by raster scanning with the CAD data. After the developing and rinsing with a developer of ZEP-RD and iso-propanol, respectively, The pattern quality was checked whether complete formation of the drawn dot arrays was done or not.

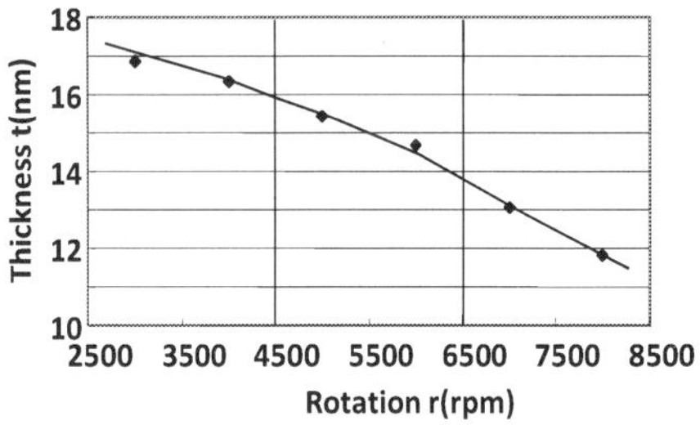

Fig. 4.3. Thicknesses of 11.8 nm to 16.9 nm controlled by the spin coating at a speed of 3000 rpm to 8000 rpm

4.3 Experimental results

4.3.1 Formation of 20 nm x 20 nm pitched dot arrays [16]

Figure 4. 4 shows the SEM images of 20 nm x 20 nm pitch dot arrays patterns drawn by 30 keV electron beam at a dosage of 16 mC/cm^2 using various resist thicknesses of 11.8 to 16.3 nm. In Figs. 4. 4(a) and (b), there are many defects such as the dots combined with neighbor dots. On the other hand, the number of defect decreases with the thickness. At the resist thickness of 13.1 nm, the 20 nm x 20 nm pitch dot array have completely been drawn as shown in Fig. 4. 4(c). Furthermore, using further thin resist film, the dot arrays appear unclearly (Fig. 4. 4(d)) because the SEM contrast becomes poor due to thin thickness. Therefore, the resist thickness of about 13 nm is very suitable for formation of 20 nm x 20 nm very fine pitch dot arrays.

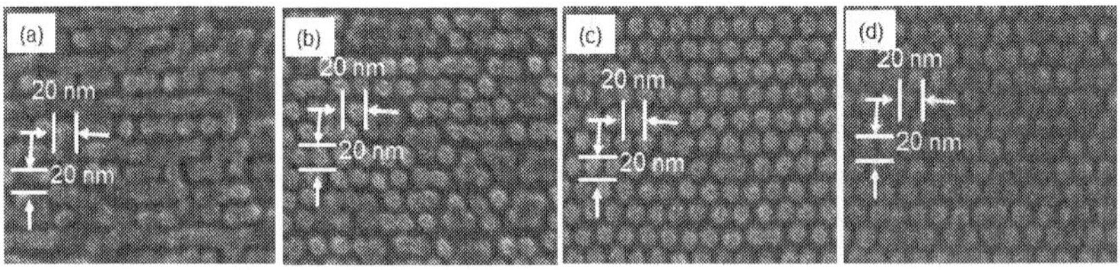

Fig. 4.4. SEM images of 20nm x 20nm fine pitch dot arrays formed by 30keV EB lithography at a dosage of 16 mC/cm^2, (a) with a resist thickness of 16.1 nm, (b) 14.7, (c) 13.1 nm and (d) 11.8 nm.

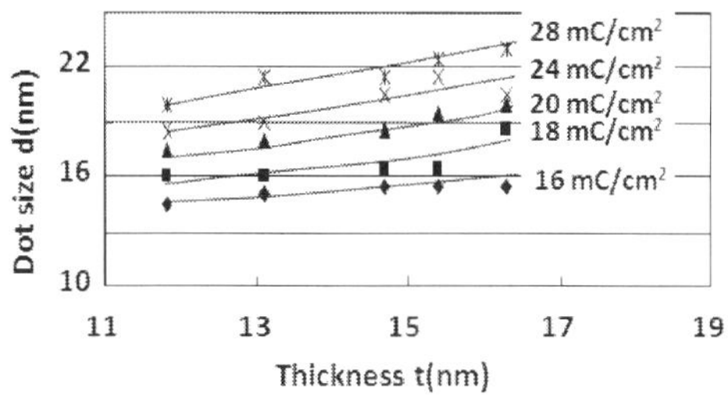

Fig. 4.5. Variations of average calixarene dot size with thicknesses for various exposure dosages.

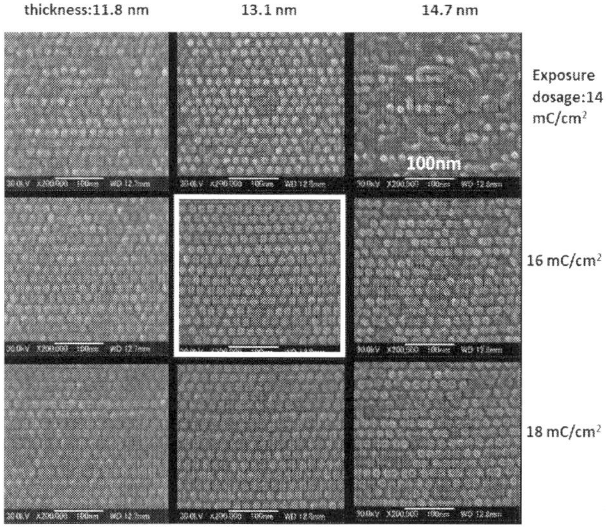

Fig. 4.6. SEM images of 20nm x 20nm pitched calixarene resist dot arrays formed by 30keV EB drawing on Si substrate (1.6 Tb/in^2) in 9 shots/dot drawing.

From Fig. 4. 4, the variations of average calixarene dot size for various thicknesses and exposure dosages were obtained as shown in Fig. 4. 5. The figure shows that the dot diameter decreases with not only thickness but also exposure dosage. Although the diameter variation with exposure dosage means that proximity effect occurs in the EB writing, the thin resist layer contributes to get small drawing probe. In order to draw the 20 nm x 20 nm pitch dot arrays pattern, we need a dot size of 15 nm at least. From Fig. 4. 5, it is neccessary to choose an exposure dosage of < 16 mC/cm^2 and a resist thickness of < 13 nm for the fine pitch arrays formation.

Figure 4. 6 shows the result of the 20 nm x 20 nm pitch dot arrays patterns drawing with a thickness of 11.8-14.7 nm at some exposure dosages. The exposure dosages were 14 mC/cm^2, 16 mC/cm^2 and 18 mC/cm^2. At a dosage of 14 mC/cm^2, there are some vacancies as defects. It may be caused by that the dosage is not enough to make the resist molecular link. When using commercial developer, the insufficient exposed resist part was solved so that the completed dots could not be formed. At a dosage of 16 mC/cm², it is enough to makes complete dots. In a case of a dosage of 18 mC/cm², the dot was combining together with neighbor dots because the dot size becomes larger than that at 16 mC/cm². Figure 4. 7 shows the variations of the dot size with exposure dosage of the 20 nm x 20 nm dot arrays pattern at a resist thickness of 13.1 nm. The fluctuation of the dot size between minimum and maximum dot sizes are about 3 nm. This is as small as that in previous section. The experimental results demonstrated that 30 keV EB drawing can form 20 nm x 20 nm very fine pitch dot arrays pattern using optimal resist thickness and exposure dosage.

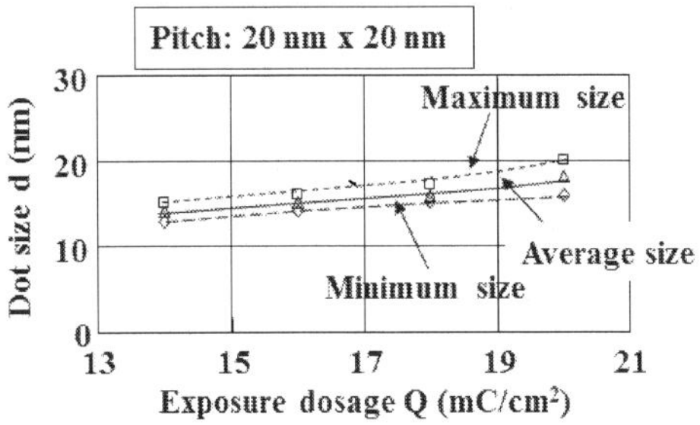

Fig. 4.7. Variation of the resist dot size measured from the 20 nm x 20 nm pitch resist dot arrays formed by EB-drawing with 30 keV incident electron.

4.3.2 Challenge to form 18 nm x 18 nm pitched dot arrays [5, 16]

(1) Dependences of resist thickness and exposure dosage on 20-nm-pitch EB drawing

Initially, we studied the effects of resist thickness and exposure dosage on 30-keV EB drawing of the dot arrays with a pitch of 20 nm x 20 nm. Figure 4. 6 indicates that the optimal thickness and exposure dosage were about 13.1 nm and 16 mC/cm², respectively. On the other hand, we did not obtain good results using a thickness of 11.8 nm, although the thickness was as thin as possible. This may be caused by insufficient contrast of SEM images due to an excessively thin resist and poor focus adjustment. Thus, it is necessary to use a thin resist film to suppress the scattering of the primary electrons in the resist. The tendency that finer dot arrays patterns can be drawn with thinner resists are demonstrated, and agrees with the simulation result. Nevertheless, there exists a critical thickness for fine evaluation of the drawn pattern as described above.

(2) Possibility of forming 18 nm pitch dot arrays pattern

To anticipate the ultimate pitch of the dot arrays using 30-keV EB drawing, we studied the EB drawing margin by EB drawing of highly packed dot arrays with pitches of 20 x 20, 25 x 25, and 30 x 30 nm² by ranging the exposure dosages between 14 and 40 mC/cm². Figure 4. 8 shows the SEM images of the 20 x 20, 25 x 25, and30 x 30 nm² pitch dot arrays patterns formed using various dosages at a constant resist thickness of 13.1 nm. The experimental results indicate that the optimal condition for drawing becomes narrow as dot packing increases. We successfully formed 20 x 20 nm² pitch dot arrays at an exposure dosage of 16 mC/cm². In contrast, the available proper exposure dosages increased to 14-22 and 14-30 mC/cm² for the pitchs of 25 x 25 and 30 x 30 nm², respectively. The results show that it is very difficult to form 20 x 20 nm² pitch dot arrays in the resist film on Si substrates even when a thinner resist film on Si substrate was used in the 30-keV-EB drawing. Nevertheless, a complete 20 nm x 20 nm pitch dot array was successfully achieved using the unique EB drawing condition for EB drawing with a resist thickness of 13.1 nm and an exposure dosage of 16 mC/cm^2 as described above. The exposure dosage is dependent on dot pattern data. In the experiment, we used the dot pattern data of 3 x 3 shots/dot. We have also indicated the exposure dosage with a unit of C/dot, as shown in Fig. 4. 9.

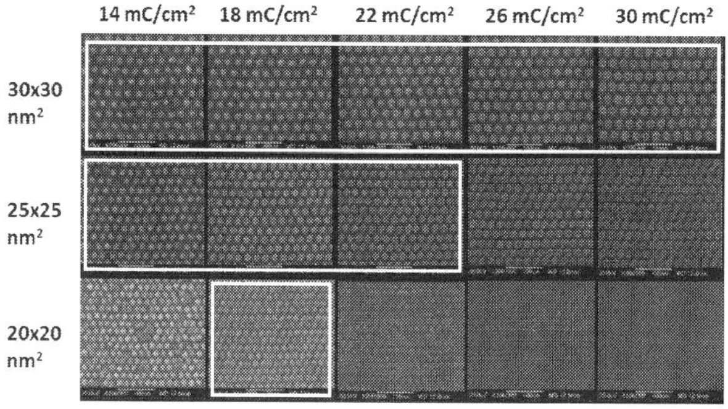

Fig. 4.8. SEM images of 20x20, 25x25, 30x30 nm^2 pitched calixarene resist dot arrays formed by 30keV EB drawing on 13.1-nm-thick resist film on Si substrate.

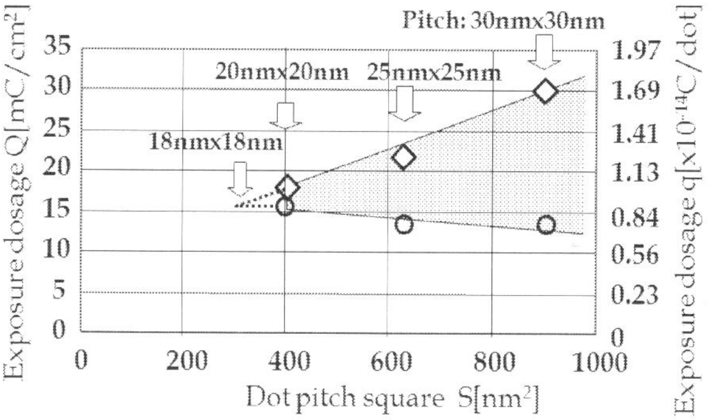

Fig. 4.9. A proper region of 30-keV-EB drawing available for very fine pitch dot arrays in a relationship between exposure dosage and unit cell area (square pitch) in 9 shots/dot drawing.

From the above results, we can obtain an available drawing margin with regard to dot packing and exposure dosage for 30-keV EB drawing. The margin indicates that a minimum pitch exists for 30-keV EB drawing of the dot array pattern. In the 30-keV EB drawing and calixarene resist system, the smallest pitch is about 18 x 18 nm^2. The dosage margin of about 6% is estimated at the pitch. This means that we need a stable EB-drawing system including the EB source, resist material and its thickness. Figure 4. 10 shows that there is the possibility of forming 18 x 18 nm^2 pitch dot arrays using EB drawing. Our results demonstrate that 18 x 18 nm^2 pitch dot arrays can be partially drawn. It is clarified that the

30-keV EB drawing has the potential to form 18 x 18 nm^2 pitch dot array patterns in thinner resist films on Si substrates by accurately selecting the optimum EB drawing conditions. For the formation of complete dot arrays pattern with 18 x 18 nm^2 pitch, stability in the EB system is also required. At least, we have to use stable electron probe current with a deviation within 6% for 2 Tbit/in^2 magnetic storage using 30-keV EB drawing.
The exposure dosages were 10-11 mC/cm^2, which are smaller than the dosage estimated in 18 x 18 nm^2 pitch drawing (Fig. 4.9). The small exposure dosages are caused by using different exposure shot numbers per square pattern as a dot. In Fig. 4. 10, we increased the number of shots/dot from 9 to 16 (4x4) for fast EB drawing. The dosage of 10 mC/cm^2 in 16 shots/dot drawing corresponds to about 18 mC/cm^2 in 9 shots/dot drawing. The 18 mC/cm^2 almost agrees with the estimated dosage in Fig. 4. 9.

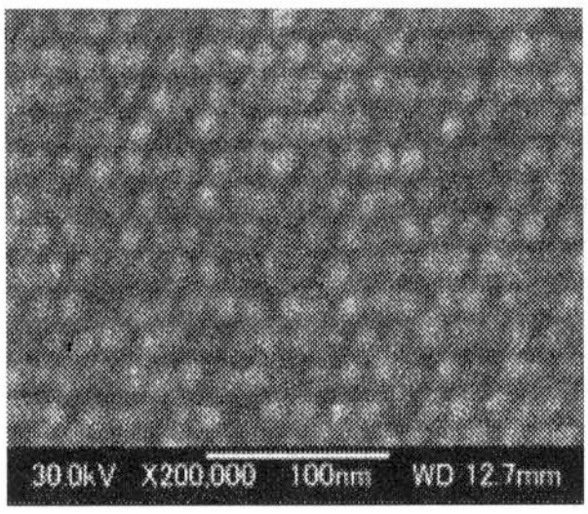

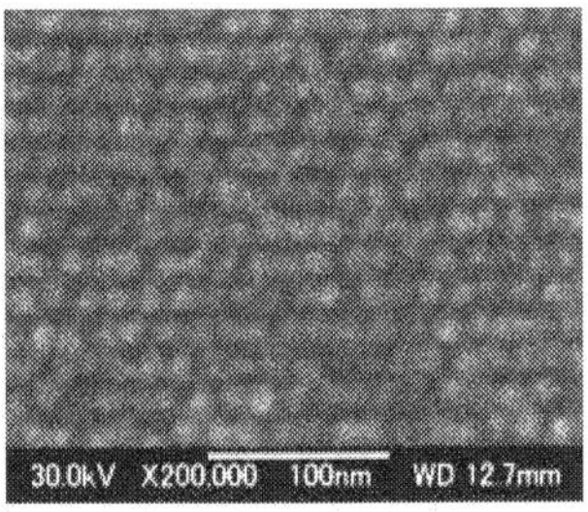

(a) 10 mC/cm^2 (b) 11 mC/cm^2

Fig. 4.10. SEM images of 18 x 18 nm^2 pitched resist dot arrays formed by 30 keV EB drawing on Si substrate (about 2.0 Tb/in^2).

4.3.3 Small proximity effect using calixarene resist [17]

The proximity effect of the EB drawing and thinner calixarene resist system is considered using the EID in the experiment and the EDD in Monte Carlo simulation.
Assuming that the EID function is defined with 2 Gaussian distributions as described in previous section, the 1st and 2nd terms in Eq. (3. 1) represent the energy depositions due to electron forward scattering (FS) and backward scattering (BS), respectively. For the miniaturization of the bit size, very small proximity effect and small σ_1 and σ_2 values are crucial. The ratio η of the total energies due to FS and BS is very important.

$$\eta = \int C_1 \exp(-\frac{r^2}{\sigma_1^2})dr \bigg/ \int C_2 \exp(-\frac{r^2}{\sigma_2^2})dr = \left(\frac{C_2}{C_1}\right)\left(\frac{\sigma_2}{\sigma_1}\right)^2 \tag{4.1}$$

The η value has to be less than 1 because the drawing energy due to FS becomes dominant for very fine dot arrays formation. On the other hand, considering EB drwing of square, the drawing energy due to BS at side and corner of the square becomes a half and a quarter of that at center, respectively, based on reciprocity principle [9]. The variation of BS drawing energy on exposure dosage at everywhere of the drawing area is suppressed to be negligible small if the η value is less than 1.
In the experiments, the exposure dosages were changed from 10 μC/cm^2 to 5 C/cm^2. We obtained the EID using calixarene as shown in Fig. 4. 11(a) [23]. When we roughly fit the EID

function to Eq. (3.1), the σ_1, σ_2, C_1/C_2, and η values are about 5 nm, 15 nm, 0.03 and 0.27, respectively. In the Monte Carlo simulation, we obtained EDD in 15-nm-thick resist on Si substrate (Fig. 4. 11(b)). When roughly fitting it to the EID function, they are about 2 nm, 10 nm, 0.02 and 0.5 nm, respectively. Although the resist materials are difficult in the experiments and simulations, the values of σ_1, σ_2, C_1/C_2 in the experiment are almost same as those in the simulation. Comparing between experimental and simulated results, the values of σ_1 in both cases agree well because the electron probe size of 2 nm and the resist molecular size of 1 nm can be considered in the EB drawing. According small σ_1 and η values, the system is very suitable for very fine dot and very fine pitch dot arrays drawing. Figures 4.12(a)-(d) show SEM images of very fine pitch resist dot arrays on Si substrate with various pitches of 20 x 20 nm^2 to 40 x 40 nm^2. The exposure dosage was 16 mC/cm^2. In these experiments, we succeeded in obtaining the highest-packed dot array pattern with a pitch of 20 x 20 nm^2 (Fig. 4. 12(a)), which corresponds to the ultrahigh recording density of about 1.61 Tb/in^2 in patterned media. The dot sizes of about 12.5 to 18 nm are changed as shown in Fig. 4. 12(e). The size fluctuation is about 2 nm and almost constant in a range of 0.4 to 1.6 Tb/in^2. This shows that the drawings are carried out in no relation with the dot pitch. Figures 4. 13(a) and (b) show SEM images of 25 x 25 nm^2 pitched resist dot arrays at side and corner of the drawing area Fig. 4.13 SEM images (a), (b) and histograms (c)-(e) of 25 nm x 25 nm pitch resist dot arrays on Si substrate at a dosage of 16 mC/cm^2, (a) at side of the written pattern area, (b) at corner, (c) at center, (d) at side and (e) at corner.at an exposure dosage of 16 mC/cm^2.
at an exposure dosage of 16 mC/cm^2. The dot sizes histograms are shown in Figs. 4. 13(c)-(e). Comparing the distributions at center, side and corner of the drawing area, the histograms and the mean values are almost same. These experimental results demonstrate the proximity effect is extremely small in this system. The system is suitable for formation of very fine pitch dot arrays with very fine dot. From the results, a curvature of corner in the drawing pattern could be sharp shape with a radius of 7-8 nm.

In addition, using 50 keV EB lithography, we can consider that the σ_1 and η values will be improved small to a twice of those in 30 keV EB lithography. This is because the forward scattering area in 50 keV EB lithography is suppressed to a twice of that in 30 keV EB lithography. Therefore, 50 keV lithography is more suitable for fine dots patterning than 30 keV EB lithography.

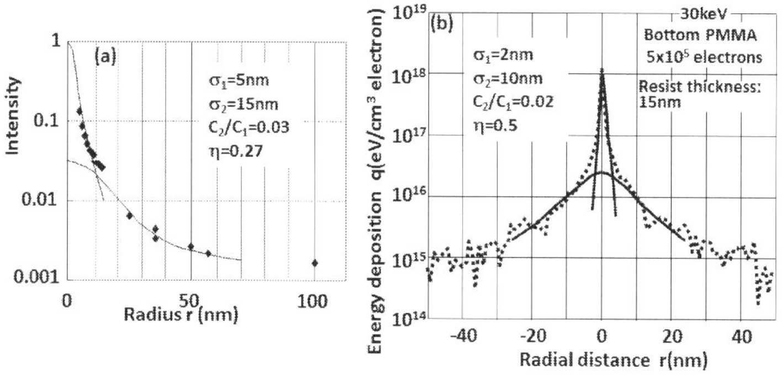

Fig. 4.11. Experimental result (a) and Monte Carlo calculation (b) of EID function and roughly fitting to EID function in 30 keV EB drawing and 15 nm thick calixarene resist on Si substrate.

The EB writing with thin calixarene resist promises to open the way toward ultrahigh-density recording at >1 Tb/in^2 and quantum devices.

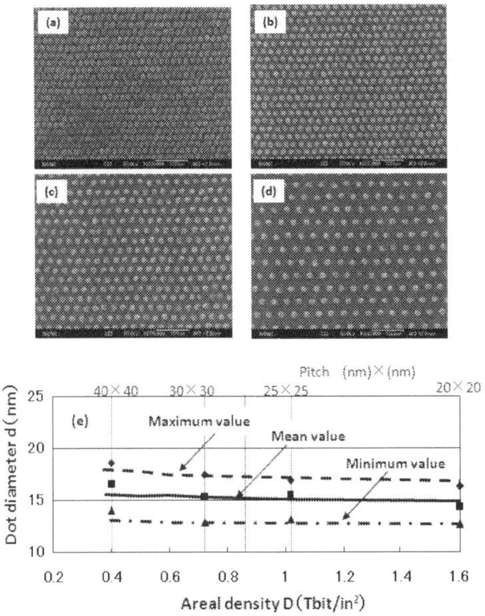

Fig. 4.12. SEM images (a)-(d) and the dot size variation (e) of very fine pitch resist dot arrays at center of the drawing area on Si substrate at a dosage of 16 mC/cm^2 with a pitch of (a) 20 nm x 20nm, (b) 25 nm x 25 nm, (c) 30 nm x 30 nm and (d) 40 nm x 40 nm.

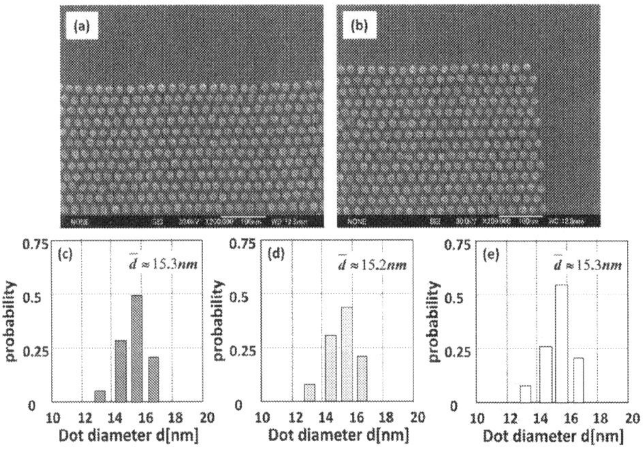

Fig. 4.13. SEM images (a), (b) and histograms (c)-(e) of 25 nm x 25 nm pitch resist dot arrays on Si substrate at a dosage of 16 mC/cm^2, (a) at side of the written pattern area, (b) at corner, (c) at center, (d) at side and (e) at corner.

5. Application of EB drawing to formation of nano-Si-dot and nano-polymer-pit arrays with a pitch of 25 nm x 25 nm using EB drawing, reactive ion etching (RIE) and nano-imprinting [23, 25]

The possibility of forming very fine pits or dots with a pitch of less than 25 nm was researched using reactive ion etching (RIE) and nano-imprinting with EB drawn pattern as a mask for the future process. We were able to fabricate ultrahighly packed dot arrays with a dot diameter of less than 15 nm and a dot pitch of 25 nm x 25 nm in negative calixarene resist using EB drawing. We also formed nan-Si dot arrays patterns by CF_4 RIE. Furthermore, pit arrays were formed in polymer film through nano-imprinting by the photo-polymer method using a Si dot arrays pattern as the master mold. We demonstrated that the EB-drawn dot arrays resist pattern is very suitable for the fabrication of Si dot arrays and pit arrays with a pitch of 25 nm x 25 nm in this polymer. The Si dot and pit diameters were less than 10 nm.

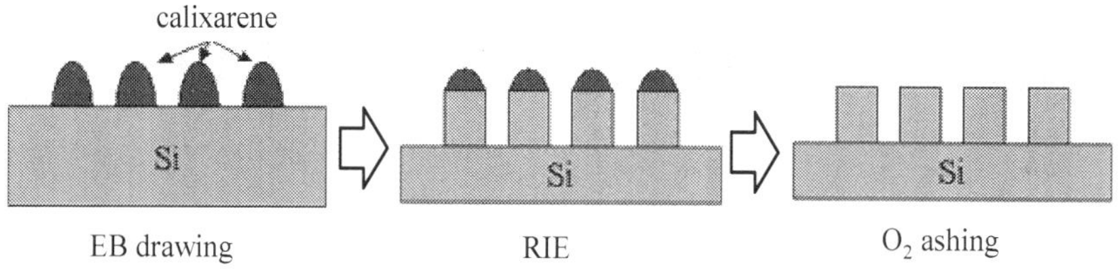

Fig. 5.1. Process flow for Si nano-dot arrays by RI etching and ashing.

5.1 Dry etching (RIE and ashing)

We tried to apply the EB drawing to dry etching process. We checked if the pattern is available for reactive ion etching (RIE) process, and we carried out CF_4-RIE using the resist patterns. We performed to do RIE of the Si substrate with the resist dot arrays pattern after post baking of the resist pattern, and to remove the remained resist by O_2 ashing. The process flow is shown in Fig. 5. 1. The experiments conditions are represented in Table 5. 1.

	RIE	Ashing
gas	CF_4	O_2
flow rate	5 sccm	10 sccm
pressure	1 mTorr	2.5 mTorr
bias	-60 V	-120 V
Etching time	0.5-2 min	10 min

Table 5.1. RIE and ashing conditions for Si dot formation and resist removal, respectively.

As the experimental results, we obtained very fine Si dot arrays with a pitch of 30 nm x 25 nm to 25 nm x 25 nm (Fig. 5. 2). The minimum diameter of the Si dot is < 10 nm, and the height is about 20 nm. Figure 5. 3 shows histograms of the EB drawn resist dot and the RIE Si dot sizes in a pitch of 25 nm x 25 nm. According RIE and ashing, the resist dot size is transferred to Si substrate. Mean dot size decreases from about 14.6 nm to about 9 nm. This may be caused by in-plane etching in RIE. On the other hand, the standard deviation increases from about 1.3 nm to about 1.5 nm. Although the cause is not clear, it may be due to the EB resist toughness or the resist pattern edge sharpness. From the etching experiments, we measured the etching rates of 7 nm/min and 10 nm/min in calixarene and Si(100) in normal direction, respectively, as shown in Fig. 5. 4. The etching rate in plane component is about 2 nm/min. The data supported the dot size decrease.

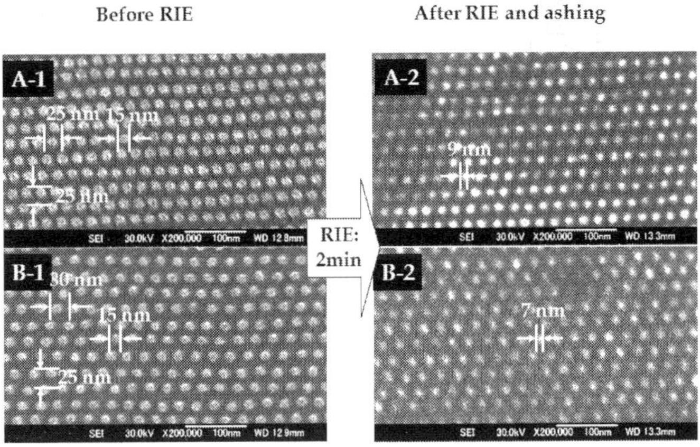

Fig. 5.2. Dot arrays patterns before and after RIE and ashing (SEM images; A: 25 nm x 25 nm in pitch and B: 25 nm x 30 nm.

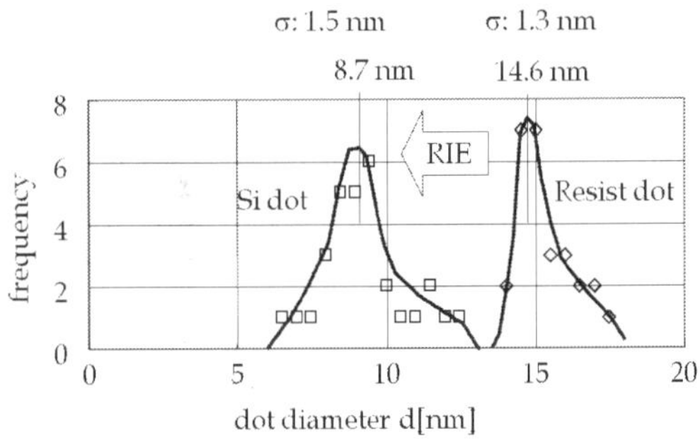

Fig. 5.3. Histograms of the resist and Si dot diameters before and after RIE and ashing, respectively, with a pitch of 25 nm x 25 nm.

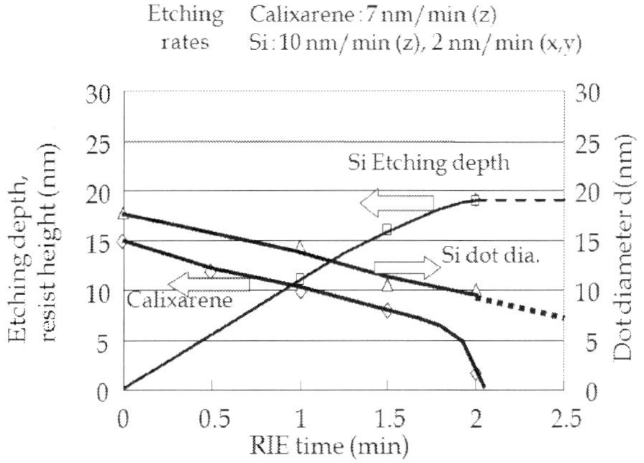

Fig. 5.4. Etching rates of Si, calixarene resist and Si dot diameter using RIE.

5.2 Nano-imprinting using UV photo polymer

By using the RIE Si dot arrays pattern as a mold, we tried to do nano-imprinting (NIP) to transfer the etched Si dot arrays pattern into UV photo polymer. Figure 5. 5 shows the nano-imprinting process flow with UV photo polymer method. After coating the polymer on the dot arrays, putting the polycarbonate substrate on the polymer and illuminating the UV light into the polymer, we obtained ultrahigh packed pit arrays in the polymer as shown in Fig. 5. 6. We could demonstrate to fabricate the 25 nm x 25 nm pitch pit arrays using EB drawing, RIE and NIP, although the polymer surfaces are in slight curvature because of deformation in the SEM observation or localized stretching in separating the polymer from the mold. From SEM images of the Si mold with 25 nm x 25 nm pitch Si dot arrays before and after NIP, we have confirmed that it is clear that the mold surface has no damage after NIP. From the SEM images of Figs. 5. 3 and 5. 6, we observed that the pit diameters are about 2 nm larger than those in the Si dots. We can consider that the polymer have a shrink property with a rate of about 8 % in the fixing process. In a detail, however, the rate of 8 % does not agree with the increase of about 2 nm against mean dot diameter of about 9 nm. We should consider the shrinkage in the future. We clarified that the very fine pitch resist dot arrays pattern written by EB drawing with calixarene are available for dry etching and nano-imprinting process. We demonstrated to form the Si nano-dot and pit arrays with a pitch of 25 nm x 25 nm using RIE and nano-imprinting with the resist pattern as a mask.

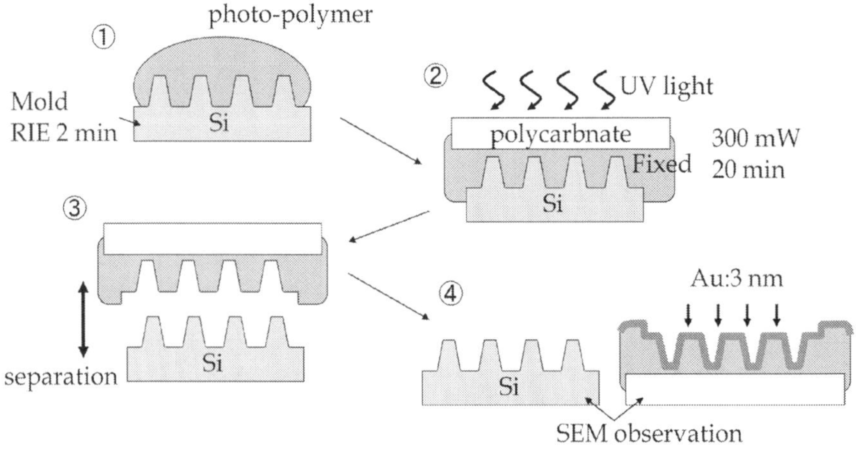

Fig. 5.5. Process flow of 2P (photo polymer) method for nano-imprinting.

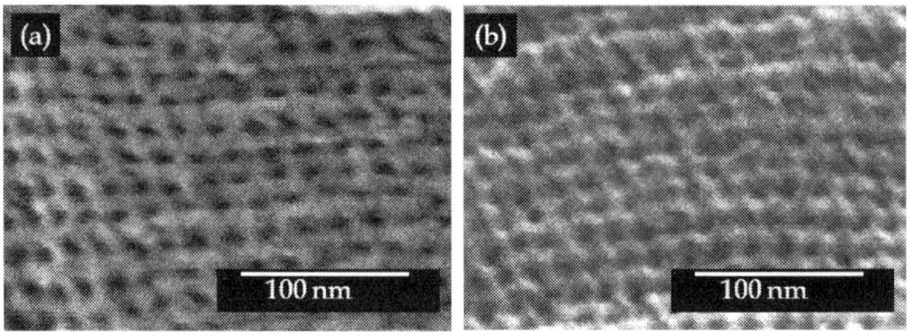

Fig. 5.6. SEM images of very fine pitch pit arrays in photo-polymer using nano-imprinting with the Si mold made by RI etched Si dot arrays pattern with a pitch of 25 nm x 25 nm drawn by 30 keV electrons with exposure dosages, (a) 28 mC/cm^2, (b) 36 mC/cm^2.

6. Conclusion

I have described an electron beam (EB) lithography using a raster drawing for fabrication of nanometer sized dot or pit arrays, theoretically and experimentally. I considered the possibility to form the nanometer-sized pitch fine dot arrays by the energy deposition distribution (EDD) calculated by the home-made Monte Carlo simulation. The experimental research is done by dependences of 2 resist materials (ZEP520 and calixarene) and thickness on the drawn dot size and pitch in EB drawing. I have used a conventional EB drawing system based on high resolution-scanning electron microscope (HR-SEM). As results using both positive and negative resists with thin thickness. I can demonstrate the possibility to form nanometer-sized dot and fine pitched dot arrays as follows.

1. The simulation shows that the EDD profile seems to be cone shape, which is very suitable for formation of nanometer-sized dots using negative resist, while it is not suitable in a case of using positive resist.

2. It is demonstrated that the calixarene negative resist is very suitable to form an ultrahigh packed bit arrays pattern, comparing with ZEP520 positive resist.
3. As the experimental results, the minimum pitch of about 18 nm x 18 nm has been demonstrated using 30-keV EB drawing with calixarene resist, while about 40 nm x 50 nm pitch has been demonstrated with ZEP520.
4. EB drawing with calixarene has extremely small proximity effect.
5. It is demonstrated that nanometer-sized Si dot arrays were fabricated by CF_4-RI etching using calixarene resist dot arrays as a mask.
6. It is also demonstrated that nanometer-sized polymer pit arrays were fabricated by nano-imprinting with 2P method using the nanometer-sized Si dot arrays as a mother pattern.

Furthermore, we try to form 15 nm x 15 nm pitched resist dot arrays using HSQ negative EB resist. We have gotten a prospect to form them by improving a developer for the resist. There are some papers to improve them using HSQ negative resist. X. Yang et al. have reported 12 nm x 12 nm or 15 nm x 15 nm pitch fine resist dot arrays using 100 keV EB drawing with hot developer of TMAH at 40 ^{o}C [6]. In addition, J. K. Yang et al. have reported 12 nm or 14 nm pitch resist line and space pattern EB-drawn using 100 keV EB drawing with salty development [27].

On the other hand, it is crucial to improve the resolution of SEM to check whether the EB-drawn pattern is complete or not. We should get high resolution observation method for very fine pitch dot arrays with a pitch of less than 15 nm x 15 nm.

7. Acknowledgements

I would like to thank Profs. H. Sone, Y. Yin, K. Itoh, Z. Mohamad and T. Tamura, and students of Ms. H. Zhang, Mr. H. Sano, Mr. M. Shirai in Hosaka labo of Gunma University for the experiments, simulation and fruitful discussion, and Mr. K. Noguchi in Dept. of Electronic Eng., Gunma University for technical supports in EB writing and resist process. This research was performed in Kiryu Ohta Toshi Area project supported by the Ministry of Monbukagaku-sho of Japan.

8. References

[1] J. A. Dohery, Solid State Technol. 22 83 (1979).
[2] E. V. Weber and R. D. Moor, J, Vac. Sci. Technol. 16, 1780 (1979).
[3] G. Varnell, D. Spiecer, J. Hebley, R. Robbins, C. Carpenter and M. Malone, J. Vac. Sci. Technol. 16, 1787 (1979).
[4] S. Hosaka, M. Ichihashi, H. Hayakawa, S. Nishi and M. Migitaka, Jpn. J. Appl. Phys. 21, 543 (1982).
[5] S. Hosaka, Y. Tanaka, M. Shirai, Z. Mohamad and Y. Yin, Jpn. J. Appl. Phys. 49, 046503 (2010).
[6] X. Yang, S. Xiao, et al., J. Vac. Sci. Technol. 25, 2202 (2007).
[7] S. Hosaka, H. Sano, K. Itoh, and H. Sone, Microelectronic Eng. 83, 792 (2006).
[8] H. Zhang, T. Tamura, Y. Yin, and S. Hosaka, Key Engineering Material, to be published.
[9] T. H. P. Chang, J. Vac. Sci. Technol. 12, 1271 (1975).
[10] W. Zhang, A. Potts, D.M. Bagnall, B.R. Davidson, Thin Solid Film 515, 3714 (2007).
[11] K. Murata and T. Matsukawa, Jpn. J. Appl. Phys. 10, 678(1971).

[12] L.M. Ren, B.Q. Chen, and Z.Y. Tan, Opto-Electronic Engineering, 29, 24 (2002).
[13] S. Hosaka, H. Sano, M. Shirai, Y. Yin, and H. Sone, Microelectronics Eng. 84, 802 (2007).
[14] J. Fujita, Y. Ohnishi, S. Manako, Y. Ochiai, E. Nomura, T. Sakamoto, and S. Matsui, Jpn. J. Appl. Phys. 36, 7769 (1997).
[15] M. Ishida, J. Fujita, T. Ogurai, Y. Ochiai, E. Ohshima, and J. Momoda, Jpn. J. Appl. Phys. 42, 3913 (2003).
[16] Z. Mohamad, M. Shirai, H. Sone, S. Hosaka, and M. Kodera, Nanotechnology 19, 125301 (2008).
[17] S. Hosaka, Z. Mohamad, M. Shirai,H. Sano, Y. Yin, A. Miyachi, and H. Sone, Appl. Phys. Express 1, 027003 (2008).
[18] C.T. Rittner, M.E. Best, B.D. Terris, IEEE Trans. Magn. 37, 1649 (2001).
[19] S. Hosaka, T. Shintani, M. Miyamoto, A. Hiritsune, M. Terao, M. Yoshida, S. Honma, and S. Kammer, J. Appl. Phys. 79, 8082 (1996).
[20] S. Hosaka. M. Ishihashi, H. Hayakawa, S. Hishi, and M. Migikata, Jpn. J. Appl. Phys. (Part 1), 21, 543 (1984).
[21] S. Ogata, M. Tada, and M. Yoneda, Appl. Opt. 33, 2032 (1994).
[22] Y. Kojima, H. Kitahara, M. Katsumura, and Y. Wada, Jpn. J. Appl. Phys. (Part 1), 37, 2137 (1998).
[23] S. Hosaka, H. Sano, M. Shirai, and H. Sone, Appl. Phys. Lett. 89, 223131 (2006).
[24] S. Hosaka, Z. Mohamad, M. Shirai, H. Sano, Y. Yin, A. Miyachi, H. Sone, Microelectronic Eng. 85, 774 (2008).
[25] S. Hosaka, H. Hayakawa, M. Ichihashi, K. Asanami, S. Nishi, and M. Migitaka, Proc. Microcircuits Eng. (Achen, 1979) p. 7.
[26] J. K. W. Yang, B. Cord, H. Duan, K. K. Berggren, J. Klingfus, S. Nam, K. Kim, M. Rooks, J. Vac. Sci. Technol. B27, 2622 (2009).

2

Focused Ion Beam Lithography

Heinz D. Wanzenboeck and Simon Waid
Vienna University of Technology – Institute for Solid State Electronics
Austria

1. Introduction

Optical lithography is the unrivalled mainstream patterning method that allows for cost-efficient, high-volume fabrication of micro- and nanoelectronic devices. Current optical photolithography allows for structures with a reproducible resolution below 32 nm. Nevertheless, alternative lithography methods coexist and excel in all cases where the requirement for a photomask is a disadvantage. Especially for low-volume fabrication of microdevices, the need for a photomask is inefficient and restricts a fast structuring, such as required for prototype device development and for the modification and repair of devices. The necessity of high-resolution masks with a price well above €10k is too cost intensive for the fabrication of single test devices. For this reason 'direct-write' approaches have emerged that are popular for several niche applications, such as mask repair and chip repair. Optical direct-write lithography and electron beam lithography are among the most prominent techniques of direct-write lithography. Less known, but highly versatile and powerful, is the ion beam lithography (IBL) method.

Optical direct-write lithography uses laser beam writers with a programmable spatial light modulator (SLM). With 500 mm^2/minute write speed and advanced 3D lithography capabilities, optical direct-write lithography is also suitable for commercial microchip fabrication. However, with a resolution of 0.6-µm minimum feature size of the photoresist pattern, optical direct-write lithography cannot be considered a nanopatterning method.

Electron beam lithography uses a focused electron beam to expose an electron beam resist. Gaussian beam tools operate with electron beams with a diameter below 1 nm so that true nanofabrication of structures is feasible. A resolution of 10 nm minimum feature size of the e-beam resist pattern has been successfully demonstrated with this method. However, special resists are required for e-beam lithography, that are compatible with the high energy of forward scattered, back-scattered and secondary electrons. A common resist for sub-50nm resolution is polymethylmetacrylate (PMMA) requiring an exposure dose above 0.2 $\mu C/\mu m^2$. For highest resolution (below 20 nm) inorganic resists such as hydrogen silsesquioxane (HSQ) or aluminium fluoride (AlF3) are used, which unfortunately require a high electron exposure dose. Hence, high-resolution electron beam lithography (EBL) is linked to long exposure times which, in combination with a single scanning beam, results in slow processing times. Therefore, this high-resolution method is only used for writing photomasks for optical projection lithography and for a limited number of high-end applications. A resolution to this dilemma may be the use of multi-beam electron tools, as are currently under development. Also electron projection lithography has been under

development but currently all development programmes for a commercial tool have been discontinued.
FIB lithography is similar to EBL, but provides more capabilities. Not only can FIB lithography (i) create a pattern in a resist layer just like EBL, but it is also capable of (ii) locally milling away atoms by physical sputtering with sub-10nm resolution (subtractive lithography), (iii) locally depositing material with sub-10nm resolution (additive lithography), (iv) local ion implantation for fabrication of an etching mask for subsequent pattern transfer and (v) direct material modification by ion-induced mixing.
The ion direct-write lithography combines the high resolution of electron beam lithography with the higher writing speed of optical laser writers. With so-called liquid metal ion sources focusing of the ion beam to a diameter down to 5 nm is feasible. Due to the higher mass of the ions the higher energy of the ion beam allows a faster exposure of resists and thus a higher processing speed. Currently, new ion sources have been developed and also ion projection systems and multi-beam systems are on the verge of commercial introduction, so that this "exotic" technique deserves more consideration for future nanofabrication.
FIB lithography is superior to EBL, as with focused ion beam (FIB) proximity, effects are negligible as no electron backscattering occurs. As a consequence, a higher resolution can be obtained with FIB as the pixel size is roughly equal to the beam spot size and no exposure occurs between pixels, hence allowing a short dwell time on each pixel. With the shorter ion range, weaker forward scattering and smaller lateral diffusion of secondary electrons, FIB lithography reaches a higher resolution than EBL with the same beam spot size. Overall, the higher resist sensitivity to ions increases the throughput in contrast to EBL.
A speciality of ion beam direct-write lithography is the possibility for resistless structuring. The application of a resist layer is not possible on non-planar samples, such as prestructured wafer surfaces or three-dimensional samples. If a resist layer can be applied, small structures are typically only feasible with ultrathin resist layers with a homogeneous thickness below 100 nm. The ion beam can also be used for direct-write implantation of direct-write milling of patterns in order to fabricate structures. The implantation of ions originating from the ion source itself can be used to fabricate locally doped hardmask layers that can be used for pattern generation in a subsequent selective etching process. This approach will be described in detail in section 4.3. The straightforward approach for pattern generation is the direct-write milling with a focused ion beam. The kinetic energy of accelerated ions may be used for physical sputtering of the substrate. With a focused Ga^+ ion beam of less than 5 nm diameter, structures with 30 nm features have been realized. This processing alternative will be described in detail in section 4.1. A sub-version of direct-write milling with an ion beam is the gas-assisted etching and the beam-induced deposition. The physical milling by the ion beam is complemented by a chemical reaction locally triggered by the energy of the ion beam. With gas-assisted etching, an etch gas is added that can react with the substrate to form a volatile etch product. With beam induced deposition a precursor gas is added that locally decomposes on those substrate areas scanned by the ion beam. From this ion beam-induced deposition, a solid material structure is formed. This is an 'additive' direct-write lithography technique.

2. Ion-solid interaction

The fundamental process of resist-based IBL is the ion-induced change of the resist. Typically, ion beam resists are used as negative resists experiencing a decrease of solubility of the ion exposed area due to ion-triggered reactions. Also with ion beam-induced etching and ion beam-induced deposition, a chemical reaction of surface species is the underlying mechanism of this structuring approach. For this reason, the ion-solid reaction shall be taken into closer examination.

Ion interaction with solid can be separated in elastic and inelastic collisions and in electronic interactions. Ion-atom as well as ion-electron collisions are typically treated as binary collisions. For the treatment of ion-atom collisions, a lower energy limit of 10 to 30 eV has to be considered. At lower energies, many body interactions are also a relevant mechanism, which is up to now widely neglected in literature due to its complexity (Eckstein, 1991).

Elastic collisions between ions and atoms of the substrate (or resist) are responsible for (i) beam broadening by scattering, (ii) amorphization of the target substrate, (iii) ion implantation into the target substrate, and (iv) sample physical sputtering. Both forward scattering and backward scattering lead to a broadening of the ion beam propagating in matter. As a practical consequence, this reduces the resolution when exposing a resist layer. Ions impinging into the substrate lead to the secondary effects of atomic mixing, which results in amorphization of crystalline samples, in the intermixing of resist and substrate at the interface and also implantation of the primary ions (often as an element) into the substrate. With photoresist, this may also lead to problems with later removal, as ion-implanted resists display a higher etch resistivity in plasma ashers. In the special case of sputtering, the substrate material is removed as a consequence of elastic collisions. The incident ions transfer their momentum to the target atoms within a collision cascade region. Atoms from the substrate surface may be ejected as a sputtered particle if it receives a kinetic energy that is sufficient to overcome the surface binding energy (SBE) of the target material. This effect is used for direct-write structuring by milling without any resist.

The ion beam may also be used to initiate chemical reactions. For this process, energy has to be converted from kinetic energy into other types of energy, such as bond dissociation energy. Such inelastic collisions involve an energy transfer either to electrons of the substrate ('electronic stopping') or an energy transfer to other nuclei or atoms of the substrate.

About two thirds of the dissipated energy is transformed into kinetic energy of so-called δ-electrons. Heavy ions dissipate their energy along their trajectories ionizing target atoms and producing free electrons. Around the ion's trajectory, secondary and tertiary ionization processes occur. Inelastic processes may lead to ionization of atoms involving also secondary electron emission. The secondary electrons are also subscribed a significant role in bond breaking mechanisms as a consequence of ion irradiation. Secondary electrons have energy between 1 and 50 eV corresponding to the energy range required to break molecular bonds (sigma and pi bonds). Other inelastic processes involve loss of kinetic energy by emission of photons including emission of x-rays, of Bremsstrahlung, or of Čerenkov radiation. Finally, heating, luminescence, shock wave or phonon excitation are other energy-loss mechanisms affecting not a single atom but rather an entire volume of the irradiated substrate.

Chemical reactions of the resist layer or of the substrate are induced by effects of inelastic collision of the primary ions. For chemical reactions of the resist layer, primarily the secondary electron-induced bond dissociation or the radical production is considered as a relevant mechanism. The low-energy secondary electrons (generated by ion-matter interaction) can expose a resist layer for lithography analogous to the secondary electron induced reactions used in EBL. Hence, electron beam resists can also be used as FIB lithography resists.

FIB lithography has the advantage of (i) a higher resolution due to the absence of proximity effects and (ii) a higher resist sensitivity. As no electron backscattering exists, the pixel size with FIB lithography is equal to the ion beam spot size and thus can be much higher than with EBL. As a primary ion can release up to 200 secondary electrons (Dietz & Sheffield, 1975), while a primary electron can release less than 2 secondary electrons (Hoyle, 1994) the exposure speed with ion lithography can be up to a factor of 100. On the other hand, FIB lithography resists suffer from a restricted exposure depth in the resist and from contamination of the resist by source ions. To circumvent larger structures resulting from the restricted exposure depth, a thin resist layer can be used, but this makes subsequent etching processes or lift-off processes more difficult. The contamination of the resist is especially problematic, if organic resists are removed by plasma ashing and the inorganic contaminations remain on the surface.

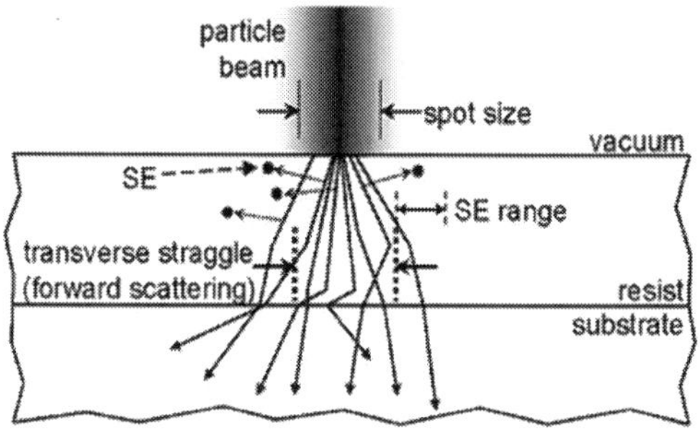

Fig. 1. Factors limiting resolution of IBL. A focused ion beam irradiates a resist layer on a substrate. The three factors limiting resolution are (i) spot size of the beam (ii) ion scattering and (iii) secondary electron emission.
Reprinted with permission from Winston D. et al., 2009. Scanning-helium-ion-beam lithography with hydrogen silsesquioxane resist. JVST B 27(6), 2702. Copyright 2009, American Vacuum Society.

Ion beam lithography has repeatedly been successfully used for exposing resist layers.
Structural modification of the resist, including chain scission, cross-linking, double-bond formation, molecular emission, changes in molecular weight distribution, and so forth are due to ion irradiation of polymers (Calcagno et al., 1992). The degradation of PMMA by proton beam irradiations for resist applications has been analyzed by Choi et al. (Choi et al., 1988). Even though the energies of the radiation sources varied considerably (up to 900 keV for H3+), they observed a 1-to-1 correspondence of loss of ester groups and generation of double bonds in the polymer chains for all radiation types.
Horiuchi et al. (Horiuchi et al., 1988) have achieved 200 nm line width in PMMA using a He+ ion beam. Using a Ga+ beam Kubena et al. (Kubena et al., 1989) could even demonstrate sub-20 nm line width in PMMA. The higher energy transmission by the ions allows for faster exposure of resists by ion beams so that also resists requiring prohibitively high electron doses with EBL can be used in IBL. Therefore, inorganic high-resolution resists such as hydrogen sesquioxane (HSQ) and aluminium fluoride can also be used. Hydrogen silsesquioxane (HSQ) is a negative-tone resist that cross-links via Si–H bond scission (Namatsu et al., 1998). The energy of a Si–H bond is roughly 3 eV and can be broken by secondary electron energy. Van Kan et al. (van Kan et al., 2006) have successfully demonstrated 22 nm line width in HSQ using a 2 MeV H_2^+ beam.

3. Ion beam equipment

For structuring with an ion beam, two complementary approaches have to be distinguished. The equipment setup differs among those approaches described in Table 1. Yet, the experimental setup always includes an ion source and an ion optical system consisting of electrostatic lenses and electrostatic deflectors. For ion optics, only ions of a specific mass and of a specific energy can be used for the focusing optics. From the use of ions for structuring 'resist-based' and 'resistless' methods can be distinguished.

Beam type	Scanning beam	Broad, collimar beam
Method	'direct-write technique'	'projection technique'
Mask	maskless (structure by pattern generator)	aperture mask

Table 1. Structuring approaches using an ion beam

3.1 Ion sources

The core component of an ion beam system is the ion source. Development of ion sources initially was motivated by mass spectrometry and ion implantation for semiconductor manufacturing. Only with the emerging resolution limits of optical lithography particle beam methods became interesting for nanostructuring. For using a focused beam, a point source is required, while broad beams can also use ion sources emitting ions over a larger area. Four basic ion source types are described below:

1. **Electron bombardment ion sources.** An electron beam is directed onto a gas. Under electron bombardment, the gas molecules in the irradiated volume become ionized. Hence, the ions are not emitted by a localized source. The resulting ion current is rather small but typically has a small ion energy spread. Due to the low currents these sources are not used for ion lithography but find their application in mass spectrometry (Dworetsky et al., 1968). This source has been used successfully to produce low-energy beams of noble gas ions, such as He+ and Ar+, without measurable contamination, as well as for H_2+ and N_2+.
2. **Gas discharge ion source.** The ions are created by plasma or by electric discharge. Typically ions are generated by capacitively coupled plasma, inductively coupled plasma or by microwave-induced plasma. An alternative are glow discharge of a gas at low pressure or spark ionization of a solid sample. As ions from gas discharge are emitted over a larger gas volume, they are not point-sources and are therefore not suitable for focused beams. However, gas discharge ion sources produce a high ion current and are therefore interesting sources for ion lithography based on the projection method. These sources are also widely used in high-energy accelerators and ion implanters for semiconductor manufacture. A widely used type is the duoplasmatron. First, gas, such as argon, is introduced into a vacuum chamber where it is charged and ionized through interactions with the free electrons from the cathode. The ionized gas and the electrons form a plasma. By acceleration through two highly charged grids, ions are accelerated and form a broad ion beam.
3. **Field ionization source.** These sources operate by desorption of ions from a sharp tip in a strong electric field. Typically, gas molecules are adsorbed on the surface of a sharp needle tip and are directly ionized in the high electrical field prevailing at the tip apex.

Due to the point-like emission of ions from a single spot, the focusing to a beam with an ultra-small diameter is feasible. The adsorbtion of gas on the tip may be enhanced by cryostatic cooling of the tip. The focused ion beam may be used for a field ion microscope and typically non-reactive ions, such as noble gas, ions are used.

4. **Liquid metal ion source (LMIS).** The LMIS operates by desorbtion of metal ions from liquid metal under a strong electrical field. Typically a thin needle or a capillary is wetted by a thin film of liquid source metal, which has been heated to the liquid state. Typically, gallium (m.p. 29,8°C) or indium (m.p. 156,6°C) or Be-Si-Au alloys ($Au_{70}Si_{15}Be_{15}$) are employed. A Taylor cone is formed under the application of a strong electric field. The force acting onto the needle due to the electric filed shapes the cone's tip to get sharper, until ions are produced by field evaporation. For emission of ions, a threshold extractor voltage (for Ga 2kV) is required. For an alloy source, an energy separator is needed to filter out one ion species. Liquid metal ion sources are particularly used in focused ion beam microscopes. The emission angle is around 30°. The angle distribution of emission current is rather uniform. Energy spread of emitted ions can be large (>15V) resulting in a large chromatic aberration.

Depending on whether a resist layer is used or resistless structuring is performed, patterning with an ion beam opens up different structuring capabilities, as shown in Table 2.

	Single focused beam	**Broad beam**
	Scanning beam ➔ maskless	Projection ➔ requires mask
Resist	Direct-write exposure of resist	Projection exposure
Resistless	Direct-write Milling or Direct-write Deposition or Direct-write Etching	Projection milling

Table 2. Resist-based and resistless patterning approaches with an ion beam

Depending on the selected source type focused beam systems and broad beam systems also have to be distinguished as depicted in Table 3.

		Focused beam	**Broad beam**
		High resolution	Low resolution
Single beam	Sequential writing → slow	Ga LMIS He-LIS	Plasma source
Multi-beam	Parallel writing → fast	Plasma source with aperture plate and projection optics	-

Table 3. System configurations with ion beam tools

3.2 Ga ion microscopes

Focused ion beam tools using a liquid metal ion source for Ga ions are currently the state of the art, because this ion beam can be made very small and therefore resembles a perfect tool for nanofabrication. The Ga ion beam can be focused below 5nm diameter, as the ion source is almost an ideal point source. The original W wire is not sharp and may have a tip radius of more than 1 μm. The sharp tip is formed by the liquid metal induced by the electric field.

This Taylor cone of liquid metal results in a very high electric field at the cone apex, which is required for field emission of the ions.

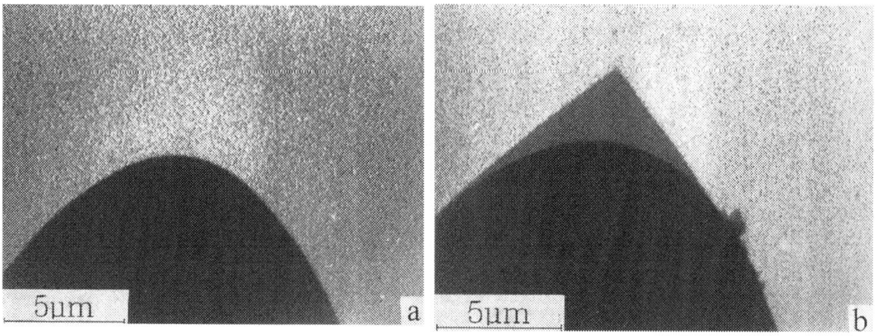

Fig. 2. Shape of the emitter tip coated with a liquid AuGe alloy (a) without Taylor cone; (b) with Taylor cone.
Reprinted with permission from Driesel W., Dietzsch C. & Muhle R., 1969. In situ observation of the tip shape of AuGe liquid alloy ion sources using a high volt transmission electron microscope. JVST B 14(5), 3367. Copyright 1969, American Vacuum Society..

The ion beam is used for imaging (Orloff et al., 1996), local implantation (Schmidt et al., 1997), physical milling (Giannuzzi & Stevie, 1999), gas-assisted etching (Utke et al., 2008), localized deposition (Matsui et al., 2000) and for exposure of resist layers (J. Melngailis, 1993) (Lee & Chung, 1998) as extensively described in literature (Tseng, 2005); (Giannuzzi & Stevie, 1999) (Jeon et al., 2010) (Tseng, 2004) and is therefore not further discussed here in detail.

3.3 He ion microscope

'Heavy' ions, such as Ga+, may displace and scatter atoms in the substrate so much that device performance suffers. The He ion beam offers a new alternative. Helium ions are more massive than electrons by over three orders of magnitude and thus diffract less around apertures. Thus, smaller apertures are possible in a helium ion column than in an electron column, and this enables a smaller spot size. The specified spot size for a Zeiss Orion Plus helium ion microscope is 0.75 nm at an accelerating voltage of 30 kV. The first commercial helium ion microscope was introduced in 2006 (B. W. Ward et al., 2006) and by now has reached a maturity so that edge resolutions of <0.35 nm have become routinely possible.
As ion source a cryogenically cooled metal needle with a tip in the shape of a three-sided pyramid is used. The He+ source can be cooled to around 80 K, which resembles a point source that can be focused. When gun temperature exceeds 95 K, this structure is unstable, meaning source size and thus brightness may become compromised at higher temperatures A small quantity of helium is admitted in the vicinity of the sharp tip in the shape of a three-sided pyramid. The apex of the pyramid consists of a set of three atoms, the trimer. The injected helium is then polarized by the large electric field and He atoms accelerate towards ionization area near the three topmost atoms. Only in this restricted area at the pyramidal tip does ionization take place. The charged He ions from one emitter atom are selected and are accelerated into the ion column to produce the focused electron beam. The ion optics should theoretically operate around unity magnification.

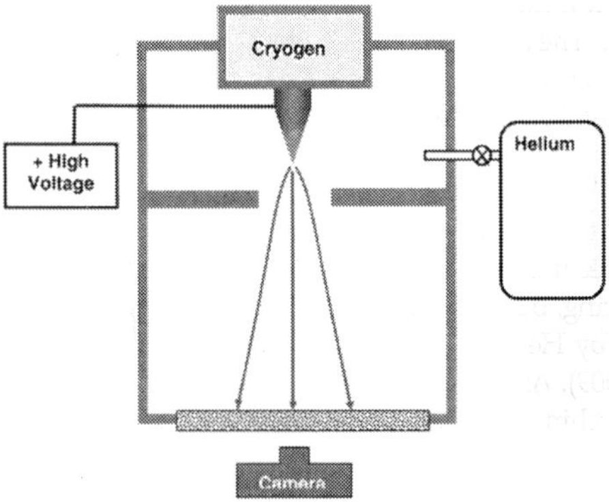

Fig. 3. Field ion microscope in its simplest form consists of a cryogenically cooled tip, biased to a high voltage. When the imaging gas is admitted, a pattern is visible on the scintillator. Reprinted with permission from Ward B.W., Notte J.A. & Economou N.P., 2006. Helium ion microscope: A new tool for nanoscale microscopy and metrology. JVST B 24(6), 2871. Copyright 2006, American Vacuum Society.

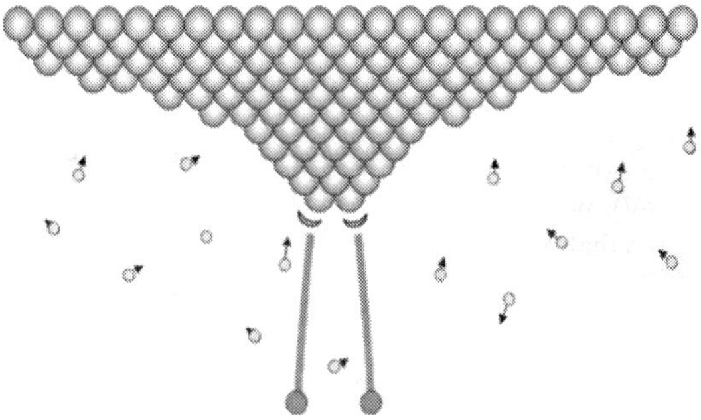

Fig. 4. Spherical tip after the atoms have been rearranged to form a three-sided pyramid. Now the ionization disks exist only over the topmost three atoms.
Reprinted with permission from Ward B.W., Notte J.A. & Economou N.P., 2006. Helium ion microscope: A new tool for nanoscale microscopy and metrology. JVST B 24(6), 2871. Copyright 2006, American Vacuum Society.

The depth of field with a He ion microscope is correspondingly five times larger as the convergence angle is typically five times smaller than an SEM. Also the diffraction curve is over two orders of magnitude smaller compared to the SEM. Consequently, an ideally focused spot may have a spot size down to 0.25 nm.
With current systems, the virtual source size is smaller than 0.25 nm while providing a brightness of approximately 10^9 Acm^{-2}. Measurements have shown an angular beam intensity of 0.5-1 μAsr^{-1} and an energy spread of 1 eV. (R. Hill & Faridur Rahman, 2010).
The interaction of the He ion beam with the sample is significantly different than with either an electron beam or a Ga+ ion beam. Versus using electrons, He ions are advantageous by the strongly reduced diffraction effect which enables a tremendously increased resolution of

smaller structures. He ion microscopy is highly suitable for imaging of insulating samples and biological samples. The scanning helium ion microscope can also be used for diffraction imaging in transmission mode (J. Notte 4th et al., 2010). This way, crystallographic information can be provided in the form of thickness fringes and dislocation images. This mode allows the recording of high-contrast images of crystalline materials and crystal defects even at modest beam energies.

Helium ion microscopy has already been successfully used for resist-based structuring and the feasibility of 6nm features has been demonstrated (Sidorkin et al. 2009). Ion beams may also be used for sample sputtering, but the light He ions have a very low yield. Yet, successful milling of graphene structures by He ion microscopes has been shown by R. Hill & Faridur Rahman (2010) and Bell et al. (2009). As with Ga ion microscopy the ion beam may also be used for gas-assisted deposition or etching. The fabrication of a W pillar with an average diameter of 50 nm grown by deposition in the He ion microscope has been demonstrated. This deposited pillar was 6.5 µm high and had a height to width aspect ratio of 130:1 (R. Hill & Faridur Rahman 2010). Also 10 nm wide nanowires have been deposited and sub-10 nm cuts in Au have been performed with a focused helium ion beam (Livengood et al. 2011)

3.4 Ne ion microscope

The development of ion microscopes with heavier noble gas ions is currently underway (Livengood et al., 2011). Utilizing neon ions extends the capabilities of high-source brightness technology. The neon ion source will also use the trimer-gas-field ion source used in the He ion microscope.

For a stable GFIS source it is necessary that other gas contaminants have lower ionization energy than the noble gas ions, otherwise ionization of contaminants might also occur. The resulting contamination of the source region would contaminate the trimer trip region. Besides helium (24.5 eV), neon (21.6 eV) is the only noble gas with an ionization energy significantly higher than that of contaminants such as O, (13.7 eV), N (14.5 eV) and CO_2 (13.8 eV). Therefore, Ar (15.8 eV) Kr and Xe (11.1 eV) are less suitable ions for this process.

With a test system the beam diameter was determined to be 1.5 nm at 28 kV, with a sputter yield around 1 atom per incident ion for Si and 4 atoms per incident ion for Cu. In comparison to a gallium FIB this is by a factor of 2x lower. In comparison to the light helium ions this is by a factor of 100x higher.

Nanomachining tests performed in a 600 nm SiO2-CDO dielectric stack and in 30 nm Cr-SiO_2 using 24 kV beam energies (1 pA beam current) have achieved smallest via a 40 nm at the mid-point and 30 nm at the base, with a depth of 240 nm (6:1 aspect ratio via). These first results indicate structure widths larger than expected and side wall profiles poorer than expected, so that further improvements have to be implemented. The benefit of a Ne ion beam will be in the field of milling, as the heavier ion species is more suitable for material modification such as ion milling or beam-induced deposition or etching.

3.5 Ion projection systems

From an industrial viewpoint, a major deficit of focused ion beam systems is the sequential scanning with a single beam resulting in a very low sample throughput. For this reason projection systems have been developed using a stencil mask (Hirscher et al., 2002). A broad helium beam is extracted from a plasma source. An ExB mass filter selects only the desired He ion species. The ion projection lithography (IPL) system uses electrostatic ion optics for

reduction printing of stencil mask patterns to a magnification factor of 4. Monitoring the position of the ion beam for correction of the projected image was achieved with a pattern lock system which consists of (i) detectors measuring the position of beamlets, (ii) transputer-based controllers and (iii) beam control elements.
An ion projection lithography system for exposure of the Shipley XP9946 resist family allowing for very high resolution of 50 nm was developed. A high pattern collapse probability was experienced at high aspect ratios. Required ion doses varied with the composition of the resist and were in the range of 1.4 down to 0.12 $\mu C/cm^2$. Alternatively sensitivity adjustment by a variation of the photo acid generator (PAG) was achieved.

3.6 Multi-beam systems

Based on an ion projection concept using a stencil mask, further development efforts of IMS Nano have brought forward an ion multi-beam system. This multi-beam system features a programmable aperture plate with integrated CMOS electronics (Hans Loeschner et al., 2010). This aperture plate is equipped with deflection electrodes and a blanking plate. A beam deflection of 300 mrad from the axis is sufficient to filter out a beamlet. By blanking through one aperture a single beam can be individually switched on and off. The produced pattern of individually switched ion beams can be demagnified leading to a 200x pattern reduction.
As ion source a broad beam generated from plasma was used. The gases ionized ranged from hydrogen H_3^+ to Argon (Koeck et al., 2010).
Using 10 keV H^+ ions, a 20 nm thick layer of the inorganic photoresist hydrogen silsesquioxane (HSQ) was exposed. For this purpose the sample was irradiated by 43.000 beams with exposure dose of 12 mC/cm2. Tetramethylammonium hydroxide (TMAH) was used for resist development. After development, an effective 15 nm half-pitch (hp) resolution in 50 nm HSQ resist could be confirmed (Hans Loeschner et al., 2010). Development in NaOH/NaCl required a 3.3x higher exposure dose but longer exposure led to reduced shot noise influence on line edge roughness (LER).

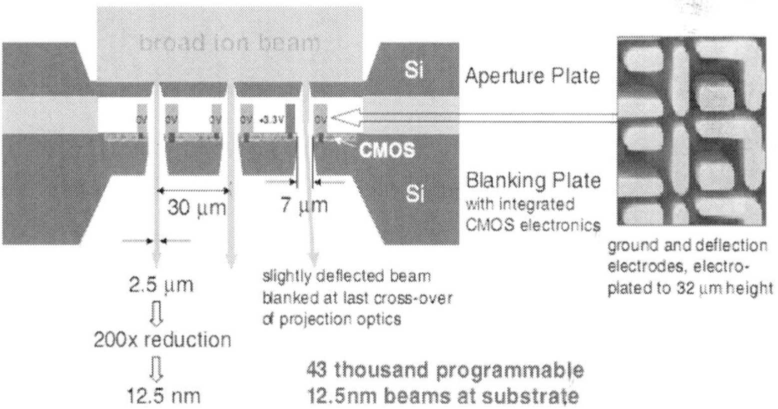

Fig. 5. Schematics of a 43k-APS unit of the IMS system providing 43 thousand programmable beams.
Reprinted with permission from Loeschner Hans, Klein C. & Platzgummer Elmar, 2010. Projection Charged Particle Nanolithography and Nanopatterning. JJAP, 49(6), 06GE01..

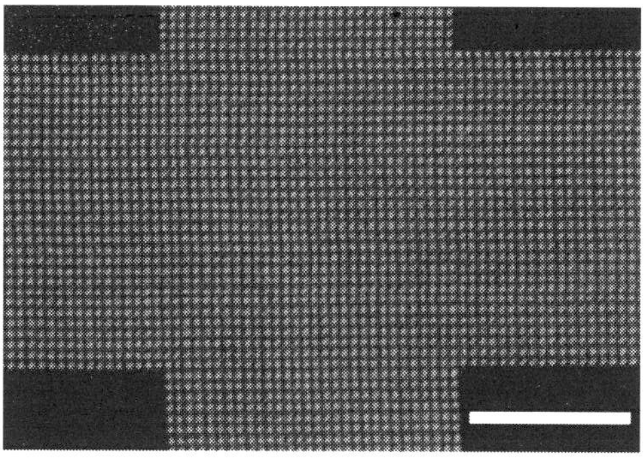

Fig. 6. SEM views of a dot array with 12.5 nm half-pitch features. HSQ was developed in NaOH/NaCl. The scale bar is 400 nm.
Reprinted from Publication Muehlberger M. et al, Nanoimprint lithography from CHARPAN Tool exposed master stamps with 12.5 nm hp, Microel. Eng. 88/8 2070, Copyright 2010, with permission from Elsevier.

This approach has been successfully used to pattern high-resolution patterns with 12.5 nm half-pitch in inorganic HSQ resist and to replicate this pattern by nanoimprint lithography (Muehlberger et al., 2011)

4. Lithography

Lithography is employed to define patterns inside a target material. Using an FIB, a multitude of processing techniques exists to achieve this goal. Similar to photolithography it can be achieved by exposing a resist material using the ion beam. However, with ions, patterns can also be defined by physically sputtering the target atoms (FIB milling), by triggering chemical reactions inside an adsorbed layer of a precursor gas (gas-assisted processing) and by ion implantation.
Among these techniques the most prominent are FIB milling and gas-assisted processing. They are employed for optical mask repair (Yasaka et al., 2008), circuit editing (CE) (Boit et al., 2008), transmission electron microscope (TEM) sample preparation and rapid prototyping (Persson et al., 2010). We will discuss these techniques only briefly since they are already well described and reviewed elsewhere (Reyntjens & Puers, 2001) (Utke et al., 2008). Instead we will focus on the less prominent FIB patterning techniques.
In the following sections we will review the work carried out on resist-based IBL and discuss the reasons for its failure as well as its chance of resurrection. Further we will present our findings on patterning using ion implantation with a focus on 3D nano patterning. Finally, we will introduce a new technique, called direct hard mask patterning (DHP) which combine the advantages of FIB milling with the speed of resist-based lithography.

4.1 FIB milling and gas-assisted processing

When an ion hits the target, surface elastic and inelastic scattering processes will take place. While inelastic processes are responsible for the generation of photons and secondary electrons, elastic scattering will transfer kinetic energy from the ion to the target atoms (Orloff et al., 2002). This kinetic energy transfer will cause the displacement of the target atoms and trigger a recoil cascade whereby kinetic energy is transferred from one atom to another by elastic scattering processes.

When the recoil cascade induced by the incident ion reaches the surface the target atoms at the surface may gather sufficient energy to leave the surface and enter the surrounding vacuum. The atoms are then either re-deposited or removed from the vacuum chamber by the pumping system. This material removal process is called milling.

How many target atoms a single ion is able to remove is largely dependent on the target material, the ions species, its energy and the angle of incidence of the ion beam. E.g. a 30keV Ga ion may eject 2 to 3 Si atoms. A 25keV Ga ion may also eject 23 Au atoms (Utke et al., 2008).

This milling process can be easily applied to almost any material and permits patterning down to the sub-100 nm regime. Due to its large material independence and high achievable resolution this process is commercially employed for transmission electron microscope (TEM) sample preparation. Due its ease of application and versatility, this process is often employed for rapid prototyping in research. Due to the high ion doses required for patterning only small areas can be patterned by FIB milling if processing times ought to remain within reasonable limits.

The energy transfer from the incident ion to the target may also be transferred to molecules adsorbed on the target surface. This may trigger a chemical reaction. Possible chemical reactions include the decomposition of adsorbed molecules and reaction of adsorbed molecules with the target atoms. This FIB-induced reaction process is illustrated in Figure 7.

The energy transfer mechanism from the ion to the adsorbed molecule is not yet fully understood. However, it is commonly agreed that it can be mainly attributed to the same recoil cascade which causes milling. Alternative explanations include secondary electron and local heating.

The decomposition products may be solid and thus deposited on the target surface, or volatile and thus be removed by the pumping system. By the choice of appropriate substances one may induce the local deposition of specific materials, e.g., metals may be deposited from appropriate metal organic precursors. This process is called gas-assisted deposition (GAD).

GAD is commercially employed for rewiring of integrated circuits in circuit editing (CE) (Boit et al., 2008), to protect the specimen surface of TEM samples during preparation by FIB milling and to correct void defects on photo masks (Boit et al., 2008).

The energy transferred to the target surface may trigger a reaction of the precursor with the target material. If the reaction product is volatile this will cause local etching of the target. This is mainly achieved by supplying light, reactive species such as halogens or halogen compound onto the surface. This process is called gas-assisted etching (GAE).

The etching efficiency of the GAE process is dependent on the target material. Thus, this can be employed to locally remove one specific material. E.g., the addition of XeF_2 will increase the removal rate of SiO_2 by a factor of nine compared to FIB milling while the removal rate of most metals (e.g., Al) will not be altered. This can be employed to remove a SiO_2 dielectric layer while mostly keeping Al interconnects intact. GAE is employed for selective removal of material in CE and for the selective removal of Cr in photo mask repair (Utke et al., 2008).

The chemical etching will also increase the removal rate compared to milling. This may be employed to increase processing speed and to minimize contamination and amorphization of the target material.

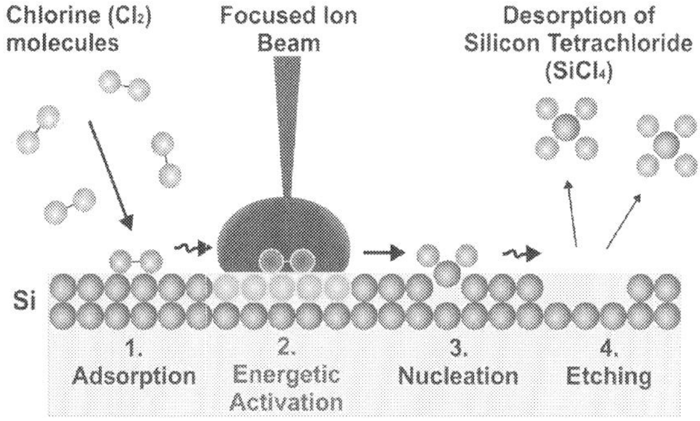

Fig. 7. Schematic illustration of the FIB-induced etching process

4.2 Resist-based lithography

The typical process flow for resist-based IBL is identical to EBL and is illustrated in Figure 8. The pattern definition is performed by the chemical modification of the resist irradiated by ions. The key elements in the process are thus the employed resist and its interaction with the beam employed for exposure.

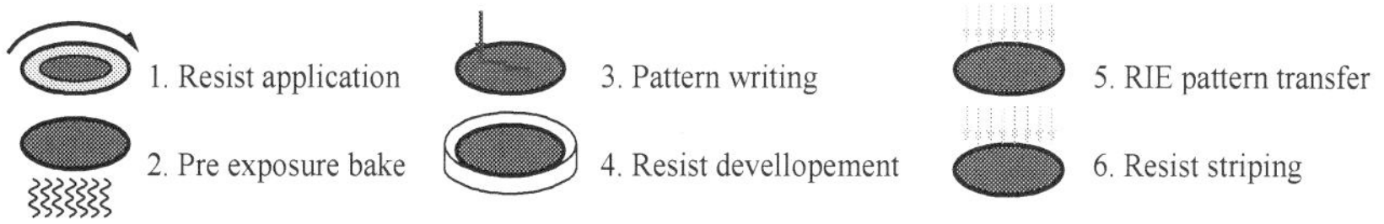

Fig. 8. Typical process flow for resist-based IBL

The resist-based IBL was developed after EBL and thus most resist materials employed in IBL were first employed for EBL and then found suitable for IBL. However, not every material suitable for EBL can be employed for IBL without restriction. Due to the different interaction of ions with resists compared to electrons, the resist properties may change significantly.

How a resist material behaves under ion beam irradiation largely depends on the form of energy deposition. Ions with a low mass to energy ratio deposit their energy mainly by electronic effects. For ions with high mass to energy, the energy deposition is mainly due to nuclear stopping (Gowa et al., 2010).

In the case of electronic energy deposition, resist materials behave similar to what is known from electron beam lithography. The energy dissipated into the resist following exposure leads to chemical damage of the polymer bonds, such as chain scission for positive resist and cross-linking in case of negative resist (Ansari et al., 2004). However, the ion resist interaction is much stronger for ions and will thus result in increased resist sensitivity. E.g., spin on glass (SOG) was found to be 500 times more sensitive to 30keV Ga ions than to electrons (Taniguchi et al., 2006).

Although most resists are more sensitive to ion irradiation than to electron irradiation, this does not imply a higher patterning speed. The practical patterning speed depends on a large number of parameters including source brightness properties of the employed optics and required resolution. Depending on the specifications, either EBL or IBL may be faster. Due to the availability of higher current densities and the variable shape beam (VSB) writers EBL usually wins this battle.

For low-energy and high-mass ions resist behavior may change severely compared to EBL. It was shown that, for low-energy Ga ions, several positive resists behave as negative resists when irradiated with a high fluency or high flux ion beam (Gowa et al., 2010). This behavior is attributed to cross-linking induced by the radicals liberated by the incident ions (Gowa et al., 2010).

At sufficient ion doses also ion implantation into the resist may be an issue, but can also be employed for patterning. Sufficiently high concentrations of Ga implanted into a resist can protect it from being developed or etched. This fact was employed in combination with a DNQ/Novolak-based resist to permit positive/negative patterning in one exposure step. Weather a feature is exposed positively or negatively is solely dependent on the ion dose (K. Arshak et al., 2004).

During IBL a significant part of the employed ions may be implanted into the substrate. This may be unacceptable in occasions where the substrate is sensitive to defects or doping, e.g., in semiconductor device fabrication. The issue may be circumvented by employing a double-layer resist system, whereby the second layer acts as an ion absorber (Hillmann, 2001).

Beside 2D patterning, resist-based 3D patterning using a FIB was also demonstrated (Hillmann, 2001) (Taniguchi et al., 2006). For this purpose, a positive resist is employed and the depth to which the resist is removed by the developer is dose dependent. Due to the absence of the proximity effect (Hillmann, 2001) this technique permits fast 3D nano patterning with high lateral resolution.

Light ions with large energies may pass thick resist materials with little deviation from their trajectory thus permitting the creation of high aspect ratio features. Due to the large impact of the ion energy and mass, the penetration depth of ions into the resist must always be considered and the resist thickness must be chosen appropriately.

At high resist sensitivities, e.g., when using chemically amplified resists (CARs) resolution and edge roughness may be limited by shot noise (Rau, 1998). Rau found that using a CAR feature printing down to an average of 7 ions was possible, but very unreliable. At an average dose of 28 ions per feature, 98% of all features were printed.

Beside classical organic resists also alternative materials are employed for FIB lithography. The change in the crystalline phase of MoO3 and WO3 (Hashimoto, 1998) was successfully used for patterning. The intermixing of the Ag2Se/GeSe2 bilayer system caused by ion bombardment proved also to be a viable patterning technique (Wagner, 1981). Self-development was shown using two materials, namely AlF3 (Gierak et al., 1997) and nitrocellulose (Harakawa, 1986).

Resist-based focused IBL was studied extensively in the 1980s (Gierak et al., 1997). FIB devices for resist-based FIB lithography were put on the market and the impact of ion irradiation on resist materials was investigated. In the end, IBL could not supersede EBL and the commercialization efforts for these specialized devices were stopped by most companies. The introduction of multi beam and high-resolution FIB systems (Elmar Platzgummer & Hans Loeschner 2009) might now lead to a renascence of resist-based FIB lithography.

4.3 3D patterning by ion implantation

During ion bombardment, ions are implanted into the irradiated substrate. Depending on their mass and energy they will come to a rest in deeper or shallower regions. If the depth in

which the ions come to rest is sufficiently narrow and if a sufficient number of ions are implanted they may form a thin layer of highly doped substrate material and locally change its chemical properties.
For the very common material system gallium on silicon a SRIM (Ziegler, 2004) simulation quickly reveals a projected range of 28 nm at 30keV ion energy. Starting from an ion dose of $2\cdot 10^{15}$ cm^{-2} (Chekurov et al., 2009), a change in the reactive ion etcher (RIE) etch speed of Si inside doped areas is noticed for fluorine-based plasmas. Only recently it was discovered that by modulating the ion dose, the time the highly doped layer is able to withstand the etching can also be modulated (Henry et al., 2010).
The dependence of the etch depth from the applied Ga dose may be employed as an effective way for 3D patterning. The process flow is illustrated in Figure 9. It consists of two steps: (i) Implantation and (ii) Pattern transfer using RIE. The key parameters for the process are: (i) The implantable Ga quantity in dependence of the scan parameters and (ii) The dependence of the etch depth on the implanted Ga quantity and on the etch parameters. For effective patterning these parameters have to be optimized.

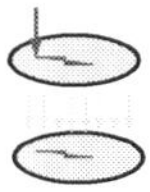

Fig. 9. Process flow

We measured the implantable Ga in dependence of the scan parameters by using EDX. Since EDX can only measure the relative Ga content, a method was needed to calculate the implanted absolute Ga quantity. We found that the dependence of the measured Ga quantity from the applied Ga quantity fitted well to the exponential function in eq. 1. Herby d_m denotes the measured Ga quantity, d_i is the applied Ga quantity and A and B are fit parameters.
Under the assumption that for low ion doses the implanted Ga dose is proportional to the applied Ga dose one can calculate the proportionality factor between measured and applied Ga dose from the fit parameters. The implanted Ga quantity in physical units D_m can thus be calculated using eq. 2.

$$d_m = Be^{Ad_i} \tag{1}$$

$$D_m = \frac{1}{AB}\, d_i \tag{2}$$

The measured Ga quantity in dependence of the applied Ga quantity for different ion energies is shown in Fig. 12. We find that at low implantation doses, the measured implantation dose is proportional to the applied Ga dose, while at higher implantation doses, the implanted Ga is also removed due to sputtering and the implanted ion dose saturates. The maximum implantable Ga quantity is dependent on the ion energy and becomes higher with increasing energy. The implantation efficiency for 30keV Ga ions is

summarized in Table 4. One will tend to maximise the implantation efficiency and thus choose sufficiently low implantation doses.
Besides the impact of the ion energy, we also measured the influence of scanning speed and ion current on the implanted dose (not shown). At doses below 100pC/μm² and 30keV ion energy we find that the effect of these parameters on the implanted Ga quantity is negligible. In practice, one will necessarily avoid higher implantation doses due to the low implantation efficiency.

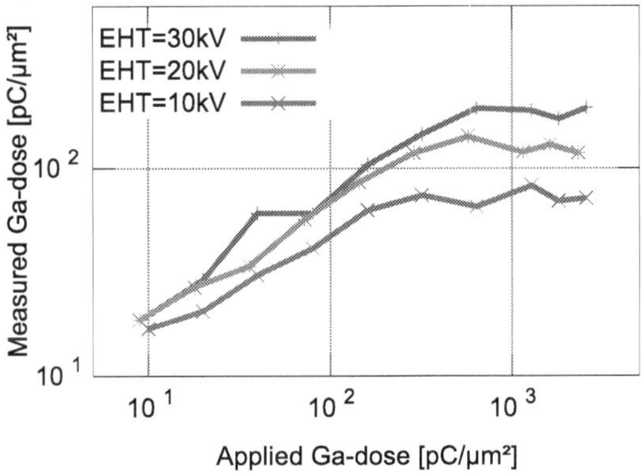

Fig. 10. Impact of the applied Ga dose on the effectively implanted ion dose. The indicated measured ion dose shown was calculated from the measured relative ion dose as described in the text.

Ion dose	Implantation efficiency
100pC/μm²	75%
200pC/μm²	60%
300pC/μm²	50%
500pC/μm²	36%

Table 4. Effect of the ion dose on the implantation efficiency at an ion energy of 30keV

For the RIE pattern transfer, the dependence of the etching depth on the applied Ga dose is of importance. We measured the dependence for three gas compositions, namely SF_6 + Ar, SF_6 + O_2 and SF_6 + $SiCl_4$. The first of these gas compositions was found to be the most interesting for 3D patterning. The SF_6 + $SiCl_4$ gas composition was found to be useful to suppress the masking effect of implanted Ga.
The resulting etch depth in dependence on the applied Ga dose for the SF_6+Ar plasma is shown in Fig. 11. We find that depending on the plasma composition the sensitivity of the etch depth on the applied Ga dose may be modulated. To optimize the overall process for speed one will tend to choose a gas composition which minimizes the required implantation dose. However, the process with the highest dose sensitivity will also exhibit the highest sensitivity against dose variations. Thus, in practice, one will have to make a trade-off between patterning speed and precise height control.

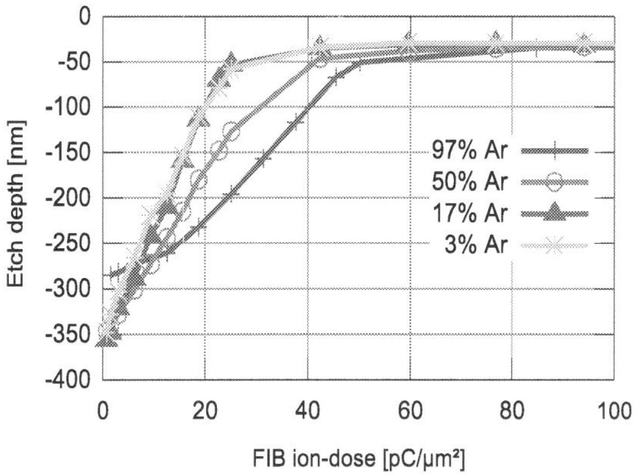

Fig. 11. Impact of the implanted ion dose on the etch depth. The etch gas was composed of Ar and SF_6.

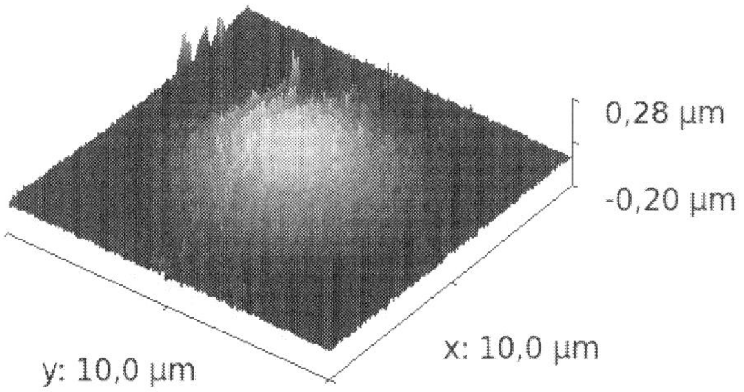

Fig. 12. AFM image of a micro lens created by ion implantation and subsequent RIE pattern transfer

Beside speed, resolution is also among the key properties of a lithographic technique. In the presented lithography technique it is mainly limited by the current distribution of the ion beam and the ion sample interaction. We find that with our Canion 31 Ga LMIS we can easily achieve line/space patterns with 50 nm HP as shown in Fig. 13.

As we have learned, in IBL the proximity effect is absent or negligible. This makes IBL very attractive for 3D nano patterning and possibly the only 3D nano patterning technique with sufficient throughput. The most popular workaround, namely EBL multilevel patterning is a good option if only a few height levels are required. However, it cannot provide real 3D patterns.

We conclude that 3D patterning by ion implantation and subsequent RIE etch is a promising patterning technique. It permits the creation of real 3D nano patterns not feasible with other methods. If combined with nano imprint lithography 3D nano patterns may be economically created and replicated.

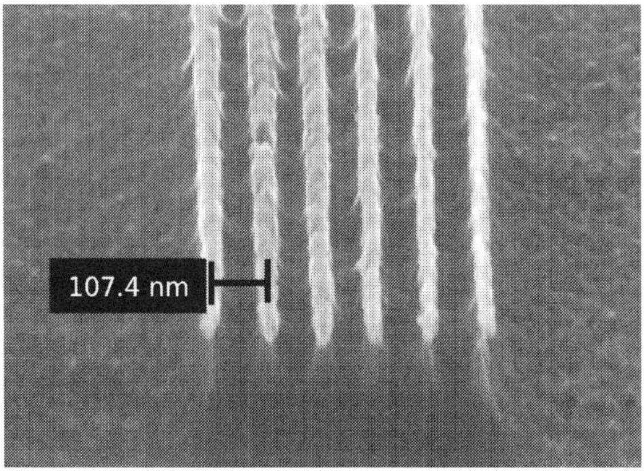

Fig. 13. 50 nm half-pitch lines/spaces. Scanning electron microscope (SEM) image of an implanted and etched lines/spaces pattern with 50 nm half-pitch.

4.4 Direct patterning of hard mask layers

Direct hard mask patterning is an alternative to resist-based patterning and direct FIB milling. To our knowledge, lithography by direct patterning of hard mask layers is a completely new technique.

Compared to resist-based methods, the employment of inorganic mask layers permits a larger flexibility in terms of patterned materials and the patterning of pre-structured substrates with relatively few patterning steps. This technique is not as flexible and straightforward as direct patterning; however, it is significantly faster. Thus, this patterning technique can be seen as a speed/complexity trade-off.

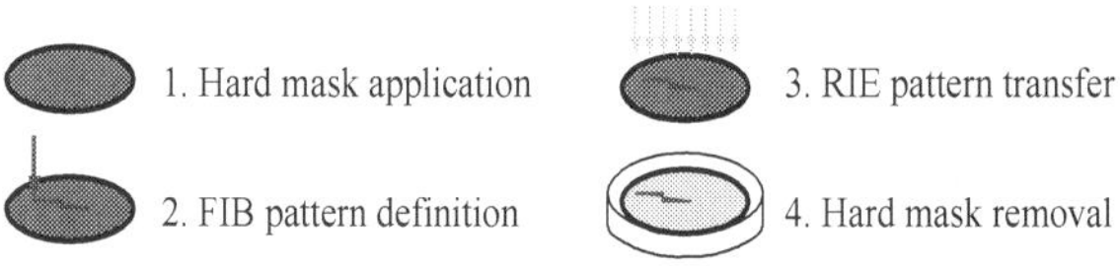

Fig. 14. Process flow for direct hard mask patterning

The direct hard mask patterning technique consists of four steps: (i) Hard mask application, (ii) Pattern definition, (iii) Pattern transfer-etch and (iv) Hard mask removal. For hard mask application standard sputtering, ALD or other methods may be employed. The hard mask removal may be performed in a dedicated etching step, e.g., by wet etching or combined with the RIE pattern transfer step.

For the pattern definition the material properties during FIB patterning are of outstanding importance. Fig. 15 shows the same pattern in two hard mask materials: AZO and Ru on Si. While the pattern in the AZO mask is clearly defined, the pattern in the Ru hard mask is heavily distorted by the surface roughening induced by the FIB milling. In this case, we believe this is due to the formation of Ru island films on the Si substrate.

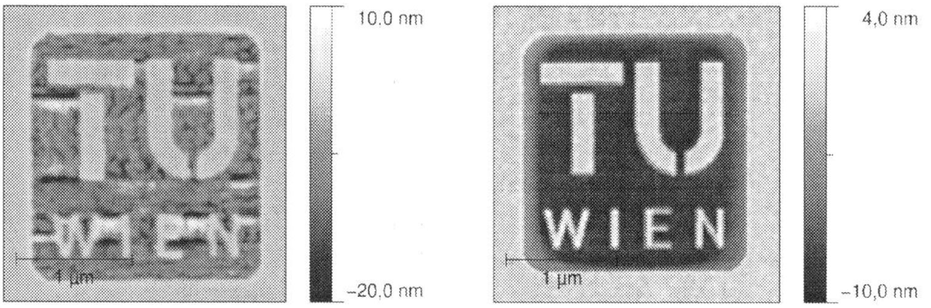

a) Ruthenium b) Aluminium doped zinc oxide (AZO)

Fig. 15. AFM images of two hard materials after pattern definition. The patterning process was identical; however, the resulting pattern quality differs significantly. The nominal hard mask thickness before patterning was 10 nm in both cases.

To quantify the impact of the FIB milling induced roughening, the roughness in dependence of the milling depth was measured for a number of materials. The resulting curves are shown in Fig. 16. Both AZO and Ta masks show excellent flatness after patterning. For Ru, the roughness is found to increase abruptly when the mask thickness approaches zero.

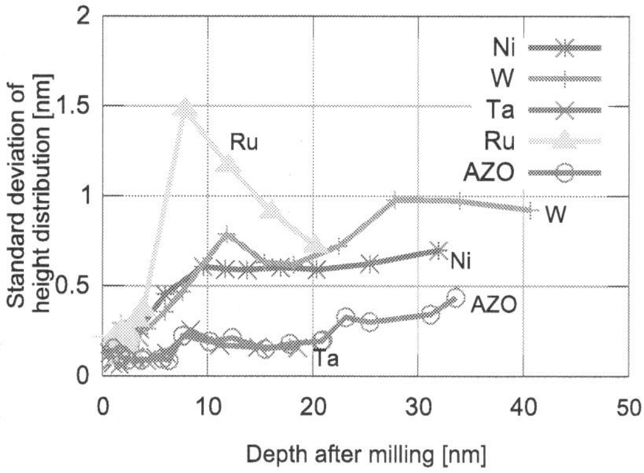

Fig. 16. Impact of ion milling on the roughness of hard mask materials. The hard mask thickness before milling was 10±2 nm in all cases.

For the RIE transfer-etch it is important to choose both the hard mask material and etching chemistry appropriately. Further, it must be considered that during FIB milling the employed ion species is implanted into the substrate and can disturb the etching process. The etching chemistry must etch the substrate material and the implanted ions but must not remove the hard mask material.

Ga implanted into an Si substrate acts as an etch stop for the routinely employed fluorine-based RIE chemistries. We found that our routinely employed Si etch recipes could not be employed for the transfer-etch due to the masking effect of the implanted Ga. Chlorine is known to be an etchant for Ga, thus we investigated the impact of the addition of a chlorine source to a well-established fluorine-based RIE recipe.

The impact of the addition of chlorine on the masking capability of Ga implanted into Si is shown in Fig. 17. While for low $SiCl_4$ concentration the Ga still shows a significant masking effect, we find that the addition of 30% $SiCl_4$ is sufficient to completely suppress the impact of the implanted Ga on the final pattern.

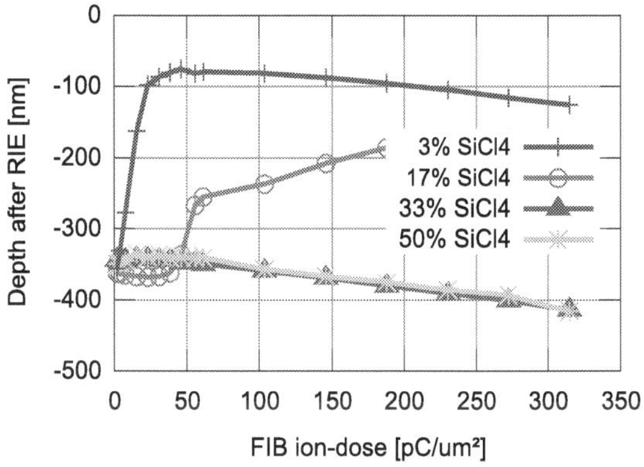

Fig. 17. Impact of the addition of $SiCl_4$ to the masking capability of Ga implanted by FIB into Si in a SF_6-based RIE process

The addition of chlorine solves the issue of Ga implantation. However, it negatively impacts the available pool of hard mask materials. Now the hard mask material must not only resist etching by a standard fluorine-based RIE process but also to the added chlorine species. As shown above, Ta exhibits excellent properties during the pattern definition process. However, it does not resist our chlorine-containing etch recipe and thus cannot be employed.

In terms of resolution, the direct hard mask patterning process compares very favourably to FIB direct milling. Since the pattern in the thin hard mask layer is transferred into a thicker layer, slopes with low steepness are imaged into slopes with higher steepness. Thus patterning close to the beam diameter becomes possible. Also the influence of the beam tails and mechanical deformation due to milling-induced strain is minimized. We found that lines down to 40 nm half pith are obtainable in a 10 nm thick Ni hard mask layer.

We conclude that the direct patterning of hard mask materials and subsequent pattern transfer by RIE can help to speed up patterning compared to direct FIB milling. We believe that this method is useful for a number of applications including patterning on uneven surfaces.

5. Conclusion

Ion beam lithography is a versatile technique with several variations of the process. Single focused ion beams of Ga ions have been successfully used for exposure of resist layers, but more common are direct milling or beam-induced deposition or etching of the material. Also implantation of the ions to pattern surfaces has been demonstrated as a powerful structuring approach. With the emergence of helium ion beams, a new tool with new structuring capabilities is on the market and allows new applications. In this work also several structuring approaches have been discussed including (i) FIB milling (ii) FIB-induced gas-assisted processes (iii) 3D patterning by ion implantation and (iv) patterning my milling of hard mask layers. Due to the versatility of these approaches an increasing amount of applications of IBL for optical systems, sensor devices, in the modification and custom-trimming of microelectronic circuitry as well as in the 'classical' fields of photomask repair and defect analysis of cross-sections may be expected. With the neon ion beam systems on the verge of commercial introduction, this is surely going to remain an exciting field of research. With the multi-beam ion systems reaching maturity, interest in IBL from the side of industrial fabrication can also be expected in the future.

6. References

Ansari, K. et al., 2004. Fabrication of high aspect ratio 100 nm metallic stamps for nanoimprint lithography using proton beam writing.

Arshak, K. et al., 2004. A novel focused-ion-beam lithography process for sub-100 nanometer technology nodes. *Superlattices and Microstructures*, 36(1-3), pp.335-343.

Bell, D.C. et al., 2009. Precision cutting and patterning of graphene with helium ions. *Nanotechnology*, 20(45), p.455301.

Boit, C. et al., 2008. Physical Techniques for Chip-Backside IC Debug in Nanotechnologies. *Design & Test of Computers, IEEE*, 25(3), pp.250-257.

Calcagno, L., Compagnini, G. & Foti, G., 1992. Structural modification of polymer films by ion irradiation. *Nuclear Inst. and Methods in Physics Research, B*, 65(1-4), pp.413-422.

Chekurov, N. et al., 2009. The fabrication of silicon nanostructures by local gallium implantation and cryogenic deep reactive ion etching. *Nanotechnology*, 20(6).

Choi, J.O. et al., 1988. Degradation of poly(methylmethacrylate) by deep ultraviolet, x-ray, electron beam, and proton beam irradiations. *Journal of Vacuum Science & Technology B: Microelectronics and Nanometer Structures*, 6(6), p.2286.

Dietz, L.A. & Sheffield, J.C., 1975. Secondary electron emission induced by 5-30-keV monatomic ions striking thin oxide films, Journal of Applied Physics, 46(10), 4361.

Driesel, W., Dietzsch, C. & Muhle, R., 1996. In situ observation of the tip shape of AuGe liquid alloy ion sources using a high voltage transmission electron microscope, JVST B, 14(5), 3367.

Dworetsky, S. et al., 1968. Electron Bombardment Ion Source for Low Energy Beams. *Review of Scientific Instruments*, 39(11), p.1721.

Eckstein, W., 1991. *Computer Simulation of Ion-Solid Interactions* 1st ed., Springer.

Giannuzzi, L.A. & Stevie, F.A., 1999. A review of focused ion beam milling techniques for TEM specimen preparation. *Micron*, 30(3), pp.197-204.

Gierak, J. et al., 1997. Focused ion beam nanolithography on AlF_3 at a 10 nm scale. *Applied Physics Letters*, 70(15), p.2049.

Gowa, T. et al., 2010. Ion beam irradiation effects on resist materials. *Journal of Photopolymer Science and Technology*, 23(3), pp.399-404.

Harakawa, K., 1986. Focused ion beam etching of resist materials. *Journal of Vacuum Science & Technology B: Microelectronics and Nanometer Structures*, 4(1), p.355.

Hashimoto, M., 1998. Application of dual-functional MoO_3/WO_3 bilayer resists to focused ion beam nanolithography. *Journal of Vacuum Science & Technology B: Microelectronics and Nanometer Structures*, 16(5), p.2767.

Henry, M.D. et al., 2010. Ga+ beam lithography for nanoscale silicon reactive ion etching. *Nanotechnology*, 21(24).

Hill, R. & Faridur Rahman, F.H.M., 2010. Advances in helium ion microscopy. *Nuclear Instruments and Methods in Physics Research Section A: Accelerators, Spectrometers, Detectors and Associated Equipment.*

Hillmann, M., 2001. *3-dimensionale Nanostrukturierung mittels fokussierter Ionenstrahllithographie,*

Hirscher, S. et al., 2002. Ion projection lithography below 70 nm: tool performance and resist process. *Microelectronic Engineering*, 61-62, pp.301-307.

Horiuchi, K., Itakura, T. & Ishikawa, H., 1988. Fine pattern lithography using a helium field ion source. *Journal of Vacuum Science & Technology B: Microelectronics and Nanometer Structures*, 6(1), p.241.

Hoyle, P., 1994. *Focused electron beam induced chemistry and its application in microelectronics.* University of Cambridge.

Jeon, J., Floresca, H.C. & Kim, M.J., 2010. Fabrication of complex three-dimensional nanostructures using focused ion beam and nanomanipulation. *Journal of Vacuum Science & Technology B: Microelectronics and Nanometer Structures*, 28(3), p.549.

van Kan, Jeroen A., Bettiol, Andrew A. & Watt, Frank, 2006. Proton Beam Writing of Three-Dimensional Nanostructures in Hydrogen Silsesquioxane. *Nano Letters*, 6(3), pp.579-582.

Koeck, A. et al., 2010. Argon ion multibeam nanopatterning of Ni–Cu inserts for injection molding. *Journal of Vacuum Science & Technology B: Microelectronics and Nanometer Structures*, 28(6), p.C6B1.

Kubena, R.L. et al., 1989. Sub-20-nm-wide line fabrication in poly(methylmethacrylate) using a Ga^+ microprobe. *Journal of Vacuum Science & Technology B: Microelectronics and Nanometer Structures*, 7(6), p.1798.

Lee, H.-Y. & Chung, H.-B., 1998. Ga+ focused-ion-beam exposure and CF4 reactive-ion-etching development of Si_3N_4 resist optimized by Monte Carlo simulation, JVST B, 16(3), 1161.

Livengood, R.H. et al., 2011. The neon gas field ion source--a first characterization of neon nanomachining properties. *Nuclear Instruments and Methods in Physics Research Section A: Accelerators, Spectrometers, Detectors and Associated Equipment*, In Press, Corrected Proof.

Loeschner, Hans, Klein, C. & Platzgummer, Elmar, 2010. Projection Charged Particle Nanolithography and Nanopatterning. *Japanese Journal of Applied Physics*, 49(6), p.06GE01.

Matsui, S. et al., 2000. Three-dimensional nanostructure fabrication by focused-ion-beam chemical vapor deposition, JVST B, 18(6), 3181.

Melngailis, J., 1993. Focused ion beam lithography. *Nuclear Inst. and Methods in Physics Research, B*, 80-81(PART 2), pp.1271-1280.

Muehlberger, M. et al., 2011. Nanoimprint lithography from CHARPAN Tool exposed master stamps with 12.5 nm hp. *Microelectronic Engineering*, In Press, Corrected Proof.

Namatsu, H. et al., 1998. Nano-patterning of a hydrogen silsesquioxane resist with reduced linewidth fluctuations. *Microelectronic Engineering*, 41-42, pp.331-334.

Notte, J., 4th et al., 2010. Diffraction imaging in a He+ ion beam scanning transmission microscope. *Microscopy and Microanalysis: The Official Journal of Microscopy Society of America, Microbeam Analysis Society, Microscopical Society of Canada*, 16(5), pp.599-603.

Orloff, J., Swanson, L.W. & Utlaut, M., 1996. Fundamental limits to imaging resolution for focused ion beams. *Journal of Vacuum Science & Technology B: Microelectronics and Nanometer Structures*, 14(6), p.3759.

Orloff, J., Swanson, L. & Utlaut, Mark, 2002. *High Resolution Focused Ion Beams: FIB and Applications* 1st ed., Springer.

Persson, A., Thornell, G. & Nguyen, H., 2010. Rapid prototyping of magnetic tunnel junctions with focused ion beam processes. *Journal of Micromechanics and Microengineering*, 20(5).

Platzgummer, Elmar & Loeschner, Hans, 2009. Charged particle nanopatterning. *Journal of Vacuum Science & Technology B: Microelectronics and Nanometer Structures*, 27(6), p.2707.

Rau, N., 1998. Shot-noise and edge roughness effects in resists patterned at 10 nm exposure. *Journal of Vacuum Science & Technology B: Microelectronics and Nanometer Structures*, 16(6), p.3784.

Reyntjens, S. & Puers, R., 2001. A review of focused ion beam applications in microsystem technology. *Journal of Micromechanics and Microengineering*, 11(4), pp.287-300.

Schmidt, B., Bischoff, L. & Teichert, J., 1997. Writing FIB implantation and subsequent anisotropic wet chemical etching for fabrication of 3D structures in silicon. *Sensors and Actuators A: Physical*, 61(1-3), pp.369-373.

Sidorkin, V. et al., 2009. Sub-10-nm nanolithography with a scanning helium beam. *Journal of Vacuum Science & Technology B: Microelectronics and Nanometer Structures*, 27(4), p.L18.

Taniguchi, J. et al., 2006. Rapid and three-dimensional nanoimprint template fabrication technology using focused ion beam lithography. *Microelectronic Engineering*, 83(4-9), pp.940-943.

Tseng, A.A., 2004. Recent developments in micromilling using focused ion beam technology. *Journal of Micromechanics and Microengineering*, 14(4), p.R15-R34.

Tseng, A.A., 2005. Recent Developments in Nanofabrication Using Focused Ion Beams. *Small*, 1(10), pp.924-939.

Utke, I., Hoffmann, P. & Melngailis, John, 2008. Gas-assisted focused electron beam and ion beam processing and fabrication. *Journal of Vacuum Science & Technology B: Microelectronics and Nanometer Structures*, 26(4), p.1197.

Wagner, A., 1981. Germanium selenide: A resist for low-energy ion beam lithography. *Journal of Vacuum Science and Technology*, 19(4), p.1363.

Ward, B.W., Notte, J.A. & Economou, N.P., 2006. Helium ion microscope: A new tool for nanoscale microscopy and metrology. *Journal of Vacuum Science & Technology B: Microelectronics and Nanometer Structures*, 24(6), p.2871.

Winston, D. et al., 2009. Scanning-helium-ion-beam lithography with hydrogen silsesquioxane resist. *Journal of Vacuum Science and Technology B: Microelectronics and Nanometer Structures*, 27(6), pp.2702-2706.

Yasaka, A. et al., 2008. Application of vector scanning in focused ion beam photomask repair system. *Journal of Vacuum Science and Technology B: Microelectronics and Nanometer Structures*, 26(6), pp.2127-2130.

Ziegler, J.F., 2004. SRIM-2003. *Nuclear Instruments and Methods in Physics Research, Section B: Beam Interactions with Materials and Atoms*, 219-220(1-4), pp.1027-1036.

3

Atom Lithography: Fabricating Arrays of Silicon Microstructures Using Self-Assembled Monolayer Resist and Metastable Helium Beam

Jianwu Zhang, Zhongping Wang and Zengming Zhang
Department of Physics, University of Science and Technology of China, Hefei
The Centre of Physical Experiments, University of Science and Technology of China, Hefei
P.R. China

1. Introduction

Lithography is a top-down technology to produce sub-micrometer feature sizes for the fundamental research and widespread applications. The conventional optical lithography has been developed to not only make the tiny electronic devices to form the basic circuits in today's computer chips, but also fabricate micro/nano electromechanical systems (MEMS/NEMS). However, the conventional photolithographic techniques have now reached their ultimate spatial resolution because of both the light diffraction and the photoresist restrictions [1]. Alternative techniques for traditional photolithography have been proposed and developed to meet the requirements of a high spatial resolution of sub-100 nm as well as the capability to fabricate an arbitrary structure and achieve mass production [1-10]. Research teams around the word have been investigating alternative techniques including extreme ultraviolet lithography, electron and ion beam lithography, X-ray lithography and atom lithography. Each technique has its own advantages and disadvantages, and no one yet knows which one will be the method of choice for the next generation lithography (NGL).

The arrays of micro- and nanostructures fabricated on silicon substrates have now attracted more attention for its interesting characteristics in applications such as photonics, electronics, optoelectronics and sensing [11-15]. The array properties can be tuned further by varying the geometry, the doping, the periodicity, and the size of the micro- and nanostructures [11,12]. These controllable and elaborate arrays are commonly fabricated by conventional optical lithography, X-ray lithography, electron-beam or ion-beam lithography [1,16-18]. Unfortunately, there are inherent limitations to pattern over large areas at nanoscale using these lithographical techniques due to the light diffraction, the long-range inter-particle interactions and the proximity effects. To overcome these limitations, it's necessary to develop new alternative techniques with a shorter wavelength and a thinner resist for traditional optical lithography.

Among the alternative techniques for traditional photolithography, atom lithography based on metastable atoms beam (MAB) and self-assembled monolayers (SAMs) has shown significant potential in fabricating arrays of micro- and nanostructures, which is a major goal in nanoscience and nanotechnology [10,19-29]. Metastable atom is one of the atom's electron is excited from ground state to metastable state. In metastable state, instead of immediately decay to ground state, the electron will stay for some time (long compare to usuall short lifetime-excited state) before it decays to its ground state. For example, the metastable helium atoms have natural lifetimes ≥ 20ms, which is much longer than its typical flight times. The energy stored in metastable noble gas atom can be used to create pattern in a SAM that act as resit. When a metastable noble gas atom strikes the SAM resist, the atom will release energy and it will go to ground state. The energy released will change the characteristic of the SAM resist in radius few angstroms from place where the atom strikes. The atom in ground state will not have any effect toward SAM resist because noble gas atoms are not chemically reactive. The locally change of characteristic of the SAM resist will enable us to do wet chemical etching. This new fabrication technique, which bridges the gap between the bottom-up chemical self-assembly techniques and the modern top-down lithography, can overcome the intrinsic resolution limitations of traditional photolithographic techniques [10,19,20]. In principle, the atom lithography based on neutral MAB and SAM ideally meets the required conditions for sub-100nm fabrication [10]. MAB with a de Broglie wavelength of less than 0.1nm eliminates diffractive resolution limitations, and an ultrathin organic SAM resist with a thickness of 1~3nm ensures the sharpness of the edge profile within the resist. Moreover, the neutral metastable atoms are insensitive to the electric and magnetic fields, and the long-range inter-particle interactions are weak. Therefore, this novel method permits the direct and large area manufacturing of micro- and nanostructures on a silicon wafer, avoiding some inherent complications of electron-beam, ion-beam or photolithography. With these unique advantages, the atom lithography with neutral MAB and high-resolution SAM resist makes it possible to achieve nanolithography and provides a potential way in manufacturing structures at a large-scale based on micro- and nanoscale features.

During the past fifteen years, considerable attention has also been given to the investigation of the mechanism of metastable atom lithography and microfabrication with different resolutions on various substrates, such as gold, silicon, silicon oxide, and mica [10,19-29]. Direct evidence of the emission of H^+ and CH_x^+ from the SAM on the Au(111) surface under the irradiation of a metastable helium atom beam (He*-MAB) at thermal energy supports the interpretation that the SAM is damaged through the interaction between the outermost surface of the SAM and the metastable atoms. This damage can cause a characteristic surface change from hydrophobic to hydrophilic or a molecular structural change from condensed to cross-linked, which consequentially affect the etching process in the fabrication of the micro- or nanostructure, resulting in the positive and negative pattern transfers, respectively. The typical micropatterning with sub-100 nm resolution on a silicon substrate covered with a SiO_2 or Au layer has been achieved successfully [10,21,22,28,29]. Most of these studies are conducted through multistep etching processes using the cover layers as intermediate masks. For example, in the He*-MAB lithography with dodecanethiol-SAM [10,21,22], a Au/Ti coating layer is used to create an intermediate mask on silicon substrate by the first etching process, and then, the intermediate mask pattern is transferred to silicon substrate by the second etching process. It is very interesting to directly transfer negative and positive patterns of SAM induced by He*-MAB into a silicon substrate without metal/oxide coating layer to fabricate arrays of micro- and nanostructures.

Recently, the He*-MAB lithography of Si with SAM, which formed directly on silicon substrate by direct covalent linkage instead of metal/oxide coating layers, has been accomplished by our research team [19,20]. The latent image formed in SAM by He*-MAB passing through a stencil is directly transferred into the underlaying Si substrate by KOH etching, which is a single step etching process. Undoubtedly, to realize and understand this direct transfer process has opened a novel way in the practical application of the atom lithography in micro- and nanofabrication of silicon, especially in arrays of micro- and nanostructures of silicon.

In this context, we summarized our experimental results obtained during the near past years in detail in the field of atom lithography, where organosilane SAMs and He*-MAB were used to pattern the surface of Si(111)/Si(110)/Si(100) wafer substrates without coating intermediate layer. The principle and procedure of atom lithography used in our experiments were introduced firstly. Then the experimental process and achievements were given. Finally, the problems and perspectives about this new technique were also discussed in brief.

2. Principle and procedure

2.1 The general principle and procedure

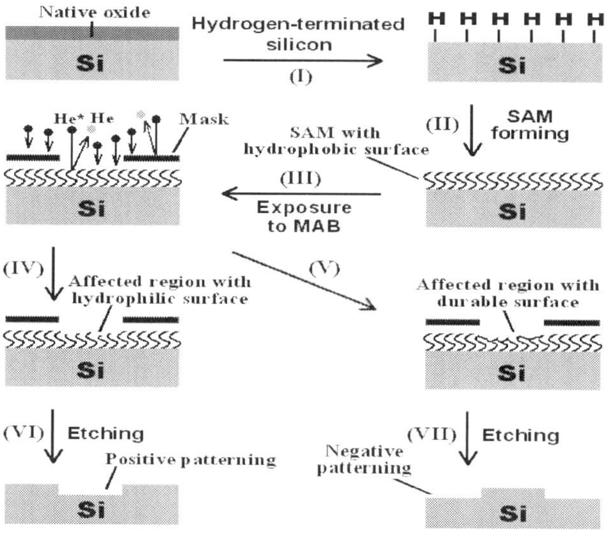

Sheme 1. Shematic illustration of the principle and procedure for the frabrication of periodic arrays of Si(111)/Si(100) by He*-MAB lithography(excited helium atom:He*; ground-state helium atom: He).

The general principle and process of atom lithography using He*-MAB and organosilane SAMs to create the arrays of silicon micro- and nanostructures on a Si substrate were illustrated in Scheme 1. By introducing a hydrogenation process into the pretreatment of silicon substrates (as described in (I) and (II) in Scheme 1), fine organosilanes SAMs was successfully formed as a resist on silicon surface under a controlled argon gas atmosphere at elevated temperature. The chemical modification of non-oxidized silicon surfaces utilizing monolayers could be achieved by neutralizing the silicon surfaces with alkyl chains through direct covalent linkage, i.e., silicon-carbon [30-32]. This provided an opportunity to directly use an ultrathin hydrophobic SAM with a thickness of 1~3nm as a super resist on silicon surfaces. The samples with the physical stencil masks were exposed to the irradiation of He*-MAB at thermal energy, and the SAM was damaged through the interaction between the outermost surface of the SAM and the metastable atoms (as described in (III) in Scheme 1). When the metastable atoms hit the outermost surface of the SAM on Si substrate, they transferred their internal energy (~20eV) to the chemisorbed hydrocarbons of SAM molecules and caused chemical changes that might result in either the formation of a durable crossing-linking polymerization material or the loss of hydrophobicity of the SAM (as indicated in (IV) and (V) in Scheme 1). After exposure, the pattern was transferred into a latent pattern in SAM. The SAM exhibited the positive-tone and negative-tone sensitivity due to largely different dosage of MAB and the length of the alkyl chains of SAM molecule. The negative pattern implied the polymerization of SAM molecules induced by more He*-MAB irradiation dose. Meanwhile, the positive pattern indicated the hydrophilic region of SAM induced by less He*-MAB irradiation dose. Furthermore, the longer alkyl chains were easier to undergo cross-linking polymerization than the shorter alkyl chains under the He*-MAB irradiation. Consequently, these could affect the etching process in the fabrication of the micro- or nanostructure (i.e., the hydrophilic exposure region was easier to etch using KOH solution than the hydrophobic unexposed region, whereas the cross-linking polymerization region was more durable to KOH etching than the hydrophobic unexposed region), resulting in the positive and negative pattern transfers (as described in (VI) and (VII) in Scheme 1), respectively.

The mechanism for the contrast inversion, i.e., the competition between the loss of hydrophobicity and the cross-linking polymerization of SAM molecules determined the polarity of the pattern transfer process, has been confirmed previously [19,20]. In our experiments, the lithographic patterns were obtained in samples with a relatively large area of about 2 cm^2. In these areas, the mask pattern was reproduced with high fidelity, and the repetition of lithographic patterns was consistent in different runs. In order to obtain better contrast patterns, we carefully optimized the experimental parameters. After etching, the resulting arrays fabricated on the silicon substrate were clearly observed under the optical microscope.

2.2 Mechanism of forming SAMs on Si substrates

The mechanism of forming SAMs on Si substrates was illustrated in details in Scheme 2. The p-type Si(111)/Si(110)/Si(100) wafers with a native oxide layer (p-type Si(111): Φ=100mm, thickness=380μm, ρ=10~20 Ω·cm; n-type Si(110): Φ=100mm, thickness=300μm, ρ=1~5 Ω·cm; p-type Si(100): Φ=100mm, thickness=300μm, ρ=10~20 Ω·cm) were used and cut into pieces of 10mm×20mm as Si substrates. The Si substrates were first rinsed using toluene, ethanol, and deionized water sequentially, then immersed into a 2vol.% aqueous HF solution for 5~10min to remove the native oxide layer and hydrogenate the silicon substrates, and finally dried. Protected from air by a flowing argon gas, the Si substrates were immersed into a 10~40mM organosilanes (octadecyltrichlorosilane, 18C-long-chain hydrocarbon molecules, ODTS; dodecyltrichlorosilane, 12C-long-chain hydrocarbon molecules, DDTS; hexycyltrichlorosilane, 6C-long-chain hydrocarbon molecules, HTS) toluene solution at room temperature for 24~50 hours, and then rinsed in toluene and dried at 100°C for 30~80min. Thus, through direct covalent linkage, i.e., silicon-carbon, fine organosilanes

SAMs were formed as a resist on silicon surface by introducing a hydrogenation process into the pretreatment of silicon wafer and a controlled argon atmosphere at elevated temperature [19,20,30].

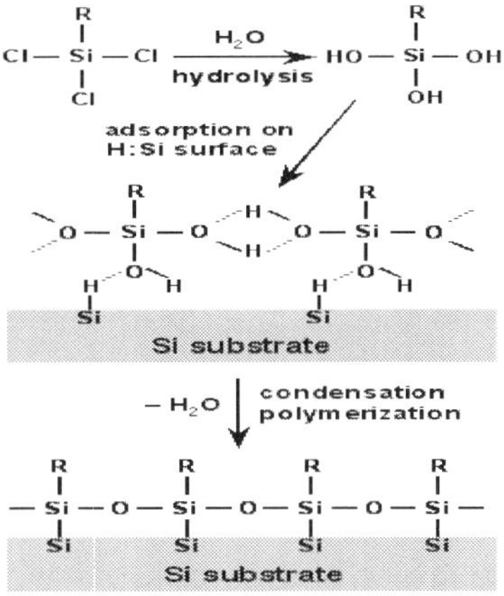

Sheme 2. The illustration of the mechanisam of forming SAMs on Si substrates.

2.3 Exposure to MAB

The experimental setup for the sample exposure to He*-MAB was depicted in Scheme 3. The Si substrate with organosilane SAMs was covered by a physical mask with TEM Cu grids. As shown in the inset of Scheme 3, six TEM grids were well distributed on a nickel sheet (with a thickness of 0.1mm and a square area of 1cm×2cm) and fixed over the opening holes with a diameter of 2mm using Ag conductive paste. The assembly was immediately inserted into the sample chamber of the MAB source (as shown in Scheme 3) and exposed to He*-MAB for 5~120 min, producing latent patterns in organosilanes SAM on Si substrates. The exposure experiments were performed with the existing apparatus depicted in Scheme 3. The He* source was operated in a dc discharge mode, and a stable pressure (~5×10^{-7} Pa for the ultimate pressure and ~3×10^{-2} Pa for the working pressure) of chambers was retained by turbo-molecule pump (T.M.P) vacuum systems. A transverse deflecting voltage of 600~800 V was applied along the beam line to eliminate ions or electrons in the beam, which otherwise damage the SAM during exposure. Through a skimmer opening of ~0.7 mm diameter, He*-MAB was directed toward a rotatable thin tantalum (Ta) sheet sample holder at a distance of 26.5 cm. Three tantalum (Ta) plates were arranged triangularly around the rotation axis (shown in the inset of Scheme 3). Two of them were used for installing samples, and the remaining one was used for monitoring the intensity of the primary beam. The thickness of the nickel sheet defined the gap between physical masks and sample surfaces (~100um), avoiding the mechanical contact that may cause possible artifacts of SAMs. Since the mask almost covered the entire sample, the metastable flux was represented by the metastable-atom-induced electron emission from the Ta plate or from the sample and mask, which was the main content of the induced current. The typical values of the measured current on both the Ta plate and sample were 100 and 50 nA, which corresponded to an electron emission rate of ~1×10^{12} and ~2×10^{12} s^{-1} cm^{-2}, respectively.

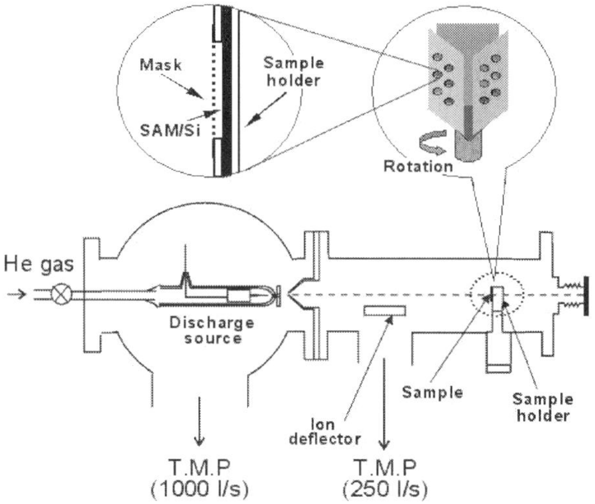

Sheme 3. Shematic setup of the sample exposure to MAB for atom lithography.

During exposure He*-MAB impacted the surface of SAM formed on silicon substrate after the beam was patterned in transverse direction by the TEM grids. Atoms with kinetic energy smaller than 1eV do not pass freely through materials. So, the physical mask (i.e., TEM grids) with thickness in the order of sub-micrometers was enough to pattern the He*-MAB. Our experimental results showed good reproducibility with a high fidelity of the patterns of the TEM copper grids and TEM holey carbon grids (shown in figure 1), as well as a good repetition of the lithographic patterns in different runs.

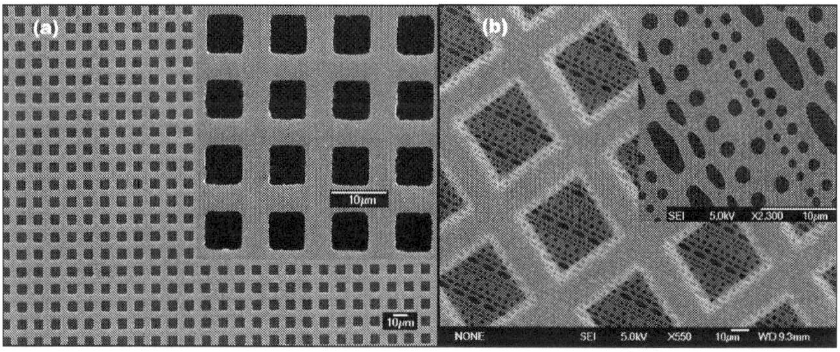

Fig. 1. (a) SEM image of TEM Cu grid used as a mask (Note: the pitch size of patterns is 12.5 um, 5.5 um for wires, and 7.0 um for spaces) and (b) SEM image of TEM holey carbon grid used as a mask (Note: The diameters of the circular holes are about 1, 1.4, and 2μm, the bar widths range from 0.5 to 4μm, the oval holes with a dimension of 8x2μm and 4x1 μm).

Two kinds of physical masks (i.e., TEM Cu grids and TEM holey carbon grids) were used in our experiments. The SEM images and parameters of the TEM grids were shown in figure 1.

The bar widths and the hole diameters define the pitch sizes of the patterns. Even though the image of a pattern was in micrometer range, a sharp edge in the tens-of-nanometers range for the resulting microstructures of silicon indicated a potential of nanosized pattern in our experiments. In this sense, actually, the spatial resolution of the patterning in our experiments was not ascribed to TEM grid mask but to the edge resolution.

2.4 Etching processing and pattern transferring

After exposure, the samples with latent patterns in SAM were taken out from the vacuum chamber and rapidly dipped into an etching solution in which a magnetic bar was used for stirring, and were etched for 5~60min at room temperature in an aqueous solution of 0.1M KOH, rinsed by deionized water, and then dried in air. By this single step KOH etching process, the latent image formed in SAM by He*-MAB passing through a stencil was directly transferred into the underlaying Si substrate (as described in (VI) and (VII) in Scheme 1).

For the wet etching of Si substrate in KOH solution, the anisotropy of etching rate of silicon have been studied in detail over the passed 20 years. Especially, in the case of the photolithography in MEMS and IC industry, the anisotropy of etching rate of silicon has been investigated completely, and the anisotropy can induce obvious differences of the spatial resolution of patterning of Si along different crystal orientation. The etching rates for the silicon wafer in the KOH solution are known to depend strongly on its orientation ($Si+2OH^{-}+2H_2O \rightarrow SiO_2(OH)_2^{2-}+2H_{2(g)}$; etching ratio: {110}>{100}>>{111}) [33]. In our experiments, the anisotropic etching process of silicon occurred in the same way, and strongly affected the patterning results including edge step resolution. The sharpest step edge resolution was only 41nm with 1 pixel and the average step edge resolution was around 100 nm.

2.5 Characterization

AFM images of the samples were taken by a multimode Nanoscope IV atomic force microscope (Veeco Metrology Group, Santa Barbara, CA), operated at ambient conditions. For all images we recorded the retrace direction of the tip using a scan angle of 0° or 90°. Substrates decorated with organosilanes SAMs were imaged in tapping mode using silicon cantilevers (NanoWorld, Neuchâtel, Switzerland) with a spring constant of 42 N m^{-1}. All images were recorded at a rate of 1.0 Hz, and with a pixel resolution of 512. SEM inspection of patterned Si substrates and the masks were carried out with a JSM-6500/SG scanning electron microscope operated at 30 keV.

3. Experimental achievements

Up to now, we have developed a process successfully, and realized a He*-MAB lithography of Si(111)/Si(110)/Si(100) wafer substrates with SAMs, where the SAMs was formed not on metal/oxide coating layers but directly on silicon substrate. The process of atom lithography using He*-MAB and organosilane SAMs to create the arrays of silicon micro- and nanostructures on a Si substrate were illustrated in Scheme 1. The first step was the formation of very fine organosilanes (ODTS, DDTS and HTS) SAMs as a resist on Si(111)/Si(110)/Si(100) surface by introducing a hydrogenation process into the pretreatment of silicon wafer and a controlled argon gas atmosphere at elevated temperature. After the development of micro stencil masks for He*-MAB exposure to generate latent patterns in SAMs was carried out, the chemical etching process to transfer the latent patterns into the underlying silicon substrate was examined and optimized. By adjusting experimental parameters (i.e., different masks, exposure time, etching time, etc.), various arrays of silicon microstructures were fabricated successfully. The resulting arrays of silicon microstructures on Si substrates were characterized by the AFM/SEM.

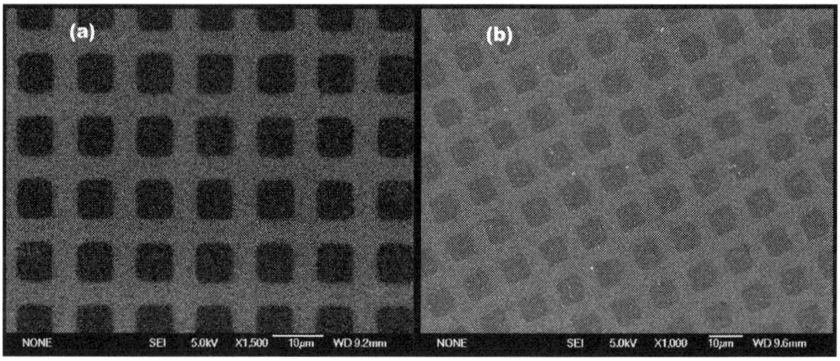

Fig. 2. (a) SEM image of negative patterning (exposure time: 60min; etching time: 5min) and (b) positive patterning (exposure time: 20min; etching time: 5min) on Si(110) substrate with the ODTS SAM and DDTS SAM, respectively. Mask: TEM Cu grid.

The SEM images of two samples were displayed in figure 2. The pattern from the stencil mask (as shown in figure 1(a)) was transferred into the Si(110) substrate, and the resulting micromesa/microwell arrays of silicon fabricated on the Si(110) substrate were very intact and clear (as shown in figure 2). The preliminary SEM observations revealed that the direct fabrication of the array on the surfaces of Si(110) substrates by He*-MAB and SAMs was viable, and our experimental results indicated that the patterning also appeared on either kind of surface of the Si(111) or Si(100) substrate with various organosilane SAM mentioned above. The main experimental results and discussions were given as following.

3.1 Arrays of Si(111) microstructures

The AFM images with the sectional analysis of the square micromesa arrays formed on the Si(111) substrate with ODTS SAM were shown in figure 3, which corresponds to negative patterning with an exposure time of 40 min and an etching time of 60 min. A square silicon micromesa arrays with a size of ~7×7 μm^2 and a nanoscale edge resolution of ~100 nm was successfully fabricated. The height of the step was ~30 nm, indicating an etching rate of 0.5 nm/min. The width of the step edge was 1-3 pixels (41~123 nm). The sharpest step edge had a width of 1 pixel (41 nm) and a height of 22 nm, indicating a steep sidewall with an aspect ratio of 2:1, which was considered to be the achievable limit in the present case. In fact, after subtracting the effects of the tip shape on the lateral resolution induced by the AFM tip with a ~10 nm radius of curvature at its end, the real step width should be smaller than the value mentioned above, and the aspect ratio should be larger. The cross sections were not flat, and the rms roughness of the top area of the micromesa and the pedestal base area were 6.2 nm and 5.8 nm, respectively. This relatively rough characteristic could be ascribed to the imperfect SAM formation (i.e., the density of SAM molecules anchored on the Si(111) surface through chemical bonds is not perfectly uniform side by side) on the silicon substrate, which induced a local fluctuation in the etching rate of transferring the latent image of the SAM resist onto the underlying silicon substrate and, consequently, produced a rough surface. Flatter surfaces was obtained using a perfect SAM and under optimal etch conditions. Another characteristic was shown in figure 3, i.e., an upward fringe around the edge of the top of the square micromesas. The etching rates for the silicon wafer in the KOH solution depended strongly on its orientation [33]. Thus, the etching parallel to the Si(111) surface proceeded much faster than that perpendicular to the Si(111) surface and undercuts the sidewalls of a micromesa capped with a durable SAM resist that withstands the etching by KOH solution, leaving a fringe around the micromesa. After the sample was dried, the fringe rolls up or down around the edge of the silicon micromesa. The upward fringe was clearly observed in the 3D image as shown in figure 3(b), suggesting the strong durability of the exposed SAM resist against chemical etching.

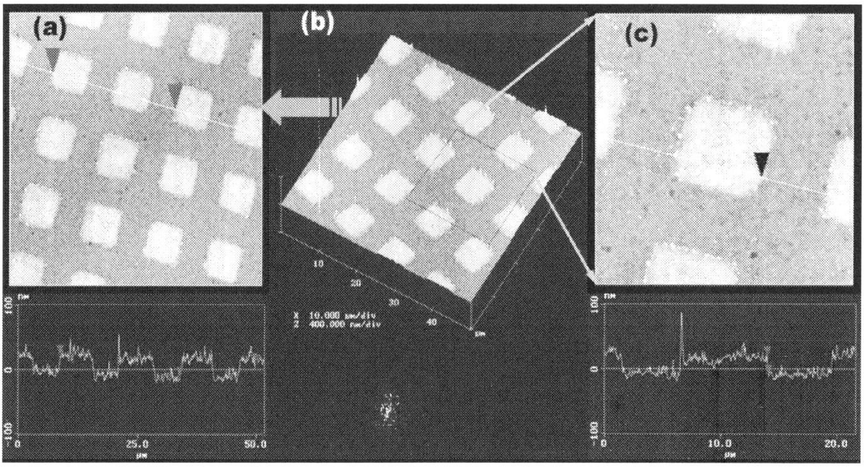

Fig. 3. Tapping-mode AFM images with sectional analysis of square silicon micromesa array created on Si(111) substrate (Negative patterning, exposure time: 40min; etching time: 60min, SAM: ODTS) (a): 50×50 μm^2; (b) 3D image; (c): 21.7×21.7 μm^2).

We also obtained a positive pattern within the exposed region with another TEM grid window near the border of the irradiation area. The border region accepted considerably sparser irradiation of He-MAB than the center of He-MAB. Consequently, this exposed border region only changed from hydrophobic to hydrophilic without cross-linking between molecules, which leaded to a weak resistance to chemical etching. Figure 4 showed AFM images and the sectional analysis of the microwell array formed on the Si(111) substrate, corresponding to the positive patterning. A square silicon microwell array with a size of ~7×7 μm^2 and a depth of ~10 nm was formed. The topography and line profile of the microwell array indicated a much rougher surface with an rms roughness of ~7 nm. In comparison with the depth of ~10 nm, this large rms value suggested that it was difficult to achieve a perfect positive pattern transfer using the ODTS SAM resist, possibly because the hydrophilic area might not be much weaker in resisting the etchant than the unexposed hydrophobic area.

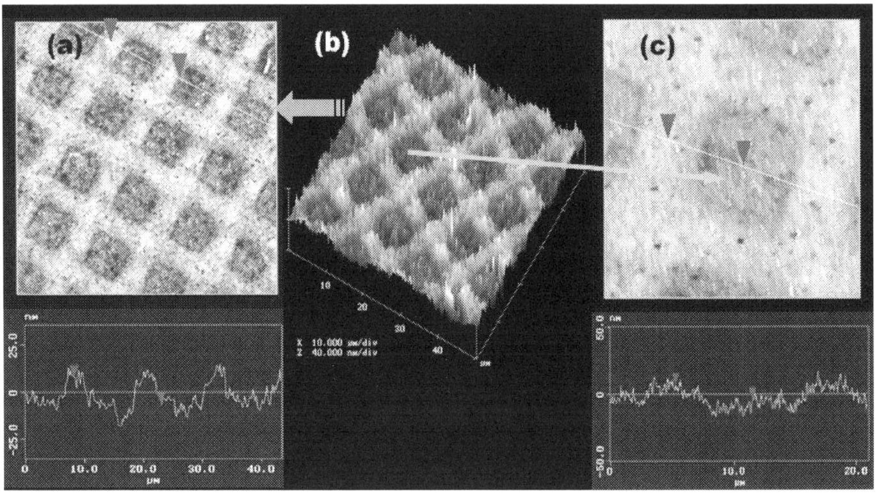

Fig. 4. Tapping-mode AFM images with sectional analysis of square silicon microwell array created on Si(111) substrate (Negative patterning, exposure time: 40min; etching time: 60min, SAM: ODTS) (a): 50×50 μm^2; (b) 3D image; (c): 21.7×21.7 μm^2).

Since the ODTS SAM resist comprised highly oriented 18C-long-chain hydrocarbon molecules chemically bonded to the silicon substrate, its surface became hydrophobic [30-32]. Previous studies revealed that, under irradiation of He-MAB with a 20 eV internal energy, C-H bond scission and C-C bond scission occured and a C=C double bond could be formed at the outermost surface of the SAM [34-37]. At low doses, breaking of the C-H bond caused H loss, leaving a C-rich film that could be oxidized in air. Consequently, the exposure region was changed from hydrophobic to hydrophilic, leading to a positive pattern [38,39]. At high doses, further degradation of the hydrocarbon molecules caused fragmentation of the carbon chain, and cross-linking polymerization between adjacent fragmented SAM molecules might occur, giving rise to the formation of a highly resistant carbonaceous layer and leading to a negative pattern [36]. In addition, the contamination resist, which was induced by hydrocarbon molecules remaining on the inner wall of the vacuum chambers, remained a concern [40-41]. When the metastable atoms impacted the surface, they transferred their internal energy to these physisorbed hydrocarbon molecules and induced a polymerization change within the exposed region. To clarify whether these resist materials could remain on the surface even after immersion in a KOH solution, a clean Si(111) sample without an SAM was used. Only a trace of negative contrast was observed, indicating a negligible contamination effect. In our negative pattern case, the exposure region of the SAM resist became etching-resistant, which might be mainly ascribed to the cross-linking polymerization of the SAM molecules.

3.2 Arrays of Si(110) microstructures

The AFM images of the square micromesa arrays formed on the Si(110) substrate with ODTS SAM was shown in figure 5, which corresponded to negative patterning with an exposure time of 20 min and an etching time of 5 min. A square silicon micromesa array with a micromesa size of ~7×7 μm^2 (e.g., the size of one mesh of TEM stencil mask) was fabricated successfully on the substrate. The average height of the step was 27 nm, indicating an etching rate at 5 nm/min. The sharpest step edge had a width of 1 pixel (41 nm) and a height of 26 nm, indicating a steep sidewall with an aspect ratio of 1:1.7. The sectional analysis shown in Figure 5(a) indicated that the rms roughness of the top area of the micromesa and the pedestal base area were 2 nm and 15 nm, respectively. The flat top area of the micromesas could be observed clearly, suggesting the strong durability of the exposed SAM resist against chemical etching. The flaw pedestal base area of the micromesa was also very conspicuous, and this rough characteristic could be attributed to the imperfect SAM formation on the silicon substrate, which induced a local fluctuation in the etching rate of transferring the latent image in SAM resist onto the underlying silicon substrate and, consequently, produced a rough surface. As described in scheme 1, the negative-tone sensitivity of the SAM might be mainly due to the cross-linking polymerization of ODTS molecules induced by the irradiation of the He*-MAB [19,20]. In addition, it was noteworthy that the etching time had a notable influence on the production of high-quality patterning. In our practical runs, the following etching times were chosen: 5, 10, and 20 min with identical exposure times. The sectional analysis of AFM images of these samples indicated that the rms roughness of top area of the micromesa in the case of overetching was bigger than that in the case of underetching.

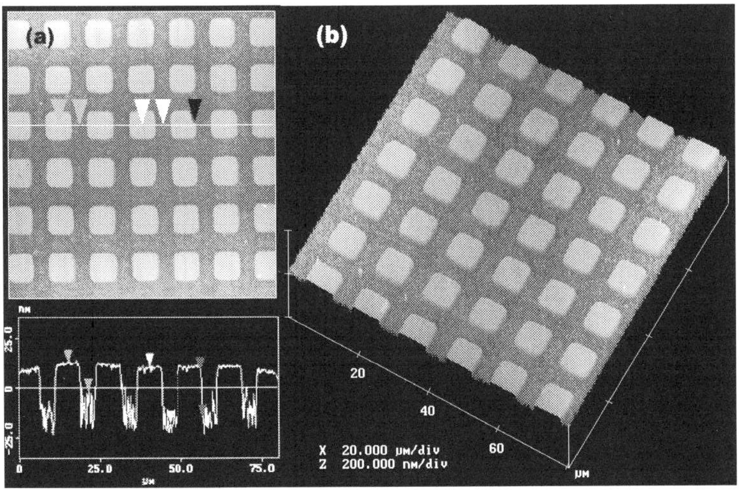

Fig. 5. Tapping-mode AFM images of periodic arrays of square silicon micromesa fabricated on Si(110) substrate with ODTS SAM (exposure time: 20min, etching time: 5min): (a) AFM image with sectional analysis of the arrays of silicon micromesas with a height of about 26 nm ; (b) 3D views of the arrays of silicon micromesa over a 80 um×80 um area and with a periodicity of 12.5 um.

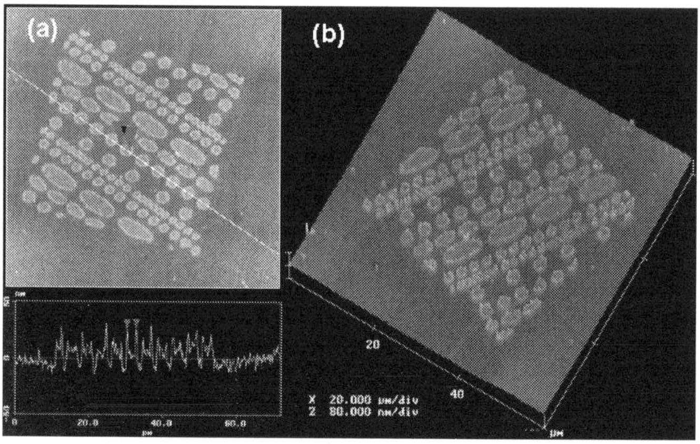

Fig. 6. Tapping-mode AFM images with sectional analysis of periodic arrays of square silicon microwell fabricated on Si(110) substrate with DDTS SAM (exposure time: 40min, etching time: 10min): (a) AFM images with sectional analysis of arrays of micromesas with a height of about 27 nm ; (b) 3D views of the arrays of silicon micromesas over a 60 um×60 um area.

When a TEM holey carbon grid was used as a mask, an interesting positive patterning result was obtained. The AFM images of the microstructures fabricated on the Si(110) substrate with DDTS SAM were displayed in figure 6, which corresponding to the positive patterning (exposure time: 30 min; etching time: 10 min). The rough micromesa with different diameters can be observed clearly in figure 6(a) and 6(b).

3.3 Arrays of Si(100) microstructures

Compared with the negative patterning shown in figure 5, a positive patterning could be obtained by adjusting the dosage of He*-MAB irradiation and the length of alkyl chains of SAM molecule. The AFM images of the square microwell arrays fabricated on the Si(100) substrate with DDTS SAM was shown in figure 7, corresponding to the positive patterning (exposure time: 20 min; etching time: 10 min). A square microwell arrays with a microwell size of ~7×7 μm^2 was fabricated on the Si(100) substrate. The depth of the microwells was about 220 nm. The sharpest wall edge had a width of 1 pixel (41 nm) and a depth of 100 nm, indicating a steep sidewall with an aspect ratio of 2.5:1. The sectional analysis shown in figure 7(a) indicated that the rms roughness of the bottom area of the microwell and the top area of the wall were 31 nm and 22 nm, respectively. The rough characteristic could also be attributed to the imperfect SAM formation on the silicon substrate, as mentioned in 3.2. The positive-tone sensitivity of the SAM should be mainly due to the loss of hydrophobicity of DDTS SAM (with shorter alkyl chains, 12C) induced by the irradiation of the metastable helium atoms. The topography and line profile of the microwells array indicated a much rougher top surface of the sidewall with an rms roughness of 22 nm. This large rms value suggested that the hydrophobic area might be much weaker in resisting the KOH etchant, and the DDTS SAM might not be a good resist for atom lithography. The corresponding 3D views of the array wae given in figure 7(b). Additionally, we also used TEM holey carbon grid as a mask to obtain a positive pattern result. The AFM images of the microstructures fabricated on the Si(100) substrate with HTS SAM was shown in figure 8, corresponding to the positive patterning. The rough microwell with different diameters can also be observed clearly in figure 8(a) and 8(b).

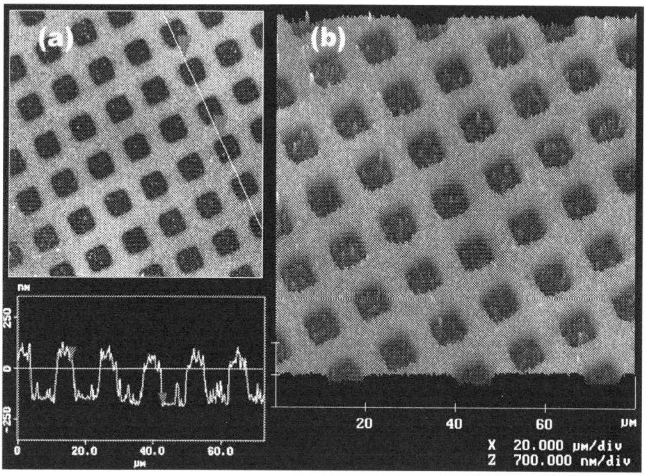

Fig. 7. Tapping-mode AFM images with sectional analysis of periodic arrays of square silicon microwell fabricated on Si(100) substrate with DDTS SAM (exposure time: 20min, etching time: 10min): (a) AFM images with sectional analysis of arrays of square microwell with a depth of about 220 nm; (b) 3D views of the arrays of silicon microwell over a 80 um×80 um area and with a periodicity of 12.5 um.

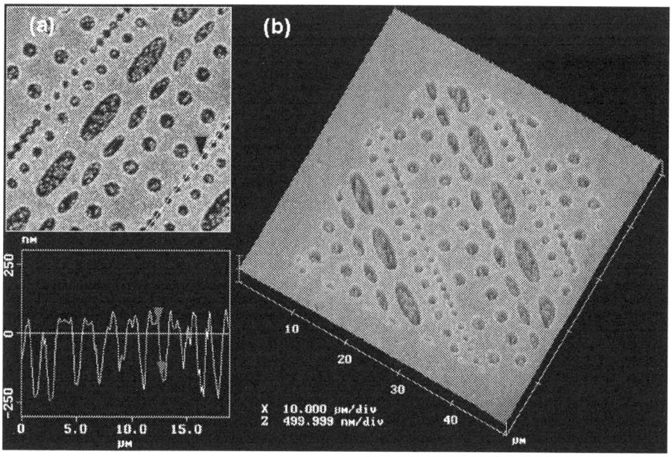

Fig. 8. Tapping-mode AFM images with sectional analysis of periodic arrays of square silicon microwell fabricated on Si(100) substrate with HTS SAM (exposure time: 30min, etching time: 10min): (a) AFM images with sectional analysis of arrays of square microwell with a depth of about 220 nm; (b) 3D views of the arrays of silicon microwell over a 50 um×50 um area.

The anisotropy of etching rate of silicon induced obvious difference of the spatial resolution of patterning of Si along different crystal orientation. In our experiments, the anisotropic etching rate seemed to improve the edge resolution, and the anisotropic etching process of silicon strongly affected the patterning results including the edge step resolution. It was easier to obtain a large aspect ratio and a better resolution for Si(110) rather than for Si(100). The sharpest step edge resolution obtained presently in Si(110) samples was only 41nm with 1 pixel and the step edge resolution of about 120 nm, indicating an aspect ratio of 3:1. In principle, the estimated value of geometrical blurring was ~240 nm with the geometry described in scheme 3, and the spatial resolution was improved by reducing the distance between the source and the sample, which was related to the divergence of the metastable helium beam. The achievable width in our experiments was much smaller than the calculated one because of the hardening effect in the process of chemical etching. If the experimental parameters in both the exposure and etching processes (e.g., the exposure time, etching time, etc.) were optimized further, higher resolution of the patterning onto the silicon substrates could be obtained.

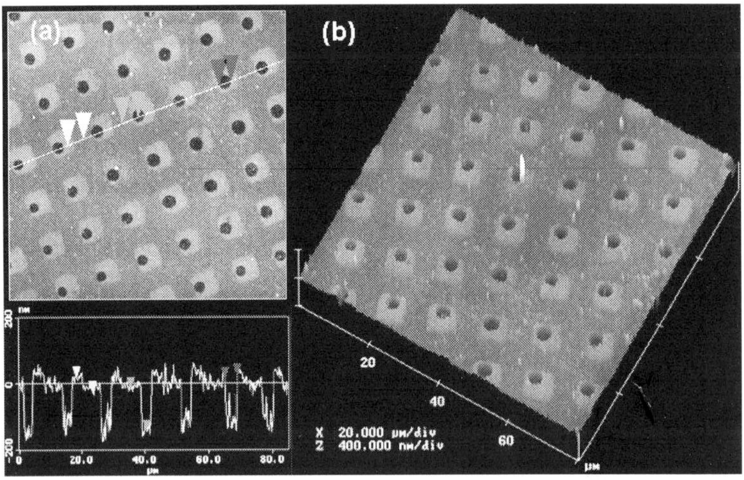

Fig. 9. Tapping-mode AFM images with sectional analysis of periodic arrays of square silicon microwell fabricated on Si(100) substrate with DDTS SAM (exposure time: 20min, etching time: 10min): (a) AFM images with sectional analysis of arrays of square microwell with a depth of about 180 nm; (b) 3D views of the arrays of silicon microwell over a 80 um×80 um area and with a periodicity of 12.5 um.

Athough the interaction between the outermost surface of the SAM and the irradiation of metstable atoms has been investigated for about ten years [34-41], the detailed change in surface chemistry of the outermost of the SAM molecules under the irradiation of helium atom beam is still difficult to predict and remains to be explored. What we demonstrated in the present study was that both the positive-tone and the negative-tone sensitivity of the SAM were related to largely different dosage of MAB and the length of the alkyl chains of SAM molecule (as illustrated in scheme 1 (IV) and (V)). When we optimized the experimental parameters, we accidentally obtained a novel patterning with both negative and positive pattern on one Si(100) substrate with DDTS SAM (as discribed in figure 9), which was very different from previous results. In this case, a novel array of square micromesas with microholes in it was fabricated successfully. The mechanism to induce this dual patterning is remained to be explored at present. This dual array also indicated some interesting flexibility of atom lithography in micro- and nanofabrication.

4. Problems and perspectives

Atom lithography based on He*-MAB and SAMs has been demonstrated to have significant potential in fabricating arrays of micro- and nanostructures. This new fabrication technique has opened a novel way in the practical application of the atom lithography in micro- and nanofabrication of silicon. Atom lithography using He*-MAB and SAMs to pattern the surface of silicon wafer without coating intermediate layer to create the arrays of silicon microstructures has been realized successfully. In order to improve the spatial resolution of the patterning, new etching method instead of KOH wet etching method need to be developed to meet the requirments of nanopaterning. At present, though the image of a pattern is in micrometers range due to TEM grids, a sharp edge in the tens-of-nanometers range indicates a potential of submicrometer-sized patterning. The new masks with nanopattern instead of present TEM masks should be used in order to generate nanostructure arrays. The mechanism of positive and negative patterning on silicon wafer has been investigated. Our results suggest that both the positive-tone and the negative-tone sensitivity of the SAM are related to largely different dosage of MAB and the length of the alkyl chains of SAM molecule. More experiments need to be carried out in order to conclude the detailed parameters to alter patterning type.

In the future, further studies on this subject can be listed as following:

1. To prepare more perfect SAMs for atom lithography;
2. To make new mask with sub-100nm pattern for patterning transfer;
3. To improve the precision to position and adjust the mask during exposing;
4. To investigate both other substrates (e.g. GaAs) and other etching processing (e.g. dry etching) for atom lithography.

5. Acknowledgement

The author gratefully acknowledges support from the National Natural Science Foundation of China (Grant No. 20777072, 91023040), the Fundamental Research Funds for the Central Universities (Grant No. WK2030020018) and would like to thank Prof. Yasushi Yamauchi, Dr. Mitsunori Kurahashi for their help with experiments and their helpful comments. The author also notes that part of the work was completed while he was at the NIMS.

6. References

[1] Xia, Y.; Rogers, J. A.; Paul, K. E. and Whitesides, G. M. (1999). Unconventional methods for fabricating and patterning nanostructures. *Chem. Rev.* 99 1823-1848.

[2] Campbell, C. J.; Smoukov, S. K. M.; Bishop, K. J.; Baker, E. and Grzybowski, B. A. (2006). Direct printing of 3D and curvilinear micrometer-sized architectures into solid substrates with sub-micrometer resolution. *Adv. Mater.* 18 2004-2008.

[3] Ahn, S. J.; Kaholek, M.; Lee, W. K.; LaMattina, B.; LaBean, T. H. and Zauscher, S. (2004). Surface-Initiated Polymerization on Nanopatterns fabricated by electron-beam lithography. *Adv. Mater.* 16 2141-2145
[4] Wu, N.; Pease, L. F. and Russel, W. B. (2006). Toward large-scale alignment of electrohydrodynamic patterning of Thin Polymer Films. *Adv. Funct. Mater.* 16 1992-1999.
[5] Palmer, A. J.; Baker, M. and Sang, R. T. (2006). Towards creation of iron nanodots using metastable atom lithography. *Nanotechnol.* 17 1166-1170.
[6] Sawada, S.; Masuda, Y.; Zhu, P. and Koumoto, K. (2006). Micropatterning of copper on a poly(ethylene terephthalate) substrate modified with a self-assembled monolayer. *Langmuir* 22 332-337.
[7] Harnett, C. K.; Satyalakshmi, K. M. and Craighead, H. G. (2000). Low-energy electron-beam patterning of amine-functionalized self-assembled monolayers. *Appl. Phys. Lett.* 76 2466-2468.
[8] Kan, J. A. V.; Bettiol, A. A. and Watt, F. (2006). Proton beam writing of 3D nanostructures in hydrogen silses quioxane. *Nano Lett.* 6 579-582.
[9] Saito, N.; Kadoya, Y.; Hayashi, K.; Sugimura, H. and Takai, O. (2003). Micropatterned 1-alkene self-assembled monolayer on hydrogen-terminiated silicon by vacuum ultraviolet lithography. *Jpn. J. Appl. Phys.* 42 2534-2537.
[10] Berggren, K. K.; Bard, A.; Wilbur, J. L.; Gillaspy, J. D.; Helg, A. G.; McClelland, J. J.; Rolston, S. L.; Phillips, W. D.; Prentiss, M. and Whitesides, G. M. (1995). Microlithography using neutral metastable atoms and self- assembled monolayers. *Science* 269 1255-1257.
[11] Dong, B.; Lu, N.; Zelsmann, M.; Kehagias, N.; Fuchs, H.; Torres, C. M. S. and Chi, L. (2006). Fabrication of high-density, large-area conducting-polymer nanostructures. *Adv. Funct. Mater.* 16 1937-1942.
[12] Jang, J. H.; Ullal, C. K.; Maldovan, M.; Gorishnyy, T.; Kooi, S.; Koh, C. Y. and Thomas, E. L. (2007). 3D micro- and nanostructures via interference lithography. *Adv. Mater.* 17 3027-3041.
[13] Henzie, J.; Lee, M. H. and Odom, T. W. (2007). Multiscale patterning of plasmonic metamaterials. *Nature Nanotech.* 2 549-554.
[14] Tao, A.; Sinsermsuksakul, P. and Yang, P. D. (2006). Tunable plasmonic lattices of silver nanocrystals. *Nature Nanotech.* 2 435-440.
[15] Lei, Y.; Cai, W. P. and Wilde, G. (2007). Highly ordered nanostructures with tunable size, shape and properties: A new way to surface nano-patterning using ultra-thin alumina masks. *Progress in Materials Science* 52 465-539.
[16] Shin, S. W.; Lee, S. G.; Lee, J.; Whang, C. N.; Lee, J. H.; Choi, I. H.; Kim, T. G. and Song, J. H. (2005). Ion-beam nano-patterning by using porous anodic alumina as a mask. *Nanotechnol.* 16 1392-1395.
[17] Tseng, A. A. (2005). Recent developments in nanofabrication using focused ion beams. *Small* 1 924-939.
[18] Carr, D. W.; Lercel, M. J.; Whelan, C. S.; Craighead, H. G.; Seshadri, K. and Allara, D. L. (1997). High-selectivity pattern transfer processes for self-assembled monolayer electron-beam resists. *J. Vac. Sci. Technol.* B 15 1446-1450.

[19] Zhang, J. W.; Kurahashi, M.; Suzuki, T.; Sun, X. and Yamauchi, Y. (2006). Microfabrication of silicon using self-assembled monolayer resist and metastable helium beam. *Jpn. J. Appl. Phys.* 45 8020-8023.

[20] Zhang, J. W. (2009). Fabricating arrays of Si(110)/Si(100) microstructures by atom lithography using organosilane self-assembled monolayers. *J. Phys.: Conf. Ser.* 188 012008

[21] Hill, S. B.; Haich, C. A.; Dunning, F. B.; Walters, G. K.; McClell, J. J.; Celotta, R. J. and Craighead, H. G. (1999). Patterning of hydrogen-passivated Si(100) using Ar(3P0,2) metastable atoms. *Appl. Phys. Lett.* 74 2239-2241.

[22] Younkin, R.; Berggren, K. K.; Johnson, K. S.; Prentiss, M.; Ralph, D. C. and Whitesides, G. M. (1997). Nanostructure fabrication in silicon using cesium to pattern a self-assembled monolayer. *Appl. Phys. Lett.* 71 1261-1263.

[23] Lu, W.; Baldwin, K. G. H.; Hoogerland, M. D.; Buckman, S. J.; Senden, T. J.; Sheridan, T. E. and Boswell, R. W. (1998). Sharp edged silicon structures generated using atom lithography with metastable helium atoms. *J. Vac. Sci. Technol.* B 16 3846-3849.

[24] O'Dwyer, C.; Gay, G.; Viaris, de Lesegno B.; Weiner, J.; Camposeo, A.; Tantussi, F.; Fuso, F.; Allegrini, M. and Arimondo, E. (2005). Atomic nanolithography patterning of submicron features: writing an organic self-assembled monolayer with cold, bright Cs atoms beams. *Nanotechnol.* 16 1536-1541.

[25] Ju, X.; Kurahashi, M.; Suzuki, T. and Yamauchi, Y. (2004). Positive and negative patterning of ethanethiol, decanethiol, and hexadecanethiol self-assembled monolayers by using a metastabl helium beam. *Thin Solid Films* 464 420-424.

[26] Wang Z. P.; Kurahashi M.; Suzuki T.; Ding Z. J. and Yamauchi Y. (2010). Silicon micro/nanofabrication using metastable helium atom beam lithography. *J. Nanosci. Nanotechnol.* 10 7443-7446.

[27] Ju, X.; Kurahashi, M.; Suzuki, T. and Yamauchi, Y. (2005). Fabrication of gold patterns with nanoscale edge by using heptanethiol self-assembled monolayer and metastable helium beam. *Appl. Surf. Sci.* 241 241-245.

[28] Brezger, B.; Schulze, T.; Drodofsky, U.; Stuhler, J.; Nowak, S.; Pfau, T. and Mlynek, J. (1997). Nanolithography with neutral chromium and helium atoms. *J. Vac. Sci. Technol.* B 15 2905-2911.

[29] Hill, S. B.; Haich, C. A.; Dunning, F. B.; Walters, G. K.; McClelland, J. J.; Celotta, R. J.; Craighead, H. G.; Han, J. and Tanenbaum, D. M. (1999). Patterning of octadecylsiloxane self-assembled monolayers on Si(100) using Ar($^3P_{0,2}$) atoms. *J. Vac. Sci. Technol.* B 17 1087-1089.

[30] Kulkarni, S. A.; Mirji, S. A.; Mandale, A. B.; Gupta, R. P. and Vijayamohanan, K. P. (2005). Growth kinetics and thermodynamic stability of octadecyltrichlorosilane self-assembled monolayer on Si (100) substrate. *Mater. Lett.* 59 3890-3895.

[31] Shirahata, N.; Masuda, Y.; Yonezawa, T. and Koumoto, K. (2004). Atomic scale flattening of organosilane self-assembled monolayer and patterned tin hydroxide thin films. Eur. Ceram. Soc. 24 427-434.

[32] Heid, S.; Effenberger, F.; Bierbaum K. and Grunze M. (1996). Self-assembled mono- and multilayers of terminally functionalized organosilyl compounds on Silicon substrates. *Langmuir* 12 2118-2120.

[33] Sato, K.; Shikida, M.; Matsushima, Y.; Yamashiro, T.; Asaumi, K.; Iriye, Y. and Yamamoto, M. (1998). Characterization of orientation-dependent etching properties of single-crystal silicon: effects of KOH concentration. *Sens. and Actuators* A 64 87-93.

[34] Yamauchi, Y.; Noro, T.; Kurahashi, M.; Suzuki, T. and Ju, X. (2005). Metastable-atom-stimulated Desorption from Dodecanethiolate Self-assembled Monolayer. *Appl. Surf. Sci.* 241 141-145.

[35] Setoyama, H.; Kera, S.; Okudaira, K. K.; Hara, M.; Harada, Y. and Ueno, N. (2003). Outermost surface reactions of molecular thin films induced by metastable-atom impacts. *Jpn. J. Appl. Phys.* 42 597-601.

[36] Thywissen, J. H.; Johnson, K. S.; Dekker, N. H.; Prentiss, M.; Wong, S. S.; Weiss, K. and Grunze, M. (1998). Metastable-atom-activated growth of an ultrathin carbonaceous resist for reactive ion etching of SiO_2 and Si_3N_4. *J. Vac. Sci. Technol.* B 16 1155-1160.

[37] Close, J. D.; Baldwin, K. G. H.; Hoffmann, K. and Quaas, N. (2000). Fragmentation of dodecanethiol molecules: application to self-assembled monolayer damage in atom lithography. *Appl. Phys.* B 70 651-655.

[38] Nowak, S.; Pfau, T. and Mlynek, J. (1996). Nanolithography with metastable helium. *Appl. Phys.* B 63 203-205.

[39] Johnson, K. S.; Thywissen, J. H.; Dekker, N. H.; Berggren, K. K.; Chu, A. P.; Younkin, R. and Prentiss, M. (1998). Localization of metastable atom beams with optical standing waves: nanolithography at the heisenberg limit. *Science* 280 1583-1586.

[40] Johnson, K. S.; Berggren, K. K.; Black, A. T.; Chu, A. P.; Dekker, N. H.; Ralph, D. C.; Thywissen, J. H.; Younkin, R.; Thinkham, M.; Prentiss, M. and Whitesides, G. M. (1996). Using neutral metastable argon atoms and contamination lithography to form nanostructure in silicon,silicon dioxide,and gold. *Appl. Phys. Lett.* 69 2773-2775.

[41] Rehse, S. J.; Glueck, A. D.; Lee, S. A.; Goulakov, A. B.; Menoni, C. S.; Ralph, D. C.; Johnson, K. S. and Prentiss, M. (1997). Nanolithography with metastable neon atoms:Enhanced rate of contamination resist formation for nanostructure fabrication. *Appl. Phys. Lett.* 71 1427-1429.

4

Character Projection Lithography for Application-Specific Integrated Circuits

Makoto Sugihara
Kyushu University
Japan

1. Introduction

In the recent fabrication of semiconductor devices, quite various devices are produced while most of them result in small production volumes. A small production volume of ICs leads to a rise of the price of an IC because the expensive investment made in its photomask set must be redeemed by passing on the price. The price of photomasks increases rapidly as the transistor integration advances. The price of photomasks has a great impact on the price of semiconductor devices.

Electron beam direct writing (EBDW) is a solution to fabricating small-lot ICs at a cheap cost. The EBDW can draw patterns onto silicon wafers masklessly or quasi-masklessly (Inanami, 2000; Pfeiffer, 1979). The throughput of the conventional EBDW equipment which adopts the variable shaped beam (VSB) method (Pfeiffer, 1978) is, however, extremely low. In the VSB method, exposed patterns are divided into a large number of small rectangular and triangular shapes to draw them as shown in the left of Fig. 1. In this figure Letter "E" is divided into four rectangles and consequently needs four "EB shots" to be drawn. The conventional VSB equipment *shoots* rectangular and triangular shapes onto silicon wafers and results in a lot of EB shots, which deteriorate the throughput of the equipment.

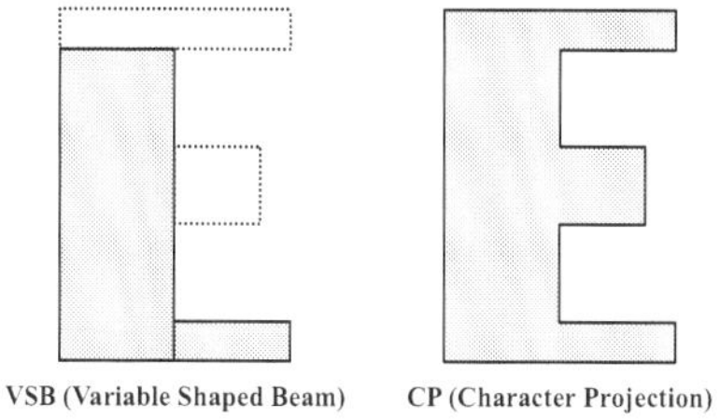

Fig. 1. VSB and CP lithographies

Character projection (CP) lithography is a promising one in which a pattern more complex than a triangle or a rectangle, called a character, is projected onto a silicon wafer with an EB shot as shown in the right of Fig. 1 (Sakitani et al., 1992; Hattori et al., 1993; Hirumi et al., 2003; Inanami et al., 2000; Inanami et al., 2003; Nakamura et al., 2006; Nakasugi et al., 2003).

The e-BEAM Corporation developed a low-energy electron beam direct writing (LEEBDW) system, which was named "EBIS" (Electron Beam Integrated System) (Inanami et al., 2000;

Inanami et al., 2003; Nakamura et al., 2006; Nakasugi et al., 2003). The system can accommodate 400 characters on a CP aperture mask and any character can be chosen at every EB shot, so that the throughput of the system can be enhanced quite effectively with the CP lithography. The projection system can also project rectangular and triangular shapes with the VSB lithography. Their system is capable of projecting patterns with both the VSB and CP lithographies.

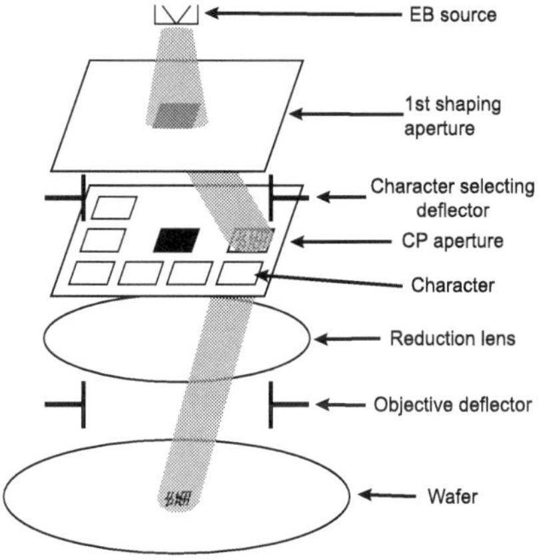

Fig. 2. Single-column-cell character projection equipment

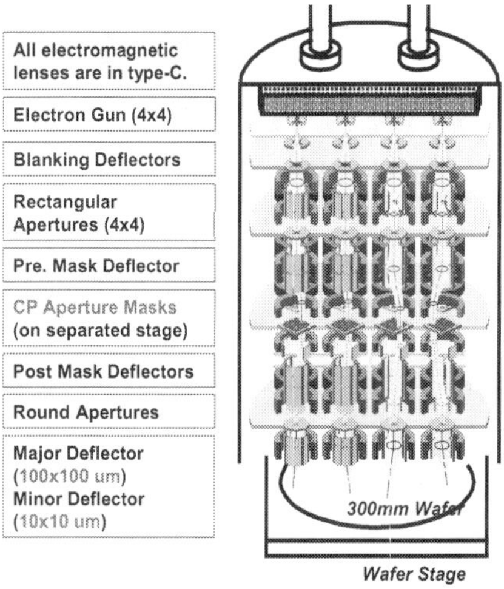

Fig. 3. An Advantest multi-column-cell character projection equipment

Yasuda et al. proposed a multi-column-cell (MCC) system, which can project multiple characters in parallel by equipping it with multiple projection mechanisms called column-cells (Yasuda et al., 2004). The motivation to develop the MCC system is to achieve higher throughput to produce ICs than sequential projection systems by parallelizing projection operations. Several ASIC design techniques were discussed (Sugihara, 2008, 2010). This chapter focuses on design techniques for single-column-cell projection equipment.

2. Cell library development for character projection equipment

Standard cell methodology is a quite popular design method to design an ASIC. The standard cell methodology exploits a cell library which is a collection of low-level logic functions, called cells, such as NAND gates, NOR gates, flip-flops, latches and buffers. From a viewpoint of character projection lithography, it is important to project as many cells as possible with character projection lithography.

Cell library development methodologies were studied for character projection lithography (Sugihara et al., 2005, 2006a, 2006c, 2007b, 2008, 2010, Inanami et al., 2000). In this section, we focus on a cell library development methodology for a single-column-cell system (Sugihara et al., 2005, 2006a, 2006c, 2007b).

Cells, which are components for IC designs, are ordinarily utilized as the basis of characters. The characters are placed in an array on a CP aperture mask as shown in Fig. 4. It accommodates several hundred characters, which are several-μm squares. The number of characters available on a single CP aperture mask is limited to a small one and not all cells in a cell library can be realized on it. For example, if there are four variations for every cell function as shown in Fig. 5, a CP aperture mask accommodates about 100 or less of cell functions at the 90 nm technology. Even if multiple CP aperture masks are utilized for placing all the cells on them, it takes forbiddingly long time to switch CP aperture masks for setting and adjusting. This chapter assumes that a single CP aperture mask is allowed to use for each layer and that the cells off the CP aperture mask are projected with the VSB lithography. It is quite important to select frequently-utilized cells to put on a CP aperture mask because a CP aperture mask is a precious resource to increase the throughput of the system.

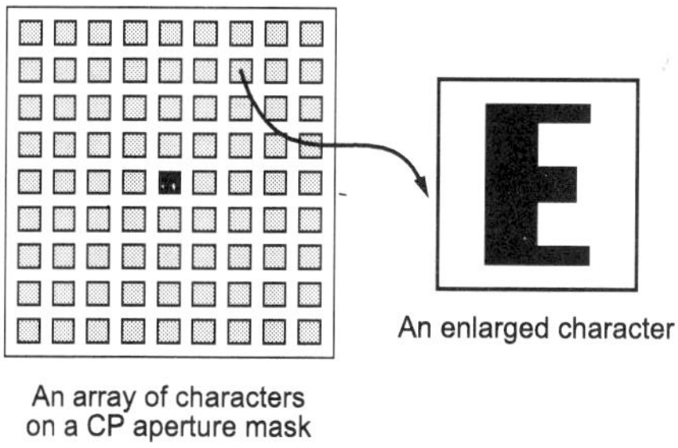

Fig. 4. A CP mask

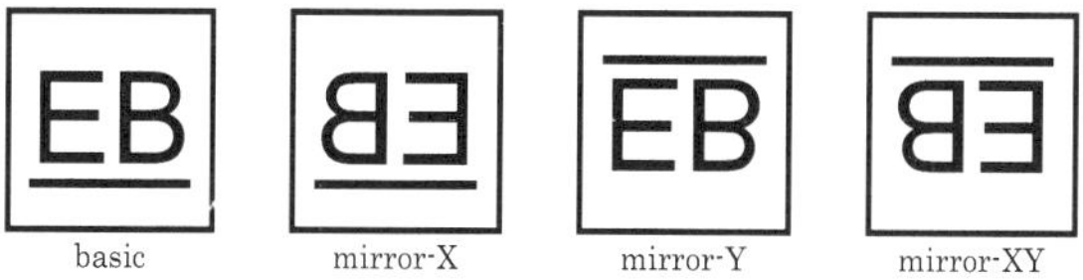

Fig. 5. Cell directions

2.1 Cell selection for CP aperture masks

In this section, a mathematical programming model is shown to select an optimal set of cells which are placed on a CP aperture mask so that the number of EB shots to draw an entire chip is minimized.

Before we describe the mathematical programming model, we briefly describe the ILP (Integer Linear Programming) to review. The ILP is a well-known way to minimize loss or maximize benefit in logistics, transportation, manufacturing and so forth (Williams, 1999). The goal of the ILP is to minimize (or maximize) a linear objective function on a set of integer variables, while satisfying a set of linear constraints. A typical ILP model can be described as follows:

$$\text{minimize: } \mathbf{Ax}$$
$$\text{subject to: } \mathbf{Bx} \leq \mathbf{C}, \text{ such that } \mathbf{x} \geq \mathbf{0} \tag{1}$$

where $\mathbf{Ax}$ is an objective function to minimize, $\mathbf{A}$ is an objective vector, $\mathbf{B}$ is a constraint matrix, $\mathbf{C}$ is a column vector of constants, and $\mathbf{x}$ is a vector of integer variables. Efficient ILP solvers are now readily available (ILOG, 2003).

A mathematical formulation is shown to select an optimal set of the cells which are placed on a CP aperture mask so that the number of EB shots to draw an entire chip is minimized We name this mathematical problem P_{A} ("A" stands for assignment). The problem P_{A} can be stated as follows:

For given C kinds of cell objects, their reference counts $r_1, r_2, \cdots, r_C$, their EB shots by the CP method, $S_{\mathrm{CP}_1}, S_{\mathrm{CP}_2}, , S_{\mathrm{CP}_C}$, their EB shots by VSB method, $S_{\mathrm{VSB}_1}, S_{\mathrm{VSB}_2}, \cdots, S_{\mathrm{VSB}_C}$, and a CP aperture mask capable of loading N_{char} characters, determine each cell's drawing method, the CP or VSB method, such that the total EB shots to draw the entire chip are minimized.

This problem is a typical combinatorial optimization problem and can be shown to be NP-hard. However, for realistic cell libraries, the sizes of the problem instances are small. The problem instances can be solved exactly using an ILP solver within short computation time.

To model this problem, consider a chip for which C kinds of cell objects are employed. Cell i appears r_i times in the chip and is drawn with either the CP or VSB methods. If Cell i is drawn with the VSB method, let the number of EB shots to draw a cell instance of Cell i be S_{VSB_i}. Likewise, If Cell i is drawn with the CP method, let the number of EB shots to draw a cell instance of Cell i be S_{CP_i}. We introduce binary variables x_i (where $1 \leq i \leq C$), which are used to determine a projection method of cells, that is the CP or VSB method. Let x_i be a binary variable defined as follows:

$$x_i = \begin{cases} 1 & \text{if cell object } i \text{ is drawn with the CP,} \\ 0 & \text{if cell object } i \text{ is drawn with the VSB.} \end{cases} \tag{2}$$

The total number of EB shots S_{A} is to draw the entire chip is given by

$$\begin{aligned} S_{\mathrm{A}}(\boldsymbol{x}) &= \text{EB shots with the CP + EB shots with the VSB} \\ &= \textstyle\sum_{i=1}^{C} S_{\mathrm{CP}_i} r_i x_i + \sum_{i=1}^{C} S_{\mathrm{VSB}_i} r_i (1 - x_i) \\ &= \textstyle\sum_{i=1}^{C} (S_{\mathrm{CP}_i} - S_{\mathrm{VSB}_i}) r_i x_i + \sum_{i=1}^{C} S_{\mathrm{VSB}_i} r_i \,. \end{aligned} \tag{3}$$

The second summation in the above equation does not include any variables so only the first summation is considered in the objective function of Eq. (1).

Depending on the size of a cell object, the number of characters to draw a cell instance of the cell object may differ from that of another. Cell i occupies c_i characters on a CP aperture mask. The area of the CP aperture makes a constraint on the number of characters. The following constraint, therefore, is introduced.

$$\sum_{i=1}^{C} c_i x_i \leq N_{\text{char}}, \tag{4}$$

where N_{char} is the maximum number of characters available on the CP aperture mask, c_i is the number of characters necessary to draw an instance of Cell i and is equivalent to S_{CP_i}.
From Eqs. (1), (2) and (3), a mathematical programming model for this problem can be formulated as follows.
Objective: Minimize $S_{\text{A}}' = \sum_{i=1}^{C}(S_{\text{CP}_i} - S_{\text{VSB}_i})r_i x_i$, subject to $\sum_{i=1}^{C} c_i x_i \leq N_{\text{char}}$, i.e. every cell adopts one drawing method, the CP or VSB in conformity with the restriction of the area of the CP aperture. The total number of EB shots is given by $S_{\text{A}}' + \sum_{i=1}^{C} S_{\text{VSB}_i} r_i$.
The above model mainly aims to develop an optimal cell library set dedicated to a product but can be applied to developing a general cell library to various products. In order to make a cell library more general among multiple products, the reference counts of cells defined as r_i $(1 \leq i \leq C)$ should be set as the reference counts of cells through the total production of the multiple products. Both dedicated and general cell libraries can be easily obtained by our proposal once the reference counts of cells r_i $(1 \leq i \leq C)$ are given. The decision to make a cell library of a product dedicated or general is made by the following two factors.

- The cost reduction by reducing the number of EB shots with a newly developed set of CP aperture masks.
- The cost increase to newly develop the set of CP aperture masks dedicated to a product.

Even if CP aperture masks are made to dedicate to a product, the total cost for the CP aperture masks is much cheaper than that for the photomask costs of the other lithography. The total amount of data to draw patterns on masks is a dominant factor in the costs of the both kinds of masks. The amount of data for photomasks is linear to the number of transistors while that for CP aperture masks to the number of cell objects on them. The developing cost of CP aperture masks is, therefore, much cheaper than that of photomasks. The CP lithography with a set of CP aperture masks is still *quasi-maskless* in terms of cost.

2.2 The impact of cell directions on area and delay

There are basically four directions of cells as cell instances are physically placed as shown in Fig. 5. A *basic* direction is literally a basis of the other directions. *Mirror-X*, *mirror-Y* and *mirror-XY* directions are horizontally-flipped, vertically-flipped and horizontally-and-vertically-flipped ones respectively. The patterns of these directions of a cell function must be distinguished from the other as different patterns on CP aperture masks.
In this section, we examine the influence of the existence of the cell directions on both area and delay time of a chip. We used a logic-synthesizable benchmark circuit which was a Z80-compatible microprocessor. We used a cell library whose feature size was 0.35 μm. The logic synthesis for the circuit was done with Synopsys Design Compiler (Synopsys, 2005). Place-and-route was done with Avant! Apollo (Avanti, 1998). Delay times for four cell directional variations are shown in Table 1. In the table, the four-bit vectors which follow "Conf. X" denote the existence of the four cell directions in the processes of place-and-route. The first bit of the vectors denotes the existence of the basic direction. If the direction is taken into

account, the number is 1, otherwise 0. Likewise, the second, third and fourth bits denote the existences of the mirror-X, mirror-Y and mirror-XY directions respectively.

Chip area	Delay times for four cell directional variations [ns], (basic, mirror-X, mirror-Y, mirror-XY)			
	Conf. 1 (1,1,1,1)	Conf. 2 (1,0,1,0)	Conf. 3 (1,1,0,0)	Conf. 4 (1,0,0,0)
799.5 × 792	N/A	N/A	N/A	N/A
810 × 808.5	7.09	6.97	N/A	N/A
819 × 808.5	7.37	7.43	N/A	N/A
829.5 × 825	7.52	-	N/A	N/A
840 × 825	7.16	-	N/A	N/A
849 × 841.5	7.24	-	N/A	N/A
859.5 × 858	7.11	-	N/A	N/A
870 × 858	-	-	7.15	N/A
879 × 874.5	-	-	7.23	7.15

Table 1. Delay times for four cell directional variations

A number in each column denotes the delay time with the least area "N/A" means that the given areas were infeasible to place and route the circuit with the place-and-route tool. For example, the least area and the delay time for the area were obtained as 810 × 808.5 and 7.09 ns respectively in Conf. 1. The least areas in Confs. 1, 2, 3, and 4 were 810 × 808.5, 810 × 808.5, 870 × 858, and 879 × 874.5 respectively. About 14% area increased when the mirror-Y and the mirror-XY were forbidden. Theoretically speaking, delay time decreases under a case in which one can use a larger place-and-route area. Delay time in a column of the table is expected to decrease downward but it did not. This is because the CAD tool is based on approximate algorithm. Comparing the two values of Conf. 2, 6.6% delay time increased while place-and-route area increased.

Conf. 1 is the configuration in which all the four cell directions are available and is supposed to be best with regard to area and delay time among the configurations because its design space includes design space of the other configurations. In other words, any layout based on Confs. 2, 3 or 4 can be realized by Conf. 1. The results which the CAD tool reported does not straightforwardly reflect this supposition because the layouts obtained by the CAD tool are approximate solutions, e.g. the delay time of Conf. 2 (6.97 ns) was shorter than that of Conf. 1 (7.09 ns) as the place-and-route area was 810 × 808.5!

Conf. 2 is the configuration in which the horizontal flippings are removed from Conf. 1. In other words, the mirror-X and mirror-XY directions are not taken into account in Conf. 2. There was no great difference of delay times among Confs. 1 and 2. Horizontal flipping seems not to be so effective to reduce delay time and area.

Conf. 3 is the configuration in which the vertical flippings (mirror-Y and mirror-XY) are removed from Conf. 1. Experimental results show that the vertical flipping of cells had little influence on delay time of the chip and it had some influence on the area. Comparing Conf. 3 with Conf. 1, about 14% area increased. This is because the gaps between cell areas were added by eliminating vertical flipping of cells and each cell area got to own its own power and ground lines.

Conf. 4 is the configuration in which any flipping cells are forbidden and only a basic direction of cells is available. Comparing Conf. 4 with Conf. 1, the differences of the delay times were insignificant while those of the areas were noticeable, that is about 14% area increase. This is because of the same reasons that the area of Conf. 3 increased. Comparing Conf. 4 with Conf. 3, the differences of their best delay times were insignificant while those of their areas were about 4.2%. Horizontal flipping had some influence on the area increase as vertical flipping was completely forbidden.

There was no great difference among delay times between the four configurations. It was experimentally confirmed that cell directions were not strongly relevant to the increase of delay time. The existence of vertical flipping of cells was relevant to increase of area. This was because gaps between cell areas come to arise and each cell area got to own its own power and ground lines.

2.3 Case study

In this section, a case study is shown for five cases to examine the relation between the number of EB shots and how to select cell objects to place on characters. The five cases are described in Table 2. We developed the cell selection software described in Section 2.1.1 with a commercial mathematical optimization engine, ILOG CPLEX 9.0 (ILOG, 2003). Every optimization process finished within a second.

	Description
Case 1	Only a basic cell direction is available. The optimal set of cells is exactly searched out by solving an ILP problem instance.
Case 2	The basic and Mirror-Y directions are available. It is assumed that the reference count of a direction of a cell function is equal to that of the other direction of the cell function. Each direction is assigned 1/2 of available characters to. This is after the fashion of Inanami's (Inanami, 2000).
Case 3	The basic and Mirror-Y directions are available. Cell objects to be placed on a CP aperture mask are exactly searched with our cell selection method.
Case 4	The four cell directions are available. It is assumed that the reference count of a direction of a cell function is equal to that of the other directions. Each direction is assigned 1/4 of available characters to. This is also after the fashion of Inanami's (Inanami, 2000)
Case 5	The four cell directions are available. Cell objects to be placed on a CP aperture mask are exactly searched with our cell selection method.

Table 2. Cell directional variations for experiments

The specification of the CP equipment for which we assumed is shown in Table 3. Two benchmark circuits were used to examine their numbers of EB shots under the five cases. The description for the benchmark circuits is shown in Table 4. Note that the cell library is from academia and comprises fewer kinds of cells than that from industry.

The maximum width and length of rectangles for VSB	3.5 μm
The width and length of characters for CP	5 μm
The number of characters on a CP aperture mask	400

Table 3. Specification of CP/VSB equipment

	Circuit 1	Circuit 2
Feature size [μm]	0.25	0.35
# cell objects in the cell library (mirroring ignored)	395	395
# mapped cell objects (mirroring ignored, one direction)	74	111
# mapped cell objects (mirroring considered, two directions)	128	191
# mapped cell objects (mirroring considered, four directions)	211	303
# cell instances	2311	3165

Table 4. Benchmark circuits

The EB shots under the five cases were sought out by solving mathematical problem instances and are shown in Table 5. In the table, the parenthesized values show the numbers of cell objects. The areas and delay times of Circuit 2 are shown in Table 6. In our experiment, areas and delay times of Circuit 1 were not examined because the benchmark circuit was not logic-synthesizable.

	Case 1	Case 2	Case 3	Case 4	Case 5
	# dirs: 1	# dirs: 2		# dirs: 4	
Circuit 1	15268 (74)	17300	16915 (128)	30925	29733 (211)
Circuit 2	51055 (111)	69760	69589 (191)	93774	91187 (303)

Table 5. EB shots and cell objects

	Case 1	Cases 2 and 3	Cases 4 and 5
Area [mm^2]	768,685.5	654,885.0	654,885.0
Delay time [ns]	7.15	6.97	7.09

Table 6. Area and delay time of Circuit 2

According to Table 5, as the number of cell objects increases, in other words, the number of cell directions increases, the number of EB shots increases. This is because the reduction of cell directions enables more cell functions to be on a CP aperture mask and to be projected with the CP. The area of Circuit 2 under Case 1 was largest among the five cases as shown in Table 6 because only a single direction, that is a basic direction, was adopted for place-and-route. This was because the gaps between cell areas came to arise and each cell area got to own its own power and ground lines. Theoretically speaking, the design with the four cell directions should be best among the five cases with regard to area and delay time. Similarly, the design with the two cell directions should be intermediate. The experimentally obtained values of areas do not reflect this supposition. The delay time of Circuit 2 under Cases 2 and 3 was found best. This is because the CAD tool returned approximate solutions of layout and happened to result against the supposition. Note that the values shown in Table 1 and Table 6 are nothing more than the ones the CAD tool reported. If a design obtained with the two cell directions is better than a design obtained with the four cell directions, the design of two-cell direction may be adopted as a design of four-cell direction.
Comparing the number of EB shots of Circuit 1 under Case 4 with that under Case 5, 3.85% reduction of the number of EB shots was achieved. The difference in the numbers of EB shots was caused by solving the problem instances exactly or approximately. The optimal sets of cells was selected exactly under Cases 3 and 5 while sets of cells was selected

approximately under Cases 2 and 4 by assigning the equal number of characters to each directions of cells.

Comparing the number of EB shots of Circuit 1 under Case 2 with that under Case 4, 44% reduction of EB shots was achieved. Likewise, comparing the number of EB shots of Circuit 2 under Case 2 with that under Case 4, about 26% reduction of EB shots was achieved. It was experimentally found that the elimination of cell directions is quite effective to reduce EB shots. It was experimentally found that the elimination of horizontal flipping reduced the much number of EB shots effectively while it has small impact on area and delay time of chips.

2.4 Conclusion

In this section, we proposed an ILP-based cell library development methodology to reduce the number of EB shots. All optimization processes finished within a second. More than 3.85% reduction of EB shots was achieved only by distinguishing between the differently mirrored cells whose functions are identical.

We examined the influence of cell directions on both area and delay time of the circuit. It was experimentally confirmed that both of the horizontal and vertical flipping of cells had little influence on delay time of chips. The horizontal flipping had little influence on area while the vertical flipping had some influence on area. This examination helps which cell direction should be implemented on CP aperture masks.

The forbiddance of horizontal flipping caused little deterioration of area while 25.6% reduction of EB shots. It was found that the forbiddance of horizontal flipping was effective to reduce the number of EB shots while it deteriorated little area and delay time of chips. The forbiddance of vertical flipping caused 13.9% increase of area while it caused less than 1% increase of delay time. The forbiddance of vertical flipping should be determined with taking a tradeoff between area and EB shots into account. For many chips of the state of the art, cells are placed so "loosely" that the deterioration of area caused by forbiddance of multi directions of cells might have less impact on area. The relation between cell directions and EB shots in design of such chips should be further examined as future work.

3. Character size optimization for higher projection throughput

Character size optimization techniques were studied for character projection (Sugihara, 2006c, 2007b, 2010). We first presented an idea to optimize a character size for higher projection throughput (Sugihara et al., 2006c). We presented a character size optimization by enumerating all possible character sizes and generating a cell library for every given character size (Sugihara et al., 2006c). Next we presented a cell library development methodology in which a character size and a set of cells were optimized at the same time (Sugihara et al., 2007b). We also presented a character size optimization technique for multi-column-cell projection equipment. In this section, we focus on the first work (Sugihara et al., 2006c) for a simple explanation.

3.1 Character size optimization problem

In Section 2, the size of characters on CP masks was given and treated as a constant because of the restriction which attributes to character projection equipment. Cells used in a design were, consequently, partitioned to fit the constant size of characters by intuition. In this section, we present a character size optimization problem P_S to minimize the number of EB shots, which affect the throughput of the CP equipment ("S" stands for size). The mathematical problem P_S can be stated as follows.

For given C kinds of cell objects, their reference counts $r_1, r_2, \cdots, r_C$, the numbers of their EB shots by the CP method, $S_{\mathrm{CP}_i}(l_{\mathrm{char}}, w_{\mathrm{char}})$, the numbers of their EB shots with the VSB method, $S_{\mathrm{VSB}_1}, S_{\mathrm{VSB}_2}, \cdots, S_{\mathrm{VSB}_C}$, the width and length of a CP mask, l_{CP} and w_{CP}, and the gap between neighboring characters, G, determine the length and width of characters, l_{char} and w_{char}, and each cell's drawing method x_i, that is the CP or the VSB method, such that the number of the total EB shots to draw the entire chip is minimized.

The number of EB shots of each cell with the CP method depends on both the size of characters and a pattern of a target layer. After the fashion of Equation (3), the total number of EB shots $S_S(\boldsymbol{x}, l_{\mathrm{char}}, w_{\mathrm{char}})$ to draw the entire chip is given by

$$\begin{aligned} S_S(\boldsymbol{x}, l_{\mathrm{char}}, w_{\mathrm{char}}) &= \text{EB shots with CP} + \text{EB shots with VSB} \\ &= \textstyle\sum_{i=1}^{C} S_{\mathrm{CP}_i}(l_{\mathrm{char}}, w_{\mathrm{char}}) r_i x_i + \sum_{i=1}^{C} S_{\mathrm{VSB}_i} r_i (1 - x_i) \\ &= \textstyle\sum_{i=1}^{C} \{S_{\mathrm{CP}_i}(l_{\mathrm{char}}, w_{\mathrm{char}}) - S_{\mathrm{VSB}_i}\} r_i x_i + \sum_{i=1}^{C} S_{\mathrm{VSB}_i} r_i \,. \end{aligned} \tag{5}$$

Depending on the size of a cell, the number of characters to draw a cell instance varies from cell to cell. The area of the CP mask makes a constraint on the number of characters. The following constraint, therefore, is introduced.

$$\textstyle\sum_{i=1}^{C} c_i(l_{\mathrm{char}}, w_{\mathrm{char}}) x_i \leq N_{\mathrm{char}}(l_{\mathrm{char}}, w_{\mathrm{char}}, l_{\mathrm{CP}}, w_{\mathrm{CP}}, G), \tag{6}$$

where $c_i(l_{\mathrm{char}}, w_{\mathrm{char}})$ is the number of characters on the CP mask for Cell i and is equal to the number of its EB shots $S_{\mathrm{CP}_i}(l_{\mathrm{char}}, w_{\mathrm{char}})$ with the CP method, $N_{\mathrm{char}}(l_{\mathrm{char}}, w_{\mathrm{char}}, l_{\mathrm{CP}}, w_{\mathrm{CP}}, G)$ is the maximum number of characters available on the CP mask when the length and width of characters are l_{char} and w_{char} respectively, the length and width of a CP mask are l_{CP} and w_{CP} respectively, and the gap between characters is G.

Equations (5) and (6) include nonlinear terms and need to be linearized if we solve this problem as an ILP problem. We did not linearize this problem but enumerate problem instances by varying both l_{char} and w_{char}. Both l_{char} and w_{char} are discrete with a narrow range. The character size enumeration results in a small number even if all problem instances are enumerated. Problem P_S can be easily solved by enumeration of Problem P_A.

3.2 Case study

We developed software to optimize the character size and select an optimal projection method for every cell so that the number of EB shots to draw an entire chip is minimized. In this development, a commercial mathematical optimization engine, ILOG CPLEX 9.0 (ILOG, 2003), was utilized. We examined the numbers of EB shots for four benchmark circuits described in Table 7. All cell libraries adopted for the four circuits include more than 300 cell functions while each circuit consists of less than 100 cell functions as a result of logic synthesis. Cells which are logically identical but are differently mirrored must be distinguished from one another on CP masks. The numbers of cell objects with or without mirroring considered are shown in the table. The numbers of cell instances are also shown in the table. Note that the size of Problems P_A and P_S are not affected by the number of cell instances but by that of cell objects. The number of cell instances affects the reference counts of cells, which are constants in the ILP model. It is the number of cell objects that is equal to the number of variables in Problem P_A and affects the size of problem instances. The experiments were based on the CP equipment which was described in Table 8. In the equipment, patterns on CP masks are demagnified to 1/5 on a wafer and all the values on the table are demagnified to 1/5 on silicon wafer. This means that a CP mask is capable of 441 characters of a 5.0 μm-square which occupies a 25.0 μm-square on CP masks. We assumed that the maximum size of an electron beam for the CP is a 10 μm-square on silicon wafers with the state-of-the-art electron beam technology.

	Circuit 1	Circuit 2	Circuit 3	Circuit 4
# mapped cell objects (mirror ignored)	39	55	74	76
# mapped cell objects (mirror considered)	118	164	211	132
# cells in the library (mirror ignored)	310	310	395	395
# cell instances	3,875	3,943	2,311	35,683
Mirror-X	Yes	Yes	Yes	Yes
Mirror-Y	Yes	Yes	Yes	No
Mirror-XY	Yes	Yes	Yes	No
# variables in subproblems	118	164	211	132
Feature size [μm]	0.35	0.35	0.25	0.25

Table 7. Benchmark circuits description

	Size [μm]
The length and width of CP masks	650
Gap between neighboring characters	5
The maximum length of characters	50

Table 8. Description for CP equipment

We examined the numbers of EB shots to draw a polysilicon gate layer of the devices under three cases shown in Table 9. In Case 1, we assumed that characters were 5.0 μm-squares. These values were given in Inanami's work (Inanami, 2000) and this size was based on the specification of the CP equipment. In Case 2, it was assumed that the size of electron beams was smaller than a 10 μm-square and the character size was optimized under the character size constraint. In Case 3, we assumed that any size of an EB was available and the character size was optimized under the character size constraint.

	Length	Width	Remarks
Case 1	5.0 [μm]	5.0 [μm]	Conventional square. This value is given from the equipment specification.
Case 2	Searched	Searched within feasible beam-size	Optimal rectangle with feasible beam-size.
Case 3	Searched	Searched within any beam-size	Optimal rectangle with ideal beam-size.

Table 9. Length and width of characters on wafer for three cases

The numbers of EB shots obtained by our experiments are shown in Table 10. Comparing Case 1 with Case 2, 72.0% of EB shots was reduced in the best. The feasible EB size was utilized in both Cases 1 and 2 and the difference between them was whether or not the character size was optimized. The gap between Cases 1 and 2 implies that the beam size of the CP equipment should be configurable for users so that the throughput of their equipment can be increased. Supposing any size of EBs can be utilized, more reduction of EB shots can be achieved as the numbers in Case 3 in Table 10 show. Comparing Case 1 with Case 3, 75.9% of EB shots was reduced in the best. Comparing Case 2 with Case 3, 39.5% of EB shots were reduced in the best. The gap between Cases 2 and 3 implies that the development effort to seek for a larger size of EBs is capable of reducing 39.5% of EB shots. The gap between Cases 2 and 3 is directive for equipment developers to determine the EB size of their equipment.

	Circuit 1	Circuit 2	Circuit 3	Circuit 4
Case 1	52,117	41,469	26,913	164,316
Case 2	23,785	21,122	7,710	46,050
Case 3	14,379	15,120	7,710	39,546

Table 10. EB shots for the four benchmark circuits under three cases

The computing platform for experiment was an Intel Pentium 4 2.4 GHz with 1 GB of main memory. Computation times to obtain the optimal character sizes under the three cases were shown in Table 11. All character size optimization processes were finished within less than two minutes. Note that computation time was not affected by the number of cell instances but by the number of cell objects. All character size optimization processes were done within practical computation time.

	Circuit 1	Circuit 2	Circuit 3	Circuit 4
Case 1	0.00	0.00	0.00	0.00
Case 2	11.86	14.29	60.55	56.28
Case 3	27.05	32.81	96.04	86.12

Table 11. Computation time to optimize a character size [s]

Fig. 6 shows the numbers of EB shots with various character sizes for Circuit 3. The figure shows that there exists a minimal point of EB shots. The number of EB shots increases rapidly from the minimal point to the point where the size of characters is smaller while it increases gradually from the minimal point to the point where the size of characters is larger. The number of EB shots rises and falls sharply in the neighborhood of the minimal point. The experimental results show that the character size optimization was quite effective to reduce the number of EB shots and to enhance the throughput of the CP equipment.

3.3 Conclusion

We proposed the character size optimization technique for improving the throughput of the CP equipment by defining a mathematical problem as an ILP problem. We also showed some experimental results by solving mathematical program instances for several benchmark circuits. According to our experiment, 72.0% reduction of EB shots was achieved with a feasible EB size in the best, comparing with the conventional and intuitional character sizing. It was experimentally found that our character size optimization technique was so

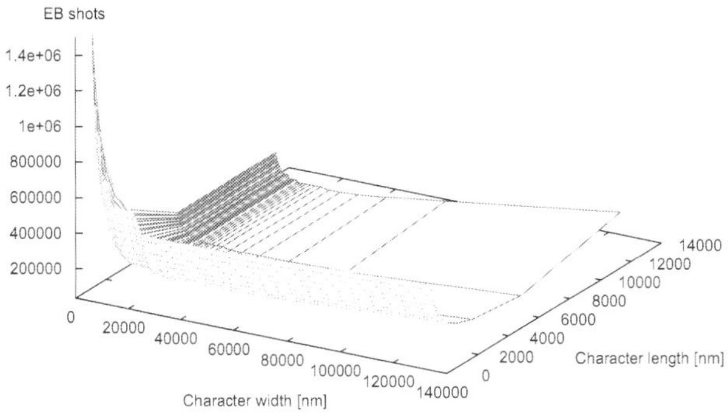

Fig. 6. The number of EB shots for Circuit 3

effective to reduce the number of EB shots. The experimental results imply that the CP equipment should be capable of modifying the size of characters by customers' demands after it is shipped out to them.

Our systematic character size optimization scheme can achieve the lower number of EB shots and can enhance the throughput of the CP equipment. The throughput enhancement of the CP equipment causes the higher production volume of semiconductors devices at a lower cost. It consequently accelerates the application of semiconductor devices to various industrial fields even if their production volumes are small. Likewise, the throughput enhancement of the CP equipment promises a lower cost in developing photomasks if the CP method is utilized in developing photomasks in the future.

4. Technology mapping technique for character projection equipment

Technology mapping techniques were discussed for character projection lithography of a single-column-cell system (Sugihara et al., 2006b, 2007c). A technology mapping technique was also proposed for a multi-column-cell system (Sugihara et al., 2007a). This section mainly discusses a technology mapping technique which reduces projection time of a single-column-cell system (Sugihara et al., 2006b, 2007c).

4.1 Review on technology mapping

In the most popular paradigm for logic synthesis, after a technology independent optimization of a set of logic equations, the result is mapped into a feasible circuit which is optimal with respect to area and satisfies a maximum critical-path delay. In this paradigm, the role of *technology mapping* is to finish the synthesis of the circuit by performing the final gate selection from a particular cell library. The role of technology mapping is the actual cell choice to implement the equations – for example, choosing the fastest cells along the critical path, and using the most area-efficient combination of cells off the critical path (Hachtel, 1996).

A set of base functions is chosen such as a two-input NAND-gate and an inverter. The logic equations are optimized in a technology-independent manner and are then converted into a graph where each node is restricted to one of the base functions. This graph is called the *subject graph*. The logic function for each library gate is also represented by a graph where each node is restricted to one of the base functions. Each graph for a library gate is called a *pattern graph*. For any given logic function there are many different representations of the function using the base function set. Therefore, each library gate is represented by many different pattern graphs (Hachtel, 1996).

Let us see some examples of technology mapping for the following Boolean network:

$$F = \overline{(d+e)\cdot(a\cdot(b+h)+c)+f\cdot g\cdot h}. \qquad (7)$$

The Boolean network and its subject graph are shown as Fig. 7 and Fig. 8. Note that a two-input NAND-gate and an inverter are chosen as the base functions. Fig. 9 and Fig. 10 show differently technology-mapped circuits for the original one.

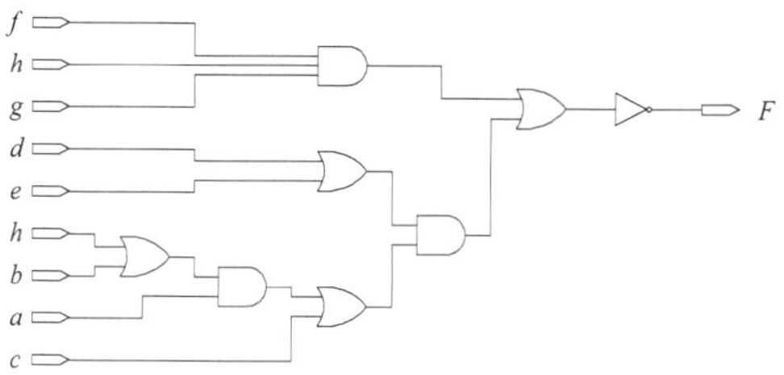

Fig. 7. A Boolean network

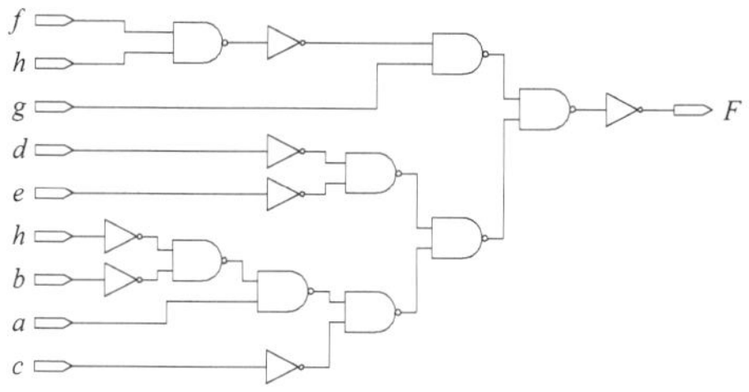

Fig. 8. A subject graph

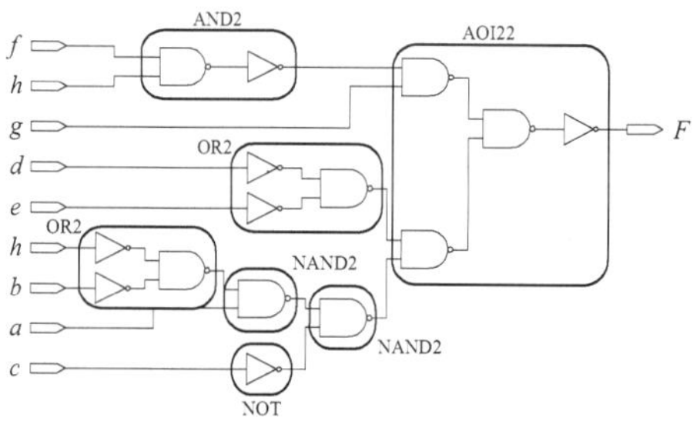

Fig. 9. A technology mapping result

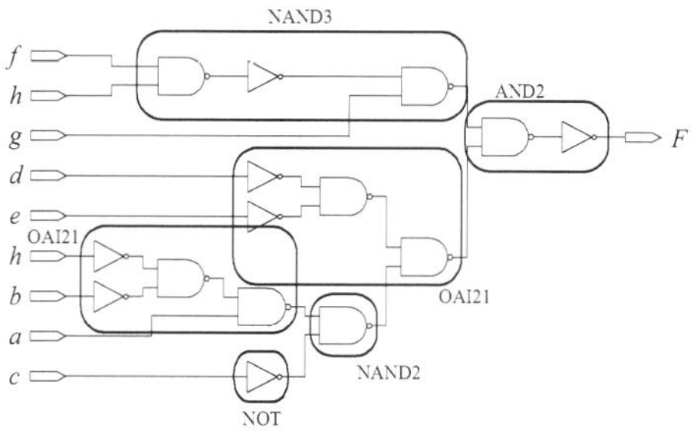

Fig. 10. Another technology mapping result

4.2 Linear combination of area and the number of EB shots

The conventional technology mapping does not consider the number of EB shots for the CP lithography because most ICs are fabricated with photolithography. The price of a photomask set is getting unaffordable and the CP lithography will be utilized gradually. This section discusses how to treat the number of EB shots to project an entire IC in the process of technology mapping.

Now let us discuss the objectives in technology mapping. The area of a circuit is simply represented with the sum of areas of the cell instances which are utilized for realizing the function the designers want and is shown as follows:

$$\mathrm{Area}_{\mathrm{all}} = \textstyle\sum_i \mathrm{Ref}_i \cdot \mathrm{Area}_i, \tag{8}$$

where $\mathrm{Area}_{\mathrm{all}}$, Ref_i, and Area_i are the total area of the IC, the reference count of Cell Function i, and the area of an instance of Cell i respectively. Likewise, the number of EB shots to project the entire IC is shown as follows:

$$\mathrm{Shots}_{\mathrm{all}} = \textstyle\sum_i \mathrm{Ref}_i \cdot \mathrm{Shots}_i, \tag{9}$$

where $\mathrm{Shots}_{\mathrm{all}}$ and Shots_i are the number of EB shots to project entire IC and that to project an instance of Cell Function i respectively.

In the conventional IC design, only Equation (6) is minimized under a timing constraint. In the design adopting the CP lithography, both of Equations (6) and (7) must be taken into account to fabricate small ICs within short fabrication time. We introduce the linear combination of $\mathrm{Area}_{\mathrm{all}}$ and $\mathrm{Shots}_{\mathrm{all}}$ as follows:

$$\alpha \cdot \mathrm{Area}_{\mathrm{all}} + \beta \cdot \mathrm{Shots}_{\mathrm{all}}, \tag{10}$$

where α and β are some constant values and should be chosen depending on the importance of area and EB shots. Technology-mapped circuits which are implemented with small area and a small number of EB shots can be obtained by minimizing the objective values in Equation (10) under a timing constraint.

4.3 Case study

4.3.1 Experimental setup

We have developed a design framework in which both area and the number of EB shots can be optimized in the process of technology mapping as shown in Fig. 11. The design flow starts with the generation of a cell library for the CP lithography. The cell library for the CP can be directly applied to a commercial logic synthesis tool. Once ASIC designers obtain the cell library for the CP, they can logic-synthesize their circuit in the same manner as the typical logic synthesis flow.

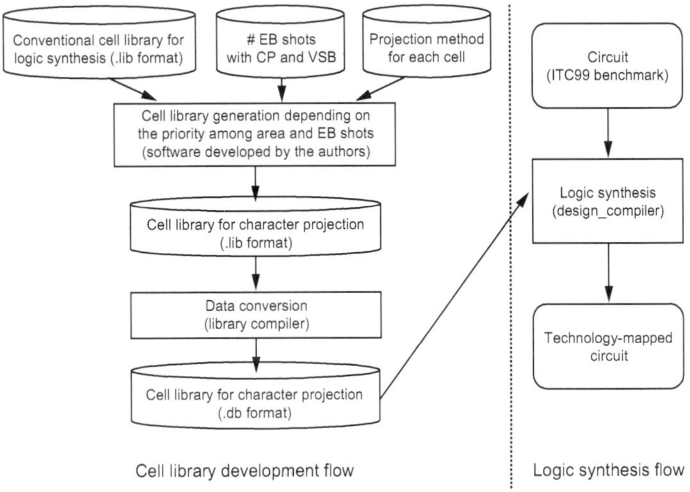

Fig. 11. Design flow in our design framework

In the design framework, a conventional cell library, the number of EB shots for every cell function, and a projection method for every cell function are required for generating a cell library specialized in the CP. The optimal projection method for every cell function can be obtained with the cell library development methodology, which we proposed in the previous work (Sugihara et al., 2005, 2006a). The simplified procedure for choosing the optimal projection method for every cell function is summarized as follows.

1. Obtain the reference counts of all cell functions, their numbers of EB shots to project a cell instance with the CP and VSB lithographies, and the number of characters available on a CP aperture mask.
2. With the above numbers, solve the mathematical problem instance shown in the previous work (Sugihara et al., 2005, 2006a). A solution to the mathematical problem instance specifies the optimal projection method for every cell functions.

To obtain the projection method for every cell functions, we have utilized a benchmark circuit, b19 from the ITC'99 benchmark circuit suite (Davidson, 1999), as a referential circuit. The benchmark circuit has been logic-synthesized to minimize its area and then the mathematical problem instance has been solved for its netlist with a commercial ILP solver (ILOG, 2003).

We adopted commercial EDA tools as possible as we could in order to reduce software development. For the design framework, we have adopted the Synopsys Design Compiler as a logic synthesis tool and the Synopsys Library Compiler as a cell library generation tool. A cell library specialized in the CP (written in the Synopsys ".lib" format) has been generated with the following three things: a conventional cell library (also written in the

Synopsys ".lib" format), the numbers of EB shots which are needed to project an instance of every cell function with the CP and VSB lithographies, and the optimal projection method of every cell function. In the cell library file format specification for the Design Compiler, the attributes designating area, delay, and power can be specified for each cell function but the number of EB shots cannot be specified. The attribute "area" has been used to represent both area and the number of EB shots by setting the linear combination of them as the attribute "area" for making the Design Compiler capable of optimizing them. And then the ".lib" file of the cell library has been converted into the Synopsys ".db" format file with the Synopsys Library Compiler.

We have utilized two cell libraries, high-performance and low-power ones, which were provided by the VLSI Design and Education Center (VDEC) at the University of Tokyo, as conventional cell libraries. The technology node of the two cell libraries was 0.35 μm. We have demagnified all patterns of the two cell libraries by a factor of 90/350 in order to simulate the number of EB shots at the 90 nm technology node. With the two conventional cell libraries, we have generated cell libraries specialized in the CP system whose specification is shown in Table 12. The numbers of cell functions, on-characters cell functions, and off-characters cell functions are described in Table 13. Note that the cell libraries were supplied from academia and the number of cell functions is smaller than commercial ones. This means that the usage of an industrial cell library increases the number of off-characters cell functions and possibly deteriorates the throughput of the projection system in exchange for more design flexibility.

The maximum width and height of rectangles for VSB	3.5 μm
The width and height of characters for CP	5.0 μm
The number of characters available on a CP aperture mask	440

Table 12. Specification of projection equipment

	# cells	# cells on characters	# cells off characters
High performance cell library	310	53	257
Low power cell library	310	101	209

Table 13. Two cell library description

4.3.2 Experimental results

We examined the numbers of EB shots to project several benchmark circuits with the two cell libraries.

First, we examined the relation between area and the number of EB shots to project each of ITC'99 benchmark circuits (Davidson, 1999). In the examination, all circuits were logic-synthesized for two objectives: area-minimization (α=1, β=0 in Equation (8)) and EB-shots-minimization (α=0, β=1 in the equation). The purpose of this examination was to obtain the minimal values of both area and EB shots of all benchmark circuits. We logic-optimized each of circuits ten times and have taken the best value of area and EB shots among the results. Fig. 12 and Fig. 13 show area and shots ratios for both area-minimized logic synthesis (α=1, β=0) and shots-minimized one (α = 0, β = 1) with the two cell libraries, respectively.

Depending on α and β, the cell library was optimized for Circuit b19. The cell library is equivalent to the original one under α=1 and β=0. The area and EB shot count are normalized with those of area-minimized circuits. Note that the numbers of EB shots shown in the figures are the ones for projecting the FEOL patterns. They do not include any EB shots to project the BEOL patterns. Shots-minimizing logic synthesis reduced 54.6% of EB shot count at the best case in exchange for 8.4% area increase. Shots-minimization logic synthesis has increased 0.0% to 24.0% of area on logic-synthesizing all circuits.

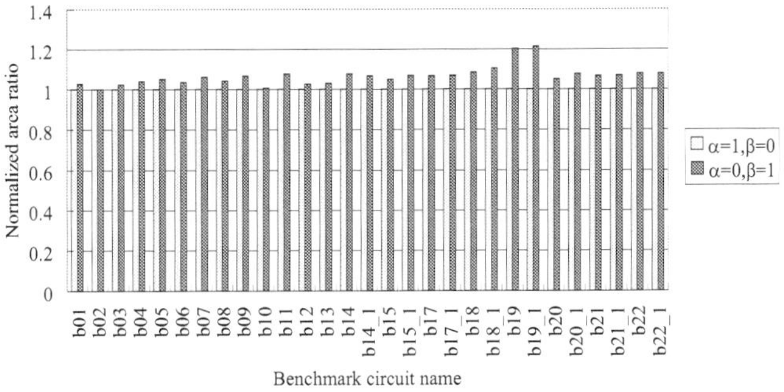

(a) Area under area- and EB shots-minimizations

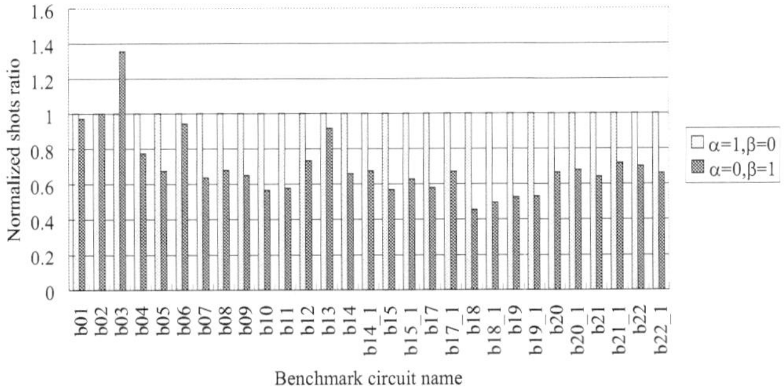

(b) The number of EB shots under area- and EB shots-minimizations

Fig. 12. Area and the number of EB shots of ITC'99 benchmark circuits with HP cell library

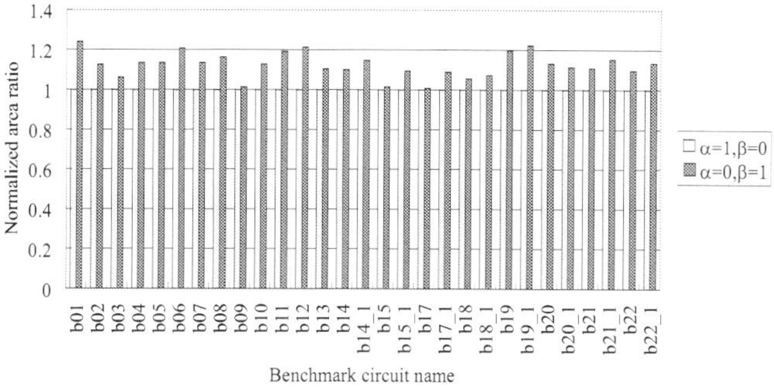

(a) Area under area- and EB shots-minimizations

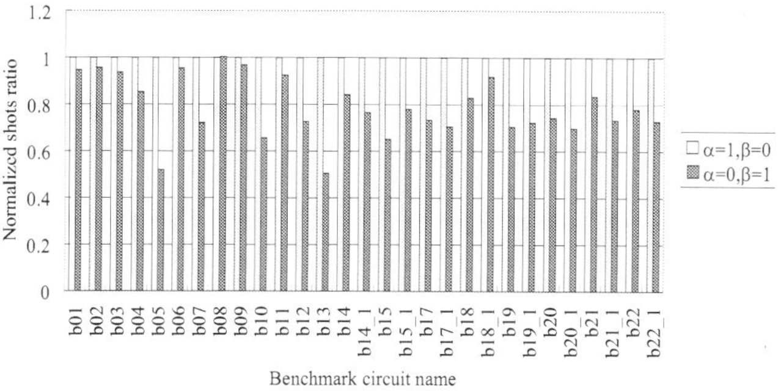

(b) The number of EB shots under area- and EB shots-minimizations

Fig. 13. Area and the number of EB shots of ITC'99 benchmark circuits with LP cell library

Changing ratio β/α, we examined the relation between area and the number of EB shots for a benchmark circuit, b18_1, as shown in Fig. 14. We show the range of both area and the number of EB shots observed in the process of ten logic optimizations in the figure in order to observe the tendency of area and the number of EB shots to change. This is because the logic synthesis tool returns quasi-optimal circuits with some variation of both area and EB shots and makes it hard to observe a consistent tendency of them. The figure shows that area tends to increase as the ratio β/α increases while the number of EB shots tends to decrease. These results show that there exists a tradeoff between area and the number of EB shots. We think that area-saving and off-characters cell functions are mapped as area is important. Such cell functions are substituted on-characters ones as the number of EB shots becomes important. ASIC designers should choose the ratio β/α according as they want to reduce area or the number of EB shots.

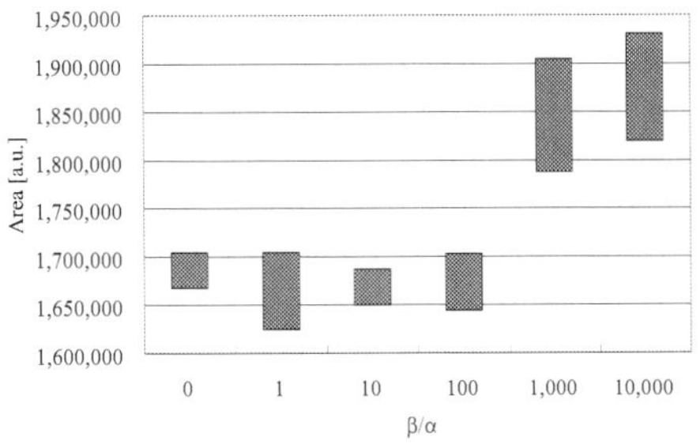

(a) Area versus β/α

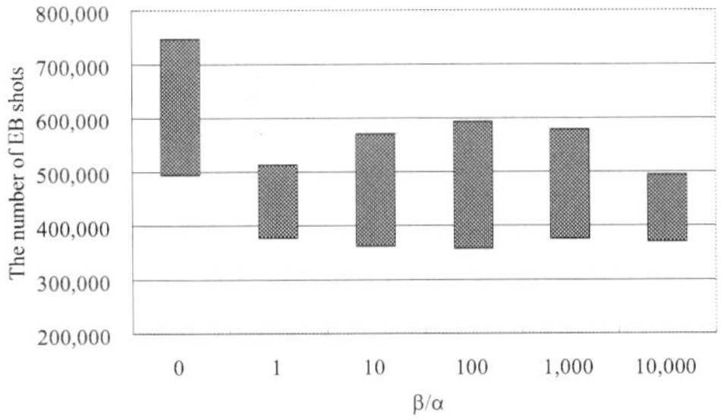

(b) The number of EB shots versus β/α

Fig. 14. Tradeoff between area and the number of EB shots

We have examined area and the number of EB shots under several timing constraints with a Z80-compatible processor. Fig. 15 and Fig. 16 show area and the number of EB shots under various timing constraints. When timing constraints were given, the effectiveness of the shots-minimization and area-minimization became less but the number of EB shots was reduced up to 26.6% nevertheless. The area increase accompanied with EB shots minimization was from 10.6% to 46.2%. These results show that area and the number of EB shots decreased as the timing constraint was loosened. We think that there exists a tradeoff between delay time and the number of EB shots (or area). A timing constraint for shorter delay time causes logic synthesis tools to choose off-characters cell functions that have higher current drivability. Such cell functions have larger transistors and require many EB shots to be projected with the VSB lithography.

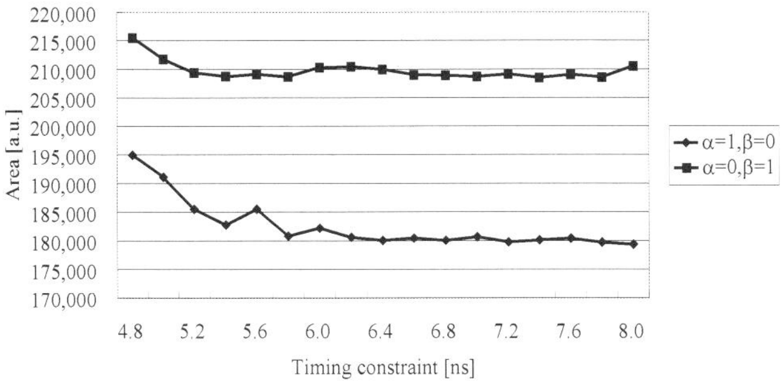

(a) Area versus timing constraint

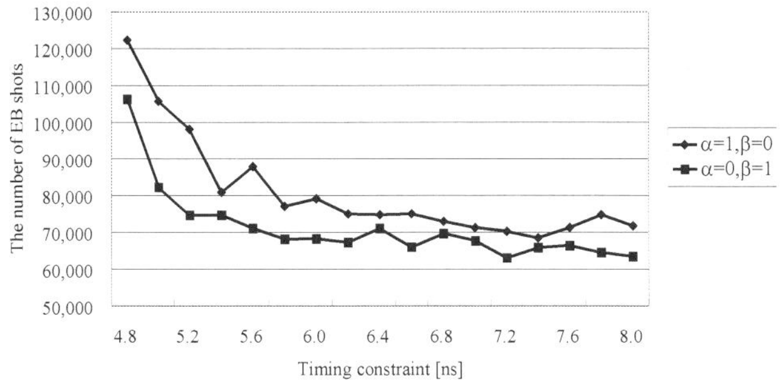

(b) The number of EB shots versus timing constraint

Fig. 15. Area and the number of EB shots with the HP cell library under timing constraints

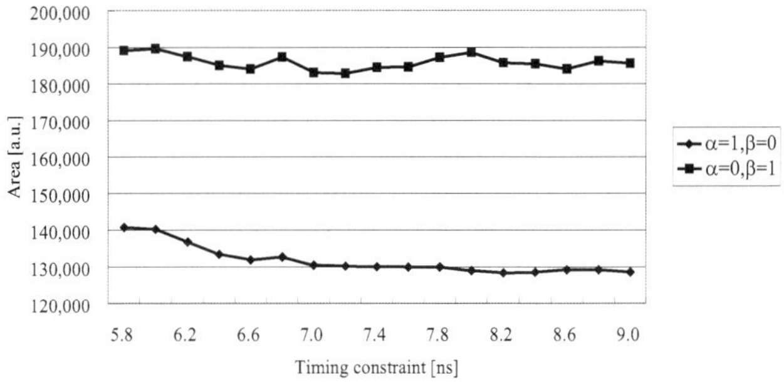

(a) Area versus timing constraint

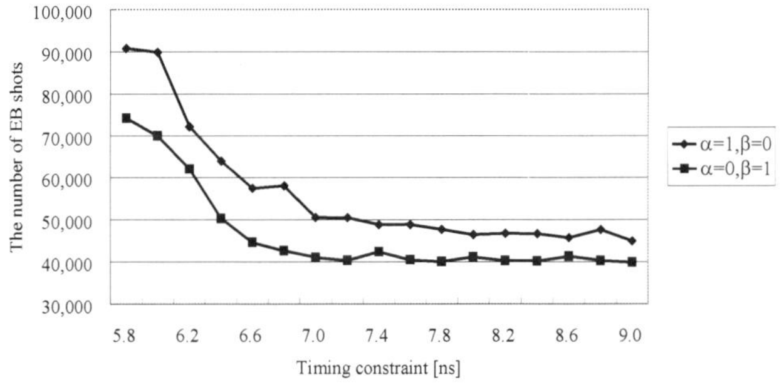

(b) The number of EB shots versus timing constraint

Fig. 16. Area and the number of EB shots with the LP cell library under timing constraints

4.4 Conclusion

Our technology mapping technique for the CP lithography achieved a 54.6% less number of EB shots with 8.4% area increase under no timing constraints than the conventional one. Our technology mapping for the CP lithography also achieved a 26.6% less number of EB shots with 41.1% area increase and without any performance degradation than the conventional one. Varying the ratio β/α, we found that there exists a tradeoff between area and the number of EB shots.

In the other experiments, tightening timing constraint increased area and the number of EB shots. It was found that there exists a tradeoff between delay time and the number of EB shots. We think that a timing constraint for shorter delay time has caused a logic synthesis tool to choose the off-characters cell functions that have higher current drivability.

Our technology mapping technique reduced the number of EB shots to project patterns for the FEOL with some area increase. It is probable that some increase of cell instances causes the number of wires to increase and does the number of EB shots for projecting the BEOL patterns to increase. A technology mapping technique should be studied for reducing the number of EB shots required for both the FEOL and the BEOL patterns as future work.

The number of cell functions which are placed on a CP aperture mask will increase as the technology node proceeds. It is probable that all cell functions of a commercial cell library, which includes more than 500 cell functions, will be placed on a single CP aperture mask at the 32 nm technology node. This indicates that the VSB lithography will be required less and less for projecting such cell functions. We suppose that our technology mapping technique is effective before the 32 nm technology node. We think that another technology mapping technique will be needed to increase throughput of MCC systems after the 32 nm technology node.

It is easy for both IC designers and equipment developers to adopt our technology mapping technique because a software approach such as our technique imposes no modification on CP equipment. This means that no additional cost is necessary to adopt our technique in their IC design.

5. Conclusion

Character projection lithography is one of promising projection methods for manufacturing application specific integrated circuits at a low cost. From the viewpoint of ASIC design, the number of EB shots, which reflects the manufacturing cost for ASICs, is reduced by (i) cell library generation and (ii) technology mapping.

Cell library generation consists of two parts: cell selection and character sizing. We presented a cell selection method (Sugihara et al., 2005, 2006a) and character sizing methods (Sugihara et al., 2006c, 2007b). Cell selection and character sizing achieved 72.0% reduction of EB shots with a feasible EB size in the best, comparing with the conventional and intuitional character sizing.

We also presented a technology mapping technique for reducing the number of EB shots in character projection lithography. Our technology mapping technique for the CP lithography has achieved a 54.6% less number of EB shots with 8.4% area increase under no timing constraints than the conventional one. Our technology mapping for the CP lithography has also achieved a 26.6% less number of EB shots with 41.1% area increase and without any performance degradation than the conventional one. Our experiments suggested that there exists a tradeoff between area and the number of EB shots.

6. Acknowledgment

This work was supported by the VLSI Design and Education Center (VDEC), the University of Tokyo in collaboration with Synopsys, Inc.

7. References

Avant!, Inc. (1998). Apollo User Guide.

Davidson, S. (1999). Notes on ITC'99 benchmarks, http://www.cerc.utexas.edu/itc99-benchmarks/bendoc1.html

Hachtel, G. D. (1996). Logic synthesis and verification algorithms, Kluwer Academic Publishers

Hattori, K.; Yoshikawa, R.; Wada, H.; Kusakabe, H.; Yamaguchi, T.; Magoshi, S.; Miyagaki, A.; Yamasaki, S.; Takigawa, T.; Kanoh, M.; Nishimura, S.; Housai, H. & Hashimoto, S. (1993). Electron-beam direct writing system EX-8D employing character projection exposure method. *Journal of Vacuum Science & Technology B*, Vol.11, Issue 6, (November 1993), pp. 2346-2351, ISSN 0734-211X

Hirumi, J.; Kuriyama, K.; Yoshioka, N.; Yoshikawa, R.; Hojo, Y.; Matsuzaka, T.; Tanaka, K. & Hoga, M. (2003). Shot number analysis at 65 nm node mask writing using VSB writer. *Proceedings of SPIE*, pp. 2346-2351, ISBN 9780819449962, Yokohama, Japan, April 2003

ILOG, Inc. (2003). CPLEX 9.0 reference manual

Inanami, R; Magoshi, S.; Kousai, S.; Hamada, M.; Takayanagi, T.; Sugihara, K.; Okumura, K & Kuroda, T. (2000). Throughput enhancement strategy of maskless electron beam direct writing for logic device. *Technical Digest of IEEE International Electron Device Meeting*, pp. 833-836, ISBN 978-0780364387, San Francisco, CA, USA, December 2000

Inanami, R.; Magoshi, S.; Kousai, S.; Ando, A.; Nakasugi, T.; Mori, I.; Sugihara, K. & Miura, A. (2003). Maskless lithography: estimation of the number of shots for each layer in a logic device with character-projection-type low-energy electron-beam direct writing system. *Proceedings of SPIE*, pp. 1043-1050, ISBN 9780819448422, Santa Clara, CA, USA, June 2003

Nakamura, F.; Watanabe, K.; Kinoshita, H.; Shinozaki, H.; Kojima, Y.; Morita, S.; Noguchi, K.; Yamaguchi, N.; Isokawa, H.; Kushitani, K.; Satoh, T.; Koshiba, T.; Oota, T.; Nakasugi, T. & Mizuno, H. (2006). A character projection low energy electron beam direct writing system for device of small production lot with a variety of design. *Proceedings of SPIE*, pp. 624-631, ISBN 9780819461940, San Jose, CA, USA, February 2006

Nakasugi, T.; Ando, A.; Inanami, R.; Sasaki, N.; Ota, T.; Nagano, O.; Yamazaki, Y.; Sugihara, K.; Mori, I.; Miyoshi, M.; Okumura, K. & Miura, A. (2003). Maskless lithography: a low-energy electron-beam direct writing system with a common CP aperture and the recent progress. *Proceedings of SPIE*, pp. 1051-1058, ISBN 9780819448422, Santa Clara, CA, USA, June 2003

Pfeiffer, H. C. (1978). Variable spot shaping for electron beam lithography. *Journal of Vacuum Science & Technology*, Vol.15, No.3, (May 1978), pp. 887-890, ISSN 0022-5355

Pfeiffer, H. C. (1979). Recent advances in electron-beam lithography for the high-volume production of VLSI devices, *IEEE Transactions on Electron Devices*, Vol.26, Issue 4, (April 1979), pp. 663-674, ISSN 0018-9383

Sakitani, Y.; Yoda, H.; Todokoro, H.; Shibata, Y.; Yamazaki, T.; Ohbitu, K.; Saitou, N.; Moriyama, S.; Okazaki, S.; Matsuoka, G.; Murai, F. & Okumura, M. (1992). Electron-beam cell-projection lithography system. *Journal of Vacuum Science & Technology B*, Vol.10, No.6, (November 1992), pp. 2759-2763, ISSN 0734-211X

Sugihara, M.; Takata, T.; Nakamura, K.; Inanami, R.; Hayashi, H.; Kishimoto, K.; Hasebe, T.; Kawano, Y.; Matsunaga, Y.; Murakami, K. & Okumura, K. (2005). Cell library development methodology for throughput enhancement of electron beam direct writing systems, *Proceedings of International Symposium on System-on-Chip*, pp. 137-140, ISBN 9780780392946, Tampere, Finland, November 2005

Sugihara, M.; Takata, T.; Nakamura, K.; Inanami, R.; Hayashi, H.; Kishimoto, K.; Hasebe, T.; Kawano, Y.; Matsunaga, Y.; Murakami, K. & Okumura, K. (2006a). Cell library development methodology for throughput enhancement of character projection equipment. *IEICE Transactions on Electronics*, Vol.E89-C, No.3, (March 2006), pp. 377-383, ISSN 0916-8524

Sugihara, M.; Takata, T.; Nakamura, K.; Inanami, R.; Hayashi, H.; Kishimoto, K.; Hasebe, T.; Kawano, Y.; Matsunaga, Y.; Murakami, K. & Okumura, K. (2006b). Technology mapping technique for throughput enhancement of character projection equipment. *Proceedings of SPIE*, pp. 71-82, ISBN 9780819462084, San Jose, CA, USA, February 2006

Sugihara, M.; Takata, T.; Nakamura, Y.; Matsunaga, Y. & Murakami, K. (2006c). A character size optimization technique for throughput enhancement of character projection lithography, *Proceedings of International Symposium on Circuits and Systems*, pp. 2561-2564, ISBN 9780780393899, Island of Kos, Greece, May 2006

Sugihara, M.; Matsunaga, Y. & Murakami, K. (2007a). Technology mapping technique for enhancing throughput of multi-column-cell systems, *Proc. SPIE*, Vol. 6517, San Jose, CA, USA, February 2007

Sugihara, M.; Matsunaga, Y. & Murakami, K2010. (2007b). Character size optimization for higher throughput of character projection lithography, *IPSJ Journal*, Vol.48, No.48, (May 2007), pp. 1888-1897, ISSN 0387-5806

Sugihara, M.; Nakamura, K.; Matsunaga, Y. & Murakami, K. (2007c). Technology mapping technique for increasing throughput of character projection lithography. *IEICE Transactions on Electronics*, Vol.E90-C, No.5, (May 2007), pp. 1012-1020, ISSN 0916-8524

Sugihara, M. (2008). Character projection mask set optimization for enhancing throughput of MCC projection systems. *IEICE Transactions on Fundamentals of Electrnonics, Communications and Computer Sciences*, Vol.E91-A, No.12, (December 2008), pp. 3451-3460, ISSN 0916-8508

Sugihara, M. (2010). Character-Size Optimization for Reducing the Number of EB Shots of MCC Lithographic Systems. *IEICE Transactions on Electronics*, Vol.E93-C, No.5, (May 2010), pp. 631-639, ISSN 0916-8524

Synopsys, Inc. (2005). http://www.synopsys.com/, May 2005

Williams, H. P. (1999). Model Building in Mathematical Programming

Yasuda, H.; Haraguchi, T. & Yamada, A. (2004). A proposal for an MCC (multi-column cell with lotus root lens) system to be used as a mask-making e-beam tool. *Proceedings of SPIE*, pp. 911-921, ISBN 9780819455130, Monterery, CA, USA, December 2004

5

Transform-Based Lossless Image Compression Algorithm for Electron Beam Direct Write Lithography Systems

Jeehong Yang[1] and Serap A. Savari[2]
[1]*University of Michigan, Ann Arbor*
[2]*Texas A&M University*
USA

1. Introduction

Conventional photolithography systems use physical masks which are expensive and difficult to create and cannot be used forever. Electron Beam Direct Write (EBDW) lithography systems are a noteworthy alternative which do not need physical masks [Chokshi et al. (1999)]. As shown in Figure 1 they rely on an array of lithography writers to directly write a mask image on a photo-resist coated wafer using electron beams. EBDW systems are attractive for a few reasons: First, their flexibility is advantageous in processes requiring the rapid prototyping of chips. Second, they are known to reduce fabrication costs [Lin (2009)]. Third, they are well suited for Next-Generation Lithography (NGL) because they are able to produce circuits with smaller features than state-of-the-art photolithography systems. Finally, since the mask images are electronically controlled EBDW systems could be improved by software. Our focus here will be on this last point.

EBDW is not at this time used in many circuit fabrication processes because it is much slower than physical mask lithography systems. One current focus of research to address the throughput problem is massively-parallel electron beam lithography. Some of the research groups/companies which are developing such systems include KLA-Tencor [Petric et al. (2009)], IMS [Klein et al. (2009)], and MAPPER [Wieland et al. (2009)].

Chokshi et al. (1999) proposed a maskless lithography system using a bank of 80,000 lithography writers running in parallel at 24 MHz. Dai & Zakhor (2006) pointed out that this lithography system can achieve the conventional photolithography throughput of one wafer layer per minute, but layout image data is often several hundred terabits per wafer and therefore data delivery becomes an important issue. Dai & Zakhor (2006) proposed using a data delivery system with a lossless image compression component which is illustrated in Figure 2. They hold compressed layout images in storage disks and transmit the compressed data to the processor memory board. This kind of EBDW lithography system can achieve higher throughput if the decoder embedded within the lithography writer can sufficiently rapidly recover the original images from the compressed files.

Dai (2008) discussed two constraints on this type of system: 1) the compression ratio should be at least (Transfer rate of Decoder to Writer / Transfer rate of Memory to Decoder), and 2) the decoding algorithm has to be simple enough to be implemented as a small add-on within the maskless lithography writer. Therefore the decoder must operate with little memory.

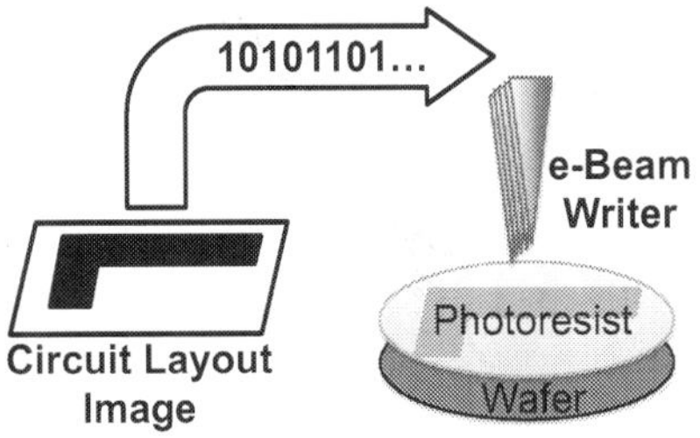

Fig. 1. EBDW Lithography

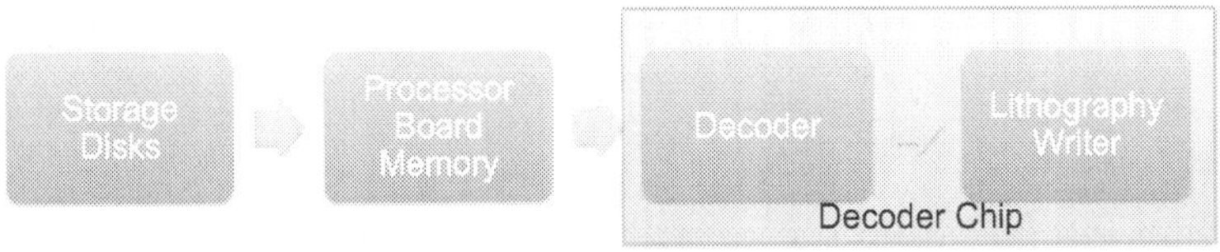

Fig. 2. Data Delivery for an EBDW Lithography System introduced in Dai & Zakhor (2006)

Dai & Zakhor (2006) reported that the layout images of control logic circuits are often irregular while the layout images of memory cells frequently contain repeated patterns. Their first algorithm *C4* attempts to handle the varying characteristics of layout images by using context prediction and finding repeated regions within an image. Liu et al. (2007) later proposed *Block C4*, which significantly reduces the encoding complexity.

Based on the framework of Dai & Zakhor (2006) and Liu et al. (2007), Yang & Savari (2010) improved the compression algorithm via a corner-based representation of the Manhattan polygons. Their initial algorithm *Corner* achieves higher compression rates than *Block C4* on an irregular circuit. Yang & Savari (2011) recently proposed an improvement *Corner2* which simplifies the corner transformation to deal with the irregular parts of the layout images and also uses a frequent pattern replacement scheme to deal with the repeated parts. Their experimental results indicate that their approach is often more efficient than the context prediction method used in *C4* and *Block C4*.

In this paper we extend the work of Yang & Savari (2011) to gray-level images to better address the issue of handling proximity correction for EBDW systems and show that we obtain better compression performance and faster encoding/decoding than *C4* and *Block C4*. Hence our work can be used to solve the data delivery problem of EBDW lithography systems with smaller features. Moreover, since our decoding speed is faster than *C4* and *Block C4* we can improve the throughput of the EBDW lithography system.

2. The compression algorithm

2.1 Overview

Layout image data is commonly cached in GDSII [Rubin (1987)] or OASIS [Chen et al. (2004)] formats. GDSII and OASIS describe circuit features such as polygons and lines by their corner points [see Rubin (1987) and Reich et al. (2003)]. GDSII and OASIS formatted data are far more compact than the uncompressed image of a circuit layer. Therefore GDSII and OASIS initially seem to be well-suited for this application, but the problem is that EBDW writers operate directly on pixel bit streams and GDSII and OASIS layout representations must therefore be converted into layout images before the process begins. The conversion process involves 1)

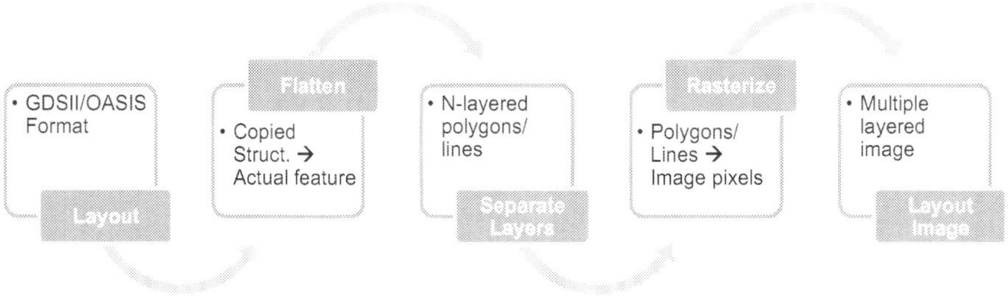

Fig. 3. Preparing Layout Images from a Circuit Layout - Rasterizing Process

removing hierarchical structures by replacing all of the copied parts with actual features, 2) placing the circuit features such as polygons and lines into the correct layers of the circuit, and 3) rasterizing (see Figure 3). This conversion process often lasts hours or even days using a complex computer system with large memory and cannot be executed by the decoder chip.

The final rasterizing step consists of two parts: a) it produces a binary image on a finer grid, and b) the binary image is processed in blocks to generate a gray-level image. In the second step the input binary string is partitioned into $m \times m$ pixel blocks. For each block the number of filled pixels are computed and normalized/quantized to the corresponding gray level. When this gray-level image is transmitted to the EBDW lithography system the lithography writer interprets the gray level (or pixel intensity) as an exposure dose which is controlled by exposing the corresponding region multiple times with an electron beam. Through this process the printed layout pattern becomes more robust to the electron beam proximity effect making better quality circuits.

Our approach is motivated by the compactness of the GDSII/OASIS format and uses corner representation. However, we bypass the complex flattening and rasterizing processes and instead work with a simple decoding process. Yang & Savari (2011) considered some of these ideas for binary images which handle the proximity correction by rasterizing the input binary image on a fine enough grid. Here we will extend these ideas to gray-level images on a coarser grid.

Figure 4 summarizes the components of the compression algorithm. We begin by applying a corner transformation to the image like the one in *Corner2* [Yang & Savari (2011)]. However, unlike *Corner2* this transformation outputs two streams: a "corner stream" and an "intensity stream". The corner stream is a binary stream which locates the polygon corners[1] and the intensity stream is a stream of pixel (corner/edge) intensities. Each stream is input to a separate entropy coding scheme which outputs a compressed bit stream. The corner stream is compressed using a combination of run length encoding [Golomb (1966)], end-of-block coding, and arithmetic coding [Moffat et al. (1998)]. The intensity stream is compressed using end-of-block coding and then compressed by LZ77 [Ziv & Lempel (1977)] and Huffman coding [Huffman (1952)].

In Section 2.2 we will first describe the corner transform process which outputs the corner stream and the intensity stream. In Section 2.3 we will describe the final entropy coding process of the corner stream, and in Section 2.4 we will describe how the intensity value is compressed.

[1] This is not actually a corner, but a horizontal/vertical transition point as explained in Subsection 2.2.

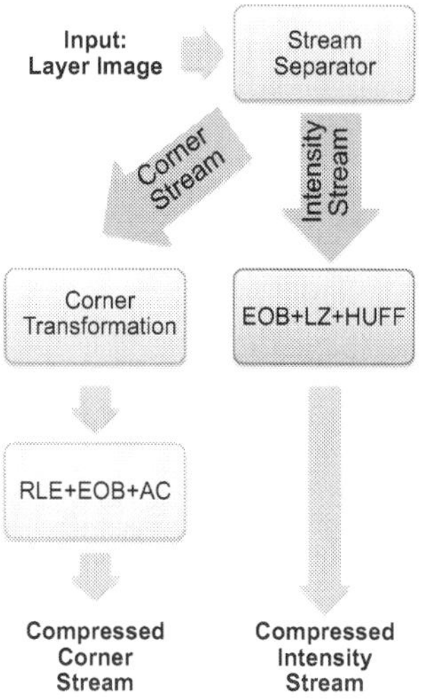

Fig. 4. Compression Process Overview

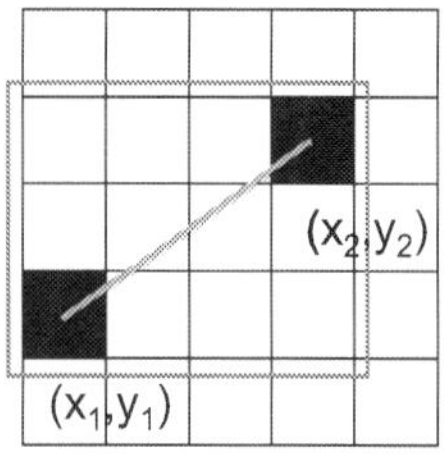

Fig. 5. Required decoder memory (red) to reconstruct a line (blue) from (x_1, y_1) to (x_2, y_2).

2.2 Corner transformation

The GDSII/OASIS representation of a structurally flattened single layer describes the layout polygons by their corner points. This representation requires large decoder memory since the decoder needs to access a memory block of size $(|x_1 - x_2| + 1) \times (|y_1 - y_2| + 1)$ for the encoder to connect an arbitrary pair of points (x_1, y_1) and (x_2, y_2) as in Figure 5. Therefore this representation is infeasible for our application.

However, the rasterizing process becomes much less complex if the angle of a contour line is constrained to a small set. Yang & Savari (2010) took advantage of horizontal and vertical contour lines and decomposed an arbitrary polygon into a collection of Manhattan polygons, i.e., polygons with right angle corners. This approach is effective because most components of circuit layouts are produced using CAD tools which design the circuit in a rectilinear space, and the non-Manhattan parts can also be described by Manhattan components.

In this framework, the decoder scans the image in raster order, i.e., each row in order from left to right. When the decoder processes a corner it must determine whether it should reconstruct a horizontal and/or a vertical line. Observe that a corner is either the beginning of a line going to the right and/or down or the end of a line. Yang & Savari (2010) assigned each pixel one of five possible values – 'not corner,' 'right,' 'right and down,' 'down,' and 'stop.'

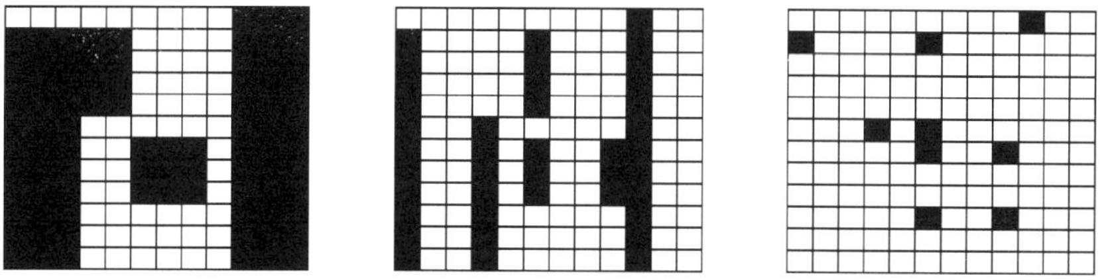

(a) Original Image (b) Horizontal coding of (a) (c) Vertical coding of (b)

Fig. 6. 2-Symbol Corner Transformation

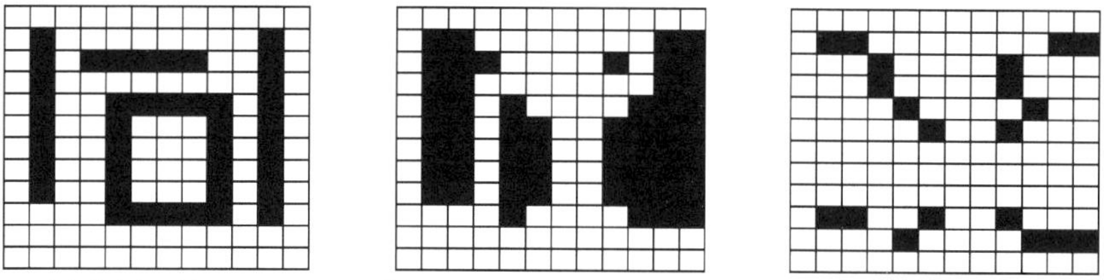

(a) Original Image (b) Horizontal coding of (a) (c) Vertical coding of (b)

Fig. 7. Handling width-1 lines

Yang & Savari (2011) more recently observed that a row (or a column) of the original binary layout image consists of alternating runs of 1s (fill) and runs of 0s (empty). Therefore it is more efficient to encode pixels where there are transitions from 0 to 1 (or 1 to 0) using symbol "1" and to encode the other places using symbol "0." Observe, as in Figure 6(b), that after applying this encoding in the horizontal direction to a collection of Manhattan polygons that the output consists of alternating runs of 1s and 0s in the vertical direction. To increase the compression we repeat this encoding in the other direction to obtain the final corner image. In the binary corner transformation the final encoded image is binary and the "1"-pixels give information about the corners of the polygons. To describe the algorithm we begin with a two-step transformation process and then shorten it to a one-step procedure which requires less memory during the encoding process and is faster than the two-step transformation process.

The two-step transformation process begins with a horizontal encoding step in which we process each row from left to right. For each row, the encoder sets the (imaginary) pixel value to the left of the leftmost pixel to 0 (not filled). If the value of the current pixel differs from the preceding one we represent it with a "1" and otherwise with a "0." The second step inputs the intermediate encoded result to the vertical encoding process in which each column is processed from top to bottom. In the specification of the algorithms x denotes the column index $[1, \cdots, C]$ of the image and y represents the row index $[1, \cdots, R]$. Algorithm 1 illustrates this process.

Line 13 of Algorithm 1 constructs $\texttt{OUT}(x, y) = 1$ only if $\texttt{TEMP}(x, y) \neq \texttt{TEMP}(x, y-1)$. That is, $\texttt{OUT}(x, y) = 1$ only if $\texttt{TEMP}(x, y) = 1$ and $\texttt{TEMP}(x, y-1) = 0$ or if $\texttt{TEMP}(x, y) = 0$ and $\texttt{TEMP}(x, y-1) = 1$. Since $\texttt{TEMP}(x, y) = 1$ only if $\texttt{IN}(x-1, y) \neq \texttt{IN}(x, y)$ as in Line 5, we can shorten the corner transform process to Algorithm 2, which has no need for intermediate memory since pixel (x, y) is processed in terms of the input pixels $(x-1, y)$, $(x, y-1)$, and $(x-1, y-1)$. Algorithm 2 is much faster than Algorithm 1. Finally, Figure 7 illustrates how the transformation handles width-1 lines.

Algorithm 1 Transformation : Two-Step Algorithm

```
Input: Binary layer image IN ∈ {0,1}^(C·R)
Output: Corner image OUT ∈ {0,1}^(C·R)
Intermediate: Temporary image TEMP ∈ {0,1}^(C·R)
  {Horizontal Encoding}
  Initialize TEMP(x,y) = 0, ∀x,y.
  for y = 1 to R do
      for x = 1 to C do
          if IN(x,y) ≠ IN(x − 1,y) then
              TEMP(x,y) = 1.
          end if
      end for
  end for
  {Vertical Encoding}
  Initialize OUT(x,y) = 0, ∀x,y.
  for x = 1 to C do
      for y = 1 to R do
          if TEMP(x,y) ≠ TEMP(x,y − 1) then
              OUT(x,y) = 1.
          end if
      end for
  end for
```

Algorithm 2 Transformation : One-Step Algorithm

```
Input: Binary layer image IN ∈ {0,1}^(C·R)
Output: Corner image OUT ∈ {0,1}^(C·R)
  Initialize OUT(x,y) = 0, ∀x,y.
  for y = 1 to R do
      for x = 1 to C do
          if IN(x − 1,y − 1) = IN(x,y − 1) and IN(x − 1,y) ≠ IN(x,y) then
              OUT(x,y) = 1
          end if
          if IN(x − 1,y − 1) ≠ IN(x,y − 1) and IN(x − 1,y) = IN(x,y) then
              OUT(x,y) = 1
          end if
      end for
  end for
```

Based on the experimental success in Yang & Savari (2010) and Yang & Savari (2011) for binary layout images it is natural to expect that a combination of the corner transformation for the outline of gray-level polygons and a separate representation for the intensity stream would outperform *Block C4*. Note that nLv-level gray images for this application have pixel intensity 0 (empty) outside the polygon outline, nLv − 1 (fully filled) inside the polygon outline, and an element of (0,nLv) along the polygon outline. Therefore we need only consider intensities along polygon corners and edges. Finally, in order to obtain the polygon outline using the corner transformation, we first have to map the gray-level image to a binary image. This is easily done by mapping all of the nonzero intensities to 1 (fill) and leaving the zero intensities (not fill) unchanged.

2.3 Entropy coding - corner stream

The corner stream typically contains long runs of zeroes and is therefore well-suited to compression algorithms like run length encoding [Golomb (1966)] and end-of-block (EOB) coding. Because the corner transformed image is a sparse binary image, if read in raster order (as we read) the string would consists of ones and runs of zeroes. During the compression process, the transitional corners (ones) of the transformed image are written unchanged, but each run of zeroes is described by its run length via an M-ary representation which we next describe. Define the new symbols "2", "3", $\cdots$, "M+1" to respectively represent the base-M symbols "0_M", "1_M", $\cdots$, "$(M-1)_M$". For example, if the transformed stream was "1 00000 00000 1 00000 0000 1 00000 00000 000" and $M = 3$, then the encoding of the stream is "1 323 1 322 1 333" because the run length are 10 (=101_3), 9 (=100_3), and 13 (=111_3), and 2/3/4 to respectively represent $0_3/1_3/2_3$.
We find that the addition of EOB coding helps represent the corner stream more efficiently. When the polygons are aligned and start/end at the same rows of the image the resulting runs of zeroes could be longer than a multiple of the row width. Although this could be handled by choosing M sufficiently large the memory requirements for the encoding and decoding of the final M-ary representation via arithmetic coding [Moffat et al. (1998)] for further compression requires a choice of M as small as possible in our restricted decoder memory setting.
We observe that it is effective to divide each line into k blocks of length L, and we define a new EOB symbol "X". If a run of zeroes appears at the end of a block we represent that run using an end-of-block symbol X instead of an M-ary representation. Hence the encoding for a line of zeroes is k X's instead of approximately $\log_M(kL)$ symbols. For the previous example, if M=2, $k = 5$, and $L = 7$, then the transformed stream "1000000 0000100 0000000 1000000 0000000" is described as "1X 3221X X 1X X," where 2/3 (=$0_2/1_2$) is used for the binary representations of runs of zeroes.
We find that EOB coding results in long runs of "X"s and it is useful to employ an N-ary run length encoding to these runs. For the previous example, if $M = N = 2$, $k = 5$, and $L = 7$, then the next description of the string is "1 4 3221 5 1 5," where 2/3 (or 4/5) handles the binary representation of runs of zeroes (or "X"s).
Finally, we compress the preceding stream using the version of arithmetic coding offered by Witten et al. (1987), and the decoder in this case requires four bytes per alphabet symbol. Since we used $M + N + 1$ symbols[2], $4(M + N + 1)$ bytes were used for arithmetic decoding.

2.4 Entropy coding - intensity stream

The corner stream contains no intensity information. Since we are applying row-by-row decompression (from left to right), the intensity values have to be given in that order. The intensity values that we require are for corner pixels and pixels on the edges. As we have mentioned earlier in Section 2.2, the pixels outside the polygons will have 0 intensity (empty) and pixels inside the polygon boundaries will have $\texttt{nLv} - 1$ intensity (fully filled).
To obtain better prediction we could apply linear prediction along the neighboring pixels as is done in *Block C4*. However, this approach requires the full information of the previous row which translates to decoder memory. Therefore we instead apply EOB encoding to the pixels corresponding to horizontal/vertical edges because the pixel intensity along an edge seldom

[2] M symbols are used for runs of zeroes, N symbols are used for runs of "X"s, and 1 is used for the transitional corners.

changes unless oblique lines are used. We encode the intensity stream as in Algorithm 3. Note that in the algorithm ρ is the length of the intensity stream which is determined at the end of the encoding process.

Algorithm 3 Intensity Stream Encoding

Input: Gray layer image $\texttt{IN} \in \{0, \cdots, \texttt{nLv} - 1\}^{C \cdot R}$
Input: Binary layer image $\texttt{BIN} \in \{0, 1\}^{C \cdot R}$
Output: Intensity stream $\texttt{OUT} \in \{0, \cdots \texttt{nLv} - 1\}^{\rho}$
1: Initialize $\rho = 0$.
2: **for** $y = 1$ **to** R **do**
3:   **for** $x = 1$ **to** C **do**
4:     **if** (x, y) is a corner pixel **then**
5:       $\rho = \rho + 1, \texttt{OUT}(\rho) = \texttt{IN}(x, y)$.
6:     **else if** (x, y) is a horizontal edge pixel with corners at $(x - 1, y)$ and $(x + \alpha, y)$ **then**
7:       **if** $\texttt{IN}(i, y)$ has the same value for $i \in [x, x + \alpha)$ **then**
8:         $\rho = \rho + 1, \texttt{OUT}(\rho) = \texttt{IN}(x, y)$.
9:         $\rho = \rho + 1, \texttt{OUT}(\rho) = \epsilon$.
10:         $x = x + \alpha - 1$.
11:       **else**
12:         **for** $i = x$ **to** $x + \alpha - 1$ **do**
13:           $\rho = \rho + 1, \texttt{OUT}(\rho) = \texttt{IN}(i, y)$
14:         **end for**
15:         $x = i$.
16:       **end if**
17:     **else if** (x, y) is a vertical edge pixel with corners at $(x, y - 1)$ and $(x, y + \beta)$ **then**
18:       **if** $\texttt{IN}(x, j)$ has the same value for $j \in [y, y + \beta)$ **then**
19:         $\rho = \rho + 1, \texttt{OUT}(\rho) = \texttt{IN}(x, y)$.
20:         $\rho = \rho + 1, \texttt{OUT}(\rho) = \epsilon$.
21:         **for** $j = y$ **to** $y + \beta - 1$ **do**
22:           $\texttt{IN}(x, j) = 0$
23:         **end for**
24:       **else**
25:         $\rho = \rho + 1, \texttt{OUT}(\rho) = \texttt{IN}(x, y)$
26:       **end if**
27:     **else if** $\texttt{IN}(x, y) > 0$ and (x, y) is a vertical edge pixel **then**
28:       $\rho = \rho + 1, \texttt{OUT}(\rho) = \texttt{IN}(x, y)$.
29:     **end if**
30:   **end for**
31: **end for**

If the current pixel corresponds to a corner (Lines 4-5), the intensity is represented as is. If the current pixel corresponds to a horizontal edge pixel (Lines 6-16) which starts from the left pixel, check the run of that intensity. If the horizontal edge pixel has constant pixel intensity throughout the entire edge, represent the intensity value followed by an end symbol ϵ and skip to the ending corner pixel (Lines 7-10). Otherwise, write the entire edge intensity as is (Lines 11-15). Similarly, if the current pixel corresponds to a vertical edge (Lines 17-29) which starts from the upper pixel determine whether or not the pixel intensities are fixed throughout the vertical edge. If they are constant then represent the intensity value followed by the end symbol ϵ (Lines 18-20) and reset the intensity values for the following rows (Lines 21-23) so that they are not processed in Lines 27-29. Otherwise, write the intensity value as is and proceed (Lines 24-26). Finally, the remaining vertical edge pixel intensities are written in Lines 27-29.

After the entire intensity stream has been processed, compress the output stream using LZ77 and Huffman coding. The LZ77 algorithm by Ziv & Lempel (1977) compresses the stream by finding matches from the previously processed data. When a pattern is repeated within the search region, it could be encoded using a short codeword. Huffman coding is used at the end of LZ77 to represent the LZ77 stream more efficiently. The combination of LZ77 and Huffman coding is widely used in a number of compression algorithms such as *gzip*. We used *zlib* [zlib (2010)] to implement it. The compression rates depend on the size of the LZ77 search region and the dictionary for the Huffman code. Because of the decoder memory restrictions we chose an encoder needing only 2,048 bytes of memory for the dictionary. 2,048 bytes is slightly less than the memory used to describe an entire row of our benchmark circuit. However, since we were applying this only to the intensity stream we were able to match more rows than *Block C4*.

3. Decoder

The decoder consists of an intensity stream decoder and a corner stream decoder as in Figure 8. The intensity stream decoder is actually an entropy decoder which can be decomposed into a Huffman decoder and an LZ77 decoder. The corner stream decoder consists of an entropy decoder which consists of an arithmetic decoder, a run length decoder, an end-of-block decoder, and a corner transform decoder which reconstructs the polygons from the entropy decoder output. The corner transform decoder utilizes the output of corner stream entropy decoder to reconstruct the polygon outlines and uses the output of the intensity stream decoder to reconstruct the polygon pixel intensity.

The entire process works on a row-by-row fashion. Since each part of the decoding procedure (arithmetic decoding, run length decoding, end-of-block decoding, inverse corner transformation, LZ77 decoding, and Huffman decoding) is simple and works with restricted decoder memory, the entire decoder can be implemented in hardware. Note that the most complex part will be the arithmetic decoder which is widely implemented in microcircuits [Peon et al. (1997)], and the other parts are comprised of simple branch, copy, and computation operations as we will see in the following subsection.

3.1 Intensity stream decoder

Decompressing the intensity stream is straightforward. We apply LZ77 and Huffman decoding to obtain the ϵ-coded intensity stream. As we have mentioned in Section 2.4, the decoder requires 2,048 bytes of memory to decode the LZ77 and Huffman codes. The ϵ-coded intensity stream is passed on to the corner transform decoder for the final reconstruction. Note that the decoder does not decompress the entire compressed intensity stream at once but rather decompresses some number of ϵ-coded intensity symbols at the request of the corner transform decoder. The detailed decompression of the ϵ-coded intensity stream will be discussed at the end of the next subsection.

3.2 Corner stream decoder - corner transform decoder

As we have mentioned earlier, the corner stream decoder consists of an entropy decoder and a corner transform decoder. The entropy decoder reverses the procedure of the entropy encoder of Section 2.3. It first reconstructs the run length and end-of-block encoded stream using the

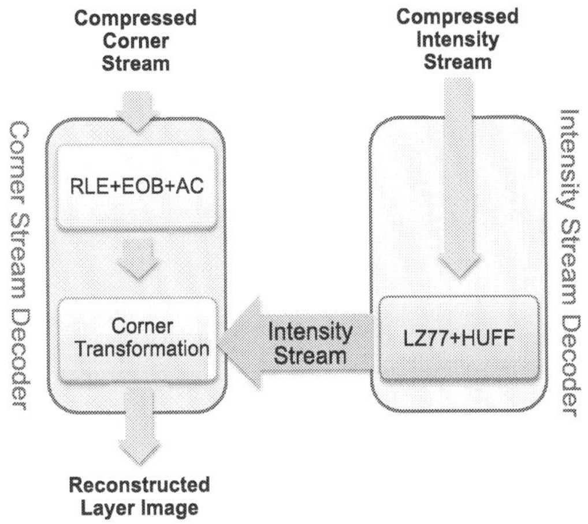

Fig. 8. Decoder Overview: Note that the decompressed corner stream is input into the image reconstructor in a row-by-row fashion and the intensity stream is actually ϵ-coded as in Section 2.4.

arithmetic decoder. Then, depending on the symbol, runs of zeroes (symbols $0_M, \cdots, (M-1)_M$), runs of EOBs (symbols $0_N, \cdots, (N-1)_N$), or the corners (symbol 1) are reconstructed. Finally, the output of the entropy decoder is a binary corner image. In this section we will focus on the operation of the corner transform decoder and why it can run in a row-by-row fashion. This feature makes our approach well-suited to the restricted memory available to an EBDW writer. In our corner transform decoder we use a row buffer `BUFF` to hold the status of the previous (decoded) row. It stores a binary representation of the status of each pixel and therefore consumes *width* bits of memory. "0" denotes 'no transition' while "1" denotes the 'transition' which delineates the starting/ending point of a vertical line. Moreover, since we need the polygon boundaries - corners and horizontal/vertical edge pixels - we use another row buffer `CNR` to hold the boundary status of the previous row. This also requires *width* bits of memory.

Since the algorithm is long, we have split its description into two parts, namely Algorithms 4 and 5. The input to the algorithm is the corner image, and the algorithm outputs the binary layer image and the corner map which shows whether a pixel in the binary layer image is outside all polygons (O), inside a polygon boundary (I), a corner (C), a horizontal edge pixel (H), or a vertical edge pixel (V) which will be used to reconstruct the pixel intensity along with the intensity stream decoder.

The first part of the algorithm, illustrated in Algorithm 4, shows how the buffers are used to pass previous row information to the current row so that the decoding process could be applied in a row-by-row fashion. Lines 5-7 process the binary image buffer `BUFF`. If the buffer is filled then the corresponding pixel is part of a vertical edge and it is filled. Lines 8-15 process the corner map buffer `CNR`. If the corresponding `CNR` pixel does not form a run, then the corresponding pixel above it was a vertical edge pixel (Line 10). The starting and ending points of the runs of 1s are interpreted as the corners (Line 12) and the other pixels in between are translated as horizontal edge pixels (Line 14).

Algorithm 4 Inverse Corner Transformation - Part I

Input: Corner image $\texttt{IN} \in \{0,1\}^{C \cdot R}$
Output: Binary layer image $\texttt{BIN} \in \{0,1\}^{C \cdot R}$
Output: Corner map $\texttt{OUT} \in \{O, I, C, H, V\}^{C \cdot R}$
Intermediate: Buffer for image layer $\texttt{BUFF} \in \{0,1\}^{R}$
Intermediate: Buffer for corner map $\texttt{CNR} \in \{0,1\}^{R}$
1: Initialize $\texttt{BUFF}(x) = \texttt{CNR}(x) = 0, \forall x$.
2: Initialize $\texttt{BIN}(x,y) = 0$ and $\texttt{OUT}(x,y) = O, \forall x, y$.
3: **for** $y = 1$ **to** R **do**
4:     **for** $x = 1$ **to** C **do**
5:         **if** $\texttt{BUFF}(x) = 1$ **then**
6:             $\texttt{BIN}(x,y) = 1$
7:         **end if**
8:         **if** $\texttt{CNR}(x) = 1$ **then**
9:             **if** $\texttt{CNR}(x-1) = 0$ and $\texttt{CNR}(x+1) = 0$ **then**
10:                 $\texttt{OUT}(x, y-1) = V$.
11:             **else if** $\texttt{CNR}(x-1) = 0$ or $\texttt{CNR}(x+1) = 0$ **then**
12:                 $\texttt{OUT}(x, y-1) = C$.
13:             **else**
14:                 $\texttt{OUT}(x, y-1) = H$.
15:             **end if**
16:         **end if**
17:     **end for**
18: **end for**

The second part of the algorithm is shown in Algorithm 5. Note that line 18 of Algorithm 4 and line 1 of Algorithm 5 should be removed when they are implemented; we inserted them to make the loops complete in each part. Lines 5-22 handle the processing of a transitional corner. If $\texttt{BUFF}(x) = 0$, then the transitional corner starts (or ends) a new horizontal edge at row y. Since the entire algorithm is designed to output the result of the previous row $y-1$, we store that horizontal edge in CNR to process it during Algorithm 4. If instead $\texttt{BUFF}(x) = 1$, then the transitional corner starts (or ends) the removal of the pre-existing horizontal edge from the buffer; i.e., the corresponding horizontal edge on the previous row was actually the horizontal boundary.

Similarly, lines 23-33 determine whether the corresponding empty corner pixel is a vertical edge pixel or an interior pixel of a polygon. Finally, note that the output OUT is always processed as a function of row $y-1$, the data that is stored in BUFF and CNR, and row y of IN. Since the buffers treat the inter-row dependencies the entire algorithm can be applied in a row-by-row fashion.

We next utilize the intensity stream to reconstruct the polygon pixel intensities. As we have mentioned earlier, the pixels in the interior of each polygon have full intensity and the pixels not within any polygon are empty (not filled). We use Algorithm 6 to reconstruct the intensity stream of the polygon boundaries. Note that the corner map CNR is the output of the inverse corner transform. In this algorithm, buffer BUFF is used to handle the vertical edge reconstruction.

Algorithm 6 is similar to the previous algorithm in that it utilizes a row buffer to handle the information that has been processed in the previous row. Lines 5-6 deal with the interior polygon pixels, lines 7-9 deal with corner pixels, lines 10-20 reconstruct the horizontal edges,

Algorithm 5 Inverse Corner Transformation - Part II
Run α means $\texttt{BIN}(x,y) = \texttt{BIN}(x,y) \bigoplus \texttt{Fill}, \texttt{BUFF}(x) = \texttt{BUFF}(x) \bigoplus \texttt{Fill}, x = x+1$

```
1: for y = 1 to R do
2:    Fill = 0.
3:    Initialize CNR(x) = 0, ∀x.
4:    for x = 1 to C do
5:       if IN(x,y) = 1 then
6:          Fill = Fill ⊕ 1
7:          if BUFF(x) = 0 then
8:             Run α
9:             while IN(x,y) = 0 do
10:               CNR(x) = 1
11:               Run α
12:            end while
13:         else
14:            OUT(x,y − 1) = C.
15:            Run α
16:            while IN(x,y) = 0 do
17:               OUT(x,y − 1) = H
18:               Run α
19:            end while
20:            OUT(x − 1,y − 1) = C.
21:         end if
22:         Fill = Fill ⊕ 1
23:      else
24:         if BUFF(x) = 1 then
25:            OUT(x,y − 1) = V
26:            Run α
27:            while BUFF(x) = 1 do
28:               OUT(x,y − 1) = I
29:               Run α
30:            end while
31:            OUT(x − 1,y − 1) = V
32:         end if
33:      end if
34:   end for
35: end for
```

and lines 21-33 reconstruct the vertical edges. Because the intensity buffer `BUFF` requires a full row, *width* × `bits per pixel` bits are required for this process.

Since the decoder requires two *width*-bit buffers for corner reconstruction and one *width* × `bits per pixel`-bit buffer for intensity decoding, the total decoder memory requirement is *width* × (2 + `bits per pixel`). However, when `bits per pixel` is at least 3 the decoder memory requirement is less than that of *Block C4* which is 2 × `bits per pixel` × *width*.

Algorithm 6 Intensity Stream Decoding

```
Input: Intensity stream IN ∈ {0,···nLv − 1}^ρ
Input: Corner map CNR ∈ {O, I, C, H, V}^(C·R)
Output: Gray layer image OUT ∈ {0,··· ,nLv − 1}^(C·R)
Intermediate: Intensity buffer BUFF ∈ {0,··· ,nLv − 1}^C
Initialize t = 1.
Initialize OUT(x, y) = 0, ∀x, y.
for y = 1 to R do
   for x = 1 to C do
      if CNR(x, y) = I then
         OUT(x, y) = nLv.
      else if CNR(x, y) = C then
         OUT(x, y) = IN(x, y), t = t + 1.
         BUFF(x) = 0.
      else if CNR(x, y) = H then
         if IN(t + 1) = ϵ then
            while CNR(x, y) = H do
               OUT(ρ) = IN(t), x = x + 1.
            end while
            t = t + 2.
         else
            while CNR(x, y) = H do
               OUT(ρ) = IN(t), t = t + 1.
            end while
         end if
      else if CNR(x, y) = V then
         if BUFF(x) = 0 then
            if IN(t + 1) = ϵ then
               OUT(x, y) = IN(t).
               Set BUFF(x) = IN(t).
               t = t + 2.
            else
               OUT(x, y) = IN(t), t = t + 1.
            end if
         else
            OUT(x, y) = BUFF(x).
         end if
      end if
   end for
end for
```

4. Experimental results

We tested the algorithm on a memory circuit which has 13 layers, uses 500 nm lithography technology, and has repeated memory cell structure. The circuit was rasterized on a 1nm grid and then regrouped with block size 250×250. The intensity for each block was quantized using 32 levels. The resulting 5-bit gray level image rasterized on a 250nm grid satisfies the minimum edge spacing of 8nm. For this circuit our algorithm *CornerGray* runs on the entire

layout image, but *Block C4* [Liu et al. (2007)] has a memory shortage/failure. We therefore divided the image in a way to enable the successful application of *Block C4*.
Our code for *CornerGray* and *Block C4* is in C/C++ for the former and in C# for the latter. The experiments were conducted on a laptop computer having a 2GHz Intel Core i7 CPU and 4GB RAM. The decoder memory requirement of *CornerGray* (with parameters $M = N = 64$) was $width \times (2 + \texttt{bits per pixel})/8 + 4(M + N + 1) + 2,048 = 8,514$ bytes while that of *Block C4* was $width \times \texttt{bits per pixel} \times 0.25 + 427 = 8,927$. We require 5% less decoder memory than *Block C4*. Tables 1 and 2 indicate that the compression performance of *CornerGray* is 6% better than *Block C4* and is 189 times faster in encoding and 24 times faster in decoding.

CornerGray Result				
Layer #	File Size (bytes)	Compression Ratio (x)	Encoding Time (s)	Decoding Time (s)
Layer1	2,826	10,720	1.27	0.28
Layer2	570,855	53	1.47	0.48
Layer3	2,826	10,720	1.28	0.28
Layer4	805,477	38	1.49	0.54
Layer5	116,516	260	1.29	0.33
Layer6	116,516	260	1.29	0.33
Layer7	531,770	57	1.44	0.44
Layer8	1,730,054	18	1.84	0.80
Layer9	255,919	118	1.29	0.37
Layer10	1,233,137	25	1.68	0.66
Layer11	107,631	281	1.21	0.31
Layer12	22,092	1,371	1.22	0.29
Layer13	116,516	260	1.27	0.33
Total	5,612,135	70	18.04	5.45

Table 1. Compression Result - *CornerGray*

Block C4 Result				
Layer #	File Size (bytes)	Compression Ratio (x)	Encoding Time (s)	Decoding Time (s)
Layer1	44,436	682	153.89	6.47
Layer2	356,424	85	170.48	7.14
Layer3	44,436	682	182.02	8.11
Layer4	595,116	51	181.68	7.32
Layer5	164,728	184	201.80	7.75
Layer6	164,728	184	300.37	11.21
Layer7	894,308	34	355.59	13.95
Layer8	1,426,824	21	356.40	11.47
Layer9	643,220	47	328.52	12.42
Layer10	1,021,240	30	302.27	9.70
Layer11	346,480	87	332.92	12.50
Layer12	115,040	263	311.24	10.70
Layer13	166,744	182	238.10	9.89
Total	5,983,724	66	3,415.28	128.63

Table 2. Compression Result - *Block C4*

The input image size was $6,800 \times 7,128$ (=30,294,000 bytes) and the *CornerGray* parameters were $N = M = 64$ for all layers. The compression ratio in Table 1 and Table 2 is defined as

$$\frac{\texttt{Input File Size}}{\texttt{Compressed File Size}}.$$

However, the results show that *CornerGray* is relatively weak for handling massively repeated patterns. Among the 13 layers, *CornerGray* did not perform well (compared to *Block C4*) for Layer2, Layer4, Layer8, and Layer10. These layers contained patterns which consist of a large array and Layer8 and Layer10 in particular had complex patterns which were scattered. For these parts the complex LZ-based copying part of *Block C4* resulted in better performance. Hence, more sophisticated pattern matching is required to improve *CornerGray*.

5. Conclusion

In the previous section we saw that the algorithm *CornerGray* outperforms *Block C4* and is considerably faster. The improvement in *CornerGray* over *Block C4* is a result of different techniques. Our corner location approach is simpler than the context prediction used by *Block C4* to handle the irregular parts of layer images. However, *CornerGray* needs a better pattern handling scheme for circuits which contain massively repeated patterns. We are currently trying to generalize the frequent pattern replacement component of *Corner2* [Yang & Savari (2011)] in order to handle frequent patterns within binary layout images and expect similar compression improvements for gray level images. The decoding operations for *CornerGray* include common decompression schemes which are widely implemented in hardware as well as simple branches, compares, and memory copies for the corner transformation part. Therefore our decoder can be deployed using hardware and is an approach to the data delivery problem of maskless lithography systems.

6. Acknowledgment

The second author was supported in part from NSF Grant CCF-1017303.

7. References

Artwork (2010) GDS_RIP. *http://www.artwork.com/gdsii/gds_rip/index.htm*

Chen, Y.; Kahng, A.;Robins, G.; Zelikovsky, A. & Zheng, Y. (2004) Evaluation of the new OASIS format for layout fill compression, *Proceedings of the 2004 11th IEEE International Conference on Electronics, Circuits and Systems*, pp. 377- 382, Tel-Aviv, Israel, December 2004.

Chokshi, N.; Shroff, Y. & Oldham, W. (1999) Maskless EUV Lithography, *International Conference on Electron, Ion, and Photon Beam Technology and Nanofabrication*, Macro Island, FL, June 1999.

Dai, V. & Zakhor, A. (2006) Lossless compression of VLSI layout image data. *IEEE Trans. Image Processing*, Vol. 15, No. 9, pp. 2522-2530.

Dai, V. (2008) *Data Compression for Maskless Lithography Systems: Architecture, Algorithms, and Implementation*, Ph.D. Thesis, Department of Electrical Engineering and Computer Science, University of California, Berkeley.

Golomb, S. (1966) Run Length Encodings. *IEEE Trans. Information Theory*, Vol. 12, No.3, pp. 399-401.

Gu, A. & Zakhor, A. (2008) Lossless Compression Algorithms for Hierarchical IC Layout. *IEEE Trans. Semiconductor Manufacturing*, Vol. 21, No. 2, pp. 285-296.

Huffman, D.A. (1952) A Method for the Construction of Minimum-Redundancy Codes, *Proceedings of the I.R.E.*, pp. 1098-1102.

Klein, C.; Klikovitsa, J.; Szikszaia, L.; Platzgummera, E. & Loeschner, H. (2009), Projection Mask-Less Lithography (PML2). *Microelectronic Engineering*, Vol. 87, No. 5-8, pp. 1154-1158.

Petric, P.; Bevis, C.; Brodie, A.; Carroll, A.; Cheung, A.; Grella, L.; McCord, M.; Percy, H.; Standiford, K. & Zywno, M. (2009) REBL nanowriter: Reflective Electron Beam Lithography, *Proceedings of the SPIE 7271 - Alternative Lithographic Technologies*, pp. 727107-1 - 727107-15, San Jose, CA, March, 2009.

Kuhn, M. (2008) JBIG-KIT. *http://www.cl.cam.ac.uk/~mgk25/jbigkit/*

LibTIFF (2010) LibTIFF. *http://www.libtiff.org*

Lin, B. (2009) NGL comparable to 193-nm lithography in cost, footprint, and power consumption. *Microelectronic Engineering*, Vol 86, pp. 442-447.

Liu, H.; Dai, V.; Zakhor, A. & Nikolic, B. (2007) Reduced Complexity Compression Algorithms for Direct-Write Maskless Lithography Systems. *SPIE Journal of Microlithography, MEMS, and MOEMS*, pp. 1-12.

Wieland, M.; de Boer, G.; Ten Berge, G.; Jager, R.; van de Peut, T.; Peijster, J.; Slot, E.; Steenbrink, S.; Teepen, T.; van Veen, A. & Kampherbeek, B. (2009) MAPPER: high-throughput maskless lithography, *Proceedings of the SPIE 7271 - Alternative Lithographic Technologies*, pp. 72710O-1 - 72710O-8, San Jose, CA, March, 2009.

Moffat, A.; Neal, R. & Witten, I. (1998) Arithmetic coding revisited. *ACM Transactions on Information Systems*, Vol. 16, No. 3, pp. 256-294.

Pennebaker, W.; Mitchell, J.; Langdon, G. & Arps, R. (1988) An overview of the basic principles of the Q-coder adaptive binary arithmetic coder. *IBM Journal of research and development*, Vol. 32, No. 6, pp. 771-726.

Peon, M.; Osorio, R. & Bruguera, J. (1997) A VLSI implementation of an arithmetic coder for image compression, *Proceedings of the 23rd EUROMICRO Conference '97 New Frontiers of Information Technology*, pp. 591-598.

Reich, A.; Nakagawa, K. & Boone, R. (2003) OASIS vs. GDSII stream format efficiency, *Proceedings of SPIE 23rd Annual BACUS Symposium on Photomask Technology*, pp. 163-173, Monterey, CA, September 2003.

Rubin, S. (1987) *Computer Aids for VLSI Design, 2nd ed.*, Addison-Wesley, Boston, MA.

Witten, I.; Neal, R. & Cleary, J. (1987) Arithmetic Coding for Data Compression, *Communications of the ACM*, Vol. 30, pp. 520-540.

Yang, J. & Savari, S. (2010) A Lossless Circuit Layout Image Compression Algorithm for Maskless Lithography Systems, *Proceedings of the 2010 Data Compression Conference*, pp. 109-118, Snowbird, UT, March 2010.

Yang, J. & Savari, S. (2011) A Lossless Circuit Layout Image Compression Algorithm for Electron Beam Direct Write Lithography Systems, *Proceedings of SPIE 7970 - Advanced Lithographic Technologies III*, pp.79701U-1 - 79701U-12, San Jose, CA, March 2011.

Ziv, J. & Lempel, A. (1977) A Universal Algorithm for Sequential Data Compression. *IEEE Transactions on Information Theory*, Vol. 23, No. 3, pp. 337-343.

zlib (2010) zlib: A Massively Spiffy Yet Delicately Unobtrusive Compression Library. *http://zlib.net*

Part 2

Nanoimprint and Soft Lithography

6

Ultrafast Fabrication of Metal Nanostructures Using Pulsed Laser Melting

Bo Cui
Waterloo Institute for Nanotechnology (WIN)
University of Waterloo
Canada

1. Introduction

Nanoimprint lithography (NIL) is a mechanical molding process. Two formats of NIL are commonly used: thermal NIL and UV-curing NIL. In thermal NIL, the resist is typically a thermoplastic or thermoset polymer that becomes soft at temperatures well above its glass transition temperature, thus it can be imprinted by a rigid mold. NIL has demonstrated low cost and high throughput patterning (Chou 1996, Schift 2008) with a high resolution of sub-5 nm (Austin 2004, Hua 2004). However, due to their very high melting temperature, direct patterning of metal or silicon using an NIL-like process is very challenging.

Previously, metal structure fabrication by transfer-printing or imprinting without melting has been demonstrated. For instance, Kim et al used a hard silicon mold to press into a metal film on a substrate with a high pressure of 290 MPa that fractured the metal film at the mold pattern edge; and they then peeled off the metal and transfer-printed it to another substrate, achieving sub-micrometer resolution (Kim 2000). Yu et al fabricated Ag metal electrodes for organic light emitting devices by transfer-printing with a polydimethylsiloxane (PDMS) stamp, which peels off the portion of the metal film that is in contact with the protruded PDMS patterns (Yu 2007). However, it has a low resolution of 13 μm and a low yield that depends on the peel direction. Buzzi et al, Pang et al and Hirai et al applied ultra-high pressure of several hundred MPa to directly imprint a solid metal at room temperature (Buzzi 2008, Pang 1998, Hirai 2003). Chen et al and Chuang et al used a mold having a sharp geometry to deform or imprint (penetrate) metal thin films (<50 nm) deposited on a soft polymer bottom layer at pressure of 10-20 MPa and temperature slightly lower than the glass transition temperature of the bottom polymer (Chen 2006, Chuang 2008). The above methods suffer from poor patterning resolution and are limited to ductile metals because, during the patterning, the metals are in the hard solid phase.

In this chapter, we present a method of direct patterning of metal or silicon nanostructures using a pulsed laser that can melt the metal or silicon. Like the aforementioned method, the most prominent feature of this technique is that it is a one-step patterning process – it replaces the steps of resist patterning in lithography, subsequent pattern transfer by etching or liftoff, and resist removal all by one single simple step. In addition, as molten metal or silicon has very low viscosity (on the same order as water), this technique can pattern them within ~100 ns. It can also anneal the patterned metallic nanostructures. More importantly, the current method is not limited to ductile metals, and hard metals like nickel can be structured readily.

Various applications have been developed, and here we will present laser assisted direct imprint, wafer planarization, via-hole filling, transfer printing, and nano-tip formation. Other important applications using pulsed laser nanofabrication, such as self-perfection by liquefaction (SPEL) that removes nanostructure fabrication defects (notably line edge roughness) (Chou 2008, Xia 2010), and nano-channel fabrication using sealing by laser-melting (Xia 2008), will not be presented.

2. Fundamentals of laser-matter interaction

Fig. 1 illustrates schematically the laser-matter interaction at a wide range of power and/or energy density. The pulsed laser we used at the Nanostructure Laboratory, Princeton University, is a 308 nm XeCl excimer laser with pulse duration of 20 ns (FWHM), maximum output energy of 200 mJ per pulse, and power of 160 MW/cm^2 for a beam size of (2.5 mm)^2 if assuming a rectangular temporal pulse shape. With this laser, a power density of order of 10 MW/cm^2 is needed to melt a 200 nm-thick thin metal or semiconductor film on an insulator, whereas order 100 MW/cm^2 is necessary for substantial vaporization and/or plasma formation. In the following sections, three topics will be covered in some detail: light absorption by metals and semiconductors, transient heat transfer and temperature distribution, and vaporization, plasma formation and plasma propagation.

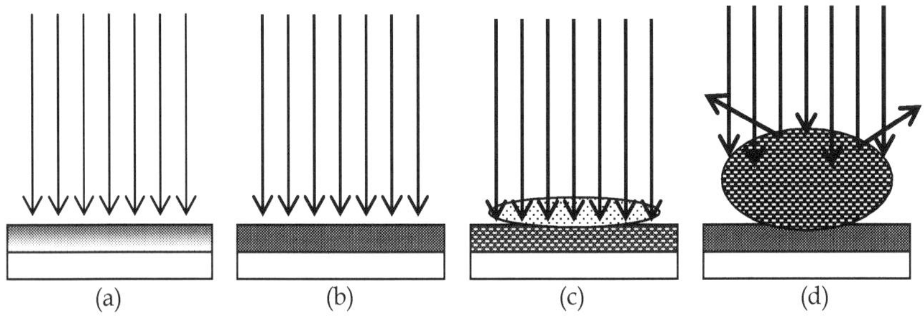

Fig. 1. Schematic laser-material interaction. (a) Absorption and heating. (b) Melting and flowing. (c) Vaporization. (d) Plasma formation in front of the target. Under certain conditions the plasma can detach from the target and propagate toward the laser beam.

2.1 Absorption of laser light

Laser light must first be absorbed in order to cause any lasting effect on a material. Electromagnetic radiation with a wavelength between UV and IR interacts exclusively with electrons, as atoms are too heavy to respond significantly to the high frequencies (>10^{13} Hz) involved. Therefore, the optical properties of matter are determined by the energy states of its valence electrons (bound or free). Generally, bound electrons can only weakly respond to the external electromagnetic wave and merely affect its phase velocity. Free electrons are able to be accelerated, i.e., to extract energy from the field. However, since the external field is periodically

changing, the oscillating electrons re-radiate their kinetic energy, unless they undergo frequent collisions with the atoms - in this case energy is transmitted to the lattice (absorbed) and the external field is weakened. Re-radiation of the energy is the cause of reflection.
For absorbing media the refractive index is a complex number with the form n=n+ik. The reflectivity at normal incidence is (n=1 for air or vacuum)

$$R = \frac{(n-1)^2 + k^2}{(n+1)^2 + k^2} \tag{1}$$

The emissivity, or the fraction of incident radiation absorbed, is then

$$\varepsilon \equiv 1 - R = \frac{4n}{(n+1)^2 + k^2} \tag{2}$$

The absorption coefficient is defined as

$$\alpha \equiv -\frac{1}{I}\frac{dI}{dz} = \frac{4\pi k}{\lambda} \tag{3}$$

where I is the intensity of the incident light (also termed irradiance, in W/cm^2). The inverse of α is referred as the absorption length. For normal incidence the power density (in W/cm^3) deposited at depth z is

$$\Phi(z) = I(1-R)\alpha e^{-\alpha \cdot z} \tag{4}$$

2.1.1 Low light intensity and temperature regime

Both n and k (thus α) are generally functions of wavelength and temperature. For a semiconductor such as silicon (see Fig. 2 (a)), the absorption coefficient α increases with decreasing wavelength λ gradually near its indirect band-gap at 1.1 eV (1.13 μm), but increases rather abruptly near its direct band-gap at 3.4 eV (0.36 μm) because no phonons are involved for direct transition. At photon energies significantly exceeding E_g, the optical response of semiconductors tends towards that of metals.
Fig. 2 (b) depicts the light-coupling parameters for Al and Au. Metals, due to their large density of free electrons close to the Fermi level, have large absorption coefficients (corresponding to absorption lengths α^{-1} around only 10 nm) over the whole spectrum. Their reflectivity is more difficult to determine than that of semiconductors. Generally, their reflectivity is high above a certain critical wavelength that is related to its electron plasma frequency ω_p given by $\sqrt{(4\pi Ne^2/m_e)} \propto \sqrt{N}$, where N is electron density and m_e is electron mass. Below the critical wavelength, which is in the UV or visible part of the spectrum, R decreases sharply. For Au and Cu, the plasma frequency is reduced (which causes their yellow color) by transition from d-band states.

2.1.2 High light intensity and/or temperature regime

At higher light intensities and/or higher temperatures, the optical properties can be modified by three mechanisms: lattice heating, free carrier generation (for semiconductors only) and excitation. At elevated temperatures, more phonons are present, leading to an increase of lattice-carrier collision. As a result, the reflectivity of most metals decreases with increasing temperature, and this effect makes metals susceptible to thermal runaway.

For semiconductors like silicon, a large number of free carriers can be generated not only by thermal excitation at high temperature and further facilitated by the band-gap narrowing, but also by photons at a rate of $I\sigma/h\nu$ (I is light intensity, σ is the absorption cross section) when $h\nu > E_g$. Fig. 3 depicts the energy diagram of the electrons (holes have similar diagram) under intense irradiation such that the carrier generation rate exceeds that of carrier relaxation rate. Due to free carrier generation, the carrier density and thus the absorption coefficient α of semiconductors increase strongly as a function of light intensity. The effect of free carriers on the optical properties is to reduce the real part and to increase the imaginary part of **n**. As a result, free carrier generation tends to decrease the reflectance for $\omega_p < \omega$ (now $\omega_p \propto \sqrt{N}$ is a variable depending on light intensity, while ω is fixed). Only for carrier density N large enough that $\omega_p > \omega$ does the reflectance begin to increase with the light intensity towards unity by further increasing ω_p. Such a rich variation of reflectivity on light intensity and time has been observed for crystalline Si irradiated by red 90 fs pulses (Shank 1983).

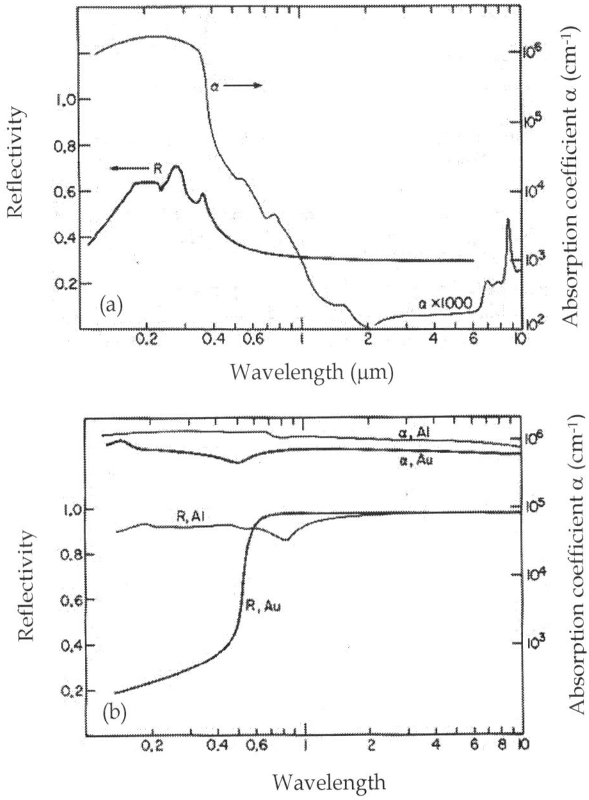

Fig. 2. Reflection and absorption coefficient as a function of wavelength for (a) silicon, and (b) Al and Au. The critical wavelength of Al is below 100 nm, so its reflectivity is close to 1 over the data range in the figure. (from Allmen, 1995)

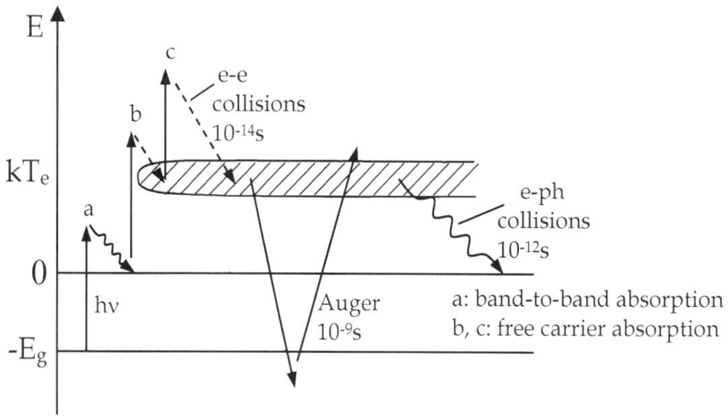

Fig. 3. Schematic energy diagram of the electrons in a semiconductor under intense irradiation. The electron temperature T_e is higher than the lattice temperature. The shaded area depicts electron plasma. (adapted from Poate 1982)

At sufficiently high light intensity, the rate of energy gain by carriers exceeds the rate of energy loss to the lattice. The result is a splitting up of carrier and lattice temperatures (see Fig. 3), and the carriers behave collectively, i.e., according to the equations governing a plasma (but this plasma behavior is not to be confused with the generation of actual plasma shown in Fig. 1(d); the plasma here is just a large number of electrons inside the conduction band of the semiconductor). The plasma may be regarded as an independent system which absorbs light and exchanges energy with the lattice. As far as coupling is concerned, the effect of the plasma is ultimately to limit the rate of local energy deposition by increasing R at high carrier density such that $\omega_p > \omega$, and by diffusing away the hot carriers from the absorbing surface to the material underneath, which effectively increases the heated volume.

Finally, Si and Ge become metallic (silicon's electrical conductivity is 30 times higher) upon melting due to bandgap collapsing, leading to a strong increase in both α and R.

2.2 Heating and melting by laser

Heating in an irradiated sample is a consequence of the balance between the deposited energy, governed by optical parameters of the sample and characteristics of the laser pulse, and the heat diffusion, determined by thermal parameters and the pulse duration. Strictly speaking, as discussed in the previous section, the primary product of absorbed laser light is not heat but the carriers' excess energy. But for simplicity, and because the our laser pulse has a relatively long duration of 20 ns, it can be assumed that the overall energy relaxation time τ_E is negligible compared to the pulse duration (τ_E~10^{-13}s for metals, ~10^{-12} – 10^{-9}s for semiconductors). With this assumption, a simple estimate of the temperature rise is possible for two limiting cases in which the optical absorption depth is small compared with the thermal diffusion length during the pulse (i.e. $\alpha^{-1} << \delta \equiv 2\sqrt{Dt}$.), or vice versa. However, for many practical cases, complex analytical or numerical evaluations must be performed. In particular, if phase transitions occur during the irradiation, only numerical solutions are available.

In the current experiment with XeCl excimer laser irradiation of semiconductors or metals, the absorption depth is much smaller than the thermal diffusion length. However, great complexities come from the multi-layered substrate/mold system bearing on its surface subwavelength patterns. Moreover, melting or even boiling was always seen in these experiments. And the parameters such as reflectivity, absorption coefficient, thermal conductivity and specific heat, are all strong functions of temperature. As a consequence, even numerical calculation will be very difficult unless significant simplifying assumptions are made. Therefore, in this section, only a rough estimation of the temperature distribution will be derived.

First, consider the case when phase change is absent. The excimer laser has a 2.5 mm beam size, which is many orders larger than the film thickness, so the heating and cooling can be approximated as a one-dimensional problem governed by the following equation:

$$\frac{\partial T}{\partial t}=\frac{\Phi(z,t)}{\rho C_p}+\frac{1}{\rho C_p}\frac{\partial}{\partial z}(\kappa\frac{\partial T}{\partial z}) \tag{5}$$

where $\Phi(z,t)$ is given by Eq. 4 that represents the source term, and ρ, C_p and κ are the density, specific heat and thermal conductivity, respectively. This equation can be solved analytically under the following assumptions: no temperature dependence of all optical and thermal parameters, rectangular temporal pulse shape of duration t_p and intensity I_0 (the actual pulse shape is close to Gaussian), very small absorption length ($\alpha^{-1}<<\delta$), and infinite sample thickness. The temperature distribution during and after the pulse is described by (Allmen 1995)

$$T(z,t)=\frac{2I_0(1-R)}{\kappa}\sqrt{Dt}ierfc[z/2\sqrt{Dt}]-[t>t_p]\sqrt{D(t-t_p)}ierfc[z/2\sqrt{D(t-t_p)}] \tag{6}$$

$$ierfc(x)=\int_x^\infty erfc(x')dx' \text{ and } erfc(x)=\frac{2}{\sqrt{\pi}}\int_x^\infty e^{-t^2}dt \tag{6'}$$

where thermal diffusivity $D\equiv\kappa/\rho C_p$ and the symbol $[t>t_p]$ equals one if $t>t_p$ and zero otherwise. The characteristic thermal diffusion length within time t is defined as $\delta\equiv2\sqrt{Dt}$. Eq. 6 is visualized in Fig. 4, which suggests the maximum temperature at $z=\delta=2\sqrt{Dt_p}$ is about 1/5 that of peak surface temperature and it is reached at about $t=2t_p$. The surface temperature at the end of the pulse is

$$T(0,t_p)=\frac{2I_0(1-R)}{\kappa}\sqrt{Dt_p/\pi}\propto\sqrt{t_p} \tag{7}$$

On the other hand, for metal or semiconductor film sandwiched between the thermal insulating mold and substrate, if the film thickness h is far less than δ, a more uniform temperature distribution is expected at the end of the pulse, and can be approximated by

$$T(z,t_p)=\frac{I_0(1-R)t_p}{C_p\rho h}\propto t_p \tag{8}$$

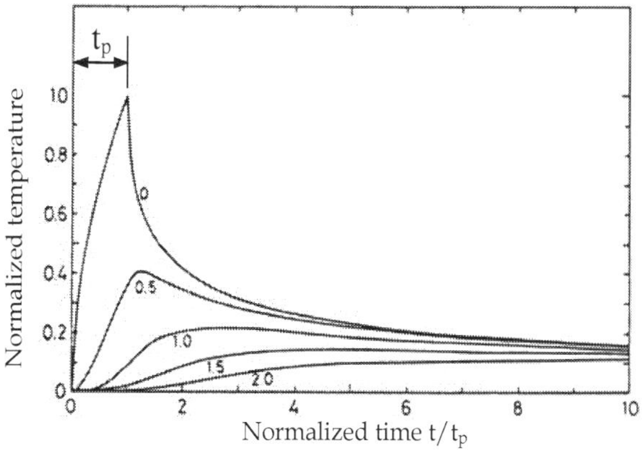

Fig. 4. Normalized temperature as a function of normalized time t/t_p at various depths z/δ ($\delta=2\sqrt{Dt_p}$, t_p is pulse duration) for a pulse of rectangular temporal shape, according to Eq. 6. Temperature is normalized to the surface temperature at the end of the pulse. (adapted from Allmen 1995)

Next, consider the case when the film is melted. The melting starts from the sample surface and propagates inside the sample until a maximum molten thickness is reached some time after the end of the pulse. The time period of melting depends on the energy density of the pulse and thermal properties of the material, and can be much longer than the pulse duration t_p. Fig. 5 sketches the temperature distribution and molten layer thickness variation with time. The latent heat ΔH_{sl} (the subscript "sl" denotes solid-liquid) for melting Al, Au, Cu, Ni and W ranges from 1/3 to 1/2 that of the energy needed to heat the materials from 300K to its melting temperature ($\Delta H_{300K\rightarrow Tsl}$), but for crystalline silicon the latent heat is 40% higher than its $\Delta H_{300K\rightarrow Tsl}$. Therefore, the film temperature is "pinned" near T_{sl} to a larger degree for silicon than for those metals. During re-solidification, the velocity of the liquid-solid interface is determined by the rate of heat extraction from the interface into the bulk, and is given by

$$v = \frac{\kappa}{\Delta H_{sl}\rho}\left(\frac{\partial T}{\partial z}\right)_i \qquad (9)$$

where $(\partial T/\partial z)_i$ is the temperature gradient just behind the interface, which is typically on the order of 1 K/nm. The resulting velocity is then on the order of several meters per second for silicon, and several tens of meters per second for metals.

2.3 Evaporation, plasma formation and plasma propagation

2.3.1 Evaporation and plasma formation

Due to their very low absorption length, the process window of irradiance for melting metals or semiconductors without surface vaporization is actually very narrow. At slightly higher laser intensity, significant material vaporization is induced and a dense vapor plume is formed in front of the target, a phenomenon known as ablation. For the our laser with 20 ns pulse duration and for film thickness around 200 nm, vaporization occurs when irradiance exceeds several tens of MW/cm². There are two distinct modes of vaporization from a liquid, namely volume vaporization or boiling, and surface vaporization. For strongly absorbing media, surface vaporization is always the case. Yet boiling would take place first for backside irradiation, where the film is deposited on a transparent support and the irradiation comes from the support side. Owing to the inverse Mott's metal-insulator transition, the neutral (not ionized) metal vapor is insulating and transparent to the laser light.

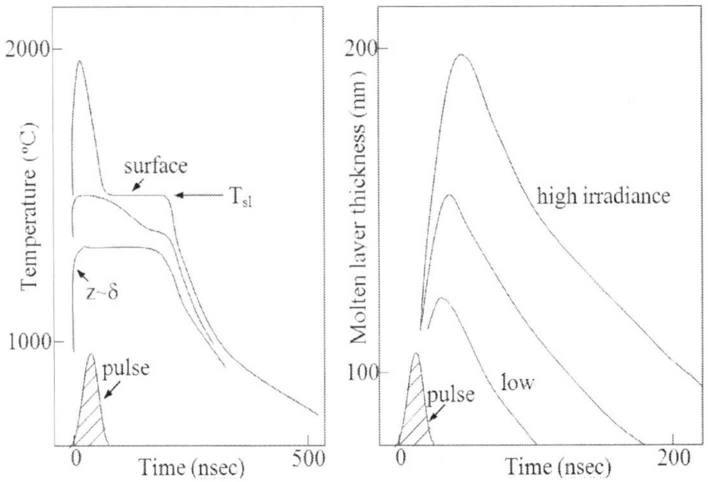

Fig. 5. Schematic time evolution of the temperature distribution (left) and the molten layer thickness (right) of a laser-irradiated sample. (sketch is partly based on the numerical simulation in Poate 1982 and Baeri 1979)

However, the regime of normal vaporization is relevant only for pulse durations longer than ~1 μs. For nanosecond pulses, unless the target is very thin and thus can be vaporized at low irradiance, the irradiance needed to substantially vaporize the target is often high enough to cause ionization and breakdown of vapor or gas (when ambient gas is present). The first breakdown mechanism is thermal ionization. The ratio of ions to neutrals for a gas in a local thermodynamic equilibrium is given by the Saha equation (Chrisey 1994):

$$\frac{n_i}{n_n} = 2.4 \times 10^{15} \frac{T^{3/2}}{n_i} e^{-E_I/kT} \tag{10}$$

where n_i and n_n are the ion and neutral densities and E_I is the first ionization energy. Partially ionized gas absorbs light by thermally excited atoms as well as by electrons and ions (inverse-Bremsstrahlung absorption) (Hughes 1975). Light absorption further heats the vapor, which leads to even more ionization and absorption. The positive feedback favors the creation of plasma (substantially ionized vapor) in front of a vaporizing target. Thermal ionization turns on at lower irradiance for longer wavelength lasers because the optical absorption (∝αI) of the plasma increases strongly with increasing wavelength. The typical plasma temperature is ~10000 K, well above the boiling point of metals; and the associated vapor pressure of the plasma is order of tens of atm. Above the cut-off electron density (1.2×10^{22} cm^{-3} for 308 nm XeCl excimer laser) where $\omega_p>\omega$, the plasma behaves essentially like a metal with large absorption coefficient and high reflectivity. Below the cut-off density, the absorption coefficient of the plasma α_p decreases strongly with increasing temperature. The second breakdown mechanism of the vapor or ambient gas is optical breakdown when the electrical field E of the light, given by $\sqrt{2I/\varepsilon_0 nc}$, reaches a critical point for avalanche ionization. The threshold irradiance I_{th} is 10^9–10^{11}W/cm^2 in *clean* air (no dust) at atmospheric pressure. In this case, initial ionization is achieved via multi-photon excitation since $h\nu<E_I$. However, for optical breakdown in front of a target, it was found that I_{th} is about two to three orders of magnitude lower than if without a target. This is mainly due to the availability of primary electrons emitted from impurities absorbed on the target, which are evaporated and ionized far below the melting point of the target material, plus the temperature increase near a target. In laser processing, optical breakdown is the chief mechanism for plasma formation.

When ambient gas is present, three different laser intensity thresholds for plasma initiation should be distinguished: surface vaporization threshold I_{vap}, metal vapor plasma threshold $I_{i,m}$, and ambient gas plasma threshold $I_{i,g}$. Generally $I_{i,g}> I_{i,m}$ because metal atoms are easier to ionize than gas atoms. But gas breakdown usually occurs first for pulsed IR lasers because I_{vap} is very high (so no metal vapor is produced for plasma formation) owing to metals' high reflectivity in the IR and rapid heat loss to the substrate. On the other hand, when a pulsed UV laser is used, metal vapor breakdown may happen first (Boulmer-Leborgne 1995).

With the formation of plasma, the plume strongly absorbs the laser light and shields the substrate from further irradiation. However, in addition to the shielding effect, there is an important energy transfer from the radiation emitted by the plasma to the surface and via heat conduction. Actually, plasma-enhanced coupling between laser and substrate often results, owing to the fact that the radiation emitted by the plasma is generally in the UV where most metals have low reflectivity.

2.3.2 Plasma propagation

At still higher irradiances depending on the wavelength, very thick (up to 10 cm) plasma can form that absorbs a significant fraction of the laser light, so laser-induced material vaporization at the target surface ceases. Then the plasma can decouple from the substrate and propagate toward the incident beam. Such plasma is often termed as laser-supported absorption wave, and its dynamic behavior is determined by the laser energy absorption within the plasma plume and the energy loss via heat conduction and plasma radiation. As the plasma leaves the target, it may fade out due to energy loss and become transparent again, which allows the substrate to be re-heated and re-vaporized that leads to a new cycle of plasma formation and propagation. If the plasma propagates with subsonic velocity, it is termed as laser-supported combustion wave (LSCW). The velocity of this wave increases with laser intensity. Plasma propagating at supersonic velocity is termed as laser-supported detonation wave (LSDW). In contrast to LSCW, LSDW is not confined to the vapor. Instead, through target surface-mediated optical breakdown, it can form in cold ambient gas even before the target has started to evaporate. Once ionized, the initial plasma tends to be strongly absorbing and can readily be sustained and propagated as an LSDW for laser irradiances of order 10^8 W/cm^2. The temperature and pressure in an LSDW can reach 10^5 K and 10^4 atm, and due to its explosive nature, a shock wave (mechanical force) is driven into both the ambient gas and the substrate material. For a more detailed discussion, refer to (Bäuerle 2000 and Allmen 1995).

3. Experimental setup

Fig. 6 shows the diagram of the setup. XeCl excimer laser with FWHM pulse duration of 20 ns was used to irradiate the samples. For 20 ns pulse duration, the characteristic thermal diffusion length ($\delta=2\sqrt{Dt_p}$) is calculated to be ~0.2 μm for SiO_2 and ~2 μm for metals and silicon. The maximum per pulse laser output energy is 200 mJ, yet 160 mJ was always set as maximum for more reproducible output fluence (energy divided by area, J/cm^2). The attenuator allows a transmission range of 10−90%. The incident laser beam has a square shape and its size, determined by the separation between the imaging doublet and the substrate, is chosen to be between 2 to 3 mm. With a 160 mJ pulse energy within 20 ns over the (2.5 mm)^2 beam area, the light intensity is calculated to be 130 MW/cm^2, high enough to cause vaporization and plasma formation in front of many metal targets. The sample stage is put either in air or in a vacuum chamber with vacuum better than 200 mTorr. The pressure between the mold and the substrate was applied by sandwiching them between two large press plates held by springs.

For laser assisted direct imprint (LADI), we used 200 nm period quartz (fused silica) grating mold that was duplicated by nanoimprint lithography, liftoff and reactive ion etching from a master mold, which was fabricated by interference lithography with uniform grating pattern over an entire 4″ wafer. The high purity quartz mold absorbs negligible laser energy at 308 nm wavelength. The substrate to be patterned was prepared by electron beam evaporation of 200 nm thick of the desired metal (Al, Ni, Au, Cu) with 5 nm thick Cr adhesion layer (the melting point of Cr is higher than the four metals studied here). The quartz mold was cut into a number of 1 mm^2 quartz pieces, which is smaller than the 2.5 × 2.5 mm^2 excimer laser spot size in order to ensure that the entire substrate area below the mold gets melted during the LADI process.

Excimer lasers are almost ideal sources for laser micro- or nano-patterning, some types of surface modifications, and thin film deposition. This is because they have high photon energies (short wavelengths), short pulse lengths (typically 10 to 40 ns), and relatively poor coherence (since an excimer laser's output is highly multimode and contains as many as 10^5 transverse modes (Bäuerle 2000)). The high photon energy of an excimer laser allows for strong optical absorption in many metals and virtually all semiconductors. The short pulse length is a prerequisite for spatially well-defined surface heating (as $\delta\propto\sqrt{t_p}$) with negligible heating or damage of the bulk substrate (see next paragraph). In addition, the short pulse produces chemically stoichiometric ablation, which is one of the key advantages for pulsed laser deposition. Finally, the poor spatial coherence of excimer-laser light diminishes interference effects. Their disadvantages include low efficiency, poor pulse-to-pulse stability (when set at 160 mJ, the output energy ranged from 140 to 180 mJ), and high operating cost.

It should be pointed out that the optimal pulse length depends on the specific application. When substrate heating is not a concern, a longer pulse is desirable for more uniform heating through the film thickness and less damage of the film surface; a longer melting time also allows larger features to be patterned. If only the surface layer needs to be melted or ablated, then a shorter pulse is favored to reduce threshold fluence and substrate heating.

All laser fluences given in this chapter are values *before* passing through the quartz window and cover. The actual laser fluences received by the films should be about 15% less due to the reflection at four quartz/air interfaces. In addition, most experiments used only one laser pulse.

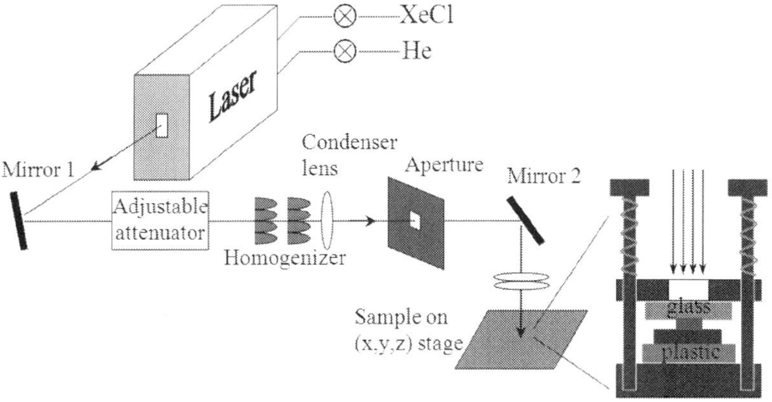

Fig. 6. Schematic setup of the excimer laser, optical components and sample stage. The laser can be used as light sources with different wavelengths by filling different rare-gas-halides into the laser gas chamber.

4. Applications of pulsed laser assisted nanofabrication

In this section we will present laser assisted direct imprint, wafer planarization, via-hole filling, transfer printing, and nano-tip formation.

4.1 Laser-assisted direct imprint (LADI)

The most straightforward application of laser assisted nanofabrication is LADI (Cui 2010, Chou 2002). In the LADI process, a single laser pulse passed through a transparent quartz mold and melted a thin layer of the substrate material, or the film material if it was different from the substrate. The characteristic time constant for electron-phonon collision to reach thermal equilibrium between excited electron gas and the lattice in metals is on the order of picoseconds, which is negligible compared to the laser pulse duration. The molten layer was then embossed by the quartz mold. Once the liquid layer re-solidified, the mold was separated from the substrate. As can be seen, the most prominent feature of LADI is that it is a one-step patterning process - it replaces the steps of resist patterning in photolithography or nanoimprint lithography, subsequent pattern transfer by etching or liftoff, and resist removal all into one single step. Besides LADI of metal or silicon, one can also imprint a polymer resist that absorbs UV light thus gets heated by the laser pulse (Xia 2003, Xia 2010). Fig. 7 shows SEM images of 200 nm period Al, Au, Cu and Ni gratings with ~100 nm line-width imprinted by LADI. Only one pulse was used to melt and imprint those metals at a fluence of 0.22, 0.53, 0.24 and 0.41 J/cm^2 for Al, Au, Cu and Ni, respectively. We found that multiple pulses up to 50 pulses had insignificant effects to the imprint results. As shown in the figure, although 100 nm features in metal were imprinted, the resolution was not near the sub-10 nm level as had been achieved in LADI of crystalline silicon (see below). This is partially due to two factors: (1) silicon expands 8.6% when re-solidified, which would help to press the un-solidified silicon to fill the void near the sharp corners; whereas Al, Au, Cu and Ni shrink 11 - 12% during re-solidification, which draws the un-solidified metals back from the sharp corners and thus results in an unfilled corner and rounded profile; and (2) the surface tension of molten silicon (0.78 N/m at its melting point) is lower than molten metals, so it is easier to imprint. We believe that here the volume shrinkage plays a major role, due to which the trench cannot be fully filled no matter how high a pressure is applied to overcome the surface tension. Besides front-side irradiation (laser beam passes from the mold to the substrate), metal can also be patterned by laser irradiation from the substrate side if the metal is deposited on a transparent substrate with a film thickness low enough (e.g. <300 nm) to be melted entirely.

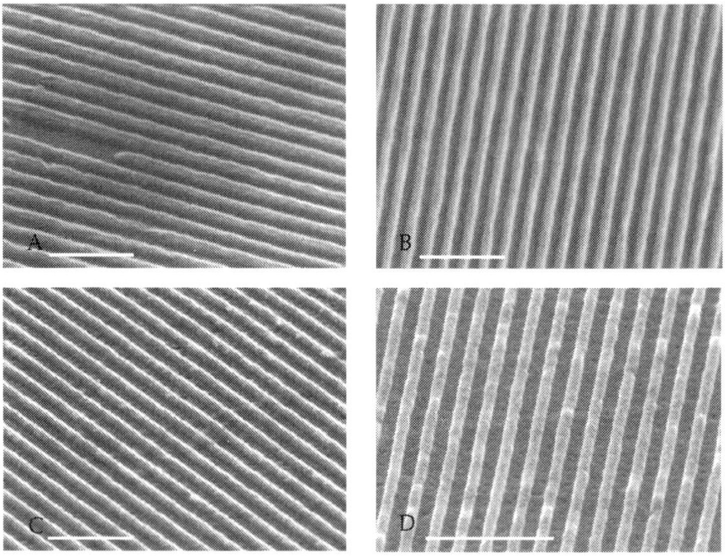

Fig. 7. SEM images of 200 nm-period metal gratings patterned by LADI. (A) Al, laser fluence 0.22 J/cm^2; (B) Au, 0.53 J/cm^2; (C) Cu, 0.24 J/cm^2; and (D) Ni, 0.41 J/cm^2. Scale bar is 1 μm.

We have also attempted to imprint tungsten. We found that it could be melted readily by a single laser pulse, but it could not be patterned with well defined gratings (yet grating pattern was still noticeable) using a quartz mold. This is likely because: (1) tungsten's melting point is 1800 K higher than that of quartz, hence quartz will be melted (softened) during LADI of W; (2) molten tungsten's surface tension is relatively high, nearly twice that of Cu, thus more difficult to imprint; (3) solid tungsten has a high Young's modulus E, three times that of Cu, leading to >3× higher thermal stress (equal to $E(\Delta L/L)$, with $\Delta L/L$ being thermal expansion between room temperature and melting point). Because of the high thermal stress, W film was found to crack after the LADI process.

LADI of single crystal silicon wafer (not thin film) is shown in Fig. 8. Due to the thermal expansion upon re-solidification that helps fill the sharp corners of the mold pattern, 10 nm fine features on the mold were faithfully duplicated into the silicon. Reflection measurement, as shown in Fig. 8c, indicates that the silicon remained in liquid state for about 220 ns, roughly 10 times the pulse duration.

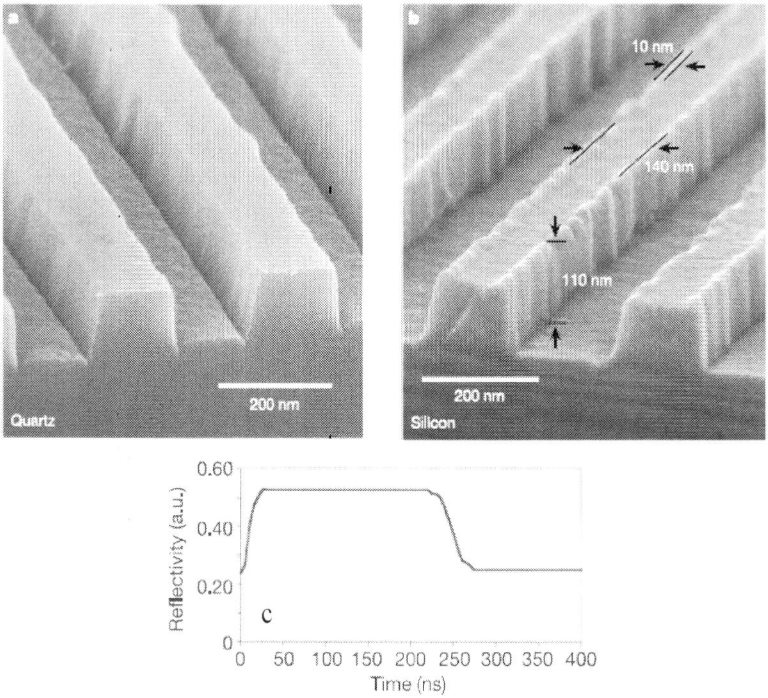

Fig. 8. LADI of single crystal silicon wafer. (a) Quartz mold. The ~10 nm wide notches were caused by the reactive ion etching trenching effect during mold fabrication; (b) Imprinted pattern in silicon showing the notches were faithfully duplicated; (c) The reflectivity of a HeNe laser beam from the silicon surface versus the time, when the silicon surface is irradiated by a single laser pulse with 1.6 J/cm^2 fluence and 20 ns pulse duration. Molten Si, becoming a metal, gives a higher reflectivity.

4.2 Laser-assisted wafer planarization

High-performance large-scale integrated circuits (ICs) require several levels of interconnect. Planarization processes which smooth and flatten the surface of an IC at various stages of fabrication are essential for high-resolution photolithography. Besides being smooth and flat, good film step-coverage without internal voids is also crucial for interconnection. Present IC interconnects are fabricated by the damascene process that consists of four major steps: patterning trenches in low-k dielectric materials by photolithography, sputtering metal plating base, copper electrical plating to fill the trench, and chemical mechanical polishing that planarizes the surface. The step-coverage of electrodeposited metal film, though better than other PVD (physical vapor deposition, including evaporation and sputtering) metal films, is not as good as CVD (chemical vapor deposition) films because trench edges plate faster than trench bottoms. This problem will become even more severe for future ICs with narrower line-widths. Therefore, it is desirable to develop a technique that can both flatten the surface and enhance the step-coverage at the same time. Laser-assisted planarization might be one answer.

The setup for planarization is the same as LADI except that now a flat un-patterned quartz "mold" is used. The process is shown in Fig. 9. First, an insulator layer, typically thermal oxide which is also a good heat insulator, was patterned by NIL and RIE. Metal was then evaporated or sputtered onto the pattern. Finally, the metal was melted by a pulsed excimer laser and thus conformed to the flat and smooth surface of the quartz mold. IC interconnect typically consists of metal lines embedded in an insulator, which can be readily achieved here by etching away the metal on top using argon ion beam etching (IBE).

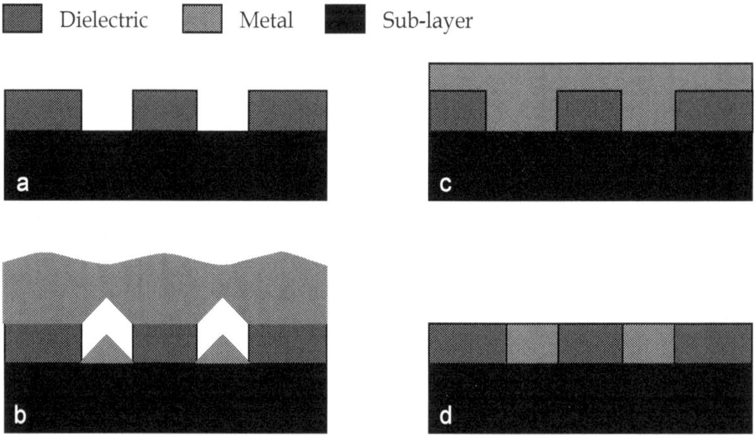

Fig. 9. Laser-assisted planarization. (a) Pattern insulator; (b) Deposit metal, here by evaporation; (c) Laser-melt and planarize. Metal on top of insulator was etched away by ion beam etching in (d).

Fig. 10 shows laser-assisted planarization of 180 nm-thick Cu evaporated onto a 200 nm-period SiO_2 grating on a silicon wafer. The laser fluence was 0.22 J/cm^2, which was close to the fluence used for LADI of Cu. As seen in Fig. 10(a), the step-coverage of evaporated Cu film was very poor. But after planarization the underlying voids were all filled with Cu (Fig. 10(b)), and the surface became very smooth with fluctuation less than 4 nm (Fig. 10(c)). Fig. 10(d) shows a 200 nm-period Cu grating embedded in an insulator (here silicon dioxide) fabricated by laser-assisted planarization and IBE removal of the Cu on top of the insulator. The laser fluence used for this sample was 70% higher than previous one, and the number of applied pulses was 150! Yet no apparent damage of Cu was found, suggesting that the number of pulses has only minor impact on laser-assisted planarization.

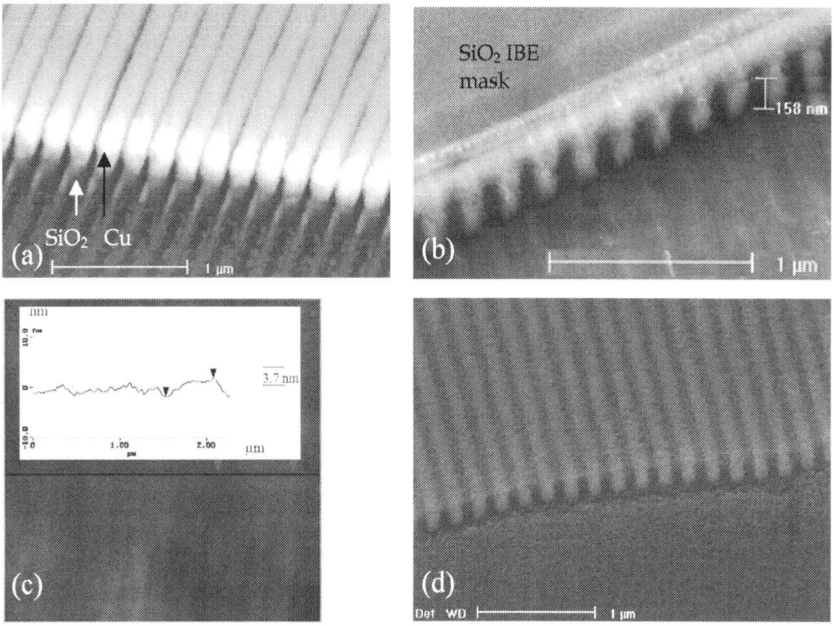

Fig. 10. Laser-assisted planarization of a 200 nm-period Cu grating on SiO_2. (a) Before planarization. (b) After planarization by 0.22 J/cm^2, shown here was a SEM image with evaporated SiO_2 IBE-mask still on top. (c) AFM image showing surface smoothness better than 4 nm. (d) Cu lines embedded in SiO_2 with flat surface fabricated after IBE removal of the Cu on top. (SEM cross-section was "cut" by IBE using evaporated SiO_2 as mask)

4.3 Laser-assisted via-hole filling

Filling via holes or trenches with a deposited material is an important step in IC manufacturing. At present, the metal interconnect for ICs is typically fabricated by the dual damascene and, for the first wiring level, tungsten CVD plug process, where the via holes and trenches are etched in a dielectric material and subsequently filled up with a metal. Clearly, any voids left in the vias or trenches caused by poor step coverage will create a serious problem for the interconnect. Though superior to PVD, the step coverage of tungsten CVD is still limited by the low volatility of the precursor gas WF_6 that leads to a low vapor pressure. The result is a mass transport limited deposition rate, and the hole opening receives faster deposition and would be closed before the holes are completely filled. Another disadvantage of CVD is its relatively high thermal budget.

Atomic layer deposition (ALD) can achieve excellent step coverage in certain conditions. It is a modified form of CVD with gas precursors introduced one at a time with pump/purge in between. So the film is deposited one atomic layer per cycle, with a typical deposition rate of order 0.5 nm/min. Not only the slow deposition rate is a manufacturing issue, but also in ALD, voids will be formed if the via holes or trenches have a sidewall with negative angle.

The process we developed to fill via-holes is very close to the laser-assisted wafer planarization process, except that the trenches were replaced with deep and high aspect ratio holes (Cui 2006). We fabricated in Si a hole array having 100 nm diameter, 500 nm depth (aspect ratio 5) and 200 nm inter-hole spacing (pitch). The hole array was patterned by nanoimprint lithography and etched by Cl_2/Ar RIE. As silicon is a good heat sink, a 15 nm thermal oxide layer was grown to reduce heat loss to the surrounding bulk Si, and to shrink the hole size at the same time. Finally, 200 nm Si or Cu was deposited on top of the hole array by e-beam evaporation. As expected and shown in Fig. 11a, evaporation has very poor step coverage, so the holes were only partially filled, with the sidewall of the holes receiving little deposition before the hole opening was closed. Fig. 11b shows via filling by Si, indicating a complete filling without voids. For via filling by Cu (Fig. 11c), the Cu plugs were found broken after wafer cut, and it is possible that the built-in stress caused by fast cooling of the liquid Cu has contributed to the Cu plug fracture. Nonetheless, we believe that most holes were completely filled to the bottom, as indicated by the Cu plug sections found on the bottom of the holes.

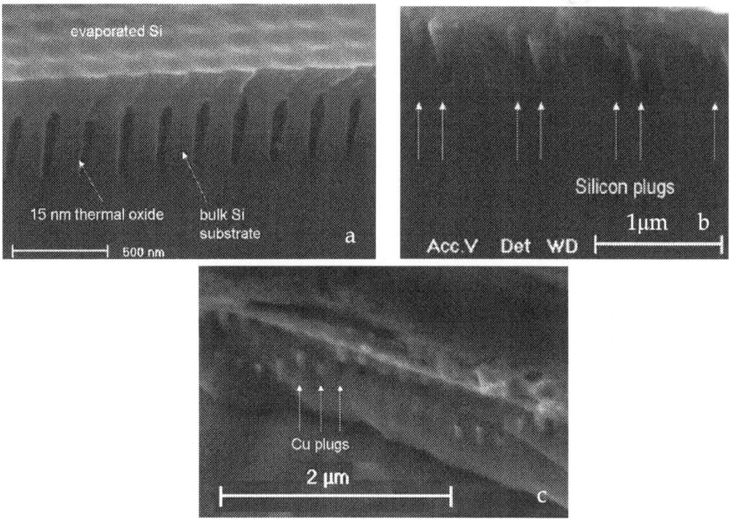

Fig. 11. (a) Holy array with 200 nm period and aspect ratio 5:1 after evaporation of filling material, here 200 nm Si. (15 nm thermal oxide has been grown to reduce heat loss to bulk Si). (b) Laser-assisted via-hole filling by α-Si, showing all holes were completely filled without void. (c) Laser-assisted via-hole filling by Cu.

4.4 Laser-assisted nanotransfer printing

Laser-assisted nanotransfer printing (LA-nTP) is a dry-contact printing process which combines into one single step several steps needed for metal nanopatterning (Le Drogoff 2006). This process shares some attributes with laser-induced forward transfer (LIFT) process (Bohandy 1986). In LIFT, a thin laser-light absorbing film is irradiated with a focused laser pulse through a transparent support, and transferred, in the form of micron-sized dots, onto a suitable substrate facing it. The pattern is created through selective deposition on the substrate by scanning the laser beam. The resolution is hence determined and limited by the laser beam size and by the gap between the two samples (which determines the lateral spreading). In the

LA-nTP process, we use a pre-patterned transparent support in intimate contact with a substrate, therefore the resolution is no longer limited by the processing parameters but is determined only by the initial feature size on the transparent support.
The LA-nTP process is shown in Fig. 12. For proof of concept, we fabricated the donor support (DS) consisting a square chromium nanodot array with 200 nm period and 35 nm height (Fig. 12A). In parallel, we fabricated the acceptor substrate (AS) by photolithography and reactive ion etching, which contained an array of microwells with a diameter of 100 μm and depth 40 nm. The two patterned surfaces were then brought into contact by sandwiching them between two press plates. After shining with a laser pulse of fluence ~0.35 J/cm^2, the Cr nanodots were transferred only onto the protruded parts of the acceptor substrate that were in intimate contact with the donor support (Fig. 12C). We also found that using higher fluences (>0.6 J/cm^2), the results obtained were similar to those observed in other LIFT studies, namely Cr dot was transferred to the protruded areas as well as the recessed areas of the acceptor substrate with an increasing amount of spatter. At even higher fluencies (>1 J/cm^2), significant chromium vaporization was observed. On the other hand, experiments using laser fluencies lower than 0.2 J/cm^2 (below the melting threshold of a 30 nm Cr thin film) failed to transfer the Cr nanopatterns.

4.5 Laser-assisted nanotip formation

Nanotips have many applications such as scanning probe microscopy (SPM) and vacuum microelectronic devices like field emission display (FED), where their performance is often limited by the tip. Carbon nanotubes (Nguyen 2005, Jeong 2006) or nanocones (Chen 2006) have been studied extensively for applications as tips or emitters owning to its fine diameter and high aspect ratio. Silicon tips with high aspect ratio can be readily fabricated by highly anisotropic dry etching (Rangelow 2001). Metal is highly conductive and mechanically strong and robust, which are favorable properties for field emitter or scanning probe. Unfortunately, it is more challenging to fabricate metal tips on a wafer due to the difficulty in etching metals. Single metal tip can be routinely fabricated by focused ion beam milling or electrochemical etching (Kim 2006). But for FED or high throughput nanolithography using SPM, tip arrays are essential, which are very difficult to fabricate by the previous two methods. Metal tip arrays with low aspect ratio were traditionally fabricated by the Spindt process where the metal is evaporated through apertures that become gradually closed by the evaporated material (Spindt 1968). Similarly yet via a different mechanism, Au tips were formed by evaporation into porous alumina (Losic 2005). Alternatively, metal tips can be replicated by physical vapor deposition or electroplating into etch holes.
We found that sharp metal tips could also be self-formed by simply melting momentarily (~100 ns) a metal film deposited on a wafer with a second wafer placed a gap (submicrometers) below, as shown in Fig. 13 (Cui 2008). In the experiment, a silicon wafer with 500 nm thermal oxide was placed below the Cr film with a gap formed by a spacer. Next, a single excimer laser pulse (fluence 0.4 J/cm^2) shone through the quartz to melt the metal film momentarily that caused it to transfer toward the lower wafer. The two wafers were separated after Cr solidification and examined by SEM that showed nano-tip formation on both surfaces. The size, shape and location of the tips were random, but more regular tips are expected if the Cr is pre-patterned into identical isolated mesas. In addition, The ends of most tips resemble a sharp cone (not flat), hence they should be broken by the tensile force due to surface tension and/or volume shrinkage upon solidification; whereas those with a flat end were broken by mechanical force when the two wafers were separated.

The quartz and SiO_2 surface was found to be rather flat after the process. Since the melting temperature of quartz is much lower than the boiling temperature of Cr, a top SiO_2 layer of order 100 nm on both wafers should have been melted by heat transfer. However, due to the six orders higher viscosity of molten SiO_2 compared to that of chromium (3.0×10^6 cP for SiO_2 versus 3.7 cP for Cr at 2673 K (Yaws 1998)), the flow of the molten SiO_2 is negligible, leading to no apparent distortion of the two wafer surfaces.

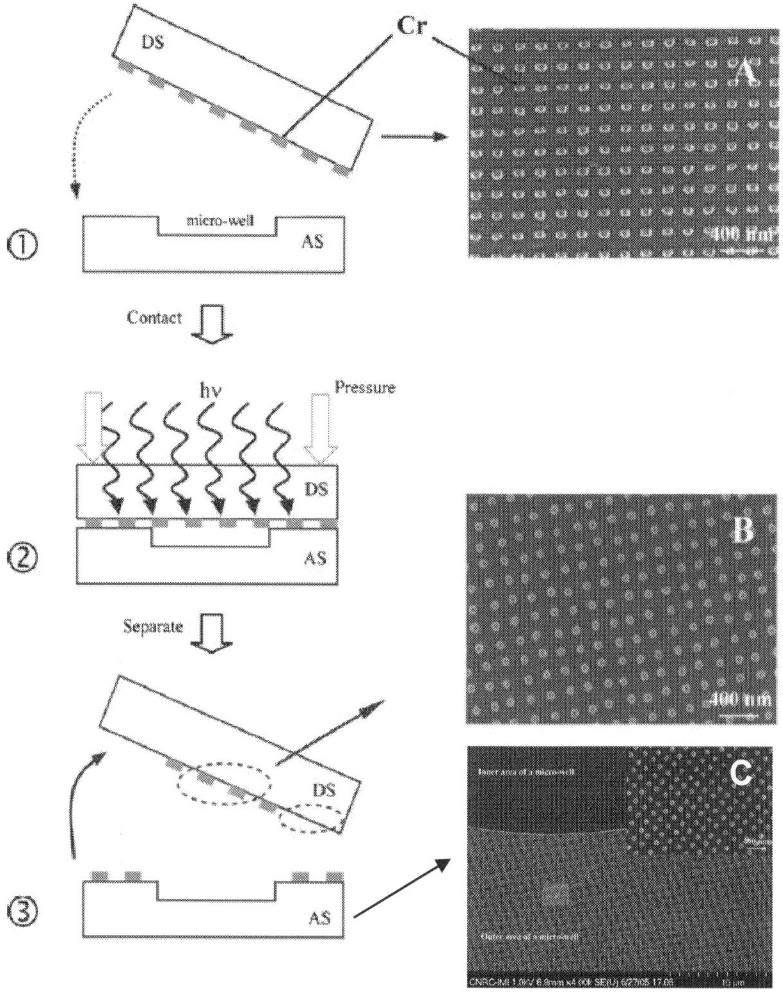

Fig. 12. (left) Schematic sequence of the fabrication process of LA-nTP (DS: donor support, AS: acceptor substrate). (right) A). Cr dot array on DS; B). Cr dot array on DS facing the recessed area of the AS; C). AS after LA-nTP showing Cr dot array was transferred from DS only to the protruded area of AS.

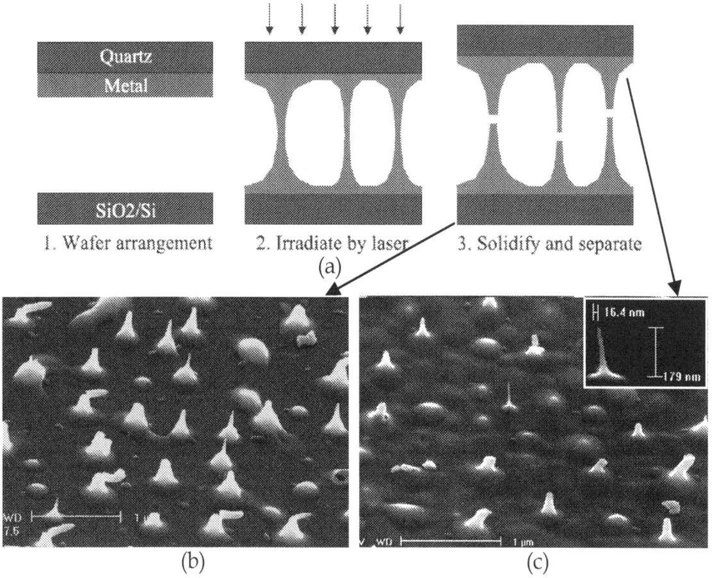

Fig. 13. (a) Schematic laser-induced metal nanotip formation process, with the metal here as 25 nm Cr. (b) Cr tips formed on the Si/SiO_2 substrate. (c) Cr tips formed on the quartz wafer. The inset is the zoom-in image of the tip at the center that has an apex of 10 nm and height of 180 nm. (This tip appears tilted due to SEM stage drift during imaging.)

The experimentally observed metal nano-tip formation phenomenon can be explained in term of an electro-hydrodynamic instability process as shown in Fig. 14. Briefly, the nano-tip formation can be roughly divided into four stages: free charge buildup, mass transfer to the lower wafer upon laser illumination, liquid pillar formation due to electrostatic attraction, and solidification and break of the pillars.

When the two wafers are brought close together, free charges are built up on both wafers due to the work function difference between Cr and SiO_2, which results in a strong electric field of order 10^5 V/cm for a 500 nm gap. Interaction between electric field and induced dipole or free charges accumulated at the interface results in an electrostatic pressure. The attractive electrostatic force sets the two molten metal films into tension subject to infinitesimal disturbances, and interfacial ripples start to develop. The periodicity of the ripple that determines the areal number density of the tips would depend on the liquid metal density, film thickness, surface tension and work function difference. The electrostatic force between two approaching peaks is further enhanced due to the locally strengthened electric field. As a result, the electrostatic force keeps on overcoming the stabilizing surface tension and viscous force and pulling the peaks further together until the peaks are bridged together to form a liquid bridge or the liquid metal solidifies. Once the peaks bridged together, the charges are neutralized and thus the electrostatic force disappears, and the bridges may break by the large surface tension of liquid metal that is about 20 times that of water. Meanwhile, solidification propagates from the two metal-wafer interfaces toward the gap center, because the wafers are heat sinks and the temperature distribution before solidification across the gap should be uniform (since the melting time is order 100 ns that is much longer than pulse duration). Nano-tips are formed at the broken bridge necks before or during solidification due to surface tension and volume shrinkage, or after the two wafers were separated.

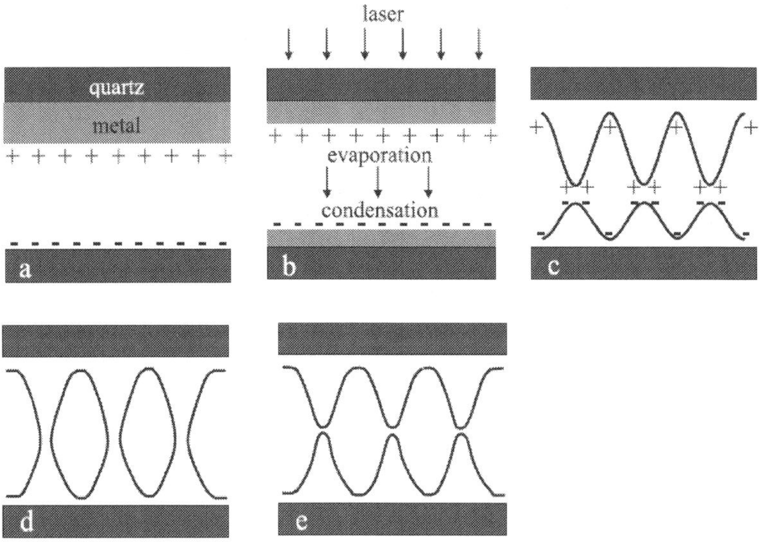

Fig. 14. Schematic of the physical mechanism behind the nano-tip formation. (a) Build-up of free charges on the two wafers. (b) Metal transfer to lower wafer by evaporation-condensation. (c) Nano-bridge formation due to attractive electrostatic force. (d) Bridging of approaching peaks and charge neutralization. (e) Break of bridges upon solidification or after separation.

5. Lower and upper limit of achievable feature size

So far, only structures with feature size of about 100 nm were described. Laser-assisted processes for much smaller and larger feature sizes will be discussed in this section, with a focus on factors that limit the achievable feature sizes of the techniques.

5.1 Resolution limit

In LADI, planarization, and via-hole filling processes, squeezing liquid metals or semiconductors into small trenches or via holes will be ultimately limited by the pulling pressure due to surface tension. Fig. 15 shows schematic drawing illustrating how much pressure is needed to fill trenches or via holes. For all cases, the pulling pressure is proportional to the surface tension constant γ and inversely proportional to feature size. Therefore, there is no fundamental resolution limit provided the applied pressure is high enough. In practice, the liquid metal and silicon generally have surface tension γ on the order of 1 N/m (for water, it is 0.076 N/m), which leads to a pressure of 200 bar to fill in 100 nm-wide trenches, or 2000 bar for 10 nm-wide trenches. This pressure is considerably higher than that used for regular thermal nanoimprint lithography (order of 10 bar). However, considerably lower applied pressure would be sufficient for patterning metals and silicon if: (1) the molten metal partly wets the SiO_2 substrate; and/or (2) the mold consists of sparse protruded patterns to create sparse recessed metal features (such as a periodic hole array having diameter ~100 nm and pitch 500-600 nm in Au for extraordinary optical transmission devices (Lesuffleur 2008)), because the effective local pressure at the protruded mold features is significantly higher than the applied pressure (applied force divided by wafer surface area).

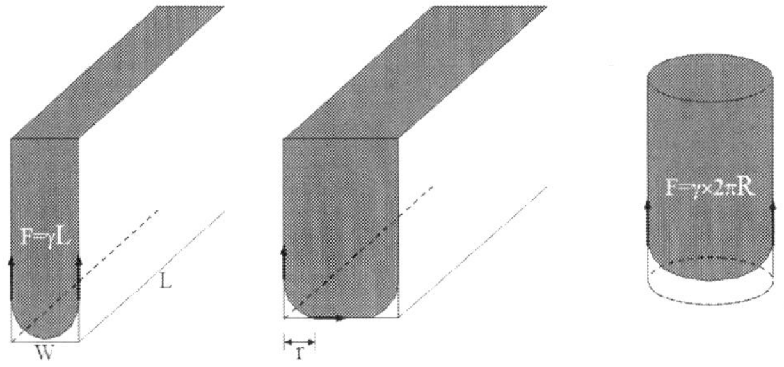

$$P = \frac{2\gamma L}{WL} = \frac{2\gamma}{W}$$
(a)

$$P = \frac{2\gamma L \cos 45^\circ}{\sqrt{2} rL} = \frac{\gamma}{r}$$
(b)

$$P = \frac{\gamma \times 2\pi R}{\pi R^2} = \frac{2\gamma}{R}$$
(c)

Fig. 15. Schematic drawing showing the calculation of minimum pressure to squeeze a liquid into (a) a trench, (b) a trench with liquid partly squeezed against the bottom, and (c) a via hole. The pressures are calculated assuming 90° contact angle and no wetting/adhesion.

5.2 Upper limit of achievable feature size

The maximum feature size that can be patterned depends on how far the molten metal can flow before re-solidification. For a rough estimation, the dynamic flowing process is modeled as in Fig. 16. It is assumed that the grating structure has a period 2L and a trench depth h_0; and the entire layer is melted at t=0 and re-solidification takes place simultaneously throughout the film thickness at time t=τ. The mold will travel a distance of $h_0/2$ and the residual molten layer thickness will decrease to $h_0/2$ when the trench is fully filled. Since L is large, the effect of surface tension can be ignored and thus the internal pressure at x=L/2 is zero. There are three forces acting on the liquid: the pressing force from the mold, the viscous force, and the inertial force.

We first consider only the effect of viscous force by assuming that a steady flow develops momentarily at t=0 (i.e. ignore inertial force). Then the maximum grating half pitch L_1 that can be patterned (trench fully filled) before the liquid metal solidifies is given by (Heyderman 2000, Cui 2006):

$$L_1 = h\sqrt{\frac{P\tau}{\mu}} \tag{11}$$

where h is the metal film thickness (200 nm), P is the applied pressure (order 100 bar), μ is the viscosity, and τ is the melting duration (~ 200 ns) that depends on laser fluence. The viscosity of molten Au and Cu is around 4 cp (only ~5× that of wafer (Yaws 1998)), leading to L_1~ 5 μm. To pattern isolated metal mesas, the liquid flows from four sides to fill a square-shaped hole in the mold, so the maximum patternable mesa size should be roughly doubled to ~10 μm, which qualitatively agrees with the experiment (17 μm). Moreover, it is interesting to see how far molten SiO_2 can flow under similar conditions. For example, its viscosity at 2200 K is 8×10^7 cp (a very tacky "liquid"), 7 orders higher than that of Au and Cu, leading to a calculated L_1 of only ~1 nm if assuming a top 200 nm SiO_2 layer is molten. This explains why the quartz mold can be used to pattern a metal at temperature slightly higher than its melting point, though the mold could not be used repeatedly for >>10 times.

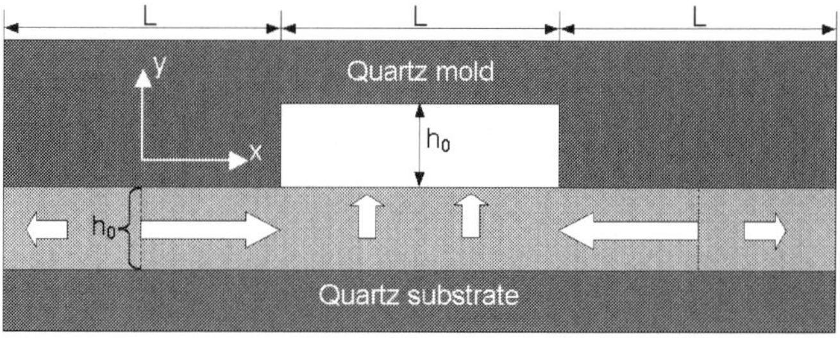

Fig. 16. Dynamics of molten metal flowing into trenches during LADI. Assume that grating line-width and trench-width are both equal to L, and trench depth and molten layer thickness are both equal to h_0.

Next, to estimate the inertial force which decides how fast the steady flow develops, it is assumed that the liquid metal is inviscid (μ = 0) for simplicity. The applied pressure is to accelerate the flow along the x-direction with acceleration given by (note that the effective pressure is doubled for a grating with equal line and trench width, and z is the direction perpendicular to the paper):

$$a = \frac{F}{m} = \frac{2P \times hz}{\rho \times hz(L/2)} = \frac{4P}{\rho L} \quad (12)$$

The velocity is then u=at at time t, and the flow-rate Q=uh=ath (ignoring the z-dimension). On the other hand, the volume of the liquid changes at a rate

$$Q = -\frac{L}{2}\frac{dh}{dt} \quad (13)$$

Therefore, solving the equation

$$\frac{4P}{\rho L} th = -\frac{L}{2}\frac{dh}{dt} \quad (14)$$

gives ($h=h_0$, $h_0/2$ at $t=0$, τ)

$$L_2 = 2\sqrt{\frac{P}{\rho \ln 2}}\tau \sim \sqrt{\frac{P}{\rho}}\tau \quad (15)$$

Thus even within a very short melting time of order 100 ns, the calculated L_2 values are order 20 μm for applied pressure of 100 bar.
The critical length can also be estimated by experiment. For example, Fig. 17a shows a backside-irradiated 200 nm-thick Ni film by a single laser pulse with fluence 0.60 J/cm² that partly blew the film off the substrate. The surface tension was the force driving the flow, and the driving pressure can be estimated by $2\gamma/h$ (h is the film thickness, and the factor 2 accounts for the top and bottom surface/interface), which was 180 bar. The SEM image indicates that Ni flowed about 4 μm before the film re-solidified. For LADI, we achieved 17 μm isolated Cu mesas with 100 nm height (Fig. 17b), but it was found that a pattern of several tens of micrometer size required multiple pulses. Therefore, the maximum patternable feature size using one laser pulse lies on the lower tens of μm, close to the theoretical prediction.

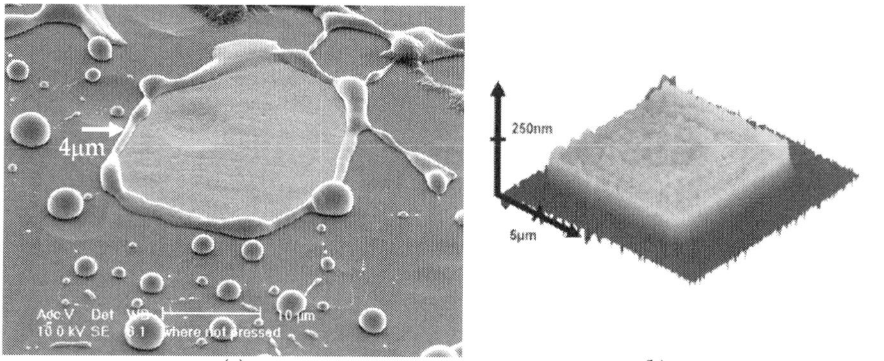

(a) (b)

Fig. 17. (a) A 200 nm-thick Ni film irradiated from backside by a single laser pulse with a fluence of 0.60 J/cm² that partly blew the film off the substrate. The liquid traveled about 4 μm before it froze. (b) AFM image of isolated Cu mesas patterned by LADI at laser fluence of 0.45 J /cm². The mesas have a length on each side of 17 μm and a height of 100 nm.

6. Conclusion

In this chapter, five nanofabrication processes using pulsed excimer laser were described. These methods share the common advantage of being orders faster than most other fabrication techniques. They are also very suitable to patterning metals, which are more difficult to process than semiconductors by conventional lithographies and etching.
The first was laser-assisted direct imprinting, which produced 200 nm-period Cu, Ni and Al gratings over about 1 mm² area within ~100 ns. LADI of W failed due to its high thermal stress. The second was wafer planarization using a flat and smooth "mold". With this technique, Cu surface was planarized by laser melting under pressure, which also squeezed the molten film to fill completely voids under the film. Cu conducting lines embedded in a dielectric matrix were created by an additional etching step. Third, a similar process could also fill 100 nm-wide and 500 nm-deep via-holes with Si and Cu. Fourth, sub-100 nm diameter Cr dots were melt-transferred to the receiving substrate using a laser-induced

nanotransfer printing. Fifth, Cr tips with apex diameter as small as 10 nm and aspect ratio up to 10:1 were achieved by laser melting and boiling through the proposed mechanism of an electro-hydrodynamic instability. Such tips could be employed as field emitters for flat panel display, or as tips for scanning probe microscopy.

For laser-assisted direct imprint, planarization, and via-hole filling, the resolution is determined by the balance between the surface tension and the applied pressure. Whereas the maximum patternable feature size is limited by the liquid viscosity and friction at the interfaces. Strong adhesion between the mold and the molten material rules out clean separation thus sets another constraint on processable materials. To scale up the process to full wafer patterning, the challenge is the availability of pulsed laser with large enough beam size and the very high pressure normally needed.

7. Acknowledgment

I would like to thank my PhD supervisor at Princeton University, Professor Stephen Y Chou, under whose guidance I carried out most of the research presented in this chapter.

8. References

Allmen, M. & Blatter, A. (1995). Laser-beam Interactions with Materials - Physical Principles and Applications, 2ed edition, Springer series in materials science.

Austin, M. D.; Ge, H.; Wu, W.; Li, M.; Yu, Z.; Wasserman, D.; Lyon, S. A. & Chou, S. Y. (2004). Fabrication of 5 nm linewidth and 14 nm pitch features by nanoimprint lithography. *Appl. Phys. Lett.* 84, 5299–301.

Baeri, P.; Campisano, S. U.; Foti, G. & Rimini, E. (1979). Melting model for pulsing-laser annealing of implanted semiconductors. *Journal of Applied Physics*, 50, 788.

Bäuerle, D. (2000). Laser Processing and Chemistry. 3rd edition, Springer-Verlag.

Bohandy, J.; Kim, B. F. & Adrian F. J. (1986). Metal deposition from a supported metal film using an excimer laser. *J. Appl. Phys.* 60, 1538.

Boulmer-Leborgne (1995). Laser-material interaction, plasma formation and applications, in Applications of Particle and Laser Beams in Materials Technology, NATO ASI Series E: Applied Sciences, Volume 283.

Buzzi, S.; Robin, F.; Callegari, V. & Loffler, J. F. (2008). Metal direct nanoimprinting for photonics. *Microelectron. Eng.* 85, 419-424.

Chen, H. L.; Chuang, S. Y.; Cheng, H. C.; Lin, C. H. & Chu, T. C. (2006). Directly patterning metal films by nanoimprint lithography with low-temperature and low-pressure. *Microelectron. Eng.* 83, 893-896.

Chen, I. C.; Chen, L. H.; Ye, X. R.; Daraio, C.; Jin, S.; Orme, C. A.; Quist, A. & Lal, R. (2006). Extremely sharp carbon nanocone probes for atomic force microscopy imaging. *Appl. Phys. Lett.* 88, 153102.

Chou, S. Y.; Krauss, P. R. & Renstrom, P. J. (1996). Imprint Lithography with 25-Nanometer Resolution. *Science*, 272, 85–7.

Chou, S. Y.; Keimel, C. & Gu, J. (2002). Ultrafast and direct imprint of nanostructures in silicon. *Nature*, 417, 835–837.

Chou, S. Y. & Xia Q. F. (2008). Improved nanofabrication through guided transient liquefaction. *Nature Nanotechnology*, 3(5), 295-300.

Chrisey, D. B. & Hubler, G. K. (editor) (1994). Pulsed Laser Deposition of Thin Films. Wiley-Interscience.

Chuang, S. Y.; Chen, H. L.; Kuo, S. S.; Lai, Y. H. & Lee, C. C. (2008). Using direct nanoimprinting to study extraordinary transmission in textured metal films. *Opt. Express*, 16(4), 2415-2422.

Cui, B. & Veres, T. (2006). Pattern replication of 100 nm to millimeter-scale features by thermal nanoimprint lithography. *Microelectron. Eng.* 83, 902-905.

Cui, B.; Wu, W.; Keimel, C. & Chou, S. Y. (2006). Filling of Via Holes by Laser-Assisted Direct Imprint (LADI). *Microelectron. Eng.*, 83(4-9), 1547-1550.

Cui, B.; Wu, L. & Chou, S. Y. (2008). Fabrication of high aspect ratio metal nano-tips by nanosecond pulse laser melting. *Nanotechnol.* 19, 345303.

Cui, B.; Keimel, C. & Chou, S. Y. (2010). Ultrafast Direct Imprint of Nanostructures in Metals by Pulsed Laser Melting. *Nanotechnol.* 21, 045303.

Heyderman, L. J.; Schift, H.; David, C.; Gobrecht, J. & Schweizer, T. (2000). Flow behaviour of thin polymer films used for hot embossing lithography. *Microelectron. Eng.* 54, 229-245.

Hirai, Y.; Ushiro, T.; Kanakugi, T. & Matsuura, T. (2003). *SPIE - Int. Soc. Opt. Eng.* 5220 74-81.

Hua, F.; Sun, Y.; Gaur, A.; Meitl, M. A.; Bilhaut, L.; Rotkina, L.; Wang, J.; Geil, P.; Shim, M. & Rogers, J. A. (2004). Polymer Imprint Lithography with Molecular-Scale Resolution. *Nano Lett.* 4, 2467–2471.

Hughes, T. P. (1975). Plasmas and Laser Light. Adam Hilger.

Jeong, T.; Heo, J.; Lee, J.; Park, S.; Jin, Y.; Kim, J. M.; Oh, T.; Park, C.; Yoo, J. B.;, Gong, B.; Lee, N. & Yu, S. (2006). Field emission from carbon nanotube emitters fabricated by the metal intermediation layer. *J. Appl. Phys.* 100, 064308.

Kim, C.; Burrows, P. E. & Forrest, S. R. (2000). Micropatterning of Organic Electronic Devices by Cold-Welding. *Science*, 288, 831–833.

Kim, P.; Kim, J. H.; Jeong, M. S.; Ko, D. K.; Lee, J. & Jeong, S. (2006). Efficient electrochemical etching method to fabricate sharp metallic tips for scanning probe microscopes. *Rev. Sci. Instrum.* 77, 103706.

Le Drogoff, B.; Cui, B. & Veres, T. (2006). Fast 3D Nanostructure Fabrication by Laser-Assisted nano-Transfer Printing. *Appl. Phys. Lett.*, 89, 113103.

Lesuffleur, A,; Im, H.; Lindquist, N. C.; Lim, K. S. & Oh, S. H. (2008). Laser-illuminated nanohole arrays for multiplex plasmonic microarray sensing. *Opt. Express,* 16, 219–24.

Losic, D.; Shapter, J. G.; Mitchell, J. G. & Voelcker, N. H. (2005). Fabrication of gold nanorod arrays by templating from porous alumina. *Nanotechnol.* 16, 2275

Nguyen, C. V.; Ye, Q. & Meyyappan, M. (2005). Carbon nanotube tips for scanning probe microscopy: fabrication and high aspect ratio nanometrology. *Meas. Sci. Technol.* 16, 2138.

Pang, S. W.; Tamamura, T.; Nakao, M.; Ozawa, A. & Masuda, H. (1998). Direct nano-printing on Al substrate using a SiC mold. *J. Vac. Sci. Technol. B*, 16, 1145–1149.

Poate, J. M. & Mayer, J. W. (editor) (1982). Laser Annealing of Semiconductors. Academic Press.

Rangelow, I. W. & Biehl, S. (2001). High aspect ratio silicon tips field emitter array. *Microelectron. Eng.* 57-58, 613.

Schift, H. (2008). Nanoimprint lithography: An old story in modern times? A review. *J. Vac. Sci. Technol. B*, 26 458–80.

Shank, C. V.; Yen, R. & Hirlimann, C. (1983). Time-resolved reflectivity measurements of femtosecond-optical-pulse induced phase-transitions in silicon. *Physical Review Letters*, 50(6), 454.

Spindt, C. A. (1968). J. Appl. Phys. 39 3504 A Thin-Film Field-Emission Cathode.

Xia, Q.; Keimel, K.; Ge, H.; Yu, Z.; Wu, W. & Chou, S. Y. (2003). Ultrafast patterning of nanostructures in polymers using laser assisted nanoimprint lithography. *Appl. Phys. Lett.* 83(21), 4417-4419.

Xia, Q.; Morton K. J.; Austin, R. H. & Chou, S. Y. (2008). Sub-10 nm Self-Enclosed Self-Limited Nanofluidic Channel Arrays, *Nano Lett.*, 8(11), 3830-3833.

Xia, Q. & Chou, S. Y. (2010). Applications of excimer laser in nanofabrication, *Appl Phys A* 98, 9–59.

Yaws, C. L. (1998). Chemical Properties Handbook: Physical, Thermodynamics, Environmental Transport, Safety & Health Related Properties for Organic & Inorganic Chemicals (McGraw-Hill Professional).

Yu, J. & Bulovic, V. (2007). Micropatterning metal electrode of organic light emitting devices using rapid polydimethylsiloxane lift-off. *Appl. Phys. Lett.* 91, 043102.

7

Soft UV Nanoimprint Lithography: A Versatile Tool for Nanostructuration at the 20nm Scale

Andrea Cattoni[1], Jing Chen[1], Dominique Decanini[1], Jian Shi [2] and Anne-Marie Haghiri-Gosnet[1]
[1]*Laboratoire de Photonique et de Nanostructures, LPN (CNRS-UPR20), Marcoussis*
[2]*Ecole Normale Supérieure de Paris, Paris*
France

1. Introduction

1.1 Why soft UV nanoimprint lithography?

Since the pioneering work of Whitesides and coworkers on microContact Printing (mCP) and Soft Lithography (Kumar & Whitesides (1993)) (Xia & Whitsides (1998)), considerable progress has been made in the last years and Soft Lithography is now a well consolidated technology utilized in academic and industrial laboratories (Rogers & Nuzzo (2005)). These printing methods use a flexible elastomer material named PDMS (poly(dimethylsiloxane)) to transfer molecules on a surface thus creating localized chemical patterns (Cerf & Vieu (2010)). The PDMS stamp inked with the desired molecules is placed in contact with the substrate to perform the molecular transfer. mCP has received large attention for biological applications since this soft transfer occurs in a gentle manner which allows the biomolecules to be transferred without any damage. In addition, this powerful technique is cheap because the soft PDMS stamp can be replicated an indefinite number of times by simply pouring the PDMS prepolymer onto a single expensive silicon master mold that contains micro or nanostructures. Since the flexibility of the elastomeric stamp ensures a perfect conformal adhesion within the substrate, thus allowing replication on large areas up to several tens of cm^2, the use of such flexible PDMS stamps was also efficiently applied to another low-cost and high-throughput manufacturing technique called Soft UV Nanoimprinting Lithography (Soft UV-NIL). This technique creates a thickness contrast by embossing thin polymeric films, highlighting the advantages of using a flexible PDMS stamp.

Historically, Nanomprint Lithography (NIL) in its original version was proposed by Stephen Chou in 1995 (Chou et al. (1995)) as an alternative technique for the embossing of high resolution patterns in thermoplastic materials. The patterning of features as small as 10 nm has been demonstrated from the beginning (Chou (1997)). This nanoimprint process, usually referred to as thermal-assisted NIL (T-NIL), is based on the use of a hard mold, namely a silicon wafer. As schematically shown in Figure 1, this hard mold containing nanoscale surface-relief features is pressed at high pressure (50-100 bar) onto a thin polymeric resist film. The resist is held some 90-100 °C above its glass-transition temperature (T_g) for few minutes to allow the flowing of the polymer in the mold nanocavities. The thin residual layer of polymer intentionally left to prevent the direct contact between the substrate and the rigid mold is then removed by reactive ion etching (RIE) to complete the resist pattern transfer. T-NIL uses a very simple experimental set-up, which results in a very short process time, from seconds to minutes.

In contrast to conventional lithography methods based on exposure and development of a resist, nanoimprint lithography is based on mechanically embossing a thin polymer film under a mold that contains micro or nanopatterns. For this reason, limitations imposed by light diffraction or beam scattering in conventional lithography techniques can be overcome. As a second advantage, NIL provides parallel processing with high throughput, which is suitable for large-scale patterning with very high resolution. Another important benefit is the low cost of the NIL equipment, which compared with the processing cost of classical lithography techniques like deep UV optical lithography or Electron Beam Lithography (EBL). Nanoimprinting lithography has rapidly received a lot of attention from both the research community and industry, so much so that MIT's 2003 Technology Review named nanoimprint lithography as one of the "ten emerging technologies that will change the world." Additionally, NIL has been added into the International Technology Roadmap for Semiconductors (ITRS) for the next 22 nm node.

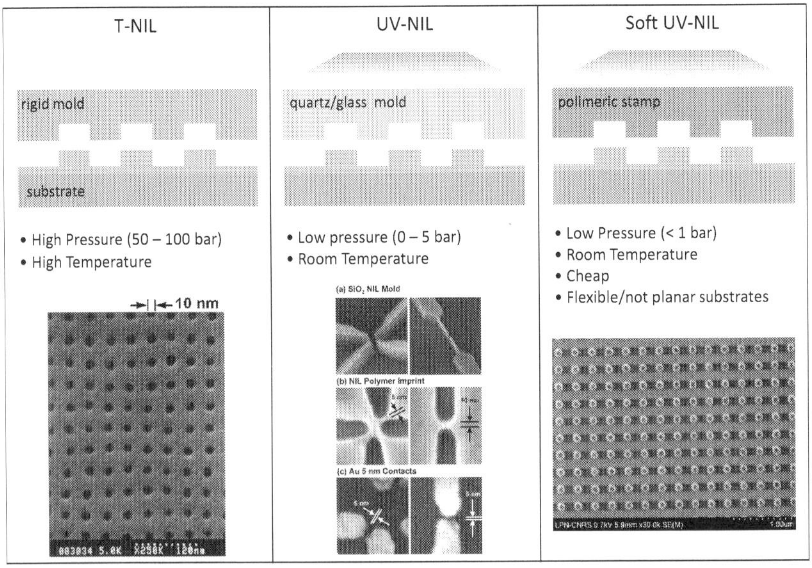

Fig. 1. Left panel: schematic of the T-NIL process proposed by S.Y. Chou in 1995 with a SEM image of 60 nm deep holes imprinted into PMMA which have a 10 nm minimum diameter and a period of 40 nm. Reprinted with permission from (Chou (1997)). Copyright 1997 American Vacuum Society. Central panel: schematic of the UV-NIL process and relative SEM image of the imprinted resist after UV curing with feature sizes as small as 5 nm. Reprinted with permission from (Austin et al. (2004)). Copyright 2004 American Vacuum Society. Right panel: SEM image of imprinted Amonil resist (AMO GmbH Aachen, Germany) of 100 nm square dots with 300 nm pitch after Soft UV NIL with PDMS stamp replicated by a Silicon master mold.

Although there are no apparent resolution limitations for T-NIL due to its purely mechanical embossing nature, this technology is unable to meet the stringent requirements of semiconductor IC manufacturing. In particular, the high pressure and temperature required during the process limit its use to few applications. UV-NIL (Haisma et al. (1996); Ruchhoeft et al. (1999)), which appeared quickly after T-NIL, is considered to be the most attractive variant for semiconductor IC manufacturing (Figure 1). With this variant, a transparent quartz mold is pressed at room temperature with a lower pressure on a liquid precursor that can be cured by UV light before the release of the mold. Industry quickly adopted UV-NIL, particularly Step and Flash Imprint Lithography (SFIL) which replicates nanostructures on a large area by patterning several fields (Bailey et al. (2002)). This industrial approach consists of dispensing a drop of the low viscosity photopolymerizable liquid resist on each field surface prior to the imprinting process. This allows for the patterning on the whole wafer area, much like a stepper lithography tool does.

As an alternative to UV-NIL, Soft UV-NIL has recently been proposed as a means to reduce the cost of master fabrication. Soft UV-NIL uses a flexible transparent stamp normally made of poly(dimethylsiloxane) (PDMS) or other flexible polymers that can be easily replicated from a single Silicon master mold. This Silicon master mold is fabricated with conventional lithography techniques, like EBL, which are more time consuming and expensive. The flexibility of the Polymeric stamp ensures contact with the surface substrate on large surfaces at low pressures, even on curved or flexible substrates. In particular, PDMS exhibits attractive properties like low interfacial free energy (~21.6 dyn/cm), chemical stability and high optical transparency. Moreover, its permeability prevents problems caused by trapped air bubbles in the resist layer when imprinting at ambient pressure. Soft UV-NIL is considered today to be the most promising variant to replicate patterns in the sub-50 nm range for mass production at low cost. Soft UV-NIL is currently being applied in a large spectrum of emerging area like Flexible Electronics, Photonics, Biotechnology and Nanomedicine.

In this paper we will discuss the ability to use PDMS based stamps, as candidates for nanoimprinting lithography in the sub-20nm scale on large surfaces and at very low pressure (< 1 atm). In terms of resolution, the replication of features with sizes as small as 2 nm has been demonstrated by using a unique carbon nanotube as a template and a modified PDMS material for the stamp (Hua et al. (2004)). These recent experiments demonstrate that PDMS properties, and specifically the density of cross-links, are important parameters that can influence the ultimate resolution. In this context, testing the resolution limits for soft UV-NIL with flexible stamps and for real applications appears very challenging.

In the frame of the European Project TERAMAGSTOR (EU funded FP7 STREP project) for the production of the next generation of perpendicular magnetic storage media with an areal density greater than 1 Tbit/inch^2, our group has been working on the feasibility to apply Soft UV-NIL for the nanofabrication of bit pattern media at ultra high resolution in the sub-10 nm scale. Our optimized approach for the fabrication of the silicon master mold using classical electron beam lithography will be presented with different fabrication techniques and resists. Due to the high cost of the master mold fabrication by EBL, we will propose new routes for the replication and/or inversion of the master mold at the whole wafer scale based on T-NIL. Once the hard mold with highly resolved patterns is obtained, the second challenge concerns the replication of the PDMS stamp. It should exhibit enough local hardness for the replication of nanometric features, and at the same time, preserve a global flexibility to ensure conformal contact at low imprinting pressures. The description of our bilayer hard-PDMS/PDMS stamp will be detailed and we will demonstrate the replication of 20 nm features under specific experimental conditions. To conclude with real application, we will present the fabrication by Soft UV NIL of plasmonic nanostructures for biosensing applications with superior optical performances directly related to their high quality and uniformity on the whole pattern area.

2. Master mold fabrication

The whole replication process of the nanostructures by Soft UV NIL is composed of three separate steps: the fabrication of the master mold, the replication by this master of the polymeric stamp and the imprinting in the UV curable resist by using this replicated polymeric stamp. Together, these steps affect the quality of the final replication in terms of resolution and line edge roughness of the nanostructures.

The first challenge is obtaining an accurate fabrication of the master mold; particular care has to be placed in order to achieve high resolution with low roughness features. Compared to other lithography techniques, the quality of the master mold is very important since the PDMS stamp will replicate any eventual imperfections with precision. These constraints can be relaxed, for instance, in the fabrication of the UV photomask for projection optical lithography, in which a demagnified image can be projected onto the photoresist (4:1). This allows for the patterning of smaller features as well as the reduction of mask damages and imperfections (Cui (2008)). For this reason, the fabrication process of such a master mold is more often based on electron beam lithography (EBL), which provides flexibility and high resolution.

The resolution limit of EBL is determined by proximity effects inside the resist, which are due to the scattering of electrons within the substrate. Working with a high energy electron beam (above 50 KeV) and with a thin resist layer reduces these proximity effects. We will compare standard processes based on a PMMA resist with direct etching, a lift-off process and a process based on a Hydrogen-silsesquioxane (HSQ) resist. Once the template is fabricated, it can be used many times, either as a mold for direct nanoimprint in T-NIL or UV-NIL, or as a master mold for the replication of polymeric stamps. In practice, both silicon and quartz templates can be damaged during these processes and it is thus highly desirable to develop a low cost solution for large area replications. Conventional methods such as EBL are time consuming and unsuitable for mass production. Other high throughput lithography techniques developed for the semiconductor industry, like immersion optical lithography and extreme UV lithography, include complicated processes and expensive equipment. To overcome this issue, we will describe a master mold replication process based on the combination of T-NIL and reactive ion etching.

2.1 EBL processes for high resolution patterns

Among the high resolution resists available for EBL, PMMA (polymethyl methacrylate) was one of the first polymeric materials to be studied for this purpose. The most prominent features of the PMMA resist are its high resolution, its wide process latitude, and its high contrast associated with its low sensitivity. Master molds have been fabricated by two EBL processes based on the PMMA resist: a conventional single-step process with PMMA exposure and direct Silicon etching (Figure 2 - Process 1(A)), and a multi-step process with PMMA exposure combined with metallic lift-off and Si reactive ion etching (Figure 2 - Process 1(B)).

In the first process, EBL exposure at 100 KeV is performed in a 430 nm thick layer of PMMA (950PMMA A5, MicroChem Corp.), spin coated on a Silicon wafer and baked at 180 °C for 30 minutes. The sample is developed in a methylisobutylketone (MIBK)/isopropanol (IPA)

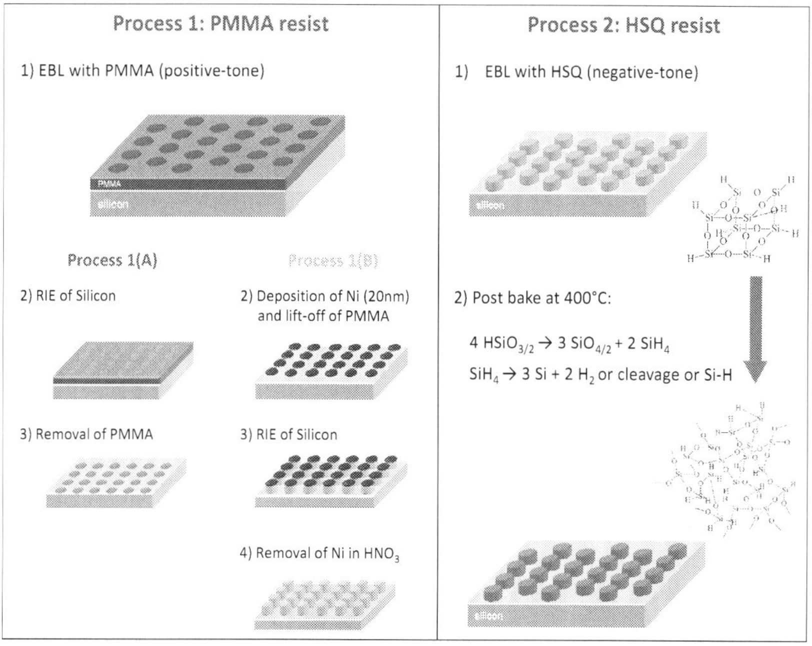

Fig. 2. Schematic of the Process 1, based on positive-tone PMMA resist and of the Process 2, based on negative-tone HSQ resist.

solution at 20 °C for 35 sec, rinsed in IPA for 10 s and then dried with a nitrogen gun. Pattern transfer in the Silicon substrate is then performed with a highly anisotropic RIE process based on $CHF_3/SF_6/O_2$, and the PMMA resist is finally removed with acetone. This process allows for the fabrication of high resolution and low sidewall roughness structures, as shown in the Figure 3 (a and b).

Since PMMA is a positive tone resist, an inverted pattern can be obtained either by exposing the complementary area, which is time consuming, or by a multi-step process with metal lift-off. In order to test the PMMA performance and the feasibility of this second process, we have carried out a challenge exposure of nanodots in a hexagonal pattern with a nominal diameter of 20 nm and a pitch of 60 nm, in a 80 nm thick PMMA resist (950PMMA A2, MicroChem Corp.). After the sample was developed as described in process 1 and lifted off of the 15 nm Ni mask, the dot patterns were transferred into the silicon wafer by reactive ion etching. The Ni mask was then removed by chemical wet etching in a HNO_3 acid solution.

The Figure 3 (c) shows a SEM image of the Si master molds after the Ni mask was removed. The inset shows a distribution graph of the area of the dots, including the standard deviation and the relative average diameter of the dots. In general, this process is suitable for the fabrication of structures up to the 100 nm scale, but it is not reproducible on sub-20 nm scale structures. This is due to the multiple intermediate steps which introduce broadening of the structures, defects and sidewall roughness. In order to improve shape uniformity and sidewall roughness, we have developed an alternative process based on negative-tone resist, Hydrogen Silsesquioxan (HSQ). HSQ is a new non-chemically amplified resist for EBL

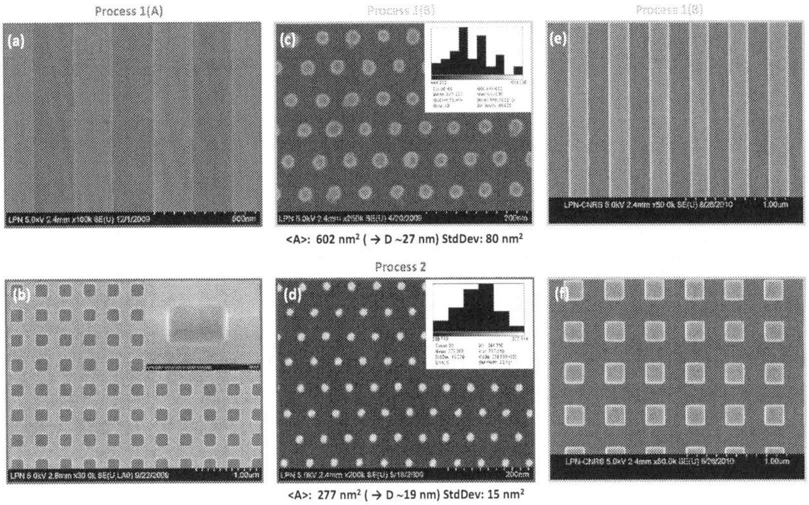

Fig. 3. SEM images recorded on Si master molds that have been processed with the PMMA process 1 (a-c) and with the HSQ process 2 (d-f). In particular, (c) and (d) show the results obtained for the processes 1(B) and the processes 2 respectively, by exposing nanodots with the same nominal diameter of 20 nm (pitch of 60 nm) in a hexagonal geometry. The correspondent dot area (A) distributions in the insets, shows that the HSQ master displays smaller dots with a better shape and area uniformity with a mean diameter (D) of 19 nm compare to the broadened mean diameter of 27 nm obtained with the multiple step process 1 (B).

which has been widely investigated in recent years (Chen et al. (2006); Falco & van Delft (20); Grigorescu & Hagen (2009); Namatsu et al. (1998)). It offers ultra-high resolution with linewidth fluctuations lower than 2 nm (Word & Adesida (2003)) and high stability and strength that can be improved by a thermal post-baking (Yang & Chen (2002)). In this second process, EBL exposure at 100 KeV is carried out on 40 nm thick HSQ (XR 1541, Dow Corning, 2% solids) spin coated on a silicon wafer. In order to avoid any change in the structure of the HSQ and possibly reduce the resist contrast, no pre-baking was conducted and the wafer was simply left for one day at room temperature to remove the solvents. The wafer is then developed in a potassium hydroxide (KOH) based solution at 20°C for 30 sec, rinsed for 2 min in deionised water, and gently dried with pure nitrogen gas. Finally, the post thermal curing of HSQ was performed at 400°C. By comparison with the multi-step fabrication process 1(B), this single step process results in a superior resolution and uniformity of the nanodots (Figure 3 (d)). The post-baking treatment induces a silica-like network redistribution of the HSQ which can be directly used, after anti-sticking treatment, as a master mold for replication of a bi-layer hard-PDMS/PDMS stamps.

2.2 Performances of the HSQ EBL process at the 10 nm scale

Even though a sub-10 nm resolution has already been achieved on a HSQ resist for isolated structures (Grigorescu & Hagen (2009)), one of the biggest challenges in nanofabrication is the patterning of high resolution and high density structures. Bit patterned media for magnetic data storage is one of the most important applications of this technology, and it was actually one of the first applications for which NIL was proposed and studied (Chou (1997)). Patterned media by means of Nanoimprinting is one of the most promising approaches pursued by global companies like Hitachi and Toshiba in order to overcome the superparamagnetic limit of current perpendicular magnetic storage media. Our group has been working within the framework of the European project "'Teramagstor"' [www.teramagstor.com], in which the main goal of the consortium is the demonstration of a Bit pattern media prototype for the next generation of disk drivers with an areal density greater than 1 Tbit/in^2 fabricated by EBL and Soft UV NIL. In contrast to writing on conventional media, where the bits can be placed everywhere on the medium, a key challenge of writing on nanodot arrays of a bit patterned media is that the write pulses must be synchronized to the dot period and a perfect displacement of the dots is compulsory. Figure 4 shows a Silicon master mold fabricated utilizing the HSQ process with 15 nm dots and a 60 nm pitch (corresponding to an areal density of 180 Gbit/in^2), and 15 nm dots and a 30 nm pitch (corresponding to an areal density of 720 Gbit/in^2), with perfect dots displacement on the whole area. At the moment, we are working on further process optimization in order to achieve 1 Tbit/in^2 which corresponds to a 27 nm pitch and will eventually demonstrate higher areal density.

Fig. 4. SEM images recorded on Si master molds fabricated by HSQ Process 2, after annealing at 420 °C of (a-b) 15 nm dots, 60 nm pitch (corresponding to an areal density of 180 Gbit/in^2) and 15 nm dots, 30 nm pitch (corresponding to an areal density of 720 Gbit/in^2). The thickness of the resist is 25-30 nm.

2.3 The whole wafer inversion based on T-NIL

We have recently proposed and developed a T-NIL based process to replicate and invert an EBL master mold (Chen et al. (2010)). The inversion by means of T-NIL allows for the fabrication of a silicon mold with inverted features, so that pillars in the original master mold become holes in the daughter mold (Figure 5). We have studied how each pattern transfer process can affect feature size, pattern shape and homogeneity. Sub-200nm nanostructures have been inverted with good reproducibility and homogeneity on field as large as $1cm^2$.

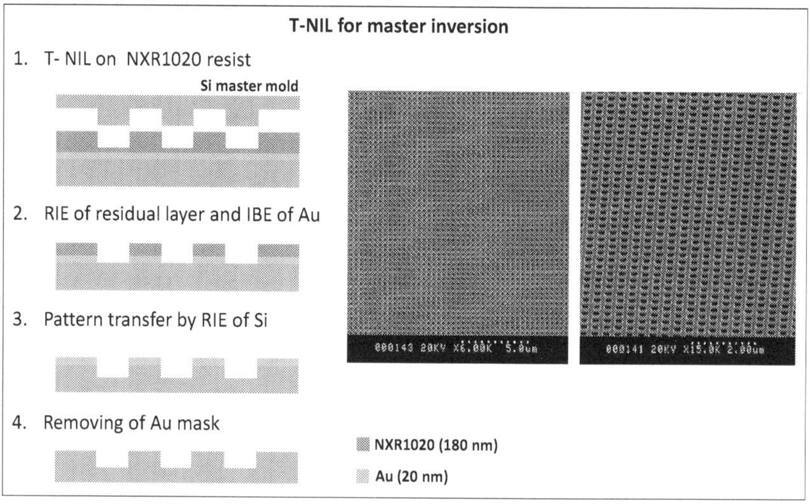

Fig. 5. (a) Schematic of master mold inversion based on T-NIL and SEM images of daughters molds

A schematic of this inversion process is presented in Figure 5. A thin gold layer (20nm) is first deposited on the silicon wafer. The thermal imprints are then performed on the NXR-1020 resist (Nanonex), spin coated at a thickness of 180 nm on Au/Si wafers. These resist films are annealed at 120°C for 10 minutes to remove the solvents. The Si master mold is gently placed in contact with the resist substrate and sandwiched between the two membranes of the NXR-2500 Nanonex, which provides optimal uniformity over the whole imprinted field. Imprinting is carried out at 130°C in two successive steps: 10 sec at 120 PSI (8 Bars) followed by 30 sec at 200 PSI (14 Bars). The imprinting tool is then cooled down below the estimated glass transition temperature of the resist, before carefully demolding the master. After removing the residual NXR1020 resist layer with O_2 plasma at a rate of 30 nm/min, the thin Au mask layer is etched by ion beam etching (IBE). The resist layer is then simply removed with acetone. Finally, the patterns are transferred into the silicon wafer with a standard Si reactive ion etching process using the Au mask, obtaining hole arrays. A trichloromethylsilane (TMCS) anti-sticking treatment is then applied to the daughter Si mold. With this process, we demonstrated that inversion is feasibile at the whole wafer scale. Figure 5 shows the SEM images of a 2 inch silicon wafer inverted daughter mold. It was patterned on a surface of 1 cm^2, with 200 nm-wide hole arrays. A good homogeneity was observed over the whole imprint field. These results clearly show the ability of the process to reach a resolution of 100 nm.

For feature sizes smaller than 100 nm, a small broadening of the pattern size is observed, even though the general shape is maintained (Chen et al. (2010)). An average broadening of 20 nm is observed for patterns (nanoline and nanohole) with feature sizes above 70 nm, while the broadening of feature sizes smaller than 50 nm become bigger (30 nm). This broadening originates from a loss of etch resistance of the NXR1020 mask resist during Ar IBE at high energy. Possible solutions to improve accuracy will be explored by using a thinner metallic underlayer or a T-NIL resist with a larger IBE resistance.

3. Fabrication of the replicated soft stamp

3.1 Polymeric stamp materials

The first unquestionable advantage of using a polymeric replica of the original master mold in the nanoimprint process is the cost reduction. As previously stated, from a single and expensive master mold, it is possible replicate a large number of polymeric stamps to use in the nanoimprint process. This prevents damages to the original master mold if it is directly used in the nanoimprint process. Moreover, the long range flexibility of the elastomeric material used for the stamp ensures the contact between the stamp and the substrate on large surfaces at low pressures (tens of bars) and on curved or flexible substrates. Several kind of polymeric materials have been tested as candidate for stamp replication including polycarbonate resins (Posognano et al. (2004)), cross-linked novolak based epoxy resin (Pfeiffer et al. (2002)), fluoropolymer materials and tetrafluoroethylene(PTFE) (Kang et al. (2006)). Nevertheless, poly(dimethylsiloxanes) (PDMS) still offers numerous attractive properties as a stamp elastomer. First of all, PDMS ensures a conformal adhesion of the stamp with the substrate on large areas without applying any pressure. To stress this point, Soft UV NIL performed with PDMS based stamps was recently renamed as UV enhanced Substrate Conformal Imprint Lithography (UV-SCIL) by Philips Research and SUSS MicroTec (Ji et al. (2010)).

Additionally, PDMS offers numerous other attractive properties: (1) its flexible backbone enables accurate replication of relief shapes in the fabrication of the patterning elements, (2) its low Young's modulus ($\sim$750 KPa) and its low surface energy enable conformal contact with the substrate without applied pressure and nondestructive release from patterned structures, (3) its good optical transparency down to a light wavelength of approximately 256 nm, (4) its commercial availability in bulk quantities at low cost. While PDMS offers some advantages, there are a number of properties inherent to PDMS which severely limit its capabilities in soft UV-NIL. First, its low Young's modulus limits the fabrication of features with high aspect ratios due to collapse, merging and buckling of the relief structures. Second, its surface energy ($\approx$ 22-25 mNm^{-1}) isn't low enough to replicate profiles with high accuracy. Third, its high elasticity and thermal expansion can lead to deformation and distortions during the fabrication. Finally, it shrinks by $\sim$1% after curing and can be readily swelled by some organic solvents. In general, long range deformations can be avoided by the introduction of a thin glass backplane which preserves a global flexibility. On the other hand, short range deformations can be avoided only by increasing the elastic modulus of PDMS, as we will discuss in deepth in the next paragraph. Therefore, compared to hot-embossing lithography (HEL) and UV-NIL which use rigid molds, it is particularly important to control the mold deformation as much as possible for the soft UV-NIL.

3.2 Basic PDMS stamp fabrication process

Simple PDMS stamps are typically replicated by first mixing two commercial PDMS components: 10:1 PDMS RTV 615 (part A) siloxane oligomer and RTV 615 (part B) cross-linking oligomers (General Electric). The mixture is then casted on the nanostructured master molds and degassed in a dessicator. A curing time of 24 h and a curing temperature of 60 °C are usually recommended in order to reduce roughness and to avoid a build up of tension due to thermal shrinkage. Longer curing times and higher temperatures allow up to twice the elastic modulus and hardness of the polymer, but can also lead to higher roughness and deformations. The stamps are left to cool to room temperature, carefully peeled off from the master mold and treated with silane based antisticking treatment to further reduce the low PDMS surface energy. These stamps are not suitable for the replication structures at the sub-100 nm scale or with a high aspect ratio because of the low elastic modulous of PDMS. To address this issue, a modified PDMS with a higher elastic modulus was already proposed 10 years ago.

3.3 Improved hard-PDMS/PDMS bilayer stamp fabrication process

In 2000, Schmid et al. (Schmid & Michel (2000)) used a modified PDMS with higher elastic modulus (hard-PDMS) to extend the range in which conventional soft lithography could be applied. The hardening of the PDMS was accomplished by decreasing the chain length of the prepolymer. Increasing the elastic modulus of PDMS allows for the replication of smaller features in the 100 nm range, but it results in poor flexibility and an increase in the brittleness of the stamp. To overcome this problem, Odom et al. (Odom et al. (2002)) proposed a composite stamp of hard-PDMS and standard PDMS which combined the advantages of a rigid layer to achieve a high resolution pattern transfer and an elastic support which enabled conformal contact even at a low imprint pressure. Another important property of hard-PDMS is the lower viscosity of its prepolymer in comparison to PDMS. PDMS prepolymer cannot completely fill up the recessed nanoareas of the master mold due to its high viscosity, and the height of the soft stamp nanostructures will thus be lower than the height of the master patterns resulting in a poor inprinted thickness contrast. The decreasing of the chain length in the preparation of the hard-PDMS prepolymer, produce a lowering of its viscosity with a consequent higher ability of hard-PDMS prepolymer to replicate with accuracy the original master mold especially for high dense and small patterns.

The importance to reduce the viscosity of the prepolymer for an accurate replication of the master mold have been reported by other two groups (Kang et al. (2006); Koo et al. (2007)). In this case the viscosity of the PDMS prepolymer was reduced with the introduction of a solvent to the mixture, together with the use of an excessive amount of modulator to delay the cross-linking. The improved stamp replication process resulted in a better more accurated imprinting process, as shown in 8.

In Figure 7 is summarized the procedure for the replication of the bi-layer hard-PDMS/PDMS stamp. The hard-PDMS prepolymer, prepared by following a recepie similar to the one proposed by Schmid (Schmid & Michel (2000)), is spin coated onto a master at 5000 rpm for 30 sec and then degassed in vacuum for 10 min (the thickness of the hard-PDMS is about 5-8 μm). A mixture of conventional PDMS (1:10) is then poured on the spin coated hard-PDMS layer curing at 60 °C for 24 hours, the bi-layer stamp can be peeled off from the master and it is treated with trichloromethylsilane (TMCS).

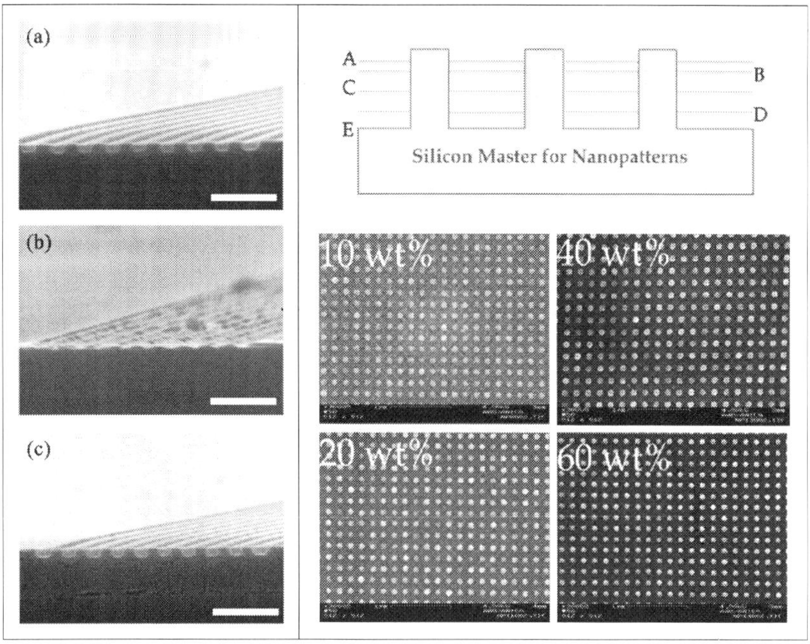

Fig. 6. Left panel: SEM images of (a) a 100 nm line of the master pattern, (b) patterned polymer with the composite stamp prepared by the method proposed by Schmid and Odom and (c) by the improved method with toluene as solvent for hard-PDMS (the scale bar is 500 nm). Reprinted with permission from (Kang et al. (2006)). Copyright 2006 IOPP Publication. Right panel: schematic of the influence of the material viscosity on the penetration depth into the master cavities for nanostructures. Exemplary SEM pictures of an 50 nm dots array imprinted with diluted PDMS molds prepared with toluene concentration of: A: 10 wt%, B: 20 wt%, C: 40 wt% and D: 60 wt% whom correspond to a pattern height increases from 10 nm (A) up to 70 nm (D). Reprinted with permission from (Koo et al. (2007)). Copyright 2007 Elsevier.

4. Soft imprinting process

4.1 Minimization of pressure and dimension control

In order to achieve an accurate pattern replication, it is important to control the eventual broadening and deformations of the replicated structures. During T-NIL and UV-NIL processes, a parallel and uniform surface contact between the hard mold and the substrate is obtained by applying high pressure, and the replicated structures preserve a high accuracy when compared to the original ones in the master mold. During the Soft UV NIL process, the flexibility of the PDMS stamp ensures a conformal adhesion with the substrate at low pressure. On the other hand, the low elastic module of PDMS can produce deformations of high aspect ratio structures when too much pressure is applied. If it is not possible to completely avoid these deformations, it is important to control and reduce them under appropriate values, depending on the specific application. Because soft UV-NIL is a very recent variant of NIL,

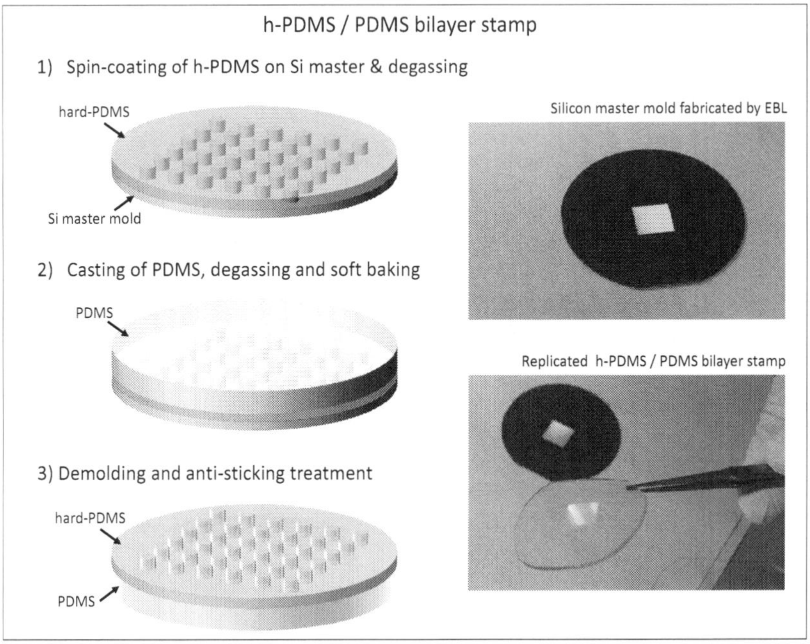

Fig. 7. Schematic of the fabrication process of a hard-PDMS/PDMS bilayer composite stamp and a pictures of a Silicon master mold and of a bilayer stamp after peeling off.

such a resolution study has been conducted by only a few groups. For instance, KarlSuss GmbH has recently demonstrated the replication of nanoholes with a diameter of 340 nm ($\pm$5%) and a minimized residual layer thickness of 36 nm in the Amonil resist with a pitch uniformity of 2 nm over a surface area of 6 inches (Ji et al. (2010)).

The pressure applied during imprinting directly influences the flow of the resist and determines the accuracy of the imprinted nanoscale structures. The first step consists of minimizing the imprinting pressure to reduce local distortions while ensuring a complete resist filling in the stamp nanocavities. We have shown that this pressure can be largely reduced to 0.7 bar (10 PSI) (Cattoni et al. (2010); Shi et al. (2009)). Combined with a UV exposure of 10 min (dose of 2 Jcm^2) at 365 nm wavelength, patterns with a high quality shape can be obtained (Figure 8(a)). To quantify the accuracy of the control dimension, we investigated the changes in pattern size on a two-inch wafer scale under this optimum pressure (10 PSI). Nanodot patterns with two different diameters (215 nm and 310 nm) have been chosen for this study. Each field size is 200 um x 200 um on the wafer. The critical dimension of the nanodot patterns on the mold and replicated in the resist were then measured using a high-resolution scanning electron microscopy (Fig. 8(b)). Black and blue points correspond to the measured diameters on the EBL master mold whereas red points correspond to measured diameter in the Amonil resist after nanoimprinting. We observe a broadening of the dot diameter, which never exceeds 10 nm, corresponding to a standard deviation of less than 5%.

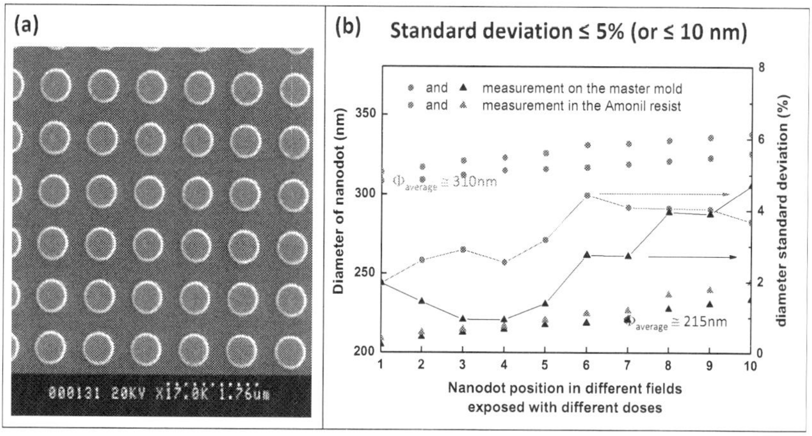

Fig. 8. Control dimension fidelity after soft UV-NIL at optimized pressure: (a) SEM image of 310 nm-diameter pillar array recorded in the Amonil resist and (b) comparison of the diameter values measured on the master mold and in the Amonil thickness contrast, with the relative calculated diameter standard deviation.

4.2 Minimization of the thickness of the resist residual layer

A second important point of the nanoimprint process concerns the removal of the residual layer of the resist. Because T-NIL and UV-NIL use rigid mold and high pressure, a thin residual layer of resist is normally left between the protrusions of the mold and the substrate. It acts as a soft cushion layer that prevents direct impact of these fragile nanostrucures and the substrate. This residual layer is normally removed by reactive ion etching (RIE) which can largely affect the original shape and size of the pattern. Soft UV NIL uses flexible stamps and this residual layer can be reduced as much as possible by simply adapting the initial resist thickness to the depth of the stamp (height of patterns). Figure 9(b) shows a SEM cross-section image of the imprinted structures at a optimized pressure of 0.7 bars, with a 25 nm thick residual layer, obtained by adjusting the initial thickness of resist. These results can be compared to the recent ones from AMO that are presented in Figure 9(a).

Finally, in order to validate the ultra-high resolution HSQ process for EBL and our improved hard-PDMS/PDMS bilayer stamp fabrication process, we have carried out Soft UV NIL in the sub-100 nm scale. Figure 9 (b-c) shows the thickness contrast in the Amonil resist after Soft UV-NIL, at 10 PSI and at room temperature, by using the bi-layer mold hard-PDMS/PDMS for 50 nm dots with a pitch of 100 nm and 20 nm dots with a pitch of 60 nm. The high and uniform contrast thinknesses obtained on the whole pattern area demonstrate the ability of Soft UV NIL with hard-PDMS/PDMS stamps to replicate nanostructures in the 20-nm scale. We believe than our current limitations in achieving higher resolution, specifically for the replication of highly dense structures, are essentially due to the replication of the bilayer stamp. It will be necessary to work on this stamp replication process, in particular by further reducing the viscosity of the hard-PDMS prepolymer and by controlling the effects of the thermal curing process, which results in a isotropic volume shrinkage of the hard-PDMS that becomes more and more important at the sub-20 nm scale.

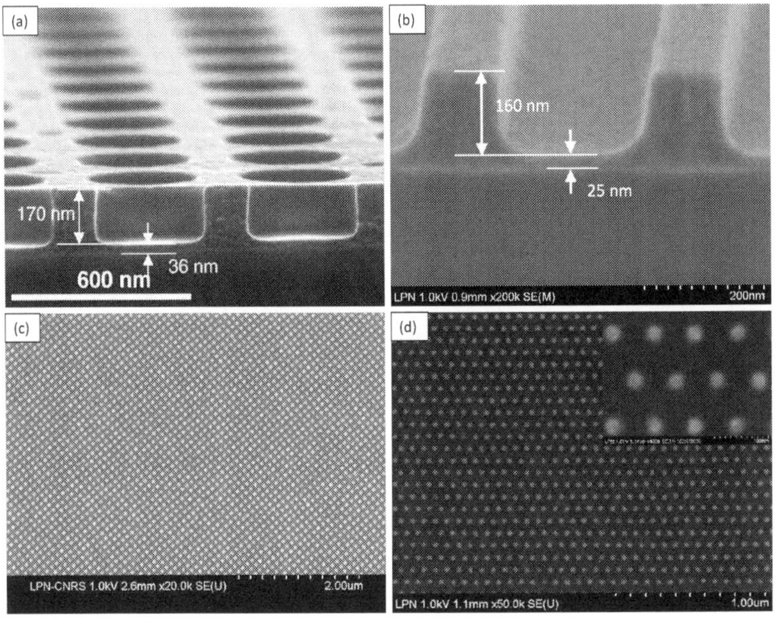

Fig. 9. SEM images of: (a) imprinted structures into AMONIL resist, the structure depth is 170 nm and the residual layer is 36 nm. Reprinted with permission from (Ji et al. (2010)). Copyright 2007 Elsevier. (b) imprinted structures into AMONIL resist, the structure depth is 160 nm and the residual layer is 25 nm. (c-d) High resolution replication at the 50 nm and at the 20 nm range in AMONIL resist.

5. Soft UV NIL for plasmonic biosensing: a real case application

Surface plasmon resonance (SPR) based sensors are a well established technology utilized for label-free bio-chemical sensing in different applications, from immunoassay and medical diagnostics, to environmental monitoring and food safety. Localized Surface Plasmon Resonance (LSPR) based sensors have different advantages and constitute a promising alternative to the SPR sensor (Stuart et al. (2005)). Because LSPR based biosensors are by design very sensitive to changes in the characteristics of nanoparticles (uniformity in nanoparticle size, shape and composition) the standard approach for the fabrication of LSPR based sensor is Electron Beam Lithography (EBL) which provide an extreme control over nanoscale size and shape of the nanoparticles thus improving sensitivity and reliability of the sensor. On the other side this is an expensive and time-consuming technique, consequently not suitable for mass production.

We have successfully realized a LSPR biosensor based on $\lambda^3/1000$ plasmonic nanocavities fabricated by Soft UV Nanoimprint Lithography on large surfaces (0,5 - 1 cm^2). These structures present nearly perfect omnidirectional absorption in the infra-red regime independently of the incident angle and light polarization and outstanding biochemical sensing performances with high refractive index sensitivity and figure of merit 10 times higher than conventional LSPR based biosensor (Cattoni et al. (2011)).

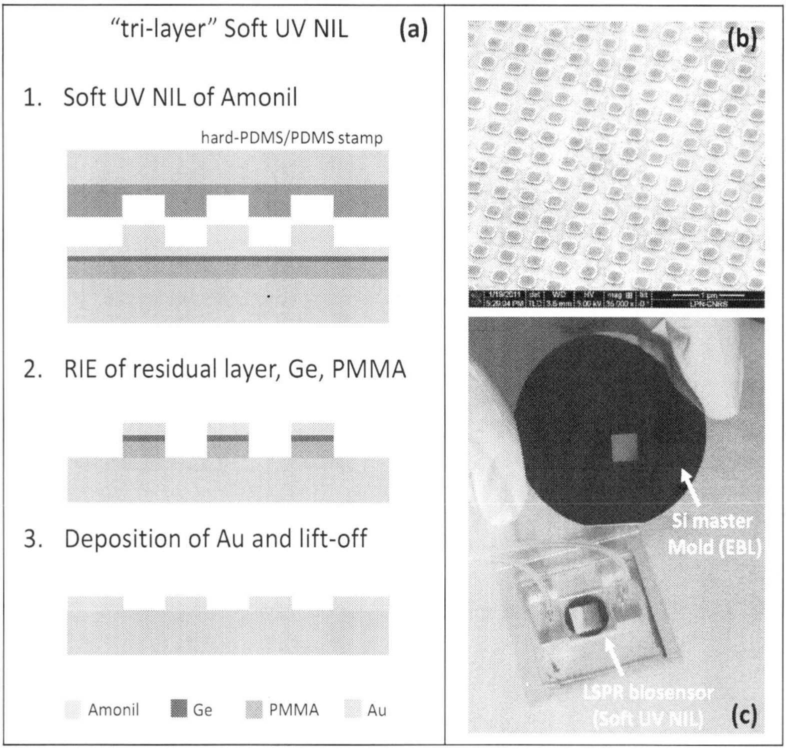

Fig. 10. (a) Schematic of the tri-layer Soft UV NIL process, (b) SEM image of square gold nanoparticle (size = 200 nm, pitch 400 nm) realized with this method and lift-off, (c) the LSPR biosensor fabricated by Soft UV NIL, integrated in a microfluidic channel with its original Silicon master mold fabricated by EBL.

The basic element of the nanocavities array (not shown) is composed by lower thick gold film acting on a glass substrate, a thin dielectric layer forming the gap of the optical antenna and an upper gold nanoparticle realized by "tri-layer" Soft UV imprint lithography and standard lift-off. In figure (a) is shown the concept of the tri-layer system on a generic substrate: typical UV NIL resists (like Amonil) are not soluble in solvents and a lift-off process is possible by using a PMMA layer under the UV NIL resist. In our process a further 10 nm Ge layer is insert between the thick PMMA layer and the thin UV NIL resist (Amonil) to improve the selectivity of the former one over the PMMA layer. After imprinting with a UV light and separation, the top layer structure is transferred into the bottom layer by a sequential reactive ion etching. The high aspect ratio tri-layer so obtained can be used directly as etching mask or for the lift-off of metals. In Figure 10 (b) is shown a SEM image of square gold nanoparticle (size = 200 nm, pitch 400 nm) realized by "tri-layer" Soft UV imprint lithography. Figure 10 (c) shows the LSPR biosensor based on $\lambda^3/1000$ plasmonic nanocavities fabricated by Soft UV Nanoimprint Lithography and integrated in a glass/PDMS/glass fluidic chamber for the optical index sensing experiments together with its original Silicon master mold fabricated by EBL.

6. Conclusion

As we have shown, the replication process of nanostructures by Soft UV NIL it is composed by three separate processes: the fabrication of the master mold, the replication of the soft polymeric stamp from this template, and the imprinting process by using this stamp. All these steps contribute equally at the quality of the final result, in terms of resolution and line edge roughness of the nanostructures. In this manuscript we have presented a detailed master mold fabrication process based on EBL. The performances of two EBL resists (PMMA and HSQ) for the replication of high density patterns (pitch=30 nm) have been discussed for ultra high resolution at the 15 nm scale. In addition a process combining T-NIL and etching has also been proposed for the cheap and fast replication of master molds. We have then detailed the replication of the polymeric stamp, based on a composite hard-PDMS/PDMS bilayer. Finally for the last third critical imprinting step, we have demonstrated the ability of Soft UV NIL to replicate nanostructures at the 20 nm scale with high uniformity at the whole pattern area. We do believe that Soft UV NIL will have in the near future an important role as a powerful and versatile tool for nanofabrication. To validate this statement, we conclude by presenting an example of a real case application, i.e. the fabrication by Soft UV NIL on large area (1 cm^2) of gold plasmonic nanostructures with outstanding optical and biosensing performances.

7. References

Austin, M. D., Ge, H. X., Wu, W., Li, M. T., Yu, Z. N., Wasserman D., Lyon, S. A., & Chou, S. Y. (2004). Fabrication of 5 nm linewidth and 14 nm pitch features by nanoimprint lithography, *Appl. Phys. Lett.* 84: 5299-5302.

Bailey, T., Johnson, S., Resnick, D., Sreenivasan, S., Ekerdt, J. & Willson., C. (2002). Step and flash imprint lithography: An efficient nanoscale printing technology, *J. Photopolymer Sci. Tech.* 15: 481–487.

Cattoni, A., Cambril, E., Decanini, D., Faini, G. & Haghiri-Gosnet, A. (2010). Soft UV-NIL at 20 nm scale using flexible bi-layer stamp casted on HSQ master mold, *Microelectron. Eng.* 87: 1015 – 1018.

Cattoni, A., Ghenuche, P., Haghiri-Gosnet, A., Decanini, D., Chen, J., Pelouard, J. L. & Collin, S. (2011). $\lambda^3/1000$ plasmonic nanocavities for biosensing fabricated by soft UV nanoimprint lithography, *submitted to Nano Letters.*

Cerf, A. & Vieu, C. (2010). *Soft Lithography, a Tool to Address Single-Objects Investigations*, InTech, chapter 22, pp. 447–456.

Chen, J., Shi, J., Cattoni, A., Decanini, D., Liu, Z., Chen, Y. & Haghiri-Gosnet, A.-M. (2010). A versatile pattern inversion process based on thermal and soft uv nanoimprint lithography techniques, *Microelectron. Eng.* 87: 899 –903.

Chen, Y., Yang, H. & Cui, Z. (2006). Effects of developing conditions on the contrast and sensitivity of hydrogen silsesquioxane, *Microelectron. Eng.* 83: 1119.

Chou, S. (1997). Patterned magnetic nanostructures and quantized magnetic disks, *PROCEEDINGS OF THE IEEE* 85: 652–671.

Chou, S. Y., Krauss, P. R. & Renstrom, P. J. (1995). Imprint of sub-25 nm vias and trenches in polymers, *Appl. Phys. Lett.* 67: 3114–3116.

Cui, Z. (2008). *Nanofabrication - Principles, Capabilities and Limits*, Springer.

Falco, C. & van Delft, J. (20). Delay-time and aging effects on contrast and sensitivity of hydrogen silsesquioxane, *J. Vac. Sci. Technol. B* 2002-2007: 2932.

Grigorescu, A. & Hagen, C. (2009). 10 nm lines and spaces written in hsq, using electron beam lithography, *Nanotechnology* 20: 292001.

Haisma, J., Verheijen, M., Heuvel, K. V. D. & Berg, J. V. D. (1996). Mold-assisted nanolithography: A process for reliable pattern replication, *J. Vac. Sci. Technol. B* 14: 4124–4129.

Hua, F., Sun, Y. G., Gaur, A., Meitl, M. A., Bilhaut, L., Rotkina, L., Wang, J. F., Geil, P., Shim, M. & Rogers, J. A. (2004). Polymer imprint lithography with molecular-scale resolution, *Nano Lett.* 4: 2467.

Ji, R., Hornung, M., Verschuuren, M. A., van de Laar, R., van Eekelen, J., Plachetka, U., Moeller, M. & Moormann, C. (2010). Uv enhanced substrate conformal imprint lithography (uv-scil) technique for photonic crystals patterning in led manufacturing, *Microelectron. Eng.* 87: 963–967.

Kang, H., Lee, J., Park, J. & Lee, H. H. (2006). An improved method of preparing composite poly(dimethylsiloxane) moulds, *Nanotechnology* 17: 197–200.

Koo, N., Bender, M., U.Plachetka, Fuchs, A., Wahlbrink, T., J.Bolten & Kurz, H. (2007). Improved mold fabrication for the definition of high quality nanopatterns by soft uv-nanoimprint lithography using diluted pdms material, *Microelectron. Eng.* 84: 904–908.

Kumar, A. & Whitesides, G. M. (1993). Features of gold having micrometer to centimeter dimensions can be formed through a combination of stamping with an elastomeric stamp and an alkanethiol "ink" followed by chemical etching, *Appl. Phys. Lett* 63: 2002–2005.

Namatsu, H., Yamaguchi, T., Nagase, M., Yamazaki, K. & Kurihara, K. (1998). Nano-patterning of a hydrogen silsesquioxane resist with reduced linewidth fluctuations, *Microelectron. Eng.* 41: 331–334.

Odom, T. W., Love, J. C., Wolfe, D. B., Paul, K. E. & Whitesides, G. M. (2002). Improved pattern transfer in soft lithography using composite stamps, *Langmuir* 18: 5314–5320.

Pfeiffer, K., Fink, M., Ahrens, G., Gruetzner, G., Reuther, F., Seekamp, J., Zankovych, S., Torres, C. S., Maximov, I., Beck, M., Graczyk, M., Montelius, L., Schulz, H., Scheer, H. C. & Steingrueber., F. (2002). Polymer stamps for nanoimprinting, *Microelectron. Eng.* 61-62: 393.

Posognano, D., D'Amone, S., Gigli, G. & Cingolani., R. (2004). Rigid organic molds for nanoimprint lithography by replica of high glass transition temperature polymers., *J. Vac. Sci. Technol. B* 22: 1759.

Rogers, J. A. & Nuzzo, R. G. (2005). Recent progress in soft lithography, *Materials Today* 8: 50–56.

Ruchhoeft, P., Colburn, M., Choi, B., Nounu, H., Johnson, S., Bailey, T., Damle, S., Stewart, M., Ekerdt, J., Sreenivasan, S. V., Wolfe, J. C. & Willson, C. G. (1999). Patterning curved surfaces: Template generation by ion beam proximity lithography and relief transfer by step and flash imprint lithography, *J. Vac. Sci. Technol. B* 17: 2965–2970.

Schmid, H. & Michel, B. (2000). Siloxane polymers for high-resolution, high-accuracy soft lithography, *Macromolecules* 33: 3042–3049.

Shi, J., Chen, J., Decanini, D., Chen, Y. & Haghiri-Gosnet, A.-M. (2009). Fabrication of metallic nanocavities by soft uv nanoimprint lithography, *Microelectron. Eng.* 86: 596–599.

Stuart, D., Haes, A., Yonzon, C., Hicks, E. & Duyne, R. V. (2005). Biological applications of localised surface plasmonic phenomenae, *IEE Proc. Nanobiotechnol.* 152: 13 – 32.

Word, M. & Adesida, I. (2003). Nanometer-period gratings in hydrogen silsesquioxane fabricated by electron beam lithography, *J. Vac. Sci. Technol. B* 21: L12.

Xia, Y. & Whitsides, G. (1998). Soft lithography, *Angew. Chem. Int. Ed* 37: 550.

Yang, C.-C. & Chen, W.-C. (2002). The structures and properties of hydrogen silsesquioxane (hsq) ?lms produced by thermal curing, *J. Mater. Chem.* 12: 1138.

8

Repairing Nanoimprint Mold Defects by Focused-Ion-Beam Etching and Deposition

Makoto Okada and Shinji Matsui
University of Hyogo
Japan

1. Introduction

Nanoimprint lithography (NIL) has been attracting attention from many industries because of its potential use in producing various nanostructure applications through a simple, low-cost, and high-throughput process. There are three primary types of NIL: thermal (T-NIL), UV-NIL, and room-temperature (RT-NIL). T-NIL has a heating and cooling process because thermoset or thermoplastic resins are usually used as T-NIL resins. When a thermoset resin is used, the mold is pressed on a substrate coated with the resin at room-temperature. During pressing, the mold and substrate are heated to harden the thermoset resin, and after cooling, the mold is separated from the substrate. It is slightly different with thermoplastic resin: in this case, the mold is pressed on a substrate coated with the resin at the resin's glass-transition temperature (Tg). The mold and substrate temperatures are then decreased and the mold is removed from the substrate. Si, SiO_2/Si, and Ni molds are usually used as T-NIL molds. UV-NIL is a room-temperature process because UV-curable resins are used as UV-NIL resins. The UV-NIL mold is pressed on the substrate coated with UV-curable resin and then the substrate is irradiated with 365-nm UV through the mold. After this irradiation, the mold is separated from the substrate. This means that UV transmissive material must be used as UV-NIL mold material. Generally, a quartz mold is used as a hard mold, and a polydimethylsiloxane (PDMS) mold is used as a soft mold. RT-NIL can be performed without heating, cooling, or UV irradiation. In this process, the sol-gel materials, such as hydrogen silsesquioxane (HSQ), spin-on-glass (SOG), and sol-gel indium tin oxide (ITO), are used as RT-NIL resins. This process requires high pressure, so Si or SiO_2/Si molds are usually used.

What type of mold to use is one of the most important factors in nanoimprint lithography because the mold must come into direct contact with the replication material and the imprinted pattern resolution depends on the mold pattern resolution. The pattern is therefore typically fabricated by electron beam (EB) lithography to obtain a high resolution pattern, thus necessitating a mold repair process with high resolution. In photolithography mask repair, focused ion beam (FIB) etching is used to remove Cr opaque defects, and FIB chemical vapour deposition (CVD), using hydrocarbon precursor gas, is used to repair clear defects. Two types of defect occur in NIL molds –protrusion and hollow defects– which correspond to the opaque and clear defects in photomasks. However, unlike photomask patterns, the NIL relief-structure patterns are formed on a substrate surface. Therefore, we must perform FIB etching and CVD directly on the substrate material to repair protrusion and hollow defects, respectively. From an economic viewpoint it is best to use existing technology, so we applied FIB etching and CVD to repair the NIL molds and then examined the resulting repair resolution.

2. Thermal nanoimprint mold repair using FIB technique

Watanabe et al. previously reported on SiO_2/Si mold repair for thermal NIL using FIB etching and CVD. We first fabricated an SiO_2/Si mold with protrusion and hollow defects by EB lithography and reactive ion etching (RIE). The scanning electron microscopy (SEM) images of the fabricated SiO_2/Si mold shown in Fig. 1 depict both the (a) protrusion and (b) hollow defects. The width and length of the protrusion defect were 210 and 130 nm, respectively, and the width of the hollow defect was 250 nm.

(a) (b)

Fig. 1. SEM images of fabricated SiO_2/Si mold with (a) protrusion and (b) hollow defects.

(a) (b)

Fig. 2. SEM images of imprinted (a) protrusion- and (b) hollow-defective lines on NEB-22.

We performed thermal nanoimprinting on NEB-22 (Sumitomo Chemical Co.) using this mold. The mold and resin were heated at 150 °C. Imprinting pressure and time were 10 MPa and 1min, respectively. The protrusion and hollow defects were clearly imprinted on the resin. We repaired these defects by FIB etching and CVD.

Figure 3(a) and (b) shows the schematic of the repair process for the protrusion and hollow defects, respectively, on the mold. We used SMI2050MS2 (SII NanoTechnology Inc.) as a FIB system. The ion source, acceleration voltage, and beam current were gallium, 30 kV, and 1 pA, respectively. The protrusion defect was removed by FIB etching. When we repaired the hollow defect, we performed FIB-CVD using phenanthrene ($C_{14}H_{10}$) as a source gas to fill in the hollow defect. Using the phenanthrene caused a diamond-like carbon (DLC) to be deposited, which upon examination we found to contain gallium. Gallium contained in DLC deposited by FIB-CVD can be evaporated by annealing at over 500 °C, but this evaporation does not occur in general thermal nanoimprinting because in such processes the temperature is usually from 100 to 200 °C. Figure 4(a) and (b) shows the SEM images of the repaired SiO_2/Si mold with protrusion and hollow defects, respectively. The protrusion defect was removed by FIB etching and the hollow defect was filled in by FIB-CVD. The etching and deposition times in this case were about 1 min and 30 sec, respectively. We then performed thermal nanoimprinting using the repaired mold on NEB-22, as shown in Fig. 5(a) and (b). The repaired lines were clearly imprinted on NEB-22. These results indicate that we can repair the protrusion and hollow defects on the thermal nanoimprint mold by FIB etching and CVD.

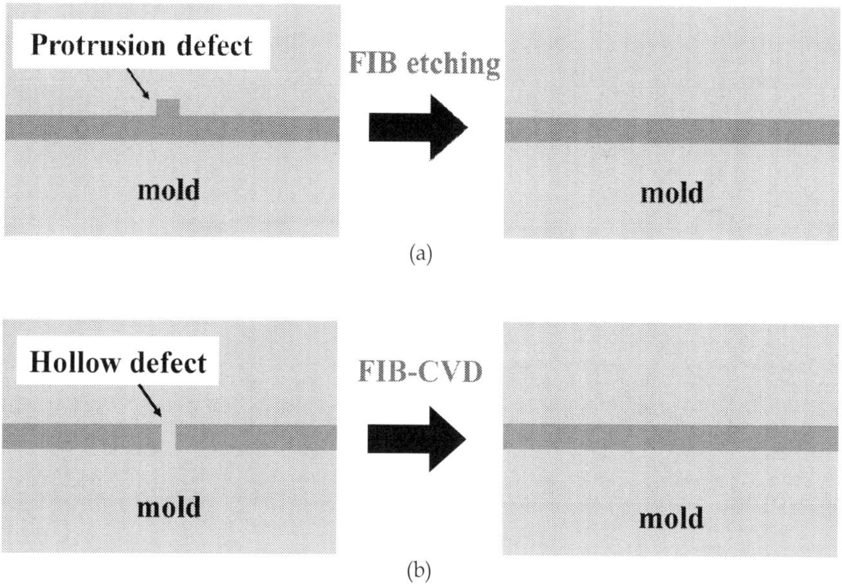

Fig. 3. Schematic of repair process of (a) protrusion and (b) hollow defects on mold.

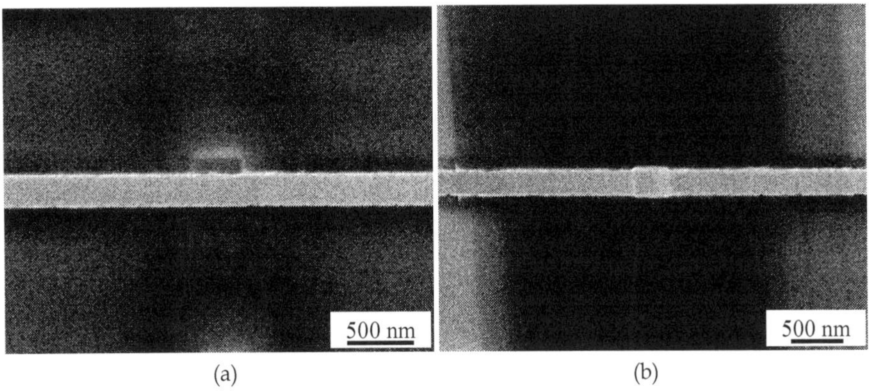

Fig. 4. SEM images of repaired SiO_2/Si mold with (a) protrusion and (b) hollow defect.

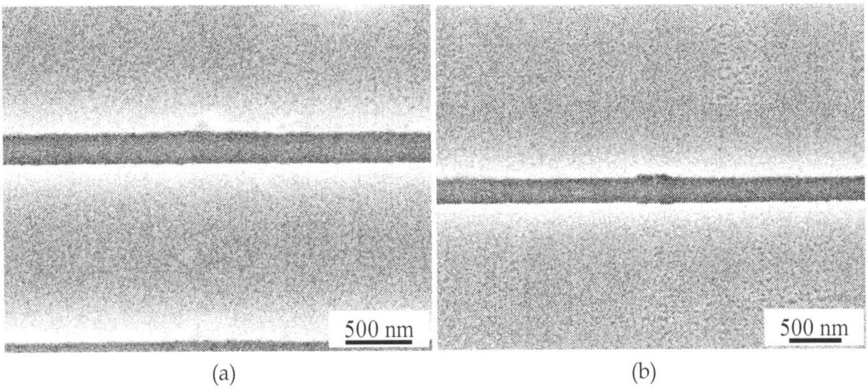

Fig. 5. SEM images of lines imprinted by T-NIL using repaired (a) protrusion- and (b) hollow-defective SiO_2/Si mold.

3. Repair of UV nanoimprint mold by using FIB etching and CVD

3.1 Characteristics of SiOx material deposited by FIB-CVD using tetraethoxysilane as source gas

In photomask repair and thermal NIL mold repair, the material deposited by FIB-CVD is DLC, which is not clear. In UV-NIL, UV-curable resin is irradiated by UV through the UV-NIL mold. Therefore, the material deposited by FIB-CVD must be transparent, meaning we cannot use DLC as the deposited material. However, we are able to deposit a SiOx material, which is transparent, by FIB-CVD using tetraethoxysilane [$Si(OC_2H_5)_4$] as a source gas. To examine whether the SiOx-deposited material has sufficient transmittance and hardness to withstand UV-NIL, we measured the atomic composition, transmittance, and hardness of SiOx film fabricated by FIB-CVD using tetraethoxysilane.

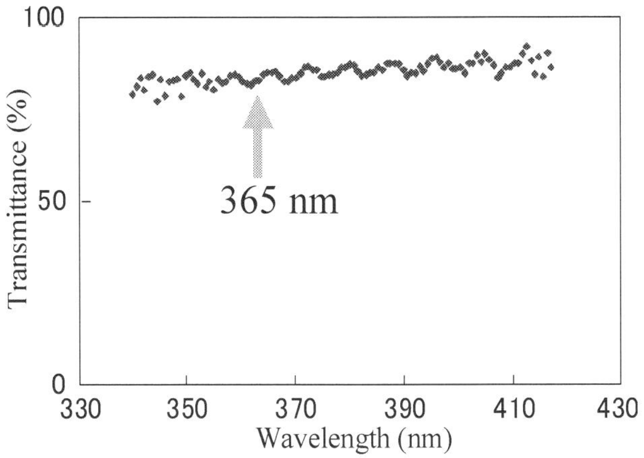

Fig. 6. Transmittance of SiOx film fabricated by FIB-CVD using tetraethoxysilane.

The atomic composition of the SiOx film measured by scanning electron microscopy-energy dispersive X-ray analysis (SEM-EDX) was 13% Si, 58% O, 9% C, and 20% Ga. A small amount of carbon is included in the SiOx film because tetraethoxysilane is composed of Si, O, H, and C. Ga is incorporated into SiOx film because Ga ion implantation is induced by the Ga ion beam used in FIB-CVD. We measured the transmittance of the 720 × 720 μm^2, 1 μm thick SiOx film using a monochromator (Hamamatsu Photonics: PMA-11) with a Xenon lamp. Figure 6 shows the results of measuring the SiOx film's transmittance. 365-nm UV is typically used in UV-NIL. The transmittance of the SiOx film was 83 % at 365 nm. This result demonstrated that the SiOx material deposited by FIB-CVD has a sufficient transmittance to carry out UV-NIL. Next, we measured the hardness of the SiOx film with a nanoindenter (Elionix: ENT-1100a). The hardness of the 150 × 150 μm^2, 1 μm thick SiOx film was 5 GPa. This result shows that the deposited SiOx material has a sufficient hardness because UV-NIL is usually performed at pressures below 1 MPa. These results clearly demonstrate that SiOx material deposited by FIB-CVD using tetraethoxysilane can be used as the repair material for a UV-NIL mold.

3.2 Necessity of antistatic treatment

When FIB is used to repair defective photomasks, the photomasks do not develop a charge because Cr metal patterns are formed on the quartz substrate. On the other hand, UV-NIL molds do develop a charge because the patterns are formed on the surface of the quartz substrate. It is therefore necessary to prevent electrical charge during the repair of a UV-NIL mold. There are two methods for controlling static. One is electron shower irradiation during FIB (Fig. 7(a)). In this method, however, the region of electron irradiation is also deposited, as in FIB-CVD. This makes it very difficult to apply an electron shower as an antistatic method. The other method is using an antistatic agent spin-coated on the UV-NIL mold (Fig. 7(b)). In this case, the deposition by FIB irradiation is possible. We used a 20-nm thick ESPACER300Z (Showa Denko) as an antistatic agent to prevent electrical charge. The FIB-irradiated part on ESPACER300Z was first etched away and then the SiOx was deposited, as shown in Fig. 7(b). ESPACER300Z can easily be washed away with water after the repair.

We evaluated the antistatic effect of ESPACER300Z on a quartz substrate. We performed FIB-CVD using tetraethoxysilane on quartz substrates both with and without an antistatic agent. Figure 8 (a) and (b) shows the scanning ion microscopy (SIM) images of the 200-nm wide patterns fabricated by FIB-CVD at 1 pA on the two substrates. The line pattern was not formed on the substrate with no antistatic agent because of substrate drift caused by electrical charge. In contrast, the line pattern was clearly formed on the substrate that did use the antistatic agent, as shown in Fig. 8(b). Next, we fabricated 3D structures by FIB-CVD on ESPACER300Z-coated quartz substrates at 1 pA and 7 pA to examine the beam current dependency. When the beam current was 1 pA, the 3D structure was successfully fabricated on the quartz substrate, as shown in Fig. 9(a). However, as shown in Fig. 9(b), thorns were formed on the edges of the 3D structure fabricated by FIB-CVD at 7 pA.

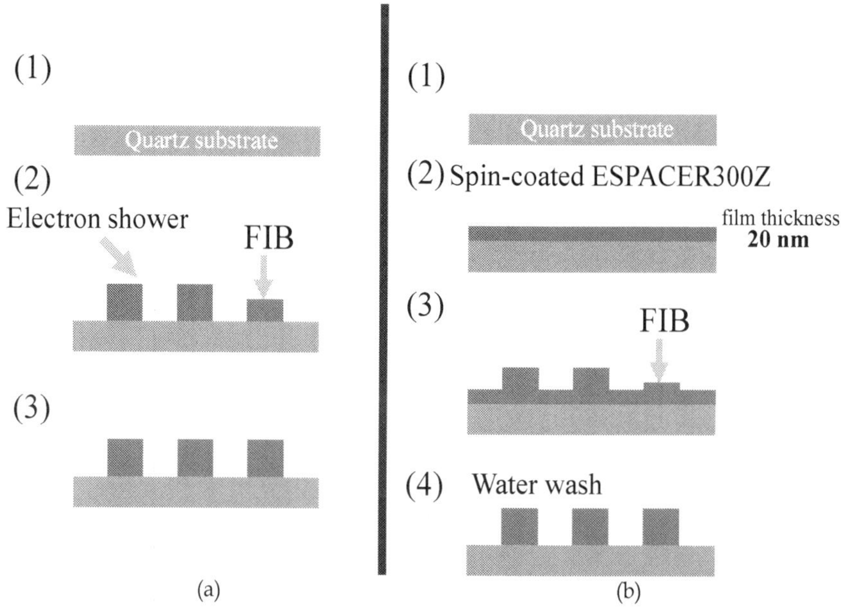

Fig. 7. Schematic of methods to control static: (a) FIB-CVD performed with an electron shower and (b) using an antistatic agent (ESPACER 300Z: Showa Denko Co.) spin-coated on the UV-NIL mold.

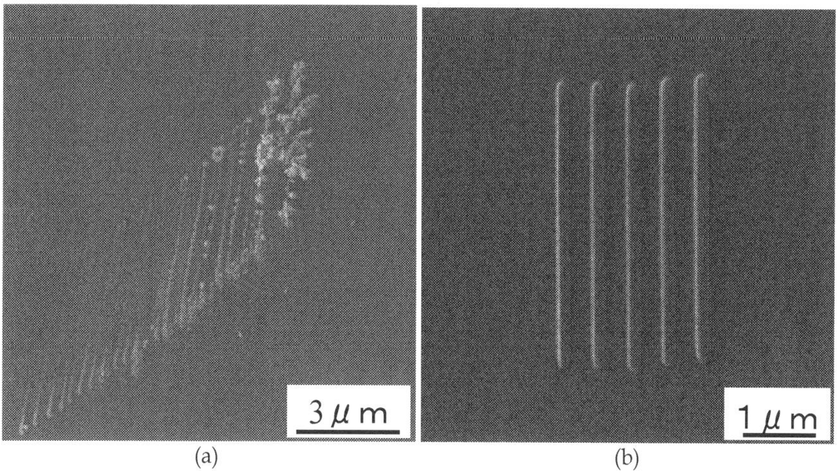

Fig. 8. SIM images of 200-nm wide patterns fabricated by FIB-CVD at 1 pA on quartz substrates without and with antistatic agent.

Fig. 9. (a) 3D structure fabricated by FIB-CVD with ESPACER300Z at 1 pA. (b) Thorns were formed on the corners of the 3D structure with ESPACER300Z at 7 pA.

To determine why the thorns were fabricated on the edges, we observed them by scanning transmission electron microscopy (STEM) and measured their atomic composition by SEM-EDX. Figure 10 shows the STEM images of the Au-coated thorns. Although the pillar fabricated by FIB-CVD has a gallium core due to Ga ion implantation, no such gallium core was observed in the thorns. According to the SEM-EDX results, the atomic composition of the thorns was 16% Si, 68% O, 13% C, and 3% Ga. In contrast, the atomic composition of the

3D structure was 21% Si, 51% O, 9% C, and 19% Ga. Ga content in the thorns was much less than that in the deposited SiOx. These results indicate that the thorn structure was caused by electric charge accumulation on the FIB-deposited region due to the increasing beam current. This makes it clear that we must use an optimum beam current to achieve hollow defect repair by FIB-CVD.

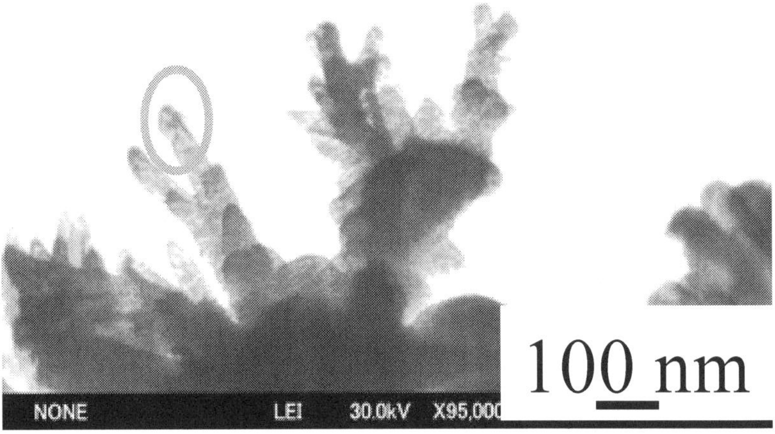

Fig. 10. STEM image of Au-coated thorns on 3D structure. The atomic content inside the circle was measured by SEM-EDX.

3.3 UV nanoimprint mold repair by FIB etching and CVD

To examine the repair resolution of the FIB etching, we fabricated a narrow line by FIB etching on a quartz mold and performed UV-NIL using this mold. The mold was pressed onto UV-curable resin (Toyo Gosei: PAK-01). The UV-NIL pressure and UV wavelength were 0.9 MPa and 365 nm, respectively. The mold was held for 90 sec during imprinting. The release agent (Daikin Industries; OPTOOL DSX: demnamsolvent = 1 : 1000 by weight) was precoated on the mold to avoid the adhesion of resin and to enable a smooth separation of the mold and the substrate. Figure 11(a) shows the imprinted line pattern using the mold on the PAK-01. The linewidth was 29 nm. Results demonstrated that FIB can etch a narrow line on the quartz substrate.

We also examined the repair resolution of FIB-CVD. First, we fabricated a narrow line by FIB etching, and then deposited it on the quartz mold by FIB-CVD. The pattern imprinted by UV-NIL using this mold is shown in Fig. 11(b). The fabricated narrow linewidth was 36 nm. These results indicate that a 30-nm defect can be repaired by FIB etching and CVD.

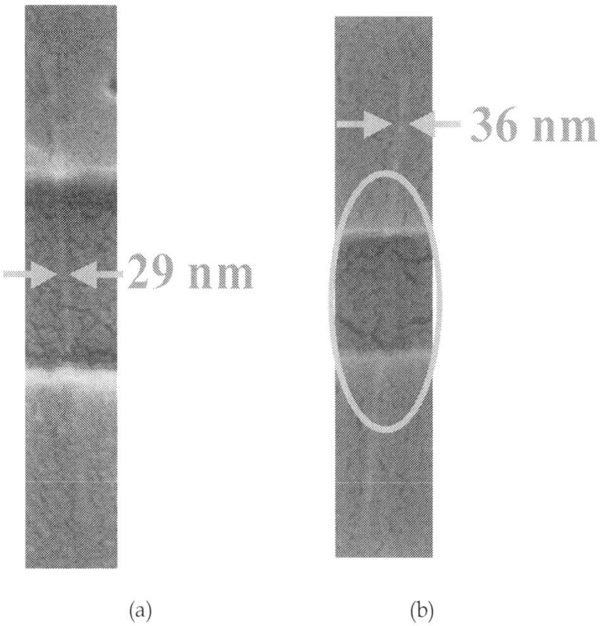

Fig. 11. SEM images of lines imprinted by UV-NIL using (a) quartz mold with narrow hollow-line fabricated by FIB etching and (b) repaired mold.

3.3.1 Repair of protrusion defects on UV-NIL mold

To repair a UV-NIL mold by 30-kV FIB etching at 1 pA, we fabricated program protrusion defects on it using a quartz substrate by EB lithography and RIE. Figure 12(a) shows the program protrusion-defective template. The protrusion width and length were 40 nm and 150 nm, respectively. Figure 12(b) shows the defective line pattern transferred by UV-NIL on the PAK-01 and as we can see, the defective line pattern was clearly imprinted on the substrate. We repaired the protrusion defects on the UV-NIL mold by FIB etching. Figure 13(a) shows the line pattern repaired by FIB etching on the mold. The repair time for one protrusion defect was about 10 sec. The protrusion defect was successfully etched away by FIB and the repaired line pattern was clearly imprinted (Fig. 13(b)).

3.3.2 Repair of hollow defects on UV-NIL mold

Figure 14(a) shows the program hollow-defective mold. The hollow-width was 60 nm. Figure 14(b) shows the imprinted line pattern using the defective mold on the substrate. The defective line pattern was clearly imprinted on the substrate. We repaired the hollow defect on the UV-NIL mold by FIB-CVD. Figure 15(a) shows the line pattern on the mold repaired by FIB-CVD using tetraethoxysilane. The repair time for one hollow defect was about 20 sec. Figure 15(b) shows the line pattern transferred using the repaired mold on the substrate. The hollow defect region was successfully deposited by FIB-CVD, and the repaired line pattern was clearly imprinted (Fig. 15(b)).

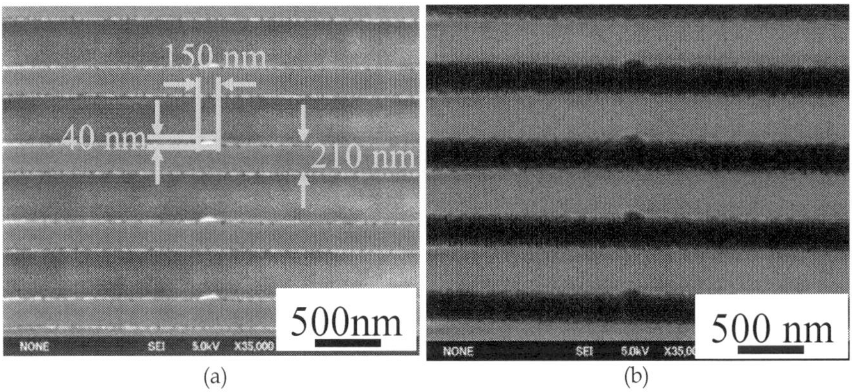

Fig. 12. SEM images of (a) program protrusion-defective quartz mold and (b) PAK-01 defective line pattern imprinted by UV-NIL using the protrusion-defective quartz mold.

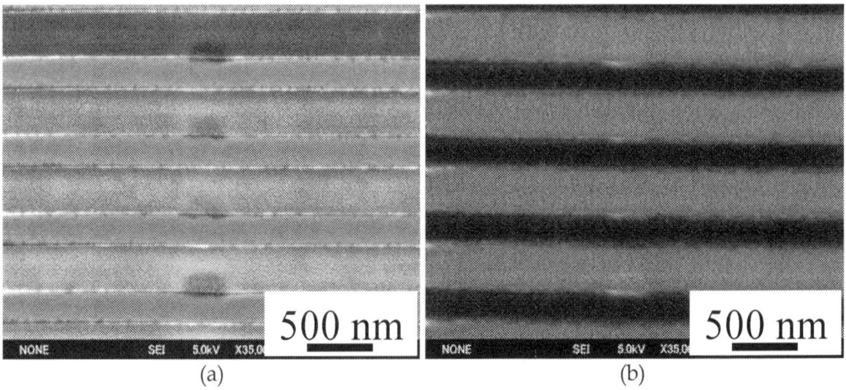

Fig. 13. SEM images of (a) line pattern in Fig. 12(a) repaired by FIB etching and (b) PAK-01 line pattern imprinted by UV-NIL using the repaired quartz mold.

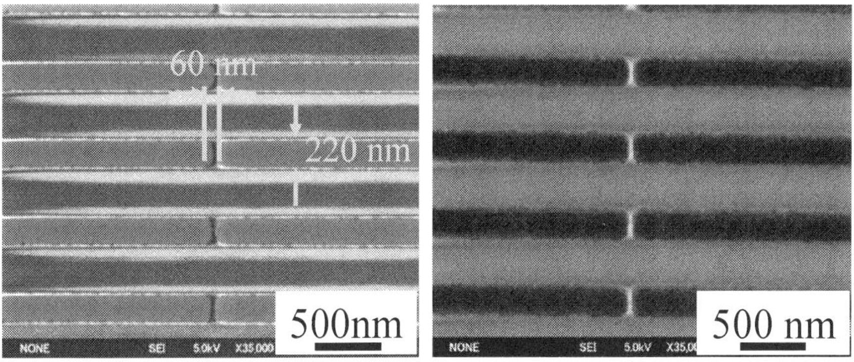

(a) (b)

Fig. 14. SEM images of (a) program hollow defective quartz mold and (b) PAK-01 defective line pattern imprinted by UV-NIL using the hollow defective quartz mold.

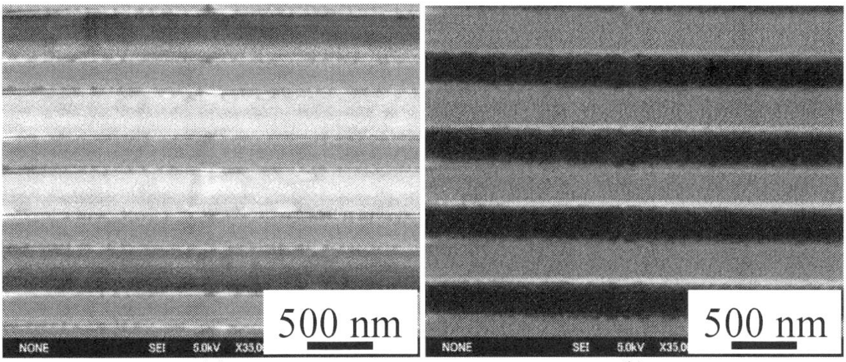

(a) (b)

Fig. 15. SEM images of (a) line pattern repaired by FIB-CVD on quartz mold and (b) PAK-01 line pattern imprinted by UV-NIL using the repaired quartz mold.

The durability of a repaired mold is crucial for applying UV-NIL to mass production. Therefore, we tested the durability of the SiOx material deposited by FIB-CVD by observing over 200 repeated uses of the mold repaired by FIB-CVD. Figure 16(a) and (b) shows the SEM images of the 100th and 200th imprinted pattern. After nanoimprintings, the pattern was still clearly imprinted on the resin. This result indicates that the deposited SiOx material has sufficient durability for repeated UV-NIL.

(a) (b)

Fig. 16. SEM images of (a) 100th and (b) 200th imprinted patterns on PAK-01 formed by UV-NIL using quartz mold repaired by FIB-CVD.

4. Conclusion

We repaired NIL molds by FIB etching and CVD. In the case of a T-NIL mold, we used phenanthrene as a source gas to repair the hollow-defect by FIB-CVD. The deposited material was DLC containing gallium. However, the gallium evaporation from DLC did not occur during T-NIL because gallium's evaporation temperature is about 500 °C. The protrusion- and hollow-defective SiO_2/Si molds were successfully repaired by FIB etching and CVD and the lines were clearly imprinted by T-NIL using the repaired mold.

In UV-NIL, UV-curable resin is irradiated by UV through a UV-NIL mold. This means that phenanthrene cannot be used as a source gas in FIB-CVD because the deposited DLC is not transparent. We therefore used SiOx material deposited by FIB-CVD using tetraethoxysilane as the source gas. We measured the transmittance and hardness of the SiOx film deposited by FIB-CVD, and results demonstrated that the deposited SiOx material has sufficient transmittance and hardness to perform UV-NIL.

To repair the defective UV-NIL mold, we must consider the effect of electrical charge because the material of UV-NIL molds is usually quartz. We proposed the use of an antistatic agent to prevent electrical charge because the antistatic agent can be spin-coated on the UV-NIL mold and is easy to wash away with water after the repair. However, failure to use an optimum beam current in FIB-CVD resulted in the fabrication of thorns. This highlights the need for an optimum beam current if we want to achieve hollow defect repair by FIB-CVD.

We repaired the defective quartz mold by FIB etching and CVD and performed UV-NIL using the repaired mold. The repaired lines were clearly imprinted on the resin. Results demonstrated that the deposited SiOx material has a sufficient durability for repeated UV-NIL.

In this work, we used gallium-FIB to repair the NIL molds. Recently, helium ion microscopy (HIM) with a subnanometer probe size has become commercially available. Apart from the obvious advantage of small probe sizes, HIM also boasts a narrow interaction volume in the substrate and the predominance of secondary electron emission. A helium ion microscope can also be used for nanofabrication. We expect the mold repair resolution to dramatically improve by using HIM.

5. References

Chou, S. Y.; Krauss, P. R. & Renstrom, P. J. (1995). Imprint of sub-25 nm vias and trenches in polymers. *Applied Physics Letter,* Vol.67, No.21, (September 1995), pp. 3114-3116, ISSN 003-6951

Chou, S. Y.; Krauss, P. R. & Renstrom, P. J. (1996). Imprint Lithography with 25-Nanometer Resolution. *Science,* Vol.272, No.5258, (April 1996), pp. 85-87, ISSN 0036-8075

Chou, S. Y.; Krauss, P. R.; Zhang, W.; Guo, L. & Zhuang, L. (1997). Sub-10 nm imprint lithography and applications. *Journal of Vacuum Science & Technology B,* Vol.15, No.6, (September 1997). pp. 2897-2904, ISSN 0734-211X

Fuchs, A.; Bender, M.; Plaschetka, U. ; Kock, L.; Wahlbrink, T.; Gottlob, H.; Efavi, J.; Moeller, M.; Schmidt, M.; Mollenhauer, T.; Moormann, C.; Lemme, M. & Kurz, H. (2006). Nanowire fin field effect transistors via UV-based nanoimprint lithography, *Journal of Vacuum Science & Technology B,* Vol.24, No.6, (November 2006). pp. 2964-2967, ISSN 0734-211X

Yoshitake, S.; Sunaoshi, H.; Yasui, K.; Kobayashi, H.; Sato, T.; Nagarekawa, O.; Thompson, E.; Schmid, G. & Resnick, J. (2007), The development of full field high resolution imprint templates, *Photomask Technology 2007 (Proceedings Volume) Proceedings of SPIE,* pp.67300E, ISBN 9780819468871 6730, Monterey, California, USA, September,2007

Schmid, G. M.; Stewart, M. D.; Wetzel, J.; Palmieri, F.; Hao, J.; Nishimura, Y.; Jen, K.; Kim, E. K.; Resnick, D. J., Liddle, J. A. & Willson, C. G. (2006), Implementation of an imprint damascene process for interconnect fabrication. *Journal of Vacuum Science & Technology B,* Vol.24, No.3, (May 2006). pp. 1283-1291, ISSN 0734-211X

Ahn, S.W.; Lee, K. D.; Kim, J. S.; Kim, S. H.; Park, J. D.; Lee, S. H. & Yoon, P. W. (2005), Fabrication of a 50 nm half-pitch wire grid polarizer using nanoimprint lithography, *Nanotechnology,* Vol.16, No.9, (July 2005). pp.1874-1877, ISSN 0957-4484

Wang, J. J.; Deng, X. G.; Liu, X. M.; Nikolov, A.; Sciortino, P.; Liu, F. & Chen, L. (2006), Ultraviolet wave plates based on monolithic integration of two fully filled and planarized nanograting layers, *Optics Letters,* Vol.31, No.12, (April 2006). pp.1893-1895, ISSN 0146-9592

Beck, M.; Persson, F.; Carlber, P.; Graczyk, M.; Maximov, I.; Ling, T. G. I. & Montelius, L. (2004), Nanoelectrochemical transducers for (bio-) chemical sensor applications fabricated by nanoimprint lithography, *Microelectronic Engneering,* Vol.73-74, (June 2004). pp.837-842, ISSN 0167-9317

Nomura, S.; Kojima, H.; Ohyabu, Y.; Kuwabara, K.; Miyauchi, A. & Uemura, T. (2005), Cell Culture on Nanopillar Sheet: Study of HeLa Cells on Nanopillar Sheet, *Japanese*

Journal of Applied Physics, Vol.44, No.37, (September 2005). pp.L1184-L1186, ISSN 0021-4922

Kim, M. S.; Kim, J. S.; Cho, J. C.; Shtein, M.; Guo, L. J. & Kim, J. (2007), Flexible conjugated polymer photovoltaic cells with controlled heterojunctions fabricated using nanoimprint lithography, *Applied Physics Letter,* Vol.90, No.12, (March 2007), pp. 123113-1-123113-3, ISSN 003-6951

Haisma, J.; Verheijen, M. & Heuvel, K. V. D. (1996), Mold-assisted nanolithography: A process for reliable pattern replication, *Journal of Vacuum Science & Technology B,* Vol.14, No.6, (August 1996). pp. 4124-4128, ISSN 0734-211X

Bailey, T.; Choi, B. J.; Colburn, M.; Meissi, M.; Shaya, S.; Ekerdt, J. G.; Sreenivasan, S. V. & Willson, C. G. (2000), Step and flash imprint lithography: Template surface treatment and defect analysis, *Journal of Vacuum Science & Technology B,* Vol.18, No.6, (September 2000). pp. 3572-3577, ISSN 0734-211X

Komuro, M.; Taniguchi, J.; Inoue, S.; Kimura, N.; Tokano, Y.; Hiroshima, H. & Matsui, S. (2000), Imprint Characteristics by Photo-Induced Solidification of Liquid Polymer, *Jpanese Journal of Applied Physics,* Vol.39, No.12B, (October 2000). pp.7075-7079, ISSN 0021-4922

Xia, Y. & Whitesides, G. M. (1998), Soft Lithography, *Annual Reviews Materials Research,* Vol.28, (August 1998). pp. 153-184, ISSN 1531-7331

Libioulle, L.; Bietsch, A.; Schmid, H.; Michel, B & Delamarche, E. (1999), Contact-Inking Stamps for Microcontact Printing of Alkanethiols on Gold, *Langmuir,* Vol.15, No.2, (December 1999). pp. 300-304, ISSN 0743-7463

Plachetka, U.; Bender, M.; Fuchs, A., Vratzov, B; Glinsner, T.; Lindner, F. & Kurz, H. (2004), Wafer scale patterning by soft UV-Nanoimprint Lithography, *Microelectronic Engneering,* Vol.73-74, (June 2004). pp.167-171, ISSN 0167-9317

Matsui, S.; Igaku, Y.; Ishigaki, H.; Fujita, J.; Ishida, N.; Ochiai, Y.; Komuro, M. & Hiroshima, H. (2001), Room temperature replication in spin on glass by nanoimprint technology, *Journal of Vacuum Science & Technology B,* Vol.19, No.6, (September 2001). pp. 2801-2805, ISSN 0734-211X

Matsui, S.; Igaku, Y.; Ishigaki, H.; Fujita, J.; Ishida, N.; Ochiai, Y.; Namatsu, H. & Komuro, M. (2003), Room-temperature nanoimprint and nanotransfer printing using hydrogen silsequioxane, *Journal of Vacuum Science & Technology B,* Vol.21, No.2, (February 2003). pp. 688-692, ISSN 0734-211X

Nakamatsu, K.; Watanabe, K.; Tone, K.; Katase, T.; Hattori, W.; Ochiai, Y.; Matsuo, T.; Sasago, M.; Namatsu, H.; Komuro, M. & Matsui, S. (2004), Bilayer Resist Method for Room-Temperature Nanoimprint Lithography, *Japanese Journal of Applied Physics,* Vol.43, No.6B, (June 2004). pp.4050-4053, ISSN 0021-4922

Nakamatsu, K; Watanabe, K.; Tone, K; Namatsu, H. & Matsui, S. (2005), Nanoimprint and nanocontact technologies using hydrogen silsesquioxane, *Journal of Vacuum Science & Technology B,* Vol.23, No.2, (March 2005). pp. 507-512, ISSN 0734-211X

Heard, P. J. & Prewett, P. D. (1990), Focused ion beam deposition of carbon for photomask repair, *Microelectronic Engneering,* Vol.11, No. 1-4 (April 1990). pp.421-425, ISSN 0167-9317

Matsui, S.; Kaito, T.; Fujita, J.; Komuro, M.; Kanda, K. & Haruyama, Y. (2000), Three-dimensional nanostructure fabrication by focused-ion-beam chemical vapor deposition, *Journal of Vacuum Science & Technology B*, Vol.18, No.6, (August 2000). pp. 3181-3184, ISSN 0734-211X

Watanabe, K.; Kanda, K.; Haruyama, Y.; Kaito, T. & Matsui, S. (2004), Nanoimprint Mold Repair by Ga^+ Focused-Ion-Beam Direct Etching, *Japanese Journal of Applied Physics*, Vol.43, No.11A, (November 2004). pp.7769-7772, ISSN 0021-4922

Fujita, J.; Ishida, M.; Ichihashi, T.; Ochiai, Y.; Kaito, T & Matsui, S. (2003), Growth of three-dimensional nano-structures using FIB-CVD and its mechanical properties, *Nuclear Instruments and Methods in Physics Research Section B: Beam Interactions with Materials and Atoms*, Vol.206, No. (May 2003). pp. 472-477, ISSN 0168-583X

Igaki, J.; Saikubo, A.; Kometani, R.; Kanda, K.; Suzuki, T.; Niihara, K. & Matsui, S. (2007), Elementary Analysis of Diamond-Like Carbon Film Formed by Focused-Ion-Beam Chemical Vapor Deposition, *Japanese Journal of Applied Physics*, Vol.46, No.12, (December 2007). pp.8003-8004, ISSN 0021-4922

Kanda, K.; Igaki, J.; Saikubo, A.; Kometani, R.; Suzuki, T.; Niihara, K.; Saito, H & Matsui, S. (2008), Effects of Annealing on Material Characteristics of Diamond-Like Carbon Film Formed by Focused-Ion-Beam Chemical Vapor Deposition, *Jpanese Journal of Applied Physics*, Vol.47, No.9, (September 2008). pp.7464-7466, ISSN 0021-4922

Kanda, K.; Okada, M.; Kang, Y.; Niibe, M.; Wada, A.; Ito, H.; Suzuki, T. & Matsui, S. (2010), Structural Changes in Diamond-Like Carbon Films Fabricated by Ga Focused-Ion-Beam-Assisted Deposition Caused by Annealing, *Japanese Journal of Applied Physics*, Vol.49, (June 2010). pp.06GH06-1-06GH06-5, ISSN 0021-4922

Edinger, K; Melingailis, J. & Orloff, J. (1998), Study of precursor gases for focused ion beam insulator deposition, *Journal of Vacuum Science & Technology B*, Vol.16, No.6, (September 1998). pp. 3311-3314, ISSN 0734-211X

Okada, M.; Nakamatsu, K.; Kanda, K.; Haruyama, Y.; Kometani, R.; Kaito, T. & Matsui, S. (2008), Examination of Focused-Ion-Beam Repair Resolution for UV-Nanoimprint Templates, *Jpanese Journal of Applied Physics*, Vol.47, Vol.6 (June 2008). pp.5160-5163, ISSN 0021-4922

Campbell, A. N.; Peterson, K. A.; Fleetwood, D. M. & Soden, J. M. (1997), Effects of focused ion beam irradiation on MOS transistors, Proceedings of International Reliability Physics Symposium, pp.72-81, ISBN 0-7803-3575-9, Piscataway, New Jersey, USA, April, 1997

Saida, Y.; Okubo, T.; Sasaki, J.; Konishi, T. & Morita, M. (2005), Evaluation of the new-type ESPACER adopted for its removal after post-exposure bake process, *Photomask and Next-Generation Lithography Mask Technology XII Proceedings of SPIE*, ISBN 0819449962, Yokohama, Kanagawa, Japan, April 2005

Ward, B. W.; Notte, J. A. & Exonomou, N. P. (2006), Helium ion microscope: A new tool for nanoscale microscopy and metrology, *Journal of Vacuum Science & Technology B*, Vol.24, No.6, (November 2006). pp. 2871-2874, ISSN 0734-211X

Alkemadea, P. F. A.; Chen, P.; Veldhoven, E. V. & Maas, M. (2010), Model for nanopillar growth by focused helium ion-beam-induced deposition, *Journal of Vacuum Science & Technology B*, Vol.28, No.6, (December 2010). pp. C6F22-C6F25, ISSN 0734-211X

Livengood, R.; Tan, S.; Greenzweig, Y.; Notte, J. & McVey, S. (2009), Subsurface damage from helium ions as a function of dose, beam energy, and dose rate, *Journal of Vacuum Science & Technology B*, Vol.27, No.6, (December 2009). pp. 3244-3249, ISSN 0734-211X

9

Improving the Light-Emitting Efficiency of GaN LEDs Using Nanoimprint Lithography

Yeeu-Chang Lee[1] and Sheng-Han Tu[2]
[1]*Department of Mechanical Engineering, Chung Yuan Christian University*
[2]*Genesis Photonics Inc.*
Taiwan

1. Introduction

A light-emitting diode (LED) is an electroluminescent device with a broad selection of emission wavelengths (colors). The unique properties of LEDs, such as compactness, low power consumption, long lifetime, and fast turn-on time have made LEDs an indispensable component in modern traffic lighting, display, car lighting, and cell phones applications. In recent years, LED usage has grown more rapidly due to the application of backlights in large-size flat panel displays, a market previously dominated by CCFL. In addition, wide speculation foresees that the next boom for LEDs could arise from the interior/exterior lighting market. As shown in U.S. energy consumption statistics conducted by the U.S. government, the energy consumption for interior/exterior lighting occupies 22%~25% of the total electrical energy produced in the U.S. Using LEDs to replace all traditional interior/exterior lighting used today can save at least an estimated amount of 20 billion U.S. dollars in annual energy costs while also significantly reducing carbon emission. Therefore, LEDs have attracted a large amount of investments from corporations in Asia, North America, and Europe, aiming to develop reliable processes to improve the production yield and optical efficiency of LEDs.

The nanoimprint lithography is a nanoscale structural formation technique with highly reproducible patterns, and is therefore suitable for LED fabrication. The typical length scale of structures for applications of LEDs ranges from a few hundred nanometers to a few micrometers, which is roughly on par with the resolution limit of the traditional optical lithography technology. Though a higher resolution is attainable using the stepper projection lithography method, the corresponding higher process cost renders the fabricated LEDs less economical. Unlike the integrated circuit, the micro- and nanostructures of an LED are often simple 2D periodic patterns. Once an imprinting mold with a high accuracy and precision is made from techniques such as the stepper projection lithography (or alternatively the electron beam, the ion beam, and the interference lithography), the LED patterns can be massively reproduced. This article discusses the role of the precision nanoimprint lithography for improving the optical efficiency of LEDs.

2. Limits and enhancements of the light-extraction efficiency of GaN LEDs

No single semiconductor material alone is capable of emitting white light. A white light LED typically consists of an appropriate mixture of (a) red, blue, and green LEDs, or (b) blue and yellow lights, where the blue LED stimulating yellow phosphor produces the yellow light. In either case, the blue LED is the main constituent of a white light LED. Most blue LEDs are made from GaN, the compound discussed in this paper.

2.1 Analysis of the light extraction efficiency

Excluding the substrate, a typical LED structure is only a few micrometers thick. The region capable of emitting light is called the active region, composed of multiple quantum wells (MQWs) of less than 1 μm in thickness. The active region can be regarded as a thin film containing a large number of point light sources radiating photons in all directions. The ratio of the number of these emitted photons to the number of electrons injected to excite the quantum wells constitutes an essential figure(also known as internal quantum efficiency) for judging the performance of an LED. Another significant figure of value concerns the number of photons out of the active region that can be extracted to the free space. This is called the light extraction efficiency, $\eta_{extraction}$, defined as follows:

$$\eta_{extraction} = \frac{\text{\# of photons emitted into free space per second}}{\text{\# of photons emitted from active region per second}}$$

The extraction efficiency is definitely not 100%. In practice, reflection at the interface between two materials with different refraction indices is unavoidable. This reflection can be considered as a loss (also known as Fresnel loss) mechanism in the LED. However, the main loss channel in the LED is caused by the total internal reflection (TIR). TIR is an optical phenomenon that occurs when the light enters from an optically dense medium to a less optically dense medium, such as when the light exits from GaN and enters the air. As the angle of incidence is greater than the critical angle, no light can be transmitted at the interface and all light is reflected. Snell's law is used to determine the critical angle:

$$n_s \sin\theta_1 = n_{air} \sin\theta_2$$

where n_s and n_{air} are the refractive indices of GaN and air, respectively. θ_1 is the angle of incidence and θ_2 is the refraction angle. When the refraction angle is greater than 90°, which forbids photons from being transmitted at the interface, the corresponding angle of incidence is the critical angle, $\theta_c = \sin^{-1}(\frac{n_{air}}{n_s})$, as shown in Figure 1- Left.

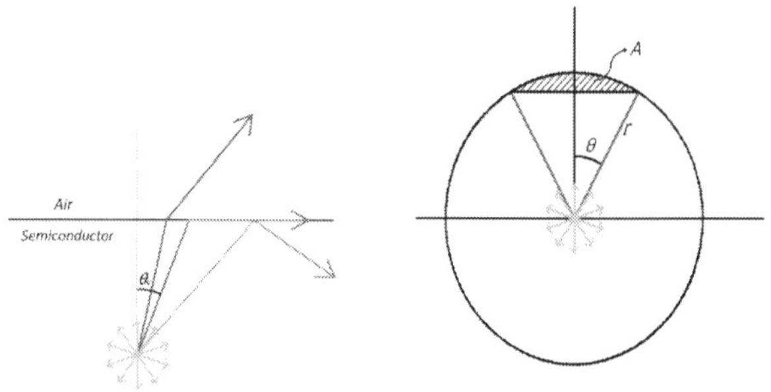

Fig. 1. (Left) The critical angle for the total internal reflection (Right) The light emission cone of an LED

For instance, the refraction index of GaN at room temperature is 2.4, corresponding to a critical angle of 24.6°. When the angle of incidence (regarding air) of the point light sources in the active region of the GaN LED is smaller than 24.6°, all emitted photons can be delivered to the free space.

In 3D space, the point light sources emit photons isotropically in spherical directions. The radiation area is the area of the sphere $4\pi r^2$, where r is the distance from the wavefront to the point light source. When the spherical wavefront reaches the GaN/air interface, only the photons within the conical region (using the critical angle as the solid angle) can escape the semiconductor (Figure1- Right). The radiation area of the escaping cone is:

$$A = \int dA = \int_{\theta=0}^{\theta_c} 2\pi r \sin\theta r d\theta = 2\pi r^2 (1 - \cos\theta_c)$$

The emitted optical power per unit area of the escaping cone is equal to the power per unit area of the emission sphere of the point source:

$$\frac{P_{escape}}{2\pi r^2 (1 - \cos\theta_c)} = \frac{P_{source}}{4\pi r^2}$$

$$P_{escape} = P_{source} \frac{2\pi r^2 (1 - \cos\theta_c)}{4\pi r^2} = P_{source} \frac{(1 - \cos\theta_c)}{2}$$

$$\frac{P_{escape}}{P_{sourse}} = \frac{(1 - \cos\theta_c)}{2}$$

By expanding the cosine term into power series, the following equation can be derived:

$$\frac{P_{escape}}{P_{sourse}} \approx \frac{1}{2}\left[1 - \left(1 - \frac{\theta_c^2}{2!} + \frac{\theta_c^4}{4!} - \cdots\right)\right]$$

For semiconductors with high refraction indices, the critical angle is small. Therefore, the high order terms in the above equation can be neglected, leading to the following:

$$\frac{P_{escape}}{P_{sourse}} \approx \frac{1}{4}\theta_c^2$$

Using GaN as an example once more, the light extraction efficiency (calculated below) from the surface of the active region to the air is only 4.61%.

$$\frac{P_{escape}}{P_{source}} \approx \frac{1}{4}\theta_c^2 = \frac{1}{4} \times \left(\frac{24.6}{180} \times \pi\right)^2 \approx 4.61\%$$

2.2 Methods to improve the light extraction efficiency

Numerous methods have been proposed to circumvent the poor light extraction efficiency imposed by the TIR effect. Modifying the geometry of the LED chip is one such method [1-3]. By shaping the sides of the LED chip into trapezoidal (or up-side-down trapezoidal) shapes, the TIR effect from the top and the sides of the chip can be released. However, the typical thickness of a blue LED chip (including the substrate) is only approximately 90 μm, rendering the shaping of the sides an extremely difficult task. Another common method to reduce the TIR effect is by roughening the sides and the surface of the LED chip[4,5], which can enhance the scattering effect. As shown in Figure 2, the light is extracted whenever the guided light reaches the roughened interface. However, the roughening technique also degrades the electrical characteristics of the LED, causing the forward voltage to rise. If the electrical issue can be resolved, the roughening technique could bea simple and reliable solution.

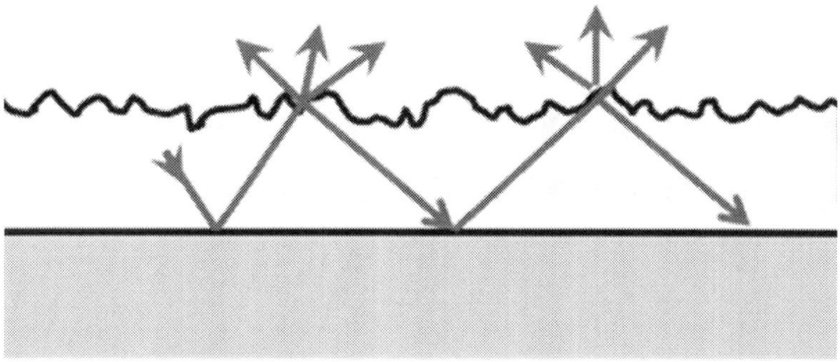

Fig. 2. Reducing the TIR by scattering

Lee et al. [6, 7] simulated the effects of the surface structures, including the size and shape effects of these structures under different packaging conditions, on the light extraction efficiency of the LED, using ray tracing. However, the ray tracing technique only applies when the size of the structures is much bigger than the wavelength of the light. For structures that are close to the size of the photon wavelength, diffraction occurs. Though the wave nature of the light can be simulated by the FDTD or the RCWA methods, the applicability of these methods to the particle-wave duality nature of the light (prominent for sub-wavelength structures) requires further evaluation.

The most commonly used substrate for growing GaN films is sapphire. However, a 16% lattice mismatch between the GaN and the sapphire exists, causing a large number of threading dislocations (10^9~10^{12} cm^{-2}) at the interface. Theoretically, these threading dislocations could extend from the epitaxial interface to the top of the p-GaN surface, leading to undesirable outcomes for the quality and lifetime of the LED chip. An approach to solving this problem requires altering the crystal growth orientation of the epitaxial film grown on etching surface structures of sapphire substrates. The interrupted and/or bent dislocations can reduce the density of the threading dislocations. In addition, these surface structures can display effects similar to the roughened surface to enhance the light extraction efficiency for the multiple reflections at the GaN/sapphire interface as shown in Figure 3, to cause guided light escape to free space while suppressing the TIR. Such substrate patterning is known as the patterned sapphire substrate (PSS) technique [8-11]. PSS is a popular technique for enhancing LED efficiency. Typical PSS patterns are on the order of micrometers, and shrinking the size of the patterns into nanoscales (also known as NPSS) is believed to significantly improve efficiency.

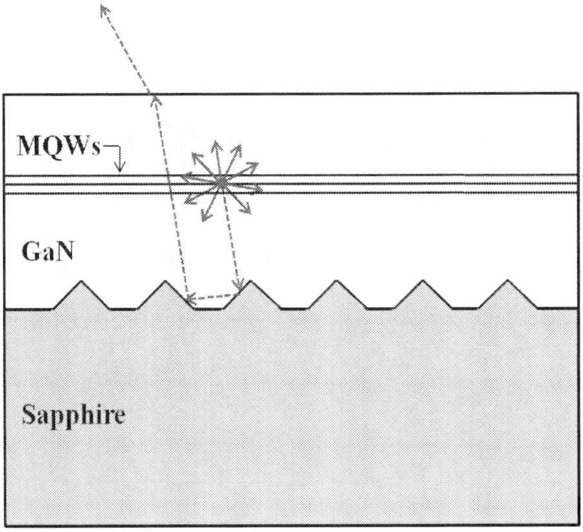

Fig. 3. Possible path for the light traveling in the patterned sapphire substrates (PSS) LED

In 1987, Eli Yablonovitch [12] and Sajeev John [13] discovered that electromagnetic waves transmission is disallowed in certain periodically patterned structures (now called photonic crystals). This phenomenon is similar to electrons in the forbidden band of an energy band diagram in solids. Much research has since focused on utilizing the photonic crystal effect. Early applications of the photonic crystal effect were in the optical communication, where the photonic bandgap was used to create a waveguide of extremely large bending angles. More recently, the photonic crystal effect has also been applied to the LED. The photonic crystal can be used to enhance the external quantum efficiency of the LED by (a) forming a photonic bandgap within the LED chip, thereby forcing the generated photons to exit the chip and/or (b) forming an efficient waveguide to couple photons to the free space. Except to enhance the light extraction efficiency, photonic crystals can also modify far-field light pattern of the LED. The far-field light pattern is related to the information transmission of traffic lights [14], fiber coupling efficiency of LED sources [15], and the color mixing homogeneity in LED displays[16], etc. Therefore, the combination of the photonic crystal and the LED has widespread application.

3. Discussions of the nanoimprint techniques

Based on the aforementioned analysis, the light-emitting efficiency of the LED can be improved by fabricating micro- or nanostructures inside or on the surface of the chip/substrate. The small dimension patterns can be fabricated via nanoimprint lithography. The advantages of nanoimprint lithography are fast production and the low cost, which satisfy industry production requirements. In addition, the method simplifies the complex optical lithography, freeing it from the diffraction limit. The nanoimprint lithography was regarded as one of the ten emerging technologies in 20003 with the potential to change the world.

Nanoimprint lithography can be distinguished in three types: hot embossing nanoimprint lithography (HE-NIL), UV-curing nanoimprint lithography (UV-NIL), and soft imprint lithography (SIL). A schematic diagram of the three imprint types is shown in Figure 4:

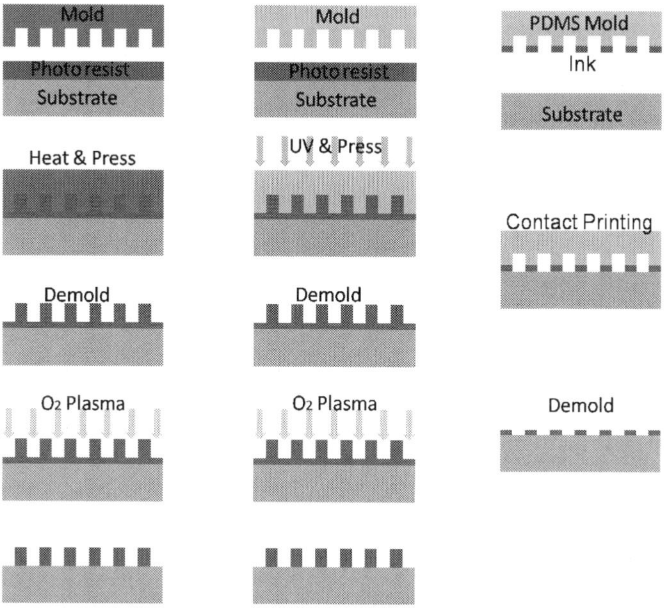

Fig. 4. Schematics of the hotembossing nanoimprint lithography, the UV-curing nanoimprint lithography, and the soft nanoimprint lithography (from left to right, respectively)

Steven Y. Chou from Princeton University developed the first hot embossing nanoimprint lithography in 1995 [17]. The electron beam lithography and etching were used to fabricate a template pattern on a silicon mold. Under a temperature of 200 °C and a pressure of 1900 psi, the silicon mold was pressed into a thermoplastic polymer called poly(methyl methacrylate) (PMMA). When the temperature was above its glass transform temperature (T_g), the PMMA became rubber-like and started to fill the pattern on the silicon mold. When the temperature dropped below T_g, the PMMA transformed into a high mechanical strength and glass-like material. Once the mold was separated from the sample, dry etching was used for cleaning the PMMA residues to complete the process. A disadvantage of the hot embossing nanoimprint lithography is that the nanostructures are fabricated under high temperature and high-pressure environments, and hence, its structure is prone to changes in temperature. Another disadvantage is the limitation of the application due to the requirement of long thermal cycle during the process.

The UV-curing nanoimprint lithography [18] uses UV-light to solidify the imprint material. A highly UV transparent and hard quartz is selected as the mold material. Quartz mold is pressed into the imprinting material to transfer the pattern into the silicon substrate at room temperature. The imprinting material is then irradiated with UV light, resulting in a crosslink of the material. After detachment of the mold, dry etching is used for cleaning the residual layer to complete the process. The UV-curing nanoimprint lithography is realized under room temperature and at a low pressure, and therefore, has the advantage of fast production and of being a simple process. However, its disadvantage is that the mold or substrate material must be UV-transparent (e.g., quartz, aluminum oxide).

In the research of soft nanoimprint lithography [19], a soft polymer such as polydimethylsiloxane (PDMS) is used as the mold to duplicate the pattern of hard mold. Acting like a stamp wetted with ink, PDMS mold is wetted with an ink made of, for instance, the self-assembly monomer (SAM). Alternatively, the substrate could also be sprayed with the imprint material. By compressing the PDMS soft mold and the substrate, the pattern of the mold is imprinted to the substrate. The soft nanoimprint lithography, similar to the UV-curing nanoimprint lithography, has the advantages of fast production, simple process, and low equipment cost. The key benefit of the soft nanoimprint lithography is its flexible mold, which makes it suitable for the non-planar imprint process, thereby having wider application. The disadvantage is that the mechanical strength of the mold is relatively weak compared to that of the others, resulting in easy wearing of the imprint patterns or the molds. To improve the imprint technique, a reversal imprint is developed. As shown in Figure 5, the aforementioned imprint techniques place the imprinting materials on the substrate surface. Conversely, the reversal imprint technique places the imprinting material on the surface of the mold before the entire material is pressed for transferring to the substrate. This procedure is similar to "planting" structures on the substrate, and the force required for the reversal imprint is comparably smaller. A 3D structure can be fabricated by stacking the structures layer-by-layer using the reversal imprint technique [20]. A successful reversal imprint requires the imprinting material to be hydrophilic to the substrate and hydrophobic to the mold, thereby enabling completion of the demolding process. Figure 5 shows the PTFE placed on the mold and the substrate preprocessed via HMDS. The attractive/repulsive surface of the materials can be analyzed by measuring the contact angle.

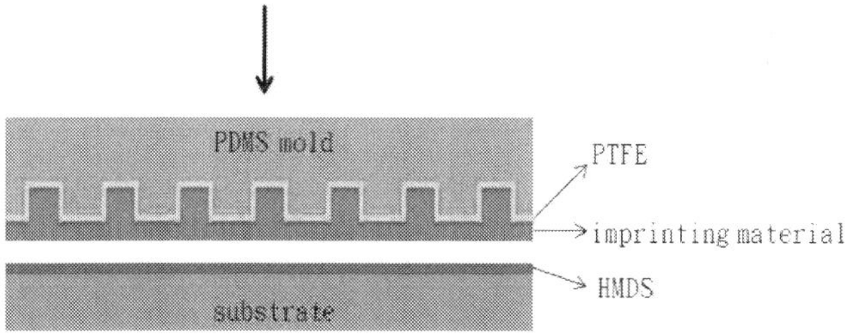

Fig. 5. Schematic diagram of the reversal imprint

Various nanoimprinting techniques have been developed to improve the abovementioned imprint processes. Among them, the roller imprint technique, originating from the roller press of the plastic sheets, is applied for nanoimprint. For example, a roller with nanostructures on its surface is rotated on a substrate composed of the thermoplastic polymer. When the temperature is set above the glass transition temperature, the thermoplastic polymer is in conjunction with the structures on the roller [21]. The nanostructure on the roller can be fabricated by the electroplating nickel mold technique,

which applies nickel film on the roller [22, 23]. If the substrate is flexible, structures can be reproduced massively via continuous roll-to-roll process. This method is suitable for the thermoplastic polymer films. The diffuser film used for current backlights of TFT-LCD panel can be fabricated via roll-to-roll process transferring of the microstructure to the PET film by roller imprint. The roller imprint, in practice, can adapt a process similar to the aforementioned imprint process. That is, PDMS soft mold is applied over the surface of the quartz roller, thereby enabling the continuous imprinting process, as shown in Figure 6. [24]. Various light sources can be used to harden the imprint material. For example, the hot embossing imprint can be heated up rapidly using an infrared light or an infrared laser [25]. The UV-hardening can be achieved via the UV light or UV laser. Using the laser has the benefit of using optical lenses to focus the beam to the ideal imprint area.
Continuous imprinting process is demonstrated using the roller imprint method. Instead of imprinting the entire surface at once, the roller and the substrate are imprinted progressively, rendering the roller imprint useful for the fabrication of structures on a large surface.

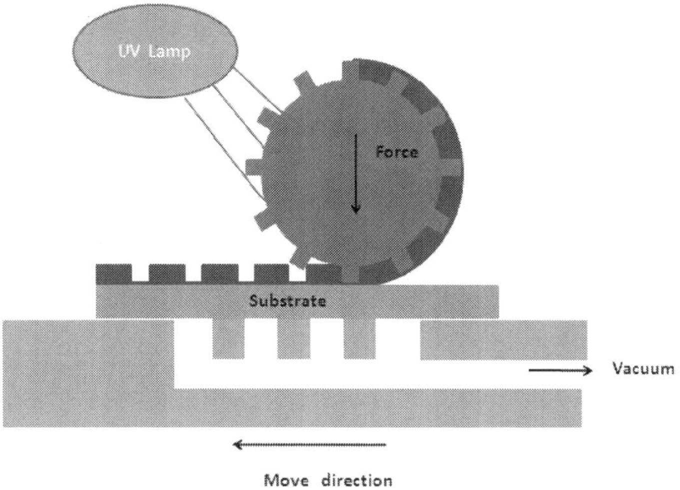

Fig. 6. Schematics of the UV-assisted roller imprint equipment

4. Improvement of the light-emitting efficiency by the imprint structure

4.1 The Effect of surface structures

As previously discussed, the light-emitting efficiency can be improved by a roughened surface, which can be achieved by the imprint of a surface structure. A uniform size and the periodicity of the structures benefit the homogeneity of the emitted light. In 2007, Chang et al. [26] applied hot embossing nanoimprint lithography to imprint the PMMA into a transparent conductive layer ITO of an LED. Inductively coupled plasma (ICP) etching was used for cleaning the process residues to expose the ITO, followed by etching in hydrochloric acid to produce periodic ITO holes, which were 90nm in depth, 0.85 μm in diameter, and 0.5μmin spacing. Compared to conventional LEDs, the periodic structure of an LED with size 350×350μm²showeda 12% improvement in the light-emitting efficiency with an input current of 20 mA. Huang et al. [27] fabricated p-GaN nanostructures on the surface of a 1×1 mm² high-power green LED chip. An SiO_2 thin film was first deposited onto p-GaN, and a silicon mold was used for the hot embossing imprint process. Reactive ion etching (RIE) was used for cleaning the residues of the imprinting material and SiO_2, to complete the transferring of the structure using SiO_2 as the resist. A p-GaN nanostructure of 75 nm depth was later produced viaICP etching. The residual SiO_2 was removed using a buffer oxidation etchant (BOE). Finally, an ITO thin film was deposited onto p-GaN. Due to high scattering effect, the light-emitting efficiency of the nanostructured LED was 48% higher than that of conventional LEDs with an input current of 350 mA. Zhou et al. [28] used the soft UV-nanoimprint lithography to fabricate nanostructures onto a 1×1 mm² p-GaN high-efficiency LED chip. A porous membrane mold with a nanohole approximately100 nm in size was first fabricated using theanodic aluminum oxide (AAO) method. A two-inch PDMS soft mold was replicated for UV-nanoimprint process. The process followed was similar to that of Huang et al. [27]. The light-emitting efficiency of the unpackaged nanostructured LED chip was 10.9% higher than that of conventional LEDs with an input current of 350 mA. The enhancement of efficiency would be 48% when the LED chip is packaged.

Other than the imprint method that imprints structures onto the LED surface via etching, Lee et al. [29] fabricated the microstructure directly onto the ITO. Figure 7 is the SEM image of the one-dimensional and two-dimensional imprint structures. The imprinting material is not a conventional polymer, but is a spin on glass (SOG), which is similar to SiO_2. The chemical and mechanical properties of this material are superior when compared to the polymer.

Fig. 7. SEM images of the one-dimensional and the two-dimensional imprint structures

Figure 8 is the schematic diagram that combines the imprint and the conventional LED fabrication process. The mesa etching (to expose the n-GaN) was first achieved usingICP. The electrode patterns were defined and deposited with p-type and n-type electrodes to fabricate conventional LED chips. After the hot embossing imprint process, the imprinting material on the electrodes was removed. Finally, the electrical and optical properties of the

LED were then measured. The results showed the forward bias of the imprinted LED is extremely similar to that of the conventional LED. Compared to the electrical property of the LED fabricated by etching above the p-GaN, the electrical property of the LED produced via this method was not damaged.

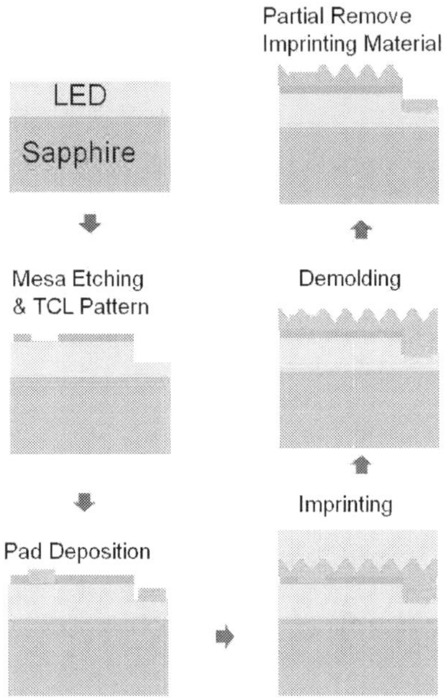

Fig. 8. Process flow of the conventional LED chip and imprint process

Regarding optical property, for a size of 300×300μm², composed of a cylinder array 3μm in diameter and 2μm in spacing of the LED chip, the efficiency enhancement is 26.2% with 20 mA current. Figure 9 is the diagram that illustrates the possible optical paths of a light incident to cylinder and pyramid structures, showing that the surface structures are beneficial to ruin total internal reflection.

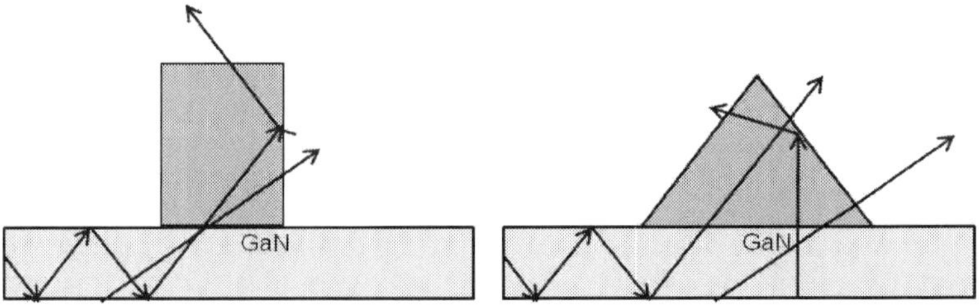

Fig. 9. The possible optical routes of an LED light incident to a micro-cylinder and to a pyramid structure

Comparing the micro-pyramid-structured LED (3 µm width and 2 µm spacing) with the nano-pyramid-structured LED (750 nm width and 500nm spacing), the light output power of the nano-pyramid-structured LED, which has a more densified structure, is superior to that of the micro-pyramid-structured LED as the input current increases (Figure 10.) [30].

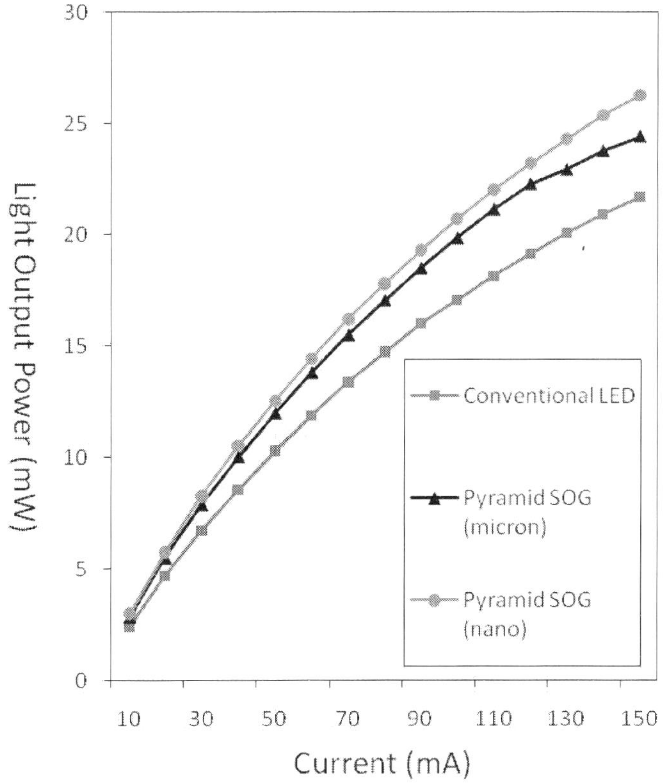

Fig. 10. A comparison of the light output powers of the microstructured, nanostructured, and conventional LEDs.

The study of the roller imprint in the application of fabricating LEDs is shown in Figure11. The light output power of the micro-cylinders with different heights increases in conjunction with the current. Theoretically, taller cylinder LED has more surface area to emit light compared to the short cylinder. However, as shown in the results, the light output power was higher for the imprint cylinder-structured LED with 3µm diameter, 2µm spacing, and 1.5µm height, compared to the cylinder-structured LED with 2.5µm or 5.0 µm height. This is because of the light absorption from the imprinting material, and thus, an optimal height of the imprint structure exists that maximizes the light emitting efficiency [24].

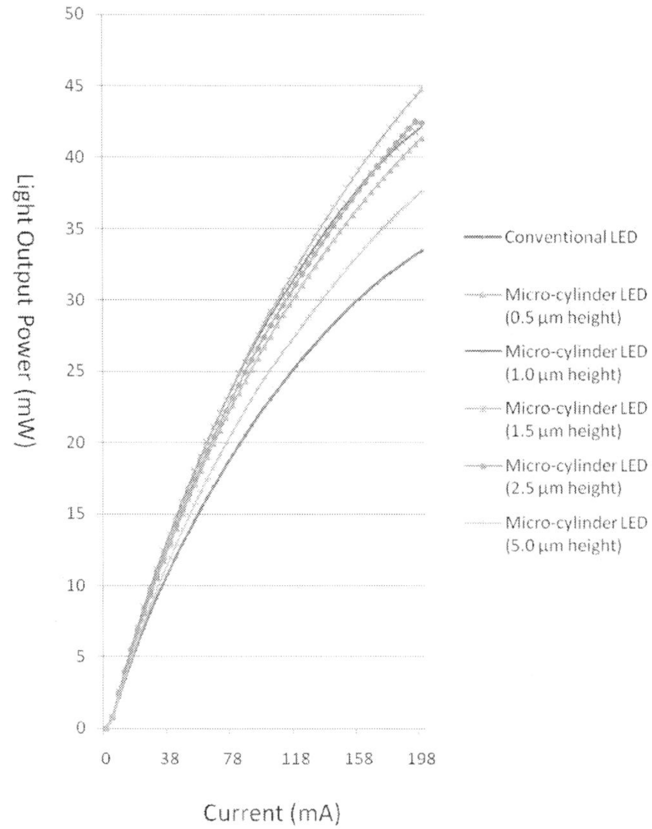

Fig. 11. A comparison between the light output power of micro-cylinder-structured LEDs of different height and conventional LEDs

4.2 The photonic crystal effect

The behavior of an electromagnetic wave traveling in a homogeneous medium can be obtained by solving the Helmholtz equation (the Helmholtz equation is derived from Maxwell's equations). The Helmholtz equation is:

$$\nabla^2 E + k^2 E = 0$$

where E is the electromagnetic wave and k is the wave number. Combined with the boundary conditions, Helmholtz equation can be solved via a harmonic solution. While solving for the equation, the wave number, frequency ($\omega = 2\pi \times v$), permittivity (ε), and permeability (μ) must satisfy below equation:

$$k^2 = \omega^2 \mu \varepsilon$$

In an ideal non-dispersive medium, the permittivity and permeability are independent of the frequency of the electromagnetic wave. Therefore, a linear relation between the wave number and the frequency exists. However, nearly all media are dispersive to a certain extent. That is, the permittivity and permeability are dependent on the frequency and lead to a nonlinear relation between the wave number and the frequency. In such a case, the group velocity and the phase velocity of the wave would not be the same.

When the medium (where the electromagnetic wave travels in) is a periodic structure in space with a periodicity comparable to the wavelength of the wave, the permittivity can also be regarded as a periodic distribution (Bloch's theorem) in space (assuming the permeability is 1). Though solving the Helmholtz equation in these periodic medium results in a harmonic solution that is similar to above equation, this solution carries the information of the periodic structure of the medium. The photonic crystal is a medium with such a periodically distributed permittivity, and the spatial periodicity of a photonic crystal is slightly smaller than the wavelength of the wave.

The solution to Helmholtz equation can be affected by the following factors: polarization of the electromagnetic wave, the interaction length between the electromagnetic wave and the photonic crystal, and the permittivity difference between the periodic structure and the substrate. Therefore, when the electromagnetic wave enters the photonic crystal, different eigen modes are formed, capable of interfering constructively or destructively to create energy bands. If the eigen modes form destructive interference in a specific direction, such a phenomenon is called the photonic band gap.

Because the existence of the photonic band gap is related to the frequency and the propagation direction of the electromagnetic wave, the photonic band gap can be used as a reflector for the large angle bending of electromagnetic wave propagation in integrated optics. Applying the photonic crystal to the LED chip can manipulate the directions of emitted photons. When the photonic crystal structure suppresses the photons with large emitting angular, the LED can be approximated a collimated light source. Conversely, when the photonic crystal structure suppresses the photons with small angular, the LED behaves akin to an expansion light source. This is the underlying principle of the far-field light pattern effect found in the photonic crystal. In addition to the far-field effect, the photonic crystal can be used to form a photonic bandgap within the active region of the LED. In this manner, photons generated by the active region are forced to escape the active region promptly, hence increasing the light-extraction efficiency [31]. If the photonic crystal is applied to the LED surface, high order diffraction lights are generated by the momentum compensation that occurred during the interaction of the photonic crystal and the guided photons. The otherwise confined photons in the LED can be coupled to the free space by the diffraction, which is another means to improve the light-extraction efficiency of the LED. Researchers at U.C. Santa Barbara conducted a series of discussions regarding the photonic crystal effect [32-35].

Analyzing the escape paths of the emitted photons from LEDs provided the observation that the chip surface's escape photons comprise 12 % of photons generated in the active region when a highly reflective coating is applied to the sapphire substrate. Approximately 22 % of the photons are confined within the sapphire substrate, while 66 % of the trapped photons are located in the chip region. Trapped photons, confined in the air/sapphire/GaN interfaces, are known as guided light. A longer interaction length between the photonic crystals and emitted photons typically results in an increased optical modulation effect on the guided light.

The significant difference (1.4) between GaN and air refraction indices makes the GaN/air interface an excellent waveguide material, which confines photons of the active region within the GaN. Photonic crystals are used to extract the guided light into the air. Considering diffraction, the guided light-extraction efficiency is dependent on the mode of the confined photons. Photons of lower order modes are confined to the center of GaN, and prevented from interacting with the photonic crystal structures on the chip's surface, which leads to inefficient photon diffraction of the lower-order modes and limits improvements to light-extraction efficiency.

A shallow photonic crystal layer on the GaN surface slightly affects the diffraction efficiency of the guided light, and therefore, the photonic crystal structure is typically 1 μm in thickness to enhance the interaction between the photonic crystal and the guided photons. However, a deep photonic crystal layer has a reduced effective refractive index, resulting in rapid evanescence of the lower-order modes in the photonic crystal. In such circumstances, the light-extraction efficiency cannot be improved. To improve the efficiency of the shallow photonic crystal layer, David et al. [32] inserted a low-refractive index material $Al_xGa_{1-x}N$ in the GaN to couple the lower-order photons trapped in the center of the GaN buffer layer to the GaN cap layer (Figure 12.). The diffraction efficiency of the lower-order photons substantially improved upon inserting the $Al_xGa_{1-x}N$.

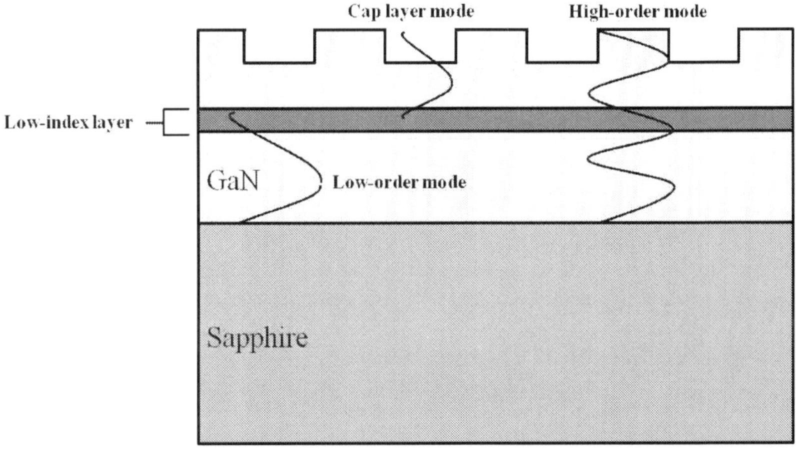

Fig. 12. Insert a low-index layer to improve mode coupling with photonics crystal

An LED is a light-emitting device comprising point light sources that emit photons in random directions. Therefore, an omnidirectional light-extraction structure reflects photons with various modes and directions from the active layer to outside the GaN, would be an ideal photonic crystal structure. To achieve the isotropic light extraction, the photonic band gap in the various directions must be of equal efficiency. Numerous photonic crystal structures designed to achieve the aforementioned objective have been proposed, including Archimedean tilings [36], Penrose lattices [37-39] and quasicrystals [40-43]. Expanding on previous discussions regarding photonic crystals, David et al. [33] used A7 Archimedean tilings as the photonic crystal layer on GaN LEDs. The A7 photonic crystal was composed of a triangular lattice and a complex basis. The unit cell is formed by 7 holes, and a is the hole interval. The lattice constant of photonics crystalis $b = a(1+\sqrt{3})$. Applying this photonic crystal structure to the surface of GaN LEDs resulted in a high order crystal, thereby diffracting the light into air, instead of onto the substrate. Diffraction occurred in the various Brillouin zone of the reciprocal lattice, demonstrating the omnidirectionality of the A7 structure.

Shields et al. [44] used the higher rotational symmetry of the photonic quasicrystals (PQCs) to achieve omnidirectionality. The designed PQCs patterns were based on square-triangular tiling using three different pitches: 450, 550, and 750 nm. Shield's experiments revealed a plateau on the far-field profile that was independent of the measurement angle. The finite-difference time-domain (FDTD) simulation also demonstrated similar behavior.

Increasing light-extraction efficiency and adjusting the emission profile can be achieved simultaneously by optimizing the coupling distance between the emitted photon and the photonic crystal. The momentum of photons parallel to the photonic crystal plate are coupled with the reciprocal lattice constant of the photonic crystal via the diffraction effect, generating photons with momentum perpendicular to the photonic crystal plate. Using an appropriate reciprocal lattice constant can increase the number of photons within the escape cone. An LED emission profile can also be adjusted using the same approach [34]. A triangular lattice photonic crystal is applied to both transparent electrodes of the LED. On the ITO side, the depth and the area of the photonic crystal is 120 nm and 400x400 μm2, respectively. On the Ni/Au side, the depth and the area of the photonic crystal is 250 nm and 500x500 μm2, respectively. The lattice constants of the photonic crystal are 185, 200, 215, and 230 nm. The experiment demonstrated that the vertical emission profile peaks occur at the second order diffraction (periodicity of photonic crystals/wavelength=0.5), and corresponds to 215 nm of lattice period/430 nm in ITO thickness, and 200 nm of lattice period/405 nm in Ni/Au thickness. Changing the lattice constant altered the vertical emission of various modes. In this sample, the extraction of cap layer mode photons into air was evident.

Methods to modulate the far-field light patterns and to improve the light-extraction efficiency of LEDs using sub-microstructures lead to that most photons confined within the LED can be coupled outside the LED, using photonic crystals. Therefore, coupling photons to the escaping cone is a major mechanism enabling photonic crystals to improve the light extraction efficiency and to adjust the far-field light pattern. Though the depth, the lattice constant and the filling factors of the photonic crystal adjust the mode of the trapped photons to enter the air lightline, the polarization and propagation directions of the photons interacting with the photonic crystals require consideration. Photons generated by the spontaneous emission process in the LED have poor temporal and spatial coherences, and therefore, the roughening effect caused by the photonic crystal should also be considered.

Numerous studies have investigated the application of nanoimprinted photonic crystals in LEDs. Examples are listed below:

Cho et al. [45] used the nanoimprint lithography and ICP etching to fabricate photonic crystal structures of various depths onto the ITO/GaN surface of a 375 μm x 330 μm blue LED chip. Upon optical efficiency optimization, the efficiency of the packaged LED can be further enhanced by 25 %, using a 20 mA current injection. Using the 3D-FDTD simulation, light-extraction efficiency increases in correlation with the etching depth of the photonic crystals. However, the light-extraction efficiency decreases when the ratio (wavelength/effective refractive index) is exceeded.

Cheng et al. [46] applied the photonic crystal [1x1 mm^2] to the green power chip and combined the omnidirectional reflector at the back of the chip. The overall efficiency increased by 88 %, with a collimated far-field light pattern.

Nao et al. [47] attached a photonic crystal in a triangular lattice (diameter=100~250 nm, height=120 nm) to the flip chip between the sapphire and the GaN, allowing the diffractive coupling of photons to enter the air from the back of the sapphire substrate. The far-field pattern was collimated, and a lobe was observed near the horizontal direction. The lobe was attributed to the intermixing of light emitted from the side and front of the chip.

Huang et al. [48] applied quasi-photonic crystals to both n- and p-GaN layers in a vertical LED. The optical efficiency exceeded that of a single photonic crystal LED.

Byeon et al. [49] created a hexagonal array of holes with a diameter of 250 to 380 nm and 600 to 900 nm of pitch as the photonic crystal layer on the ITO of a green LED, using soft imprint technology (to prevent damages on the GaN LED). The light emitting efficiency improved maximally by 25 % under a 20 mA current injection.

Khokhar et al. [50] compared e-beam lithography and nanoimprinting technology concerning the creation of photonic crystal structures. Though the e-beam lithography created a more precise and accurate pattern, it is not suitable for mass production, and while nanoimprinting technology is suitable for mass production, several issues remain. For example, the residual layer must be removed using etching, which limits the practical thickness of the photonic crystal layer, and the imprint material requires a high-etching selectivity.

The photonic crystal structure can be applied quickly and accurately to the surface and the interior of the LED via the nanoimprint lithography technique [51, 52]. However, because the electrical characteristics of the LED will be degraded using dry etching of the p-GaN, T. A. Truong et al. [53] applied sol gel titania to the substrate (soft imprint plus a 300 °C solidification process) to form titanium oxide photonic crystal structures on the LED without damaging the LED surface.

Motivated by the great potential of photonic crystal application on LED, several equipment manufacturers have devised intriguing strategies to develop this field. Molecular imprints Inc. [54] sprays the imprint material to a 6″ quartz substrate. Using a step-and-repeat exposure procedure, they progressively transfer the structure of a hard mother mold (5×5 mm^2) to the entire area of the quartz substrate, which now acts as a substitute mold. The nanoimprint of the 3″ GaN LED occurs from the quartz mold. Due to the hard imprint nature of this technique, the cleanness of the mold and the substrate play crucial roles. Suss MicroTec Lithography GmbH [55] uses a PDMS soft mold adhered to the substrate by the vacuum force before imprinting. To prevent trapping air between the mold and the substrate during imprinting, the PDMS mold is gradually released from the vacuum force until the mold and the imprinting material are held together by the capillary force. After UV exposure, the gradual vacuum force removes the mold once more. EV Group Inc. [56] also adopted PDMS molds. The difference is that the mold comprises two soft PDMS layers with different degrees of hardness. The harder layer is the outer layer that directly touches the imprinting materials, which is believed to solve the rigidity issue of a typical soft mold, and prevent the deformation of the mold. Obducat AB [57] uses non-reusable polymer stamp as the mold. The polymer stamp is fabricated by pressing a hard master stamp against the polymer film. Once the polymer stamp is used, it is disposed. Due to its single usage nature, the stamp has no lifetime issues. In addition, the flexibility of the polymer stamp allows them to be applied to nonplanar surfaces. Obducat's solution has already found a significant number of industrial users, including two LED manufacturers from Taiwan, Luxtalek and Epistar.

4.3 The Effects of NPSS

NPSS has three distinct advantages:

1. **Enhancing the light extraction efficiency**

The PSS on the micrometer scale has been experimentally demonstrated to improve the light extraction efficiency of LEDs significantly. Furthershrinking the size of PSS can increase the structural density, and hence,enhancing the light extraction efficiency even more.

2. **Simplifying the process parameters**

Changing the condition on the current micron scale PSS process usually requires careful/tedious optimization of the subsequent epitaxial process parameters. By shrinking the patterning size into nanoscales, this optimization step can be neglected, facilitating a more rapid LED growth process.

3. **Reducing the thickness of the buffer layer**

A thick buffer layer is typically necessaryto smooth out the PSSpatterned surface prior to the epitaxy process. If the size of the PSS structures is in nanoscales, the required buffer layer is thin and may be achieved by the side growth of the epitaxial layer, streamlining the LED growth process and increasing the process yield.

Huang et al. [58] created polymer nanostructures on the sapphire substrate using the thermal pressing technique. The imprint material also acted as the resist layer for the ICP etching process. A concave NPSS pattern was obtained with a periodicity of 450 nm, a diameter of 240 nm, and a depth of 165 nm. SEM measurements showed that the GaN did not fulfill the the NPSS holescompletely, causing a hole array at the GaN/NPSS interface. Due to the difference in the refraction indices between these holes and the surrounding matrix, the light originating from the MQW underwent scattering and multiple reflections at the GaN/NPSS interface. The light was bounced back to the GaN/Air interface, increasing the likelihood of light escaping the LED. Under a 20 mA electrical current injection, the NPSS was found to enhance the light extraction efficiency by as much as 1.33 times that ofconventional LEDs. Huang et. al. [59] used a similar thermal pressing technique combined with ICP etching to create a convex NPSS with a periodicity of 750 nm and a diameter of 450 nm. The height of the NPSS structures was 182 nm. Under a 20 mA electrical excitation, the $300\times300\ \mu m^2$ NPSS had 35% more light-extraction efficiency compared toconventional LEDs. If the quasi-photonic crystal effect wereconsidered, the efficiency could be further increased to48%. C.C. Kao et al. [60] used a similar process for creating NPSS to study the dependance of the light extraction on the aspect ratios of the NPSS. Their experiment showed that efficiency was higher for higher aspect ratios. The efficiency was increased from 11% to 27% for aspect ratios from 2.00 to 2.50, respectively.

Results from the micron scale PSS structures shows that increasing the aspect ratio of the PSS can also increase the light-extraction efficiency. However, the height of the ICP-etched NPSS is directly related to the size of the nanoimprinted polymer structures. Due to the high mechanical strength and the soundchemical stability, the etching selective ratio between the sapphire and the polymer resist is small, resulting in a less than desired etching depth on the sapphire substrate. Figure 13 shows the NPSS structures etched by the sulfuric acid and the phosphoric acid at high temperatures. The periodicity is 1.25 nm and the depth is 340 nm. A thin layer of silicon oxide is used as the etching resist at high temperatures. If the etching time is long, the PSS pattern transforms intocylindrical rods with amaximum height of 700 nm (Figure 14). With the ICP etching process using anNi thin film as the resist layer, an even higher cylindrical rod can be obtained due to the higher etching selective ratio between the sapphire and the Ni, as shown in Figure 15. [61].

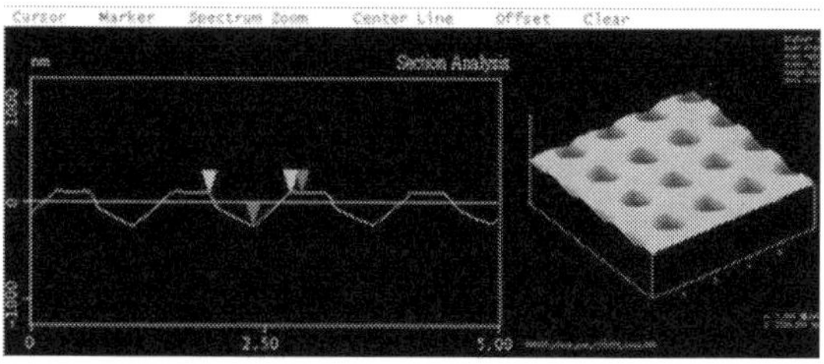

Fig. 13. AFM images of concave NPSS formed using wet etching

Fig. 14. SEM images of convex NPSS formed using wet etching

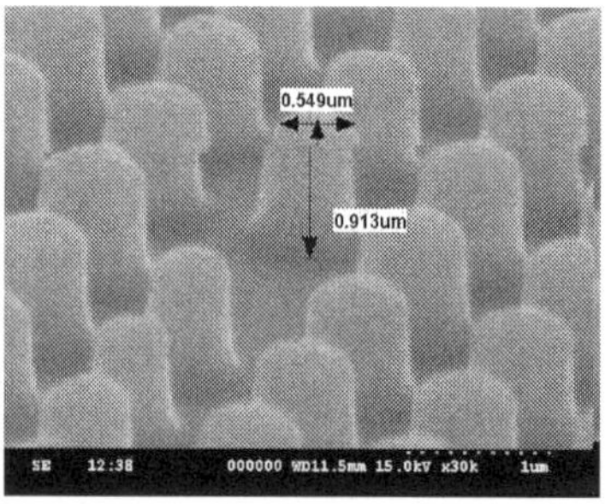

Fig. 15. SEM images of dry etched cylinder NPSS

Hsieh et al. [62] created NPSS via a thermal nanoimprint technique, usingNi as the resist layer. A anti-adhesive layer (1H, 1H, 2H, and 2H-perfluorooctyltrichlorosilane) was coated on the mold, followed by a thin Ni layer. A layer of PMMA was sprayed onto the substrate. The Ni layer on the protrudent mold was peel-off before being transferred to the PMMA, using thermal pressing. NPSS (~0.4 μm depth) was then created via ICP etching. The diameters were 0.4, 0.6, 0.8, 1.0, 2.0, and 3.0 μm, with the corresponding EL intensity of 128.4, 120.2, 109.2, 102.5, 91.6, 90.3, and 69.5 mcd, respectively. In comparison to the unpatterned substrate, the efficiency of the NPSS (0.4 μm) LED was increased by 84.7%.

5. Conclusion

Nanotechnology is typically defined as the fabrication of structures under 100 nm. The fabrication of structures between 100 nm~999 nm is referred to as sub-micron technology. However, much literature related to LEDs referring to the sub-micron process as nanotechnology exists. Therefore, this study referred to the nanoimprint as the fabrication of several hundred nanometer structures via the imprint technique, nonetheless indicating that the few hundred-nanometer fabrication process is applicable to the industry. This scale is suitable for the development of the nanoimprint technique because the limitation of optical lithography does not affect it, and the technique is low cost compared to the expensive advanced projection lithography and the electron beam lithography. The LED chip has a 0~10 μm bow and the epitaxy is usually processed under a temperature near 1000°C. Because the coefficients of thermal expansion of the substrate and the GaN epitaxial layer are different, the epitaxy process is likely to cause the deformation of the chip. Moreover, due to (a) the cleanness issues, (b) surface defects of the epitaxial layer, and (c) the emphasis on the height of the electrode after the chip process, the imprint process of the entire wafer is rendered difficult. Therefore, the use of a flexible soft mold is crucial.

This article presents the nanoimprinting technique integrated with the conventional LED fabrication process. Due to its low cost, simplicity, easy integration, and the enhancement of the optical efficiency, nanoimprinting has a widespread application in various industries. The future business aspect of nanoimprint technology is dependent on the development of manufacturing equipment suitable for LEDs fabrication.

6. References

[1] M. R. Krames, G. E. Höfler, E. I. Chen, I. H. Tan, P. Grillot, N. F. Gardner, H. C. Chui, J. W. Huang, S. A. Stockman, F. A. Kish, and M. G. Craford, "High power truncated inverted pyramid AlxGa12x.0.5In0.5P/GaP light-emitting diodes exhibiting >50% external quantum efficiency", Applied physics letters, Vol. 75, pp. 2365, 1999.

[2] C. F. Lin, Z. J. Yang, B. H. Chin, J. H. Zheng, J. J. Dai, B. C. Shieh, and C. C. Chang, "Enhanced light output power in InGaN light-emitting diodes by fabricating inclined undercut structure", Journal of The Electrochemical Society, Vol. 153, pp. 1020, 2006.

[3] C. E. Lee, Y. C. Lee, H. C. Kuo, M. R. Tsai, T. C. Lu, and S. C. Wang," High brightness GaN-based flip-chip light-emitting diodes by adopting geometric sapphire shaping structure", Semiconductor science and technology, Vol. 23, pp. 025015, 2008.

[4] C. Huh, K. S. Lee, E. J. Kang, and S. J. Park, "Improved light-output and electrical performance of InGaN-based light-emitting diode by microroughening of the p-GaN surface", Journal of Applied Physics, Vol. 93, pp.9383, 2003.

[5] T. Fujii, Y. Gao, R. Sharma, E. L. Hu, S. P. DenBaars, and S. Nakamura, "Increase in the extraction efficiency of GaN-based light-emitting diodes via surface roughening," Applied physics letters, Vol.84, pp. 6, 2004.

[6] T. X. Lee, K. F. Gao, W. T. Chien, and C. C. Sun, "Light extraction analysis of GaN-based light-emitting diodes with surface texture and/or patterned substrate", Optics Express Vol. 15 pp. 6670, 2007.

[7] T. X. Lee, K.F. Go, T. Y. Chung, and C. C. Sun, "Light extraction analysis of GaN-based LEDs", Proceeding of SPIE Vol. 6473, pp. 64731P-1, 2007.
[8] K. Tadatomo, and T. Taguchi, "High output power InGaN ultraviolet light emitting diodes fabricated on patterned substrates using metal organic vapor phase epitaxy," Japanese Journal of Applied Physics, Vol. 40, pp. L583, 2001.
[9] M. Yamada, T. Mitani, Y. Narukawa, S. Shioji, I. Niki, S. Sonobe, K. Deguchi, M. Sano, and T. Mukai, "InGaN-based near-ultraviolet and blue-light-emitting diodes with high external quantum efficiency using a patterned sapphire substrate and a mesh electrode," Japanese Journal of Applied Physics, Vol. 41, pp. L1431 ,2002.
[10] A. Bell, R. Liu, F. A. Ponce, H. Amano, I. Akasaki, and D. Cherns, "Light emission, and microstructure of Mg-doped AlGaN grown on patterned sapphire," Applied Physics Letters, Vol. 82, pp. 349, 2003.
[11] K. T. Lee, Y. C. Lee, and J. Y. Chang, "Characterization of Gallium Nitride Grown on Patterned Sapphire Substrate with Shallow U-Shaped Stripe Grooves", Journal of The Electrochemical Society, Vol. 155, pp. H673, 2008,
[12] E. Yablonovitch, "Inhibited Spontaneous Emission in Solid-State Physics and Electronics ",Physical Review Letters, Vol. 58, pp. 2059, 1987.
[13] S. John, "Strong localization of photons in certain disordered dielectric superlattices", Physical Review Letters, Vol. 58, p.2486–2489, 1987.
[14] M. Akanegawa, Y. Tanaka, and M. Nakagawa, "Basic study on traffic information system using LED traffic lights," IEEE Transactions on Intelligent Transportation Systems, Vol. 2, pp. 197, 2001.
[15] H.Ghafoori-Shiraz, "Butt coupling efficiency of long-wavelength edge-emitting LEDs to single-mode fibers", Transactions of the Institute of Electronics, Information and Communication Engineers, Section E, Vol. E70, pp. 617, 1987.
[16] Y. Tu, S. Jin, Y. Wang, and L. Dou, "Color uniformity and data simulation in high-power RGB LED modules using different LED-chips arrays," Proceeding of SPIE, Vol. 6828, pp. 682816, 2007.
[17] S. Y. Chou, P. R. Krauss and P. J. Renstrom, "Imprint of sub-25 nm vias and trenches in polymers", Applied Physics letters,Vol. 67, pp. 3114, 1995.
[18] M. Bender, M. Otto, B. Hadam, B. Vratzov, B. Spangenberg and H. Kurz, "Fabrication of nanostructures using a UV-based imprint technique", Microelectronic Engineering, Vol. 53, pp. 233, 2000.
[19] Y. Xia and G. M. Whitesides, "Soft Lithography", Annual Review of Materials Science, Vol. 28, pp. 153, 1998.
[20] 20. C. Peng and S. W. Pang, "Hybrid mold reversal imprint for three-dimensional and selective patterning", Journal of Vacuum Science and Technology B: Microelectronics and Nanometer Structures, Vol. 24, pp. 2968, 2006.
[21] H. Tan, A. Gilbertson, and S. Y. Chou, "Roller nanoimprint lithography," Journal of Vacuum Science and Technology B: Microelectronics and Nanometer Structures, Vol. 16, pp. 3926, 1998
[22] M. T. Gale, "Replication technique for diffractive optical elements", Microelectronic Engineering, Vol. 34, pp. 339, 1997.

[23] T. Makela, T. Haatainen, P. Majander, J. Ahopelto, and V. Ambertini, "Continuous double-sided roll-to-roll imprinting of polymer film," Japanese Journal of Applied Physics, Vol. 47, pp. 5142, 2008.
[24] Y. C. Lee, B. T. Chen, and T. H. Wu, "Fabrication of SU-8 microstructures by roller imprinting technique to enhance light extraction efficiency of LEDs", Proceedings of the 6th International Conference on MicroManufacturing, pp. 73, Tokyo, Mar. 7-10,2011.
[25] 25.C. H. Chen, Y. C. Lee, C. D. Chen, S. J. Lai, and S. J. Liaw, "Roller imprinting based on focus infrared heating," Proceedings of the 3rd IEEE Int. Conf. on Nano/Micro Engineered and Molecular Systems, Sanya, China, 2008.
[26] S. J. Chang, C. F. Shen, W. S. Chen, C. T. Kuo and T. K. Ko, S. C. Shei, and J. K. Sheu, "Nitride-based light emitting diodes with indium tin oxide electrode patterned by imprint lithography," Applied Physics Letters, Vol. 91, pp. 013504, 2007.
[27] H. W. Huang, C. H. Lin, C. C. Yu, B. D. Lee, C. H. Chiu, C. F. Lai, H. C. Kuo, K. M. Leung, T. C. Lu, and S. C. Wang, "Enhanced light output from a nitride-based power chip of green light-emitting diodes with nano-rough surface using nanoimprint," Nanotechnology, Vol. 19, pp. 185301, 2008.
[28] W. Zhou, G. Min, Z. Song, J. Zhang, Y. Liu and J. Zhang, "Enhanced efficiency of light emitting diodes with a nano-patterned gallium nitride surface realized by soft UV nanoimprint lithography", Nanotechnology, Vol. 21, pp. 1, 2010.
[29] Y. C. Lee, M. J. Ciou, and J. S. Huang, "Light output enhancement for nitride-based light emitting diodes via imprinting lithography using spin-on glass", Microelectronics Engineering, Vol. 87, No. 11, pp. 2211, 2010.
[30] Y. C. Lee, M. J. Ciou, and J. S. Huang, "The influence of nanoimprinting surface structures on the optical efficiency of GaN-based LEDs", IEEE Transactions on Nanotechnology, in press, 2010.
[31] S. Fan, P. R. Villeneuve and J. D. Joannopoulos, "High extraction efficiency of spontaneous emission from slabs of photonic crystals", Physical review letters, Vol. 78, pp.3294, 1997.
[32] A. David, T. Fujii, E. Matioli, R. Sharma, S. Nakamura, S. P. DenBaars, and C. Weisbuch, "GaN light-emitting diodes with Archimedean lattice photonic crystals," Applied Physics Letters Vol. 88, pp. 073510, 2006.
[33] A. David, T. Fujii, E. Matioli, R. Sharma, S. Nakamura, S. P. DenBaars, and C. Weisbuch, "Photonic-crystal GaN light-emitting diodes with tailored guided modes distribution," Applied Physics letters, Vol. 88, pp.061124, 2006
[34] K. McGroddy, A. David, E. Matioli, M. Iza, S. Nakamura, S. DenBaars, J. S. Speck, C. Weisbuch, and E. L. Hu, "Directional emission control and increased light extraction in GaN photonic crystal light emitting diodes", Applied Physics Letters, Vol. 93, pp. 103502, 2008.
[35] T. A. Truong, L. M. Campos, E. Matioli,1 I. Meinel, C. J. Hawker, C. Weisbuch, and P. M. Petroff, "Light extraction from GaN-based light emitting diode structures with a noninvasive two-dimensional photonic crystal", Applied Physics Letters,Vol.94, pp. 023101, 2009.

[36] S. David, A. Chelnokov, and J.-M. Lourtioz, "Isotropic photonic structures: archimedean-like tilings and quasi-crystals", IEEE Journal of Quantum Electronics, Vol. 37,pp. 1427, 2001.

[37] A. D. Villa, S. Enoch, G. Tayeb, V. Pierro, V. Galdi, and F. Capolino, "Band gap formation and multiple scattering in photonic quasicrystals with a penrose-type lattice", Physical Review Letters, Vol. 94,pp. 183903, 2007.

[38] S. P. Gorkhali, J. Qi, and G. P. Crawford, "Electrically switchable meso scale penro sequasi crystal structure",Applied Physics Letters, Vol. 86,pp. 011110, 2005.

[39] M. Notomi, H. Suzuki, T. Tamamura, and K. Edagawa, "Lasing action due to the two-dimensional quasiperiodicity of photonic quasicrystals with a penrose lattice", Physical Review Letters, Vol. 92, pp. 123906, 2004.

[40] Y. S. Chan, C. T. Chan, and Z. Y. Liu, "Photonic band gaps in two dimensional photonic quasicrystals", Physical Review Letters, Vol. 80,pp.956, 1998.

[41] M. E. Zoorob, M. D. B. Charlton, G. J. Parker, J. J. Baumberg, and M. C. Netti, "Complete photonic band gaps in 12-fold symmetric quasi crystals",Nature, Vol. 404,pp. 740, 2000.

[42] P. L. Hagelstein and D. R. Denison, "Nearly isotropic photonic bandgap structures in two dimensions", Optics Letters, Vol. 24, pp. 708, 1999.

[43] J. Zarbakhsh, F. Hagmann, S. F. Mingaleev, K. Busch, and K. Hinger, "Arbitrary angle waveguiding applications of two-dimensional curvilinear-lattice photoniccrystals", Applied Physics Letters, Vol. 84, pp. 4687, 2004.

[44] P. A. Shields, M. D. B. Charlton, T. Lee, M. E. Zoorob, D. W. E. Allsopp, and W. N. Wang, "Enhanced light extraction by photonic quasi-crystals in GaN blue LEDs", IEEE Journal of Selected Topics in Quantum Electronic, Vol. 15, pp. 1269, 2009.

[45] H. K. Cho, J. Jang, J. H. Choi, J. Choi,J. Kim, J. S. Lee, B. Lee, Y. Ho Choe, K. D. Lee, S. H. Kim, K. Lee, S. K. Kim, and Y. H. Lee, Light extraction enhancement from nanoimprinted photonic crystal GaN-based blue light emitting diodes", Vol. 14, Optics Express, pp.8654, 2006.

[46] B. S. Cheng, C. H. Chiu, K. J. Huang, C. F. Lai, H. C. Kuo, C. H. Lin, T. C. Lu, S. C. Wang and C. C. Yu, "Enhanced light extraction of InGaN-based green LEDs by nano-imprinted 2D photonic crystal pattern", Semiconductor Science Technology, Vol. 23,pp. 055002, 2008.

[47] Y. Naoi, M. Matsumoto, T. Tan, M. Tohno, S. Sakai, A. Fukano,and S. Tanaka, " GaN-based light emitting diodes with periodic nano-structures on the surface fabricated by nanoimprint lithography technique", Physica Status Solidi C, Vol. 7, pp. 2154, 2010.

[48] H. W. Huang, C. H. Lin, Z. K. Huang, K. Y. Lee, C. C. Yu, and H. C. Kuo, "Double photonic quasi-crystal structure effect on GaN-based vertical-injection light-emitting diodes, Japanese Journal of Applied Physics Vol. 49 pp. 022101, 2010.

[49] K. J. Byeon, H. Park, J. Y. Cho, K. Y. Yang, J. H. Baek, G. Y. Jung, and H. Lee, "Fabrication of photonic crystal structure on indium tin oxide electrode of GaN-based light-emitting diodes, Physica Status Solidi A, Vol. 208, pp. 480, 2011.

[50] A. Z. Khokhar, K. Parsons, G. Hubbard, F. Rahman, D. S. Macintyre, C. Xiong, D. Massoubre, Z. Gong, N. P. Johnson, R. M. De La Rue, I. M. Watson, E. Gu, M. D. Dawson, S. J. Abbott, M. D. B. Charlton, M. Tillin, "Nanofabrication of gallium nitride photonic crystal light-emitting diodes", Microelectronic Engineering Vol. 87, pp. 2200, 2010.
[51] K. D. Lee, S. H. Kim, J. D. Park, J. Y. Kim and S. J. Park, "Application of nanoimprint lithography to nano-optics: wire grid polarizer and photonic crystal LED", Proceedings of SPIE, Vol. 6462, pp. 64620P-1, 2007.
[52] B. S. Cheng, C. H. Chiu, K. J. Huang, C. F. Lai, H. C. Kuo, C. H. Lin, T. C. Lu, S. C. Wang, and C. C. Yu, "Enhanced light extraction of InGaN-based green LEDs by nano-imprinted 2D photonic crystal pattern", Semiconductor Science and Technology, Vol. 23, pp. 055002, 2008.
[53] T. A. Truong, L. M. Campos, E. Matioli, I. Meinel, C. J. Hawker, C. Weisbuch, and P. M. Petroff, "Light extraction from GaN-based light emitting diode structures with a noninvasive two-dimensional photonic crystal", Applied Physics Letters Vol. 94, pp. 023101, 2009.
[54] R. Hershey, G. Doyle, C. Jones, D. LaBrake, and Mike Miller, "Imprint lithography advances in LED manufacturing", Physica Status Solidi C, Vol. 4, pp. 21, 2007.
[55] R. Ji , M. Hornung, M. A. Verschuuren, R. van de Laar, J. van Eekelen, U. Plachetka, M. Moeller, and C. Moormann, "UV enhanced substrate conformal imprint lithography (UV-SCIL) technique for photonic crystals patterning in LED manufacturing", Microelectronic Engineering Vol. 87,pp. 963, 2010.
[56] M. Bender,U. Plachetka,J. Ran,A. Fuchs,B. Vratzov,H. Kurz,T. Glinsner,and F. Lindner, "High resolution lithography with PDMS molds", Journal of Vacuum Science and Technology B: Microelectronics and Nanometer Structures, Vol. 22, pp. 3229, 2004.
[57] "NanoImprint Lithography: Ready for high volume manufacturing", Euro Asia Semiconductor, v 29, pp. 15, 2007.
[58] H. W. Huang, C.H. Lin, J.K. Huang, K.Y. Lee, C.F. Lin, C.C. Yu, J.Y. Tsai, R. Hsueh, H.C. Kuo, S. C. Wang, "Investigation of GaN-based light emitting diodes with nano-hole patterned sapphire substrate (NHPSS) by nano-imprint lithography", Materials Science and Engineering B Vol. 164,pp. 76, 2009.
[59] H. W. Huang, J. K. Huang, S. Y. Kuo, K. Y. Lee, and H. C. Kuo, "High extraction efficiency GaN-based light-emitting diodes on embedded SiO2 nanorod array and nanoscale patterned sapphire substrate", Applied Physics Letters, Vol. 96, pp. 263115, 2010.
[60] C. C. Kao, Y. K. Su, C. L. Lin, and J. J. Chen, "The aspect ratio effects on the performances of GaN-based light-emitting diodes with nano patterned sapphire substrates", Applied Physics Letters, Vol. 97, pp. 023111 , 2010.

[61] H. J. Horng, "A Study of Submicron Patterned Sapphire Substrates Applied to GaN-based Light-Emitting Diode", Master thesis, Chung Yuan Christian University, Taiwan, 2010.

[62] E. D. Hsieh and Y. C. Lee, "Contact transfer and metal mask embedded lithography apply to fabricate patterned sapphire substrate", The 14th Nano and Micro-System Technology Conference, Kaohsiung, Taiwan, 2010.

10

Fabrication of Circular Grating Distributed Feedback Dye Laser by Nanoimprint Lithography

Yan Chen[1], Zhenyu Li[1], Zhaoyu Zhang[1] and Axel Scherer[2]
[1]*Shenzhen Institutes of Advanced Technology, Chinese Academy of Sciences*
[2]*California Institutes of Technology*
[1]*China*
[2]*USA*

1. Introduction

Within recent years the development of polymer dye lasers has progressed to higher levels of performance and functionality. The most attractive advantages of polymer dye lasers include low-cost processing, wide choice of emission wavelengths, and easy fabrication on flexible substrates. Several waveguide dye lasers have been studied with emission wavelengths ranging from ultraviolet to near infrared (Y. Oki et al., 2002). By simply changing the fluorophore doped in the polymer, these lasers can be used as the tunable sources for various applications, such as spectroscopy (Y. Oki et al., 2002) and fluorescence excitation source (C. Vannahme, 2011). Furthermore, microlaser array with multiwavelength emissions can be achieved (Y. Huang et al., 2010) for more applications such as compact displays and multiwavelength biosensors.

Currently the integration of miniaturized active light sources such as lasers into microfluidic systems becomes an attractive approach for biological and chemical processes (D. Psaltis et al., 2006). A majority of microfluidic systems are based on external light sources. However, the coupling of optical signals in and out of the devices, typically by optical fibers, remains one of the major challenges in integrated optics. By making on-chip light sources, we can eliminate the optics alignment, which greatly reduces the complexity of the system (E. Verpoorte, 2003). For applications in biochemical analysis in microfluidic systems, a surface emitting laser would appear to be more useful than other lasers because of its stacked substrate structure. Therefore, we choose a circular grating structure as the laser resonator design to produce low-threshold surface emitting lasing. The laser operating characteristics can be significantly improved by the two-dimensional nature of the resonator structure, and they are suitable to serve as low-threshold, surface-emitting coherent light source in microfluidic networks.

The 1-D distributed feedback (DFB) structure is a widely employed resonator geometry, and has been previously demonstrated for polymer lasers (Y. Oki et al., 2002). However, operating characteristics can be significantly improved within 2-D structures. Here, we choose a circular grating distributed feedback structure to obtain low threshold operation, a well-defined output beam, and vertical emission perpendicular to the device plane.

Although surface emitting circular grating lasers using semiconducting polymers have been previously demonstrated (Bauer et al., 2001; Turnbull et al., 2005), their lasers were fabricated by depositing the organic gain material onto prepatterned dielectric substrates, limiting the depth and the accuracy of the shape of the grating.

For better geometric control, we choose nanoimprint lithography (S. Y. Chou et al., 1996) as a direct patterning method. Nanoimprint lithography is the technique that can effectively produce nano pattern with line width below 100nm. In general, a hard mold is used to transfer patterns with high fidelity into target polymers, and this technique has become an attractive approach to define nanofabricated optical resonator structures. Conjugated polymer lasers fabricated by hot embossing have been studied (J. R. Lawrence et al., 2002), and 1-D DFB lasers based on organic oligomers using a room temperature nanoimprint method were reported (D. Pisignano et al., 2003, 2004).

The basic idea of nanoimprint lithography is to press a mold with nanostructures on its surface into a thin layer of resist on a substrate, followed by the removal of the mold. Nanoimprint is a low cost nanopatterning technology based on the mechanical deformation of a resist, and it is a high-throughput alternative to traditional serial nanolithography technologies.

The imprint step creates a thickness contrast and duplicates the nanostructures in the resist film. During the imprint process, the resist is heated to a temperature above its glass transition temperature. At this temperature, the resist, which is thermoplastic, becomes a viscous liquid and can be deformed into the shape of the mold. Therefore, this method allows the nanostructure on the mold to be faithfully transferred to the polymer substrate.

The well developed nanoimprint technology provides a convenient way of mass production and large-scale fabrication of low-cost dye laser arrays with a wide wavelength output range. It is also straightforward to build on-chip dye lasers with waveguides to replace the optical fibers necessary for the integrated optics. The miniaturized dye lasers can serve as surface emitting coherent light sources, which are very important in various biochemical applications, such as laser-induced fluorescence and spectroscopy.

In this chapter, we report the fabrication of a circular grating distributed feedback laser on dye-doped poly(methylmethacrylate) (PMMA) films (Y. Chen et al., 2007). The schematic diagram of a nanoimprinted circular grating dye laser chip is illustrated in Figure 1. The laser was fabricated on a glass substrate using a low-cost and manufacturable nanoimprint method. In this solid-state dye laser device, the laser dye is doped in the polymer forming the laser resonator, which can produce high-intensity and narrow-linewidth lasing with a well-defined output beam. With certain selected grating period, surface emission lasing with single frequency at 618 nm and a linewidth of 0.18 nm was measured from the polymer dye laser exhibiting a threshold value of 1.31 $\mu J/mm^2$. The laser operation characteristics of the circular grating resonator are improved through the high accuracy and aspect ratio nanoimprint pattern transfer. Moreover, the mold can be re-used repeatedly, providing a convenient way of mass production and large-scale fabrication of low-cost polymer dye laser arrays. The on-chip dye lasers allow the integration of coherent light sources with other microfluidic and optical functionalities, and provide possibilities for building more complete "lab-on-a-chip" systems.

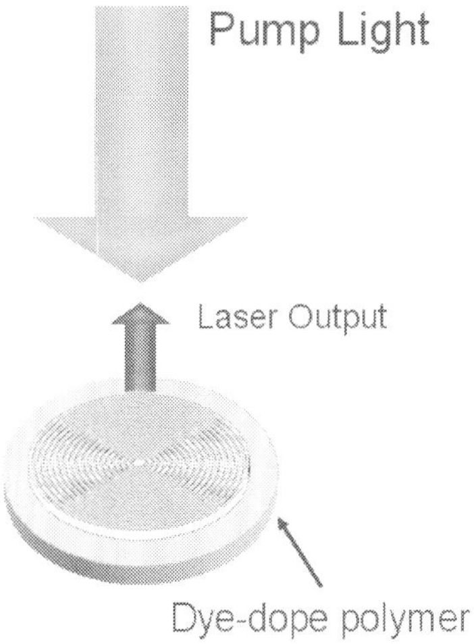

Fig. 1. Schematic diagram of a nanoimprinted circular grating dye laser chip

2. Laser cavity design

Laser (Light Amplification by Stimulated Emission of Radiation) is composed of a gain medium and a resonant optical cavity (A. E. Siegman, 1986; F.P.Schafer, 1990). The gain medium amplifies the beam by stimulated emission, and the resonant cavity provides the feedback necessary for the lasing operation. For our solid-state dye lasers, we choose the circular grating resonator as laser cavities for their unique two-dimensional nature and enhanced lasing performance.

The circular grating structure proposed (T. Erdogan, 1990) and demonstrated (T. Erdogan, 1992) by Erdogan provides a natural 2-D extension of the basic DFB structure. It allows feedback to be applied in all in-plane directions, and the second-order grating couples the emitted radiation perpendicularly out of the surface of the sample. Figure 2 shows a general design of a circular grating distributed feedback structure. A theoretical analysis of circular grating lasers is described in detail elsewhere (C. M. Wu et al., 1991; T. Erdogan et al., 1992; P. L. Greene et al., 2001; G. F. Barlow et al.,2004; G. A. Turnbull et al.,2005) predicting that only the radial propagating components define the modes in the circularly symmetric grating.

The distributed feedback scheme indicates that the gain material is directly implemented in the grating structure. The circular grating DFB structure satisfies the second-order Bragg condition, $\lambda_{Bragg} = n_{eff}\Lambda$, where λ_{Bragg} is the emission wavelength, n_{eff} is the effective index of the waveguide mode, and Λ is the grating period, with an inner cavity providing a quarter- or half-wavelength shift similar to the classical DFB case.

The second-order grating is used to obtain surface emission, because it not only couples counter-propagating radial waves (via second-order Bragg reflection), but also induces coupling of radially propagating waves into the direction normal to the grating surface (via first-order Bragg reflection). The corrugations in the grating structure provide both distributed feedback and output coupling of the guided optical mode via second-order and first-order Bragg scattering.

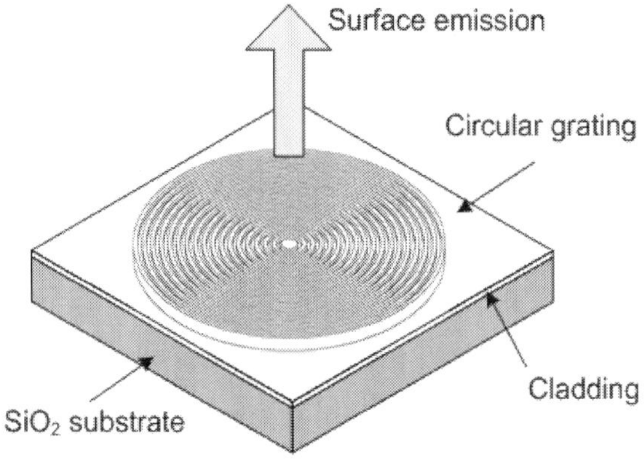

Fig. 2. General design of a circular grating distributed feedback structure

For the theoretical analysis of the grating structure, we use a transfer matrix method appropriate for description of the optical modes of circular grating microcavities. The electromagnetic modes of cylindrical multilayer structures are analyzed in terms of propagating waves, i.e., Hankel functions (A. Yariv, 1997; D. Ochoa et al., 2000). Using the transfer matrix method based 2-D cylindrical model, the spectrum information of the cavity modes can be obtained to analyze the energy confinement in the circular grating structure (A. Jebali et al. 2007).

The design parameters of the circular gratings fabricated are selected based on electromagnetic mode calculations and experimental results. A grating period of 440 nm is chosen to match the second-order Bragg condition. The center defect is a 440 nm diameter gain region. The 400 nm groove depth is defined to ensure maximum confinement, whereas the 200 μm overall diameter of the circular grating and the 50% duty cycle are used to reach the maximum coupling strength (A. Jebali et al., 2004).

3. Method of device fabrication

3.1 Materials

The materials used in the solid state dye laser chip are chosen for three layers of the device: the substrate, the cladding, and the polymer matrix. The substrate for the solid-state dye laser could be Silicon or Silicon dioxide, while the cladding material is Cytop, and the gain medium is dye doped PMMA.

The PMMA (poly(methylmethacrylate)) is a well-known highly transparent thermoplast. In our laser device, we chose PMMA to be the dye host matrix as well as the nanoimprint material. PMMA was selected as the polymer matrix because of its solubility of the dye molecules, as well as its low absorption at the wavelength for activating the dye molecule. Using PMMA in nanoimprint lithography is very common due to its small shrinkage under large changes of temperature and pressure (S. Y. Chou, et al., 1995). The mold release property of PMMA can be improved by spray coating a release agent on its surface (M. Okada, et al., 2009).

The organic laser dye we use in the laser device is Rhodamine 640 (Exciton). This laser dye has excellent stability for its large quantum efficiency and relatively long life time before bleaching. We chose Rhodamine 640 to match its absorption peak with the cavity resonant frequency. To dope the organic dye molecules into the polymer, we mixed PMMA with chlorobenzene, methyl isobutyl ketone, ethanol, and Rhodamine 640 to make a 30 mM solution.

Cytop is a low refractive index perfluoropolymer with special molecular structure. This cyclic fluoro-polymer, poly(1,1,2,4,4,5,5,6,7,7-decafluoro-3-oxa-1,6-heptadiene) is made by Asahi chemicals and used in the electronics industry. It is a hard but amorphous material with Tg ~ 108 °C. We chose Cytop as a cladding material because of its low refractive index (n=1.34). The material system of PMMA and Cytop has previously been used for commercial polymer optical fibers and simple waveguides (Y. G. Zhao et al., 2000), and using Cytop-PMMA as polymer waveguides for sensing has also been reported (B. Agnarsson et al., 2010).

3.2 Mold fabrication

The mold fabrication process is essential, it defines the laser resonator geometry, and the shape of the mold structure and surface roughness will eventually affect the laser device performance. In our experiments, silicon dioxide (SiO_2) was used as the mold material. The grating pattern was defined by electron beam lithography on a LEICA EBPG 5000+ e-beam writer. 8% 495K PMMA was spun on a SiO_2 substrate and baked for 15 min at 170 °C, which formed a 400 nm thick resist layer. The PMMA was exposed by electron beam with proximity correction. Development of patterned PMMA film was carried out in a 1:3 MIBK:IPA (methyl isobutyl ketone and isopropanol alcohol) solution for 1 min. The pattern was subsequently transferred from PMMA into SiO_2 substrate via reactive ion etching (RIE) using fluorine chemistry (CHF_3). The condition of RIE was 20 sccm, 60 mTorr of CHF_3 at 110 W for 15 min. Finally the PMMA residue was removed by sonicating the wafer in Chloroform for 2 min. The SiO_2 etching rate in the CHF_3 RIE process is 30 to 35 nm per min.

The SEM images of both the top view and the angled view of an etched SiO_2 mold of circular grating are shown in Figure 3. In this particular mold, the grating period is 440 nm, with a center defect of 440 nm and an overall diameter of 200 μm, and the trench depth is 400 nm.

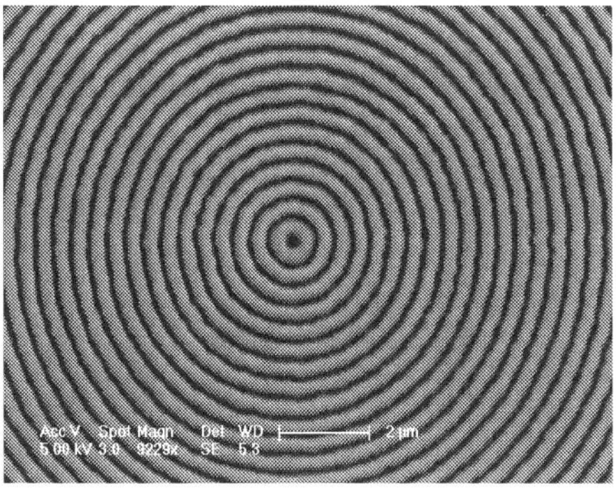

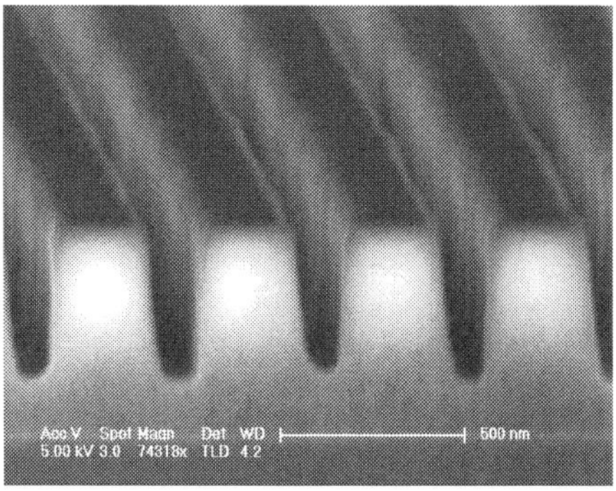

Fig. 3. The SEM images of the top view and the angled view of SiO2 mold

3.3 Laser chip fabrication

The laser chip consists of three layers, the substrate, the cladding, and the polymer matrix. PMMA was selected as the polymer matrix because of the solubility of the dye in PMMA, as well as its low optical absorption within the wavelength range for activating the dye molecules, and its excellent properties for nanoimprint lithography.

To construct the dye laser, a glass substrate (SiO_2) was spin-coated with Cytop, a low-refractive-index material (n=1.34) as the lower cladding to ensure the vertical optical confinement. After an oxygen plasma treatment to improve the adhesion of Cytop to the PMMA, dye-doped PMMA (n=1.49) was spun on top of the Cytop layer to serve as the gain medium.

The Cytop and PMMA preparation process for the nanoimprint process is summarized in Figure 4. We began the fabrication process by depositing a 5 μm thick layer of Cytop (CTL-809M, Asahi Glass) on a silicon dioxide substrate. The deposition of the Cytop was accomplished via a series of spinning and thermal curing steps to ensure flatness and uniformity over the wafer.

First, we spun the Cytop on the substrate at 1500 rpm (adhesion promoters were not necessary). Next, the Cytop was baked at 65 °C for 60 s, 95 °C for 60 s, and 180 °C for 20 min. The ramping of the bake temperature was critical in attaining flat and uniform surfaces. The spinning and baking steps were then repeated two more times, with a final bake at 180 °C for 3 hours. After the chip cooled down, an oxygen plasma treatment (Anatech SP100) of the Cytop was necessary for the adhesion of Cytop to PMMA. We exposed the oxygen plasma to Cytop at an RF power of 80 W and O_2 pressure of 200 mTorr for 30 s.

Next, dye (Rhodamine 640, Exciton)-doped PMMA (30 mM) was spin-coated on top of the Cytop layer at 500 rpm for 15 s and then 5000 rpm for 1 min. This produced a dye-doped polymer thin film with 600 nm thickness as the gain medium. A prebake at 170 °C for 2 min before the nanoimprint process ensured solvents were evaporated and improved the adhesion between the Cytop and PMMA. Then the substrate was ready for the nanoimprint process to define the laser cavity structure.

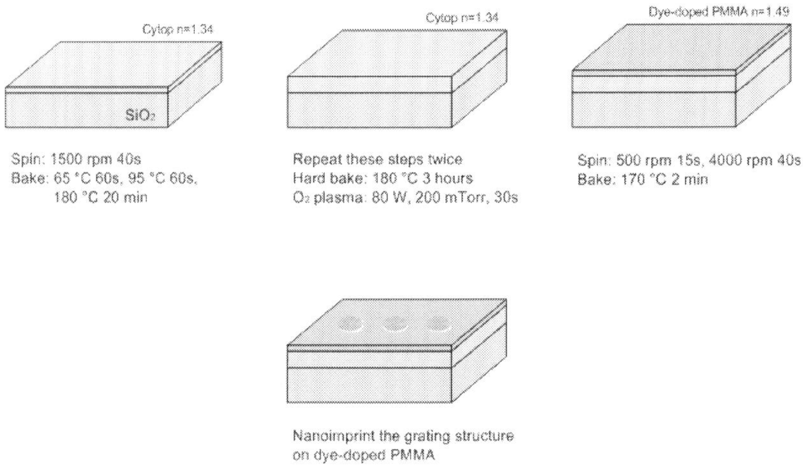

Fig. 4. The schematic procedure of the Cytop and PMMA substrate preparation process

3.4 Nanoimprint process

Nanoimprint lithography exploits the glass transition of polymers to achieve high-fidelity pattern transfer. However, degradation of the light emission efficiency of the organic materials during air exposure at high temperatures presents a challenge in nanoimprint lithography (J. Wang et al., 1999). To solve this problem, a modified nanoimprint method is used to prevent this degradation of the dye-doped PMMA film by sealing the mold and the PMMA substrate into a curable polymer during the imprinting process.

During the nanoimprint process, a mold release reagent such as 1H,1H,2H,2H-perfluorodecyl-trichlorosilane (Alfa Aesar) was also deposited on the dye from the vapor phase to reduce the resist adhesion to the mold. Then, the mold was pressed into the PMMA film by using an automatic mounting press machine (Buehler SimpliMet 1000) at a temperature of 150 °C (above PMMA's glass transition temperature) and a pressure of 1200 psi. After sample cooling, the mold could be easily separated from the patterned polymer laser chip. The nanoimprint process is schematized in Figure 5.

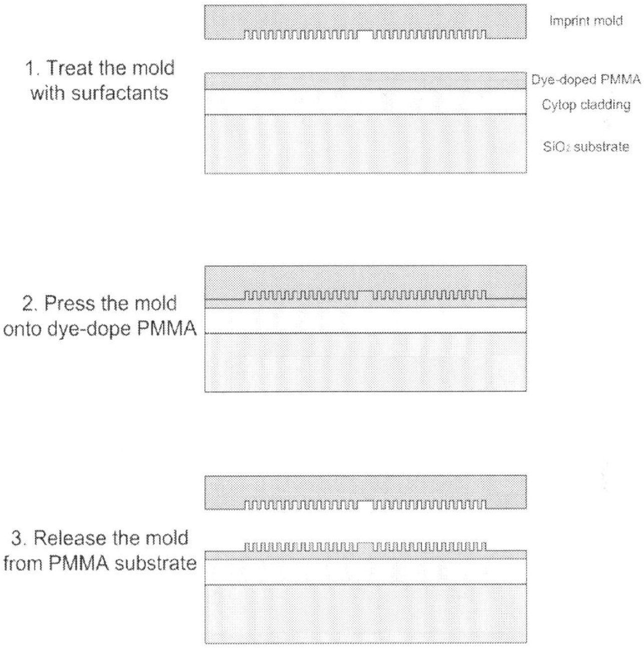

Fig. 5. The schematic nanoimprint process of circular grating polymer dye laser

Figure 6 shows the SEM images of the mold and the imprinted PMMA. From these pictures, we can observe that the structure on the SiO_2 mold is faithfully replicated on the PMMA substrate surface with high resolution. Photoluminescence spectra confirm that there is no degradation of the luminescence performance of the polymer. Compared to various methods of defining nanostructures such as Extreme UV and E-beam lithography, the modified nanoimprint lithography is a suitable method for fabricating dye laser resonator structures, since it will not cause the degradation of fluorophores doped in the polymer matrix. Also nanoimprint lithography is considered a low cost fabrication technique, enabling the mass production of dye laser array devices using a single master mold.

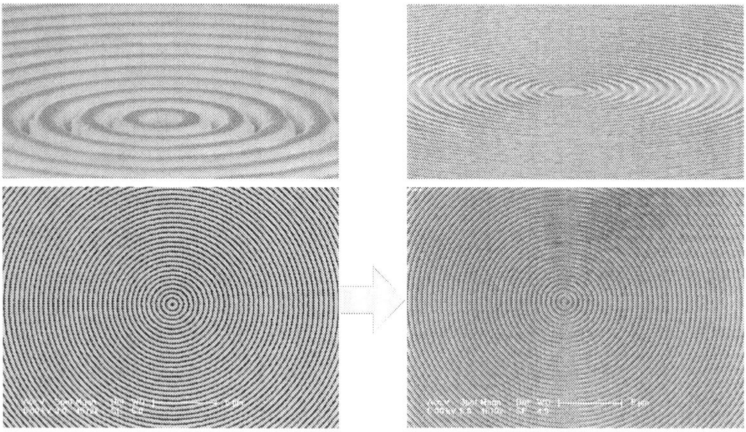

(a) (b)

Fig. 6. SEM images of (a) the SiO_2 mold and (b) the imprinted PMMA film.

4. Results and discussion

The measurement setup of the polymer dye laser chip is shown in Figure 7. The polymer laser chip was optically pumped with 6 ns Q-switched Nd:YAG laser pulses at 532 nm wavelength, focused through a 20X objective to the top side of the chip. A 10X microscope objective was used to collect the emission from the bottom side of the chip and deliver it to a fiber coupled CCD-array-based spectrometer with 0.1 nm resolution (Ocean Optics HR4000).

Fig. 7. The measurement setup of polymer dye lasers

A typical single-frequency lasing spectrum of the dye laser chip is shown in Figure 8. The lasing wavelength is 618.52 nm, and the measured linewidth is 0.18 nm. Lasing occurs near the Bragg resonance, determined by the equation $m\lambda_{Bragg} = 2n_{eff}\Lambda$, where $m = 2$ is the order of diffraction, n_{eff} is the effective refractive index of the propagation mode, and Λ is the grating period. The linewidth near threshold is measured as 0.20 nm, which results in a cavity quality factor (Q) of over 3000. The measured lasing from the solid-state dye laser shows that a high intensity, narrow linewidth, well-defined output beam is achieved by the circular grating resonator. Different lasing wavelength output can be obtained by changing the dye molecule doped in the polymer or varying the period of the grating structure.

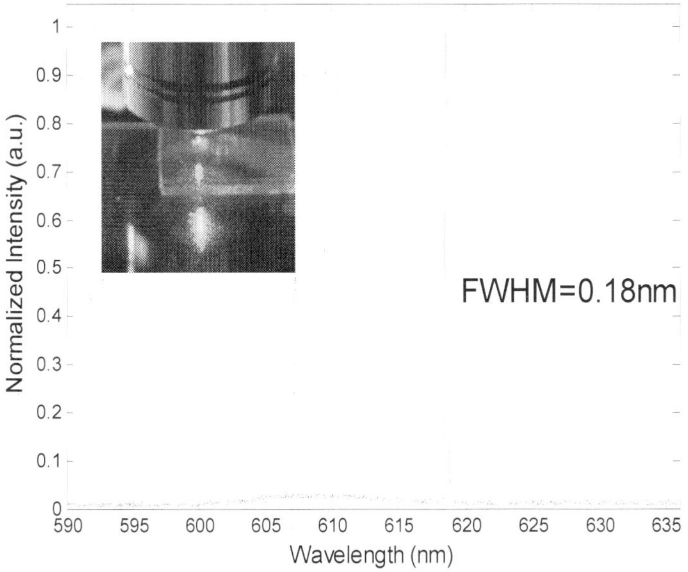

Fig. 8. Nanoimprinted circular grating DFB dye laser spectrum. The measured linewidth is 0.18 nm. Inset: Polymer laser chip excited by Nd:YAG 532 nm laser pulse.

Figure 9 shows the variation of the output laser power as a function of absorbed pump energy. With the absorbed threshold energy of 41.3 nJ, the threshold pump fluence is estimated to be 1.31 $\mu J/mm^2$. This pump intensity is well within the reach of commercial high power blue laser diodes , enabling a self-contained Laser diode pumped device. The polymer laser is pumped from the surface of the chip and the lasing emission is collected from the back side of the chip. The transparency of the substrate, the size and geometry of the laser cavity, and the low threshold match well with the output beams of high power LEDs and Laser diodes. Therefore the replication-molded circular grating geometry represents a very promising structure for the construction of compact LED or Laser diode pumped portable dye lasers.

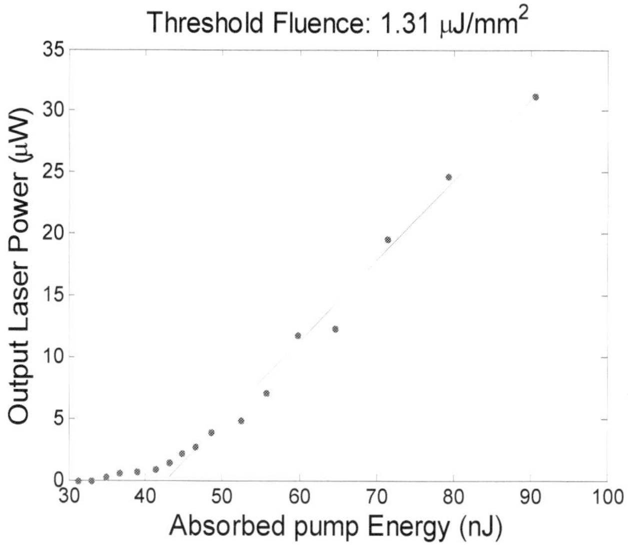

Fig. 9. The output laser power vs. the absorbed pump energy curve. The threshold pump fluence is 1.31 μJ/mm^2

Figure 10 (a) represents the far-field image of the emission pattern recorded by a CCD camera, and Figure 10 (b) shows the far-field radiation patterns of the laser passing through a linear polarizer with different orientation angles. The laser is expected to be azimuthally polarized (R. H. Jordan et al., 1997), as illustrated in the polarization patterns. The azimuthal polarization also results in a zero electrical field (a dark spot) at the center of the laser (T. Erdogan et al., 1992). The Polarization studies of circularly symmetric beams verified theoretical predictions that these beams are azimuthally polarized. In the lasing process, many spatial modes can be excited with their mode thresholds very close to each other (T. Erdogan et al., 1992). The fundamental mode is normally the favored one, because higher order modes do not overlap well with the gain region.

We observe decreases in the laser emission with increasing exposure time. This result is consistent with previous studies on polymer DFB structures (G. Heliotis et al, 2004). The lifetime of polymer dye laser can last over 10^6 shots of pump laser pulse, and if the characterization of the laser device is carried out under vacuum to inhibit photo-oxidation, the lifetime can be further extended (P. Del Carro et al., 2006). Because of the low cost of materials and fabrication, replication molded devices are disposable and may not require a long lifetime. With the mass production capability, nanoimprinted solid-state dye lasers are suitable for disposable light sources for integration in microsystems.

The integration of solid-state dye laser with microfluidic platform is important. Because of its stacked substrate structure, the alignment of surface emitting dye lasers with microfluidic channels would be straight forward. Since optofluidic dye lasers also have great advantages in microfluidics integration, many on-chip liquid dye lasers with distributed feedback structure have been demonstrated (Z. Y. Li et al., 2006; M. Gersborg-Hansen and A. Kristensen, 2006; S. Balslev and A. Kristensen, 2005) by soft lithography (Y. N. Xia and G. M. Whitesides, 1998).

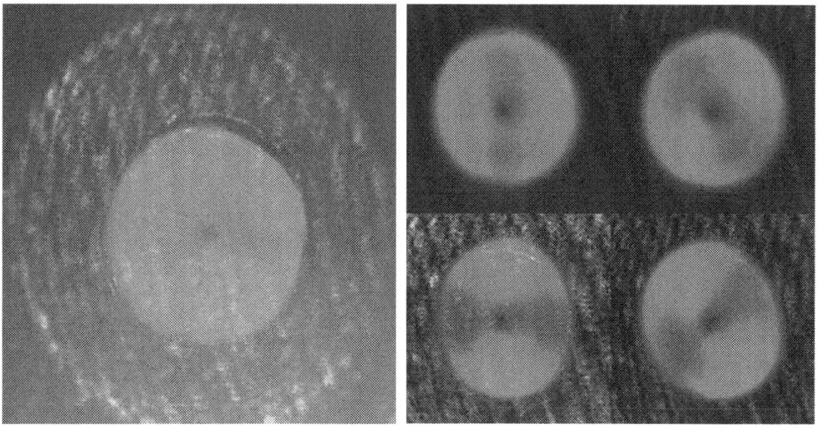

Fig. 10. (a) Far-field image of the emission pattern recorded by a CCD camera. (b) Circular grating DFB laser far-field radiation patterns through a linear polarizer with different orientation angles. The laser emits an azimuthally polarized, well-confined circular beam.

There are many unique properties of the miniaturized liquid dye lasers in microfluidic systems, the mixing and circulating capabilities enable various ways to tune the laser wavelength (J. C. Galas et al., 2005; Z. Y. Li et al., 2006; M. Gersborg-Hansen and A. Kristensen, 2007). Based on this idea, we can make an optofluidic version of the circular grating dye laser, in which the laser dye is dissolved in an organic solvent and flowed through a microfluidic channel with laser resonator embedded (Y. Chen et al., 2009). The flexibility and versatility of microfluidic fabrication enables the large-scale integration of laser arrays in compact devices with more functionality, which allows us to constantly change the dye to increase the device lifetime and to tune the wavelength.

There are still several issues with the current scheme of the solid-state dye laser. Its relatively high excitation power requires a pulsed Nd:YAG laser as the pumping source, and the reduction of pumping threshold is a very challenging problem. Recent studies show that if the cavity length can be reduced to the order of several micrometers, the optical pumping by a low-power light source, such as a laser diode, can be realized (H. Sakata, and H. Takeuchi, 2008; Y. Yang et al., 2008). The lifetime of the device can also be largely increased by optimizing the organic component of the polymer and reducing the lasing threshold. For the structure of the laser device, surface roughness will greatly affect the quality factor and lasing performance. Further improvements of the anti-adhesive properties of the stamp and the optimization of etching parameters will contribute to future devices.

5. Conclusion

In summary, we have described the fabrication of a surface emitting polymer dye laser with circular grating distributed feedback structure using nanoimprint lithography. We have

achieved excitation thresholds as low as 1.31 μJ/mm^2 and FWHM linewidths of 0.18 nm. The technique described here enables the fabrication of low cost, high quality and mass producible laser arrays, which may be deployed as compact and inexpensive coherent light sources for lab-on-a-chip applications such as sensing and spectroscopy.
Future work will be focused on improving the laser cavity quality factor (Q) values with better electromagnetic design, optimizing the dye concentration, and fabricating smoother surfaces. One of the future research directions is to use novel soft lithography technique to develop optofluidic dye lasers based on circular grating geometry, in order to realize dye solution circulating and wavelength tuning. Another direction is to use conductive polymer as the gain medium to enable electrically pumped laser scheme. The ultimate goal is to reduce the lasing threshold to enable the use of LEDs or laser diodes as integrated and inexpensive pump sources for on-chip polymer lasers, and integrate the laser chip into microfluidics for further development of complete "lab-on-a-chip" systems.

6. Acknowledgment

This work is supported by the Caltech DARPA Center for Optofluidic Integration under Contract No. HR0011-04-1-0032 and by the Boeing Corporation.

7. References

Y. Oki, K. Aso, D. Zuo, N. J. Vasa, and M. Maeda, (2002), Wide-wavelength-range operation of a distributed-feedback dye laser with a plastic waveguide," Japanese Journal of Applied Physics Part 1-Regular Papers Short Notes & Review Papers, vol. 41, pp. (Nov 2002), 6370-6374

Y. Oki, S. Miyamoto, M. Maeda, and N. J. Vasa, (2002), Multiwavelength distributed-feedback dye laser array and its application to spectroscopy, Optics Letters, vol. 27, (Jul 2002), pp. 1220-1222

C. Vannahme, S. Klinkhammer, U. Lemmer, and T. Mappes, (2011), Plastic lab-on-a-chip for fluorescence excitation with integrated organic semiconductor lasers, Opt. Express vol. 19, (Apr 2011), pp. 8179-8186

Y. Huang, and S.-T. Wu, (2010), Multi-wavelength laser from dye-doped cholesteric polymer films, Opt. Express, vol. 18, (2010), pp. 27697-27702 (2010)

D. Psaltis, S. R. Quake, and C. H. Yang, (2006), Developing optofluidic technology through the fusion of microfluidics and optics, Nature, vol. 442, (Jul 2006), pp. 381-386

E. Verpoorte, (2003), Chip vision - optics for microchips," Lab on a Chip, vol. 3, (2003), pp. 42n-52n

Y. Oki, T. Yoshiura, Y. Chisaki, and M. Maeda, (2002), Fabrication of a distributed-feedback dye laser with a grating structure in its plastic waveguide, Applied Optics, vol. 41, (Aug 2002), pp. 5030-5035

C. Bauer, H. Giessen, B. Schnabel, E. B. Kley, C. Schmitt, U. Scherf, and R. F. Mahrt, (2001), A surface-emitting circular grating polymer laser, Advanced Materials, vol. 13, (Aug 2001), pp. 1161

G. A. Turnbull, A. Carleton, G. F. Barlow, A. Tahraouhi, T. F. Krauss, K. A. Shore, and I. D. W. Samuel, (2005), Influence of grating characteristics on the operation of circular-grating distributed-feedback polymer lasers, Journal of Applied Physics, (Jul 2005), vol. 98, pp.

S. Y. Chou, P. R. Krauss, and P. J. Renstrom, (1996), Nanoimprint lithography, Journal of Vacuum Science & Technology B, vol. 14, (Nov-Dec 1996), pp. 4129-4133

J. R. Lawrence, P. Andrew, W. L. Barnes, M. Buck, G. A. Turnbull, and I. D. W. Samuel, (2002), Optical properties of a light-emitting polymer directly patterned by soft lithography," Applied Physics Letters, vol. 81, (Sep 2002), pp. 1955-1957

D. Pisignano, L. Persano, P. Visconti, R. Cingolani, G. Gigli, G. Barbarella, and L. Favaretto, (2003), Oligomer-based organic distributed feedback lasers by room-temperature nanoimprint lithography, Applied Physics Letters, vol. 83, (Sep 2003), pp. 2545-2547

D. Pisignano, M. F. Raganato, L. Persano, G. Gigli, P. Visconti, G. Barbarella, L. Favaretto, M. Zambianchi, and R. Cingolani, (2004), The luminescence quantum yield of organic one-dimensional periodic nanostructures, Nanotechnology, vol. 15, (Aug 2004), pp. 953-957

Y. Chen, Z. Li, Z. Zhang, D. Psaltis, and A. Scherer, (2007), Nanoimprinted circular grating distributed feedback dye laser, Appl Phys Lett 91, (Jul 2007), 051109

A. E. Siegman, (1986), Lasers. Mill Valley: University Science Books

F.P.Schafer, (1990), Dye Lasers, 3rd ed. New York.

T. Erdogan and D. G. Hall, (1990), Circularly Symmetrical Distributed Feedback Semiconductor-Laser - an Analysis, Journal of Applied Physics, vol. 68, (Aug 1990), pp. 1435-1444

C. M. Wu, T. Makino, J. Glinski, R. Maciejko, and S. I. Najafi, (1991), Self-Consistent Coupled-Wave Theory for Circular Gratings on Planar Dielectric Wave-Guides, Journal of Lightwave Technology, vol. 9, (Oct 1991), pp. 1264-1277

A. Yariv, (1997), Optical Electronics in Modern Communications, 5th ed. New York: Oxford University Press

D. Ochoa, R. Houdre, M. Ilegems, H. Benisty, T. F. Krauss, and C. J. M. Smith, (2000), Diffraction of cylindrical Bragg reflectors surrounding an in-plane semiconductor microcavity," Physical Review B, vol. 61, (Feb 2000), pp. 4806-4812

A. Jebali, D. Erni, S. Gulde, R. F. Mahrt, and W. Bachtold, (2007), Analytical calculation of the Q factor for circular-grating microcavities, Journal of the Optical Society of America B-Optical Physics, vol. 24, (Apr 2007), pp. 906-915

S. Y. Chou, P. R. Krauss, and P. J. Renstrom, (1995), Imprint of Sub-25 Nm Vias and Trenches in Polymers, Applied Physics Letters, vol. 67, (Nov 1995), pp. 3114-3116

M. Okada, M. Iwasa, K.-i. Nakamatsu, K. Kanda, Y. Haruyama, and S. Matsui, (2009), Nanoimprinting using release-agent-coated resins, Microelectronic Engineering, vol. 86, (Jan 2009), pp. 673-675

Y. G. Zhao, W. K. Lu, Y. Ma, S. S. Kim, S. T. Ho, and T. J. Marks, (2000), Polymer waveguides useful over a very wide wavelength range from the ultraviolet to infrared, Applied Physics Letters, vol. 77, (Nov 2000), pp. 2961-2963

B. Agnarsson, J. Halldorsson, N. Arnfinnsdottir, S. Ingthorsson, T. Gudjonsson, and K. Leosson, (2010), Fabrication of planar polymer waveguides for evanescent-wave sensing in aqueous environments, Microelectron. Eng., vol 87, (May 2010), pp. 56-61

T. Erdogan, O. King, G. W. Wicks, D. G. Hall, E. H. Anderson, and M. J. Rooks, (1992), Circularly Symmetrical Operation of a Concentric-Circle-Grating, Surface-Emitting, Algaas/Gaas Quantum-Well Semiconductor-Laser, Applied Physics Letters, vol. 60, (Apr 1992), pp. 1921-1923

T. Erdogan and D. G. Hall, (1992), Circularly Symmetrical Distributed Feedback Laser - Coupled Mode Treatment of Te Vector-Fields, IEEE Journal of Quantum Electronics, vol. 28, (Mar 1992), pp. 612-623,

P. L. Greene and D. G. Hall, (2001), Effects of radiation on circular-grating DFB lasers - Part I: Coupled-mode equations, IEEE Journal of Quantum Electronics, vol. 37, (Mar 2001), pp. 353-364

P. L. Greene and D. G. Hall, (2001), Effects of radiation on circular-grating DFB lasers - Part II: Device and pump-beam parameters, IEEE Journal of Quantum Electronics, vol. 37, (Mar 2001), pp. 365-371

G. F. Barlow, A. Shore, G. A. Turnbull, and I. D. W. Samuel, (2004), Design and analysis of a low-threshold polymer circular-grating distributed-feedback laser, Journal of the Optical Society of America B-Optical Physics, vol. 21, (Dec 2004), pp. 2142-2150

G. A. Turnbull, A. Carleton, A. Tahraouhi, T. F. Krauss, I. D. W. Samuel, G. F. Barlow, and K. A. Shore, (2005), Effect of gain localization in circular-grating distributed feedback lasers," Applied Physics Letters, (Nov 2005), vol. 87, pp. -

A. Jebali, R. F. Mahrt, N. Moll, D. Erni, C. Bauer, G. L. Bona, and W. Bachtold, (2004), Lasing in organic circular grating structures, Journal of Applied Physics, vol. 96, (Sep 2004), pp. 3043-3049

J. Wang, X. Y. Sun, L. Chen, and S. Y. Chou, (1999), Direct nanoimprint of submicron organic light-emitting structures, Applied Physics Letters, vol. 75, (Nov 1999), pp. 2767-2769

R. H. Jordan, D. G. Hall, O. King, G. Wicks, and S. Rishton, (1997), Lasing behavior of circular grating surface-emitting semiconductor lasers, Journal of the Optical Society of America B-Optical Physics, (Feb 1997), vol. 14, pp. 449-453

G. Heliotis, R. Xia, D. D. C. Bradley, G. A. Turnbull, I. D. W. Samuel, P. Andrew, and W. L. Barnes, (2004), Two-dimensional distributed feedback lasers using a broadband, red polyfluorene gain medium, Journal of Applied Physics, vol. 96, (Dec 2004), pp. 6959-6965

P. Del Carro, A. Camposeo, R. Stabile, E. Mele, L. Persano, R. Cingolani, and D. Pisignano, (2006), Near-infrared imprinted distributed feedback lasers," Applied Physics Letters, vol. 89, (Nov 2006), pp. -

Z. Y. Li, Z. Y. Zhang, T. Emery, A. Scherer, and D. Psaltis, (2006), Single mode optofluidic distributed feedback dye laser, Optics Express, vol. 14, (Jan 2006), pp. 696-701

M. Gersborg-Hansen and A. Kristensen, (2006), Optofluidic third order distributed feedback dye laser, Applied Physics Letters, vol. 89, (Sep 2006), pp.

S. Balslev and A. Kristensen, (2005), Microfluidic single-mode laser using high-order Bragg grating and antiguiding segments, Optics Express, vol. 13, (Jan 2005). pp. 344-351

Y. N. Xia and G. M. Whitesides, (1998), Soft lithography, Annual Review of Materials Science, vol. 28, (1998), pp. 153-184

J. C. Galas, J. Torres, M. Belotti, Q. Kou, and Y. Chen, (2005), Microfluidic tunable dye laser with integrated mixer and ring resonator, Applied Physics Letters, vol. 86, (Jun 2005), pp. -

Z. Y. Li, Z. Y. Zhang, A. Scherer, and D. Psaltis, (2006), Mechanically tunable optofluidic distributed feedback dye laser, Optics Express, vol. 14, (Oct 2006), pp. 10494-10499

M. Gersborg-Hansen and A. Kristensen, (2007), Tunability of optofluidic distributed feedback dye lasers, Optics Express, vol. 15, (Jan 2007), pp. 137-142

D. V. Vezenov, B. T. Mayers, R. S. Conroy, G. M. Whitesides, P. T. Snee, Y. Chan, D. G. Nocera, and M. G. Bawendi, (2005), A low-threshold, high-efficiency microfluidic waveguide laser, Journal of the American Chemical Society, vol. 127, (Jul 2005), pp. 8952-8953

Y. Chen, Z. Y. Li, M. D. Henry, and A. Scherer, (2009), Optofluidic circular grating distributed feedback dye laser," Appl Phys Lett 95, (Jul 2009), 031109

H. Sakata, and H. Takeuchi, (2008), Diode-pumped polymeric dye lasers operating at a pump power level of 10 mW, Applied Physics Letters 92, (Mar 2008), 113310

Y. Yang, G. A. Turnbull, and I. D. W. Samuel, (2008), Hybrid optoelectronics: A polymer laser pumped by a nitride light-emitting diode, Applied Physics Letters 92, (Apr 2008), 163306

11

Application of Nanoimprint Lithography to Distributed Feedback Laser Diodes

Masaki Yanagisawa
Sumitomo Electric Industries, Ltd.
Japan

1. Introduction

The recent growth of information-communication facilities such as the Internet, mobile telecommunication and video-on-demand services has led to explosion of worldwide communication traffic, which increase demand for faster and denser communication infrastructures including optical communication networks. Distributed feedback laser diodes (DFB LDs) have been widely used as optical sources in networks because of their high selectivity and stability of wavelength. Although they had limitedly been used in long-haul and high-speed network at one time, they are now increasingly required in metro and end-user fields because of the increasing traffic. Thus, necessity for inexpensive DFB LDs increases rapidly.

The characteristics of a DFB LD with uniform (constant period) gratings depend on the grating phase at the cleaved facet (Matsuoka et al., 1984). The variation of characteristics with the facet phase is a serious issue in view of productivity and usability of the LDs. One of the most effective ways of reducing the facet phase effect is to adopt phase-shifted gratings instead of uniform gratings (Kaden et al., 1992). The uniformities of the LD characteristics such as mode-stability and output power are improved by adopting phase-shifted gratings, thus the yield of LDs increase and their production cost is effectively reduced.

In general, there are various fabrication methods for diffraction gratings of DFB LDs, for example, interference exposure, electron beam lithography (EBL), and optical projection exposure. Interference exposure cannot feasibly be used for fabricating phase-shifted gratings, because it exclusively generates exposure patterns with a uniform bright-and-dark period. Although EBL has sufficient resolution to be used for phase-shifted gratings, exceedingly expensive apparatus is necessary for volume production with sufficient throughput. For the optical projection method, a forefront stepper having sufficient resolution for gratings is also expensive, and the cost is too high for fabricating DFB LDs, of which production volume is relatively small compared to that of such semiconductor devices as LSIs.

Nanoimprint lithography (NIL) has been studied by many organizations since the middle of the 1990s. Chou et al. indicated that sub-10-nm features could be formed by imprint, which started the era of NIL technology (Chou et al., 1995). A novel method of NIL using a UV-curable resin was introduced by Haisma et al (Haisma et al., 1996), and Bailey et al.

demonstrated a step-and-repeat imprint method named SFIL (Bailey et al., 2000). NIL is a simple method applicable to forming fine patterns smaller than 100 nanometers, so it is studied as a candidate for a next-generation fabrication technique in many application fields such as storage devices, displays, optical devices, MEMS, semiconductor and so on.

Since 2004, we have investigated the use of NIL for fabricating diffraction gratings, because the new technology have been considered as an attractive solution to the above issues concerning the fabrication of phase-shifted gratings because of its high resolution, throughput, and low cost. In 2009, we have reported that our NIL method have been successfully applied to fabricating quarter-wavelength shifted DFB LDs (Yanagisawa et al., 2009). To the best of our knowledge, it is the unprecedented demonstration of NIL application to DFB LDs in wafer-scale fabrication having the potential of near-future volume production.

2. Fabrication process

Our fabrication method is based on our conventional production process of DFB LDs except for the formation of diffraction gratings by NIL, which is a strong point because we make the best use of the mature process and minimize the risk of unpredictable difficulties induced by adopting new methods.

From an early stage of this study, we have used UV-NIL rather than other imprint methods such as thermal NIL and soft lithography, because UV-NIL has advantages of high over-lay alignment accuracy, low imprint pressure and high throughput. Over-lay alignment is especially important in view of mixing and matching the NIL method with the conventional LD fabrication. Imprint pressure is desirable as low as possible, because it directly influences crystal damage of a substrate as described later.

2.1 Mold design

One of the major issues of applying NIL process to DFB LDs is preparation of molds (templates). In general, NIL molds are fabricated by utilizing the forefront method such as electron-beam lithography (EBL), whose throughput is much lower than other optical lithography methods. Thus, long delivery time and high cost for mold fabrication are serious problems. DFB LDs generally used in optical networks vary in emission wavelengths from 1.3 μm to 1.6 μm, thus corresponding periods of diffraction gratings accordingly vary from 200 nm to 250 nm. If a mold were designed with containing the sole type of grating period, we would have to prepare a large number of molds for covering all wavelength range. It would lead to insurmountably high initial costs for DFB LD fabrication.

We have developed a novel NIL method with a new concept of mold design named "VARI-mold." In the new process, one mold can be used for fabricating various types of DFB LDs with different wavelengths. The prepared mold contains more than 9000 grating patterns in an imprint field, consisting of about 900 subfields, and each subfield contains more than 10 different types of grating patterns with the period from 194 nm to 248 nm exemplified in Fig. 1. As described in the next section, various types of LDs are fabricated by utilizing one VARI-mold, leading to reducing the cost for mold preparation.

The molds used in this study are fabricated by a reticle fabrication method and an etching procedure.

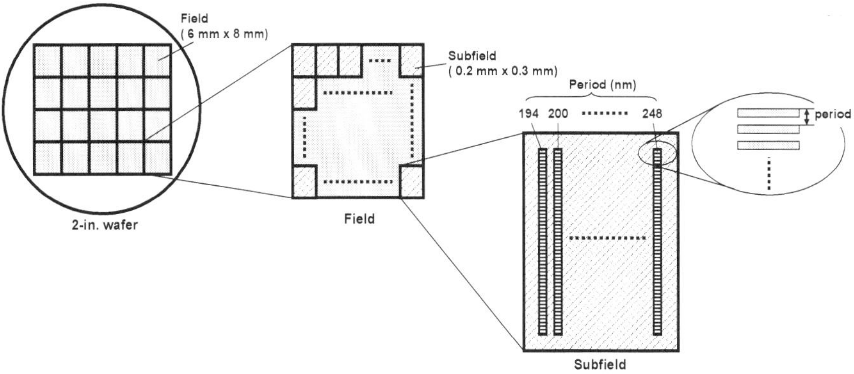

Fig. 1. Schematic structure of a "VARI-mold"

2.2 Wafer fabrication

In general, commercially available compound semiconductor substrates used for optical devices, such as GaAs, InP, and GaN, have large undulations compared with Si substrates widely used for electronic devices such as LSIs and memories. For example, 2-in. diameter InP substrates, which are generally used for fabricating DFB LDs, typically have over 3 μm in total thickness variation. If such undulating substrates are applied to NIL, the mold probably come in contact with a limited portion of the substrate, in which case severe nonuniformity of the residual layer thickness will lead to large variations of the transferred figures in the imprinted area. Thus, we have formed diffraction gratings by using a reverse-tone nanoimprint in order to suppress the variation of residual layer thickness resulting from the undulation of substrates (Miller et al., 2005).

A schematic structure and a layer structure of the LD are shown in Fig. 2. We have prepared a 2-in. InP wafer with epitaxial layers including a grating layer, an active layer, a buffer layer and a lower cladding layer grown by metalorganic vapor phase epitaxy (MOVPE).

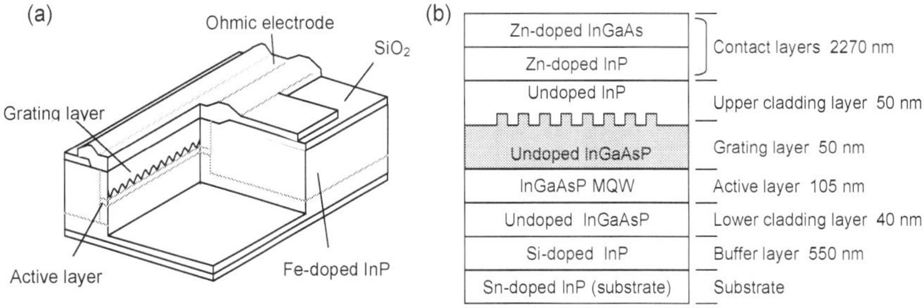

Fig. 2. Schematic structure of a DFB LD (a) and crystal layer structure of the prepared wafer (b).

The fabrication procedure is shown in Fig. 3. First, a 50-nm-thick SiN film is deposited on the wafer by plasma-enhanced chemical vapor deposition (PE-CVD). Next, a primer material is spin-coated in order to increase adhesion between UV-curable resin and the SiN film. Then, UV-NIL using a VARI-mold is conducted to form grating patterns in the UV-curable resin layer (Figs. 3(a) and 3(b)). In this study, imprinting was performed in 16 fields on the wafer with a step-and-repeat equipment. The imprint pressure before UV exposure is approximately 0.1 MPa, and the exposure time is 20 seconds. After the imprint, Si-containing resin is spin-coated in order to cover and planarize the grating corrugations (Fig. 3(c)). The thickness of the planarization layer is approximately 200 nm. Subsequently, the planarization layer is etched by reactive ion etching (RIE) until the tops of the corrugations are revealed (Fig. 3(d)). After that, the revealed layer is selectively etched and penetrated until the SiN masks are revealed (Fig. 3(e)). This penetration etching is one of the essential techniques of this process, so detailed in next section. The formed resin patterns are used as masks for the subsequent etching, transferring the grating patterns to the SiN film (Fig. 3(f)). Next, the resin layers are removed by O_2 plasma etching. After that, we use inductively coupled plasma RIE (ICP-RIE) with CH_4 / H_2 gas for etching of the crystal layer (Fig. 3(g)). Finally, the SiN masks are stripped by a wet chemical process using HF solution, and the diffraction grating structure is achieved. After the formation of the gratings, an upper cladding layer and contact layers are formed on the grating layer by MOVPE. The contact layers consist of InP and InGaAs layers. Then, stripe patterns of SiO_2 are formed on the contact layer by using chemical vapor deposition (CVD) and conventional photolithography method in order to define the cavities of the DFB LDs. In this step, cavity stripes are overlaid to the grating patterns having a specific period corresponding to the required wavelength of LDs. For example, when we would fabricate DFB LDs with the wavelength of 1310 nm, stripe patterns had to be aligned onto 200 nm-period gratings (Fig. 4). As a matter of course, different types of LDs with various wavelengths can be achieved simultaneously on a wafer provided that we adjust the alignment of the cavity stripe layer in each imprint field. The stripe patterns of SiO_2 are used as masks for subsequent crystal etching by ICP-RIE with CH_4 / H_2 gas. In this etching step, all unused grating patterns (excepting the selected one under the stripe) are removed. After that, Fe-doped InP is selectively grown onto the etched area as an insulating layer by MOVPE. Subsequently, a SiO_2 film is deposited as a passivation layer, in which contact holes are formed by selective etching by RIE. Finally, metal electrodes are formed by high-vacuum evaporation and a lift-off method.

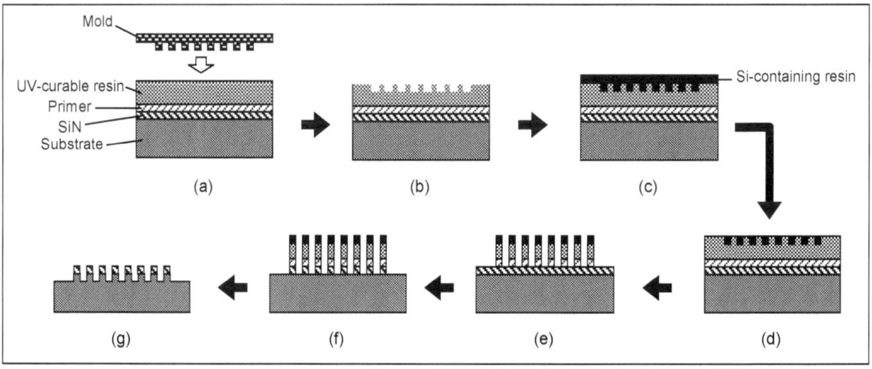

Fig. 3. Fabrication process of diffraction gratings.

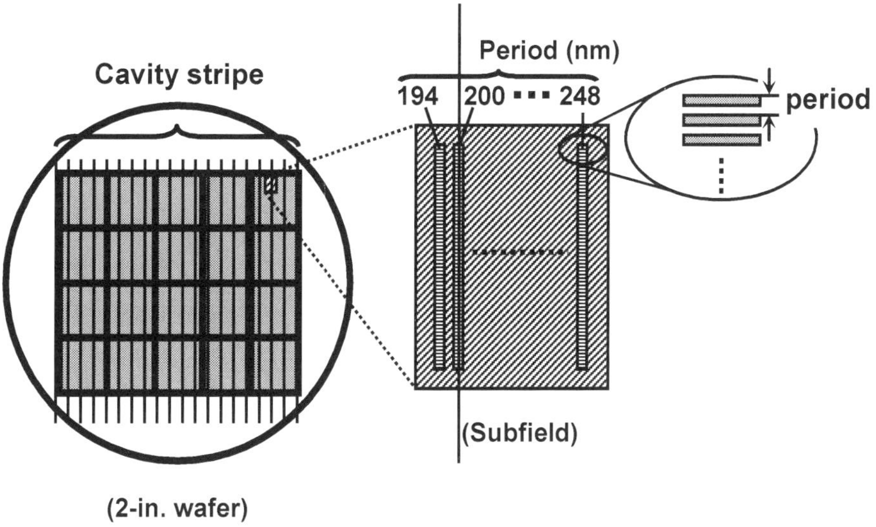

Fig. 4. Concept of cavity stripe delineation.

2.3 Etching process

As mentioned above, the resin etching is one of the key techniques of our fabrication process. Nonuniformity of the linewidth of corrugations after the penetration etching leads to inhomogeneity of the grating figures (Fig. 3(e)), resulting in yield reduction of DFB LDs.

We have developed the etching method using a low-temperature ICP-RIE system in order to achieve highly-uniform and highly-repeatable grating fabrication (Tsuji et al., 2011). The etching method is characterized by the etching gas and the low-temperature substrate stage. Oxygen and nitrogen are used as the etching gas, and the substrate stage is controlled with the temperature from 260 K to 270 K. The both features contributes to suppress the undercut of the UV-curable resin during the penetration etching, resulting from the sidewall effect produced by the optimized plasma condition and substrate temperature (Kure et al., 1991; Kinoshita et al., 1999). The optimized etching condition is shown in Table 1, and a cross-sectional view of the grating after the penetration etching observed by a scanning electron microscope (SEM) is shown in Fig. 5.

Gas flow rate	N_2	9 sccm
	O_2	1 sccm
Bias power		70 W
ICP power		250 W
Substrate temperature		263 K

Table 1. Etching condition of the penetration etching.

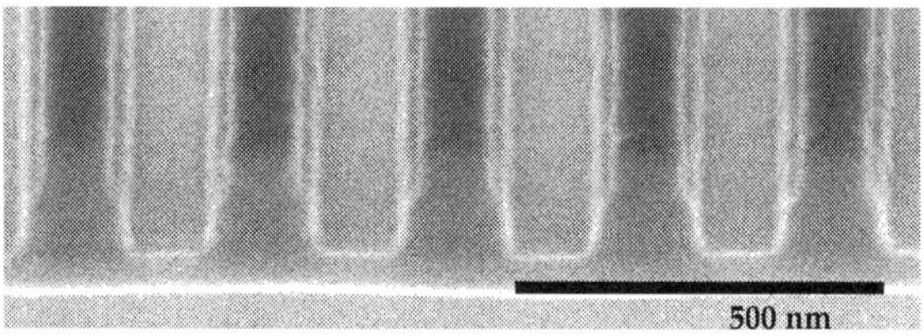

Fig. 5. SEM image of the diffraction grating after the penetration etching.

We have evaluated the linewidth uniformity of the corrugations within 6 wafers. Figure 6 is the histogram of the linewidth, indicating the standard deviation is less than 4 nm.

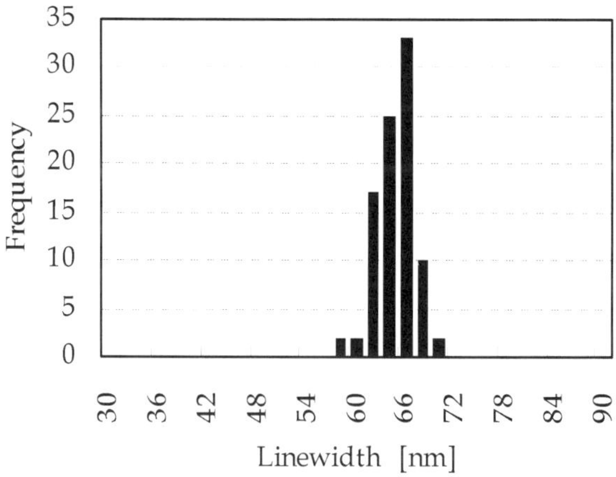

Fig. 6. Histogram of the corrugation linewidth within 6 wafers after the penetration etching.

3. Results

When we use NIL for fabricating DFB LDs, we have paid attention to mechanical damage in epitaxial layers by imprint pressure, because compound semiconductor crystals used for LDs are easily damaged by mechanical stress, leading to severe deterioration in long-term reliability of device characteristics (Fukuda et al., 1985). In this study, photoluminescence (PL) method have been used for evaluation of the crystal damage, and simple 2-D simulation is conducted as qualitative analysis of distribution of the PL intensity.

Furthermore, we have also focused on the accuracy of the grating period, because it dominantly influences the emission wavelength which is the essential parameter of LD characteristics. The emission wavelength directly impacts on the yield of DFB LDs, so its repeatability is required to be sufficiently high for volume production. We have verified the accuracy of the grating period by measuring diffraction angles described below.

Finally, we have fabricated quarter-wavelength shifted DFB LDs by utilizing the new fabrication process combining NIL and the conventional LD process, and evaluated their characteristics including long-term reliability.

3.1 Influence of imprint pressure

3.1.1 Photoluminescence intensity

In order to evaluate the mechanical damage in semiconductor crystal induced by imprinting pressure, we have investigated deterioration of PL intensities from the epitaxial layers. We prepared two indium-phosphide substrates with epitaxial layers and a blank (with no patterns) mold. We compared photoluminescence intensities between the two samples: imprinted with UV-curable resin between the mold and the substrate, and imprinted without resin. The field size of the blank mold used here is 10 mm x 10 mm. The evaluation results of PL intensities are shown in Fig. 7. Field size of the blank mold used here is 10 mm x 10 mm. Imprinting pressure is 0.8 MPa for both samples. The sample without resin shows evident deterioration of PL intensities indicated in dark (green-like) colors in Fig. 7. The deteriorations of intensity are found mainly in the edge of the imprinted area. This means that the imprinting pressure concentrates near the edge of the mold. On the other hand, no evident deterioration is found in the sample with resin. These results indicate that the resin functions such as a cushion to prevent severe damage in epitaxial layers by imprinting pressure.

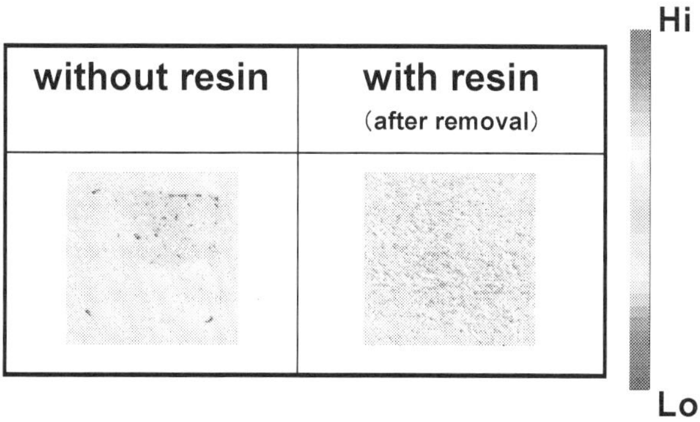

Fig. 7. PL intensities of the epitaxial layers after imprints. The field size of the imprint is 10 mm x 10 mm.

3.1.2 Simulation

We have calculated how resin between a mold and a substrate influences on mechanical stress induced in the substrate. We used a simple model for the finite element method (FEM) as described in Table 2 and Fig. 8. In this study, UV-curable resin is to be considered as liquid having nonlinear viscoelasticity; however, elastomer is substituted for resin because we do not have an adequate tool for dynamic simulation of viscoelastic material. Thickness and the Young's modulus of the elastomer are assumed of 50 nm and 1000 MPa, respectively. Imprinting pressure is 1×10^7 MPa. The elastomer and the other materials are connected with common FEM nodes at the boundaries.

Figure 8 shows two-dimensional distribution of von Mises stress in molds and substrates. For the sample without elastomer, mechanical stress is concentrated in periphery of the edge of the mold [Fig. 8(a)]. On the other hand, when the elastomer is supposed between the mold and the substrate, concentration of stress on the surface of the substrate is clearly suppressed [Fig. 8(b)]. These results are qualitatively consistent with the evaluation using PL described in above section.

		Young's modulus [MPa]	Poisson ratio
Mold	SiO_2	70000	0.17
Substrate	InP	60700	0.36
Liquid	Elastomer	1000	0.49

Table 2. Mechanical properties used for the stress simulation.

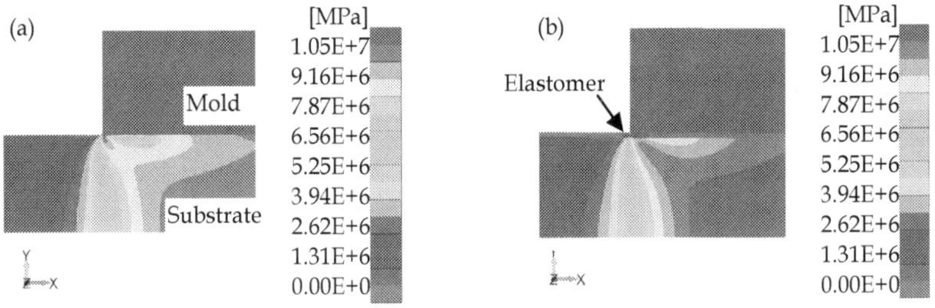

Fig. 8. Distribution of von Mises stress without (a) and with (b) elastomer between a mold and a substrate.

3.2 Grating figure

A SEM image of the fabricated diffraction gratings is shown in Fig. 9. It demonstrates that the line edge roughness (LER) of the gratings is markedly low, indicating that the LER of the grating corrugations in the master mold is sufficiently suppressed and that the master patterns are precisely transferred to the substrate by the imprinting and the subsequent etching process. The depth of the diffraction gratings was measured to be approximately 15 nm by an atomic force microscope.

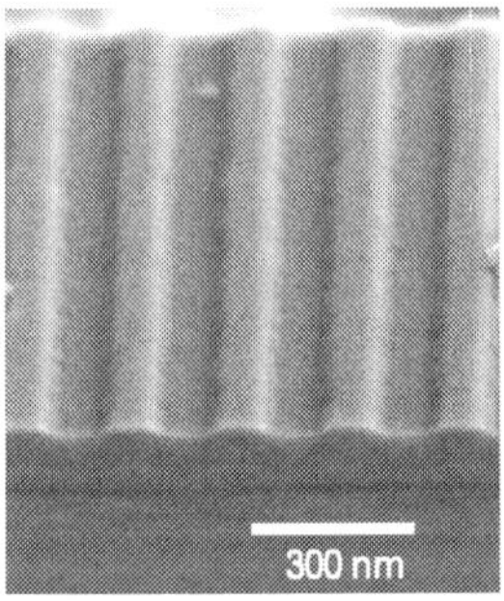

Fig. 9. SEM image of diffraction gratings.

3.3 Reproducibility of the grating period

The diffraction gratings with the period from 194 to 204 nm have been formed on a 2-in. InP wafer. The grating periods have been verified by measuring the diffraction angles of the transferred grating patterns. The wavelength of the incident light is 364.8 nm, and the measurement error of the grating period is estimated less than 0.03 nm. Figure 10 shows a correlation between the grating periods of the mold and those of the transferred patterns. The difference between both of the values is less than 1%, and the wafer-to-wafer variation of the periods is less than 0.2 nm through the 6 wafers. This shows that the NIL process has high reproducibility in transferring grating patterns.

3.4 Device characteristics

We have evaluated DFB LDs fabricated by our novel process in order to verify the basic characteristics and their uniformity in 2-in. wafer. The nominal wavelength of the measured LDs is 1490 nm, and the corresponding grating period is approximately 232 nm (excluding Fig. 12).

Figure 11 shows the dependence of the optical output and the slope efficiency on the supplied current for a typical phase-shifted DFB LD fabricated in this study. The threshold current and the slope efficiency at room temperature were measured to be 8 mA and 0.28 W/A, respectively, which are comparable to those of typical phase-shifted DFB LDs fabricated by utilizing the conventional EBL process.

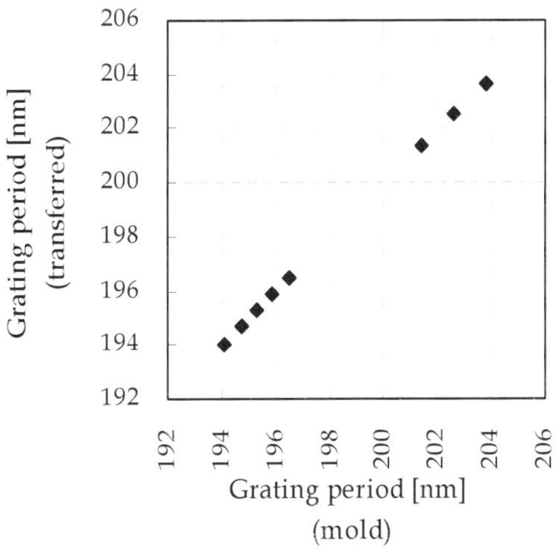

Fig. 10. Correlation between the designed (horizontal) and the measured (vertical) grating periods.

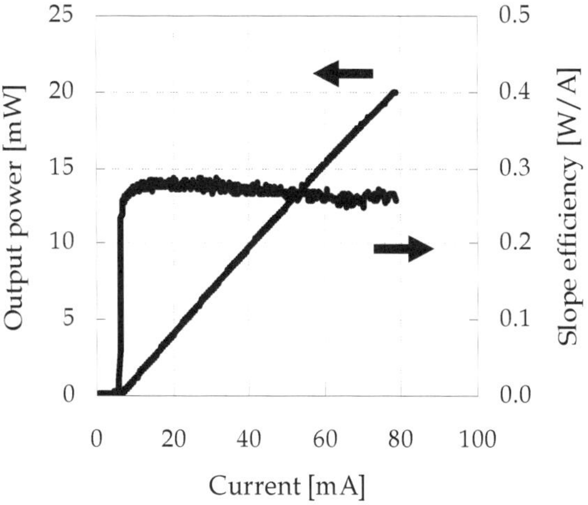

Fig. 11. Supplied current vs optical output and slope efficiency.

Figure 12 shows the oscillation spectrum of a phase-shifted DFB LD. The resolution of the wavelength is 0.02 nm. This demonstrates that the peak wavelength corresponds to the Bragg wavelength at the center of the stopband, indicating that the phase-shifted gratings function properly.

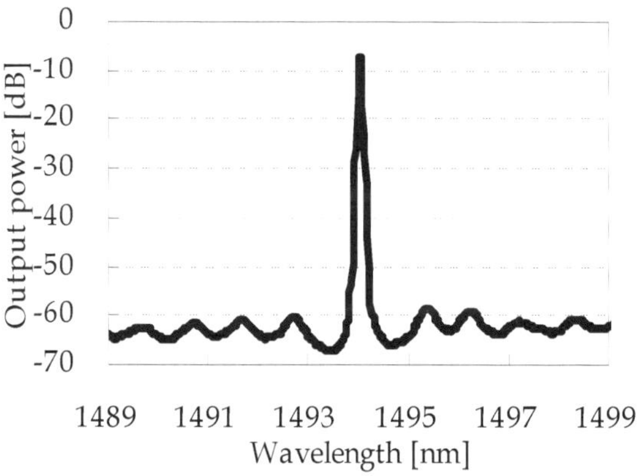

Fig. 12. Oscillation spectrum of the DFB LD.

Figure 13 shows the oscillation spectra of DFB LDs with the grating period of 195 nm and 202 nm, which are simultaneously fabricated on a wafer. This demonstrates that the peak wavelengths correspond to the respective Bragg wavelengths corresponding to each grating period.

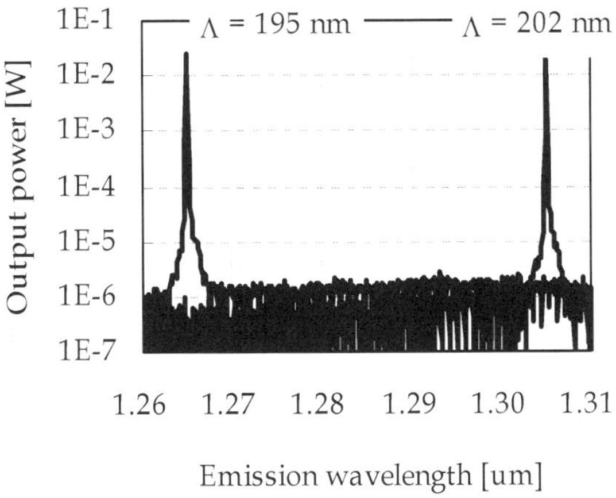

Fig. 13. Oscillation spectra of the DFB LDs with the grating period (Λ) of 195 nm and 202 nm.

We have compared the side-mode suppression ratio (SMSR) of phase-shifted LDs fabricated by NIL with those fabricated by EBL. SMSR is one of the parameters indicating the stability of the single-mode emission of DFB LDs. More than 300 LDs randomly sampled from respective 2-in. wafers have been evaluated. Histograms of SMSR for both types of LD are shown in Fig. 14, which demonstrate that they show comparable variations in SMSR.
These results indicate that DFB LDs fabricated here by utilizing NIL have comparable characteristics and their uniformities to ones by conventional EBL process.

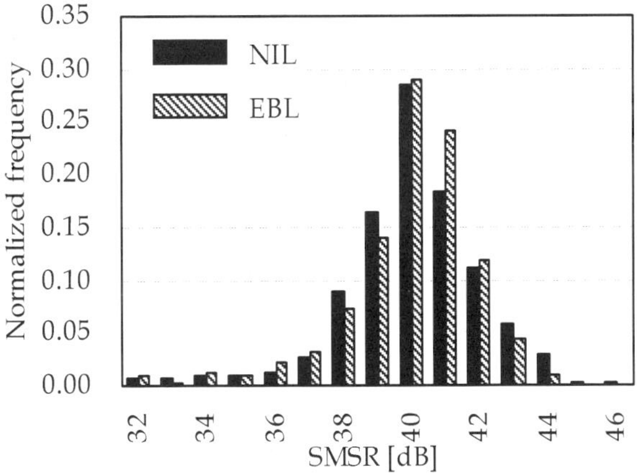

Fig. 14. Histograms of SMSR for phase-shifted LDs fabricated by NIL and by EBL. The standard deviations of SMSR are almost the same, 2.0 and 1.8, respectively.

3.5 Reliability

We have investigated long-term reliability of fabricated DFB LDs. Figure 15 shows the time-dependent change in the threshold current of phase-shifted LDs with the output power of 10 mW at the ambient temperature of 358 K. The number of samples is 18. The change in the threshold current after 5000 hours is less than ±1%, indicating that the phase-shifted DFB LDs fabricated in this study have high stability in lasing characteristics.

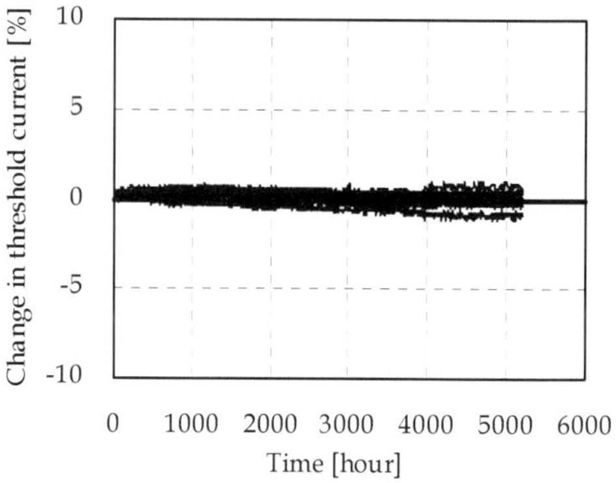

Fig. 15. Time-dependent change in the operation current of the DFB LDs with the output power of 10 mW at the ambient temperature of 358 K.

4. Conclusion

We have successfully demonstrated fabrication of phase-shifted DFB LDs by utilizing NIL process, which have comparable characteristics and their uniformities to ones fabricated by conventional EBL process. Fabricated DFB LDs have shown high stability of characteristics in long-term reliability test. We have also demonstrated the feasibility of the VARI-mold, which can be used for the fabrication of DFB LDs with various wavelengths, indicating that we can drastically reduce the cost of molds in our mass production phase in the near future. Considering the results above, we conclude that NIL is a promising candidate of the production technique for phase-shifted DFB LDs featuring low cost and high throughput.

NIL is expected to be used as a fabrication process for many applications. However, there are still some difficulties with its use as a mass-production process, for example, defects and low throughput in patterned media, defects and poor alignment accuracy in semiconductor lithography, and the necessary of increasing field size and throughput in displays. Although those difficulties may be common to the fabrication of DFB LDs, they would not be insurmountable problems. Even if defects in the imprinted pattern influence the yield of LDs, failed chips could be easily rejected because each dye is as small as approximately 300 μm. There is no need for a larger field size than a 3-in.-diameter circle, because the diameter of compound semiconductor substrates used for LDs is 3 inches or smaller. Regarding alignment accuracy, an error of up to approximately 5 μm is acceptable. Even if throughput is limited to less than one wafer per hour, NIL would still have a higher throughput than EBL.

As described above, NIL is an effective and promising method for fabricating phase-shifted DFB LDs and is expected to have the advantage of mass-production capability in the near future. We conclude that NIL has high potential for fabricating DFB LDs, and we expect that NIL will be used for fabricating various optical devices consisting of nanostructures.

5. Acknowledgment

We would like to appreciate Dr. Taniguchi and his laboratory members at Tokyo University of Science for cooperative works and fruitful discussion.

6. References

Bailey, T.; Choi, B. J.; Colburn, M.; Meissl, M.; Shaya, S.; Ekerdt, J. G.; Sreenivasan, S. V. & Willson, C. G. (2000). Step and flash imprint lithography: Template surface treatment and defect analysis, *J. Vac. Sci. Technol.*, B 18, pp. 3572-3577, ISSN 1071-1023

Chou, S. Y.; Krauss, P. R. & Renstrom, P. J. (1995). Imprint of sub-25 nm vias and trenches in polymers, *Appl. Phys. Lett.*, Vol. 67, pp. 3114-3116, ISSN 1077-3118

Fukuda, M. & Iwane, G. (1985). Degradation of active region in InGaAsP/InP buried heterostructure lasers, *J. Appl. Phys.*, Vol. 58, pp. 2932-2936, ISSN 0021-8979

Haisma, J.; Verheijen, M.; Heuvel, K. and Berg, J. (1996). Mold-assisted nanolithography: A process for reliable pattern replication, *J. Vac. Sci. Technol.*, B 14, pp. 4124-4128, ISSN 1071-1023

Kaden, C.; Griesinger, U.; Schweitzer, H.; Pilkuhn, M. H. & Stath N. (1992). Fabrication of nonconventional distributed feedback lasers with variable grating periods and phase shifts by electron beam lithography, *J. Vac. Sci. Technol.*, B 10, pp. 2970-2973, ISSN 1071-1023

Kinoshita, H.; Yamauchi, A. & Sawai, M. (1999). High-resolution resist etching for quartermicron lithography using O_2/N_2 supermagnetron plasma, *J. Vac. Sci. Technol.*, B 17, pp. 109-112, ISSN 1071-1023

Kure, T.; Kawakami, H.; Tachi, S. & Enami, H. (1991). Low-Temperature Etching for Deep-Submicron Trilayer Resist, *Jpn. J. Appl. Phys.*, 30, pp. 1562-1566, ISSN 1347-4065

Matsuoka, T.; Nagai, H.; Noguchi, Y.; Suzuki, Y. & Kawaguchi, Y. (1984). Effect of the grating phase at the cleaved facet on DFB laser properties, *Jpn. J. Appl. Phys.* 23, pp. L138-L140, ISSN 1347-4065

Miller, M.; Doyle, G.; Stacey, N.; Xu, F.; Sreenivasan, S. V.; Watts, M. & LaBrake, D. L. (2005). Fabrication of nanometer sized features on non-flat substrates using a nano-imprint lithography process, *Proc. SPIE 5751*, pp. 994-1002

Tsuji, Y.; Yanagisawa, M.; Yoshinaga, H.; Inoue, N. & Nomaguchi, T. (2011). Highly uniform fabrication of diffraction gratings for distributed feedback laser diodes by nanoimprint lithography, *Jpn. J. Appl. Phys.* 50, 06GK06, ISSN 1347-4065

Yanagisawa, M.; Tsuji, Y.; Yoshinaga, H.; Kono, N. & Hiratsuka, K. (2009). Application of Nanoimprint Lithography to Fabrication of Distributed Feedback Laser Diodes, *Jpn. J. Appl. Phys.*, Vol. 48, 06FH11, ISSN 1347-4065

Yanagisawa, M.; Tsuji, Y.; Yoshinaga, H.; Kono, N. & Hiratsuka, K. (2009). Evaluation of nanoimprint lithography as a fabrication process of phase-shifted diffraction gratings of distributed feedback laser diodes, *J. Vac. Sci. Technol.*, B 27, pp. 2776-2780, ISSN 1071-1023

Yanagisawa, M.; Tsuji, Y.; Yoshinaga, H.; Kouno, N. & Hiratsuka, K. (2011). Wafer-level fabrication of distributed feedback laser diodes by utilizing UV nanoimprint lithography, *Proc. SPIE*, 7970, pp. 797014, ISSN 0277-786X

12

Guided-Mode Resonance Filters Fabricated with Soft Lithography

Kyu J. Lee[1], Jungho Jin[2], Byeong-Soo Bae[2] and Robert Magnusson[1]
[1]*Department of Electrical Engineering, University of Texas at Arlington, Texas*
[2]*Department of Materials Science and Engineering,*
Korea Advanced Institute of Science and Technology, Daejeon
[1]*USA*
[2]*Republic of Korea*

1. Introduction

Methods of microscale and nanoscale patterning can be applied to fabricate a variety of optical devices. Periodic layered structures are found in integrated optics, communication systems, spectroscopy, lasers, and in many other important optical systems. Diffractive optical elements and photonic crystals consist of fine periodic patterns affecting the spectrum, polarization, phase, and amplitude of light. Often, holographic interferometry, or direct electron-beam patterning, is used to define the periodic structure. As an alternative method, soft lithography is effective for fabricating and transferring periodic patterns and structures as reported in recent papers (Xia & Whitesides, 1998) (Xia et al., 1999) (Schmid & Michel, 2000) (Odom et al., 2002) (T.-W. Lee et al., 2005) (Tang et al., 2003) (Rogers & Nuzzo, 2005) (D.-H. Lee et al., 2007). Accordingly, using methods and processes associated with soft lithography, new narrow-band resonant optical filters fabricated in hybrimer compounds are presented.

In this work, a new material system for fabricating the resonant optical filters is employed. Hybrimers are typical organic-inorganic hybrid materials fabricated using a sol-gel process (Choi et al., 2005) (Kim et al., 2006) (Kim et al., 2005). Hybrimers have several advantageous properties, including high modulus, low surface tension, low shrinkage, and high etching resistance. In particular, they have excellent optical properties including high transparency (>90% in the visible region), controllable refractive indices, low optical loss (<0.2 dB/cm), low birefringence ($\sim 10^{-4}$), and low viscosity compared to common ultraviolet (UV)-curable polymers (Kim et al., 2005) (T.-H. Lee et al., 2006). These materials possess thermal stability beyond 300 °C. The versatile properties of hybrimers offer new options for practical applications related to micro-optical devices. In the case of the fluorinated hybrimer used in this work, an organoalkoxysilane precursor functionalized with a perfluoroalkyl chain is used in the sol-gel reaction to lower the surface tension of the final compound. Hybrimers qualify both as molds and as resists in nanoimprint lithography (Kim et al., 2006). Significantly, there is no additional chemical treatment needed to release the mold due to the presence of fluorine molecules in the hybrimer compound.

We present guided-mode resonance (GMR) filters fabricated by soft lithography with hybrimer materials. The term GMR refers to a rapid variation in the intensities of the electromagnetic fields in a periodic waveguide, or photonic crystal slab, as the wavelength or the angle of incidence of the excitation light varies around their resonance values. A resonance occurs when incident light is phase-matched to a leaky guided mode allowed by the waveguide-grating structure (Magnusson & Wang, 1992). Numerous potentially useful devices based on resonant waveguide modes have been theoretically predicted and experimentally verified (Magnusson & Wang, 1992) (Avrutsky & Sychugov, 1989) (Peng & Morris, 1996) (Ding & Magnusson, 2004) (K.J. Lee & Magnusson, 2011)(Sharon et al., 1996) (Brundrett et al., 1998) (Priambodo et al., 2003) (Liu et al., 1998) (K.J. Lee et al., 2008). However, these devices were designed to work with conventional materials and processes. Therefore, additionally, we provide example fabrication and characterization of GMR filters made by soft lithography. As these resonant elements are highly sensitive to parametric variations, it is important to develop methods for their reliable fabrication. Thus, we provide a fabrication process that is consistent and simple, employing an elastomeric mold and a UV-curable organic-inorganic hybrid material. A particular fabricated device exhibits measured spectra showing ~81% reflectance and ~8% transmittance at a resonance wavelength of 1538 nm. The filter's linewidth is ~4.5 nm, and the sideband reflectance is ~5%. Experimental and theoretical results are in good agreement. We conclude that soft lithography combined with hybrimer media is an advantageous methodology for fabricating resonant photonic devices.

2. Fabrication of a surface-relief structure with optical polymers

Before hybrimers were considered UV-curable materials for surface-relief structures, commercially available optical prepolymers such as J-91, SK-9 (Summers Optical), and NOA-73 (Norland Products Inc.) were applied in soft lithography. Table 1 shows the main properties of these materials (Summers) (Norland).

Product name	Material	Refractive index	Viscosity (cps)
J-91	polyurethane	1.55	250 ~ 300
SK-9	polyacrylate	1.49	80 ~ 100
NOA-73	polyurethane	1.56	130

Table 1. The properties of three commercial optical prepolymers

2.1 Fabrication process of GMR filters by microtransfer molding

Soft lithography uses soft elastomeric materials to make patterns and structures without the use of complicated and expensive facilities that conventional photolithography uses. Therefore, soft lithography has been shown to be a simple and cost-effective method for pattern transfer. Xia and Whitesides reported several soft lithography methods including replica molding (REM), microtransfer molding (μTM), micromolding in capillaries (MIMIC), and solvent-assisted micromolding (SAMIM) (Xia & Whitesides, 1998). Of the four methods, we choose to conduct preliminary fabrication using the μTM method. REM duplicates the structure with a thick supporting layer, which is not desired for GMR device implementation. MIMIC requires very low-viscosity materials and it cannot get capillary filling with the optical prepolymers in Table 1. SAMIM is a method which applies a solvent that can dissolve or soften the material to form the structure. It is also not proper for our application.

Fig. 1 summarizes the procedure for fabrication of the GMR device using a μTM method. The first step of the fabrication is making a master template, which has the requisite grating structure on the surface of a silicon wafer. The silicon grating structure can be made by photoresist spin-coating, holographic recording, photoresist development, and plasma etching. These conventional fabrication process steps using photolithography are described, for example, in (Priambodo et al., 2003). However, for the results reported here, two types of commercial holographic gratings (Newport Co.) are used as master templates. One has 556 nm grating period (1800 grooves/mm) and ~170 nm grating depth for developing GMR devices operating in the near-infrared region (~850 nm wavelength). Another grating has 1111-nm grating period (900 grooves/mm) and ~340 nm grating depth for the communication band (~1550 nm wavelength). These gratings have sinusoidal profiles.

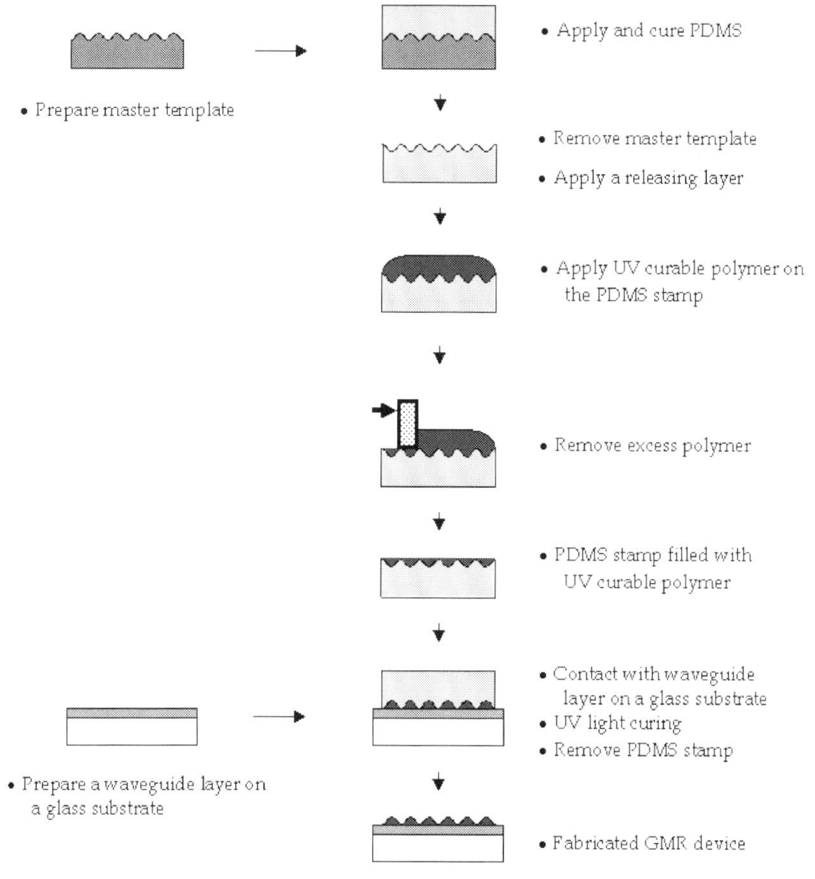

Fig. 1. Schematic fabrication process of a GMR device by μTM

As an elastomeric mold, polydimethylsiloxane (PDMS) is commonly used in soft lithography. Sylgard 184 silicone elastomer from Dow Corning, which is commonly used, is applied. Sylgard 184 silicone elastomer is mixed in a 10:1 ratio of base and curing agent and its mixture is degassed in a vacuum to eliminate bubbles. The prepolymer of Sylgard 184 is poured onto the top of the master template and cured at room temperature for 48 hours. The result is a mold with the negative replica of the grating on its surface. Atomic force microscope (AFM) images of the master template and the replica are shown in Fig. 2. These AFM images are obtained by Asylum MFP-3D AFM (Asylum Research).

A few drops of a UV-curable optical prepolymer are applied on the patterned surface of the elastomeric mold. The excess prepolymer is scraped off using a flat PDMS block, such that the grooves of the PDMS mold are filled with UV-curable prepolymer. Next, the filled PDMS mold is put in contact with a silicon nitride (Si_3N_4) thin-film on a glass substrate. As a waveguide layer, the Si_3N_4 thin-film is prepared by sputtering; this layer can also be replaced by a high-refractive index polymer material, such as a titanium dioxide (TiO_2)-rich film made by spin-coating (Wang et al., 2005). The prepolymer is cured using a UV lamp (central wavelength λ = 365 nm). Patterned grating structures are obtained on the Si_3N_4 film after the PDMS mold is peeled off.

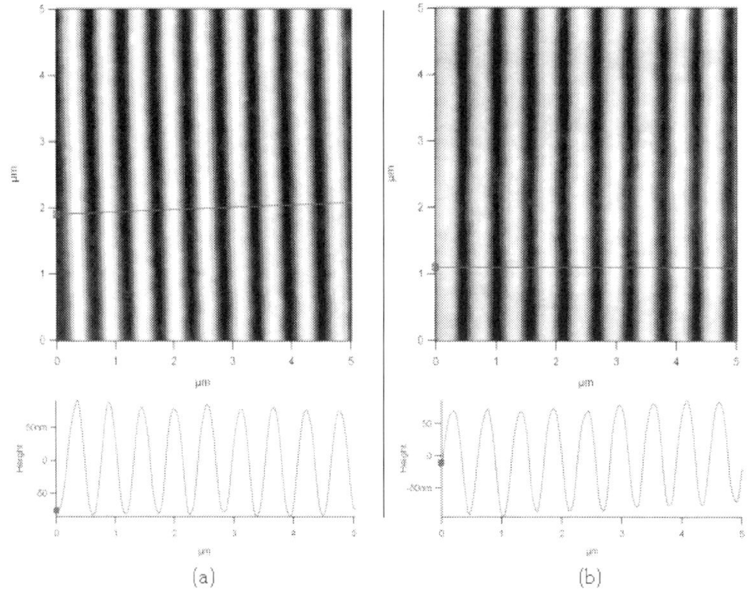

Fig. 2. AFM images of (a) the master template and (b) the replica. The size of the scanned area is 5 μm x 5 μm

2.2 Discussion of the fabrication process

Fig. 2 shows AFM images of the master template and the PDMS stamp. Both images show approximately sinusoidal profiles and a grating period of 556 nm. The average depths of the grating layer for the master template and the PDMS stamp are ~170 nm and ~168 nm, respectively. Odom et al. demonstrated that soft lithography techniques are useful for the patterning on the size scale of 500 nm and larger. However, the application of these methods is limited in the sub-100-nm range because of the low modulus (2.0 N/mm^2) of PDMS (Odom et al., 2002). During the process, deformations such as roof collapse and lateral collapse can occur in the PDMS stamp and details of deformations are discussed in (Schmid & Michel, 2000) (Odom et al., 2002) (T.-W. Lee et al., 2005). This deformation is also found in our processes reported here. For example, Fig. 3 shows AFM images of the deformed area. This deformation is more evident if there are higher aspect ratios. Fig. 4 shows AFM images of a PDMS stamp having a 520-nm grating period and 220-nm grating depth, which is prepared by photoresist grating. Pairings of grating lines are easily observed in this image. Delamarche et al. discussed the stability of lines molded in PDMS with different mold thicknesses and showed the limitation of shape formation in PDMS without deformation (Delamarche et al., 1997). They also proposed that surface treatment with 1% sodium dodecylsulfate in water and a heptane rinse could provide the recuperation of the paired lines. Even though their method is applied to this stamp, it shows little improvement.

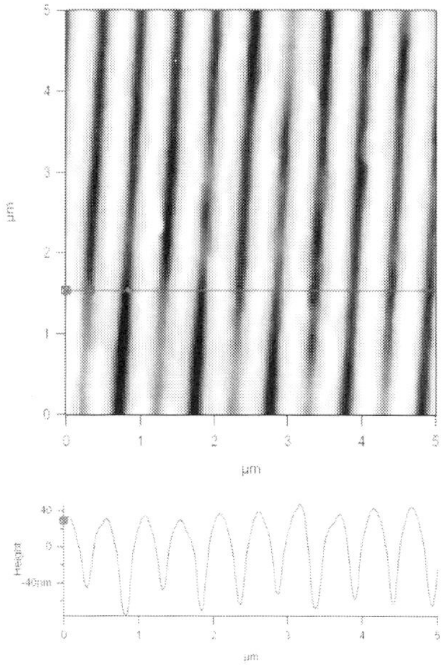

Fig. 3. AFM image of a deformed area on a PDMS stamp. The size of the scanned area is 5 µm x 5 µm

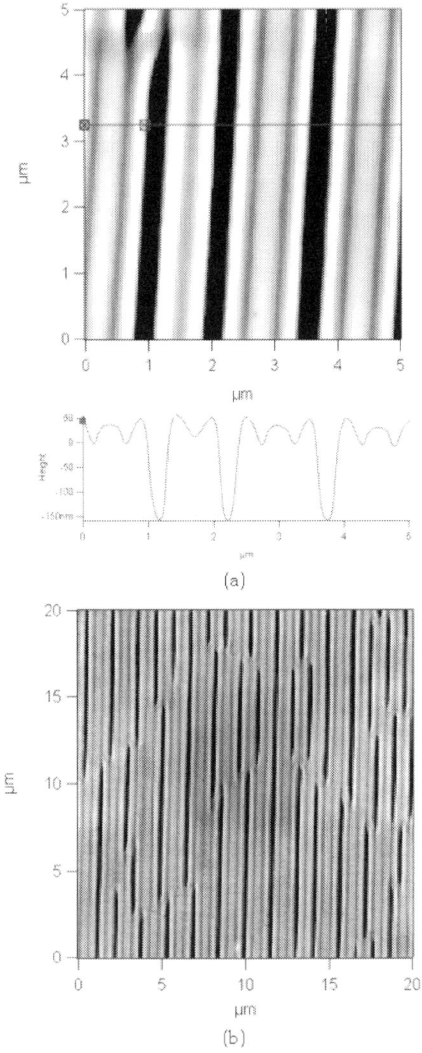

Fig. 4. AFM images of a PDMS stamp having a 520-nm grating period and 220-nm grating depth showing grating-line pairing. The size of the scanned areas are (a) 5 μm × 5 μm and (b) 20 μm × 20 μm

Schmid et al. formulated an alternative siloxane polymer, called hard-PDMS (h-PDMS), which has a higher modulus (~9 N/mm^2) than that of Sylgard 184 silicone elastomer (Schmid & Michel, 2000) (Odom et al., 2002). Odom et al. and Lee et al. demonstrated improved results of surface structure formation by using h-PDMS compared with Sylgard 184 PDMS (Odom et al., 2002) (T.-W. Lee et al., 2005). In our research, we apply h-PDMS as described in Section 3.

In addition, the UV-curable polymers can stick to the PDMS stamp and this causes deformation of the structure during the process. To solve this problem, silane treatment is widely used for the surface of the PDMS stamp and acts as a releasing layer. Tang et al. reported that the modification of PDMS stamps by an adsorbed monolayer of bovine serum albumin (BSA) can provide distortion-free separation between the PDMS stamp and molded gel from the stamp (Tang et al., 2003). In our process, we find that the cured polymer does not attach well to the waveguide but tends to remain on the PDMS stamp after separation. Therefore, we apply the BSA treatment to improve the release. The PDMS stamp is immersed in a 3% solution of aminopropyldimethylethoxysilane (Gelest, SIA0603.0) in methanol for 2 hours. The modified PDMS is next rinsed in phosphate buffer saline (PBS, Fisher Scientific, BP2438-4) solution for 90 seconds. A solution of BSA (100 mg/ml Fisher Scientific, NC9806065) is applied over the patterned surface of the PDMS for 60 seconds. The treated PDMS is then rinsed in a PBS solution for 90 seconds to remove any unbounded BSA. After this BSA treatment, it is possible to successfully get the structure to adhere to the waveguide layer with no remaining polymer in the PDMS stamp.

2.3 Fabrication results

Fig. 5 shows the experimental spectral response of a fabricated GMR device using the NOA-73 prepolymer. A tunable Ti:Sapphire laser pumped by an Argon-ion laser is used to measure its spectral response. The resonance wavelength (maximum point of reflectance) is at λ = 863.6 nm and the full-width at half-maximum (FWHM) linewidth is 4.0 nm. The reflectance at resonance is about 33%. This relatively low efficiency is attributed to the loss of the Si_3N_4 thin-film (measured extinction coefficient, k ~ 10^{-3}) and imperfections on the grating surface. Even though the efficiency of the fabricated device is low, it can be used in resonant sensor devices because high-efficiency is not essential in GMR sensor application and the key quantity is the resonance wavelength shift. Fig. 6 shows AFM images of the GMR device fabricated with NOA-73. The cross-sectional view shows that the thickness of the grating layer is not uniform. This thickness non-uniformity is due to the scraping process applied to remove excess prepolymer by hand.

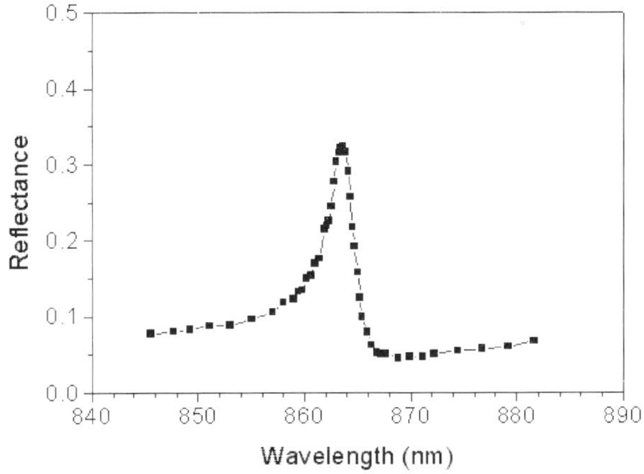

Fig. 5. Experimental spectral response of a fabricated GMR filter for transverse-electric (TE) polarization

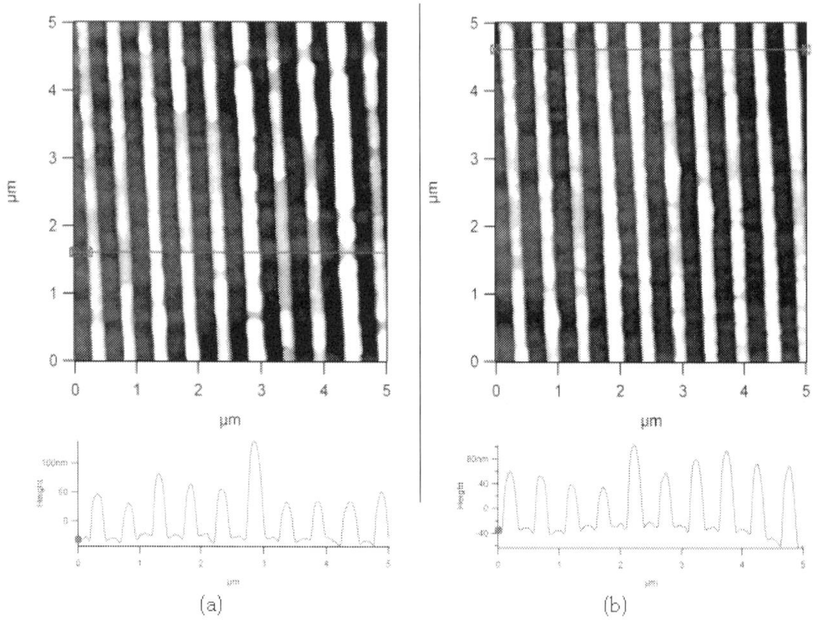

Fig. 6. AFM images of fabricated GMR devices with NOA-73 at two different locations. The size of the scanned area is 5 µm × 5 µm

The structure of GMR filters fabricated by J-91 and SK-9 is not much different from that of the NOA-73 filter as shown in Fig. 6. The improved µTM method with a machine-controlled system was reported by (J.-H. Lee et al., 2005). They used a dragging speed of ~30 µm/sec and a metal blade, which was controlled by mechanical actuators for scraping.

3. Improved fabrication of GMR filters using hybrimers and MIMIC

As described in section 1, hybrimers have many advantages, especially low viscosity, which can be lowered to ~8 cps. This low-viscosity material is applied with the MIMIC method. The combination of hybrimers and MIMIC to fabricate GMR filters is described in this section. It is shown to provide much-improved GMR devices relative to the devices reported above.

3.1 Fabrication process and characterization

Fig. 7 shows a schematic procedure for fabrication of a GMR device using the MIMIC method. As described in section 2.1, MIMIC is one of several soft lithography methods proposed by Whitesides and co-workers, and it is simple to apply (Xia & Whitesides, 1998) (Xia et al., 1999). The first step of the fabrication is preparing a master template, which has the grating structure on the surface of a silicon wafer or a glass substrate. In this section, a commercial holographic grating, which has a 1111-nm grating period and ~340-nm grating depth with a sinusoidal profile is used as a master template.

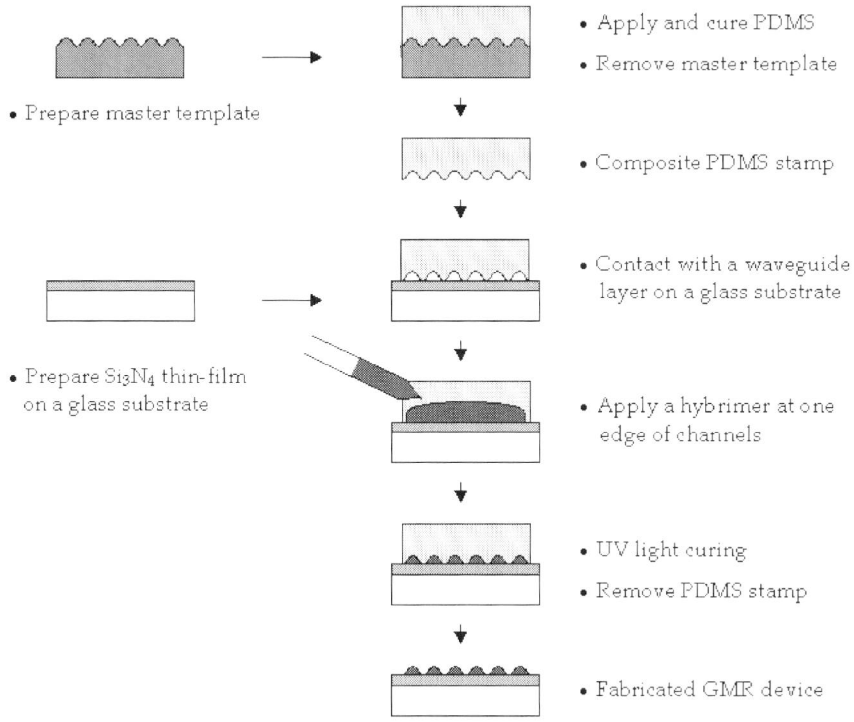

Fig. 7. Schematic fabrication process of a GMR device using the MIMIC method

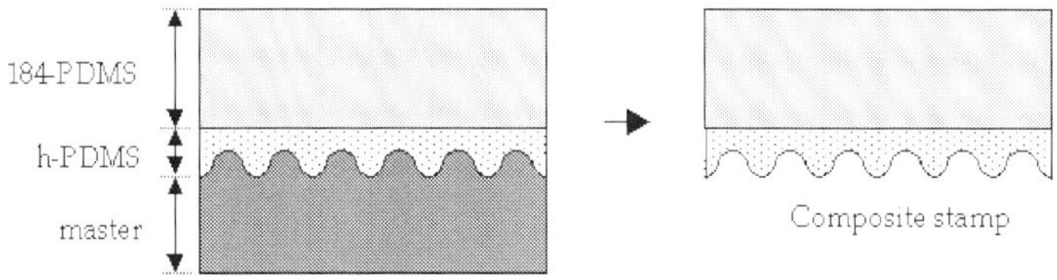

Fig. 8. Configuration of the composite stamp. The h-PDMS is spin-coated on the master template followed by pouring Sylgard 184 PDMS

Composite polymeric stamps as elastomeric molds are applied to achieve high-quality patterning. The composite stamps consist of two parts as shown in Fig. 8; one part is h-PDMS, which has different mechanical properties than the Sylgard 184 silicone elastomer. The other part is Sylgard 184 PDMS to support h-PDMS and maintain the flexibility of the stamp. The h-PDMS is prepared as follows (Odom et al., 2002): a vinyl PDMS prepolymer (3.4 gram, Gelest Inc., VDT-731), platinum (Pt) catalyst (18 μL, Gelest Inc., SIP6831.1), and modulator (50 μL, Sigma-Aldrich, Product No. 87927) are mixed and degassed in a vacuum for 5 minutes. A hydrosilane prepolymer (1.0 gram, Gelest Inc., HMS-301) is added into the mixture and stirred gently to avoid bubbles. Within 5 minutes after stirring, this h-PDMS prepolymer is spin-coated (1000 rpm, 40 sec.) on a commercial holographic grating and cured in an oven at 60 °C for 30 minutes. Then a prepolymer of Sylgard 184 silicon elastomer is poured on the h-PDMS layer and is cured in an oven again at 60 °C for 3 hours. Therefore, a pattern with a negative replica of the master template is formed on the h-PDMS surface. AFM images of the surface of the master template and the composite stamp are shown in Fig. 9. We obtain faithful replication and a high-quality, scatter-free surface.

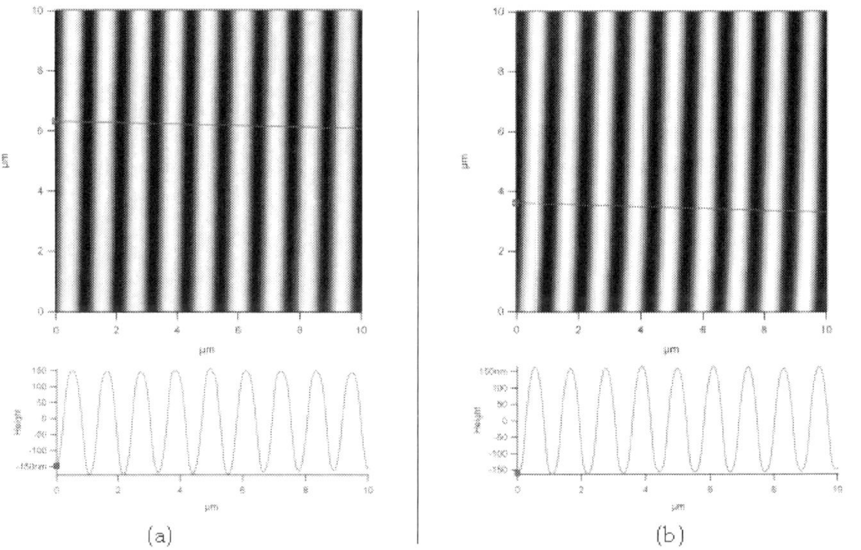

Fig. 9. AFM images of (a) the master template and (b) the composite stamp replica. The size of the scanned area is 10 μm × 10 μm

To realize a GMR filter, the 10mm × 10mm composite mold is placed in contact with a Si_3N_4 thin film on a glass substrate. The Si_3N_4 film, prepared by plasma-enhanced chemical vapor deposition (PECVD, Surface Technology Systems), serves as the waveguide layer of the GMR device. A diluted hybrimer is prepared to obtain a lower viscosity prepolymer. A few drops of the UV-curable hybrimer are applied at one edge of the patterned surface of the composite mold as indicated in Fig. 7. The applied hybrimer spreads through the channels, which are formed by contact between the patterned mold and the thin-film layer on the substrate. Then the composite mold set is put into a vacuum chamber and is allowed to remain in low vacuum (~450 Torr) for 12 hours so that the channels fill with the hybrimer prepolymer by capillary force. Subsequently, the hybrimer in the h-PDMS channel is cured using a UV lamp (central wavelength λ = 365 nm). The surface-relief type grating structure remains on the Si_3N_4 film after the composite mold is peeled off. This easy-release process is associated with the hybrimers medium.
A tunable laser source (HP 8168F) is used to measure the spectral response. The angle of incidence (θ_{in}) is set at 10° to locate the resonance wavelength of the GMR device within the operating spectral range of the laser source. Next, the reflected and transmitted power are measured in the wavelength range of 1450 nm to 1590 nm, in 0.5 nm steps. A polarizer is used to set the polarization state.

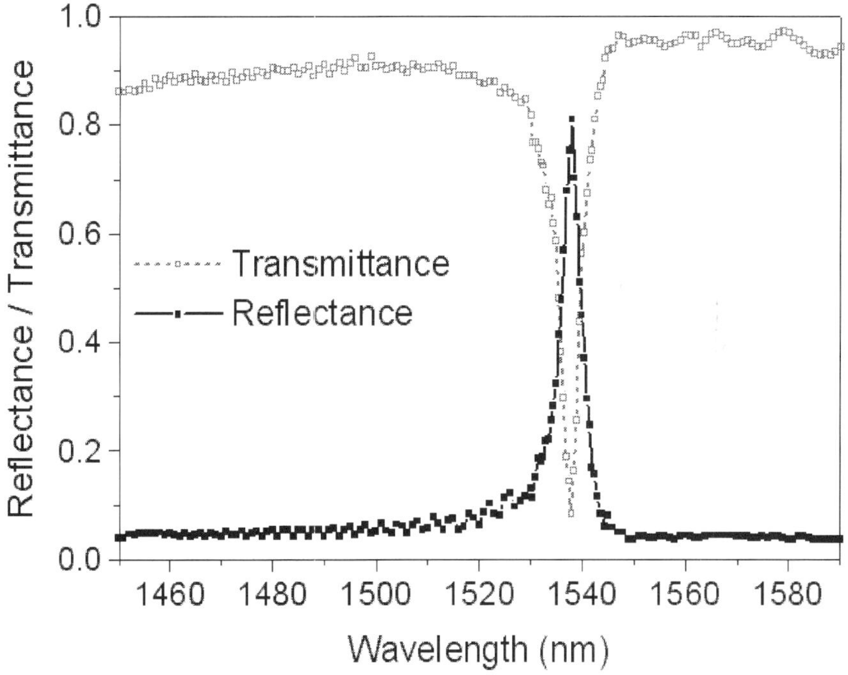

Fig. 10. Experimental spectral response of a fabricated GMR filter for TE polarization

3.2 Results and discussion

Fig. 10 shows the experimental spectral response of the fabricated device. The resonance wavelength (maximum point of reflectance or minimum point of transmittance) is λ = 1538 nm, and the reflectance at resonance is ~81%. The FWHM linewidth is ~4.5 nm, and the sideband reflectance is ~5%. The transmittance at resonance is ~8%. These results pertain to the TE polarization state of the input light.

Fig. 11 shows the calculated spectral response of the fabricated GMR device whose model is shown as an inset. The calculations are performed with a computer code that we wrote based on rigorous coupled-wave theory (Gaylord & Moharam, 1985). The device parameters used in Fig. 11 correspond to the experimental values used in the fabrication. The calculated resonance wavelength is λ = 1536 nm at θ_{in} = 10.0°, and the FWHM linewidth is 4.4 nm. The calculated reflectance at resonance is ~82%. This non-100% reflection is due to the presence of a higher-order transmitted wave at resonance as shown in Fig. 11 and noted as T_{+1}.

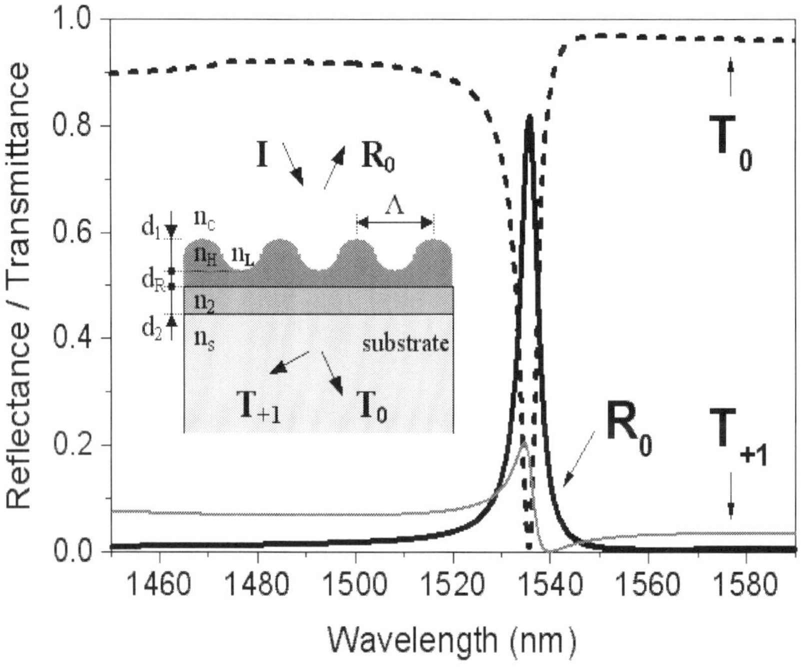

Fig. 11. Calculated spectral filter response with parameters corresponding to those of the fabricated filter for TE polarization. The parameters are as follows: thicknesses, d_1 = 333 nm, d_R = 0 nm (No residue), d_2 = 250 nm; refractive indices n_H = 1.51, n_L = 1.00, n_2 = 1.87, n_c = 1.00, n_s = 1.50; grating period Λ = 1111 nm; incident angle θ_{in} = 10°

Generally, for this device class, there is good agreement between theory and experiment as shown, for example, in (Priambodo et al., 2003) and by the results presented here. As quantified in (Kemme et al., 2003), small variations in the device parameters can shift the location of the resonance peak significantly. A residual layer, shown in the inset in Fig. 11 with thickness d_R, can have significant effects on the spectrum and central filter wavelength. By numerical modeling, the thickness of the residual layer can be estimated. This layer, in this work, is an unwanted layer arising during the soft lithography process. Theoretical calculation shows that the resonance wavelength shifts and the FWHM linewidth decreases as the thickness of the residual layer increases as shown in Fig. 12. By comparing the resonance wavelength and linewidth of the calculated data with the measured data, the thickness of the residual layer is estimated to be ~11 nm.

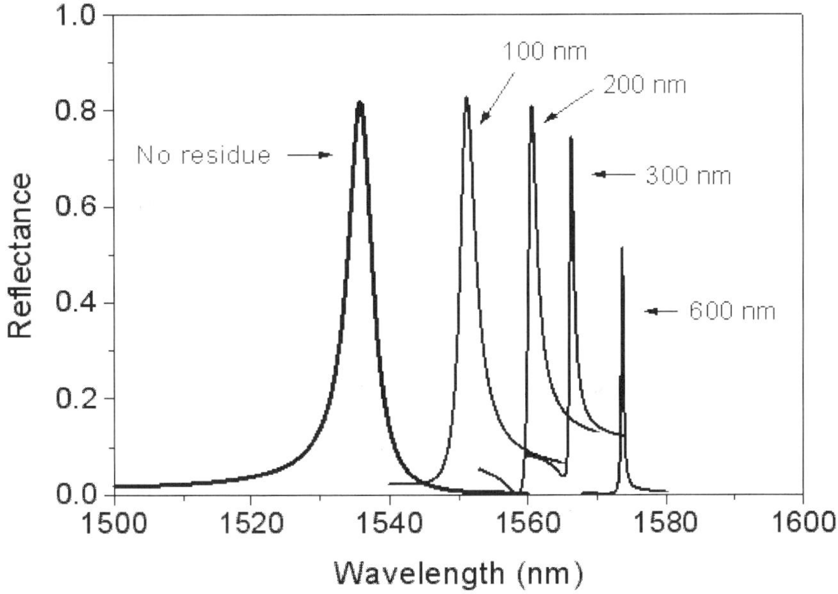

Fig. 12. Calculated spectral response of the fabricated GMR filter with varying thickness of the residual layer (d_R)

Fig. 13 shows a cross-sectional view of the fabricated GMR filter obtained with a scanning electron microscope (SEM, FEI Company, Quanta 3D FEG) indicating no apparent residual layer for the fabricated filter. Moreover, the SEM confirms the parameters used in the theoretical calculations in Fig. 11.

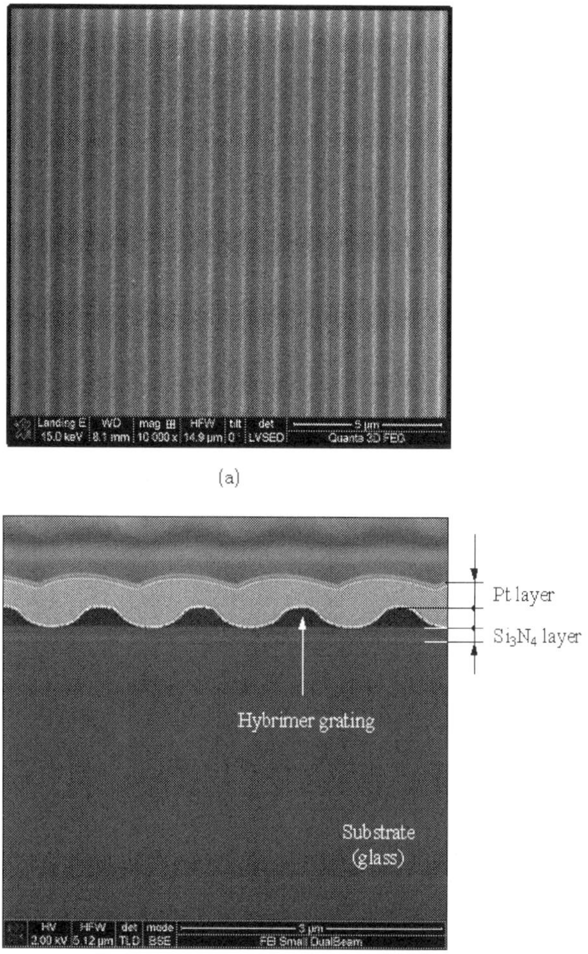

(a)

(b)

Fig. 13. SEM images of the fabricated GMR device. (a) Top down view, Magnification = 10,000. The size of the image is ~15 µm. (b) Cross-sectional view, Magnification = 25,000. Note that the white part on top is a platinum (Pt) layer for protection during ion-beam sectioning

4. Conclusion

We presented fabrication and characterization of guided-mode resonant (GMR) filters made by soft lithography. As these resonant elements are highly sensitive to parametric variations, it is important to develop methods for their reliable, repeatable fabrication. Thus, a fabrication process that is consistent and simple, employing an elastomeric mold and a UV-curable organic-inorganic hybrid material is provided. By combining MIMIC, a hybrimer, and an h-PDMS mold, a photopolymer grating structure is readily fabricated. Measured spectra show ~81% reflectance and ~8% transmittance at a resonance wavelength of 1538 nm. The filter's linewidth is ~4.5 nm, and the sideband reflectance is ~5%. Experimental and theoretical results are in good agreement. The use of hybrimer media yields improved processes and high-quality photonic devices.

5. Acknowledgments

This work was supported in part by the National Science Foundation (NSF) under grant ECCS-0925774 and by the UT System Texas Nanoelectronics Research Superiority Award funded by the State of Texas Emerging Technology Fund. The authors gratefully acknowledge the NanoPort Applications Team at FEI Company for providing the results shown in Fig. 13. The authors thank Prof. B. Huey from The Institute of Materials Science at the University of Connecticut for providing access to AFM facilities. This work was performed in part at the Center for Nanoscale Systems (CNS), a member of the National Nanotechnology Infrastructure Network (NNIN), which is supported by the NSF under award ECS-0335765. CNS is part of the Faculty of Arts and Sciences at Harvard University. Additional support was provided by the Texas Instruments Distinguished University Chair in Nanoelectronics endowment.

6. References

Avrutsky I.A. & Sychugov, V.A. (1989). Reflection of a beam of finite size from a corrugated waveguide. *J. Mod. Opt.*, Vol. 36, No. 11, (1989), pp. 1527-1539, ISSN 1362-3044

Brundrett, D.L.; Gaylord, T.K. & Glytsis, E.N. (1998). Polarizing mirror/absorber for visible wavelengths based on a silicon subwavelength grating: design and fabrication. *Appl. Opt.*, Vol. 37, No. 13, (May 1998), pp. 2534-2541

Choi, D.-G.; Jeong, J.-H.; Sim, Y.-S.; Lee, E.-S.; Kim, W.-S. & Bae, B.-S. (2005). Fluorinated organic-inorganic hybrid mold as a new stamp for nanoimprint and soft lithography. *Langmuir*, Vol. 21, No. 21, (October 2005), pp. 9390-9392

Delamarche, E.; Schmid, H.; Michel, B. & Biebuyck, H. (1997). Stability of molded polydimethylsiloxane microstructures. *Advanced Materials*, Vol. 9, No. 9, (1997), pp. 741-746, ISSN 0935-9648

Ding, Y. & Magnusson, R. Resonant leaky-mode spectral-band engineering and device applications. *Optics Express, V*ol. 12, No. 23, (2004), pp. 5661–5674

Gaylord, T.K. & Moharam, M.G. (1985). Analysis and applications of optical diffraction by gratings. *Proc. IEEE*, Vol. 73, No. 5, (May 1985), pp. 894-937, ISSN 0018-9219

Kemme, S.A.; Peters, D.W.; Wendt, J.R.; Carter, T.R. & Samora, S. Integration and tolerance issues for resonant subwavelength gratings," *Proc. of the SPIE Gradient Index, Miniature, and Diffractive Optical Systems III*, Vol. 5177, pp. 1-8, San Diego, California, USA, August 2003.

Kim, W.-S.; Kim, K.-S.; Eo, Y.-J.; Yoon, K.B. & Bae, B.-S. (2005). Synthesis of fluorinated hybrid material for UV embossing of a large core optical waveguide structure. *J. Mater. Chem.*, Vol. 15, No. 4, (January 2005), pp. 465-469

Kim, W.-S.; Choi, D.-G. & Bae, B.-S. (2006). Ultraviolet-nanoimprint of 40 nm scale patterns using functionally modified fluorinated hybrid materials. *Nanotechnology*, Vol. 17, No. 13, (July 2006), pp. 3319-3324

Lee, D.-H.; Park, K.-H.; Hong, L.-Y. & Kim, D.-P. (2007). SiCN ceramic patterns fabricated by soft lithography techniques. *Sens. Actuators A: Physical*, Vol. 135, No. 2, (April 2007), pp. 895-901

Lee, J.-H.; Kim, C.-H.; Ho, K.-M. & Constant, K. (2005). Two-polymer microtransfer molding for high layered microstructures. *Advanced Materials*, Vol. 17, No. 20, (October 2005), pp. 2481-2485

Lee, K.J.; LaComb, R.; Britton, B.; Shokooh-Saremi, M.; Silva, H.; Donkor, E.; Ding, Y. & Magnusson, R. (2008). Silicon-layer guided-mode resonance polarizer with 40-nm bandwidth. *IEEE Photonics Tech. Lett.*, Vol. 20, No. 22, (November 2008), pp. 1857-1859, ISSN 1041-1135

Lee, K.J. & Magnusson, R. Single-layer resonant high reflector in TE polarization: Theory and experiment. *IEEE Photonics Journal*, Vol. 3, No. 1, (February 2011), pp. 123–129

Lee, T.-H.; Kim, J.H. & Bae, B.-S. (2006). Synthesis of colorless imide hybrid nanocomposites using amine functionalized oligosiloxane nano-building clusters. *J. Mater. Chem.*, Vol. 16, No. 17, (May 2006), pp. 1657-1664

Lee, T.-W.; Mitrofanov, O. & Hsu, J.W.P. (2005). Pattern-transfer fidelity in soft lithography: the role of pattern density and aspect ratio. *Advanced Functional Materials*, Vol. 15, No. 10, (October 2005), pp. 1683-1688

Liu, Z.S.; Tibuleac, S.; Shin, D.; Young, P.P. & Magnusson, R. (1998). High-efficiency guided-mode resonance filter. *Opt. Lett.*, Vol. 23, No. 19, (October 1998), pp. 1556-1558

Magnusson, R. & Wang, S.S. (1992). New principle for optical filters. *Appl. Phys. Lett.*, Vol. 61, No. 9, (August 1992), pp. 1022-1024, ISSN 0003-6951

Norland Products Inc., n.d. Available from: <http://www.norlandprod.com>

Odom, T.W.; Love, J.C.; Wolfe, D.B.; Paul, K.E. & Whitesides, G. W. (2002). Improved pattern transfer in soft lithography using composite stamps. *Langmuir*, Vol. 18, No. 13, (June 2002), pp. 5314-5320

Peng, S. & Morris, G.M. (1996). Experimental demonstration of resonant anomalies in diffraction from two-dimensional gratings. *Opt. Lett.*, Vol. 21, No. 8, (April 1996), pp. 549-551

Priambodo, P.S.; Maldonado, T.A. & Magnusson, R. (2003). Fabrication and characterization of high-quality waveguide-mode resonant optical filters. *Appl. Phys. Lett.*, Vol. 83, No. 16, (October 2003), pp. 3248-3250

Rogers J.A. & Nuzzo, R.G. (2005). Recent progress in soft lithography. *Materials Today*, Vol. 8, No. 2, (February 2005), pp. 50-56

Schmid H. & Michel, B. (2000). Siloxane polymers for high-resolution, high-accuracy soft lithography. *Macromolecules*, Vol. 33, No. 8, (April 2000), pp. 3042-3049

Sharon, A.; Rosenblatt, D. & Friesem, A.A. (1996). Narrow spectral bandwidths with grating waveguide structures. *Appl. Phys. Lett.*, Vol. 69, No. 27, (December 1996), pp. 4154-4156, ISSN 0003-6951

Summers Optical, n.d. Available from: <http://www.optical-cement.com>

Tang, M.D.; Golden, A.P. & Tien, J. (2003). Molding of three-dimensional microstructures of gels. *J. Am. Chem. Soc.*, Vol. 125, No. 43, (October 2003), pp. 12988-12989

Wang, Y.; Flaim, T.; Mercado, R.; Fowler, S.; Holmes, D. & Planje, C. (2005). Hybrid high refractive index polymer coatings, *Proc. of the SPIE Organic Photonic Materials and Devices VII*, Vol. 5724, pp. 42-49, San Jose, California, USA, April 2005

Xia, Y. &Whitesides, G.M. (1998). Soft lithography. *Angewandte Chemie International Edition*, Vol. 37, No. 5, (March 1998), pp. 550-575

Xia, Y.; Rogers, J.A.; Paul, K.E. & Whitesides, G.M. (1999). Unconventional methods for fabricating and patterning nanostructures. *Chem. Rev.*, Vol. 99, No. 7, (July 1999), pp. 1823-1848

Part 3

Interference, Two-Photon, UV and X-Ray Lithography

13

DUV Interferometry for Micro and Nanopatterned Surfaces

Olivier Soppera et al.*
Institut de Sciences des Matériaux de Mulhouse, CNRS LRC 7228, Mulhouse, France

1. Introduction

Recent developments in nanoscience and nanotechnology were strongly supported by advances in nanofabrication. Controlled patterning of nanostructured materials has become increasingly important because of the ever-decreasing dimensions of various devices, including those used in electronics, optics, photonics, biology, electrochemistry, and electromechanics (Henzie et al., 2004; Fan et al., 2006). Today, the production of structures with typical dimension in the 1 to 100 nm range with engineered physical and chemical properties is challenging.
Different nano-fabrication techniques have been reported in the literature (Nie & Kumacheva, 2008). Recent examples include optical lithography (Cotton et al., 2009), electron beam lithography (Gonsalves et al. 2009), X-ray lithography (Im et al., 2009), laser writing (Soppera et al., 2008), scanning probe techniques (including optical near-field lithography (El Ahrach et al., 2007), pen nanolithography (Cai & Ocko, 2005), dip-pen lithography (Christman et al., 2009), nanoshaving (Seo & Borguet, 2006) and thermal scribing (Lee et al., 2008)), microcontact printing (Huh et al., 2009), micro-phase separation of block copolymers (Greater et al., 2007), dewetting (Yoon et al., 2008), nanoimprint lithography (He et al., 2009) or electrochemical nanopatterning (Jegadesan et al., 2006). The major remark is that the size of the achievable patterns is strongly dependent of the technique used and can vary between the micrometer to the sub-10 nanometre length scale. This point is a serious limitation when different length scales are needed. Furthermore, in most cases these techniques suffer from different material requirements and limited dimensions of the patterned surface.
In this context, interferometric lithography appears of high interest when periodical patterns are needed. Indeed, interferometric techniques can be considered as massively parallel nanofabrication techniques since patterns can be obtained over large area within a single exposure. Moreover, the recourse to wavelength in the Deep-UV range (DUV corresponds

*Ali Dirani[1,2], Fabrice Stehlin[1], Hassan Ridaoui[1,3], Arnaud Spangenberg[1], Fernand Wieder[1] and Vincent Roucoules[1]
[1]*Institut de Sciences des Matériaux de Mulhouse, CNRS LRC 7228, Mulhouse, France*
[2]*Université Catholique de Louvain, Division of Bio- and Soft Matter, Belgique*
[3]*CEA, Laboratoire d'Innovation pour les Technologies des Energies Nouvelles et les nanomatériaux / DEHT/ LPCE, Grenoble, France*

to λ<300 nm) allows producing periodic nanostructures with typical dimension down to several tens of nm.

The aim of this chapter is to review some recent works about DUV interferometric lithography nanofabrication. In the first part, a brief introduction to interferometric lithography will allow illustrating its interest and main applications. The second part will be dedicated to applications with organic materials (polymers) that have been widely used for micro and nanopatterning with such a technique. However, organic materials present some inherent limitations that have justified many efforts during the last years to developed inorganic materials prepared by sol-gel technique. This will be the topic of the third part of this chapter.

2. DUV interferometry

2.1 Introduction

The needs for simple, fast and versatile techniques for micro- and nanomachining have accounted for many works during the last years. The most active sector in this field has been the microelectronics industry. Researches in this area were essentially motivated by finding new solutions to follow the trend towards a constant decrease of the size of the transistors as stated in the "Moore's law" (Moore, 1965). To reach these objectives, the recourse to deeper wavelengths has been proposed. Today, the most used wavelength for chips manufacturing is 193 nm provided by ArF excimer lasers. In this case, the use of short wavelength allows decreasing the limitations linked to diffraction. Patterns with dimensions as small as 22 nm can be produced by industrial machines (steppers) on 300 mm diameter Silicon wafers. However, these high performance lithographic tools are restricted to production use and complex patterns due to their very high cost.

For these reasons, there is a parallel need of less sophisticated setups but able to produce sub-100 nm structures on relatively wide surfaces. For some applications, the structures do not need to be complex and can be limited to a regular replication of simple shapes like dots or lines. For the fabrication of such periodical nanostructures, interferometric techniques fulfil most of the needed requirements.

Interferometry consists in combining two or more monochromatic, coherent and polarized beams to produce an interference pattern. This technique is sometimes also called holography. The spatially controlled irradiation is used to induce a local modification of the material that can be either a photocrosslinking or an ablation due to different molecular mechanisms.

The period p is given by the following relation:

$$p = \frac{\lambda}{2 \cdot n \cdot \sin\theta} \tag{1}$$

Where λ is the wavelength, and n is the refractive index of the medium in which the two beams are recombined and θ is the half-angle between the two beams.

The recourse to short wavelengths (248 nm, 193 nm) allows producing periods as low as 100 nm in air (Charley et al., 2006), meaning that this route is competitive in terms of resolution with most advanced photolithography industrial tools. In comparison, the alternative to mask projection imaging is limited to gratings with periods > 1 µm (Mihailov & Gower, 1994). Interferometric techniques are much more adapted for sub-micron structuring.

Interferometry in the DUV range has been enabled by the development of excimer lasers that have two main advantages: first, the use of short wavelengths is an effective way to provide the requested resolution since the period is directly proportional to the wavelength. Secondly, DUV wavelength permits direct writing via photoinduced processes provoked by high-energy photons. Examples of suitable materials for such wavelengths will be given in the following sections.

On the experimental point of view, one of the difficulties for interferometry in the DUV range is due to the low coherence of available DUV lasers. As an example, typical coherence of ArF lasers is limited to a few hundreds of microns, which justifies efforts to develop specific experimental setup for short wavelengths (Figure 1):

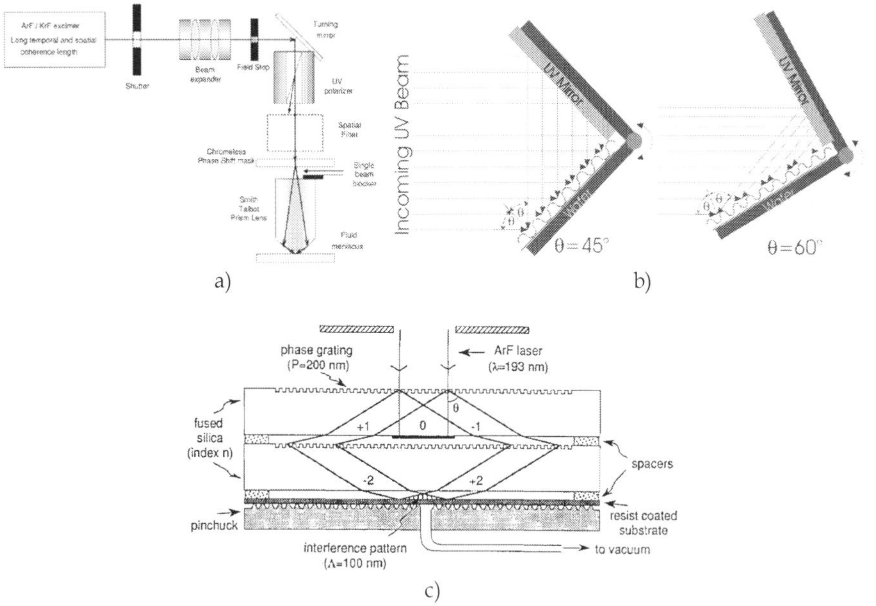

Fig. 1. Example of interferometric lithography configuration: a) Talbot prism (Bourov et al., 2004), b) Lloyd setup (Raub & Brueck, 2003) and c) achromatic holographic configuration (Yen et al., 1992)

- Holographic grating formation has been proposed (Askins et al., 1992; Archambault et al., 1993). However, in all cases, it has been necessary to increase the spatial coherence and decrease the spectral emission band width of excimer lasers, which considerably increase the laser system complexity and demonstrations have been mostly done with 248 nm lasers.
- Several systems based on Compact size Talbot prism have been developed to fit with the different excimer wavelengths (193 nm, 248 nm), with periods down to 90 nm (Bourov et al., 2004). The reduced dimensions of the system are compatible with coherence requirements linked to excimer lasers and such configuration is compatible with immersion.

- Fresnel corner cube mirror with 213 nm Nd-YAG (5ω) has been proposed. The interest of this configuration is to be compatible with immersion. However, it is limited in laser sources since it requires spatial coherence higher than what is available with excimer lasers (Raub &Brueck, 2003).
- A fourth approach relies on the use of diffraction element designed to generate two coherent beams. This was the most widely used method since it is relatively simple and compatible with the low coherence of the excimer gratings (spectral line widths of 1 nm and spatial coherence size < 300 µm). This last route has been studied from theoretical and practical point of view. Among other interesting works, specific studies have investigated the effect of the 0th order (non-diffracted beam in the direction of the incident beam) on the spatial light distribution (Xiong et al., 1999; Dyer et al., 1995). It is shown that even small contributions from orders other than the ±1 can produce significant deviations from the distribution expected for the ideal two-beam cases and it is thus to be considered with particular care.

One very elegant way to cope with this problem has been proposed by Yen et al (Yen et al., 1992). It consists in using an achromatic configuration. In this case, two matched fused silica phase gratings were employed and, the demonstration of 100 nm period grating was achieved using an ArF excimer laser. High contrast fringes were obtained with a depth-of-focus compatible with practical applications. Here, the 0th order can be physically blocked providing a perfect sinusoidal light pattern on the photosensitive resin. The advantages of the achromatic configuration are obvious for laser sources with limited coherence. Since then, this technique has been used by several research teams. Recently, Bourgin et al. (Bourgin et al., 2009) have proposed an integrated solution providing the 2 gratings on the same substrate, which simplifies the alignments.

Significant progresses in resolution have also been achieved using immersion technique. Using an immersion fluid between the phase mask and the sample, it is possible to increase the numerical aperture and thus decrease the period (see Eq. 1). The most widely used fluid for immersion is water since water is transparent at 193 nm. High refractive index fluids have demonstrated their interest to reach resolutions as low as 32 nm HP corresponding to the 65 nm node in microelectronics (Santillan et al., 2006).

2.2 Applications of periodical structures

Most of the applications of periodical micro- or nanostructures are in the field of optics and photonics. Indeed, such structures with periods in the range, or under, the wavelength of light exhibit strong interaction with light with specific effects.

It is not possible to mention here all the applications of gratings that could be provided by means of DUV laser interferometry. The most important ones are probably linked to spectroscopy, especially for high resolution spectrometers for astronomy (Heilmann et al., 2004), low-loss polarisers, grating for laser pulses shortening, motion sensors, displays (Braun, 2002), microlasers (Wegmann, 1998; Schon, 2000), white light processing like anti-reflective (Gombert et al., 2004) or diffusing surfaces (Menez et al., 2008), and in the solar cell technology (light concentrators (Karp et al., 2010), and Sub-Wavelength Gratings SWG (Y. Kanamori et al., 2005)).

Beside optical applications, new applications have emerged from the spectacular properties of patterned surface when the size of patterns is reaching the nanometre scale. These properties can be superhydrophobicity, interaction with biofilms, nanotribology, etc...

Finally, the development of photoresists for microelectronics applications has also accounted largely for the success of DUV interferometric lithography. This point will be developed below.

2.3 3D micro and nanostructures

There are specific interests in developing 3D periodical structures. One of the most important applications is the fabrication of photonics crystals (PC). PCs are crystalline materials where the refractive index is periodically modulated on a length scale comparable to the light wavelength of interest. Interference of the light waves scattered from the dielectric lattice (i.e., Bragg scattering) leads to omnidirectional stop bands or photonic band gaps (PBGs), which are analogous to the electronic energy band gaps in a semiconductor. (Joannopoulos et al., 1995; Lin et al., 1998). PCs potentially offer revolutionary advances in the next-generation microphotonic devices and the integration of existing optoelectronic devices, including integrated optical circuits, lasers, sensing, spectroscopy, and pulse shaping.

In recent years there has been a considerable effort to develop novel methods for mass production of 3D PCs with controlled size, symmetry, and defect(s) on a large-scale basis (Moon & Yang, 2009). Among others techniques, interferometric techniques appear very interesting since they are massively parallel techniques of microfabrication (unlike 2-photon fabrication), and they are potentially free of random defects (unlike self-assembly techniques). Several experimental configurations have been developed, including multibeam interference (Figure 2. Campbell et al., 2000; Yang et al., 2002) and mask interference lithography (Jeon et al., 2004).

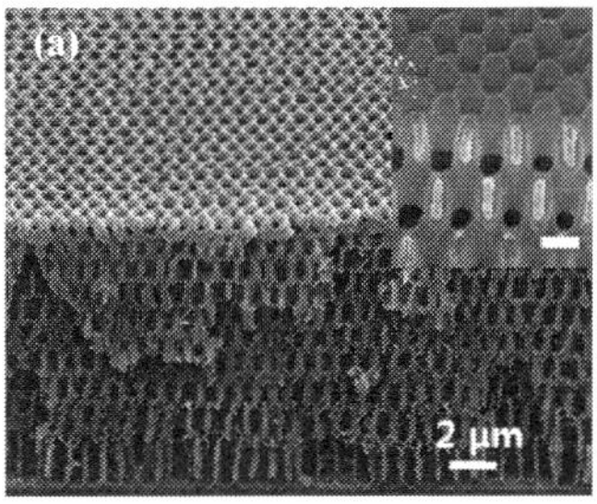

Fig. 2. Left) Holographic lithography process using an umbrella-like four beam setup, forming diamond- like interference patterns (Moon & Yang, 2009); Right) Structure and optical reflectance of 3D hydrogel PCs via holographic lithography (Kang et al., 2008).

Many different materials have been proposed for the fabrication of 3D structures. Most of these materials are sensitive in the UV or visible range of wavelengths (Moon & Yang, 2009) since the requirements in terms of period are not targeting the highest resolutions. Recently, a 3D "woodpile" structure with 1.55μm lattice constant and a 2mm-by-2mm pattern area was demonstrated using DUV wavelengths (Yao et al., 2008).

2.4 Wide surface micro and nanopatterning

Applications of nanopatterned substrates in practical application in optics or biology require the generation of nanopatterns over wide surfaces. Such requirements are specially needed for applications in displays, light concentrators for solar cells, displacement sensors or compression of high power laser pulses (Figure 3). The extension of nanostructures over meter square area has generated many efforts. The Fraunhofer Institute fur Solare Energiesysteme (ISE, Freiburg en Brisgau) developed a holographic tool (Holotool) based on a Mach-Zender configuration with an irradiation surface greater than 1 m^2. The Lawrence Livermore National Laboratory, in Livermore (Califormia, USA), has developed also an interference lithographic tool compatible with substrates as wide as 80 cm. In both case, the requirements of environmental conditions stability (temperature, mechanical vibrations, etc...) are extremely severe and despite sophisticated monitoring and correctives devices, the resolutions are limited to periods greater than 200 nm.

A very interesting alternative has been proposed at Massachusetts Institute of Technology, named Scanning Beam Interference Lithography (SBIL). The principle consists in generating a small area interferometric pattern and then, scans the surface to cover a wide substrate. The main difficulty relies on insuring a controlled displacement at the nanoscale of the writing interferometric head over 1 m^2. This is achieved thanks to the development of a sophisticated interferometric displacement sensor. This technique allowed producing 900 mm x 500 mm gratings.

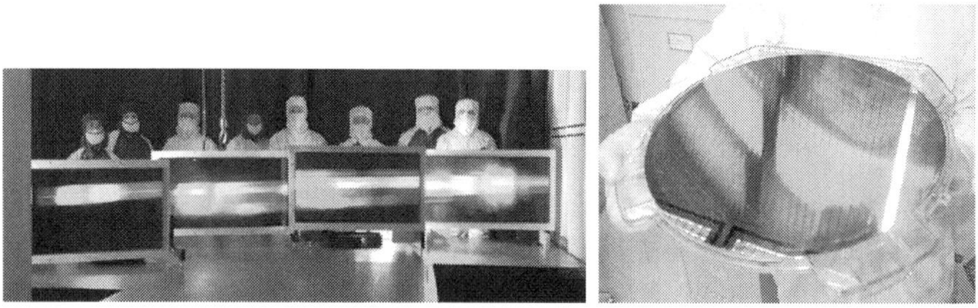

Fig. 3. Left) Multilayer dielectric diffraction gratings produced for NIF's Advanced Radiographic Capability petawatt laser have record size, damage resistance and efficieny (www.lasers.llnl.gov) and right) a 300 mm-diameter silicon wafer patterned with a 400 nm-period grating by the Nanoruler (http://snl.mit.edu/).

3. Polymer-based materials for DUV interferometry lithography

3.1 Chemically amplified photoresists for applications in microelectronics

Polymers are probably the most widely used materials for nanopatterning and many different strategies have been developed to achieve a spatial control of the material deposition down to the nanoscale.

As already mentioned previously, Deep-UV (DUV) photolithography has become the current technique used in the industry of microelectronics for production of sub-micron structures. In this field, Chemically Amplified Photoresists (CAR) are the predominant materials used in the fabrication of nanoscale structures with 193 nm photolithography (Bowden & Turner, 1988; Macdonald et al., 1994). Patterns are defined chemically through the production of acid in areas exposed to DUV light (Figure 4) and an acid-catalyzed deprotection reaction that changes the solubility of the reacted material in an aqueous base solution (Ito, 2005). The patterning process is complex since many composition and process parameters have an impact on the photolithographic performances of the photoresist. Among those parameters, those linked to the resist materials are considered as the most critical. Three factors are essential to consider: resolution limit, sensitivity and line-width roughness (LWR). Resolution limit is the most important criterion since the new platforms of resists should be able to address the challenging next lithographic nodes under 45 nm. However, sensitivity is a parameter of importance for practical applications due to the necessity to achieve short exposure times. Line width roughness (LWR) and line edge roughness (LER), measuring the deviation from an atomically smooth surface have also become new parameters preventing the feasibility of smaller feature sizes (Yoshimura et al., 1993).

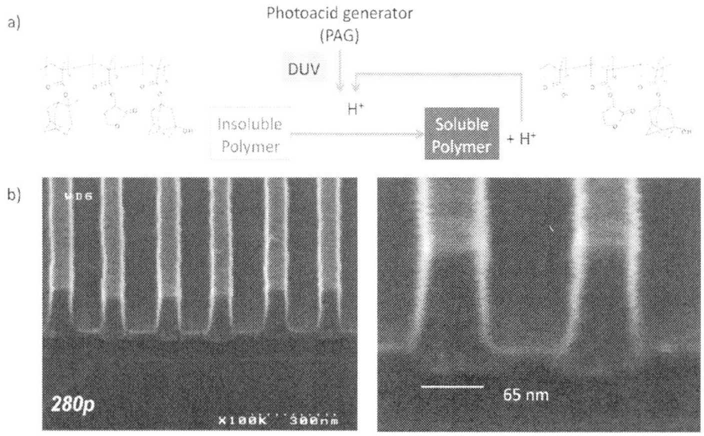

Fig. 4. a) Molecular mechanism of photoinduced modification of CAR photoresists (positive tone resist). b) SEM images of typical samples prepared by DUV interferometric lithography.

These material requirements for the next generations of DUV lithography have justified the recent efforts to develop innovative photoresists. The cost of industrial lithography tools destined at microelectronics applications (few tens of M$) is hardly compatible with time-consuming and potentially contaminant resist development experiments. This is one of the reasons explaining the success of DUV interferometric lithography for developing new photoresists since this tool is relatively easy to install with a reasonable cost, and however, it provides resolutions in the range of the most advanced DUV industrial lithographic tools (few tens of nm). Moreover, immersion lithography or double patterning can be proceeded.

The nature of the polymer significantly contributes to all aspects of resist characteristics and performance. Most of the polymers used as resists are linear copolymers or terpolymers synthesized by the free radical polymerization technique (Ito, 2005; Kang et al., 2006). Such technique has the advantage to be relatively simple but the main drawback is a limited control of the polymer chain structure.

Advanced polymer synthesis strategies like Atom Transfer Radical Polymerization (ATRP) were recently proposed to achieve a better control of the polymer structure, with both linear and hyperbranched structure and for a large variety of monomers (Xia & Matyjaszewski, 2001;

Chochos et al., 2009). Interestingly, a correlation can be obtained between the polymer molecular weight, the chemical composition of the polymer and its polydispersity and its performance in photolithography, for linear and branched polymers, demonstrating the interest of a precise of the polymer molecular structure for nanolithography. (Ridaoui et al., 2010).

3.2 Molecular glass photoresists

Amorphous molecular glasses were recently proposed as a promising class of photoresists matrixes in addition to the traditional polymeric materials. They are amorphous materials with low molecular weight and designed for DUV, Extreme-UV (EUV) or e-beam lithography (Tsuchiya et al., 2005). Low molecular weight materials can form a stable glass above room temperature and they offer several advantages over traditional linear polymers as patterning feature size decreases. Compared to polymeric resists, molecular glasses provide better control of molecular structure and a broader range of building blocks. The recourse to molecular elementary building blocks allows reducing the variations in line width roughness (LWR) and line edge roughness (LER). In addition, the small uniform molecular size offers excellent processability, flexibility, transparency and uniform dissolution properties based on elemental composition.

Many different molecular structures have been proposed recently. Among other works, Figure 5 illustrates 2 examples. First example is based on an Adamantane architecture (Tanaka & Ober, 2006). Adamantane structure brings both high transparency at 193 nm wavelength and high etch resistance by the cage structure. The second example is a 3-component system composed of a monomer, a crosslinker and a photoacid generator.

Upon irradiation, photoacid is generated in the exposed region. Cross-linking reactions between the TMMGU cross-linker and the hydroxy groups on the monomer are catalyzed by the acid generated during the post-exposure baking period. The resulting cross-linked oligomers are insoluble in aqueous base, thus providing the solubility switch required for development. This system is capable of producing 60 nm line/space patterns (Yang et al., 2006).

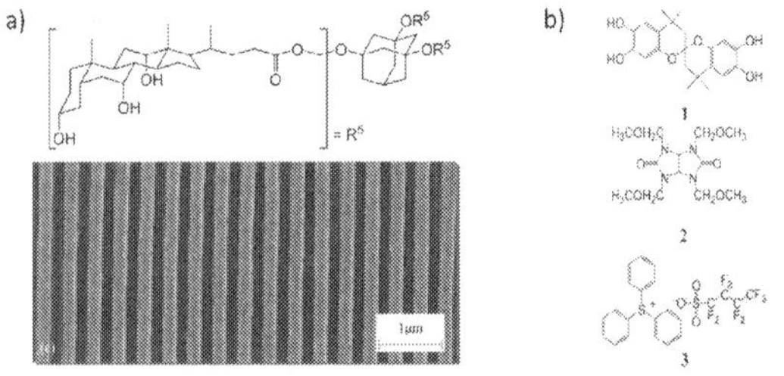

Fig. 5. a) Adamantane based molecular glass photoresist (from Tanaka & Ober, 2006) and b) Components of negative-tone molecular glass resist: monomer (1), TMMGU cross-linker (2), and photoacid generator (3) (Yang et al., 2006).

3.3 Plasma polymer for micro and nanopatterned surfaces

The examples given in the previous parts are in relation with applications in microelectronics that are obviously the most important applications. Besides the fabrication of nanostructures by lithography, the generation of functional materials with chemical control at the nanoscale has drawn considerable interest during the last years.

One of the most important objectives is to propose fast and easy processes to generate hydrophilic/hydrophobic surfaces with nanoscale lateral resolution possibly coupled with topography generation. Such surfaces are of importance for controlled deposition of nano-objects, for applications in biology, or sensors.
In this context, polymers deposited by plasma polymerization have proved to be relevant since they present the following inherent advantages: i) the plasmachemical surface functionalization step is substrate-independent (Boening, 1988), ii) the plasma polymer thin film provides a good adhesion with most of the substrates (Roucoules et al., 2007), iii) the surface density of immobilized molecular species can be finely tuned by varying the pulsed plasma duty cycle (Teare et al., 2002; Oye et al., 2003) and iv) the plasma polymerization step is easily scaled up to industrial dimension (Yasuda & Matsuzawa, 2005).
It was recently demonstrated that Deep-UV lithography could be used to generate topography patterns at the surface of maleic anhydride-based plasma polymers with typical dimension down to 75 nm (Soppera et al., 2008). Macroscopic spectroscopic characterization demonstrated that the surface chemistry was affected by DUV-irradiation (Figure 6). The choice of maleic anhydride-based polymer was guided by the reactivity of anhydride moieties that allows the introduction of further functionalities and/or immobilization of bioactive molecules using different binding strategies. These options are very valuable for the preparation of model substrates for fundamental studies on biointerfacial phenomena as well as for the controlled surface modification of a great variety of bulk substrates.

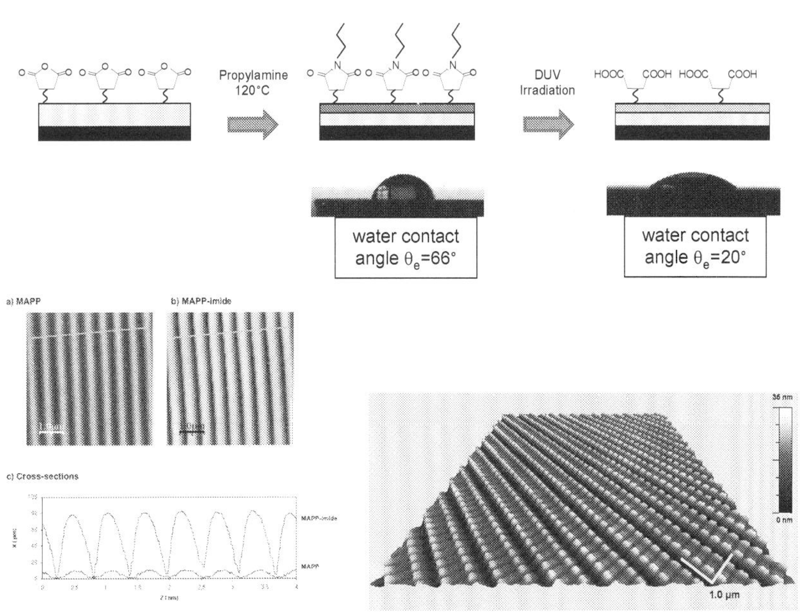

Fig. 6. Plasma polymer thin film preparation and surface chemistry characterization (Soppera et al., 2008; Dirani et al., 2010)

It was also demonstrated that the DUV patterning allows creating topographic nanopatterns associated to a precise tuning of the local surface chemistry (Dirani et al, 2010a; Dirani et al., 2010b; Soppera et al., 2008). The control of the surface chemistry contrast was achieved by a new method using Atomic Force Microscopy in Pulsed Force Mode with plasma polymer functionalized tips (Figure 7).

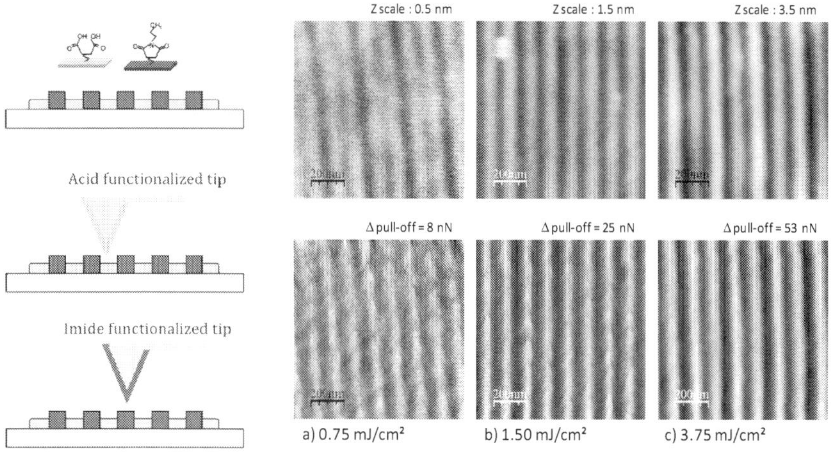

Fig. 7. PFM images of the imide plasma polymer surface for increasing DUV dose. The period of pattern was 150 nm. An imide terminated tip was used for surface probing in air. First raw is the topography image. Last raw shows the local pull-off force. Doses of a) 0.75 mJ/cm^2, b) 1.50 mJ/cm^2 and c) 3.75 mJ/cm^2 were used (Dirani et al., 2010)

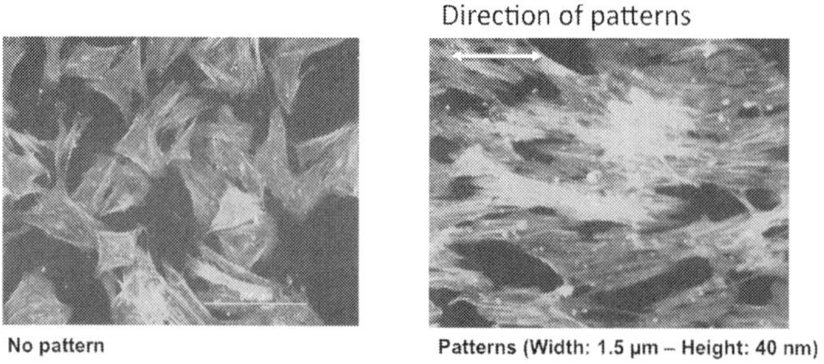

Fig. 8. Comparison of human osteoprogenitor cells development on non-patterned (left) and patterned (right) plasma polymer surface (Ploux et al., 2009).

These chemically and topographically patterned surfaces have high potential as model surfaces for studying cell and bacteria responses to surface chemistry and surface topography. Biological experiments were conducted on patterned maleic anhydride plasma polymer thin films using human osteoprogenitor cells and Escherichia coli K12 (Ploux et al., 2009). Proliferation and orientation of cells and bacteria were analyzed and discussed according to the size and the chemistry of the features. This work showed interesting opposite behavior of bacteria compared to eukaryotic cells, in response to the surface chemistry and to the surface topography. This result may be particularly useful on medical implants design.

4. Inorganic materials for DUV interferometry lithography

4.1 Introduction

Despite many advantages listed below, organic-based polymers have specific limitations like poor mechanical properties, low refractive index and thus, there is a major interest to develop non-organic materials suitable for nanofabrication. In this context, metal oxides present many advantages. As an example, Zirconium dioxide (ZrO_2) has unique properties such as high refractive index, wide optical band gap, low absorption and dispersion in the visible and near-infrared spectral regions, as well as high chemical and thermal stabilities. Other metal oxides like TiO_2, HfO_2, ZnO have many applications in fields like photocatalysis, photovoltaic, displays, biology, optics, and photonics... (Lebeau & Innocenzi, 2010) and require convenient techniques for nanopatterning.

The chemical sol-gel process is probably the most attractive route to elaborate inorganic thin films due to its powerful control on the structural and properties of films, low-cost, and abilities to deal with substrates of large area and/or complex shape (Judenstein & Sanchez, 1996). It is a versatile process that can be adapted for a wide range of oxides (Si, Zr, Ti, Hf, Zn) using simple deposition techniques (dip-coating, spin-coating) and relatively low curing temperatures (few hundreds of °C). Basically, sol-gel chemistry is based on the succession of hydrolysis and condensation reactions as follow:

Hydrolysis

$$M(OR)_4 + H_2O \;->\; HO\text{-}M(OR)_3 + R\text{-}OH \tag{2}$$

$$M(OR)_4 + 4\,H_2O \;->\; M(OH)_4 + 4\,R\text{-}OH \tag{3}$$

Condensation

$$(OR)_3\text{-}M\text{-}OH + HO\text{-}M\text{-}(OR)_3 \;->\; [(OR)_3M\text{-}O\text{-}M(OR)_3] + H\text{-}O\text{-}H \tag{4}$$

$$(OR)_3\text{-}M\text{-}OR + HO\text{-}M\text{-}(OR)_3 \;->\; [(OR)_3M\text{-}O\text{-}M(OR)_3] + R\text{-}OH \tag{5}$$

Where M stands for different elements (Si, Zr, Ti...)

Usually the crosslinking of the layer is achieved by thermal process. In this case, the process is not compatible with photopatterning at the micro or the nanoscale. Several strategies have been developed to achieve patterning on sol-gel deposited thin metal oxide layers. Some examples for ZrO_2 are given in (Thomas, 1994; Belleville et al., 2000; Belleville et al., 2003; Zhang et al., 2000; Tian et al. 2005). Usually they rely on complex multistep processes, such as photolithographic patterning and chemical etching or lift-off. These methods often use sacrificial masking materials and the transfer step induces a loss of resolution.

There is thus a major challenge to develop direct means for nanopatterning. One way consists in proposing hybrid materials based on precursors combining an inorganic function and a photopolymerizable one (Soppera et al., 2003; Matejka et al., 2004). The main limitation of this route is to finally provide a hybrid material with a relatively high proportion of organic part.

4.2 Nanopatterning of hybrid precursors by DUV interferometry

In order to obtain inorganic nanostructures, sol-gel-based spin-coatable precursor of ZrO2 that is not only amenable to direct-write using DUV lithography but is also capable of providing nanoscale resolution was proposed. (Ridaoui et al., 2010)

The crosslinking photoreaction relies on a direct photolysis of the complexed Zr atom. The interaction between UV light and metal alkoxide complex has been already described : after light absorption, charge transfer complexes can be created and they can induce a photolysis of the ligand (Versace et al., 2008). For Titanium alkoxides, it has been proved that Ti-oxo complex gave rise to a decarboxylation reaction (Hundiecker reaction (Soppera et al., 2001)). In the present case, we can assume that the same kind of mechanism occurs, leading to the production of reactive Zr species that can react on free alkoxides functions to create a tridimensional ZrO_2 network (Figure 9).

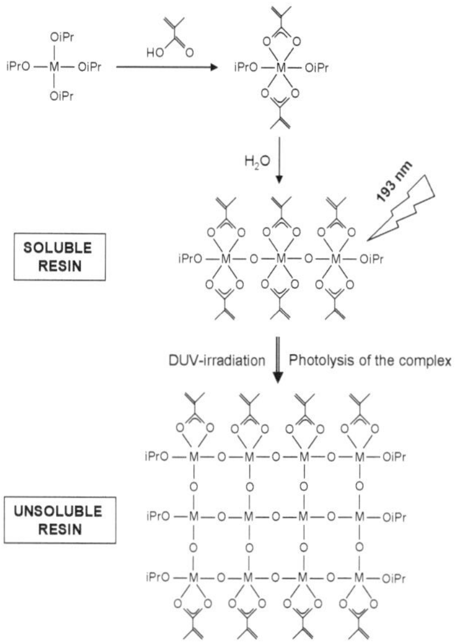

Fig. 9. Schematic representation of the material preparation and modification under DUV irradiation of the negative tone inorganic resist. M is a transition metal (Zr, Ti...). As shown in this scheme, the DUV irradiation results in condensation of the partially condensed metal alkoxide precursors resulting in a modification of solubility of the thin film (Ridaoui et al., 2010)

The main interest of this negative tone resist relies on the possibility to remove the organic part and obtain ZrO_2 after thermal treatment. The AFM scan of the sample is plotted in Figure 10. It can be observed that the patterns remained perfectly defined after the thermal treatment, opening a very convenient way to produce micro or nanostructures. Applications of such nanostructures with high refractive index are expected in the field of optics and photonics. Many applications of such technology are also expected in all applications fields in which robust nanostructures with inertness towards chemical, temperature, and pressure are needed such as photovoltaic, photocatalysis or biology.

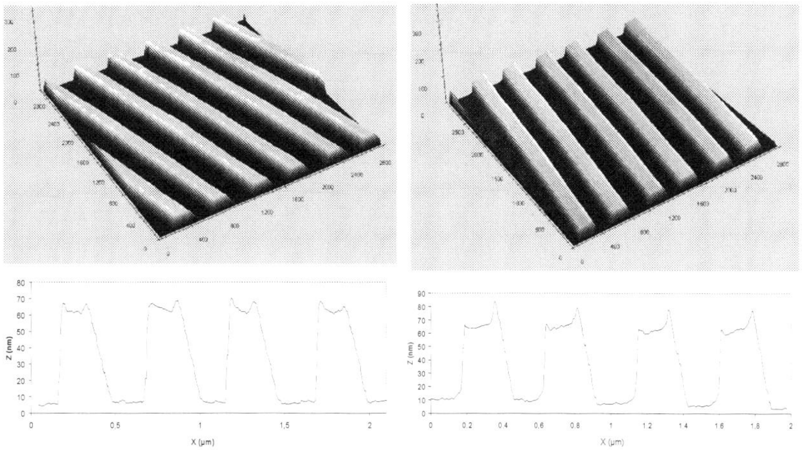

Fig. 10. AFM image of a periodic patterns written on the Zr sol-gel film. Period was 500 nm. Left) represent the grating before thermal heating, and Right) is a grating after thermal heating (600°C, 2 hours).

5. Conclusion

The development of large-area, high-resolution nanostructures is a challenging problem that must be addressed for applications in high performance nanoscale devices, such as nanoelectronics, optics, microfluidics, organic solar cell, display devices and biosensing devices. Today's challenges are not limited to the resolution issue but many others aspects are to be considered. In particular, there is a growing need for simple processes enabling integration of functional materials.

With this regards, DUV interferometric lithography techniques are still of high interest, as illustrated in this chapter. The recourse to advanced optical setup and immersion allow creating patterns with typical dimensions much smaller than 100 nm on the basis of many different materials. This photon-based technique is thus competitive in terms of resolution with other advanced nanofabrication techniques and because it is a massively parallel technology to produce nanoparts on a large substrate, it is well-complementary to e-beam, ion-beam or nanoimprint.

6. Acknowledgment

Agence Nationale pour la Recherche (ANR - Projects NANORUGO, HOLOSENSE and NIR-OPTICS), CNRS and Région Alsace are gratefully acknowledged for financial supports

7. References

Archambault, J.-L.; Reekie L. & Russell, P.St.J. (1993). 100% reflectivity Bragg reflectors produced in optical fibres by single excimer laser pulses. *Electron. Lett.*, Vol.29, No.5, (March 1993), pp. 453-455

Askins, G.; Tsai, T.-E.; Williams, G.M.; Putnam, M.A.; Bashkansky M. & Friebele E.J. (1992). Fiber Bragg reflectors prepared by a single excimer pulse. *Optics Lett.*, Vol.17, No.11, (1992), pp. 833-835

Belleville, P.; Bonnin, C. & Priotton, J.J. (2000). Room-Temperature Mirror Preparation Using Sol-Gel Chemistry and Laminar-Flow Coating Technique. *J. Sol–gel Sci. Technol,* Vol.19, No.1-3, (2000), pp. 223-226

Belleville, P.; Prené, P. ; Bonnin, C. ; Beaurain, L. ; Montouillout, Y. & Lavastre, E. (2003). *Proc. SPIE,* 5250, 196

Boening, H.V. (1988). Fundamental of plasma chemistry and technology, Technomic Publishing Company, Inc., 1988, p. 75.

Bourgin, Y.; Bakkali, S.; Jourlin, Y.; Tonchev, S. & Parriaux, O. (2009). Monolithic double-grating phase mask for large-period highly coherent grating printing. *Optics Letters,* Vol.34, No.24, (2009), pp. 3800-3802

Bourov, A.; Fan, Y.; Cropanese, F.C.; Lafferty, N.V.; Zavyalova, L.V.; Kang., H.; Smith, B.W. (2004). Immersion microlithography at 193 nm with a Talbot prism interferometer. *Proc. SPIE,* vol. 5377, (2004), p. 1573.

Bowden, M. J. & Turner, S. R. (1988) Electronic and photonic applications of polymers. *ACS Adv. Chem. Ser.*, Washington (1988) 218, 109.

Braun D., (2002). Semiconducting polymer LEDs. *Materials Today,* Vol.5, No.6, (June 2002), pp. 32-39

Cai Y. & Ocko B. (2005). Electro Pen Nanolithography. *J. Am. Chem. Soc., Vol.*127, No.46, (October 2009), pp. 16287-16291

Campbell, M.; Sharp, D. N.; Harrison, M. T.; Denning, R. G. & Turberfield, A. J. (2000). Fabrication of photonic crystals for the visible spectrum by holographic lithography. *Nature,* Vol.404, (March 2000), pp. 53-56

Charley, A.L.; Lagrange, A.; Lartigue, O.; Bandelier, P.; Derouard, M & Schiavone, P. (2006). Liquid Immersion Lithography at 193 nm Using a High-NA Achromatic Interferometer, *Proc. of SPIE* Vol. 6154 61541Z

Chochos, C. L.; Ismailova, E.; Brochon, C.; Leclerc, N.; Tiron, R.; Sourd, C.; Bandelier, P.; Foucher, J.; Ridaoui, H.; Dirani, A.; Soppera, O.; Perret, D.; Brault, C.; Serra, C. A. & Hadziioannou, G. (2009). Hyperbranched Polymers for Photolithographic Applications : Towards Understanding the Relationship between Chemical Structure of Polymer Resin and Lithographic Performances. *Advanced Materials, Vol.*21, No.10-11, (2009), pp. 1121-1125

Christman K.L., Schopf E., Broyer R.M., Li R.C., Chen Y. & Maynard H.D. (2009). Positioning Multiple Proteins at the Nanoscale with Electron Beam Cross-Linked Functional Polymers. *J. Am. Chem. Soc.* Vol.131, No.2, (December 2008), pp. 521-527

Cotton D. V., Fell C.J. & Dastoor P.C. (2009). Fabricating semi-conducting polymer photonic structures via near-field scanning optical lithography. *Synthetic Metals.* Vol.159, No.5-6, (March 2009), pp. 456-461

Dirani, A.; Roucoules, V.; Haidara, H. & Soppera, O. (2010). Plasma polymer tailoring of the topography and chemistry of surfaces at the nanoscale. *Langmuir,* Vol.26, No.22, (October 2010), pp. 17532-17539

Dirani, A.; Wieder, F.; Roucoules, V.; Airoudj, A. & Soppera, O. (2010). Nanopatterning of Plasma Polymer Thin Films by ArF Photolithography: Impact of Polymer Structure on Patterning Properties. *Plasma Processes and Polymers,* Vol. 7, No.7, (February 2010), pp.571-581

Dyer, P.E.; Farley, R.J. & Giedl, R. (1995). Analysis of grating formation with excimer laser irradiated phase masks. *Opt. Comm.* Vol.115, No.3-4, (August 2005), pp 327-334

El Ahrach H., Bachelot R., Vial A., Lerondel G., Plain J., Royer P. & Soppera O. (2007). Spectral Degeneracy Breaking of the Plasmon Resonance of Single Metal Nanoparticles by Nanoscale Near-Field Photopolymerization. *Phys. Rev. Lett.,* Vol.98, No.10, (March 2007), pp. 107402-1-107402-2

Fan H. J., Werner P. & Zacharias M. (2006). Semiconductor Nanowires: From Self-Organization to Patterned Growth. *Small.* Vol.2, No.700, (March 2006), pp. 700-717

Gombert A., Benedikt B., Christopher B., Peter N., Jörg M., Wolfgang H. & Michael N., (2004), Replicated microstructures with optical functions in solar and display applications (Proceedings Paper). *Optical Engineering,* Vol.43, No.22, (December 2003), pp. 2525-2533

Gonsalves K.E., Wang M., Lee C.T., Yueh W., Tapia-Tapia M., Batina N. & Henderson C.L. (2009). Novel chemically amplified resists incorporating anionic photoacid generator functional groups for sub-50-nm half-pitch lithography. *J. Mater. Chem.* Vol.19, (March 2009), pp. 2797-2802

Graeter S.V., Huang J., Perschmann N., Lopez-Garcia M., Kessler H., Ding J. & Spatz J.P. (2007). Mimicking Cellular Environments by Nanostructured Soft Interfaces. *Nano Letters,* Vol.7, No.5, (March 2007), pp. 1413-1418

He X., Winkel J. & Huck W.T. (2009). Nanopatterning via Pressure-Induced Instabilities in Heilmann, R.K.; Akilian, M.; Chang, C.H.; Fleming, R.; JOO, C.; Montoya, J.C.; Torkmann, A. & Schattenburg, M.L. (2004). MIT CSR IAP l-Lectures series Frontiers of Astronomy, Astrophysics and Space Sciences.

Henzie J., Barton J. E., Stender C. L. & Odom T. W. (2006). Large-area nanoscale patterning: chemistry meets fabrication. *Accounts of Chemical Research.* Vol.39, No.4, (January 2006), pp. 249-257

Huh J.W., Jeong J. W., Lee J.W., Shin S.I., Kwon J.H., Choi J., Yoon H.G., Cho G.I., You I.K., Kang S.Y. & Ju B.K. (2009). Carbon nanotube and conducting polymer dual-layered films fabricated by microcontact printing. *Appl. Phys. Lett.* Vol.94, (June 2009), pp. 223311-1-223311-2

Im S.G., Kim B.S., Tenhaeff W.E., Hammond P.T. & Gleason K.K. (2009). A directly patternable click-active polymer film via initiated chemical vapor deposition (iCVD). *Thin Solid Films.* Vol.517, No12, (January 2009), pp. 3606-3611

Ito H., (2005). Chemical amplification resists for microlithography. *Advances in Polymer Science,* Vol.172, No.276, (2005), pp. 37-245.

Jegadesan S., Sindhu S., Advincula R.C. & Valiyaveettil S. (2006). Direct Electrochemical Nanopatterning of Polycarbazole Monomer and Precursor Polymer Films: Ambient Formation of Thermally Stable Conducting Nanopatterns. *Langmuir,* Vol.22, (November 2005), pp. 780-786

Jeon, S.; Menard, E.; Park, J. U.; Maria, J.; Meitl, M.; Zaumseil, J. & Rogers, J. A. (2004). Three-Dimensional Nanofabrication with Rubber Stamps and Conformable Photomasks. *Adv. Mater.*, Vol.16, No.15, (August 2004), pp. 1369-1373

Joannopoulos, J. D.; Meade, R. D. & Winn, J. N. (1995). Photonic Crystals; Princeton University Press: Princeton, NJ, 1995.

Judeinstein, P. & Sanchez, C. (1996). Hybrid Organic-Inorganic Materials : A Land of Multidisciplinarity. *J. Mater. Chem.*, Vol.6, No.4, (1996), pp. 511-525

Kamigaito M., Ando T. & Sawamoto M., (2001). Metal-Catalyzed Living Radical Polymerization. *Chem. Rev.*, Vol.101, No.12, (December 2001), pp. 3689-3746

Kanamori Y., Roy E., Chen Y., (2005). Antireflection sub-wavelength gratings fabricated by spin-coating replication. *Microelectronic Engineering*, Vol.78-79, (March 2005), pp. 287-293

Kang S. H., Prabhu V. M., Vogt B. D., Lin E. K., Wu W.-L. & Turnquest K., (2006). Effect of copolymer composition on acid-catalyzed deprotection reaction kinetics in model photoresists. *Polymer*, Vol.47, No.18, (August 2006), pp. 6293-6302

Kang, J. H.; Moon, J. H.; Lee, S. K.; Park, S. G.; Jang, S. G.; Yang, S. & Yang, S. M. (2008). Thermoresponsive Hydrogel Photonic Crystals by Three-Dimensional Holographic Lithograph. *Adv. Mater.*, Vol.20, No.16, (August 2008), pp. 3061-3065

Karp, Jason H., Tremblay, Eric J., Ford & Joseph E., (2010). Planar micro-optic solar concentrator. *Optics Express*, Vol.18, No.2, (January 2010), pp. 1122-1133

Lebeau, B & Innocenzi, P. (2011). Hybrid materials for optics and photonic. *Chemical Society Review*, Vol.40, No.2, (2011), pp. 886-906

Lee W.K., Whitman L.J., Lee J., King W.P. & Sheehan P.E. (2008). The nanopatterning of a stimulus-responsive polymer by thermal dip-pen nanolithography. *Soft Matter*, Vol.4, No.9, (2008), pp. 1844-1847

Lin, S. Y.; Fleming, J. G.; Hetherington, D. L.; Smith, B. K.; Biswas, R.; Ho, K. M.; Sigalas, M. M.; Zubrzycki, W.; Kurtz, S. R. & Bur, J (1998). A three-dimensional photonic crystal operating at infrared wavelengths. *Nature*, Vol.394, (July 1998), pp. 251-253

Macdonald S. A., Wilson C. G. & Fréchet J. M. J., (1994). Chemical Amplification in High-Resolution Imaging Systems. *Acc. Chem. Res.* Vol. 27, No.6, (June 1994), pp. 151-158

Matejka, L.; Strachota, A.; Plestil, J.; Whelan, P.; Steinhart, M. & Slouf, M. (2004). Epoxy Networks Reinforced with Polyhedral Oligomeric Silsesquioxanes (POSS). Structure and Morphology. *Macromolecules*, Vol.37, No.25, (November 2004), pp. 9449-9456

Menez, L.; Mulet, J.P.; Lalanne, P. (2008). Patent WO/2008/145895 Light Scattering Network

Mihailov, S.J.; & Gower, M.C. (1994). Recording of efficient high-order Bragg reflectors in optical fibres by mask image projection and single pulse exposure with an excimer laser. *Electron. Lett.*, Vol.30, No.9, (April 1994), pp. 707-709

Moon, J.H. & Yang, S. (2009). Chemical Aspects of Three-Dimensional Photonic Crystals. *Chem. Rev.*, Vol.110, No.1, (August 2009), pp. 547-574

Moore G. E. (1965). Cramming More Components Onto Integrated Circuits. *Electronics*, Vol.38, No.8, (April 1965, pp. 114-117

Nie Z. & Kumacheva E. (2008). Patterning surfaces with functional polymers. *Nature Mater.*Vol.7, pp. 277-290

Oye, G. ; Roucoules, V. ; Oates, L.J.; Caameron, A.M.; Cameron, N.R.; Steel, P.G.; Badyal, J.P.S.; Davis, B.G.; Coe, D. M. & Cox, R.A. (2003). Plasmachemical Amine Functionalization of Porous Polystyrene Spheres: The Importance of Particle Size. *J. Phys. Chem. B.*, Vol.107, No.15, (2003), pp. 3496-3499

Ploux, L.; Anselme, K.; Dirani, A.; Ponche, A.; Soppera, O. & Roucoules, V. (2009). Opposite Responses of Cells and Bacteria to Micro/Nanopatterned Surfaces Prepared by

Pulsed Plasma Polymerization and UV-Irradiation. *Langmuir,* Vol. 25, No.14, (June 2009), pp. 8161-8169
Raub, A.K. & Brueck, S.R.J. (2003). Deep UV immersion interferometric lithography. *Proc of SPIE* 2003, 5040, 667.
Ridaoui, H.; Dirani, A.; Soppera, O.; Ismailova, E.; Brochon, C.; Schlatter, G.; Hadziioannou, G.; Tiron, R.; Bandelier, P. & Sourd, C. (2010). Chemically amplified photoresists for 193-nm photolithography: Effect of molecular structure and photonic parameters on photopatterning. *J. Polym. Sci. Part A: Polym. Chem.*", Vol.48, No.6, (February 2010), pp. 1271-1277
Ridaoui, H.; Wieder, F.; Ponche, A. & Soppera, O. (2010). Direct ArF laser photopatterning of metal oxide nanostructures prepared by the sol–gel route. *Nanotechnology,* Vol.21, No.6, (January 2010), pp. 065303
Roucoules, V.; Ponche, A.; Siffer, F.; Ergurrolla, U. & Vallat, M. F. (2007). Strengthening the Junction Between EPDM and Aluminium Substrate via Plasma Polymerisation. *J. Adhes.* Vol. 83, No.10, (2007), pp. 875-895
Santillan, J.; Otoguro, A.; Itani, T.; Fujii, K.; Kagayama, A.; Nakano, T.; Nakayama, N.; Tamatani, H. & Fukuda, S. (2006). A study of 193-nm immersion lithography using novel high refractive index fluids. *Microelectronics Engineering,* Vol.83, No.4-9, (September 2006), pp. 651-654
Schon J.H., Kloc C., Dodabalapur A. & Batlogg B., (2000). An Organic Solid State Injection Laser. *Science,* Vol.289, No.5479, (July 2000), pp. 599-601
Seo K. & Borguet E. (2006). Nanolithographic Write, Read, and Erase via Reversible Nanotemplated Nanostructure Electrodeposition on Alkanethiol-Modified Au(111) in an Aqueous Solution. *Langmuir,* Vol.22, No.4, January 2006), pp 1388-1391
Soppera O., Jradi S. & Lougnot D.J. (2008). Photopolymerization with microscale resolution: Influence of the physico-chemical and photonic parameters. *J. Polym. Sci. Part A: Polym. Chem.,* Vol.46, No.11, (April 2008), pp. 3783-3794
Soppera, O. & Croutxe-Barghorn C. (2003). Real-time Fourier transform infrared study of free-radical UV-induced polymerization of hybrid sol–gel. I. Effect of silicate backbone on photopolymerization kinetics. *J. Polym. Sci., Part A : Polym. Chem.,* Vol.41, No.5, (January 2003), pp. 716-724
Soppera, O.; Croutxe-Barghorn & C.; Lougnot, D.J. (2001). New insights into photoinduced processes in hybrid sol–gel glasses containing modified titanium alkoxides. *New J. Chem.,* Vol.25, No.8, (2001), pp. 1006-1014
Soppera, O.; Dirani, A.; Ponche, A. & Roucoules, V. (2008). Nanopatterning of plasma polymer reactive surfaces by DUV interferometry. *Nanotechnology,* Vol.19, No. 39, (August 2008), pp 395304-395312
Tanaka, S. & Ober, C.K. (2003). Adamantane-based Molecular Glass Resist for 193nm Lithography *Proc. of SPIE* 2006, 6153.
Teare, D.O.H.; Spanos, C.; Ridley, P.; Kinmond, E.J.; Roucoules, V. & Badyal, J.P.S. (2002). Pulsed plasma deposition of super-hydrophobic nanospheres. *Chem. Mater.* Vol. 14, No.11, (2002), pp. 4566-4571.
He, X.; Winkel, J.; Huck, W. Nanopatterning via Pressure-Induced Instabilities in Thin Polymer Films. *Adv. Mater.* Vol. 21, (March 2009), pp. 2083-2087
Thomas I.M. (1994). Preparation of dielectric highly reflective (HR) mirrors from colloidal oxide suspensions containing organic polymer binders. *Proc. of SPIE* 2288, 50

Tian, G.L.; Huang, J.B.; Wang, T.; He, H.B. & Shao, J.D. (2005). Microstructure and laser-induced damage threshold of ZrO_2 coatings dependence on annealing temperature. *Appl. Surf. Sci.*, Vol.239, No.2, (July 2004), pp. 201-208

Tsuchiya, K.; Chang, S. W.; Felix, N. M.; Ueda, M. & Ober, C. K. (2005). Lithography Based on Molecular Glasses. *Journal of Photopolymer Science and Technology*, Vol.18, No.3, (May 2005), pp. 431-434

Versace, D.L.; Soppera, O.; Lalevée, J. & Croutxe-Barghorn, C. (2008). Influence of zirconium propoxide on the radical induced photopolymerisation of hybrid sol-gel materials. *New J. Chem*, Vol.32, No.12, (2008), pp. 2270-2278

Wegmann G., Giessen H., Greiner A. & Mahrt R.F., (1998). Laser emission from a solid conjugated polymer : Gain, tunability, and coherence. *Phys. Rev. B*, Vol.57, No.8, (February 1998), pp. R4218-R4221

Xia J. & Matyjaszewski K., (2001). Atom Transfer Radical Polymerization. *Chem. Rev.*, Vol.101, No.9, (February 2001), pp. 2921-2990

Xiong, Z.; Peng, G.D.; Wu, B. & Chu, P.L. J. (1999). Effects of the zeroth-order diffraction of a phase mask on Bragg gratings. *Lightwave Techn*, Vol.17, No.11, (November 2002), pp. 2361-2365

Yang, D.; Chang, S.W. & Ober, C.K (2006). Molecular glass photoresists for advanced lithography. *J. Mater. Chem.*, Vol.16, No.18, (2006), pp. 1693-1696

Yang, S.; Megens, M.; Aizenberg, J.; Wiltzius, P.; Chaikin, P. M. & Russel, W. B. (2002). Creating Periodic Three-Dimensional Structures by Multibeam Interference of Visible Laser. *Chem. Mater.* Vol.14, No.7, (June 2002), pp. 2831-2833

Yao, P.; Qiu, L.; Shi, S.; Schneider, G.J.; Prather, D.W.; Sharkawy, A. & Kelmelis, E. (2008). Fabrication of 3D polymer photonic crystals for near-IR applications. *Proc. SPIE*, Vol.6883, (January 2008), pp. 688313-688319-9

Yasuda, H. & Matsuzawa, Y. (2005). Economical Advantages of Low-Pressure Plasma Polymerization Coating. *Plasma Process. Polym.*, Vol. 2, No.6, (July 2005), pp. 507-512

Yasuda, H. (1985). Plasma Polymerization; Academic Press: London, 1985.

Yen, A.; Anderson, E.H.; Ghanbari, R.A.; Schattenburg, M.L. & Smith, H.I.; Achromatic holographic configuration for 100-nm-period lithography. *Applied Optics* Vol.31, No.22, (August 1992), pp. 4540-4545

Yoon B., Acharya H., Lee G., Kim H-C., Huh J. & Park C. (2008). Nanopatterning of thin polymer films by controlled dewetting on a topographic pre-patter. *Soft Matter*, Vol.4, No.7, (2008), pp. 1467-1472

Yoshimura T., Shiraishi H., Yamamoto J. & Okazaki S. (1993). Correlation of Nano Edge Roughness in Resist Patterns with Base Polymers. *Jpn. J. Appl. Phys.* Vol. 32,(August 2003), pp. 6065-6070

Zhang, Q.Y.; Shen, J.; Wang, J.; Wu, G.M. & Chen, L.Y. (2000). Sol-gel derived ZrO_2–SiO_2 highly reflective coatings. *International Journal of Inorganic Materials*, Vol.2, No.4, (April 2000), pp. 319-323

14

Ultrashort Pulsed Lasers – Efficient Tools for Materials Micro-Processing

Marian Zamfirescu[1,2], Magdalena Ulmeanu[1], Alina Bunea[2], Gheorghe Sajin[2] and Razvan Dabu[1]
[1]*National Institute for Laser, Plasma and Radiation Physics,*
[2]*National Institute for Microtechnology,*
Romania

1. Introduction

Fabrication technologies on the micro and nanometer scale are becoming more and more important from the viewpoint of industrial applications, for example: high-resolution lithography for manufacturing high-density recording media, high-resolution displays, and high sensitivity biomolecule sensor arrays. Electron-beam lithography, interferometric lithography, electron-beam evaporation of constituent materials and lift-off procedures were used to fabricate negative refractive-index metamaterials at the near-infrared (NIR) wavelengths (Zhang et al., 2005; Dolling et al., 2006; Dolling et al., 2007). Due to its capability for large-area fabrication, conventional photo-lithography is a wide-spread fabrication technology. Because the minimum feature size is limited by optical diffraction, sub-micrometer structures can be created by deep ultra-violet (DUV) at hundred-nanometer wavelengths or extreme ultra-violet (EUV) at ten-nanometer wavelengths lithography techniques (Gwyn et al., 1998). Because these technologies require complex vacuum optics, their cost remains prohibitive.

Ultrashort pulsed lasers, particularly femtosecond lasers, could offer an alternative to currently used micro and nanostructuring methods (Nishiyama et al., 2008; Qi et al., 2009; Allsop et al., 2010; Sugioka et al., 2010). High peak power can be reached for femtosecond pulses at relatively low energy per pulse. Due to the high radiation intensity, nonlinear effects dominate the interaction of tightly focused femtosecond laser beams with materials.

Pulsed laser micromachining involves the removal of material through the ablation process which consists in some consecutive physical processes: laser energy absorption, material heating, material expelling, and material cooling (Liu et al., 1997; Stuart et al., 1996).

The first step in laser ablation is the absorption of laser energy by the target material. The absorption mechanism depends on laser intensity (laser fluence and pulsewidth) and can be accomplished by linear and nonlinear processes. For opaque materials at laser radiation wavelength, linear absorption is the main mechanism at long pulsewidths with low intensity, whereas the nonlinear absorption becomes dominant at ultrashort pulsewidths with high intensity. For transparent materials, absorption comes from nonlinear processes through laser-induced optical breakdown. It is a process where a normally transparent material is first transformed in absorbing plasma by avalanche ionization and multiphoton ionization. At relatively low laser intensity, the avalanche ionization process in a transparent dielectric material is seeded by free electrons coming from material impurities, thermal or linear optical ionization of shallow energy levels of inclusions. Free electrons can absorb laser energy through the inverse Bremsstrahlung process, when they collide with the bound electrons and the material lattice. Seed electrons can be accelerated at energy exceeding the ionization potential of the bound electrons. The collisions of seed electrons with bound electrons give rise to an avalanche ionization process growing exponentially from the initial very low seed electrons density. When plasma with a critical density is created, the transparent material is broken down and becomes absorbing at laser radiation wavelength. If the laser field strength is very high, as in the case of femtosecond-pulse laser-matter interaction, bound electrons can be directly ionized through multiphoton absorption. For longer laser pulses (microsecond-nanosecond pulsewidths) where the field strength is lower, the multiphoton ionization contribution is negligible and laser-induced breakdown is dominated by avalanche ionization. For ultrashort pulses, multiphoton ionization determines the breakdown threshold behavior: only when the laser intensity exceeds a certain threshold the plasma density grows to the critical value where irreversible breakdown takes place. The avalanche ionization fluence breakdown threshold exhibits large fluctuations due to statistical variations of the number of seed electrons already present in the material. Multiphoton ionization can directly generate free electrons, it is not related to the impurity seed electrons. Consequently, the laser induced breakdown threshold at ultrashort pulsewidths becomes more precise.

The material is heated up by the electrons energy transfer to the ions and the lattice. The amount of laser energy transfer during the laser pulse depends on the laser pulse duration and the energy coupling coefficient. Due to the thermal gradient, the absorbed energy can leave the laser focal volume by heat conduction and a larger volume is heated. For absorbing materials, such as metals and semiconductors, the laser pulse energy is absorbed in a surface layer whose thickness is given by the skin penetration depth, $l_s = (\alpha)^{-1}$, where α is the absorption coefficient. The heat penetration depth due to the thermal conduction is given by the diffusion length, $l_d = \sqrt{D\tau_l}$, where D is the heat diffusion coefficient and τ_l is the laser pulsewidth. For microsecond and nanosecond laser pulses, $l_d > l_s$, and the volume of the heated material, hence the temperature, is determined by the heat diffusion length. Therefore, for long pulses, the fluence breakdown threshold varies with the laser pulsewidth as $F_{th} \propto \sqrt{\tau_l}$. If the laser pulsewidth decreases, the diffusion length decreases too. For laser pulses in the femtoseconds range, $l_d < l_s$, the skin penetration depth determines the heated volume during the laser pulse. As a result, the breakdown threshold becomes independent of the pulsewidth.

For long pulses, a larger volume of material is heated and melted, but only a small layer of material reaches the vaporization temperature. Material removal is accomplished through melt expulsion driven by the vapor pressure and the recoil of the light pressure. This physical process involving fluid dynamics and vapor conditions is an instable one. For micro-processing applications, the resolidification of the melt after the ablation can lead to very irregular shapes of the holes or cuts. On the other hand, for ultrashort pulses, the laser energy is deposited in a thin layer with a thickness limited by the skin penetration depth. The localized energy heats the material very quickly at high temperatures to the vapor phase with high kinetic energy. The material removal takes place mainly by direct vaporization. Because most of the heated material reaches the vaporization temperature, the resulting melt layer thickness is much smaller than in case of long pulses. After the laser pulse irradiation is finished, there is a rapid cooling due to the steep temperature gradient. Because little liquid is involved, the ablation process becomes highly precise in comparison to the long pulse case. By precision micromachining with femtosecond laser pulses, feature sizes with submicrometer resolution can be obtained (Chimmalgi et al., 2003).

Micro- and nanostructures can be created also through some nonlinear optical processes that occur when the interaction of high intensity femtosecond laser radiation with materials is performed under the ablation threshold. By tightly focusing the femtosecond laser pulses into the bulk of transparent materials, 3D micro-structures can be produced by a permanent refractive-index modification inside a small focal volume (Will et al., 2002; Osellame et al., 2003; Vega et al., 2005). Laser energy is deposited in this volume by multi-photon absorption and avalanche ionization. The photogenerated hot electron plasma rapidly transfers its energy to the lattice giving rise to high temperatures and pressures. The nonelastic thermo-mechanical stress created in the focal region produces a local densification with an increase of refractive index over a micrometer-sized volume of the material. This index gradient allows one to fabricate complicated photonic structures in many transparent materials like fused silica (Will et al., 2002; Shah et al., 2005), silicate and phosphate glasses (Homoelle et al., 1999; Ams et al., 2005), chalcogenide thin films (Zoubir et al., 2004), sapphire (Wortmann et al., 2008), poly(methyl methacrylate) (Sowa et al., 2006). Various microstructures contributing to developments in integrated optics, optical communications, and optical data storage have been obtained by refractive-index patterning: optical waveguides (Will et al., 2002; Osellame et al., 2003; Zoubir et al., 2004; Ams et al., 2005; Sowa et al., 2006; Allsop et al., 2010), beam splitters (Homoelle et al., 1999), micro-channels (Wortmann et al., 2008), directional couplers (Streltsov & Borelli, 2001), three-dimensional data storage (Glezer et al., 1996), diffraction gratings (Takeshima et al., 2004), photonic crystals (Takeshima et al., 2005).

In case of irradiation of UV-photopolymerizable materials with high-intensity tightly focused NIR femtosecond laser pulses, a nonlinear optical process of multi-photon absorption can take place. In such photopolymers, the two-photon absorption of NIR femtosecond laser pulses induces photochemical reactions and then photopolymerization, just like in the case of a single UV photon absorption (Maruo et al., 1997; Witzgall et al., 1998). Unlike the UV-radiation photo-polymerization by single photon absorption, a high-probability of two-photon absorption occurs only in a very tiny volume of material near the center of the focused laser spot, leaving behind photo-polymerized patterns with micrometer or sub-micrometer dimensions. Direct laser writing by multiphoton polymerization of photoresists has emerged as a technique for the rapid, cheap and flexible fabrication of 3D structures with a resolution beyond the diffraction limit, leading to advanced applications in telecommunication, photonics, metamaterials and biomedicine (Deubel et al., 2004; Rill et al., 2008; Ovsianikov et al., 2009; Farsari & Chichkov, 2009).

In this chapter, we present some experimental results in the field of femtosecond laser micromachining and micro/nanostructuring of both opaque and transparent materials. Three-dimensional (3D) micro-structuring of negative photo-resists by two-photon photo-polymerization (TPP) using low energy high repetition rate NIR femtosecond lasers is described in the Section 2 of the paper. Fabrication of GHz range composite right/left handed (CRLH) devices by a technology that combines classical photo-lithography and femtosecond laser processing is described in the Section 3. Near-field laser lithography reaching a resolution beyond the diffraction limit through the interaction of ultra-short pulsed laser beams with small size particles is presented in Section 4. In the Section 5, a micro-printing method based on femtosecond laser induced forward transfer (LIFT) of various materials deposited in thin films is demonstrated.

2. 3D structuring by two-photon photo-polymerization (TPP)

Building a 3D micro-structure was always a challenge, since micrometer and submicrometer resolutions are hardly obtained on complex arrangements using various materials. Applications such as microfluidics, micro- and nanoelectronics, integrated optics, motivated the development of manufacturing techniques. Among these, micro-milling, micro-moulding, electro-discharge machining, imprinting lithography, X-ray and UV lithography, electron and ion-beam lithography, chemical and plasma etching, laser rapid-prototyping, direct laser writing (DLW), are the most used microfabrication technologies (Frassila, 2010). However, none of these techniques satisfies simultaneously all the requirements in terms of spatial resolution, high-aspect-ratio of the structures, or ability of processing a large variety of materials and complex designs. Combinations of different techniques are sometimes the way for overcoming some technological bottlenecks. For example, X-ray lithography combined with electroforming and/or moulding allows microfabrication on metals or plastics with resolution below 100 nm and aspect-ratio up to 100 or even higher (Becker et al., 1986). However, the resulting structures are not fully 3D designs. These are more 2D structures with a finite dimension in Z direction, the so called 2½D structures. Despite of good resolution of the method in XY directions, such structures have limited complexity.

3D-stacking approach was recently used in combination with thin film deposition, lithographic and selective etching techniques for fabrication of 3D integrated circuits such as microprocessors and memories (Pavlidis et al., 2009). It requires complex and very expensive machineries justified only by very high mass production. From commercial point of view, for custom applications or limited series, these technologies are not the most competitive ones.

The laser processing of materials always offered a cheap and reliable solution for microfabrication. The lasers become an omnipresent tool, from thin film deposition, to laser processing by ablation, or laser spectroscopy characterization. Concerning the fabrication of full 3D structures, the laser rapid prototyping or stereolithography have been developed by C.W. Hull in 1986. A 3D model can be created inside an UV-curable material. The UV-laser irradiates a photosensitive material layer by layer following a certain path. The size of the 3D model is typically of the order of few-mm to hundreds of mm in each direction, with resolution of tens of micrometers.

Since the femtosecond lasers were developed, the approach of rapid prototyping inspired the development of micro-stereolithography (Maruo et al., 1997). The nonlinear effect such as two-photon absorption of NIR radiation is easily achieved in photoresists which usually absorb the UV radiation. When the material is transparent to NIR photons, the laser can be focused deeply in the volume of the material. At the tightly focused spot of a femtosecond nJ laser pulse, the laser intensity is high enough to exceed the threshold of the nonlinear absorption. Permanent physical and chemical modifications of the material take place in a small volume, deep inside of the transparent material. If the laser fluence is kept low enough, small features can be created with resolution down to tens of nm's (Farsari et al., 2005; Tan et al., 2007). By following the rapid-prototyping algorithms, a 3D model can be created layer by layer inside the material.

A large series of negative photoresists, among them SU-8, organically modified ceramics (Ormocer), or organically modified silicate (Ormosil), have the maximum of the absorption band in the UV-blue spectral range. In such photopolymers, the two-photon absorption of NIR Ti:sapphire femtosecond laser radiation induces photochemical reactions and then photo-polymerization, just like in the case of a single UV photon absorption. In contrast with the single photon processing, the two-photon absorption occurs in a very tiny volume of material, near the centre of the focused spot, allowing creation of features with resolution below 1 micrometer.

2.1 Experimental set-up

The laser workstation for material processing consists in the following main parts: the femtosecond laser, the beam delivery and focusing optics, and the scanning mechanics. In our experiments we use a set-up compatible with different type of laser structuring methods, by laser ablation as well as by TPP. The set-up consists in a modular microscope for laser writing built to be coupled with laser beams at different radiation wavelengths, depending on the experimental requirements. For applications such as laser ablation, requiring high pulse energy from tens nano-Joules up to micro-Joules, a laser amplifier CPA2101 system (Clark-MXR) was employed. This laser emits femtosecond pulses of as much as 0.6 mJ energy, 200 fs pulse duration at 775 nm wavelength and 2 KHz repetition rate. In experiments requiring high repetition rate, such as TPP, a 75 MHz repetition rate femtosecond laser oscillator Synergy Pro (Femtolasers) was coupled with the processing microscope. The oscillator delivers laser pulses of 5 nJ pulse energy, 10 fs duration at 790 nm central wavelength, with spectral bandwidth of 100 nm. The beam delivery optics is interchangeable and can be easily replaced with optics adapted for other working wavelengths, such as harmonics of fundamental wavelength.

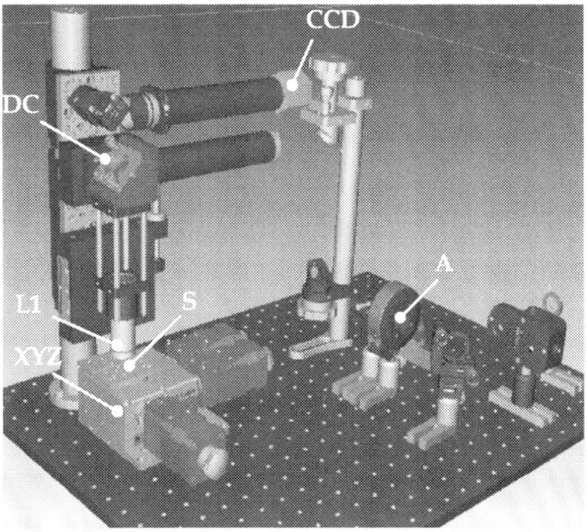

Fig. 1. Experimental set-up for laser processing. A - Attenuator with half waveplate and Glan Polarizer; DC - Dichroic mirror; L1 - Focusing lens; S - Sample; *XYZ* -Translation stage; CCD - Video camera.

The laser workstation has a modular configuration as shown in figure 1. The main modules are: the attenuation module, the focusing optics, the *XYZ* sample translation stages, and the visualisation system with video camera.

The laser energy can be continuously attenuated by a motorised half waveplate placed in front of a Glan polarizer providing a 300:1 extinction rate. In the case of extremely short femtosecond pulses, a reflective polarizer is used for avoiding the temporal stretching laser pulses due to the dispersion introduced by bulk material. A dielectric mirror reflects the laser beam at 800 nm to the focusing optics and transmits the visible radiation to the visualization system.

For beam focusing, different microscope objectives or lenses with a wide range of numerical aperture, adapted to a specific application, are used. The same focusing objective is used for visualization. The focusing optics is chosen depending on the desired resolution of the processed sample or the depth of the structure to be obtained. Submicrometer resolutions were obtained with a 100x Mitutoyo microscope objective with 0.5 numerical aperture and long working distance of 12 mm, allowing the processing deeply in the volume of a transparent material. With such objective, the focused laser spot has a minimal diameter of about 1 μm, and a confocal parameter of about 2 μm.

In case of 100-nm bandwidth pulses, the positive group delay dispersion (GDD) introduced by the beam delivery optics was partially compensated by an optical temporal compressor formed by a pair of SF_{11} prisms. By changing the distance between the two prisms, the negative dispersion in the compressor is adjusted and laser pulses with less than 40-fs duration were delivered to the processed materials.

For processing the sample according to a computed design, a *XYZ* translation (Nanocube - Thorlabs) with motorized stages and piezo drivers is used. The stage has a total travel range of 4x4x4 mm^3 with hundreds of nm accuracy. The embedded piezo stage has 20 μm travel range per each axis and accuracy down to 5 nm. For longer travels, linear stages with 50 mm maximum travel are used. The maximum translation speed is 2 mm/s.

The sample focusing is done by the visualization system with CCD and a 200 mm tube lens. The resolution of the visualization system is better than 1 μm when a 100x microscope objective is mounted.

Dedicated software was realized for controlling the laser processing of samples. Common structures such as parallel lines, grids for 2D geometries, alphanumerical characters are included in a predefined library. Also, 3D predefined periodical structures such as vertical pillars in rectangular, hexagonal, or woodpile symmetries can be fabricated.

2.2 Laser processing protocol of photosensitive materials

For a reproducible fabrication of a required design a good control of processing protocol is necessary. The main processing steps are: the substrate preparation; deposition of the photosensitive material; laser irradiation; development and final curing of the polymerized structures.

2.2.1 Substrate preparation for photoresist deposition

A critical step is the initial treatment of the substrate in order to provide a good adhesion of the polymerized structure. The frequently used treatments are cleaning by reactive plasma etching, chemical corrosion, or a simple ultrasonic bath in solvents followed by thermal dehumidification, depending on the substrate nature (silicon wafer, glass, fused silica). A promoter film deposition is sometime a good solution to increase adhesion, especially in the case of glass substrate with reduced adherence. Producing photoresists with enhanced adhesion is also desirable, since the treatment of the substrate is simplified. All the following processing steps are preferable to deploy in a controlled environment since the dust can easily stick to the photoresist film, compromising the final geometry.

2.2.2 Photoresist films deposition and treatment

The commonly used technique to deposit films of non-polymerised resin on the substrate is the spin-coating. The thickness of the resulting photoresist film can be controlled by the viscosity of the initial monomer and the spinning speed. It can vary from tens of micrometers down to few micrometers for spinning speeds varying typically from 500 rot/min up to 6000 rot/min. If structures with height of the order of hundreds of micrometer are required, the sample has to be prepared differently. A drop of resin is placed between two transparent slides separated by spacers with fixed width of hundreds of micrometers up to 1-2 mm. Thicknesses below 1 μm are also possible for specific applications. In such case, resins with low viscosity are spin-coated at speed up to 10.000-12.000 rot/min. However, the uniformity of the film thickness is difficult to be controlled on the entire surface.

After deposition, the processing protocol is followed by a specific treatment for each type of photoresists. Since the viscosity of the material is controlled by concentration of the solvent present in the liquid resin, the solvent has to be removed prior to laser treatment of the material. An insufficient removal of solvent will result in soft and deformed final microstructure, with low adherence to the substrate. For example, in the case of SU-8, the deposited film resist has to be baked for 15-30 minutes at 95°C. This is the so called Hard Backing step. The baking time depends on the photoresist thickness. The increase of the Hard Backing temperature should in principle accelerate the removal of the solvent. However, a fraction of solvent always remains in the volume of the material, and increasing the Hard Backing temperature causes a gradient of the solvent concentration and distortion of the microstructure. Low bake temperature for longer time can provide a constant evaporation rate, and finally low and uniform concentration of the remaining solvent in the resulting solid film resin (Liu et al., 2005).

Other materials such as Ormocers and Ormosil can be directly processed without any intermediate thermal treatment. Such photoresists remains liquid till are laser irradiated, then the cross-linking process transforms the resin into a solid. However, the removal of solvent before laser processing will significantly improve the fabrication of the microstructures, especially in terms of aspect ratio, adhesion to the substrate and smoothness of the surfaces.

2.2.3 Laser irradiation of the photoresists

Laser exposure conditions, such as laser fluence and scanning speed should be established prior to fabrication of complex structures. The map of processing parameters for a commercial Ormocore, a derivate of Ormocer photoresist, is shown in figure 2. Different scanning speeds and laser fluences are tested. The scanning speed were varied from 0.1 to 0.8 mm/s, and the laser fluence from Fo to 2Fo, with Fo = 0.28 J/cm^2. The structures from figure 2 were obtained by focusing the laser beam at 2 KHz repetition rate by a 75 mm lens with a 25 μm beam diameter at the beam waist. In order to determine the optimal laser irradiation dose, structures of *XY* parallel lines, with 50-μm period on each direction, were produced under different exposure conditions. For low speed and high laser fluence, the polymer is over exposed and damaged. For example, in case of scanning speed of 0.1 mm/s, the laser fluence of 1.5-2 Fo was over the damage threshold. With a laser spot diameter of 25 μm and a laser repetition rate of 2 KHz, we can estimate about 520 laser pulses per each spot contributing to the polymerization process. Under these conditions, the irradiation dose is high enough to damage the material by producing bubbles in the photoresist. The structure optical quality is compromised.

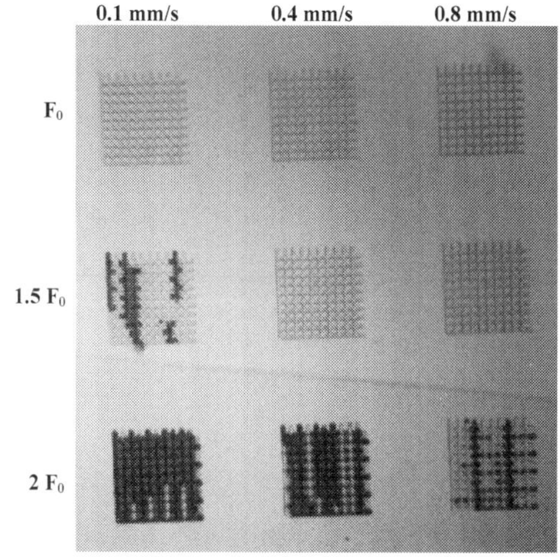

Fig. 2. Processing map for the determination of optimal laser irradiation parameters.

At 1.5 Fo laser fluence, but much higher scanning speed of 0.8 mm/s, only 65 laser pulses contribute to the photopolymerization process. Such irradiation parameters produce smooth structures with good optical quality, transparent to the visible light. Similar structures were obtained for Fo fluence and 0.1 mm/s processing speed. Decreasing the laser fluence to Fo, thinner structures can be obtained. From practical point of view, higher scanning speed is preferred because of shorter processing times. However, the accuracy of the fabricated structures is better at lower scanning speeds. The optimum parameters have to be a compromise between the total processing time and the processing quality.

2.2.4 Development

After laser exposure, the non irradiated material is removed by rinsing the sample in a specific solvent, depending on the nature of the photoresist. In the case of SU-8 photopolymer, an intermediate step of post-exposure bake is needed, before rinsing, for accelerating the process of cross-linking of molecules and then polymerization. For other materials such as Ormocers, Ormosil, or chalcogenic glasses, this step is not required, and the processing protocol is simplified. In the case 3D microstructuring of liquid resist, the UV curing of the structures after development is preferred for strengthening the processed microstructure.

2.3 Micro-rapid-prototyping for 3D microstructuring in photoresists

The classical photolithography techniques are usually limited to the fabrication of 2D structures. Also, several technological steps are required in order to complete a final design: mask fabrication, exposure, selective corrosion, etc. The laser direct-writing method provides a more simplified protocol for fabrication of different devices. Moreover, it can be successfully applied to the fabrication of 3D structures, using the rapid prototyping approach (Kawata et al., 2001). As in the case of macroscale fabrication, the micro-rapid-prototyping supposes the reconstruction of a 3D structure, using the laser imprinting in bulk material, from a computer generated solid by CAD software.

In our laser processing workstation, for 3D structuring of transparent materials, the software allows the import of complex designs as standard format STL files, commonly used in rapid prototyping. The solid is numerically divided in series of sections as in the figure 3b. Each layer is laser imprinted inside the photosensitive material by translating the focused femtosecond laser beam relative to the sample, accordingly to a calculated path. The two-photon or multiphoton absorption of the femtosecond radiation will induce photochemical modification and densification of the material at the focused spot, as described in the previous section. By incrementing the position in Z direction, a solid shape will result. The irradiated material is transparent to the laser wavelength and allows deep focusing of the beam inside the volume of the material. The formation of high aspect ratio 3D microstructures is possible.

A high-aspect-ratio structure produced by TPP in Ormoclad photoresist is shown in figure 3c. The 3D structure is a miniaturised "Endless Column", a sculpture created by the Romanian artist Constantin Brâncuşi (1876-1957). The microstructure is 600 μm in height, and is formed by 4 octahedron-shaped modules ended by a half module. Each module is constructed by squared layers with maximum size of 100x100 μm^2. In order to obtain a good overlap between layers the increment in Z direction was 2 μm, while the vortex size in Z direction was about 7 μm.

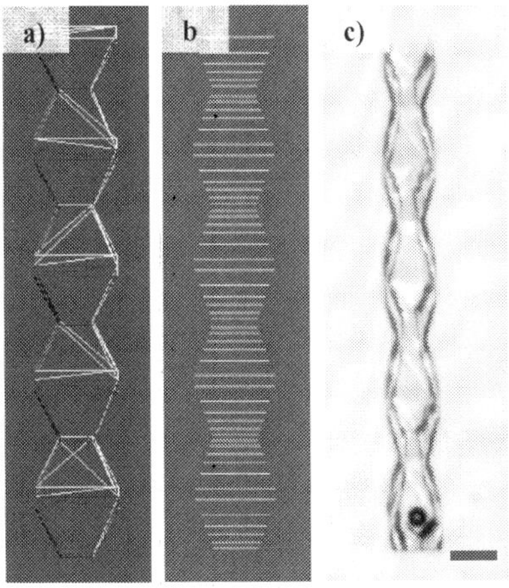

Fig. 3. Description of 3D stereolithography algorithm: a) The design in STL format; b) Computed slices of the 3D design; c) High aspect ratio 3D structure fabricated in Ormoclad. Scale bar: 100 μm. The solid structure is formed by TPP in layer by layer approach.

The laser irradiation was done by the femtosecond laser beam at 75 MHz repetition rate, with pulse duration less than 40 fs, and laser beam power of 60 mW. The scanning speed was 0.5 mm/s. The focusing optics was a 100x microscope objective with 0.5 numerical aperture.

After irradiation, the non-polymerised resin was removed by OrmoDev, a solvent specific for Ormocers. The remained 3D model was gently rinsed in isopropyl alcohol. As observed in the figure 3c, the miniaturised Endless Column has a hollow shape. At the bottom part of the structure a bead is trapped. Its high-aspect-ratio shape demonstrates the capability of our experimental set-up for fabricating complex structures with applicability in microfluidics. The same microfabrication approach can be involved in applications such as photonics, integrated optics, or tissue engineering. Some structures demonstrating the application of TPP are shown in figure 4. Biocompatible scaffolds were fabricated by photopolymerisation of Ormosil as support for culture of live cells (Matei et al., 2010). After deposition of the material on glass substrate, a 2D grid is realized by TPP (figure 4a). The grid is formed by polymerised lines in *X* and *Y* direction, spaced at 100 μm, with thickness of about 10 μm, and highness of about 50 μm. After appropriate microbial decontamination of the sample, epithelial human cells MRC-5 were grown on the polymerised scaffold. As shown in figure 4b, the cells attach on the grid setting each on a square element. Such biocompatible scaffolds are of big interest in biomedical applications such as BioMEMS, tissue engineering, and medical implants (Weiß et al., 2009).

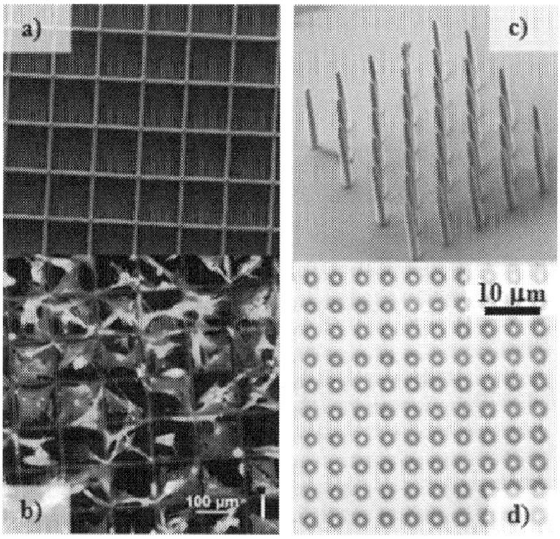

Fig. 4. Microstructures fabricated by TPP: (a) Scaffold for tissue engineering fabricated in ORMOSIL. (b) Stem cells deposited on the polymerised structure. Periodical photonic structures in SU-8 photoresist in hexagonal (c) and rectangular (d) arrangements.

Photonic structures can be also produced by TPP in photopolymers such as SU-8. Columnar structures of 20 μm height and 2 μm diameter were obtained by scanning the laser beam in *Z* direction (figure 4b and 4c). The lattice constant of the structure is 5 μm in both *X*, *Y* directions. Such a periodical structure has a photonic bandgap near 10 μm wavelength. Bandgaps at visible spectral range are also possible for photopolymerised photonic crystals with submicrometre lattice constant and different arrangements, such as woodpile structures (Deubel et al., 2004).

2.4 The limits of TPP microstructuring method

The quality of a 3D structure is characterized by some parameters, such as adhesion to the substrate, shrinkage of the obtained geometry compared to the desired design, limited aspect ratio, and limited minimal feature size. The limitations of the TPP technique are mainly given by intrinsic physical and chemical properties of the photoresist and the laser focusing capability. For overcoming them, new polymers have to be developed with better performances such as low shrinkage and good adhesion on commons substrates such as glass or silica. Also, the behaviour of the photoresist and its solvent strongly affect the aspect-ratio of the structure. Some materials become very soft during development, and the structure falls down to the substrate after removing the sample from the solvent. Beside these, the limitations due to the processing equipments are mainly related to the processing resolution. Limited smoothness of the structures is given by the non uniform movement of the translation stages. Another limitation of the rapid prototyping method is the total processing time required to fabricate a structure. Since the rapid prototyping is a laser scanning technique, this is a time consuming method requiring tens of minutes up to few hours in order to complete a complex 3D structure. The solution to overcome this disadvantage is to use lasers with very high repetition rate and also fast scanning mechanics such as galvano-scanners. However, a compromise between writing speed and accuracy has to be done, the piezoelectric stages remaining the most accurate positioning systems used in 3D DLW.

3. Femtosecond laser microprocessing of composite right/left handed (CRLH) metamaterials for millimeter wave devices

3.1 Metamaterial CRLH millimeter wave devices

Metamaterials (MTM) are propagation media presenting simultaneously both negative permittivities ($\varepsilon < 0$) and negative permeabilities ($\mu < 0$). These media, with unusual properties not readily available in nature, are called Left Handed Materials (LHM). For a LHM, the triade formed by the electric field, magnetic field and Poynting vector of an electromagnetic field propagating through this media has a "left hand" orientation, different from common materials where this orientation is a "right hand" one. As a result, LHM support propagation of an electromagnetic wave where the group velocity is anti-parallel to the phase velocity. This phenomenon associates with a negative refraction index.

Although presented as a theoretical curiosity since 1968 (Veselago, 1968), practical applications of "left handed" media appeared 30 years later when the first experimental investigations were made (Pendry, 1999; Shelby et al., 2001). Since that time, a great variety of media with metamaterial characteristics and subsequent applications from microwave to the visible domain were developed.

In the microwave and millimeter wave frequency domain, the main conventional propagation media are the transmission lines (microstrips, coplanar waveguides). The right handed form of these transmission line structures may be assimilated to a large enough number of cascaded cells, each cell being made of *series inductor – parallel capacitor*. For this type of line, called **R**ight **H**anded **T**ransmission **L**ine (RH-TL), the group velocity and the phase velocity have positive values.

The left handed transmission lines are made by cascading a large enough number of cells, each cell composed of *series capacitor – parallel inductor*. For this type of line, called **L**eft **H**anded **T**ransmission **L**ine (LH-TL), the group velocity has positive values while the phase velocity has, in some conditions, negative values. In practice LH-TLs are made as artificial lines using lumped or distributed circuit elements. Due to the parasitic inductances and capacitances associated with series capacitors and parallel inductors respectively, the line configuration will present, actually, cells of type *series LC – parallel LC*, the result being a composite kind of line named **C**omposite **R**ight / **L**eft **H**anded Transmission Line (CRLH-TL).

CRLH structures act as LH-TLs at frequencies where the guided wavelength is larger than the cell size and as RH-TLs at high frequencies, where the guided wavelength is smaller than the cell size. Between these frequency domains there is a forbidden frequency band where the propagation is cut. These are so called unbalanced CRLH structures. If this forbidden band is reduced to a single point, a balanced CRLH structure is obtained and this is one of the conditions for designing some CRLH devices for microwaves and millimeter waves.

Based on the CRLH structures, a large class of microwave and millimeter wave devices (MMW) has been developed (Caloz & Itoh, 2006). Among such devices, made on the basis of metamaterial approach, various types of directional couplers (Caloz et al., 2004; Wang et al., 2006), filter configurations (Li et al., 2007; Liu et al., 2009), antennas (Sajin et al., 2007; Seongmin et al., 2008; Ziolkowski et al., 2009; Basharin & Balabukha, 2009; Eggermont et al., 2009) were reported. The metamaterial CRLH-TL approach allows substantial space reduction compared to the standard devices. Moreover, combined with other metamaterial devices and circuits, it offers the possibility to develop a new and different kind of microwave and mm-wave circuitry.

Complex monolithically integrated circuits can be fabricated using coplanar waveguide (CPW) configurations on semiconductor substrates. The design of the MTM microwave and millimeter wave devices is based on CRLH-TL elementary cells. A CPW CRLH elementary cell with distributed parameters is usually composed of series interdigital capacitors and parallel inductive stubs to the ground. The schematic layout of an elementary CPW CRLH cell is shown in figure 5a, while the equivalent circuit is given in figure 5b.

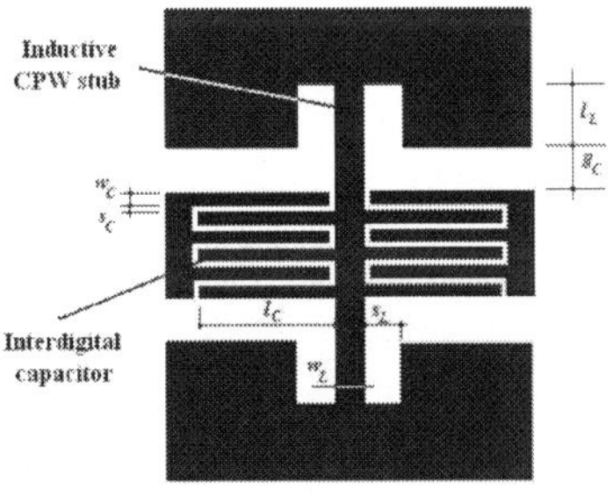

(a)

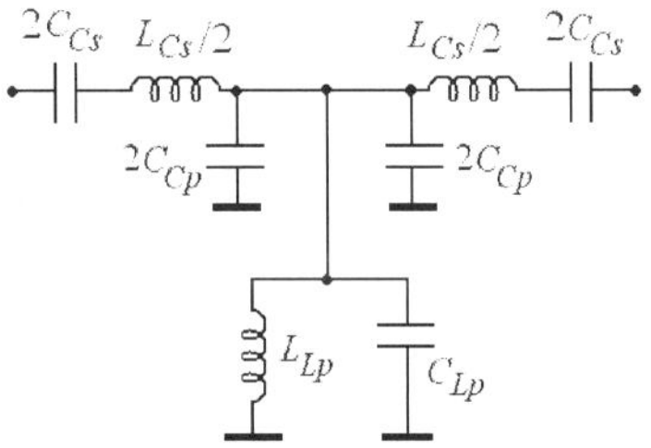

(b)

Fig. 5. Layout of an elementary CRLH cell (a) and the equivalent circuit (b).

C_{Cs} are the series connected interdigital capacitors and L_{Lp} is the ground connected inductor. At high frequencies, L_{Cs} (parasitic inductance of the interdigital capacitor), C_{Cp} (parasitic capacitance to the ground of the interdigital capacitor), and C_{Lp} (parasitic capacitance of the inductive stub) become significant circuit elements. The CRLH-TL elementary cell configuration is determined by the working frequency and the materials used. The values of capacitances and inductances forming the CRLH cell (see figure 5b) were obtained with appropriate computer software (Microwave Office, HFSS). These programs also give the geometrical dimensions of the CRLH cell, as they are presented in figure 5a, where: w_C, l_C, and s_C are the width, length and distance between two digits of the interdigital capacitor, respectively; w_L and s_L are the width and distance to the ground of the inductive CPW stub, respectively; g_C is the distance to the ground planes of the CPW line.

At high frequencies (ten-hundred GHz), the cell geometrical details reach the limits of classical photolithography and the laser ablation becomes a valuable technique to process such MMW devices.

3.2 Zeroth-order resonant CRLH antenna on silicon substrate

This antenna consists of an array of CRLH cells. Each cell is made of two series connected CPW interdigital capacitors and two parallel conected short-ended CPW inductive transmission lines. The layout of an elementary CRLH cell and its equivalent circuit are presented in figure 5. The designed antenna is a zeroth-order resonance device configuration consisting of an open-ended array of CRLH cells. Using CPW transmission lines, the circuit area of the CRLH antenna could be much smaller compared to the microstrip antenna where microstrip lines are used.

For an open-ended CRLH antenna, the zeroth-order resonance is at the frequency $f_{sh} = 1/[2\pi(L_{Lp}C_{Lp})^{1/2}]$ which is the parallel resonance due to the two CPW short-ended inductive transmission lines. Also, there are resonance frequencies corresponding to the right-hand (RH) and the left-hand (LH) CRLH behaviour (Caloz & Itoh, 2006). For the operating frequency of the zeroth-order antenna, f_{sh} , $\beta = 0$ where β is the equivalent phase constant of the CRLH cell, this frequency being the highest one for the LH frequency range.

A zeroth-order resonance CPW CRLH antenna working at 28 GHz frequency was designed. It consists of three resonant T - shaped CRLH cells processed on a high resistivity silicon wafer substrate. The conditions and mathematical relations for the CRLH cells design are presented in literature (Caloz & Itoh, 2006; Sajin et al., 2007). For future integration in a more complex circuit, the CRLH cell was designed on silicon substrate, using CPW transmission lines. The substrate was a 500 μm thick silicon wafer ($\varepsilon_{r,Si} = 11.9$) of 5 kΩcm resistivity, covered with 1 μm SiO_2 layer ($\varepsilon_{r,SiO2} = 4.7$) grown by thermal oxidation. The Si wafer was plated by a sputtering process with a metallic layer of 3000 Å Au / 500 Å Cr. The backside of the silicon wafer was not metalized. The calculated dimensions of the interdigital capacitors and inductive stubs for these CRLH cells are the following: s_C = 5 μm, l_C = 250 μm, w_C = 10 μm, w_L = 42 μm, s_L = 10 μm, l_L = 212 μm, g_C = 65 μm. Each capacitor has 10 digits. The antenna access line has 3400 μm length with a geometry computed to match the 50 Ω characteristic impedance of the measuring system.

The processing technology applied to obtain the antenna structure consists of two steps by combining two technologies: photolithography and direct laser writing. In the first step, the Au/Cr metallization was removed from the large areas of the structure by standard wet photolithography. Since the laser structures writing by ablation is a time consuming scanning method, the lithography was preferred for large size area processing in order to obtain the main layout of the future device. Also, the removal of the metallic film by laser ablation from large area could result in a high quantity of ablated material re-deposited on the sample surface, contaminating the active area of the device. In the second step, for the fine details of the interdigital capacitor, the precise laser ablation was used.

The samples were laser ablated by tightly focusing the femtosecond laser pulses with 200 fs duration, 775 nm wavelength, tens of nJ pulse energy, and 2-kHz repetition rate. A 20x microscope objective, 0.4 NA, was used for the Au/Cr layer ablation, obtaining interdigital spaces of 5 μm, as designed. The 2D structures were generated according to a computer controlled algorithm by precisely translating the sample with a resolution below 1 μm.

An optical microscopy image of the active part of the antenna structure is presented in figure 6a, where the grounded lines forming the inductive stubs and the interdigital capacitors, processed by laser ablation, can be observed. A Scanning Electron Microscopy (SEM) image showing details of the antenna interdigital capacitors is presented in figure 6b.

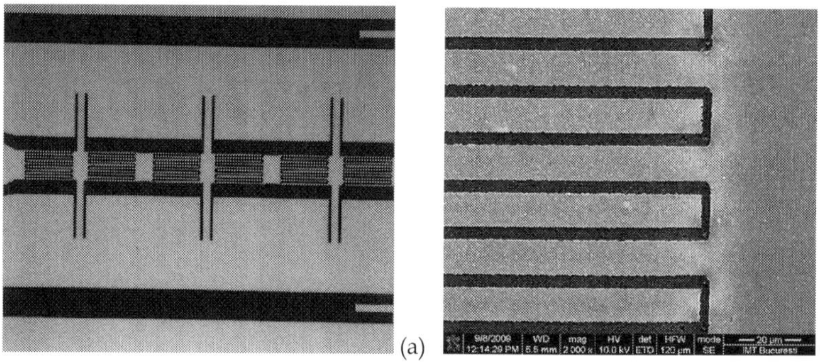

(a) (b)

Fig. 6. (a) Optical microscopy image of the active part of the CRLH antenna. (b) SEM image showing a detail of the interdigital capacitor made by femtosecond laser processing.

Many electrical properties such as gain, return loss, voltage standing wave ratio, reflection coefficient, and amplifier stability of microwave and mm-wave networks may be expressed using S-parameters (Kurokawa, 1965). In the context of S-parameters, scattering refers to which way the traveling currents and voltages in a transmission line are affected when a discontinuity is inserted. These losses appear when the incident wave meets an impedance which differs from the line's characteristic impedance.

The input return loss (RL) parameter of the CRLH antenna is presented in figure 7. It represents a scalar measure of how close the actual input impedance of the device is to the nominal system impedance value and, expressed in decibels, is given by RL_{in} (dB) = $|20\times\log_{10}|S_{11}||$ where, by definition, $|S_{11}|$, is equivalent to the reflected voltage magnitude divided by the incident voltage magnitude.

For a frequency domain between 28.01 GHz and 29.66 GHz, an input return loss having values RL < -15 dB can be observed. The minimum value of RL = -27.05 dB was obtained at 28.78 GHz. It represents a very good matching of the antenna structure to the 50 Ω characteristic impedance of the millimeter wave measurement system and to the most usual microwave and millimeter devices.

In order to measure the radiation characteristic, antenna chips were mounted on specialized test fixtures. The radiation characteristic was measured at 28.7 GHz frequency where the antenna return losses have a minimum. The measuring setup allows the rotation of the CRLH emitting antenna both in transversal and longitudinal planes.

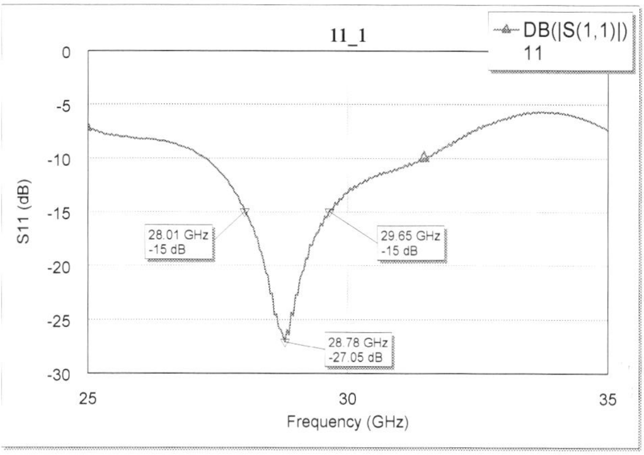

Fig. 7. Return loss of the CRLH antenna for a frequency sweep between 25-35 GHz

The radiation characteristic rated to the maximum power in the transversal plane of the CRLH antenna structure is shown in figure 8a, while the characteristic in the longitudinal plane is presented in figure 8b. The -3 dB beam-width of the radiation characteristic is about 37 degrees in the transversal plane and 25 degrees in the longitudinal plane. The longitudinal radiation characteristic has the maximum at the angle of ~ 24 degrees.

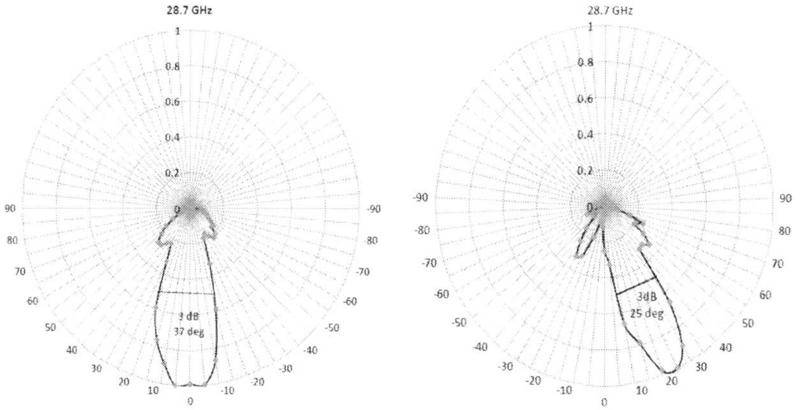

(a) (b)

Fig. 8. Radiation characteristic in the transversal plane (a) and in the longitudinal plane (b) for the measured CRLH antenna

The antenna gain, which relates the intensity of an antenna radiation in a given direction to the intensity that could be produced by a hypothetical ideal antenna that radiates isotropically (equally in all directions), was computed using the De Friis formula (Balanis, 1997). The gain was nearly constant in the 28 ÷ 29 GHz frequency band, with a maximum value of 2.99 dB at 28.6 GHz. The area occupied by the antenna is 2.15 × 0.6 mm^2. There is a considerable size reduction compared to a standard $\lambda/2$ patch antenna.

3.3 Millimeter wave CRLH band-pass filter and directional coupler

The band-pass filter was made by cascading a number of identical CRLH cells. The directional coupler consists of two coupled CRLH artificial lines, each composed of two identical cascaded CRLH cells. Both band-pass filter and directional coupler (figure 9a,b) were microprocessed by the two-steps technology described in section 3.2.

The geometrical dimensions of the interdigital capacitor and inductive CPW stub of the CRLH band-pass filter at 50 GHz working frequency were calculated: s_C = 5 µm, l_C = 250 µm, w_C = 10 µm, g_C = 65 µm, the number of capacitor digits = 10, w_L = 42 µm, s_L = 10 µm, and l_L = 212 µm (see figure 5a). Unlike the antenna CRLH cell, the filter CRLH cell has only one inductive CPW stub. A silicon substrate plated with 2000 Å Au/500 Å Cr was used.

The working parameters of the band-pass filter structure were measured in the 40-60 GHz frequency range. The S_{11} parameter evaluation show return loss values < -15 dB in a frequency range of 46.32 GHz ÷ 55.4 GHz, whereas losses in the same frequency range, given by S_{21} parameter, are around 6 ÷ 7 dB (Sajin et al., 2010a).

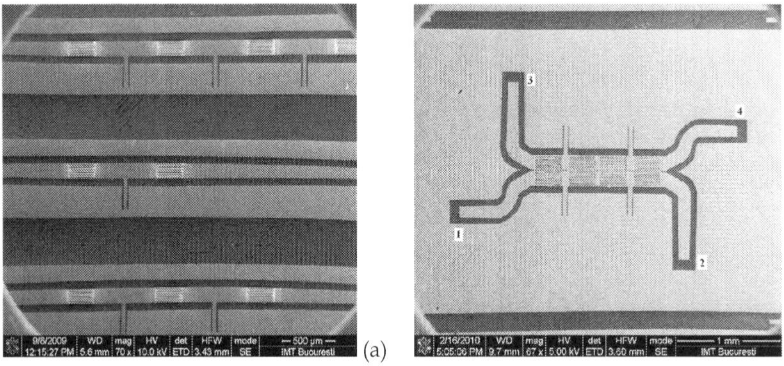

(a) (b)

Fig. 9. SEM images of CRLH band pass filter (a) and directional coupler (b) structures microprocessed by laser ablation.

For the CRLH directional coupler the port 1 is the input port, while the ports 2, 3 and 4 are the transmission, coupled, and isolated ports, respectively (figure 9b). Measured return losses (S_{11}) were better than -20 dB for a frequency domain between 24.01 GHz ÷ 38.11 GHz, whereas the isolation (S_{41}) is greater than 30 dB in a large frequency range, exceeding the domain 20-40 GHz (Sajin et al., 2010b).

4. Enhanced near-field laser lithography

The current trend towards sub-micrometer structures requires new methods and technologies of surface structuring. The traditional masking approach in optical lithography

is limited to a minimal resolvable feature size given by the diffraction limit. Optical near-fields were explored for their ability to localize optical energy to length scales smaller than half-wavelength. This localization was achieved for ultrasensitive detection (Fischer, 1986) and for high-resolution optical microscopy and spectroscopy (Novotny & Stranick, 2006). Near-field optics research essentially determined the advance of nano-optics (Novotny & Hecht, 2006), single molecule spectroscopy (Xie & Trautman, 1998), and nanoplasmonics (Barnes et al., 2003). Micron and sub-micron patterning was performed by "direct writing" where the laser light is just projected onto a sample via a direct-contact mask or by the interference of laser beams. Another technique is based on scanning near field optical microscopy (SNOM). Here, the light is coupled into the tip of a solid or hollow fiber. By positioning the substrate within the near field of the fiber tip, one can produce patterns with widths that are beyond the diffraction limit. Structures with lateral dimension less than 30 nm, well below the radiation wavelength, could be produced underneath the tip (Gorbunov & Pompe, 1994). This technique has been employed for nanolithography, ablation, material etching and local reduction of oxides.
Spherical particles can act as spherical lenses and therefore increase the laser intensity if their diameter is bigger than the laser wavelength. Laser-induced submicron patterning of surfaces has been demonstrated by means of two-dimensional colloidal lattices of microspheres that are formed by self-assembly (Piglmayer et al., 2002). If the diameter of spherical particles is of the order of magnitude of the radiation wavelength, according with the Mie solution to Maxwell's equations (Mie, 1908), the optical field enhancement occurs quite near laser irradiated particles. Electric field intensity distributions were calculated with the finite-difference time-domain (FDTD) method using simulation software (RSoft Design Group). For colloidal sub-wavelength particles (700 nm diameter of the particle, 532 nm wavelength radiation) placed on a glass substrate, the electric field can be greatly enhanced (by a factor of ~ 8) in the near-surface field region under particles as shown in the simulation from figure 10a (Ulmeanu et al., 2009a). The enhancement value quoted above is a theoretical estimation and the actual field strength enhancement may differ due to various influencing factors, such as surface roughness and oxidation of the thin film layer.

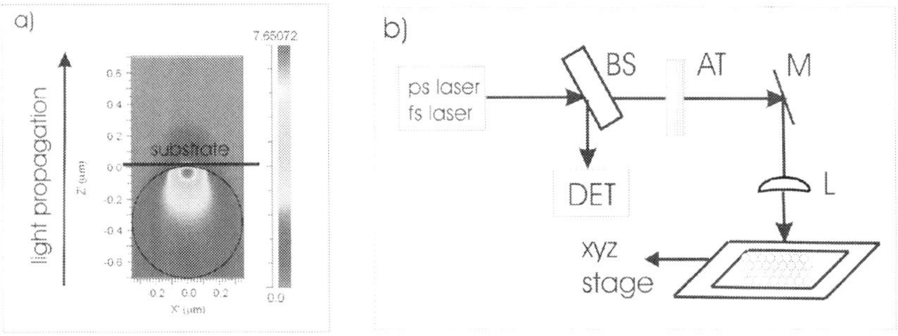

Fig. 10. (a) Field intensity enhancement of a 700 nm colloidal particle sitting on a glass substrate in a free space for λ = 532 nm (b) Experimental setup (BS - beam splitter, AT - attenuator, L - convergent lens, DET - pyroelectric detector, M - high reflectivity mirror.

We demonstrated surface patterning in the enhanced near-field by scanning a quasi-Gaussian laser beam through a self-assembled monolayer of colloidal particles onto different substrates:

glass and metallic thin films of nanometers thickness like Ag, Au, Co, Cu thin films as well as Co/Cu/Co multilayer structures (Ulmeanu et al., 2009b). Single pulse laser ablation of monolayer thin films and multilayer structures was performed by irradiating with a Nd:YAG picosecond laser (400-ps pulsewidth at λ =532 nm) and a Ti:sapphire femtosecond laser system (200-fs pulsewidth at λ = 775 nm). The experimental setup is shown in figure 10b.
The laser beam was focused on the colloidal particles using a 75 mm convex lens corrected for reduced spherical aberrations. The irradiating energy was adjusted through neutral attenuators. For the control of the irradiation process, a certain fraction of the incident laser energy is split off by a glass splitter and measured with a pyroelectric detector.
The sample was mounted on a precision three-axis xyz stage and was aligned for normal incidence of the laser beam. The ablation spots were produced with single pulse shots at different laser energies. All experiments were performed under ambient conditions. The irradiated area of the sample was imaged using Scanning Electron Microscpy (SEM) and Atomic Force Microscopy (AFM) to obtain the surface maps and crater profiles. From our investigations, it is obvious that, using 2D colloidal masks, arrays with regularly arranged holes can be created through a parallel process on the whole surface (Ulmeanu et al., 2009a). The shape of the holes depends on the laser fluence and thermophysical properties of the surfaces involved in the experiments (figure 11).

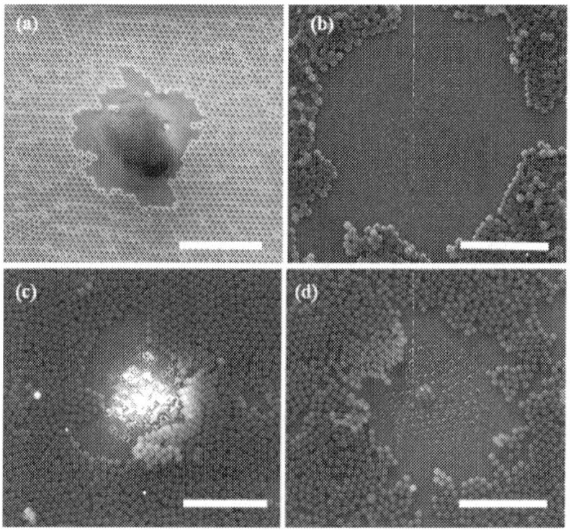

Fig. 11. SEM images for the near-field experiments at fluences above the threshold level: (a) glass substrate (F_{peak}= 7.3 J/cm^2), (b) Co thin film (F_{peak} = 0.82 J/cm^2), (c) Ag thin film (F_{peak} = 0.18 J/cm^2) and (d) Au thin film (F_{peak} = 0.17 J/cm^2). Scale bar: 10 μm.

For the case of glass near-field processing, we have obtained regular nano-holes, with a uniform distribution of depths and heights at the same fluence value (figure 11a). The smallest diameter of the holes, according to the SEM images, was 100 nm. For the case of Ag thin film (figure 11c), bumps and nanoholes were visible away from the peak fluence location, but not in a regular structure like in the case of the Au thin film (figure 11d) and Co thin film (figure 11b). Despite of similar ablation thresholds of Ag and Au thin films, due to a high thermal diffusivity coefficient and low adherence of the thin film on glass substrate, the near-field experiments gave a non-uniform pattern in the case of the Ag thin film surface. The discrepancy between Ag and Au thin film may potentially be due also to appreciable surface roughness of the thermally evaporated Ag thin film, that has a great impact on the optical near-field distribution.

The near-field processing in the case of Au thin film indicated different patterns produced by a single shot laser: from bowl type structure to bumps depending on the fluence, as in AFM investigations presented in the figure 12a,b. For the Co thin film near-field ablation, we were able to determine the dependence of the holes diameter versus the depth at different fluences, the holes having shapes from bowl type to an almost rectangular shape with flat bottom.

By comparing near-field processing on Co and Au thin films (figure 12c), we put into evidence holes with different diameters (*d*) and heights (*z*) at fluences nearby the thresholds: d_{Au} = 145 nm, z_{Au} = 14 nm (Au thin film), d_{Co} = 250 nm, z_{Co} = 60 nm (Co thin film).

For a multilayer Co(3 nm)/Cu(6 nm)/Co(20 nm) material, we investigated a method of preferential laser processing of a specific layer by changing the laser fluence instead of selecting different laser wavelengths. Figure 13a represents a schematic multilayer structure. The heterostructure composition determined by Sputtered Neutral Mass Spectroscopy (SNMS) measurements is shown in figure 13b. We demonstrated the formation of ordered areas of nano-holes in a multilayer structure using single femtosecond laser pulse irradiation (200 fs, λ = 775 nm) on 700 nm diameter colloidal particles (Ulmeanu et al., 2010). The ablation threshold for the multilayer structure in the enhanced near-field regime was measured to be 2.8 x 10^{-2} J/cm^2.

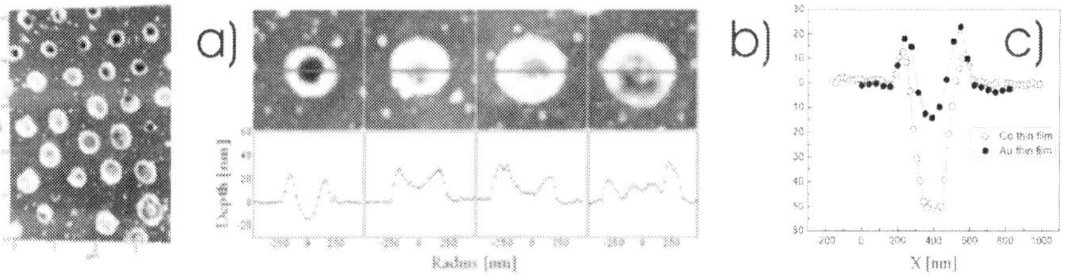

Fig. 12. (a) 3D AFM image; (b) scan lines for the structures produced by a single shot laser in the case of the near-field experiments for Au thin film; (c) depth profiles of the holes for Au and Co thin films.

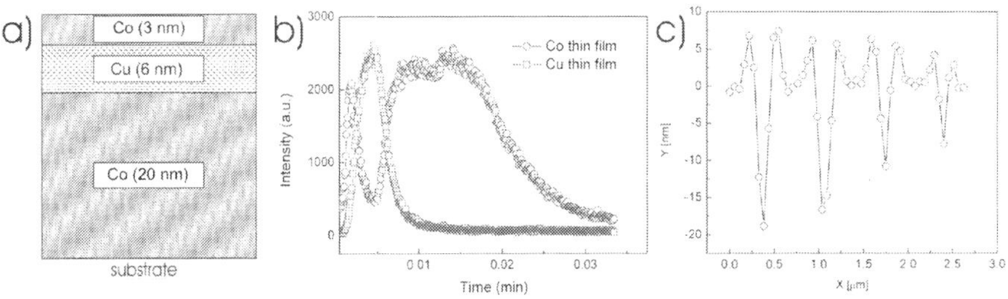

Fig. 13. (a) Schematic representation of the multilayer structure; (b) heterostructure composition determined by the SNMS measurements; (c) holes depths and widths distribution for the Co(3nm)/Cu(6nm)/Co(20 nm) multilayer structure (peak laser fluence, F_{las} = 0.04 J/cm^2).

The dependence of the holes diameter versus the depth at different fluences, obtained by AFM measurements, is showed in the figure 13c. As expected, with the decrease of the laser energy from the central zone to the edges of the irradiated area, the depth of the holes is decreasing. In the case of multilayer structures, the hole depth decrease depending on the laser fluence corresponds to the ablation of the first, second or third layer. This type of planar metal/dielectric interfaces with a selective distribution of layers can open new perspective in the excitation of propagating surface plasmons and, consequently, in creating transducers for sensing of biomolecular recognition reactions.

5. Femtosecond laser induced forward transfer

Laser Induced Forward Transfer (LIFT) represents a challenging technique to the conventional etching microfabrication techniques. It becomes particularly interesting when a very small quantity of material has to be deposited on a substrate. Many kinds of materials such as metallic films, semiconductors, polymers, or even biological material can be transferred (Thomas et al, 2007; Sanz et al, 2010; Colina et al, 2005). The material to be transferred is initially deposited in thin films on a transparent substrate named donor substrate, or "ribbon" (transparent at laser radiation used for LIFT process). Usually, but not necessarily, a very thin metallic layer is deposited as buffer between the donor substrate and the film to be transferred. The donor sample, is placed at a short distance, parallel to another acceptor substrate (virtually any material). The donor film is backward irradiated with a pulsed laser, like in figure 14.

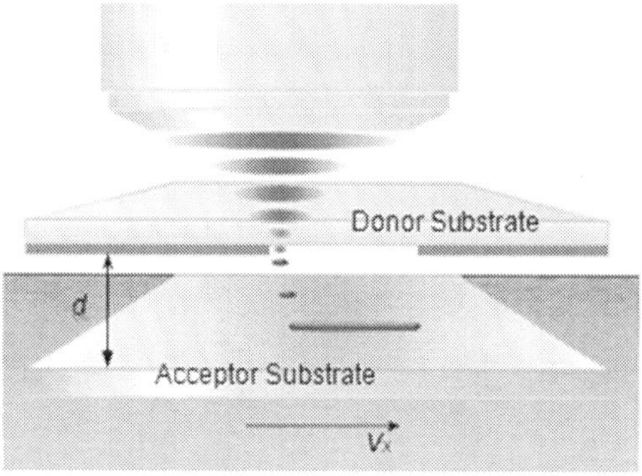

Fig. 14. The schematic of LIFT principle.

The laser is focused on the donor thin film at the interface with the donor substrate. Then, a small amount of buffer material is ablated and transformed in gaseous faze. This gas expands pushing forward the rest of the material which is projected to the acceptor substrate. If the parameters are correctly chosen, the ejected material is deposited on the acceptor's surface. The role of the buffer layer is only to protect the material to be transferred and is used especially in the case of organic materials susceptible to be affected by a direct exposure to the laser beam. Otherwise, in absence of a buffer layer, the material itself can be vaporized at the interface with the donor substrate, the pressure of the created gas transferring a small quantity of material from a substrate to another.

In LIFT experiments some parameters, like distance *d* between the donor film and acceptor substrate, or laser fluency, have to be investigated in order to find the optimal processing conditions for the deposition of a certain material. In our LIFT experiments, we demonstrated the transfer of a polymer material, an Ormocer photoresist, using our laser processing workstation. The polymer layer was directly deposited by spin coating on a glass substrate, without any buffer layer. The distance between donor and acceptor was fixed at 15 μm. Series of 5x5 pixels were created by single pulses, shot by shot. The laser source was the Clark MXR CPA-2101 laser, with 200 fs pulse duration and 775 nm wavelength, externally triggered for single shot experiments. The sample was translated from a pixel to another by a computer controlled translation stage. The distance between pixels was 50 μm. The laser was focused to the donor layer by a 75 mm focusing lens with about 25 μm focus spot diameter. The energy per pulse was varied from 2.5 to 7.5 μJ.
Optical images of the structures, as transferred to the acceptor substrate at different pulse energies, are shown in figure 15. The quality of the obtained structures strongly depends on the pulse energy. At the highest pulse energy used, non uniform droplets results, sparse on the donor surface. Decreasing the pulse energy the transferred droplets remain well defined.

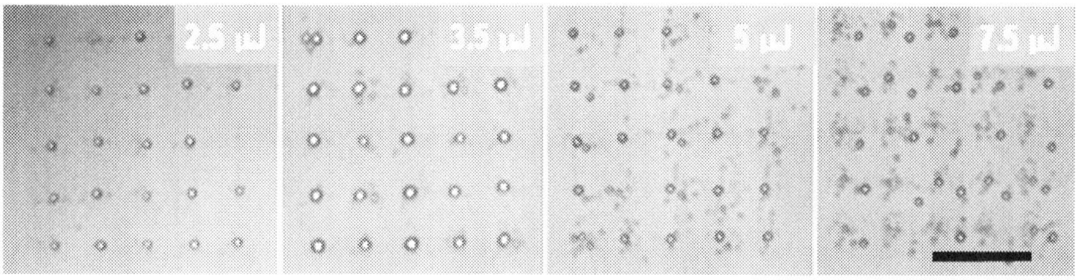

Fig. 15. LIFT generated microstructures at different laser energy. Scale bar: 100 μm.

The smallest size of the droplets obtained in these experimental conditions was about 2 μm. Smaller structures, such as nanodroplets, can be also transferred (Banks et al., 2006) and even an entire microstructure or a microdevice could be deposited by LIFT (Piqué et al., 2006). This technique can be efficiently used as a microprinting method.

6. Conclusions

Various experimental techniques for materials micro-processing based on ultra-short pulsed lasers are presented.
Direct femtosecond laser writing technique by two-photon photo-polymerization was used to produce microstructures in the volume of transparent materials. By this technique, photonic devices such as photonic crystals, optical couplers, diffractive elements, 3D structures for microfluidics, scaffolds for tissue engineering, and other MEMS can be fabricated.
Femtosecond laser ablation was used to produce 2D microstructures on different materials surface. Electronic devices, based on CRLH transmission lines having metamaterials characteristics in the tens-GHz frequency range, such as band-pass filters, antennas, and directional couplers were manufactured by combined photo-lithography and femtosecond laser ablation techniques. A microprinting method, based on laser induced forward transfer

of a small quantity of ablated material from a donor substrate to an acceptor substrate, was demonstrated.

Using self-assembled monolayers of colloidal micro-spheres on glass and metallic thin film substrates, nano-patterning was obtained by ultrashort pulse laser ablation in the near-surface enhanced laser field irradiation regime. Formation of ordered areas of nano-holes in a multilayer structure was demonstrated by selective near-field laser ablation.

Ultrashort pulsed laser micro/nanostructures writing technique could be considered as an alternative to the classical techniques based on photo-lithography. New micro/nano-processing technologies can be developed by combining femtosecond laser writing techniques with already proved lithography techniques.

7. Acknowledgment

M. Zamfirescu acknowledges the support of the Sectoral Operational Programme Human Resources Development (SOP HRD), financed from European Social Fund and by Romanian Government under the contract number POSDRU/89/1.5/S/63700. M. Ulmeanu acknowledges the financial support of the European Social Fund POSDRU 2007-2013 through the contract POSDRU/89/1.5/S/60746.

8. References

Allsop, T.; Dubov, M.; Mezentzev, V. & Bennion, I. (2010). Inscription and characterization of waveguides written into borosilicate glass by a high-repetition-rate femtosecond laser at 800 nm, *Applied Optics*, Vol. 49, No. 10, (1 April 2010), pp. 1938-1950, ISSN 0003-6935

Ams, M.; Marshall, G. D.; Spence, D. J. & Withford, M. J. (2005). Slit beam method for femtosecond laser direct-write fabrication of symmetric waveguides in bulk glasses, *Optics Express*, Vol. 13, No. 15, (July 2005), pp. 5676-5681, eISSN 1094-4087

Balanis, C.A. (1997). *Antenna Theory. Analysis and design*, John Wiley & Sons, Inc., ISBN: 0-471-59268-4, New York, Chichester, Brisbane, Toronto, Singapore

Banks, D. P.; Grivas, C.; Mills, J. D.; Eason, R. W. and Zergioti, I. (2006). Nanodroplets deposited in microarrays by femtosecond Ti:sapphire laser-induced forward transfer, *Appl. Phys. Lett.*, Vol. 89, No. 19, (November 2006) pp. 193107, ISSN 1077-3118

Barnes, W. L.; Dereux, A. & Ebbesen, T. W. (2003), Surface plasmon subwavelength optics. *Nature*, Vol. 424, (August 2003), pp. 824-830, ISSN 0028-0836

Basharin, A.; & Balabukha, N. (2009). The Radiation of Antennas Based on Metamaterial Waveguides, *Proceedings of 3rd International Congress on Advanced Electromagnetic Materials in Microwaves and Optics, Metamaterials 2009*, pp.224 - 226, ISBN 978-0-9551179-6-1, 30th Aug. - 04th Sept. 2009, London, UK

Becker, EW.; Ehrfeld W.; Hagmann, P.; Maner, A & Munchmeyer, D. (1986). Fabrication of microstructures with high aspect ratio and great structural heights by synchrotron radiation lithography, galvanoformung and plastic moulding

(LIGA process). *Microelectronic Eng.*, Vol. 4, No.1, (May 1986), pp. 35–36, ISSN 0167-931

Caloz, C. & Itoh, T. (2006). *Electromagnetic metamaterials: Transmission line theory and microwave applications*, John Wiley & Sons, Inc., ISBN-10: 0-471-66985-7. ISBN-13: 978-0-47166985-2, Hoboken, New Jersey

Caloz, C.; Sanada, A. & Itoh, T. (2004). A novel composite right-/left-handed coupled-line directional coupler with arbitrary coupling level and broad bandwidth, *IEEE Trans. on Microwave Theory and Techniques*, Vol. 52, March 2004, pp. 980-992, ISSN 0018-9480

Chimmalgi, A.; Choi, T. Y.; Grigoropoulos, C. P. & Komvopoulos, K. (2003). Femtosecond laser aperturless near-field nanomachining of metals assisted by scanning probe microscopy, *Applied Physics Letters*, Vol. 82, No. 8, (24 Febraury 2003), pp. 1146-1148, ISSN 0003-6951

Colina, M.; Serra, P.; Fernández-Pradas, J.M.; Sevilla, L. & Morenza, J.L. (2005). *Biosensors and Bioelectronics*, Vol. 20, No. 8, (February 2005), pp. 1638–1642, ISSN 0956-5663

Deubel, M.; Von Freymann, G.; Wegener, M.; Pereira, S.; Busch, K. & Soukoulis, C. M. (2004). *Nature Materials*, Vol. 3, No. 7, (July 2004), pp. 444-447, ISSN 1476-1122

Dolling, G.; Enkrich, C.; Wegener, M.; Soukoulis, C. M. & Linden, S. (2006). Low-loss negative-index metamaterial at telecommunication wavelengths, *Optics Letters*, Vol. 31, No. 12, (June 15, 2006), pp. 1800-1802, ISSN 0146-9592

Dolling, G.; Wegener, M.; Soukoulis, C. M. & Linden, S. (2007). Negative-index metamaterial at 780 nm wavelengths, *Optics Letters*, Vol. 32, No. 1, (January 1, 2007), pp. 53-55, ISSN 0146-9592

Eggermont, S.; Platteborze, R. & Huynen, I. (2009). Analysis of Radiation in a Metamaterial Leaky Wave Antenna Based on Complementary Split Ring Resonator, *Proceedings of 3rd International Congress on Advanced Electromagnetic Materials in Microwaves and Optics, Metamaterials 2009*, pp.102 - 104, ISBN 978-0-9551179-6-1, 30th Aug. - 04th Sept. 2009, London, UK

Farsari, M.; Filippidis, G. & Fotakis, C. (2005). Fabrication of three-dimensional structures by three-photon polymerization, *Optics Letters* , Vol. 30, No. 23, (luna N, 2005) pp. 3180-3182, ISSN 0146-9592

Farsari, M. & Chichkov, B. N. (2009). Two-photon fabrication, *Nature Photonics*, Vol. 3, No. 8, (August 2008), pp. 450-452, ISSN 1749-4885

Fischer, U. Ch. (1986). Submicrometer aperture in a thin metal film as a probe of its microenvironment through enhanced light scattering and fluorescence. *Journal of the Optical Society of America B*, Vol. 3, No. 10, (October, 1986), pp. 1239-1224, ISSN 0740-3224

Frassila, Sami. (2010). Introduction to Microfabrication. Second Edition, John Wiley & Sons Ltd., ISBN 978-0-470-74983-8

Glezer, E. N.; Milosavljevic, M.; Huang, L.; Finlay, R. J.; Her, T. H.; Callan, J. P. & Mazur, E. (1996). Three-dimensional optical storage inside transparent materials, *Optics Letters*, Vol. 21, No. 24, (December 15, 1996), pp. 2023-2025, ISSN 0146-9592

Gorbunov, A. A. & Pompe, W. (1994). Thin film nanoprocessing by laser/STM combination. *Physica Status Solidi A*, Vol. 145, No. 2, (October 2004), pp. 333-338, ISSN 0031-8965

Gwyn, C. W.; Stulen, R.; Sweeney, D. & Attttwood, D. (1998). Extreme ultraviolet lithography, *Journal of Vacuum Science & Technology B*, Vol. 16, No. 6, (November 1998), pp. 3142-3149, ISSN 0734-211X

Homoelle, D.; Wielandy, S.; Gaeta, A. L.; Borelli, N. F. & Smith, C. (1999). Infrared photosensitivity in silica glasses exposed to femtosecond laser pulses, *Optics Letters*, Vol. 24, No. 18, (September 15, 1999), pp. 1311-1313, ISSN 0146-9592

Kawata S., Sun H.-B., Tanaka T., and Takada K. (2001). Finer features for functional microdevices, *Nature* , Vol. 412, No. 6848, (August 16, 2001) pp. 697-698, ISSN: 0028-0836

Kurokawa, K. (1965). Power Waves and the Scattering Matrix, *IEEE Trans. on MTT,* Vol.13, March 1965, pp194 - 202, ISSN: 0018-9480

Li, C.; Liu, K.-Y. & Li, F. (2007). Composite right/left-handed coplanar waveguide band-pass filter using capacitively-coupled zeroth-order resonators, *Appl. Phys. Lett.* Vol. A 87, pp. 317–319, ISSN 0003-6951

Liu, X.; Du, D. & Mourou, G. (1997). Laser Ablation and Micromachining with Ultrashort Laser Pulses, *IEEE Journal of Quantum Electronics*, Vol. 33, No. 10, (October 1997), pp. 1706-1716, ISSN 0018-9197

Liu, C. & Menzel, W. (2009), On the Relation between Composite Right-/Left-Handed Transmission Lines and Chebyshev Filters, *International Journal of Microwave Science and Technology,* Volume 2009, Article ID 184895, DOI:10.1155/2009/184895

Liu, G.; Tian, Y. and Kan., Y. (2005). Fabrication of high-aspect-ratio microstructures using SU8 photoresist, Microsystem Technologies, Vol. 11, No. 3-4, (November 6, 2003), pp.343–346, ISSN: 1432-1858

Maruo, S.; Nakamura, O. & Kawata, S. (1997). Three-dimensional microfabrication with two-photon-absorbed photoplymerization, *Optics Letters,* Vol. 22, No. 2, (January 15, 1997), pp. 132-134, ISSN 0146-9592

Matei, A.; Zamfirescu, M.; Jipa, F.; Luculescu, C.;Dinescu, M.; Buruiana, EC.; Buruiana, T.;Sima, L.E. & Petrescu, S.M. (2010). Two Photon Polymerization of Ormosils, *AIP Conference Proceedings, International Symposium on High Power Laser Ablation 2010,* Vol. 1278, No. 1, (October 2010), pp. 843-851, ISSN 0094-243X

Menzel, W. (1979). A new traveling-wave antenna in microstrip, *Arch. Electron. Uebertrag. Tech.,* Vol. 33, 1979, pp. 137 - 140.

Mie, G. (1908). Beitrage zur Optik trueber Medien, speziell kolloidaler Metallosungen. *Annalen der Physik,* Vol. 330, No. 3, pp. 377-445, ISSN 1521-3889

Nishiyama, H.; Mizoshiri, M.; Kawahara, T.; Nishii, J. & Hirata, Y. (2008). SiO_2- based nonplanar structures fabricated using femtosecond laser lithography, *Optics Express,* Vol. 16, No. 22, (27 October 2009), pp. 17288-17294, eISSN 1094-4087

Novotny, L. & Stranick, S. J. (2006). Near-field optical microscopy and spectroscopy with pointed probes. *Annual Review of Physical Chemistry,* Vol. 57, (May 2006), pp. *303-331,* ISSN 0066-426X

Novotny, L. & Hecht, B. (2006). *Principles of Nano-Optics*, Cambridge University Press, ISBN 978-0-521-83224-3, New York

Osellame, R.; Taccheo, S.; Marangoni, M.; Ramponi, R.; Laporta P.; Polli, D.; De Silvestri, S. & Cerullo, G. (2003). Femtosecond writing of active optical waveguides with astigmatically shaped beams, *Journal of the Optical Society of America B*, Vol. 20, No. 7, (July 2003), pp. 1559-1567, ISSN 0740-3224

Ovsianikov, A.; Xiao, S. Z.; Farsari, M.; Vamvakaki, M.; Fotakis, C. & Chichkov, B. N. (2009). Shrinkage of microstructures produced by two-photon polymerization of Zr-based hybrid photosensitive materials, *Optics Express*, Vol. 17, No. 4, (February 16, 2009), pp. 2143-2148, eISSN 1094-4087

Pavlidis, V. F. and Friedman, E. G. (2009). Three-Dimensional Integrated Circuit Design, Morgan Kaufmann, Elsevier Inc., ISBN 978-0-12-374343-5.

Pendry, J.B. (1999). Magnetism from conductors and enhanced nonlinear phenomena. *IEEE Trans. on MTT*, Vol.47, nr.11, Nov. 1999, pp. 2075-2084, ISSN: 0018-9480

Piglmayer, K.; Denk, R. & Bauerle, D. (2002). Laser-induced surface patterning by means of microspheres. *Applied Physics Letters*, Vol. 80, No. 25, (June 2005), pp. 4693-4695, ISSN 0003-6951

Piqué, A.; Mathews, S.A.; Pratap, B.; Auyeung, R.C.Y.; Karns, B.J. & Lakeou, S. (2006) Embedding electronic circuits by laser direct-write, *Microelectronic Engineering*, Vol. 83, No. 11-12, (November-December 2006) pp. 2527–2533, ISSN 0167-9317

Qi, L.; Nishii, K. & Nambai, Y. (2009). Regular subwavelength surface structures induced by femtosecond laser pulses on stainless steel, *Optics Letters*, Vol. 34, No. 12, (June 15, 2009), pp. 1846-1848, ISSN 0146-9592

Rill, M. S.; Plet, C.; Thiel,, M.; Staude, I.; Von Freymenn, G.; Linden, S. & Wegener, M. (2008). Photonic metamaterials by direct laser writing and silver chemical vapour deposition, *Nature Materials*, Vol. 7, No. 7, (July 2008), pp. 543-546, ISSN 1476-1122

Sajin G.; Simion S.; Craciunoiu, F. & Marcelli, R. (2007). Silicon supported microwave zeroth-order resonance antenna on metamaterial approach, *Proceedings of Asia-Pacific Microwave Conference, APMC 2007*, pp. 221 - 224, ISBN 1-4244-0748-6, Bangkok, Thailanda, December 2007

Sajin G.; Simion S.; Craciunoiu Fl.; Bunea, A. C.; Dinescu, A.; Zamfirescu, M.; Neagu, L. & Dabu, R. (2010a). Millimeter Wave Metamaterial Band-Pass Filter by Femtosecond Laser Ablation, *Proceedings of European Microwave Conference, EuMC*, ISBN 978-2-87487-015-6, pp. 1393–1396, Paris, France, September 26 - October 01, 2010.

Sajin, G.; Simion, S.; Craciunoiu, Fl.; Bunea, A. C.; Muller A.A. & Dinescu, A. (2010b). Metamaterial Millimeter Wave Directional Coupler on silicon substrate, *Proc. of 33rd International Semiconductor Conference, CAS*, pp. 269–272, ISBN 978-14244-5781-6; ISSN 1545-827X, Sinaia, Romania, October 11 - 13, 2010.

Sanz, M.; Walczak. M.; Oujja, M.; Domingo, C.; Klini, A.; Papadopoulou, E.L.; Fotakis, C. & Castillejo, M. (2010), Femtosecond laser deposition of TiO2 by laser induced forward transfer, *Thin Solid Films*, Vol. 518, No. 19, (July 2010) pp. 5525–5529, ISSN 0040-6090

Seongmin, P.; Baik, J. W.; Cho, S. H. & Kim, Y. S. (2008). A metamaterial-based symmetrical periodic antenna with efficiency enhancement, *Proceedings of Asia-Pacific Microwave Conference,* Paper A3-49, ISBN 978-1-4244-2642-3, Hong Kong, China, December 2008.

Shelby R.A.; Smith D.R. & Schultz S. (2001). Experimental verification of a negative index of refraction, *Science,* vol. 292, (April 2001), pp. 77–79, ISSN: 0036-8075

Sowa, S.; Watanabe, W.; Tamaki, T.; Nishii, J. & Itoh, K. (2005). Symmetric waveguides in poly(methyl methacrylate) fabricated by femtosecond laser pulses, *Optics Express,* Vol. 14, No. 1, (9 January 2006), pp. 291-297, eISSN 1094-4087

Streltsov, A. M. & Borrelli, N. F. (2001). Fabrication and analysis of a directional coupler written in glass by nanojoule femtosecond laser pulses, *Optics Letters,* Vol. 21, No. 1, (January 1, 2001), pp. 2023-2025, ISSN 0146-9592

Stuart, B. C.; Feit, M. D.; Herman, S.; Rubenchick, A. M.; Shore, B. W. & Perry, M. D. (1996). Optical ablation by high-power short-pulse lasers, *Journal of the Optical Society of America B,* Vol. 13, No. 2, (February 1996), pp. 459-468, ISSN 0740-3224

Sugioka, K.; Meunier, M. & Pique, A. (2010). *Laser Precision Microfabrication,* Springer Series in Materials Science, Vol. 135, Springer_Verlag Berlin Heidelberg, ISBN 978-3-642-10523-4

Takeshima, N.; Kuroiwa, Y.; Narita, Y.; Tanaka, S. & Hirao, K. (2004). Fabrication of a periodic structure with a high refractive-index difference by femtosecond laser pulses, *Optics Express,* Vol.12, No. 17, (23 August 2004), pp. 4019-4024, eISSN 1094-4087

Takeshima, N.; Nartita, Y.; Nagata, T. & Tanaka, S. (2005). Fabrication of photonic crystals in ZnS-doped glass, *Optics Letters,* Vol. 30, No. 5, (March 1, 2005), ISSN 0146-9592

Tan, D.; Li ,Y.; Qi, F.; Yang, H.; Gong Q.; Dong, X. & Duan X. (2007). Reduction in feature size of two-photon polymerization using SCR500, Appl. Phys. Lett. Vol. 90, (January 13, 2007) pp. 071106 1-3 , ISSN 1077-3118

Thomas, B.; Alloncle, A. P.; Delaporte, P.; Sentis, M.; Sanaur, S.; Barret, M. & Collot, P. (2007). Experimental investigations of laser-induced forward transfer process of organic thin films, *App. Surf. Sci.*, Vol. 254, No. 4, (December 2007) pp. 1206-1210, ISSN 0169-4332

Ulmeanu, M.; Zamfirescu, M.; Rusen, L.; Luculescu, C.; Moldovan, A. ; Stratan, A. & Dabu, R. (2009a). Structuring by field enhancement of glass, Ag, Au, and Co thin films using short pulse laser ablation. *Journal of Applied Physics,* Vol. 106, No.11, (December 2009) pp.114908 - 114908-8, ISSN 0021-8979

Ulmeanu, M.; Zamfirescu, M. & Medianu, R., (2009b). Self-assembly of colloidal particles on different surfaces. *Colloids and Surfaces A: Physicochemical and Engineering Aspects,* Vol. 338, No. 1-3, (April 2009), pp. 87-92, ISSN 0927-7757

Ulmeanu, M.; Filipescu, M.; Scarisoreanu, N.D.; Georgescu, G.; Rusen, L. & Zamfirescu, M. (2010). Selective removal and patterning of a Co/Cu/Co trilayer created by femtosecond laser processing. *Applied Physics A,* DOI:10.1007/s00339-010-6119-9, ISSN 1432-0630

Vega, F.; Armengol, J.; Diez-Blanco, V.; Siegel, J.; Solis, J.; Barcones, B.; Perez-Rodriguez, A. & Loza-Alvarez, P. (2005). Mechanisms of refractive index modification during femtosecond laser writing of waveguides in alkaline lead-oxide silicate glass, *Applied Physics Letters,* Vol. 87, 021109 (2005), ISSN 0003-6951

Veselago, V. G. (1968). The electrodynamics of substances with simultaneously negative values of ε and μ, *Sov. Physics – Usp.,* vol.47, January-February 1968, pp. 509-514

Wang, Y. Z.; Zhang, Y. W.; He, L.; Liu, F.-Q.; Li, H. Q. & Chen, H. (2006). Discussion on coupling mechanism of asymmetric CRLH/RH TL coupler, *Journal of Zhejiang University - Science A,* Vol. 7, January 2006, pp. 95-98, ISSN 1673-565X

Weiß, T.; Hildebrand. G.; Schade, R. & Liefeith K. (2009). Two-Photon polymerization for microfabrication of three-dimensional scaffolds for tissue engineering application, Engineering in Life Sciences Vol. 9, No. 5, (October 2009), pp. 384–390, ISSN 1618-2863

Will, M.; Nolte, S.; Chichkov, B. & Tunnermann, A. (2002). Optical properties of waveguides fabricated in fused silica by femtosecond laser pulses, *Applied Optics,* Vol. 41, No. 21, (July 2002), pp. 4360-4364, ISSN 1559-128X

Witzgall, G.; Vrijen, R.; Yablonovitch, E.; Doan, V. & Schwartz, B. J. (1998). Single-shot two-photon exposure of commercial photoiresisit for the production of three-dimensional structures, *Optics Letters,* Vol. 23, No. 22, (November 15, 1998), ISSN 0146-9592

Wortmann, D.; Gottmann, J.; Brandt, N. & Horn-Solle, H. (2008). Micro- and nanostructures inside sapphire by fs-laser irradiation and selective etching, *Optics Express,* Vol. 16, No. 3, (4 February 2008), pp. 1517-1522, eISSN 1094-4087

Xie, X. S. & Trautman, J. K. (1998). Optical studies of single molecules at room temperature. *Annual Review of Physical Chemistry,* Vol. 49, (October 1998), pp. 441–480, ISSN 0066-426X

Yuanyuan, S.; Yewen, Z.; Fuqiang, L.; Li, H.; Hongqiang, L. & Hong, C. (2007). A novel filter based on zeroth-order resonance by means of CRLH transmission line, *Microwave and Optical Technology Letters,* Vol. 49, Issue 5, pp. 1015 - 1018, ISSN: 0895-2477

Yu, C. C.; Huang, M. H.; Lin, L. K. & Chang, Y. T. (2008). A compact antenna based on metamaterial for WiMAX, *Proceedings of Asia-Pacific Microwave Conference,APMC 2008,* Paper J2-05, ISBN 978-1-4244-2642-3, Hong Kong, China, December 2008

Zhang, S.; Fan, W.; Panoiu, N. C.; Malloy, K. J.; Osgood, R. M. & Brueck, S. R. J. (2005). Experimental Demonstration of Near-Infrared Negative-Index Metamaterials, *Physical Review Letters,* Vol. 95, 137404 (23 September 2005), ISSN 0031-9007

Ziolkowski, R.W.; Peng, J. & Lin, C. C. (2009). Electrically Small Metamaterial-Inspired Antennas: The Next Generation, *Proceedings of 3rd International Congress on Advanced Electromagnetic Materials in Microwaves and Optics, Metamaterials*

2009, pp.44 - 46, ISBN 978-0-9551179-6-1, 30th Aug. - 04th Sept. 2009, London, UK

Zoubir, A.; Richardson, M.; Rivero, C.; Schulte, A.; Lopez, C.; Richardson, K; Ho, N & Vallee, R. (2004). Direct femtosecond laser writing of waveguides in As_2S_3 thin films, *Optics Letters,* Vol. 29, No. 7, (April 1, 2004), pp. 748-750, ISSN 0146-9592

15

Laser-Based Lithography for Polymeric Nanocomposite Structures

Athanassia Athanassiou, Despina Fragouli,
Francesca Villafiorita Monteleone, Athanasios Milionis,
Fabrizio Spano, Ilker Bayer and Roberto Cingolani
Center for Biomolecular Nanotechnologies @UNILE,
Istituto Italiano di Tecnologia (IIT), Via Barsanti, 73010 Arnesano, Lecce
Italy

1. Introduction

In recent years, the synthesis of polymeric nanocomposites has gained much interest in the scientific community thanks to their unique capability to combine the properties of the host polymer matrices, such as toughness, elasticity, processability, solubility, thermal stability, etc, with those of inorganic nanoparticles (NPs), such as hardness, chemical resistance, optical and electronic properties. Among a variety of nanofillers, semiconductor and metallic NPs are extensively studied and used, because of their unique properties especially in the nanoscale. In this work we deal with polymeric nanocomposites incorporating various nanofillers, each one of them having extremely attractive properties for technological applications. In particular we focus on titanium dioxide (TiO_2) NPs due to their unique reversible wettability, iron oxide (Fe_2O_3) NPs due to their superparamagnetic nature, gold (Au) NPs due to their enhanced electronic conductivity, and cadmium sulphide (CdS) NPs due to their tuned photoemission in the quantum size regime.

Nanocomposite materials are usually prepared by simple blending of the nanosized inorganic components into polymer solutions. Nevertheless, this method quite often leads to formation of aggregates or macroscopic phase separation, since the control of the dispersion and distribution of the nanofillers within the organic matrices is impossible. To obtain a higher compatibility between the filler and the host polymeric material, and achieve coatings with high content of inorganic particles, the use of polymerization techniques is widely adopted (Fouassier 1995, Decker et al 2005, Lee et al 2006, Wang & Ni 2008). From this point of view, several methods can be considered, depending on the type of monomers and nanomaterials, such as bulk polymerization, photoinitiated polymerization, emulsion polymerization, in situ thermal polymerization, or copolymerization in solution. Among them, the use of ultraviolet (UV) light in combination with proper photoinitiators to produce polymeric nanocomposite films is one of the most rapid and effective method, the main advantage being the creation of well defined patterned structures.

Patterning of nanocomposites using UV polymerization, the so-called UV photolithography is ideal for the direct incorporation of nanocomposites into specific parts of systems and devices. Photolithographically prepared nanocomposite structures can be used for the selective deposition of molecules, which have specific affinity to the photopolymerized material, for the creation of molecular sandwiches, for cells deposition, for microfluidic devices, sensors, photoemitting parts in devices, etc. Herein, we present two different UV light-based lithography techniques for formation of nanocomposite patterns that consist of polymers incorporating inorganic nanofillers. The latter are either colloidal NPs produced by chemical synthesis, or NPs that are formed directly into the polymer matrices starting from an appropriate precursor, using laser light irradiation.

In particular the photolithography presented in this work follows two approaches:

1. UV laser photopolymerization: the nanocomposite solutions preexist in the form of colloidal NPs mixed with monomers and photocuring agent and laser light beams induce the patterned structures. Three systems are demonstrated: PMMA with TiO_2 NPs, PMMA with Fe_2O_3 NPs , SU-8 with Fe_2O_3 NPs. In the last two systems we can induce further nanopatterning into the patterned polymers structures using an external magnetic field to align the nanofillers into nanowires (NWs).
2. UV laser light-induced formation of NPs into polymers: the starting system is a polymeric film incorporating light-sensitive precursors of NPs and laser light irradiation induces the formation of NPs in specific locations in the polymer matrix. Three systems are demonstrated: Chitosan with Au NPs, TOPAS with CdS NPs, PMMA with CdS NPs.

The patterns of nanocomposites produced with the above mentioned lithography methods are categorized according to their properties as follows: patterns with tunable surface properties, magnetic patterns, conducting patterns, patterns with tunable emission. Such patternable composite materials deal with most of the present technologies, including bioengineering and medical instrumentation, packaging, electrical enclosures, sensors, actuators and energy.

2. Lithography using laser photopolymerization

Lithographic patterning using UV light in combination with monomers and appropriate photoinitiators is a very attractive technology due to low energy consumption, room temperature operation, rapid curing, spatial control, possibility of curing structures on heat-sensitive substrates etc. (Andrzejewska 2001). The use of the specific technique for lithography of nanocomposites, permits to realize materials with homogeneous dispersion of the inorganic phase, but also to simultaneously create patterned surfaces in an easy and single step. (Sun et al 2008)

The photopolymerization technique involves UV/visible light absorption from the photoinitiators forming radical species (*initiation reaction*), which are responsible for the initiation of the photopolymerization process due to their addition to the monomer molecules (*propagation reaction*). The chain growth termination occurs upon annihilation of the radical centres due to radical-radical recombination (*termination reaction*). Among different light sources, lasers are most widely used when lithography is involved due to their unique characteristics that lead to small, highly controlled, sharp and precise patterns. In the presented work a pulsed laser was used permitting high energy concentrations in very short times, and allowing polymerization to develop to a high extent in the dark periods between consecutive pulses (Van Herk 2000).

2.1 TiO_2-based patterned nanocomposites

In the development of polymeric nanocomposites, semiconductor oxides are extensively studied and applied as nanofillers due to their unique electromagnetic, mechanical,

chemical and numerous other properties. In particular, among them titanium dioxide (TiO_2) is possibly the most widely used, thanks to two exceptional properties: the photocatalytic activity and the reversible wettability, both activated upon UV irradiation (Fujishima 2000, Nakajima 2000, 2001, Wang 1997, 1998, 1999, Sakai, 1998, 2001, 2003).
In this section we focus on the reversible wettability properties of TiO_2 upon excitation with UV laser light. Specifically, nanorods (NRs) of TiO_2 are used as the basic building blocks of photolithographically patterned nanocomposite materials with functional and responsive surfaces, which show UV-sensitive wettability. It has been demonstrated that upon UV irradiation of TiO_2, oxygen vacancies are created on its surface resulting into the conversion of Ti^{4+} into Ti^{3+} sites. These sites are favourable for dissociative adsorption of atmospheric water molecules, leading to the formation of a highly hydroxylated (hence, hydrophilic) surface. This procedure is reversible, since upon long term storage (few months) under ambient dark conditions, or for an accelerated process, upon thermal treatment, vacuum storage or visible irradiation (Wang et al., 1999, Nakajima et al., 2001, Sakai et al., 1998), the adsorbed hydroxyl groups can be removed and eventually replaced by ambient oxygen, allowing the initial hydrophobicity to be recovered (Caputo et al., 2008). Taking advantage of the above described mechanism we present herein how the formed TiO_2/polymer nanocomposite patterns can serve as paths for the directional movements of water drops onto them.

2.1.1 Photopolymerized patterns of PMMA/TiO_2

Solutions of MMA monomer, TiO_2 NRs, and photoinitiator were prepared in toluene to obtain concentrations of 94 wt.%, 5 wt.% and 1 wt.%, respectively. The photoinitiator was the IRGACURE®1700. It is formed by two distinct molecules, a phosphine oxide derivative and a hydroxyl alkyl phenyl ketone, which acts as co-initiator. Upon UV irradiation, electrons and hydrogen atoms are transferred from the co-initiator molecules to the photoinitiator, generating radicals that initiate the polymerization of the methyl methacrylate (MMA) monomers. All the solutions were stirred and left under dark for several minutes to equilibrate. Glass and silicon substrates were washed with isopropanol and subsequently with acetone and dried with nitrogen. About 200 μL of each solution were initially spin coated on both glass and silicon substrates at 1000 rpm for 20 s. Next, about 40 μL of the same solution were drop casted onto each sample.
To obtain photopolymerization, the third harmonic of a pulsed Nd:YAG laser (Quanta-Ray GCR-190, Spectra Physics, energy density = 10.5 $mJ \cdot cm^{-2}$, $\lambda = 355$ nm, pulse duration = 6 ns, repetition rate = 10 Hz) was used. In particular, the previously casted samples were irradiated through aluminum masks characterized by different patterns of mm-dimensions. After the photopolymerization, each sample was washed 3 times with methanol to remove unreacted monomer and photoinitiator, and then dried in ambient dark conditions. In Figure 1 are demonstrated characteristic patterns created upon UV laser irradiation of the drop casted solution through a photomask of two parallel lines in the mm range, demonstrating the potentiality of the specific lithography technique.
A closer look of the surfaces of the photolithographically produced nanocomposite patterns shown in Figure 1 was performed using lateral force Atomic Force Microscopy (AFM). The obtained AFM images demonstrate that NRs are apparent onto the surface of the photopolymerized MMA-TiO_2 nanocomposite films. In general, they appear quite aggregated, however it is occasionally possible to distinguish also single NRs laying parallel to the substrate (Figure 2a). In contrast, the photopolymerized coatings without TiO_2 NRs appear flat and homogeneous under the AFM (Figure 2b).

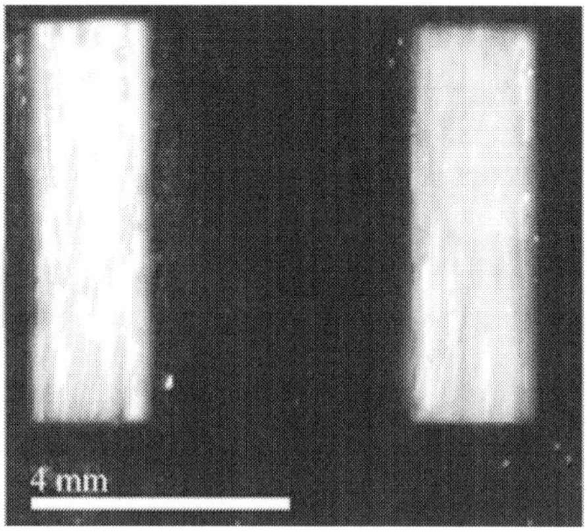

Fig. 1. Photolithographically created lines of mm-dimensions of PMMA/TiO_2 NRs nanocomposites.

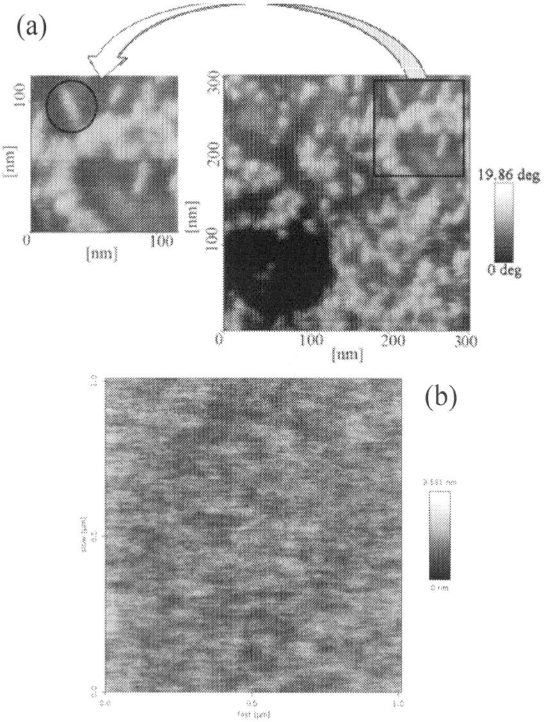

Fig. 2. AFM images of (a) photolithographically produced nanocomposite patterns demonstrating NRs onto their surface, and of (b) photopolymerized coatings of PMMA without TiO_2 NRs.

In order to enhance the hydrophilicity of the TiO_2 NRs exposed on the surface of the photopolymerized nanocomposites, the prepared patterns were irradiated for diverse time intervals with a pulsed Nd:YAG laser at 355 nm (energy density = 7 $mJ \cdot cm^{-2}$, repetition rate = 10 Hz, pulse duration ~ 4-6 ns). Increasing irradiation time up to 90 min under the specific experimental conditions, increases the hydrophilicity of the nanocomposites' surfaces (Villafiorita Monteleone et al 2010). The complete recovery of the initial wettability of the films was achieved by placing them in vacuum at a pressure of $3 \cdot 10^{-3}$ mbar for 48 hours. We take advantage of the controlled wettability changes to realize liquid flow paths, irradiating adjacent surface areas of the photolithographically produced lines of nanocomposites with increasing time, and creating wettability gradients along their surfaces, essential for liquid droplet motion. An example is demonstrated in Figure 3, where is shown a X×Y = **6x1** mm^2 laser photopolymerized nanocomposite film, where in order to achieve the drop movement along the X-axis, a first area of **XxY** = **1x1** mm^2 was left without irradiation, the next **XxY** = **1x1** mm^2 area was irradiated for 20 min and two adjacent areas of **XxY** = **2**×**1** mm^2 each, were irradiated for 40 and 90 min, respectively. The spontaneous water drop movement occurs as shown at the right hand side of Figure 3 (Figure 3c) because the following criteria essential for the movement are fulfilled: (1) The contact angle hysteresis ($\Delta\theta_h$) of the rear area, irradiated for less time must be smaller than the difference between the advancing contact angles of the two adjacent areas (nominated $\Delta\theta$ in Figure 3). (2) The side of each irradiated area along the direction of the movement should be smaller than the diameter of the drop on

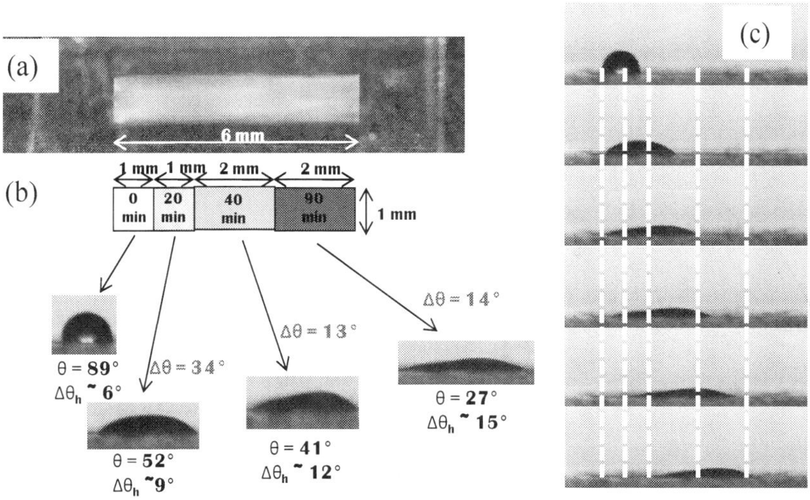

Fig. 3. (a) Photolithographically created line of dimensions 6×1 mm^2, of PMMA/TiO_2 NRs nanocomposite. (b) Graphical representation of the dimensions of the adjacent irradiated areas together with the irradiation times on each area. The pictures of the water drops obtained on each area with their water contact angle and hysteresis values and the difference between the advancing contact angles of the two adjacent areas are also shown. (c) Side view photographs of a water droplet that is moving on the PMMA/TiO_2 surface with UV-induced gradient hydrophilicity. The hydrophilicity increase is in the direction of the drop movement, from left to right.

the specific area, so that the front edge of the drop is always in contact with a more hydrophilic area than the back edge (Villafiorita Monteleone et al., 2010). This kind of samples are very versatile and can be used as described above or incorporated in more complicated systems and devices, such as microfluidics, labs on chip etc.

2.2 γ-Fe_2O_3-based patterned nanocomposites

Magnetic particles are useful in different biomedical applications, such as magnetically induced hyperthermia of malignant tumors, contrast enhancement agents for magnetic resonance imaging (MRI), tissue repair (Gupta et al 2005, Pankhurst et al 2003), and delivery of drugs (Voltairas et al 2002) or nucleic acids (Mair et al 2009, Buerli et al 2007), as they can be manipulated via external magnetic fields to improve transport in biological systems. They are also used in the fabrication of photonic crystals (Ding et al 2009), as nanowire contacts in electronic devices (Bangar et al, 2009), and as device components in microfluidics (Kavcic et al 2009; Fahrni et al 2009). Additionally, there has been growing interest in the use of magnetic particles in the fabrication of nanomotors and nanomachines (Ghosh & Fischer, 2009).

Among all magnetic NPs, iron oxides are possibly the most frequently used, due to their high magnetic moment, chemical stability, low toxicity, biocompatibility, easy and economical synthetic procedures. Iron oxides exist in many forms in nature, with magnetite (Fe_3O_4), maghemite (γ-Fe_2O_3) and hematite (α-Fe_2O_3) being the most common (Cornell & Schwertmann 2003). Magnetite is ferrimagnetic at room temperature (Cornell & Schwertmann 2003), and usually particles smaller than ≈6 nm are superparamagnetic at room temperature, although their magnetic properties depend strongly on the methods used in their synthesis (Kado 2008, Zhao et al 2007). At room temperature, maghemite is ferrimagnetic while particles smaller than ≈10 nm are superparamagnetic (Neuberger et al 2005). Aggregation of ultrafine maghemite particles sometimes leads to their magnetic coupling and ordering of their magnetic moment, which is termed superferromagnetism (Cornell & Schwertmann 2003). Hematite is paramagnetic at temperatures above 956 K. At room temperature, it is weakly ferromagnetic and undergoes a phase transition at 260 K to an antiferromagnetic state. The magnetic behavior of hematite depends also on crystallinity, particle size and on the extent of cation substitution (Cornell & Schwertmann 2003).

The magnetic properties of iron oxides have been exploited in a broad range of applications including magnetic seals and inks, magnetic recording media, catalysts, ferrofluids, contrast agents for magnetic resonance imaging, therapeutic agents for cancer treatment, and numerous others (Azhar Uddin et al 2008, Cui et al 2006, dos Santos Coelho et al 2008). These applications demand nanomaterials of specific sizes, shapes, surface characteristics, and magnetic properties, so the specific type of iron oxide is chosen every time depending on the field of use. On the top, iron oxide in polymer matrices, forming functional nanocomposites, pave the way to novel plastic devices for gas and vapor sensing, actuation, molecular separation, electromagnetic wave absorption, nonlinear optical systems, and photovoltaic solar cells (Kaushik et al 2009, Merkel et al 2002, Huo et al 2009, Long 2005). Furthermore, the magnetic properties of iron oxide can be used in future generations of electronic, magnetic, and photonic devices for information storage or magnetic imaging (Weller et al 2000, Wang et al 2004).

For enhanced performances and occasionally directional, nanocomposite systems require high density of oriented anisotropic magnetic nanostructures, like NWs, in polymer matrices. Indeed, the realization of arrays of NWs is a new interesting solution to obtain novel collective properties, different from those of isolated NPs (Tang and Kotov, 2005).

One-dimensional magnetic NWs can be produced by the assembly of isotropic magnetic NPs, under external magnetic field (MF). This is an attractive technique for the fabrication of NWs, due to its simplicity and at the same time, high effectiveness. In this perspective, several studies have recently demonstrated the possibility of producing oriented magnetic nanocomposites through the dispersion of magnetic NPs in polymer or prepolymer solutions, and subsequent evaporation or polymerization under a weak magnetic field (Park et al 2007, Jestin et al 2008, Fragouli & Buonsanti et al 2010, Fragouli & Torre et al 2010).
Here, we present photolithographically realized patterned nanocomposites of PMMA or SU-8 polymers which incorporate magnetic NWs, formed starting from spherical iron oxide (γ-Fe_2O_3) colloidal NPs. Indeed, applying a homogeneous magnetic field produced by two magnets to the nanocomposites solutions, NWs are formed, which are aligned along the magnetic field lines. We demonstrate that the photolithography process does not affect the NPs alignment, and, more importantly, that it allows the creation of polymeric patterns with magnetic properties in specific areas.

2.2.1 Photopolymerized patterns of PMMA/Fe_2O_3

Solutions of methyl methacrylate (MMA) monomer mixed with colloidal γ-Fe_2O_3 NPs in the presence of a photoinitiator IRGACURE®1700 were prepared in chloroform at concentrations of 89.5 wt.%, 10 wt.% and 0.5 wt.%, respectively. The Fe_2O_3 NPs were previously prepared in chloroform, by modifying a wet-chemical synthetic approach previously reported (Sun et al 2004), in order to obtain hydrophobic-capped Fe_2O_3 spherical particles with a mean diameter of 10 nm. Oleic acid, oleylamine, and hexadecane-1,2-diol were used as capping molecules of the produced NPs. All solutions were stirred and left in the dark for few minutes to equilibrate. 200 µL of each solution were spin-coated at 1000 rpm for 20 s, and subsequently ~40 µL of the same solution were drop-casted onto glass substrates. For the alignment of the nanoparticles, the system was subjected to a homogeneous magnetic field (~160 mT), produced by two permanent magnets, applied parallel to the substrate during the deposition, evaporation and photopolymerization processes.
The patterned PMMA/Fe_2O_3 nanocomposites were obtained by irradiating all samples with the third harmonic of a pulsed Nd:YAG laser (Quanta-Ray GCR 190, Spectra Physics) with an energy density of 10.5 $mJ \cdot cm^{-2}$ (λ=355 nm, pulse duration =4-6 ns, repetition rate=10 Hz) for 90 minutes, using aluminum photomasks with different geometries. After the photopolymerization, each sample was washed three times with methanol to remove unreacted monomer and photoinitiator and then dried under ambient dark conditions for 2 days to achieve complete solvent evaporation. In Figure 4 are demonstrated characteristic patterns created upon irradiation of the spin coated and subsequently drop casted MMA/Fe_2O_3 NPs solution through a photomask of two parallel lines in the mm range.
During the photopolymerization, a homogeneous magnetic field produced by two permanent magnets was applied to the samples, at saturated chloroform atmosphere as shown in Figure 5. All samples prepared with or without the application of the magnetic field were studied under an optical microscope. Figure 6 presents samples with 10 wt. % concentration of NPs. The analysis reveals that in the samples created without the application of the magnetic field (Figures 6a and 6b), the NPs form aggregates randomly distributed in the film. The application of a magnetic field produced by two magnets leads

to NPs aligned along the magnetic field lines, forming parallel NWs (Figure 6c and 6d). By these two figures is demonstrated that the formation of magnetic NWs when the samples are placed under the magnetic field occurs both before and after the photopolymerization procedure, with the latter leaving unaffected the geometry of the wires. In particular, Figure 6d shows that the area of the sample that was covered by the mask (bottom part of the picture) is not polymerized and thus, after washing with methanol, the composite is removed and no aligned structures are left onto the substrate.

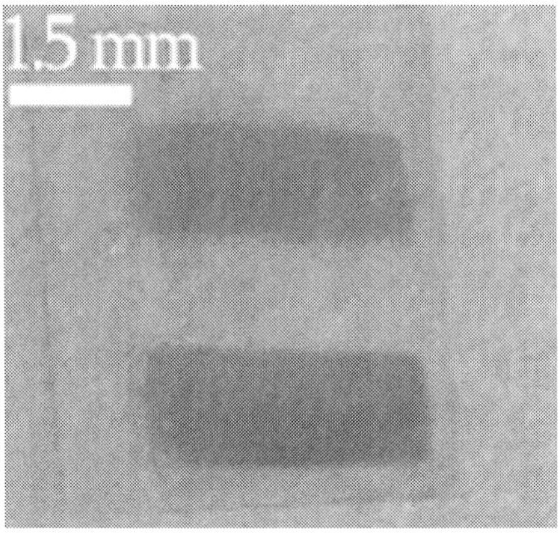

Fig. 4. Photolithographically created lines of mm-dimensions of PMMA/γ-Fe_2O_3 NPs nanocomposites.

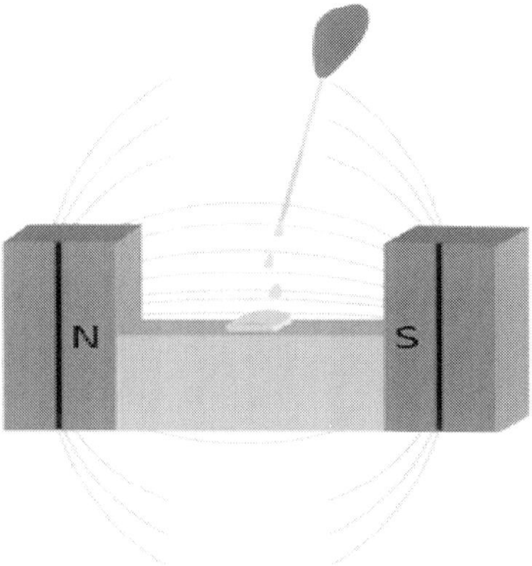

Fig. 5. Schematization of the experimental set-up where a homogeneous magnetic field is applied parallel to the samples.

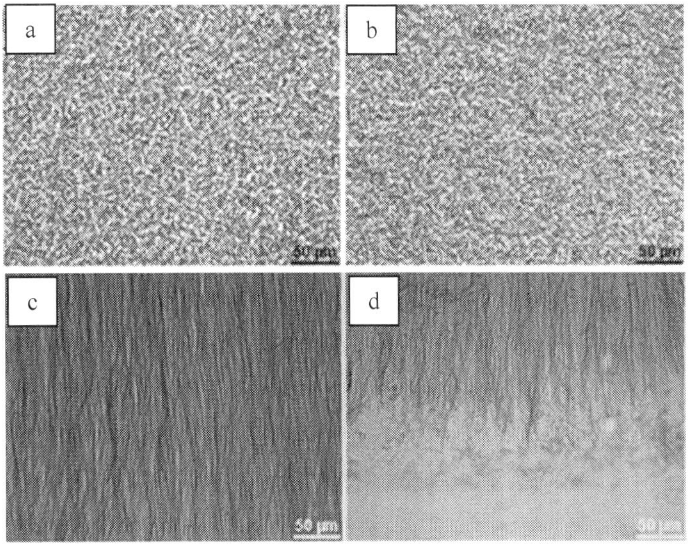

Fig. 6. Optical microscopy images of PMMA/Fe_2O_3 films formed without the application of a magnetic field (a) before and (b) after the photopolymerization, and with the application of a magnetic field (c) before and (d) after the photopolymerization and washing. In the panel (d) is demonstrated the edge of the photopolymerized film after the washing.

A detailed topographic study of the photolithographically produced patterns with or without the application of the magnetic field was conducted using AFM (Figures 7a-7d). Figures 7a and 7b show the topography of samples produced in the absence of a magnetic field. The nanofillers of these samples do not show any particular alignment, as expected, but rather organize themselves in order to form aggregates of different sizes distributed all over the surface, in agreement with the results of the optical microscopy examination. On the contrary, the AFM images of the samples realized under the magnetic field produced by two magnets show the presence of parallel structures aligned along the direction of the external magnetic field (Figures 7c and 7d). In particular, parallel lines with thickness of ~1 µm and length of tens of microns clearly appear both in the 2D and in the 3D topography images. The images also demonstrate the presence of bigger aggregated structures with width around 5 µm, which can be attributed to aligned NWs that in some areas of the sample get very closely packed, due to the high NPs concentration. It is very likely that the NWs formation inside the photopatterned polymer are facilitated by the initial organization of the NPs into bigger clusters, as demonstrated already in Figures 7a and 7b. These clusters exhibit higher magnetic moments than the single Fe_2O_3 NPs, resulting in an increased response to the external magnetic field that leads eventually to chain formation (Lalatonne et al 2004).

The formation of magnetic NWs inside and on the surface of polymeric patterned structures opens up the possibility of various applications of these systems, related to the oriented growth and patterning of molecules bound on the NWs, and to the effect of this binding on the magnetic properties of the NWs. In particular, specially designed structures can be created where various biological molecules such as DNA, proteins, or cells can be bound on the NWs. Such structures can be extremely useful in biological applications for the formation of biological sensors, and can be also incorporated in molecular recognition devices

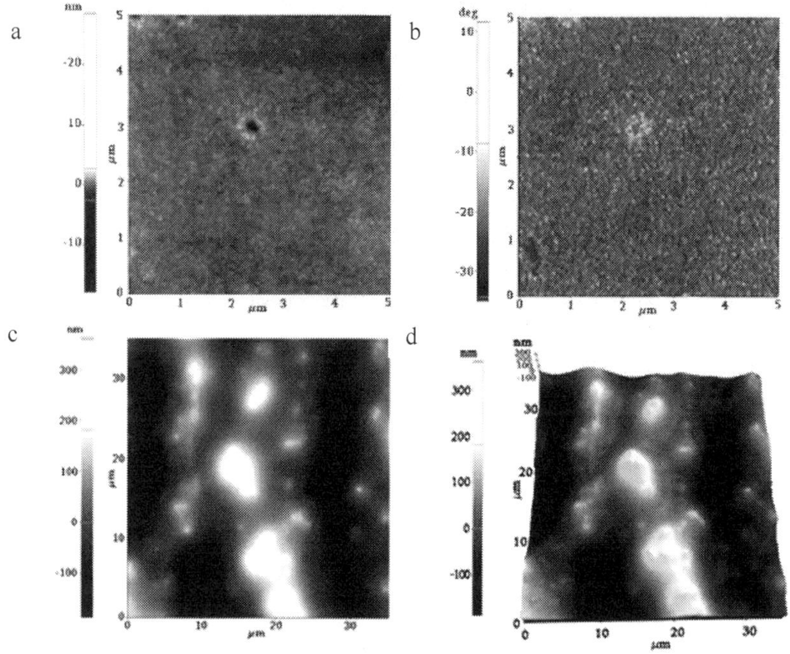

Fig. 7. AFM images of PMMA/Fe_2O_3 films realized with or without a magnetic field. The images show the topography (a) and the phase (b) of a sample realized in the absence of a magnetic field, and the 2D (c) and 3D (d) topography images of a sample realized under the magnetic field of two magnets. In particular, the 2D and 3D topography images of sample areas with parallel lines show the existence of aligned structures on the surface of the sample.

2.2.2 Photopolymerized patterns of SU-8/Fe_2O_3

SU-8 is a commercial biocompatible epoxy-based negative photoresist that is suitable for the microfabrication of high aspect-ratio (>20) structures. When exposed to UV light, its molecular chains crosslink causing the polymerization of the material. Here the SU-8 3050 (Microchem) was used for the fabrication of the SU-8 pillars. Fe_2O_3 NPs of diameter 20 nm in choloroform solution were prepared, by modifying the wet-chemical synthetic approach previously reported by Sun et al 2004. Size control was achieved by varying the concentration of the precursor and the surfactant-to-precursor molar ratio.

For the fabrication of SU-8 magnetic pillars the following steps were performed: The existing cyclopentanone solvent was evaporated from the SU-8 3050 resin. Subsequently, SU-8 was dissolved in chloroform (1:5 wt.) by sonication for 10 min and then stirring. Then, the Fe_2O_3 NPs (2 wt. %) were slowly added and mixed with the SU-8 solution under sonication. After obtaining a homogeneous solution, the solvent was slowly evaporated under nitrogen flow, until it was obtained a nanocomposite solution viscous enough to form a thick coating. The solution was then spin-coated on a silicon wafer and subsequently left under a magnetic field (400 mT) with a vertical direction with respect to the substrate, resulting in the formation of wire like magnetic structures. These magnetic wires induce magnetic anisotropy so for specific magnetic field orientations the magnetic response is higher (Fragouli & Buonsanti et al 2010). After, the sample was irradiated with a UV lamp mask aligner for 70 sec with energy of 25 mW. Masks of square shaped pillar patterns were used, with 42 μm side and different interpillar distances varying from 14 to 77 μm. The post-bake was for 1 min at 65° C and for 4 min at 95° C. Finally, the samples were developed with SU-8 developer (15min) and rinsed with isopropanol with subsequent drying under nitrogen flow. The fabricated nanocomposite pillars formed in this way, had a square side of 42 μm and 40 μm high as shown in the image recorded with Scanning Electron Microscopy (SEM) (Figure 8).

Fig. 8. SEM image of the SU-8/iron oxide nanocomposite pillar structures.

The AFM image of figure 9 demonstrates a part of the surface of a nanocomposite pillar in which the NPs are aligned vertically to the substrate. The nanoroughness shown is due to the protruding edges of the formed NWs, demonstrating that the diameters of the NWs are around 100 nm.

The magnetic alignment of the iron oxide NPs inside the SU-8 matrix creates a magnetic anisotropy that enhances the possibility for the pillar structures to respond to an external magnetic field. The responsivity of the magnetic pillars to an external magnetic field was checked using water drops placed onto the pillars. In particular, the pillars were inclined in order to observe a difference between the right and the left part of the drops and a static magnet of maximum field strength 500 mT, was moved towards the sample with a direction parallel to it. The test was performed onto magnetic pillars with interpillar distance 14 μm. The water contact angle on these pillars is about 130°, while the values of the right and the left contact angle after the tilting of the sample are shown in table 1. After few minutes it was observed a change in the shape of the drop when the magnet was moving towards the drop, while it was recovering when the magnet was far from the substrate. This reversible difference in the shape of the drop indicates the tendency of the drop to move, which is connected with the reversible movement of the magnetic pillars responding to the external magnetic field.

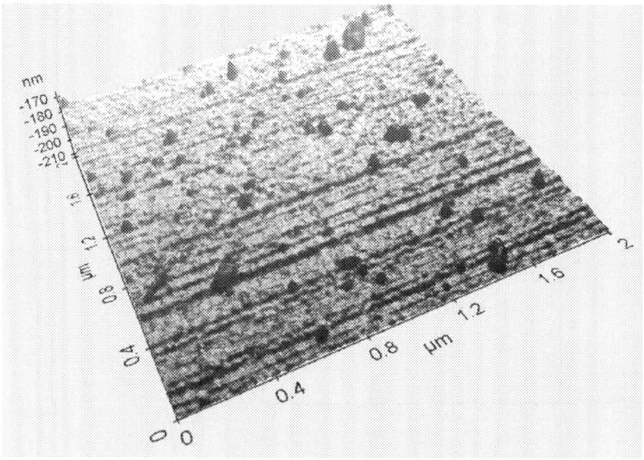

Fig. 9. AFM topography of the nanocomposite SU-8 mixed with iron oxide NPs, vertically aligned under the application of an external magnetic field.

Right Part (°)	Left Part (°)	Drop Images
127	130	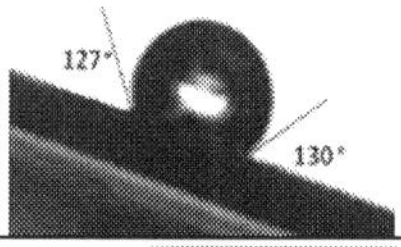
122	135	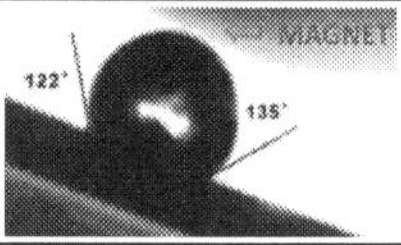

Table 1. Contact angle values of the left and right part of the water drop when the magnet is away from the substrate (top raw) and when the magnet in coming closer (down raw). At the right column are demonstrated the two examined frames.

3. Lithography using light sensitive precursors of nanofillers

In situ synthesis of NPs into their matrices is a very powerful lithographic technique because the properties of the nanocomposites can be controlled and tailored locally in a unique way. In this section we will present the laser-induced formation of NPs directly into the polymer matrices following two main strategies. The first one is based on the use of chloroauric acid salt, which is a gold (Au) precursor, introduced in transparent polymer films. Herein, chitosan (CTO) is used as a matrix, which is a very promising polymer due to its biocompatibility. In CTO polymer, Au NPs can be generated by UV irradiation (Miyama & Yonezawa 2004). Using photomasks and UV laser beam we obtain the creation of gold NPs in precise areas of the polymeric film, turning the insulating polymer into electron conducting material. This process allows us to localize and design accurately surface patterns and moreover to tune the metallic particle size in the range of nanoscale by varying the laser irradiation time and energy.

Another strategy that we follow for the in situ creation of nanofillers into the polymer matrix is the UV laser irradiation of polymer films containing cadmium thiolate precursors. This method results in the spatially selective formation of cadmium sulphide (CdS) crystalline NPs in the host matrix, through a macroscopically non-destructive procedure for the matrix. Using a pulse by pulse approach, we accomplished the formation of NPs with gradually increasing dimensions, and consequently the progressive change of the emission characteristics of the formed nanocomposites. The optimized combination of irradiation wavelength with polymer matrix gives patterned nanocomposite materials incorporating nanocrystals of high quality, ready to be used in various optical applications.

3.1 Au precursor-based nanocomposite patterns

The use of polymeric nanocomposite materials is expanding to a huge range of applications since they combine the flexibility, easy processability and low cost of the polymers with the unique properties of the nanofillers. On the other hand the intrinsic insulating characteristic of the polymers limit the possibilities of using polymeric-based systems in devices where electronic conductivity is desired, like sensors, miniaturized electronic chips, etc. In order to overcome this limitation, the use of metallic NPs that exhibit very high electronic conductivity as nanofillers is proved to be a successive strategy (Gelves 2006, 2011, Huang et al 2009).

Our strategy is based on the use of the Au precursor, chloroauric acid salt ($HAuCl_4$), introduced in a transparent polymer film by immersion, and the generation of Au particles in specific areas by means of laser irradiation. The lithographically produced nanocomposite areas have tailored properties, dependent on the density and size of the produced Au NPs. The possibility to produce Au-polymeric nanocomposite materials with enhanced electrical properties, in combination with the spatial control of the specific property by introducing in situ the nanofillers in the desired areas increases enormously the potentiality of such systems in a wide variety of applications.

3.1.1 Spatially controlled in situ formation of Au NPs in chitosan

Chitosan (CTO) is a natural biodegradable and biocompatible polysaccharide polymer derived from chitin, a linear chain of acetylglucosamine groups, extracted from crustaceans shells and the cell walls of many fungi. CTO is fiberlike and is obtained by the deacetylation process of the natural chitin, a process that gives rise to amine groups which can be used for further functionalization (Yi et al 2005, Luther et al 2005, Su et al 2005, Zhou et al 2006, Zangmeister et al 2006). CTO is becoming widely used due to its potential polysaccharide resource and properties as non-toxicity, excellent processability, adsorption properties, hydrogel behavior, electrospinning, etc (Guibal 2005, Nirmala et al 2011). Its chemical structure is illustrated in figure 10.

CTO has a hydrogel nature resulting in the tendency to absorb ambient moisture or liquids. In this work, the process was optimized and used to introduce gold precursor in CTO polymer thin films. By controlling the immersion time of the polymer film in gold precursor solution, the absorption of gold precursors is highly controlled. The CTO used in this work was purchased from Sigma Aldrich with a degree of deacetylation about 80%. Various concentrations of CTO polymer solutions (0.5%, 1% and 2% wt.) are prepared in acetic acid. CTO polymeric films on glass substrates are obtained by drop-casting or spin-coating. The use of high CTO polymer concentration allows the formation of CTO hydrogel films able to absorb gold precursor crystals behaving as a "gold precursor reservoir". The gold precursor

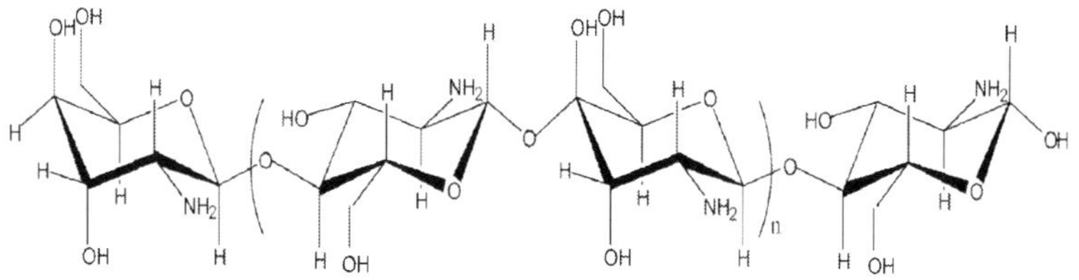

Fig. 10. Chemical structure of chitosan.

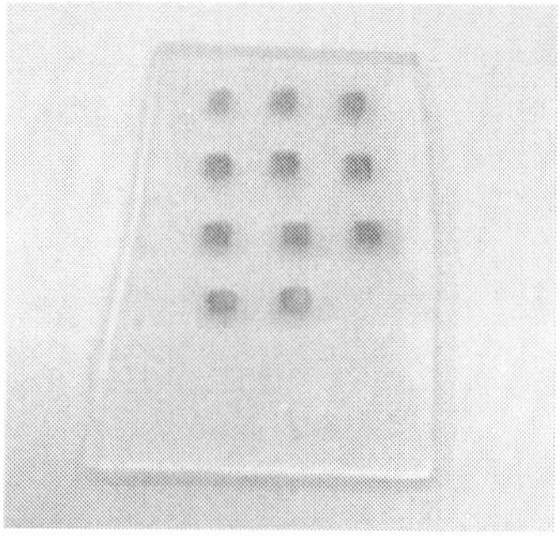

Fig. 11. Lithographic production of Au NPs at adjacent areas (dark spots) of a CTO film incorporating Au precursor, irradiated with increasing times.

used is a chloroauric acid salt (Mw ($HAuCl_4$)=339.5g/mol) and is dissolved in distilled water by providing two solutions (0.01M and 0.02M). Irradiation of the salt embedded in CTO with UV light induces photoreduction of metallic ions ($AuCl_4^-$) into metal atoms (Duff et al 1993), clusters, aggregated metal clusters and eventually gold NPs, as described in equation 1. The Au NPs were obtained by irradiating the samples with the third harmonic of a pulsed Nd:YAG laser (λ=355 nm, pulse duration =4-6 ns, repetition rate=10 Hz, Quanta-Ray GCR 190, Spectra Physics) with fluence in the range of 0.5 to 5.0 J·cm^{-2} through photomasks. The results of the lithographic production of Au NPs at adjacent areas of a representative sample irradiated with increasing irradiation times is shown in Figure 11.

$$\begin{aligned} &RNH_3{+}AuCl_4^- \rightarrow Au^0 + RNH_3{+}Cl^- + 3Cl \\ &Au^0 \rightarrow Au_n\,(\text{cluster}) \rightarrow Au_p\,(\text{particles}) \end{aligned} \quad (1)$$

For the fluence range used in this work the irradiation time needed for the creation of the NPs is very short, in the range of few seconds, and after 2-3 minutes of irradiation the Au/CTO films appear destroyed. The generation of Au NPs at the surface of the CTO-Au polymeric film in function of the irradiation time is illustrated in the AFM images of Figure 12, for a laser fluence of 1 J cm^{-2}. In particular, Fig. 12a illustrates a non-irradiated area with a very smooth surface. Fig. 12b demonstrates an area irradiated for 30 sec. At this stage, the photolysis reaction results in the formation of few Au NPs with low density and sub-micron size. In Fig. 12c is demonstrated the surface of an area after its irradiation for 75 sec. In this figure the Au NPs appear much denser and a clear reduction of their size is also demonstrated. The size of the smallest Au NPs that appear onto the surface of the sample is around 20 nm. Under these experimental conditions, the distribution of Au NPs becomes almost homogeneous after 90 sec and they form a sort of continuous film on the surface of the CTO film. It can also be noted that the form of the NPs seems to change from undefined geometrical structures to spherical shapes after prolonged irradiation.

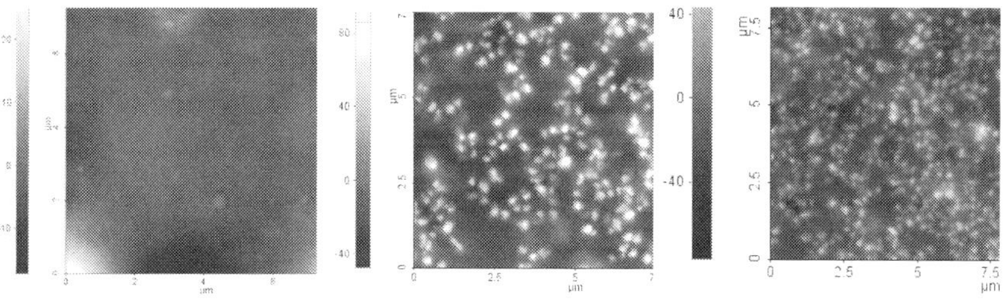

a b c

Fig. 12. Evolution of the formation of the Au NPs onto the CTO surface upon increasing irradiation time: (a) non-irradiated area, (b) area irradiated for 30 sec, (c) area irradiated for 75 sec. The density of the NPs increases and their size decreases with the irradiation time.

In the literature (Miyama & Yonezawa 2004) is mentioned that the growth of Au NPs occurs by aggregation of the photolytically formed Au atoms and clusters (equation 1). Nevertheless, the aggregation mechanism is not confirmed by the AFM images presented in figure 12, where is demonstrated that increasing irradiation time causes the decrease and not the increase of the size of the formed particles. The mechanism occurring in our case seems to be closer to the one reported on the formation of metal NPs by laser ablation in water solutions, where the reduced size of the NPs with increasing irradiation time is explained by their fragmentation due to self-absorption of laser pulses (Mafune´ et al 2000, Shukla & Seal 1999, Videla et al 2010, Kadossov & Burghaus 2010, Kabashin & Meunier 2003). Indeed, the Scanning Electron Microscopy (SEM) images obtained on the same sample irradiated at two different areas for very few seconds (≈5) and 30 sec, respectively, using fluence of 1 J cm^{-2}, demonstrate that the first particles that appear onto the CTO surface are quite big (in the micrometer scale) and then they fragment into smaller ones (Figure 13). A closer look into the NPs clearly demonstrates that some of them are already fragmented into smaller pieces. Due to the irradiation conditions used in our experiments the aggregation of Au atoms and clusters into small particles seems to occur already in the very first seconds of the procedure and then the fragmentation mechanism becomes predominant. It is also interesting to mention that the fragmentation mechanism as described in literature is linked with an increase of local temperature due to the absorption of the laser photons by the Au NPs, which may increase the mobility of the latter towards the surface of the CTO polymeric films.

Finally, in figure 14 is demonstrated the intensity of the electronic current that is conducted through a CTO/Au NPs sample as a function of the laser irradiation time, for a laser fluence of ~5 $J{\cdot}cm^{-2}$. It is clear that the density increase of the formed Au NPs and the interconnection between them, previously demonstrated in figures 12 and 13, has an effect on the electrons mobility into the samples. Indeed, the current passing through the samples increases almost linearly with the laser irradiation time. The enhancement of the electrical conductivity in specific areas of nanocomposites by laser induced lithography, can be used for a wide range of applications such as implantable gas sensors, liquid sensors for robotics and nanocircuits.

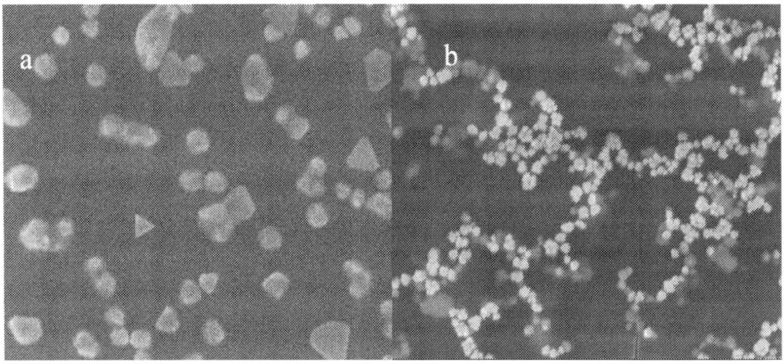

Fig. 13. SEM images showing the fragmentation of the Au NPs with increasing irradiation time: (a) area irradiated for 5 sec, (b) area irradiated for 30 sec. The density of the NPs increases and their size decreases with the irradiation time.

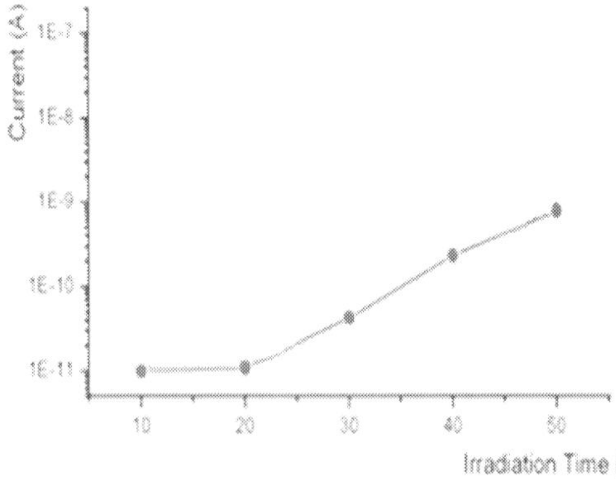

Fig. 14. Evolution of the current intensity in function of the irradiation time for laser fluence ~5 J·cm^{-2}.

3.2 CdS precursor-based nanocomposite patterns

Semiconductor nanocrystals (NCs) embedded into polymeric matrices can be exploited in several technological applications, taking advantage of the unique photophysical characteristics of the former due to the quantum confinement effect. In particular, the size-dependent optical properties of the NCs, such as, high emission quantum yields, narrow emission bands, and tunable emission/absorption spectra, have been the topic of many recent research works (Xia et al 2008, Medintz et al, 2005, Bruchez et al 1998, Michalet et al 2005). On the top, the careful selection of the polymer matrices can lead to highly processable nanocomposite materials with increased stability.

For different applications, the lithographically patterned formation of well-dispersed NCs into polymers is highly requested, since it provides spatially selective tailoring of specific properties of the nanocomposites. Indeed, on one hand a good dispersion of the NCs optimize their quantum size effect, meaning the control of their emission properties. On the other hand, the localization of the NCs in specific sites of the polymer provides the possibility of the direct incorporation of the nanocomposites in various advanced technological devices, such as sensors, biological chips, photoemission devices, etc. In this respect our lithography approach involves the localized in-situ formation of NCs inside polymer matrices by UV laser irradiation of polymer-precursor films. We focus our study on the *in situ* localized formation of NCs of one very promising II–VI semiconductor, the CdS, by the use of pulsed UV laser irradiation of a polymer film incorporating the photosensitive metal sulphide precursor.

3.2.1 Spatially controlled in situ formation of CdS NCs in polymers

Cadmium bis-dodecanthiolate $Cd(SC_{12}H_{25})_2$, (C12), is a photosensitive metal precursor that was mixed with the polymer poly-methylmethacrylate (PMMA) or TOPAS®, a thermoplastic cyclo-olefin copolymer consisting of ethylene and norbonene units, transparent in the visible, in order to produce the CdS NCs after laser irradiation. In particular, 20 wt.% of the metal thiolate precursors was mixed with 80 wt. % of polymer, and then diluted in toluene. The solutions, after being sonicated for 30 min, were cast in Petri capsules. The polymer-precursor films formed after the evaporation of toluene had thickness ~200 μm.

For the *in situ* formation of the NCs in the polymer matrix, the films were irradiated through photomasks of different shapes with pulses of Nd:YAG laser (Quanta-Ray PRO-290-30, Spectra Physics) operating at the fourth harmonic, (wavelength 266 nm, pulse duration 8 ns, and repetition rate 2 Hz). The wavelength 266 nm was chosen since the metal precursor exhibits at it enhanced absorption, as demonstrate in figure 15. Both TOPAS and PMMA polymers have intriguing physical properties for a number of applications, taking advantage at the same time of the in situ photoinduced CdS nanocomposite micropatterns with accurate control of the NCs size upon UV irradiation.

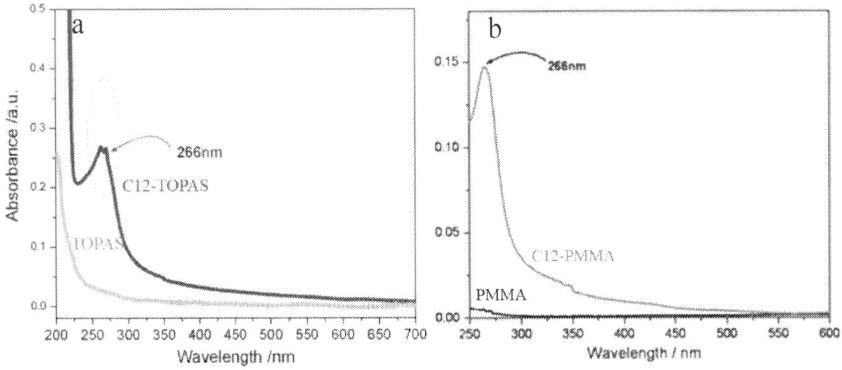

Fig. 15. Normalized absorption spectra of the precursor C12 mixed with the polymers (a), and of the pure TOPAS and PMMA (b).

Upon laser irradiation through photomasks the CdS NCs are localized exclusively in the irradiated area, making possible the lithographic patterning of the samples. A characteristic micro pattern of 3 lines is illustrated in figure 16. For the specific pattern the lithographic technique that we used was not involving photomasks but it was done by focusing the laser beam on the C12-TOPAS sample, which was fixed onto a motorized stage moving at constant velocity. The spot size of the beam was ~0.2 × 0.1 mm^2, and the speed of the motor about 0.2 mm·s^{-1}. The lines are fairly clear and their width (240 μm) is constant along the irradiation path. The total number of pulses in each spot area is about 100 and is enough to form CdS NCs, as demonstrated by their emission spectra shown in figure 16. The bright areas correspond to CdS NCs in the bulk region, since the spectrum has the characteristic emission peak at 506 nm, while the dark areas, not being irradiated, have no fluorescence emission in the studied spectral region.

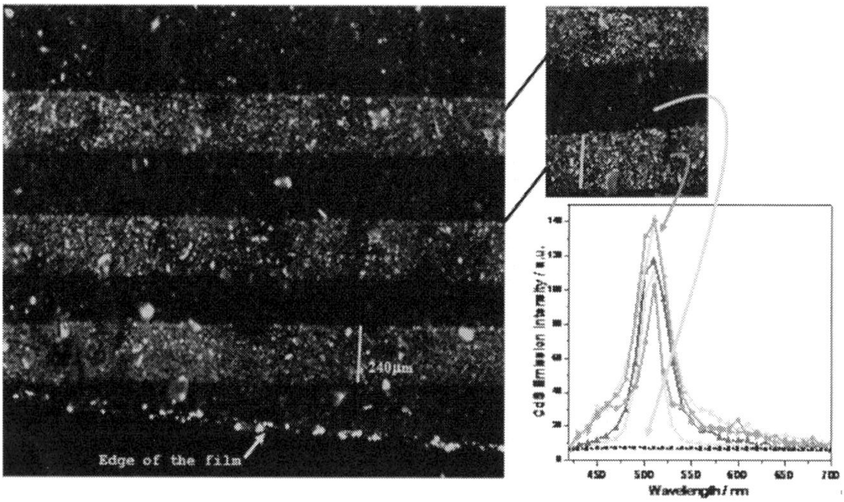

Fig. 16. Fluorescence image of a patterned sample of cadmium thiolate precursor-TOPAS polymer excited with a 405 nm diode laser after irradiation at 266 nm. The narrow fluorescence spectra of the formed CdS NCs are also shown.

The formed CdS NCs following an increasing number of laser pulses are characterized by spatially resolved photoluminescence measurements using a confocal microscope. Indeed, in figure 18 are presented the fluorescence spectra from selected areas of the films irradiated with various laser pulses at a laser fluence of 25 mJ·cm^{-2}. After 6 pulses the emission peak of the formed NCs is close to λ = 440 nm. After successive laser pulses the emission is shifted towards higher wavelengths with the peak reaching 506 nm above 40 laser pulses, which coincides with the emission of the bulk CdS material. It is clearly demonstrated that the luminescence of the samples changes dramatically after the first irradiation pulses, while an increase of the incident pulses causes a red shift to the emission, which is attributed to the increasing dimensions of the CdS NCs. Therefore, the UV irradiation of metal precursor-TOPAS polymer films results in the formation of CdS NCs, with dimensions extending from the quantum size effect range to the bulk, depending on the number of the incident laser pulses. This, in combination with the spatial control of the formation of these NCs thanks to the presented lithographic technique opens the way for the incorporation of such systems in complex multicomponent devides.

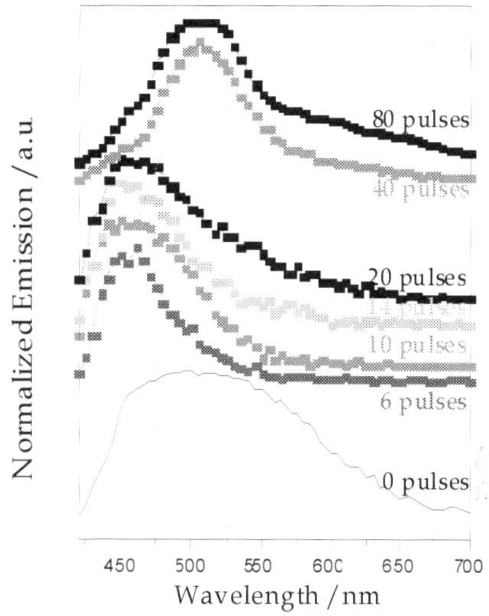

Fig. 17. The change of the fluorescence spectra of C12-TOPAS film with an increasing number of incident laser pulses.

The lithographic formation and size tuning of CdS NCs with increasing UV irradiation time of metal precursors occurs also in PMMA, in the same way as in TOPAS. However, intense deteriorations (broadening) are observed in the emission spectra of the NCs formed in PMMA compared to TOPAS upon increasing number of pulses. Figure 18 shows the fluorescence image of a C12-PMMA film irradiated with 80 laser pulses of fluence F=20 $mJ\cdot cm^{-2}$, at λ=266 nm, and the corresponding emission taken from various areas of the film using a confocal microscope. The emission spectra show the existence of few CdS NCs emitting at the bulk region (NCs diameter > 7 nm) with an emission peak close to 500 nm and FWHM ~32 nm (Figure 18b), while many areas exhibit a very broad emission with the peak around 550 nm (Figure 18b, blue line), representative of the trap states emission formed on the surface of the NCs. Indeed, the quality of the semiconductor NCs is generally studied by their emission characteristics verifying that the broader the emission spectra the higher the number of the trap states on their surfaces (Athanassiou et al 2007, Fragouli et al 2009, 2010, Wang et al 2004, Antoun et al 2007, Wu et al 2000, Khanna et al 2007). After irradiation with the same number of laser pulses but increased laser fluence (F=50 $mJ\cdot cm^{-2}$, 80 pulses) the characteristic emission of the NCs in PMMA matrix is no more evident in the irradiated area, while the trap state emission is dominant (Figure 18c). Therefore, when PMMA matrix is used the laser irradiation at 266 nm results in the formation of NCs with trap states on their surface, an evidence that becomes clearer as the incident fluence increases. The comparison between the emission spectra of the CdS NCs formed in the TOPAS and in the PMMA matrix shows that they are very narrow and characteristic of CdS in the first case while they are broad and red shifted in the second case. This finding demonstrates that the trap states on the NCs surfaces formed with laser lithography, and thus, the optical quality of the NCs strongly depends on the surrounding matrix.

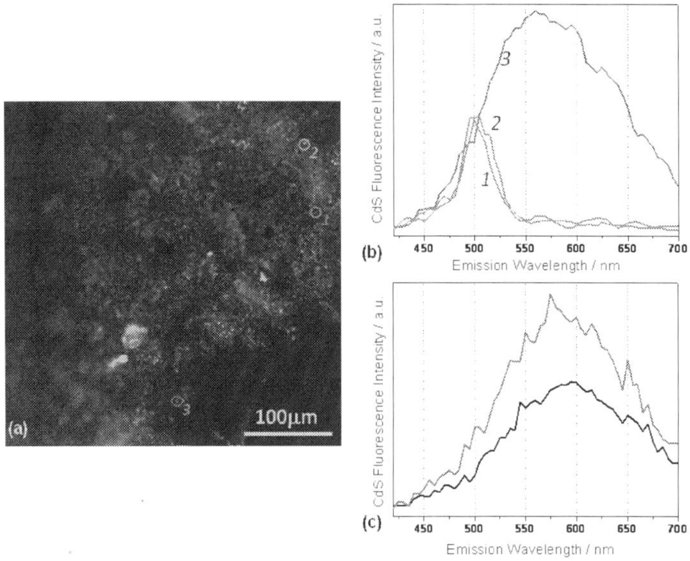

Fig. 18. (a) Confocal microscopy fluorescence image of the C12-PMMA film after irradiation at 266 nm with 80 incident pulses and fluence F=20 mJ·cm^{-2}. (b) Emission spectra of selected areas of the image; each curve corresponds to the marked area of (a) with the same colour and number. (c) Emission spectra of selected areas of the C12-PMMA film after irradiation at 266 nm with 80 incident pulses and higher fluence F=50 mJ·cm^{-2}.

4. Conclusion

In this work are demonstrated two different lithography techniques based on the use of pulsed UV laser light for the creation of well defined polymeric nanocomposite structures. The presented techniques open up new ways for nanocomposite film patterning with well dispersed and occasionally size controlled NCs. The first technique is the UV laser photopolymerization lithography, where UV laser pulses are used for the polymerization and patterning of nanocomposite solutions of preformed colloidal NPs mixed with monomers and photoinitiators. In the second lithography technique UV laser pulses induce the in situ formation of NPs in specific locations in the polymer matrices, starting from polymeric film incorporating light-sensitive precursors of NPs. Using both techniques we present patterned structures of various nanocomposite systems, such as PMMA with TiO_2 NPs, PMMA and SU-8 with Fe_2O_3 NPs aligned into NWs under external magnetic field, CTO with Au NPs, TOPAS and PMMA with size tuned CdS NPs. The formed nanocomposite structures can exhibit controlled and enhanced surface, optical, electronically conducting and magnetic properties, in an easy and fast way, promising for the development of nanodevices used in biotechnology, optoelectronics, microfluidics and other advanced technological areas.

5. Acknowledgment

The authors would like to thank Dr Rafaella Buonsanti, Dr Gianvito Caputo, Dr P. D Cozzoli, Dr Anna Maria Laera, Professor Leander Tapfer who provided us with diverse colloidal nanofillers or nanocomposite samples and for the fruitful discussions and suggestion that helped in the realization of this research work.

6. References

Andrzejewska, E. (2001). Photopolymerization kinetics of multifunctional monomers. *Progress in Polymer Science*, Vol. 26, No. 4, (May 2001), pp. 605-665, ISSN 0079-6700

Antoun, T.; Brayner, R.; Alterary, S.; Fiévet, F.; Chehimi, M.; Yassar, A. (2007). Facile Synthesis of Oligothiophene-Capped CdS Nanoparticles. *European Journal of Inorganic Chemistry*, Vol. 2007, No. 9, (March 2007), pp. 1275-1284, ISSN 1099-0682

Athanassiou, A.; Cingolani, R.; Tsiranidou, E.; Fotakis, C.; Laera, A. M.; Piscopiello, E.; Tapfer, L. (2007). Photon-induced formation of CdS nanocrystals in selected areas of polymer matrices. *Applied Physics Letters*, Vol. 91, No. 15, (October 2007), 153108, ISSN 1077-3118

Athanassiou, A.; Blasi, L.; De Giorgi, M.; Caputo, G.; Fragouli, D.; Tsiranidou, E.; Laera, A. M.; Tapfer, L.; Cingolani, R. (2010). Comparison between laser-induced nucleation of ZnS and CdS nanocrystals directly into polymer matrices. *Polymer Composites*, Vol. 31, No. 6, (June 2010), pp. 1075–1083, ISSN 1548-0569

Athanassiou, A.; Cingolani, R.; Caputo, G.; Cortese, B.; Gigli, G.; Nobile, C.; Cozzoli, P. D. (2010). Reversible wettability of hybrid organic/inorganic surfaces of systems upon light irradiation/storage cycles. *Internal Journal of Nanomanufacturing*, Vol. 6, No. 1/2/3/4, pp. 312-323, ISSN 1746-9406

Azhar Uddin, M.; Tsuda, H.; Wu, S.; Sasaoka, E. (2008). Catalytic decomposition of biomass tars with iron oxide catalysts. *Fuel*, Vol. 87, No. 4-5, (April 2008), pp. 451-459, ISSN 0016-2361

Bangar, M. A.; Hangarter, C. M.; Yoo, B.; Rheem, Y.; Chen, W.; Mulchandani, A.; Myung, N. V. (2009). Magnetically Assembled Multisegmented Nanowires and Their Applications. *Electroanalysis*, Vol. 21, No. 1, (January 2009), pp. 61–67, ISSN 1521-4109

Bruchez, M. Jr.; Moronne, M.; Gin, P.; Weiss, S.; Alivisatos, A. P. (1998) Semiconductor Nanocrystals as Fluorescent Biological Labels. *Science*, Vol. 281, No. 25, (September 1998), pp. 2013-2016, ISSN 1095-9203

Buerli, T.; Pellegrino, C.; Baer, K.; Lardi-Studler, B.; Chudotvorova, I.; Fritschy, J.-M.; Medina, I.; Fuhrer, C. (2007). Efficient transfection of DNA or shRNA vectors into neurons using magnetofection. *Nature Protocols*, Vol. 2, No. 12, (December 2007), pp. 3090–3101, ISSN 1750-2799

Caputo, G.; Nobile, C.; Kipp, T.; Blasi, L.; Grillo, V.; Carlino, E.; Manna, L.; Cingolani, R.; Cozzoli, P. D.; Athanassiou, A. (2008). Reversible Wettability Changes in Colloidal TiO2 Nanorod Thin-Film Coatings under Selective UV Laser Irradiation. *Journal of Physical Chemistry C*, Vol. 112, No. 3, (January 2008), pp. 701-714, ISSN 1932-7455

Caputo, G.; Cortese, B.; Nobile, C.; Salerno, M.; Cingolani, R.; Gigli, G.; Cozzoli, P. D.; Athanassiou, A. (2009). Reversibly Light-Switchable Wettability of Hybrid

Organic/Inorganic Surfaces With Dual Micro-/Nanoscale Roughness. *Advanced Functional Materials*, Vol. 19, No. 8, (April 2009), pp. 1149–1157, ISSN 1616-3028

Caputo, G.; Cingolani, R.; Cozzoli, P. D.; Athanassiou, A. (2009). Wettability conversion of colloidal TiO_2 nanocrystal thin films with UV-switchable hydrophilicity. *Physical Chemistry Chemical Physics*, Vol. 11, No. 19, pp. 3692 – 3700, ISSN 1463-9076

Caputo, G.; Nobile, C.; Buonsanti, R.; Kipp, T.; Manna, L.; Cingolani, R.; Cozzoli, P. D.; Athanassiou, A. (2008). Determination of surface properties of various substrates using TiO_2 nanorod coatings with tunable characteristics. *Journal of Material Science*, Vol. 43, No. 10, (May 2008), pp. 3474-3480, ISSN 1573-4803

Cornell, R. M.; Schwertmann, U. (2003). *The Iron Oxides: Structure, Properties, Reactions, Occurrences and Uses*, second ed. Wiley-VCH, Weinheim

Cui, X.J.; Antonietti, M.; Yu, S.H. (2006). Structural Effects of Iron Oxide Nanoparticles and Iron Ions on the Hydrothermal Carbonization of Starch and Rice Carbohydrates. *Small*, Vol. 2, No. 6, (June 2006), pp. 756-759, ISSN 1613-6829

Decker, C.; Keller, L.; Zahouily, K.; Benfarhi, S. (2005). Synthesis of nanocomposite polymers by UV-radiation curing. *Polymer*, Vol. 46, No. 17, (August 2005), pp. 6640-6648, ISSN 1521-4095

Ding, T.; Song, K.; Clays, K.; Tung, C.-H. (2009). Fabrication of 3D Photonic Crystals of Ellipsoids: Convective Self-Assembly in Magnetic Field. *Advanced Materials*, Vol. 21, No. 19, (May 2009), pp. 1936–1940, ISSN 1521-4095

dos Santos Coelho, F.; Ardisson, J.D.; Moura, F.C.C.; Lago, R.M.; Murad, E.; Fabris, J.D. (2008). Potential application of highly reactive Fe(0)/Fe_3O_4 composites for the reduction of Cr(VI) environmental contaminants. *Chemosphere*, Vol. 71, No 1, (March 2008), pp. 90-96, ISSN 0045-6535

Duff, D. G.; Baiker, A.; Edwards, P. P. (1993). A new hydrosol of gold clusters. 1. Formation and particle size variation. *Langmuir*, Vol. 9, No. 9, (September 1993), pp. 2301-2309, ISSN 1520-5827

Fahrni, F.; Prins, M. W. J.; van Ijzendoorn, L. J. (2009). Magnetization and actuation of polymeric microstructures with magnetic nanoparticles for application in microfluidics. *Journal of Magnetism and Magnetic Materials*, Vol. 321, No. 12, (June 2009), pp. 1843–1850, ISSN 0304-8853

Fouassier, J. P. (1995) Photoiniatiation, Photopolymerization and Photocuring: Fundamentals and Applications, Hanser/gardner Publications Inc., Cincinnati 1995.

Fragouli, D.; Resta, V.; Pompa, P. P.; Laera, A. M.; Caputo, G.; Tapfer, L.; Cingolani, R.; Athanassiou, A. (2009). Patterned Structures of In-Situ Size Controlled CdS Nanocrystals in a Polymer Matrix Under UV Irradiation. *Nanotechnology*, Vol. 20, No. 15, (April 2009), 155302, ISSN 1361-6528

Fragouli, D.; Laera, A. M.; Pompa, P. P.; Caputo, G.; Resta, V.; Allione, M.; Tapfer, L.; Cingolani, R.; Athanassiou, A. (2009). Localized formation and size tuning of CdS nanocrystals upon irradiation of metal precursors embedded in polymer matrices. *Microelectronic Engineering*, Vol. 86, No. 4-5, (April-June 2009), pp. 816–819, ISSN 0167-9317.

Fragouli D.; Laera, A. M.; Caputo, G.; Resta, V.; Pompa, P. P.; Tapfer, L.; Cingolani, R., Athanassiou, A. (2010). The Effect of Polymer Matrices in the In-Situ CdS Formation Under UV Irradiation of Precursor-Polymer Films. *Journal of Nanoscience and Nanotechnology*, Vol. 10, No. 2, (February 2010), pp. 1267-1272, ISSN 1361-6528

Fragouli D., Buonsanti, R.; Bertoni, G.; Sangregorio, C.; Innocenti, C.; Falqui, A.; Gatteschi, D.; Cozzoli, P. D.; Athanassiou, A.; Cingolani, R. (2010). Dynamical Formation of

Spatially Localized Arrays of Aligned Nanowires in Plastic Films with Magnetic Anisotropy. *ACS NANO*, Vol. 4, No. 4, (April 2010), pp. 1873-1878, ISSN 1936-086X

Fragouli D., P.P. Pompa, M. Kalyva, G. Caputo, L. Tapfer, R. Cingolani, A. Athanassiou, (2010). The Effect of Irradiation Wavelength on the Quality of CdS Nanocrystals Formed Directly into PMMA Matrix. *Journal of Physical Chemistry C*, Vol. 114, No. 33, (August 2010), pp. 13985-13990, ISSN 1932-7455

Fragouli, D.; Torre, B.; Bertoni, G.; Buonsanti, R.; Cingolani, R.; Athanassiou, A. (2010). Formation and microscopic investigation of iron oxide aligned NWs into polymeric nanocomposite films. *Microscopy Research and Technique*, Vol. 73, No. 10, (October 2010), pp. 952-958, ISSN 1097-0029

Fujishima, A.; Rao Tata, N.; Tryk, D. A. (2000). TiO_2 photocatalysts and diamond electrodes. *Electrochimica Acta*, Vol. 45, No. 28, (October 2000), pp. 4683-4690, ISSN 0013-4686

Gelves G. A.; Lin B.; Sundararaj U.; Haber J. A. (2006). Low Electrical Percolation Threshold of Silver and Copper Nanowires in Polystyrene Composites. *Advanced Functional Materials*, Vol. 16, No. 18, (December 2006), pp. 2423-2430, ISSN 1616-3028

Gelves, G. A.; Al-Saleh M. H.; Sundararaj U. (2011). Highly electrically conductive and high performance EMI shielding nanowire/polymer nanocomposites by miscible mixing and precipitation. *Journal of Material Chemistry*, Vol. 21, No. 3, pp. 829-836, ISSN 1573-4803

Ghosh, A.; Fischer, P. (2009). Controlled Propulsion of Artificial Magnetic Nanostructured Propellers. *Nano Letters*, Vol. 9, No. 6, (June 2009), pp. 2243-2245, ISSN 1530-6992

Guibal, E. (2005). Heterogeneous catalysis on chitosan-based materials: a review. *Progress Polymer Science*, Vol. 30, No. 1, (January 2005), pp. 71-109, ISSN 0079-6700

Gupta, A. K.; Gupta, M. (2005). Synthesis and surface engineering of iron oxide nanoparticles for biomedical applications. *Biomaterials*, Vol. 26, No. 18, (June 2005), 3995-4021, ISSN 0142-9612

Huang, X.; Kim, C; Jiang, P; Yin, Y; Li, Z. (2009). Influence of aluminum nanoparticle surface treatment on the electrical properties of polyethylene composites. *Journal of Applied Physics*, Vol. 105, No. 1, (Janury 2009), 014105, ISSN 1089-7550

Huo, J.; Wang, L.; Yu, H. (2009). Polymeric nanocomposites for electromagnetic wave absorption. *Journal of Material Science*, Vol. 44, No. 15, (August 2009), pp. 3917-3927, ISSN 1573-4803

Jestin, J.; Cousin, F.; Dubois, I.; Ménager, C.; Schweins, R.; Oberdisse, J.; Boué, F. (2008). Anisotropic Reinforcement of Nanocomposites Tuned by Magnetic Orientation of the Filler Network. *Advanced Materials*, Vol. 20, No. 13, (July 2008), pp. 2533-2540, ISSN 1521-4095

Kabashin A. V.; Meunier M.. (2003). Synthesis of colloidal nanoparticles during femtosecond laser ablation of gold in water. *Journal of Applied Physics*, Vol. 94, No. 12, (December 2003), 7941, ISSN 1089-7550

Kado, T. (2008). Structural and magnetic properties of magnetite-containing epitaxial iron oxide films grown on MgO(001) substrates. *Journal of Applied Physics*, Vol. 103, No. 4 (February 2008), 043902, ISSN 1089-7550

38 Kadossov, E.; Burghaus, U. (2010). Adsorption Dynamics of CO on Silica Supported Gold Clusters: Cluster Size Effects in Molecular Beam Scattering Experiments. *Catalysis Letters*, Vol. 134, No. 3-4, (February 2010), pp. 228-232, ISSN 1572-879X

Kaushik, A.; Solanki, P. R.; Ansaria, A. A.; Sumanaa, G.; Ahmadb, S.; Malhotraa, B. D. (2009). Iron oxide-chitosan nanobiocomposite for urea sensor. *Sensors and Actuators B*, Vol. 138, No. 2, (May 2009), pp. 572-580, ISSN 0925-4005

Kavcic, B.; Babic, D.; Osterman, N.; Podobnik, B.; Poberaj, I. (2009). Magnetically actuated microrotors with individual pumping speed and direction control. *Applied Physics Letters*, Vol. 95, No. 2, (July 2009), 023504, ISSN 1077-3118

Khanna P. K.; Singh, N. (2007). Light Emitting CdS Quantum Dots in PMMA: Synthesis and Optical Studies. *Journal of Luminescence*, Vol. 127, No. 2, (December 2007), pp. 474-482, ISSN 0022-2313

Lalatonne, Y.; Richardi, J.; Pileni, M. P. (2004). Van der Waals versus dipolar forces controlling mesoscopic organizations of magnetic nanocrystals. *Nature Materials*, Vol. 3, No. 2, (February 2004), pp. 121-125, ISSN 1476-4660

Lee, B.-H.; Choi, J.-H.; Kim, H.-J. (2006). Coating performance and characteristics for UV-curable aliphatic urethane acrylate coatings containing norrish type I photoinitiators. *Journal of Coatings Technology and Research*, Vol. 3, No. 3, (July 2006), pp. 221-229, ISSN 1935-3804

Long, Y. (2005). Electrical and magnetic properties of polyaniline/Fe_3O_4 nanostructures. *Physica B: Condensed Matter*, Vol. 370, No. 1-4, (December 2005), 121-130, ISSN 0921-4526

Luther, M.; Parry, J.; Moore, J.; Meng, J.; Zhang, Y.; Cheng, Z.; Yu, L. (2007). Inhibitory effect of Chardonnay and black raspberry seed extracts on lipid oxidation in fish oil and their radical scavenging and antimicrobial properties. *Food Chemistry*, Vol. 104, No. 3, pp. 1065–1073, ISSN 0308-8146

Mafuné, F.; Kohno, J.; Takeda, Y.; Kondow, T.; Sawabe, H. (2000). Formation and Size Control of Silver Nanoparticles by Laser Ablation in Aqueous Solution. *Journal of Physical Chemistry B*, Vol. 104, No. 39, (October 2000), pp. 9111-9117, ISSN 1520-5207

Mair, L.; Ford, K.; Alam, M. R.; Kole, R.; Fisher, M.; Superfine, R. (2009). Size-Uniform 200 nm Particles: Fabrication and Application to Magnetofection. *Journal of Biomedical Nanotechnology*, Vol. 5, No. 2, (April 2009), pp. 182–191, ISSN 1550-7041

Medintz, I. L.; Tetsuouyeda, H.; Goldman, E. R.; Mattoussi, H. (2005). Quantum Dot Bioconjugates for Imaging, Labelling and Sensing. *Nature Materials*, Vol. 4, No. 6, (June 2005), pp. 435-446, ISSN 1476-4660

Merkel, T. C.; Freeman, B. D.; Spontak, R. J.; He, Z.; Pinnau, I.; Meakin, P.; Hill, A. J. (2002). Ultrapermeable, Reverse-Selective Nanocomposite Membranes. *Science*, Vol. 296, (April-June 2002), pp. 519-522, ISSN 1095-9203

Michalet, X.; Pinaud, F. F.; Bentolila, L. A.; Tsay, J. M.; Doose, S.; Li, J. J.; Sundaresan, G.; Wu, A. M.; Gambhir, S. S.; Weiss, S. (2005). Quantum Dots for Live Cells, in Vivo Imaging, and Diagnostics. *Science*, Vol. 307, (October-December), pp. 538-544, ISSN 1095-9203

Miyama, T.; Yonezawa, Y. (2004). Aggregation of Photolytic Gold Nanoparticles at the Surface of Chitosan Films. *Langmuir*, Vol. 20, No. 14, (July 2004), pp. 5918-5923, ISSN 1520-5827

Myroshnychenko V., J. Rodriguez-Fernandez, I. Pastoriza-Santos, A. M. Funston, C. Novo, P. Mulvaney, L. M. Liz-Marzan, F. J. Garcia de Abajo, (2008). Modelling the optical response of gold nanoparticles. *Chemical Society Reviews*, Vol. 37, No. 9, pp. 1792–1805, ISSN 1460-4744

Nakajima, A.; Koizumi, S.-I.; Watanabe, T.; Hashimoto, K. (2001). Effect of repeated photo-illumination on the wettability conversion of titanium dioxide. *Journal of Photochemistry and Photobiology A*, Vol. 146, No. 1-2, (December 2001), pp. 129-132, ISSN 1010-6030

Nakajima, A.; Koizumi, S.-I.; Watanabe, T.; Hashimoto, K. (2000). Photoinduced Amphiphilic Surface on Polycrystalline Anatase TiO_2 Thin Films. *Langmuir*, Vol. 16, No. 17, (August 2000), pp. 7048-7050, ISSN 1520-5827

Neuberger, T. ; Schopf, B.; Hofmann, H.; Hofmann, M.; von Rechenberg, B. (2005). Superparamagnetic nanoparticles for biomedical applications: Possibilities and limitations of a new drug delivery system. *Journal of Magnetism and Magnetic Materials*, Vol. 293, No. 1, (May 2005), pp. 483-496, ISSN 0304-8853

Nirmala, R.; Navamathavan, R.; El-Newehy, M. H.; Kim, H. Y. (2011). Preparation and electrical characterization of polyamide-6/chitosan composite nanofibers via electrospinning. *Materials Letters*, Vol. 65, No. 3, (February 2011), pp. 493–496, ISSN 0167-577X

Pankhurst, Q. A.; Connolly, J.; Jones, S. K., Dobson, J. (2003). Applications of magnetic nanoparticles in biomedicine. *Journal of Physics D*, Vol. 36, No. 13, (July 2003), pp. R167-R181, ISSN 1361-6463

Park, J. I.; Jun, Y. W.; Choi, J. S.; Cheon, J. (2007). Highly crystalline anisotropic superstructures *via* magnetic field induced nanoparticle assembly. *Chemical Communications*, Vol. 2007, No. 47, 5001–5003, ISSN 1364-548X

Sakai, N.; Fujishima, A.; Watanabe, T.; Hashimoto, K. (2001). Enhancement of the Photoinduced Hydrophilic Conversion Rate of TiO_2 Film Electrode Surfaces by Anodic Polarization. *Journal of Physical Chemistry B*, Vol. 105, No. 15, (April 2001), pp. 3023-3026, ISSN 1520-5207

Sakai, N.; Fujishima, A.; Watanabe, T.; Hashimoto, K. (2003). Quantitative Evaluation of the Photoinduced Hydrophilic Conversion Properties of TiO_2 Thin Film Surfaces by the Reciprocal of Contact Angle. *Journal of Physical Chemistry B*, Vol. 107, No. 4, (January 2003), pp. 1028-1035, ISSN 1520-5207

Sakai, N.; Wang, R.; Fujishima, A.; Watanabe, T.; Hashimoto, K. (1998). Effect of Ultrasonic Treatment on Highly Hydrophilic TiO_2 Surfaces. *Langmuir*, Vol. 14, No. 20, (September 1998), pp. 5918-5920, ISSN 1520-5827

Shukla S.; Seal, S. (1999). Cluster size effect observed for gold nanoparticles synthesized by sol-gel technique as studied by X-ray photoelectron spectroscopy. *NanoStructured Materials*, Vol. 11, No. 8, (November 1999), pp. 1181–1193, ISSN 0965-9773

Su, L.; Yin, J.-J.; Charles, D.; Zhou, K.; Moore, J.; Yu, L. (2005). Total phenolic contents, chelating capacities, and radical-scavenging properties of black peppercorn, nutmeg, rosehip, cinnamon and oregano leaf. *Food Chemistry*, Vol. 100, No. 3, pp. 990–997, ISSN 0308-8146

Sun Z.-B.; Dong X.-Z.; Chen W.-Q.; Nakanishi S.; Duan X.-M.; Kawata S. (2008). Multicolor Polymer Nanocomposites: In Situ Synthesis and Fabrication of 3D Microstructures. *Advanced Materials*, Vol. 20, No. 5, (March 2008), pp. 914–919, ISSN 1521-4095

Sun, S.; Zeng, H.; Robinson, D. B.; Raoux, S.; Rice, P. M.; Wang, S. X.; Li, G. (2004). Monodisperse MFe_2O_4 (M = Fe, Co, Mn) Nanoparticles. *Journal of American Chemical Society*, Vol. 126, No. 1, (January 2004), pp. 273–279, ISSN 1520-5126

Tang, Z.; Kotov, N. A. (2005). One-Dimensional Assemblies of Nanoparticles: Preparation, Properties, and Promise. *Advanced Materials*, Vol. 17, No. 8, (April 2005), pp. 951–962, ISSN 1521-4095

Van Herk, A. M. (2000). Pulsed initiation polymerization as a means of obtaining propagation rate coefficients in free-radical polymerizations. II Review up to 2000. *Macromolecular Theory and Simulations*, Vol. 9, No. 8, (November 2000), pp. 433-441, ISSN 1521-3919

Videla, F. A.; Torchia, G. A.; Schinca, D. C.; Scaffardi, L. B.; Moreno, P.; Méndez, C.; Giovanetti, L. J.; Ramallo Lopez, J. M.; Roso, L. (2010). Analysis of the main optical mechanisms responsible for fragmentation of gold nanoparticles by femtosecond laser radiation. *Journal of Applied Physics*, Vol. 107, No. 11, (June 2010), 114308, ISSN 1089-7550

Villafiorita-Monteleone, F.; Caputo, G.; Canale, C.; Cozzoli, P. D.; Cingolani, R.; Fragouli, D.; Athanassiou, A. (2010). Light-Controlled Directional Liquid Drop Movement on TiO2 Nanorods-Based Nanocomposite Photopatterns. *Langmuir*, Vol. 26, No. 23, (December 2010), pp. 18557–18563, ISSN 1520-5827

Voltairas, P. A.; Fotiadis, D. I.; Michalis, L. K. (2002). Hydrodynamics of magnetic drug targeting. *Journal of Biomechanics*, Vol. 35, No. 6, (June 2002), pp. 813–821, ISSN 0021-9290

Wang, C.-W.; Moffit, M. G. (2004). Surface-Tunable Photoluminescence from Block Copolymer-Stabilized Cadmium Sulfide Quantum Dots. *Langmuir*, Vol. 20, No. 26, (December 2004), pp. 11784-11796, ISSN 1520-5827

Wang, J.; Ni, X. (2008). Interfacial structure of poly(methyl methacrylate)/TiO_2 nanocomposites prepared through photocatalytic polymerization. *Journal of Applied Polymer Science*, Vol. 108, No. 6, (June 2008), pp. 3552-3558, ISSN 1097-4628

Wang, M.; Singh, H.; Hatton, T.A.; Rutledge, G. C. (2004). Field-responsive superparamagnetic composite nanofibers by electrospinning. *Polymer*, Vol. 45, No. 16, (July 2004), pp. 5505-5514, ISSN 0032-3861

Wang, R.; Hashimoto, K.; Fujishima, A.; Chikuni, M.; Kojima, E.; Kitamura, A.; Shimohigoshi, M.; Watanabe, T. (1997). Light-induced amphiphilic surfaces. *Nature*, Vol. 388, No. 6641, (July 1997), pp. 431, ISSN 1476-4687

Wang, R.; Hashimoto, K.; Fujishima, A.; Chikuni, M.; Kojima, E.; Kitamura, A.; Shimohigoshi, M.; Watanabe, T. (1998). Photogeneration of Highly Amphiphilic TiO_2 Surfaces. *Advanced Materials*, Vol. 10, No. 2, (January 1998), pp. 135-138, ISSN 1521-4095

Wang, R.; Sakai, N.; Fujishima, A.; Watanabe, T.; Hashimoto, K. (1999). Studies of Surface Wettability Conversion on TiO_2 Single-Crystal Surfaces. *Journal of Physical Chemistry B*, Vol. 103, No. 12, (March 1999), pp. 2188-2194, ISSN 1520-5207

Weller, D.; Doerner, M. F. (2000). Extremely High-Density Longitudinal Magnetic Recording Media. *Annual Review Materials Science*, Vol. 30, pp. 611-644, ISSN 0084-6600

Wu, F.; Zhang, J. Z.; Kho, R.; Mehra, R. K. (2000) Radiative and Nonradiative Lifetimes of Band Edge States and Deep Trap States of CdS Nanoparticles Determined by Time-Correlated Single Photon Counting. *Chemical Physical Letters*, Vol. 330, No. 3-4, (November 2000), 237-242, ISSN 0009-2614

Xia, H.; Peng, J.; Liu, K.; Li, C.; Fang, Y. (2008). Preparation and Gas Sensing Properties of Novel CdS-Supramolecular Organogel Hybrid Films. *Journal of Physics D*, Vol. 41, No. 10, (May 2008), 105405, ISSN 1361-6463

Yi, H.; Wu, L.-Q.; Bentley, W. E.; Ghodssi, R.; Rubloff, G. W.; Culver, J. N.; Payne, G. F. (2005). Biofabrication with chitosan. *Biomacromolecules*, Vol. 6, No. 6, (November 2005), pp. 2881–2894, ISSN 1525-7797

Zangmeister, R. A.; Park, J. J.; Rubloff, G. W.; Tarlov, M. J. (2006). Electrochemical study of chitosan films deposited from solution at reducing potentials. *Electrochimica Acta*, Vol. 51, No. 25, (July 2006), pp. 5324–5333, ISSN 0013-4686

Zhao, X.; Xu, M.; Qin, L.; Xiao, J.Q. (2007). Microstructure and magnetic properties of magnetite thin films prepared by reactive sputtering. *Journal of Applied Physics*, Vol. 102, No. 11, (December 2007), 113913, ISSN 1089-7550

Zhou, K.; Yin, J.-J.; Yu, L. (2006). ESR determination of the reactions between selected phenolic acids and free radicals or transition metals. *Food Chemistry*, Vol. 95, No. 3, (April 2006), pp. 446–457, ISSN 0308-8146

16

Fabrication of 3-D Structures Utilizing Synchrotron Radiation Lithography

Mitsuhiro Horade and Susumu Sugiyama
Ritsumeikan University
Japan

1. Introduction

Microfabrication of high-aspect-ratio or three-dimensional (3-D) structures is critical for the production of various components for micro electro mechanical systems (MEMS). The term "three-dimensional structure" refers to a structure with a free-form surface or sloped sidewall. This article describes the fabrication of 3-D microstructures using synchrotron radiation (SR) lithography. SR lithography technology is one component of the LIGA process, and it is also called X-ray lithography.

MEMS devices have attracted a great deal of attention, and further studies are needed to realize their full potential. Among fabrication technologies, microfabrication, developed using a semiconductor process, is in high demand. Recently, the demand for MEMS devices has diversified, and microfabrication technologies for the production of high-aspect-ratio and 3-D structures are required to meet this demand.

Microfabrication technologies for the production of high-aspect-ratio structures include deep reactive-ion etching (D-RIE) and deep X-ray lithography in the LIGA process utilizing SR light. In the former, because SR light is highly directional, it is possible to fabricate a structure with a thickness of several hundred to one thousand micrometers. Moreover, because SR light contains X-ray (short wavelength) regions, it is possible to transfer patterns that are ≤ 1 μm (diffraction during exposure does not occur readily). Therefore, SR lithography has been used as a fabrication technology for high-aspect-ratio structures. It is possible to fabricate high-aspect-ratio structures using D-RIE. However, because a patterned, indented sidewall called a scallop is formed due to the nature of the process mechanism, it is difficult to fabricate structures with smooth sidewall surfaces. On the other hand, it is possible to fabricate structures with smooth sidewall surfaces using SR lithography, which is discussed in more detail in Chapter 2.

In the field of 3-D microfabrication, techniques such as KOH anisotropic etching of silicon and laser machining have been employed (Tsukada et al., 2005). However, 3-D fabrication using SR lithography was recently achieved, and results have already been reported (Horade & Sugiyama, 2009; Lee & Lee, 2003; Matsuzuka et al., 2005; Mekaru et al., 2007; Sugiyama et al., 2004; Tabata et al., 2000). Nanoscale 3-D microfabrication technology using SR lithography can be used to fabricate high-aspect-ratio structures by exploiting the properties of SR, and free-form structures with inclined sidewall surfaces can be fabricated. Additionally, this article describes 3-D polytetrafluoroethylene (PTFE) microstructures fabricated by SR ablation. Because PTFE is a remarkable material, there are high expectations regarding its application in various devices. PTFE fabrication by SR ablation is discussed in detail in Chapter 4.

2. SR lithography

This section outlines the LIGA process, the fabrication mechanism of SR lithography, and optimum experimental conditions. To achieve high-accuracy microfabrication, optimum exposure and development conditions were determined using both experimental and theoretical values. Energy distribution, etching rate, control of processing depth, and micro-loading effects are described.

2.1 LIGA process

The LIGA process fabricates microstructures as components of MEMS (Figure 1). As shown in Figure 1, resist materials are first exposed using a soft X-ray source from an SR light through the mask, and the exposed resists are then developed. The exposure and development processes are discussed in more detail in Sections 2.2 and 2.3. Next, metallic moulds, such as Ni, are fabricated using an electroforming technique. Resist is then removed from the metallic moulds. In this process, it is possible to remove the resist under pressure as well as dissolve the resist using wet etching. Finally, microstructures of various materials are fabricated by moulding. The LIGA process can be applied to a wide range of materials, including plastics, metal, and ceramics, when the electroforming technique and moulding process are utilized.

This technology was developed in the 1980s by a group of researchers led by Becker and Ehrfeld of the Kernforschungszentrum Karlsruhe (KfK). LIGA is a German acronym consisting of the initial letters of three processes: lithographie (lithography), galvanoformung (electroforming), and abformung (moulding).

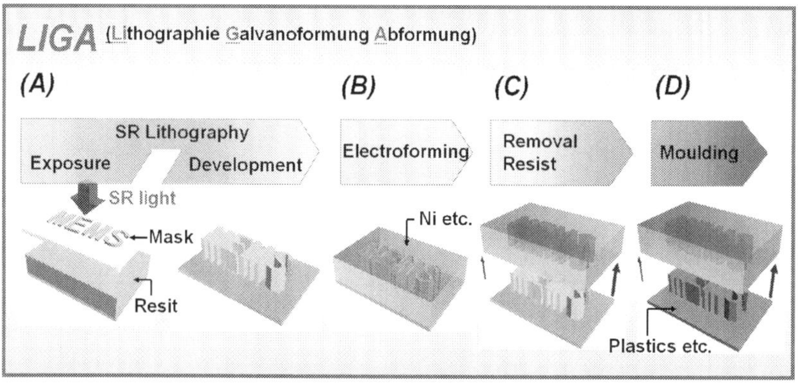

Fig. 1. Process flow of the LIGA process; (A) SR lithography; (B) electroforming; (C) removal resist; and (D) moulding

Next, we describe the characteristics of the LIGA process. Because SR light is highly directional, it is possible to fabricate a structure with a thickness of several hundred to one thousand micrometers. Moreover, because SR light contains X-ray (short wavelength)

regions, it is possible to transfer patterns that are ≤ 1 μm (diffraction during exposure does not occur readily). Therefore, SR lithography has been used as a fabrication technology for high-aspect-ratio microstructures. Fabrication of Ni structures with a line width of 2 μm and an aspect ratio of 100 or a line width of 0.2 μm and an aspect ratio of 75 have been reported (Kato et al., 2007; Kondo et al., 2000; Ueno et al., 2000). Although it is also possible to transfer patterns that are ≤ 1 μm using electron beam lithography, it has the disadvantage of a long exposure time. On the other hand, because the production of large volumes is possible using electroforming and moulding in the LIGA process, it is a superior technology in terms of time and cost.

2.2 Exposure

2.2.1 Light source

In SR lithography, the use of ultra-bright and highly directional SR light sources provides perfect conditions for fabricating structures with the required thickness. Although SR light is spectrally continuous and includes a wide wavelength range, wavelengths of 0.2 to 0.5 nm are most suitable for SR lithography because they reduce the spread of light by Fresnel diffraction in the long-wavelength domain and the generation of secondary electrons inside resists in the short-wavelength domain, enhancing resolution.

Experiments described in this article utilized the superconductivity compact SR source "AURORA" at the SR Centre of Ritsumeikan University in Japan (Figure 2). The properties of SR at AURORA include a wavelength range from 0.15 nm to visible light and an applied electron energy and maximum storage current of 575 MeV and 300 mA, respectively. This light source was adapted for our studies; there are 16 beam lines, 4 of which are used in SR lithography. The light from AURORA penetrates two 200-μm beryllium (Be) windows and, within the exposure chamber, uses light with a 0.15- to 0.95-nm wavelength domain. The outline of the beam line is shown in Figure 3. For beam line number 13 (Bl-13), the distance from the light source to the sample is 3.388 m. The exposure environment in the chamber was helium (He) gas at 1 atm to prevent the attenuation of X-rays by N_2 or O_2 gases and to prevent damage to the mask or resist by heat generated during exposure. Figure 4 shows the wavelength and photon density after penetration of the two 200-μm Be windows; the peak wavelength was 0.37 nm.

Fig. 2. Superconductivity compact SR source "AURORA" at the SR Centre of Ritsumeikan University in Japan

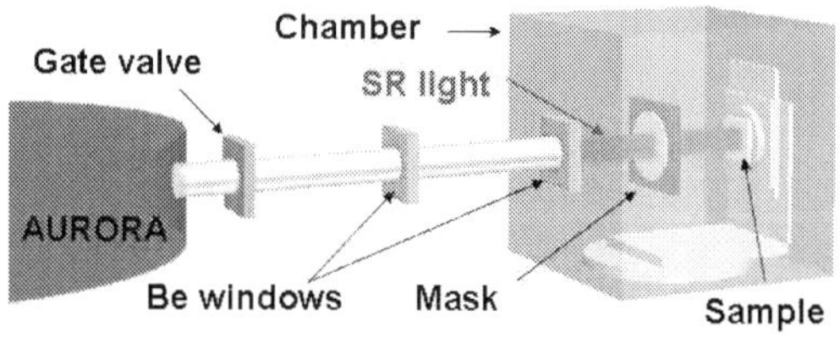

Fig. 3. The outline of the beam line; since wavelengths of 0.2 to 0.5 nm are most suitable for SR lithography, the light from AURORA penetrates two 200-μm Be windows and, within the exposure chamber, uses light with a 0.15- to 0.95-nm wavelength domain

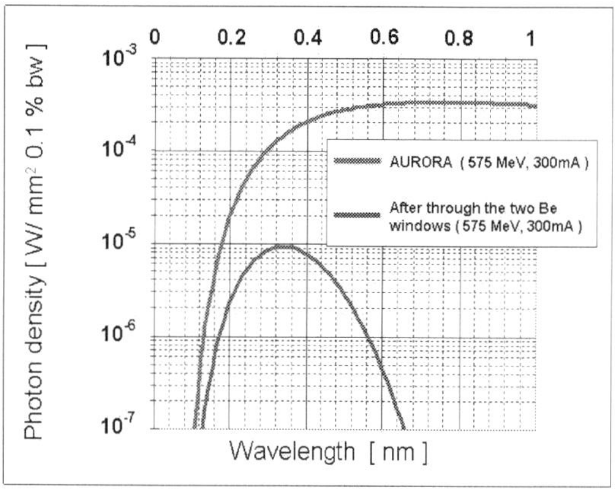

Fig. 4. Relationship between wavelength and photon density after penetration of the two 200-μm Be windows

2.2.2 X-ray mask

Masks used in SR lithography consist of an X-ray absorber, a high-permeability membrane that supports it, and a frame that forms the entire mask. Figure 5A shows the typical structure of an X-ray mask. It is necessary for an X-ray mask to have high contrast to fabricate structures with the required thickness (Singleton & Detemple, 2003; Suzuki & Sugiyama, 1997). As shown in Figure 5B, the mask contrast indicates the ratio of energy through the "membrane only" to energy through the "membrane and absorber."

The primary requirements of X-ray mask membranes include (1) high X-ray transmission, (2) moderate tensile stress, high Young's modulus and high mechanical strength, (3) strong X-ray exposure resistance, and (4) a low rate of thermal expansion. To meet these requirements, Ti, Be, Si_3N_4, SiC, and polyimide are typically used as membrane materials. Although Si_3N_4 and SiC are becoming mainstream, the production cost of X-ray masks based on these materials is high. Thus, polymer materials with low production costs, such as polyimide, are often used as membranes.

The primary requirements for X-ray absorbers of X-ray masks include (1) high X-ray absorptance, (2) a low rate of thermal expansion and no membrane strain, (3) high processing accuracy, and (4) the ability to fabricate high-aspect-ratio structures. To meet these requirements, heavy metals such as Au, Cu, W, and Ta are typically used as X-ray absorber materials. In general, X-ray absorptance increases with increasing atomic number. Additionally, it is possible to fabricate Au or Cu using electroforming. Because the absorptance of Cu in the X-ray region is lower than that of Au, a Cu absorber must be thicker and have a higher aspect ratio to achieve the same contrast. Therefore, fabrication of Cu is difficult. On the other hand, it is possible to pattern W or Ta using RIE.

Experiments described in this article utilized an X-ray mask consisting of a polyimide membrane with a thickness of 50 μm and an Au absorber with a thickness of 3 μm. The X-ray mask was from Optnics Precision Co., Ltd. The linear expansion coefficient of polyimide is very low compared with other organic compounds and is close to that of metals; therefore, thermal expansion produces low strain when polyimide combines with a metal absorber.

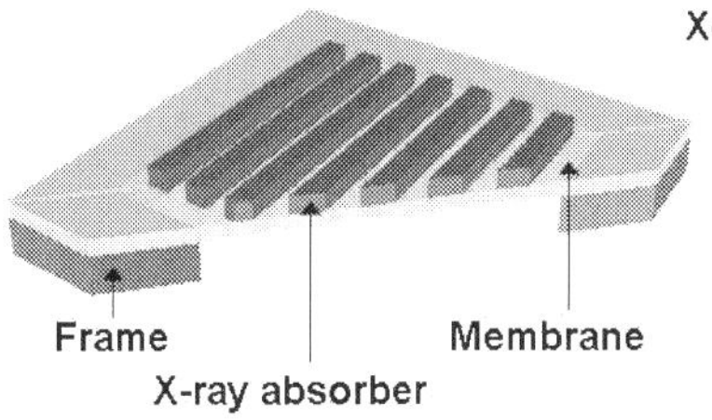

(A)

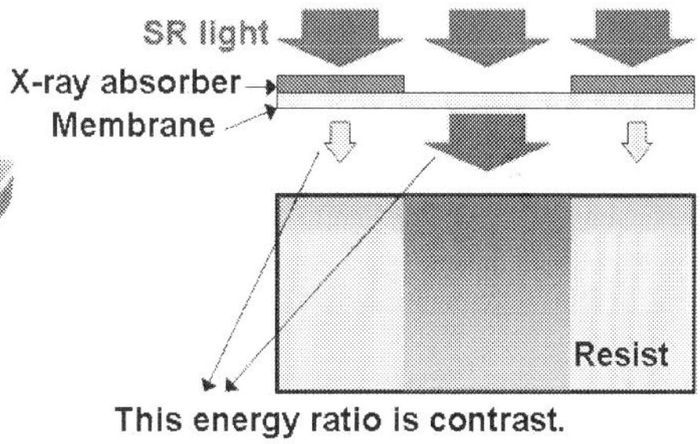

(B)

Fig. 5. X-ray mask; (A) typical structure of an X-ray mask; and (B) the mask contrast indicates the ratio of energy through the "membrane only" to energy through the "membrane and absorber"

2.2.3 Resist

When some types of polymer materials are exposed to light, the exposed areas undergo a photochemical reaction, and molecular structures are changed. There are three primary types of photochemical reactions: cross-linking, polarity change, and main-chain breaking. It is possible to design a wide variety of resist materials using these various reaction mechanisms. Additionally, there are two major classes of resist materials: positive resists, in which non-exposed areas are not dissolved after development, and negative resists, in which exposed areas are not dissolved after development.

In general, polymer materials that have high sensitivity and X-ray resolution are suitable as X-ray resists. Most polymer materials used as electron beam resists are not exposed to visible or ultraviolet (UV) light. Therefore, X-ray resists are often used as electron beam resists. Further requirements of X-ray resists include a lack of exposure to small amounts of X-rays through the absorber for fabricating high-aspect-ratio structures and high mechanical strength to endure prolonged exposure and electroforming. To meet these requirements, polymethylmethacrylate (PMMA) is typically used as the X-ray resist.

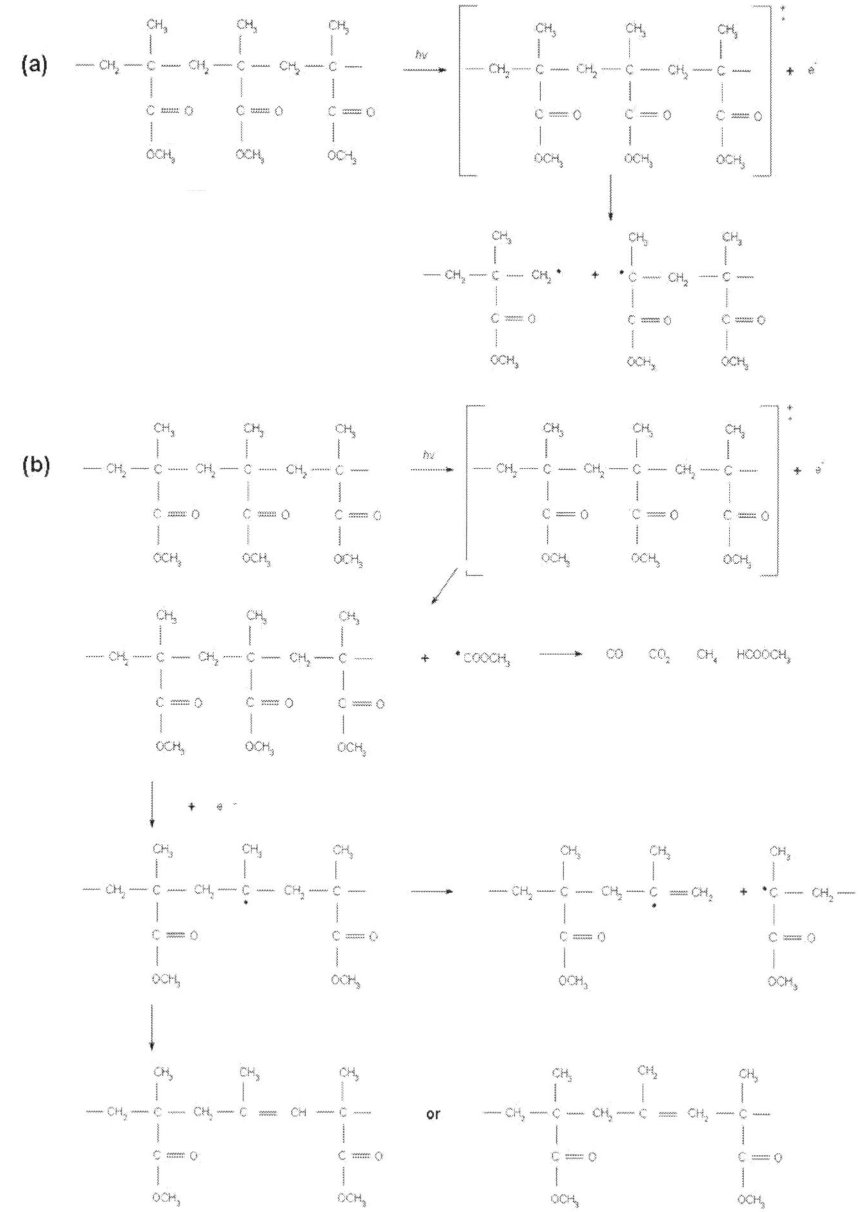

Fig. 6. Photodegradation reaction mechanisms; (A) main-chain breaking without side-chain cleavage; and (B) main-chain breaking via side-chain cleavage

Although the sensitivity of PMMA is low, its resolution is very high (50 Å), and reproducibility of fine structures for moulding can be enhanced by further processing. PMMA is a material in the acrylic plastic group that has a broad range of applications, including lenses, instrument windows, signboards, displays, etc. Among the aforementioned photochemical reactions, PMMA preferentially causes main-chain breaking. If main-chain breaking occurs, molecular mass decreases, and solubility in the developer increases. PMMA has a distinguished transparency and weatherability in many polymer materials. Additionally, because PMMA also has good formability, its mechanical properties are also good. Because PMMA is a malleable optical material, it is possible to use fabricated microstructures directly as optical devices. Experiments described in this article utilized PMMA sheets provided by Nitto Jushi Kogyo Co., Ltd. Figure 6 shows photodegradation reaction mechanisms (Schmal et al., 1996). Figure 6A shows main-chain breaking without side-chain cleavage, and Figure 6B shows main-chain breaking via side-chain cleavage.

2.2.4 X-ray absorbed energy

This section describes absorbed energy in the PMMA resist after exposure. To achieve highly accurate microfabrication or 3-D fabrication, it is necessary to calculate the absorbed energy distribution in the resist. To calculate the absorbed energy in the vertical direction, the following factors must be considered: the photon density spectrum of SR light, attenuation by distance from the light source to the resist, divergence of light, and attenuation when light is transmitted through Be windows. If the spectrum of synchrotron orbital radiation is set to $P_s(\lambda)$, the absorption of the resist is given by the following formula.

$$\Delta P_R(\lambda) = -d/dz\{P_S(\lambda)\cdot T_w(\lambda)\cdot T_M(\lambda)\cdot \exp(-\mu(\lambda)z)\} \quad (1)$$

$T_W(\lambda)$ is the transmissivity in the constituent factor of a beam line until it reaches the front of a mask, such as the Be window and exposure to atmospheric gas, $T_M(\lambda)$ is the transmissivity of the membrane of a mask, $exp(-\mu(\lambda)x)$ is the attenuation in depth x to the inside of the resist, and $\mu(\lambda)$ is an absorbing coefficient of the resist. Moreover, the exposure energy I integrates with the synchrotron orbital radiation spectrum in the wavelength range (0.15–0.95 nm).

$$I = \int_{0.15}^{0.95} P_S(\lambda)\cdot d\lambda \quad (2)$$

The exposure energy after polyimide membrane (50 μm) penetration, I_{Poly}, computed using the conditions described in Section 2.2.2, is given by the following formula.

$$I_{Poly} = \int_{0.15}^{0.95} S_{poly}(\lambda)\cdot d\lambda = 1.587\times 10^2 \text{ [mJ/sec mA mm}^2\text{]} \quad (3)$$

The exposure energy density is given per second and milliamp; thus, if the dosage is applied under the experimental conditions, the arbitrary exposure energy per unit area [mJ/mm^2] can be calculated. The penetration energy spectrum $S_{PMMA}(x)$ in arbitrary depth x inside PMMA is given by Formula 4. Additionally, the exposure energy at an arbitrary depth x in PMMA $E_{PMMA}(x)$ is given by Formula 5.

$$S_{PMMA}(x) = S_{Poly}(\lambda)\times \exp(-\mu_{PMMA}(\lambda)x_{PMMA}) \quad (4)$$

$$E_{PMMA}(x) = \int_{0.15}^{0.95} S_{PMMA}(x) \cdot d\lambda \quad (5)$$

The penetration energy spectrum $S_{PMMA}(x)$ was differentiated by the depth direction, and the absorbed energy spectrum $S_{ABS}(x)$ of PMMA was calculated.

$$S_{ABS}(x) = d/dx\ S_{PMMA}(x) = S_{poly}\{\exp[-\mu_{PMMA}(\lambda)(x_{PMMA} + \Delta x)] - \exp(-\mu_{PMMA}(\lambda)x_{PMMA})\} \quad (6)$$

Therefore, the absorbed energy $E_{ABS}(x)$ at the arbitrary depth x is given by the following formula.

$$E_{ABS}(x) = \int_{0.15}^{0.95} S_{ABS}(x) \cdot d\lambda = \int_{0.15}^{0.95} d/dx S_{PMMA}(x) \cdot d\lambda = E_{PMMA}(x + \Delta x) d\lambda - E_{PMMA}(x) \cdot d\lambda \quad (7)$$

The amount of absorbed energy in PMMA after SR was transmitted through the mask was calculated (Figure 7). The amount of absorbed energy [J/s·mA·mm³] to a depth x [μm] when SR penetrated the membrane (50-μm-thick polyimide) was approximated as $F_{poly}(x)$, and the amount of absorbed energy after SR penetrated the membrane (polyimide) and the absorber (3-μm-thick Au) was approximated as $F_{poly+Au}(x)$.

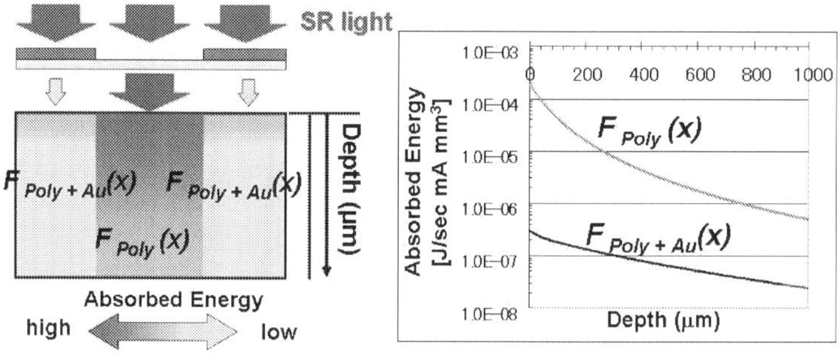

Fig. 7. Amount of absorbed energy in PMMA after SR was transmitted through the mask was calculated

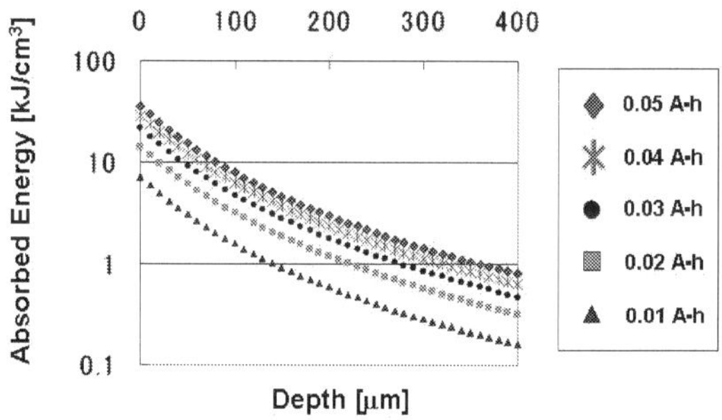

Fig. 8. Relationship between depth and absorbed energy

We multiplied $F_{poly}(x)$ and $F_{poly+Au}(x)$ by the dosage, an exposure condition, and the amount of absorbed energy that was actually exposed was calculated. Because the dose was 1 A·h = 3,600,000 mA·s, multiplication of $F_{poly}(x)$ and $F_{poly+Au}(x)$ by the dosage (converted to mA·s) results in a unit of [kJ/cm^3]. The relationship between depth and absorbed energy can be calculated from $F_{poly}(x) - F_{poly+Au}(x)$ (Figure 8).

2.3 Development

An important consideration of resist development is the etching rate ratio of the exposed area to the non-exposed area. The developer must select a ratio that decreases the etching rate of the non-exposed area and increases the etching rate of the exposed area. Experiments described in this article utilized GG developer (60 vol% 2-(2-butoxy-ethoxy) ethanol; 20 vol% tetra-hydro-1, 4-oxazine; 5 vol% 2-amino-ethanol-1; and 15 vol% water). Next, stopper liquid (80 vol% 2-(2-butoxy-ethoxy) ethanol and 20 vol% water) was used for 10 min, followed by rinsing with water for another 10 min. All processes were performed at exactly 37°C. PMMA was exposed, and its molecular mass decreased due to a photochemical reaction. Although the molecular mass of PMMA is usually between 10^5 and 10^6 g/mol, it began to dissolve in GG developer when its molecular mass reached 10^4 g/mol.

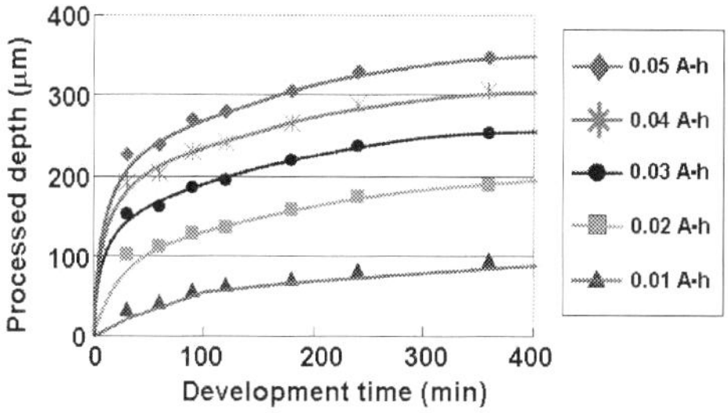

Fig. 9. Relationship between development time and processed depth

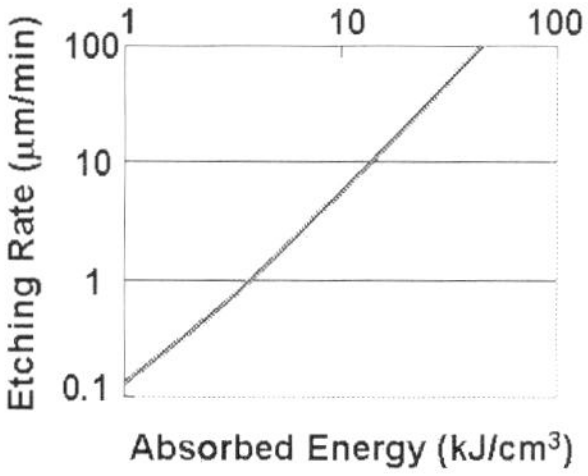

Fig. 10. Relationship between the absorbed energy and the etching rate

The relationship between development time and processed depth was investigated experimentally (Figure 9). The relationship between the absorbed energy and the etching rate is shown Figure 10. Etching rate is proportional to the amount of absorbed energy. In general, when the development temperature was high, the etching rate, and thus the processing depth, increased. Figure 11 shows the relationship between dosage and processed depth at a development time of 180 min. Figures 9 and 11 show the experimental values and formula approximations. The processing depth is determined by the dosage and development time, as shown in Figures 9 and 11. Thus, it is possible to determine the processing depth using these approximation formulas.

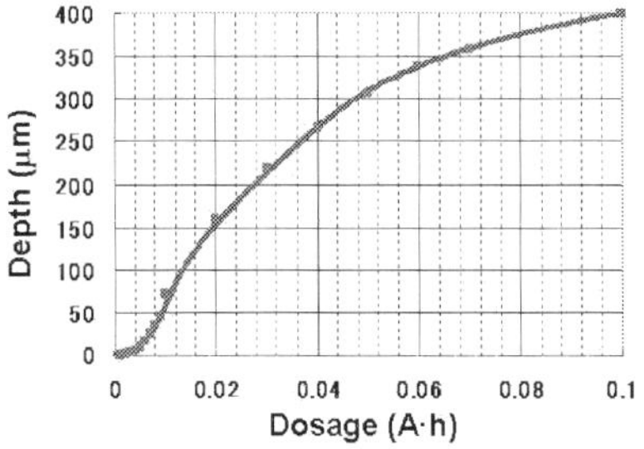

Fig. 11. Relationship between dosage and processed depth at a development time of 180 min

2.4 Approaches to high-accuracy microfabrication

This section describes important parameters that are necessary for high-accuracy microfabrication.

2.4.1 Resolution

The effect of Fresnel diffraction on resolution is determined by wavelength and the gap between the mask and resist surface, whereas the effect of secondary electrons on resolution is determined by wavelength (Figure 12). To transfer a narrow line pattern, it is necessary to narrow the gap between the mask and the resist surface.

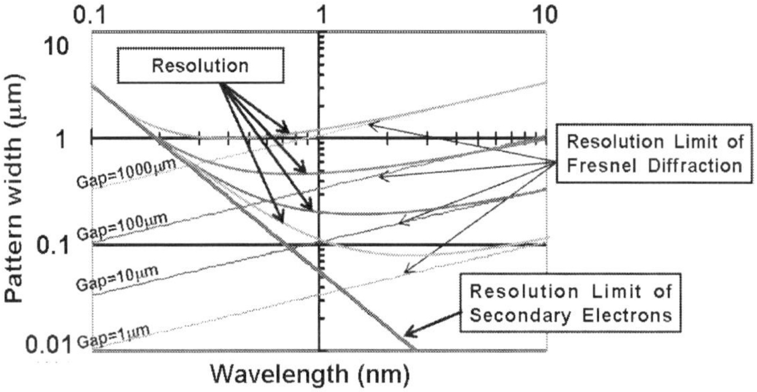

Fig. 12. Relationship between wavelength and pattern width

2.4.2 Optimum experimental conditions

As mentioned in Section 2.3, the processing depth is determined by the dosage and development time and increases with increasing development temperature. For example, to increase processing depth, even when the dosage is small, the development time or temperature must increase. On the other hand, when the dosage is high, the development time can be short. However, to achieve high-accuracy microfabrication, it is necessary to determine optimum experimental conditions.

When the dosage is too high, bubbles often form inside the resist, especially near the surface. Figure 13 shows a microscopic photograph of damage to the PMMA sheet after exposure. Figure 13B shows exposed microchannel patterns. As seen in the figure, bubbles spread not only to the exposed area but also to the non-exposed area. We confirmed that bubbles influenced the structure after development. Because bubbles often form when the amount of absorbed energy exceeds 80 kJ/cm^3, it is necessary to select a dosage that does not exceed this "surface-damage energy amount."

As mentioned in Section 2.3, the molecular mass of PMMA must be below a certain value in order for PMMA to dissolve in developer. The minimum energy required for dissolution is called the "development-limit energy amount." Although the development limit energy amount varies in the literature, it is approximately 1 kJ/cm^3 based on previous experiments. Variation in the literature values is due to differences in the molecular mass of the PMMA. As the depth from the surface increases, the absorbed energy decreases exponentially; thus, it is necessary to consider not only the surface-damage energy amount but also the development-limit energy amount to determine the optimum dosage.

Next, we discuss the development temperature. As mentioned in Section 2.3, the etching rate increases with increasing development temperature. Because of the higher development temperature and faster etching rate, the processing depth may become large in a short amount of time. However, when the development temperature is too high, the etching rate increases, but the development-limit energy amount decreases. Thus, it is necessary to select the optimum development temperature (Fujinawa et al., 2006). If the development-limit energy amount is too low, PMMA resist develops not only in the exposed area, but also in the non-exposed areas (through the absorber). Therefore, sloped-sidewall structures are fabricated. On the other hand, when the development temperature is low, the processing depth is low, and a long development time is required. Because PMMA swells in water, a long development time is a disadvantage. For these reasons, the development process was performed at 37°C.

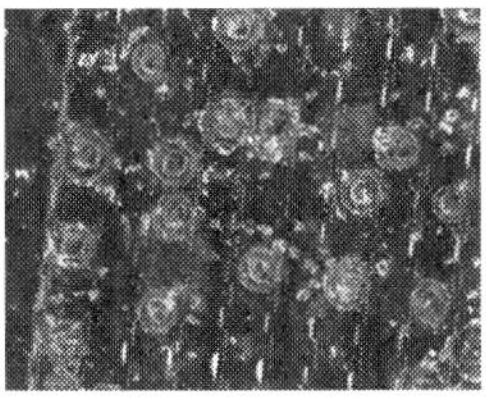

(A)

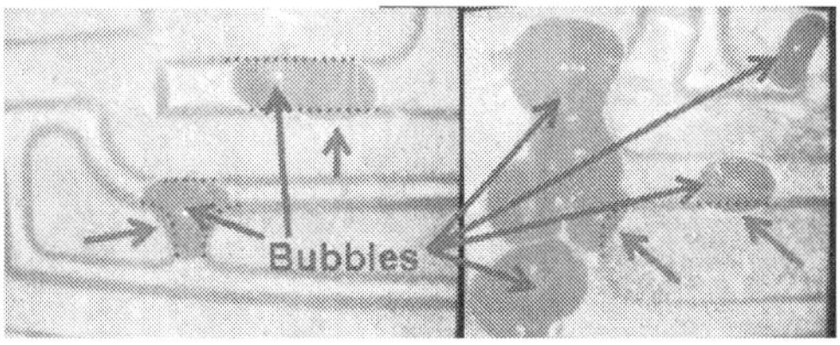

(B)

Fig. 13. Microscope photographs of damaged to the PMMA sheet after exposure

2.4.3 Micro-loading effect

In this section, the influence of the micro-loading effect is described. The micro-loading effect is a phenomenon that leads to different processing depths depending on the line width. If the line width is narrow, circulation of the developer worsens, it is difficult to supply new developer to the bottom, and dissolved PMMA tends to remain near the bottom. Figure 14 shows the relationship between line width and processing depth using dosages of 0.03 and 0.05 A·h. Plots show experimental values for line widths of 10, 15, 20, and 25 μm, and lines show the processing depth of the domain where the line is wide. Figure 15 shows the percentage of processing depths that reached the domain where the line is wide. As shown in Figures 14 and 15, processing depth decreased with decreasing line width. However, higher percentages of processing depth where the line is wide correspond to longer development time. That is, although developer circulation was reduced, dissolution progressed slowly. When the dosage was low, the percentage was high. Additionally, the processing depth was low, and thus the circulation of the developer was good. However, as shown in Figure 14B and C, when the line width was too narrow, dissolution did not progress slowly. Thus, when narrow line patterns are transferred, it is necessary to consider the influence of the micro-loading effect.

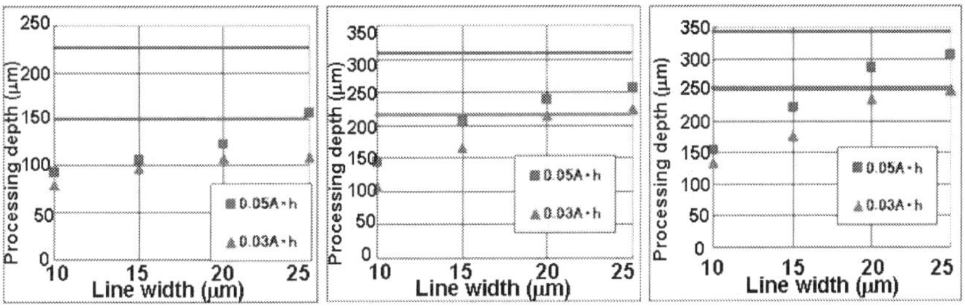

(A) (B) (C)

Fig. 14. Relationship between line width and processing depth; (A) is 30 min; (B) is 180 min; and (C) is 360 min

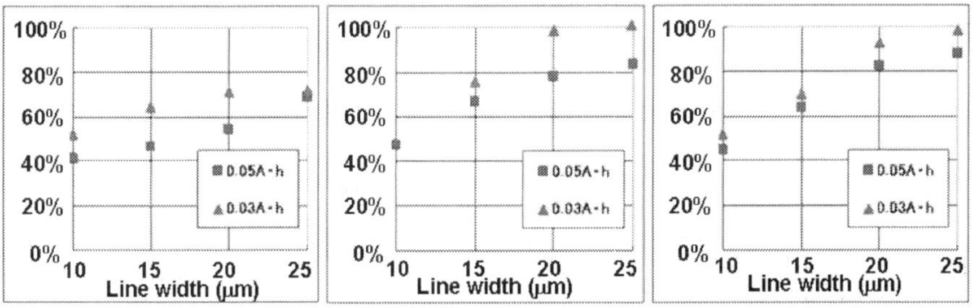

(A) (B) (C)

Fig. 15. Percentage of processing depths that reached the domain where the line is wide; (A) is 30 min; (B) is 180 min; and (C) is 360 min

3. 3-D fabrication method

This section describes the fabrication of 3-D structures using SR lithography. In this article, two fabrication methods are described in detail: the plain-pattern to cross-section transfer

(PCT) technique and the pixels exposure technique. When the PCT technique is used, it is possible to fabricate a microneedle array structure (Khumpuang et al., 2006). When the pixels exposure technique is used, it is possible to fabricate a 3-D structure with a complex surface, such as an asymmetrical structure.

3.1 Advancement in the LIGA process

As mentioned in Section 2.3, a characteristic of SR lithography is that it is possible to fabricate a thick structure, and the processing depth can be controlled by dosage. Thus, it was expected that SR characteristics could be applied to the fabrication technology of 3-D structures with free-form surfaces or sloped sidewalls. Additionally, studies on 3-D processing methods using the LIGA process have been reported previously. "SLIGA" adds a sacrificial layer process to the LIGA process. Although the traditional LIGA process can only fabricate a mechanical component fixed to the substrate, the SLIGA technique enables fabrication involving sensors with moving parts (Ruzzu et al., 1998). For example, fabrication of a high-aspect-ratio microactuator was reported. However, this processing technique produces structures with sidewalls that are vertical to the resist surface. To control the inclination angle of the sidewall, the "skew exposure technique" is used to incline resists for X-rays during exposure (Tsuei et al., 1998; Ehrfeld et al., 1999). However, the structures that can be fabricated are limited, and fabrication of structures with free-form surfaces is not possible. Recently, methods that give energy distributions to the resist surface, such as grayscale mask UV lithography, have been reported. If complicated energy distributions can be given to resist surfaces, it will be possible to fabricate complicated arbitrary 3-D structures.

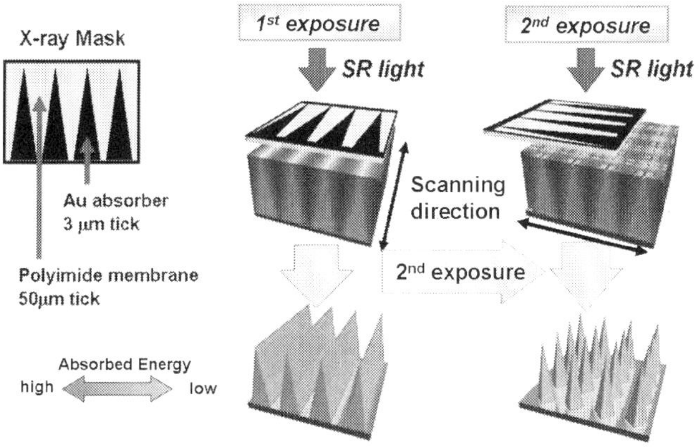

Fig. 16. PCT technique, black color shows X-ray absorber and corn color shows membrane, energy amount is shown by shading, the energy distribution is deposited in the resist by scanning, and a more complex energy distribution can be given by rotating the mask 90°

3.2 PCT technique

Figure 16 shows the PCT technique. The energy distribution is deposited in the resist by scanning, as shown in the figure. Three-dimensional structures were fabricated by subsequent development. Additionally, a more complex energy distribution can be given by rotating the mask 90°. For example, a needle shape can be fabricated using the PCT method. When the PCT technique is used, a shape similar to that of the mask absorber pattern is expected to be fabricated. Therefore, if the exposure distribution onto the resist is expected to form a curved shape or sloped-sidewall structure whose cross-section is similar to the mask absorber pattern as the target shape. As shown in Figure 16, when the PCT technique is used, it is possible to fabricate a microneedle array structure, as indicated by the diagram below. If this method is used, it is possible to fabricate both a microneedle and microlens array. Fabrication results of 3-D structures are shown in Figure 17.

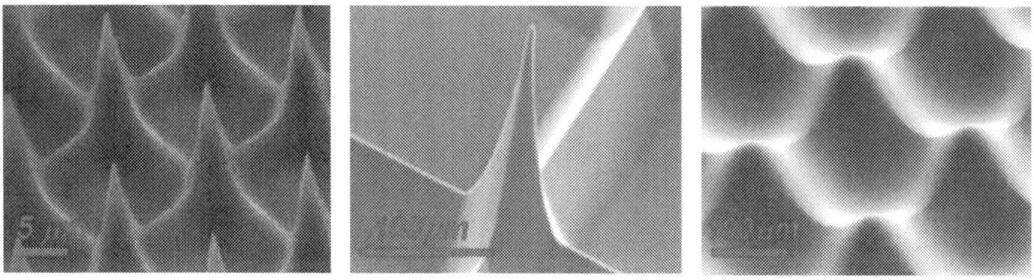

(A) (B) (C)

Fig. 17. Fabrication results of 3-D structures

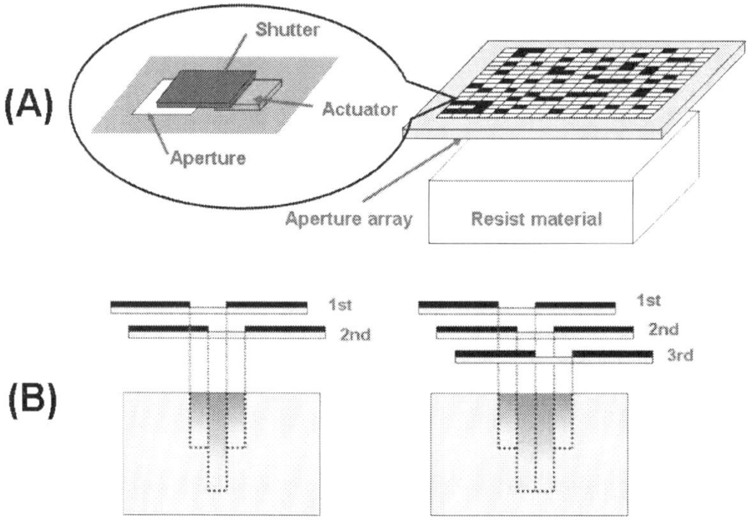

Fig. 18. Pixels exposure technique; (A) SR light is shaped by the aperture, and the amount of exposure energy is controlled by opening and closing the aperture with an actuator; and (B) an ideal mosaicked exposure energy by two-axis scanning of the PMMA resist

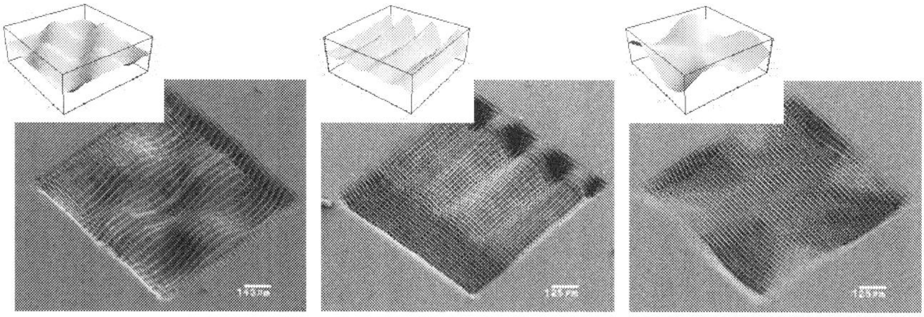

Fig. 19. The target forms and fabrication structures

3.3 Pixels exposure technique

With the pixels exposure technique, it is possible to fabricate arbitrary 3-D microstructures using a mask with an appropriate pattern. In this technique, SR light is shaped by the aperture, and the amount of exposure energy is controlled by opening and closing the aperture with an actuator (Figure 18A). This technique applies the energy distribution in a mosaicked shape, as shown in the figure. Because the array was difficult to fabricate, we fabricated a 20-μm-thick, 75- × 80-μm single aperture made of Ni to distribute an ideal mosaicked exposure energy by two-axis scanning of the PMMA resist (Figure 18B). Because fabrication masks were not needed and it is suitable for rapid prototyping, fabrication time and cost were reduced. When the PMMA resist is exposed to a mosaicked energy distribution with the same aperture size, a smooth free-form surface cannot be fabricated. Thus, because the resolution of PMMA is very high, if even a few adjoining pixels overlap or there is a gap between pixels, channels or pillars appear on the boundary surface. Based on these basic studies, fabrication of structures with smooth free-form surfaces was achieved using an exposure method in which the pixels were overlapped beforehand. Moreover, an algorithm to determine the amount of exposure energy while taking overlap into consideration was written for the target form to fabricate a 3-D structure that had an arbitrary shape. The target forms and fabrication structures are shown in Figure 19.

4. PTFE fabrication by SR ablation

This section describes 3-D PTFE microstructures fabricated by SR ablation. Advantages of ablation technology and the mechanism of fabrication are described.

4.1 PTFE

PTFE is a fluoroplastic material known for its excellent material characteristics, including insulation against high voltages, resistance to chemicals and creep, and high thermal stability. PTFE was discovered in 1938 by Dr. Roy J. Plunkett, who was a researcher of E.I. du Pont de Nemours and Company, a commercial reality in 1946. PTFE is best known by the DuPont brand name Teflon®. In addition to its primary use in non-stick frying pans and filters, PTFE is also used in a broad range of fields, such as those involving household articles, OA equipment, semiconductors, and cars. Despite its long (> 60-year) history, new uses for PTFE are developed continuously because of its outstanding characteristics. Because PTFE has excellent material characteristics, it is expected to be applicable to MEMS.

4.2 Microfabrication of PTFE

PTFE microfabrication is difficult to achieve. It is impossible to fabricate this material through wet etching with chemicals (acids and alkali) used in numerous microfabrication techniques, such as lithography, because of its excellent chemical resistance. Additionally, when the temperature is increased above the melting point of PTFE (327°C), the viscosity becomes too high for moulding. Moreover, laser ablation, which has been widely used in recent years as an effective tool for direct microfabrication, is also difficult. Because the first absorption band of PTFE is near 160 nm, where little light is absorbed by the UV to infrared (IR) domain, ablation processing using a laser with this domain is not possible. Most laser processing of PTFE is not laser ablation but rather thermal processing, which causes a deformation in structure surfaces. Therefore, lasers with narrower wavelengths, such as vacuum ultraviolet (VUV) pulsed-lasers or ultrashort pulsed-lasers, are used for microfabrication of PTFE. However, the aspect ratio of the structures that can be fabricated is very small, generally <1. For these reasons, laser ablation was developed, replacing lasers with SR light.

4.3 SR ablation

In 1996, a technique for direct etching of PTFE using an SR light source was reported (Katoh & Zhang, 1996). Unlike X-ray lithography, direct etching was performed entirely with dry etching and did not use wet etching. With this technology, it is also possible to combine the electroforming technique and moulding process, as in the LIGA process. This technology is referred to as the TIEGA™ (Teflon® Included Etching Galvanic Forming) process. Figure 20 shows the process flow of the TIEGA™ process. The etching rate of PTFE using SR ablation is very high, 10–100 μm/min. Moreover, because the developing process is not necessary, structure collapse is not caused by sticking effects. Sticking is problematic in microfabrication utilizing wet etching, such as lithography. Although fabrication of high-aspect-ratio structures is also possible, surface roughness of sidewalls is undesirable, and the taper angle is large compared with that of the LIGA process.

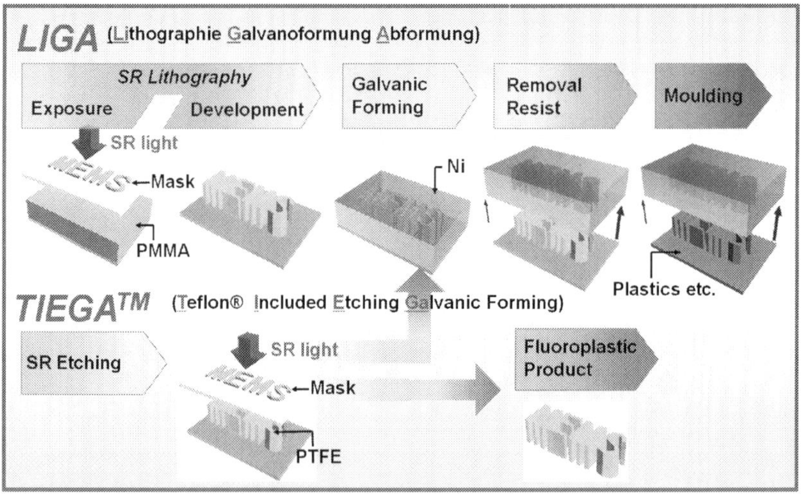

Fig. 20. Process flow of the TIEGA™ process

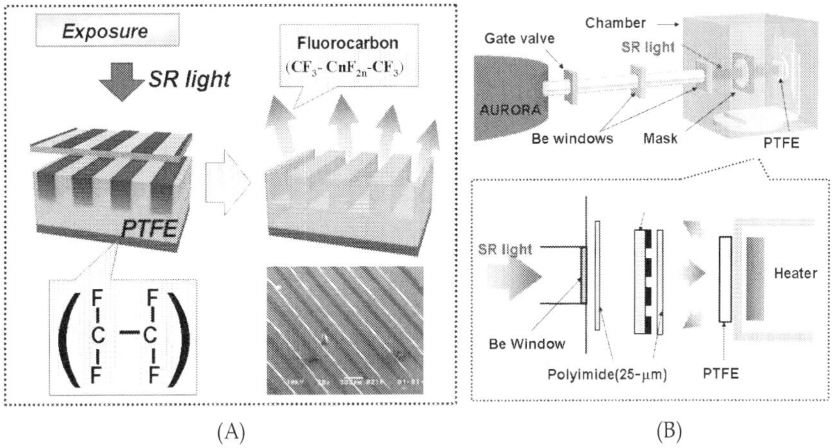

Fig. 21. (A) is processing mechanism; and (B) is experimental system

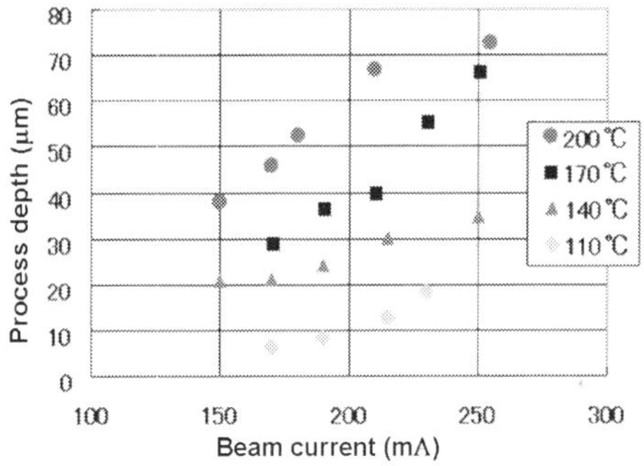

Fig. 22. Relation between the beam current and processing depth

We used a 1-mm-thick product from Yodogawa Hu-tech Co., Ltd. for PTFE. The processing mechanism is outlined in Figure 21A. When PTFE is exposed to an SR light source, a photochemical reaction occurs in which the main chain of the material decomposes, forming a fluorocarbon gas (CF_3-C_nF_{2n}-CF_3), and exposed parts are etched. The etching rate generally increases when PTFE is etched in a vacuum chamber and heated during the SR ablation fabrication process. The secession rate of fluorocarbon gas increases if the PTFE temperature is excessively high at this time. Therefore, a vacuum atmosphere at 10^{-5} torr was used, and a substrate heater was installed in the chamber. To prevent the mask and Be window from becoming contaminated by fluorocarbon gas, a 25-μm polyimide film was placed on each (Figure 21B). The processing depth was checked after an exposure of 30 min, and the PTFE surface temperature was increased to 110°C, 140°C, 170°C, and 200°C in succession. Figure 22 shows the relationship between the beam current and processing depth. In general, the TIEGA™ process utilizes white light. However, because this experimental system was designed for use with the TIEGA™ process as well as the LIGA process, PTFE was only exposed to the X-ray domain. Therefore, the etching rate and processing depth were low.

4.4 3-D Fabrication of PTFE

Because the PCT technique successfully provided PMMA 3-D structures, we adapted the PCT technique to expose PTFE while the stage was in motion. Fabrication of 3-D PTFE microstructures using other techniques has also been reported, including 3-D PMMA structures (Nakamura and Tabata, 2006; Nishi et al., 1999). Figure 23 shows 3-D PTFE structures fabricated using the PCT technique with a triangular mask pattern. Figure 23A shows a structure in which the PTFE surface temperature was 140°C, and Figure 23B shows a structure in which the PTFE surface temperature was 200°C. As shown in the figure, higher PTFE temperatures correspond to enhanced surface roughness. Figure 24 shows the target forms and 3-D PTFE structures fabricated using the PCT technique with other mask patterns.

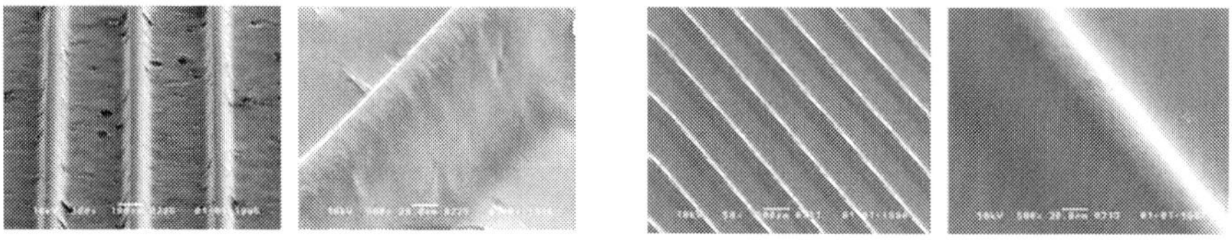

(A) (B)

Fig. 23. 3-D PTFE structures fabricated using the PCT technique with a triangular mask pattern; (A) PTFE surface temperature was 140°C; and (B) PTFE surface temperature was 200°C

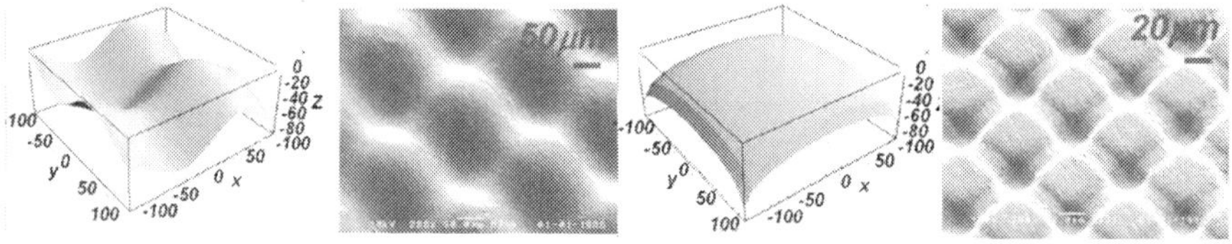

(A) (B)

Fig. 24. Target forms and 3-D PTFE structures fabricated using the PCT technique with other mask patterns; (A) is utilizing sine-curved mask pattern; and (B) is utilizing semicircular mask pattern

5. Conclusions

MEMS devices have attracted much attention, and further studies are needed to realize their full potential. In fabrication technology, an elemental technology, microfabrication, developed primarily using a semiconductor process, is in high demand. Recently, MEMS devices have diversified, and microfabrication technologies for the production of high-aspect-ratio and 3-D structures are in demand. This article described the fabrication of 3-D microstructures utilizing SR lithography.

Chapter 1 (Introduction) described the field of fabrication technologies of high-aspect-ratio structures and 3-D structures and explained the purpose of SR lithography. Chapter 2 (SR Lithography) provided an outline of the LIGA process, fabrication mechanism of SR lithography, processing-depth control, and optimum experimental conditions. Sections 2.2 (Exposure) and 2.3 (Development) described the mechanism of SR lithography and the requirements of the light source, the X-ray mask, and the resist materials. Section 2.4 (Approaches to High-Accuracy Microfabrication) described the resolution, optimum experimental conditions, and micro-loading effect based on both experimental and theoretical values. To achieve high-accuracy microfabrication, all of these are important. Chapter 3 (3-D Fabrication Method) described the fabrication method of 3-D structures utilizing SR lithography. Two fabrication techniques were described in detail: the PCT technique and the pixels exposure technique. In this chapter, the fabrication process and results, as well as the mechanism of 3-D fabrication, were described. Chapter 4 (PTFE Fabrication by SR Ablation) described the 3-D PTFE microstructures fabricated by SR ablation. Advantages of ablation technology and the fabrication mechanism were described. We expect that this research will contribute to elemental technologies in the field of MEMS, the achievement of high functionality, and to the development of high-performance devices that, so far, have been difficult to realize.

6. Acknowledgment

We would like to express our sincere gratitude to our supervisor, Prof. Toshiaki Ohta who is head of SR Center of Ritsumeikan University.
We would like to thank Dr. Hiroshi Ueno, Dr. Sommawan Khumpuang, Mr. Kazuya Fujioka, Mr. Shinya Fujinawa, and Mr. Shunsuke Kajita for providing us with useful data.
We would like to thank Dr. Yasukazu Yamamoto and Mr. Hiroyuki Ikeda for their technical assistance.

7. References

Ehrfeld, W.; Wood, R.L.; Hessel, V.; Lowe, H.; Schulz, C ; Weber, L (1998), Materials of LIGA technology, *Microsystem Technologies*, Springer, Vol.5, Num.3, pp.105-112, ISSN 0946-7076

Fujinawa, S.; Kato, F.; Sugiyama, S (2006), Development of fabrication process for shape-control of three-dimensional submicron structure by synchrotron radiation lithography, *Proceedings of SPIE - The International Society for Optical Engineering*, Vol.6037, pp.331-338,, ISSN 0277-786X, Brisbane, Australia, Dec.11-15,2005

Horade, M. & Sugiyama, S (2010), Study on fabrication of 3-D microstructures by synchrotron radiation based on pixels exposed lithography, *Microsystem Technologies*, Springer, Vol.16, Num.8-9, pp.1331-1338, ISSN 0946-7076

Kato, F.; Fujinawa, S.; Li, Y.G.; Sugiyama, S (2007), Fabrication of a spiral microcoil using a 3D-LIGA process, *Microsystem Technologies*, Springer, Vol.13, Num.3-4, pp.221-225, ISSN 0946-7076

Kato, T. & Zhang, Y (1998), High aspect ratio micromachining by synchrotron radiation direct photo-etching, *Microsystem Technologies*, Springer, Vol.4, Num.3, pp.135-138, ISSN 0946-7076

Kondo, R.; Takimoto, S.; Suzuki,K.; Sugiyama, S (2000), High aspect ratio electrostatic micro actuators using LIGA process, *Microsystem Technologies*, Springer, Vol.6, Num.6, pp.218-221, ISSN 0946-7076

Kumpuang, S.; Horade, M.; Fujioka, K.; Sugiyama, S (2006), Microneedle fabrication using the plane pattern to cross-section transfer method, *Smart materials and structures*, Vol.15, Institute of Physics, pp.600-606, ISSN 0964-1726

Lee, K.C & Lee, S.S (2003), 3D fabrication using deep X-ray mask with integrated micro-actuator, *Proceedings of 16th IEEE International conference on Micro Electro Mechanical Systems (MEMS 2003)*, pp.558-561, ISBN 0-7803-7744-3, Kyoto, Japan, Jan 19-23, 2003

Matsuzuka, N.; Hirai, Y.; Tabata, O (2004), A novel fabrication process of 3-D microstructures by double exposure in standard deep X-ray lithography, *Proceedings of 17th IEEE International conference on Micro Electro Mechanical Systems (MEMS 2004)*, pp.681-684, ISBN 0-7803-8265-X, Maastricht, Netherlands, Jan 25-29, 2004

Mekaru, H.; Kusumi, S.; Sato, N.; Shimizu, M.; Yamashita, M.; Shimada,O.; Hattori, T (2007), Fabrication of a spiral microcoil using a 3D-LIGA process, *Microsystem Technologies*, Springer, Vol.13, Num.3-4, pp.393-402, ISSN 0946-7076

Nakamura, Y. & Tabata, O (2006), Moving Mask Direct Photo-Etching (M2DPE) for 3D Micromachining of Polytetrafluoroethylene, *IEEJ Transactions on Sensors and Micromachines*, Vol.126, Num.9, pp.499-503, ISSN 1347-5525

Nishi, N.; Katoh, T.; Ueno, H.; Kohish, S.; Sugiyama, S (1999), 3-Dimensional Micromachining of PTFE Using Synchrotron Radiation Direct Photo-Etching, *Proceedings of the 1999 International Symposium on Micromechatronics and Human Science*, Vol.5148, pp.93-98, ISBN 0-7803-5790-6, Nagoya, Japan, Nov 23-26, 1999

Ruzzu, A.; Fahrenberg, J.; Heckele, M.; Schaller, T (1998), Fabrication of a spiral microcoil using a 3D-LIGA process, *Microsystem Technologies*, Springer, Vol.4, Num.3, pp.128-131, ISSN 0946-7076

Schmalz, O.; Hess, M.; Kosfeld, R (1996), Structural changes in poly(methyl methacrylate) during deep-etch X-ray synchroton radiation litography, *Die Angewandte makromolekulare Chemie*, Num.239, pp.63-78, ISSN 0003-3146

Singleton, L& Detemple, P (2003), Mask technologies for deep x-ray LIGA, *Proceedings of SPIE - The International Society for Optical Engineering*, Vol.5148, pp.210-217, ISSN 0277-786X

Sugiyama, S.; Khumpuang, S.; Kawaguchi, G. (2004), Plain-pattern to Cross-section Transfer (PCT) Technique for Deep X-ray Lithography and Applications, *Journal of Micromechanics and Microengineering*, Vol.14, pp. 1399-1404, ISSN : 0960-1317

Suzuki, K. & Sugiyama, S (1997), Silicon IC compatible LIGA mask-fabrication process, *Microsystem Technologies*, Springer, Springer, Vol.4, Num.1, pp.7-11, ISSN 0946-7076

Tabata, O.; Matsuzuka, N.; Yamaji, T.; Uemura, S.; Yamamoto, K (2002), 3D fabrication by moving mask deep X-ray lithography (M2DXL) with multiple stages, *Proceedings of 15th IEEE International conference on Micro Electro Mechanical Systems (MEMS 2002)*, pp.180-183, ISBN 0-7803-7185-2, Las Vegas, USA, Jan 20-24, 2002

Tsukada, N.; Nakao, T. & Higuchi, T. (2005), 3D microstructures fabrication of paraffin or plastics with laser heating, *Proceedings of 18th IEEE International conference on Micro Electro Mechanical Systems (MEMS 2005)*, pp.576-579, ISBN 0-7803-8732-5, Miami, USA,Jan 30- Feb 3, 2005

Ueno, H.; Zhang, Y.; Nishi, N.; Sugiyama, S (2000), Fabrication of sub-micron structures for MEMS using deep X-ray lithography, *Microsystem Technologies*, Springer, Vol.6, Num.6, pp.210-213, ISSN 0946-7076

17

Emerging Maskless Nanolithography Based on Novel Diffraction Gratings

Guanxiao Cheng[1,2,3], Yong Yang[4], Chao Hu[2,3,5], Ping Xu[1], Helun Song[6], Tingwen Xing[4] and Max Q.-H. Meng[2,3]

[1]*College of Electronic Science and Technology, Shenzhen University, Shenzhen,*
[2]*Shenzhen Institutes of Advanced Technology, Chinese Academy of Sciences, Shenzhen,*
[3]*The Chinese University of Hong Kong, Hong Kong,*
[4]*Institute of Optics and Electronics, Chinese Academy of Sciences, Chengdu,*
[5]*Ningbo Institute of Technology, Zhejiang University, Ningbo,*
[6]*Suzhou Institute of Nano-tech and Nano-bionics, Chinese Academy of Sciences, Suzhou,*
China

1. Introduction

The development of lithographic technology that has been used in semiconductor electronics has led to systems that put a premium on spatial resolution, throughput and reliability, regardless of cost and flexibility. According to Abbe's theory, the spatial resolution can be improved by using either shorter wavelength or higher numerical aperture (*NA*). Although the semiconductor industry has made significant progress in increasing the lithography resolution in the past decades, further improvement of the resolution by accessing shorter wavelengths is facing critical challenges due to the availability of optical materials with suitable refractive index. The expansion of nanoscale science and engineering will require flexible, high spatial resolution, and low-cost nanolithographic techniques and systems other than those employed in the semiconductor industry, for reasons of both cost and limited flexibility. The research on Emerging Maskless Nanolithography Based on Novel Diffraction Gratings presented in this chapter is a step in the direction of providing affordable, highly flexible nanolithography.

In this chapter, we present two cases for maskless nanolithography employed novel diffraction optics elements as objective lens to focus illumination light. The diffractive objective lens (DOL) operates by the principle of diffraction other than conventional objective lens functioning by refraction. DOL can be designed to operate at any wavelength while refractive elements are constrained at short wavelengths by material transmission properties. DOL are thinner, and can be fabricated by planar techniques that are reliable and low cost.

DOL, in a form of photon sieves originally used in x-ray microscopy and achieved 6nm resolution, is possible to extend the use of diffractive elements down to the limits of nanolithography. Recently, we present the scheme of photon sieve array X-ray maskless nanolithography (PSAL) to fabricate novel nanometer devices (Cheng et al., 2006, 2007a, 2007b, 2008). The lithographic principle is shown in figure 1. Firstly, each of the photon sieve array focuses incident X-ray into a diffraction-limited on-axis spot on the surface of a photoresist-coated substrate, the X-ray intensity of each spot is modulated by means of a spatial light modulator. Then, patterns of arbitrary geometry are written in a dot matrix fashion while the photoresist-coated substrate on a precision stepping stage is exposed to the properly modulated X-ray.

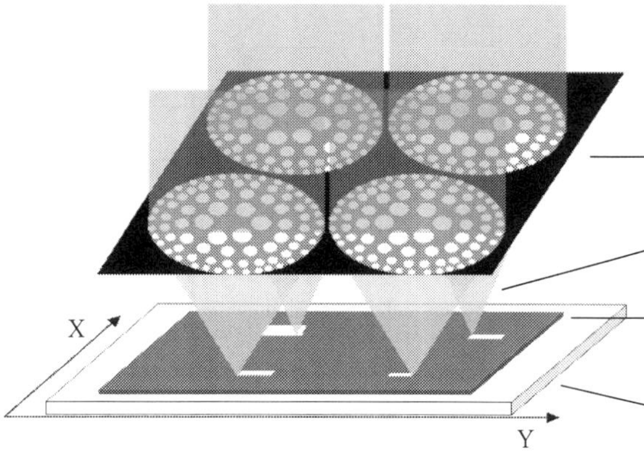

Fig. 1. Schematic of PSAL. An array of photon sieves focuses incident X-ray into a matrix of spots on the substrate coated photoresist. Patterns of arbitrary geometry are recorded while the stepping stage is driven.

In combination with the synchrotron light sources, PSAL can offer a new lithographic tool for high-resolution X-ray nanolithography in physical and nanoscale sciences. The PSAL lithographic system will be further discussed in detail on the synchrotron radiation light, resolution limits, depth of focus, etc. The design, fabrication, and experimental characteristic of X-ray photon sieve will also be illustrated by numerical analyse and experimental results. According to Fresnel-Kirchhoff diffraction theory, the diffractive field of photon sieve is described by means of the discrete fast Fourier transform algorithm. The approaches to enhancing imaging resolution of photon sieve are presented in detail. The related contents include the calculation of point spread function, the suppression of side lobes, the imaging bandwidth, the physical limit of resolution, and the diffraction efficiency. Imaging properties of photon sieve are analyzed on the basis of precise test and shown that photon sieve is a kind of diffractive optical element modulating either amplitude or phase and thus suffers from chromatic aberration or low diffraction efficiency. Hybrid lens consisting of both refractive optical surfaces and photon sieve are suggested to correct the chromatic aberration. Phase-photon sieve technology and surface plasmon polaritons technology are promising approaches to improve the diffraction efficiency and spatial resolution.

DOL, in another form of superlens consisted of nano-filmed noble metals on which the evanescent field is strongly enhanced using the resonant excitation of surface plasmons that can be excited at given conditions, is also possible to extend the use of diffractive elements down to the limits of nanolithography (Yang et al., 2007a, 2007b, 2009). The high-resolution plasmonic nanolithography has been investigated by using optical proximity exposure in the evanescent near field in nano-filmed noble metals. Sub-diffraction-limited feature size can be resolved by using I-line illumination exposure. Compared with the model of original superlens, we separated the superlens 100nm away from the substrate, under the illumination of I-line light, the initial simulation shows that the sub-diffraction-limited feature as small as 60nm line width with 120nm pitch can be clearly resolved without hard contact between the substrate and the superlens. This is shown in figure 2. By proper design of the materials and the parameters of nano-filmed layers, better resolution can be realized.

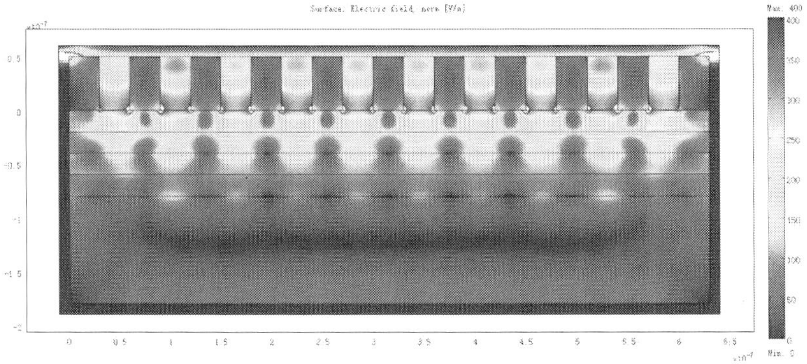

Fig. 2. Electric field distribution of dual-layered heterostructure for CD=30nm.

2. Photon sieve array X-ray maskless nanolithography

Lithography has been the key technology in the semiconductor industry. Beyond the semiconductor industry, lithography has also been widely employed in such technological fields as microoptics, nanophotonics, MOEMS, nanotechnology, etc. In order to achieve the minimum feature size, the development of lithography has resulted in setups with high throughput and reliability regardless of cost and flexibility.

For nanoscale science and engineering, however, the lower cost and higher flexibility of lithography must be considered. Fortunately, maskless lithography can meet the requirements of nanotechnology. There are various forms of maskless lithography that include scanning electron-beam lithography, focused ion-beam lithography, multiaxis electron-beam lithography, interference lithography, maskless optical-projection lithography, scanning probe lithography, zone plate array lithography (ZPAL) (Menon et al., 2005), etc. It is a very complex process of selecting an optimum lithography tool that requires knowledge and experience in several disciplines including physics, chemistry, electronics, device design, manufacturing, processing, cost and marketing. Although the selection strategy consists of many aspects, the technical aspect is the dominant item because the tool has to be technically performable. Menon et al. showed the feasibility of ZPAL operating at a wavelength of 400nm and its potential for the fabrication of novel devices.

Photon sieve is a novel diffractive optical element which consists of a great number of pinholes distributed appropriately over the Fresnel zones for the focusing and imaging of soft X-rays (Kipp et al., 2001). Photon sieve has advantages of the diameter of pinholes beyond the limitation of the corresponding Fresnel zone width and the minimum background in the focal plane, which is shown in figure 3. The focal spot produced by a Fresnel zone plate is surrounded by rings of intensity (secondary maxima) that blur the images. For a zone plate, each ring contributes equally to the amplitude at the focus. This contribution drops abruptly to zero beyond the outermost ring which leads to strong intensity oscillations in the diffraction pattern (top row). With a photon sieve, the number of pinholes per ring can be readily adjusted to yield a smooth transition which minimizes the secondary maxima (bottom row). This provides sharper images. Note that diffraction intensities are plotted on a log scale (right column). Furthermore, photon sieve can be fabricated on a single surface without any supporting struts required unlike the Fresnel zone plate.

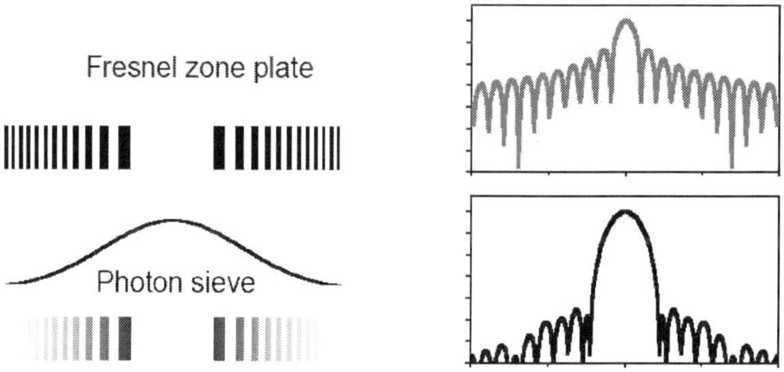

Fig. 3. Schematic of photon sieve suppressing side lobes.

In this section, we present the scheme of maskless nanolithography using a large array of photon sieves, each of which focuses incident X-ray into a diffraction-limited on-axis spot on the surface of a photoresist-coated substrate (PSAL). The X-ray intensity of each spot is modulated by means of a spatial light modulator. Patterns of arbitrary geometry are written in a dot matrix fashion while the photoresist-coated substrate on a precision stepping stage is exposed to the properly modulated X-ray. In combination with the synchrotron light sources, PSAL can offer a new tool for high-resolution X-ray nanolithography in physical and nano sciences. Lastly, the design and fabrication of photon sieve are illustrated with a low-*NA* amplitude-photon sieve fabricated on a chrome-coated quartz plate using laser-beam lithographic process.

2.1 Scheme of photon sieve array X-ray maskless nanolithography

In PSAL system operating at wavelength of 0.5~2nm synchrotron light sources radiated, each of a large array of photon sieves focuses incident X-ray into a diffraction-limited on-axis nanoscale spot on the substrate coated photoresist. Patterns of arbitrary geometry are exposed in a dot matrix fashion while the substrate on a stepping stage is precisely driven in two dimensions according to the computer program. A Schematic of PSAL is shown in Figure 1. PSAL provides a promising approach for the fabrication of nanometric device and integrated circuit (IC). There are emphases on light source, resolution, depth of focus and process latitude in both nanodevice and the IC areas. In the section, we will go into detail the four characteristics.

2.2 Source of X-ray

PSAL utilizes very short wavelength synchrotron radiation source emitted. Synchrotron radiation arises from the changing direction of a high energy relativistic beam of electrons caused to run a circular modified elliptical path. The key characteristics of synchrotron radiation light are high brightness, very small divergence, extremely narrow bandwidth, and so on. In the case of synchrotron radiation, the operating wavelength ranged from 0.5nm to 2nm is very suitable for nanolithographic research.

2.3 Resolution

Resolution limits are caused by the diffraction effects and photoelectron spreading. On one hand, the minimum diffraction-limited resolution of a PSAL system is determined by

$$R_{\min} = k_1 \lambda / NA \tag{1}$$

where λ is the exposure wavelength of the incident light, k_1 is a factor between 0.5 and 1. The shortest exposure wavelength contemplated is 0.5nm. NA of approximately 0.8 is expected. As a result, the minimum feature size R_{min} for such a system may approach 0.3nm. Therefore, PSAL is very promising for the fabrication of nanodevices.

On the other hand, the incident x-ray photon usually transforms all its energy into a single high energy photoelectron in the incident solid. Both a secondary electron shower and the chemical alterations in photoresist necessary for exposure are caused by this primary photoelectron. Therefore, resolution limits arise also from the range of this photoelectron.

2.4 Depth of focus

The depth of focus (DOF) of a PSAL system is given by

$$DOF = \pm k_2 \lambda / NA^2 \tag{2}$$

where k_2 is determined by the lithography process in use, ranging from 0.5 to 1. DOF defines the maximum tolerable displacement of the image plane from its ideal position. The approach to optically polishing imaging surfaces tends to make lithography easier by eliminating depth-of-focus problems. For example, a photon sieve of NA = 0.8 and k_2 = 0.6, operating at 2 nm, the DOF is $\pm 1.9nm$.

2.5 Process latitude

If tiny variations in exposure and development conditions cause large varieties in the exposed pattern, the process is worthless. Process latitude is defined that the variety of some critical process parameters (such as line width, edge acuity, etc.) concerning the processing conditions (such as exposure time, development time, etc.). Ideally, a process whose critical parameters change as little as possible for a given change in process condition is desired.

2.6 Design of photon sieve

Photon sieve can produce effects on their own depending on what the optical engineer is trying to achieve for a particular design goal. In situations where size or weight are critical, photon sieves bring substantial advantages. The flexibility in constructing wavefronts gives an optical designer the ability to use photon sieves fabricated microscopic structures on flat surfaces as converging lenses all by themselves. Acted as a diffractive optical element, photon sieves can also be used to correct image aberrations and color aberrations in a lens in a manner similar to the use of aspheric surfaces and additional refractive components. For example, the hybrid lens, consisting of a photon sieves with a conventional refractive lens, can provide the required color correction.

One useful method of the approaches to designing photon sieve is the paraxially individual far-field model and the nonparaxial model (Cao & Jahns, 2002, 2003). The former derived from the paraxial Fresnel diffraction is valid for the low-NA photon sieve. The latter based on the nonparaxial expression for the far field of individual pinholes and the linear superposition principle is valid for high-*NA* photon sieve, such as the photon sieves operating in X-ray region.

The key methods designing photon sieve are to optimize positions and radii of pinholes. Kipp et al. pointed out that photon sieve consists of a large number of pinholes distributed appropriately over the zones of a Fresnel zone plate pattern. The size of the focused spot is approximately equal to the width of the smallest zone in the underlying zone plate pattern. The diffracted field of a single pinhole can be expressed as

$$U \propto KJ_1(0.5\pi K) \tag{3}$$

where K is the ratio of pinhole diameter to the corresponding zone width and J_1 is the first order Bessel function of the first kind. The maximum modulus of U appears when K is approximately equal to 1.5, 3.5, 5.5, etc, as shown in figure 4. The parameter K may be utilized to relax the fabrication requirements for the photon sieve considerably as compared with a zone plate of the same *NA*.

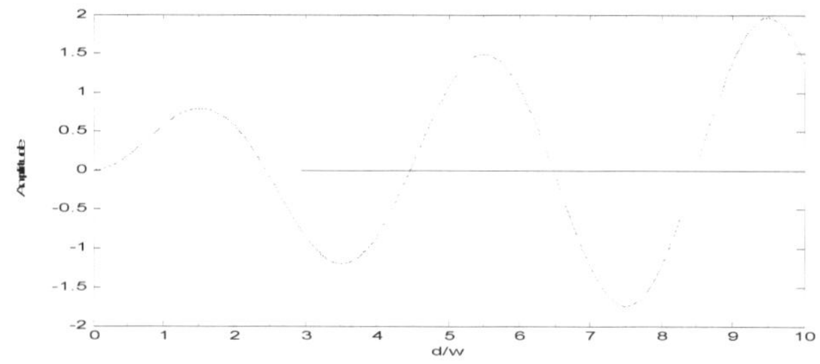

Fig. 4. The diffracted field of a single pinhole.

The point-spread function (PSF) of a traditional Fresnel zone plates generally has side-lobes of noticeable strength. The side-lobes blur the image obtained by the zone plates. The photon sieve may create a PSF that has lower side-lobes at the expense of a slightly wider main lobe than the zone plates. Moreover, an additional benefit of the photon sieve is that it breaks the periodicity that is inherent in the zone plates, thus strongly attenuating the higher diffracted orders. For an amplitude-photon sieve, apodization may be achieved by varying the transmissivity of its pupil gradually, i.e., the number of its pinholes, as a function of radial distance so as to follow approximately the function form of a Gaussian distribution

$$G(r) = \exp(-0.5r^2/\sigma^2) \tag{4}$$

where r is the radius at the center of the underlying zone pattern and is the parameter that controls the strength of apodization. The apodization parameter may be adjusted to achieve the optimum desired intensity distribution at the focal plane (Cheng et al., 2010, 2011).

We calculated the PSF intensity distribution of photon sieves with different apodization parameters at focal plane compared with a zone plates. Figure 5(a) shows a plot of the pupil intensity transmission of a zone plate with rectangular transmission window and two photon sieves with Gaussian transmission windows $(\sigma = 1, 0.5)$, respectively. In fact, the rectangular transmission pupil can also be derived from the Gaussian transmission pupil when $\sigma = \infty$. Figure 5(b) shows cross sections of the intensity PSF for the three cases. The logarithm of intensity is plotted vertically in order to emphasize the side-lobes, and the intensity normalization is proportional to the total integrated intensity passed by the pupil in each case. It may be seen that the Gaussian apodization significantly suppresses the side-lobes at the expense of slightly broadening the main lobe. The relative attenuation of side-lobes of the zone plate and photon sieves is -13.2dB, -16.7dB, -32.8dB, respectively. The full width at half maximum (FWHM) of these main lobes is 0.05, 0.05, 0.07, respectively. Therefore, a less value of σ produces a narrower window and a less strength of side-lobes at the expense of a somewhat wider main lobe.

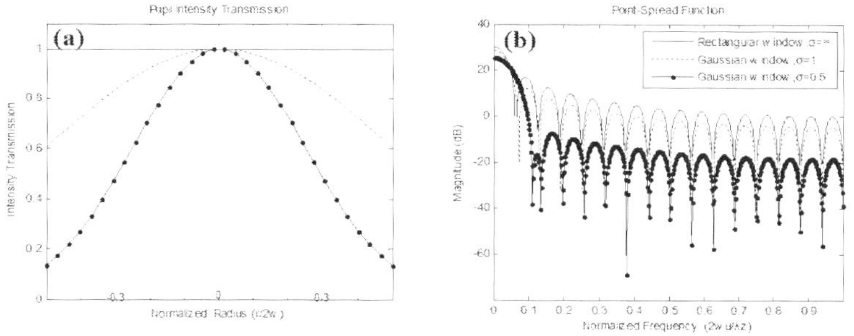

Fig. 5. Comparison of calculated intensity distributions of photon sieves and zone plate with different apodization windows. (a) Transmission windows in pupil plane for a zone plate (solid line) with a rectangular apodization window and two different photon sieves with Gaussian apodization window: σ=1 (dotted line), σ=0.5 (solid-dot line), respectively. (b) Transverse intensity point-spread function distribution in focal plane for the corresponding zone plate and photon sieves.

We designed an amplitude-photon sieve. The diffraction element had a diameter of 30mm and a focus length of 175 mm for an operating wavelength 632.8 nm. The pinholes were distributed over zones of order ranged from 2 to 1000. The minimum diameter of pinhole was 5.6 μm. We give the simulation results of pinhole positions distributed over the zones of order from 2 to 100 and the point spread function of the former 10 zones, which are shown in Figure 6.

A Gaussian apodized X-rays amplitude-photon sieve at a wavelength of 0.5nm (20mm focal length, 2mm pupil diameter, 0.05 *NA*) is simulated based on the scalar diffraction theory. The minimum feature size and total number of pinholes is 7.5nm, 14215227, respectively. The spatial resolution of 6.1nm can be obtained. Figure 7 shows the distribution of central part of the photon sieve, which the ratio of pinhole diameter to underlying zone width is chosen as 1.5.

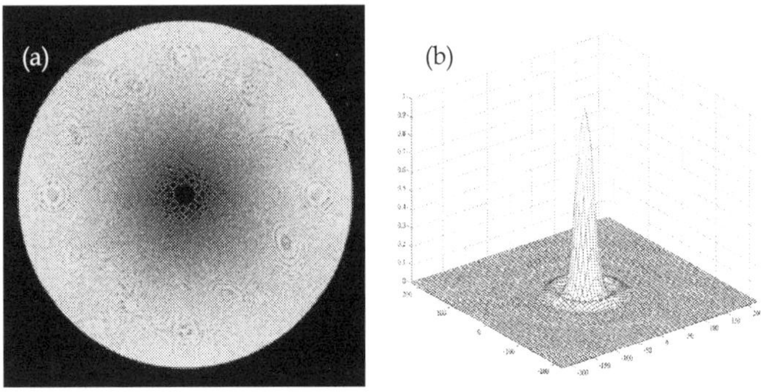

Fig. 6. Simulated pinholes distribution and PSF of a photon sieve.

Fig. 7. Pinholes distribution of an apodized X-rays photon sieve.

The diffraction efficiency scales as the square of the transmission area. A photon sieve transmits only 15~30% of the incident light because of the Gaussian apodization whereas an amplitude-zone plate has a transmission of 50%. Therefore, the first-order diffraction efficiency of a photon sieve is lower than that of a zone plate of equal *NA* by a factor of 10. However, the photon sieve is an attractive alternative to conventional zone plates for the X-rays focusing and imaging elements in the situation using the brilliant X-rays from synchrotron light sources as illumination, the diffraction efficiency is not a very important consideration but the side-lobes suppression and the fabrication ease are relatively more important.

The process of designing an amplitude-photon sieve is shown as follows. Firstly, the operating wavelength, diameter and focal length of photon sieve are given according to the purpose of imaging or focusing. Secondly, the number of corresponding Fresnel zones is calculated. Lastly, the optimum radius and coordinate of each pinhole in each corresponding zone are determined, and then the data of all pinholes are stored in the same file in order to create the photon sieve.

2.7 Fabrication and experiment

In order to verify the feasibility of the above methods, we fabricated a Gaussian apodized amplitude-photon sieve (633nm design wavelength, 175mm focal length, 30mm pupil diameter, 0.08 *NA*) on chrome-coated quartz plate using laser-beam lithographic techniques. The resolution image was recorded at 627nm peak-wavelength and 20nm spectral half-width produced by a light-emitting diode. Figure 8 demonstrates the pinholes distribution of the central part of the Gaussian apodized amplitude-photon sieve and the ratio of pinhole diameter to underlying zone width is chosen as 1.5.

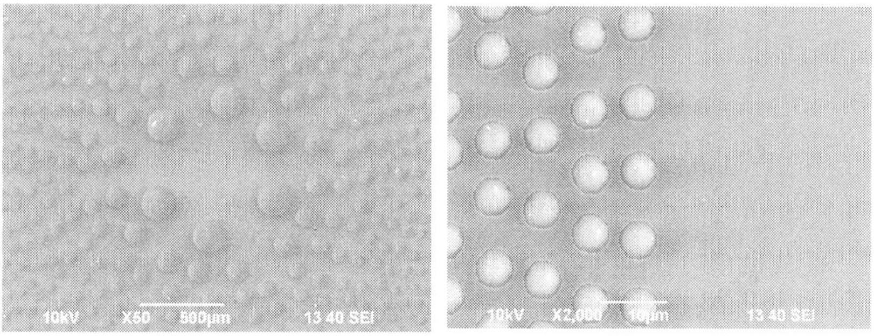

(a) (b)

Fig. 8. Scanning electron microscope images of an amplitude-photon sieve. (a)Central portion, (b)the outermost portion.

For the purpose of testing the imaging property, a beam of even illumination was produced by a light-emitting diode (LED, with peak-wavelength 627nm and spectral half-width 20nm) and a ground-glass diffuser. The beam transmitted through a WT1005-62 resolution test target was then incident upon the photon sieve. The resolution images were recorded by an intensified charge coupled device (ICCD). Figure 9 shows that good agreement between experimental and theoretical results concerning reduction of side-lobes but the resolution of 5.9um obtained is somewhat lower than the expected resolution limit of 4.8um because of the wavelength difference and chromatic aberration.

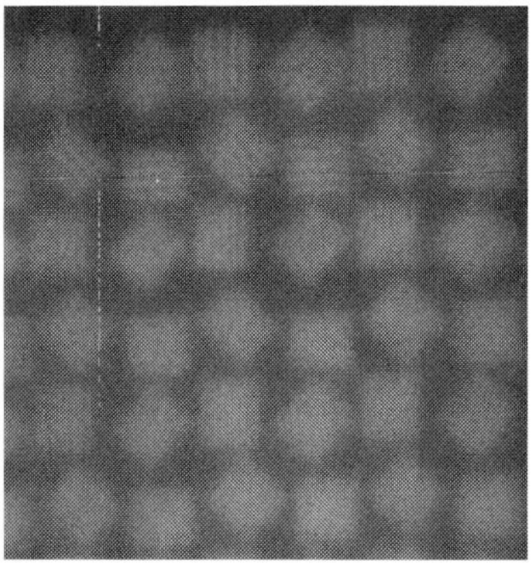

Fig. 9. Resolution test target imaged with an apodized photon sieve at 627 nm.

In brief, the experiment mentioned above demonstrates the feasibility of using photon sieves as the focusing optical element. We believe in that PSAL will be used as one of the promising tools in the nanometric device and special IC areas for the purpose of high resolution regardless of low yield.

3. Nanolithography in the evanescent near field by using nano-filmed noble metal layers

The sharpness of object can't be resolved by conventional lens due to the limitation by the wavelength of illumination light. J. B. Pendry had predicted that a slab of negative refractive index material has the power to focus all Fourier components of a 2D image. The super resolution of the negative index materials using silver layer, which was called 'superlens', can reconstruct the image of a pattern with line width of 40nm (Fang, et al., 2005). They made mask, silver slab and photoresist integrity in Fang's experiment, which likes the traditional contact exposure of lithography. It is not practical in real application by Fang's method of nanolithography because each wafer needs its respective mask. The experiments of super resolution using silver slab was reported, and the line width with one fifth of illumination wavelength can be successfully resolved by the silver slab (Blaikie et al., 2006).
Though Blaikie's experiment made mask and silver integrity, they separated photoresist from silver slab. This kind of configuration still had limitation in practical application. In order to investigate the influence of distance between mask and noble metal slab on imaging, we designed a separated 'superlens' with silver slab 100nm away from mask. We analyzed the distribution of optical field by Finite Difference Time Domain (FDTD). The results show that the images of object can be reconstructed by the structure.

3.1 Nanolithography method

The exposure method of near-field nanolithography that was proposed by this paper is illustrated in Figure 10. A UV transparent substrate with refractive index of n = 1.6 (@365nm) is used for supporting the mask. The object layer with line width of 60nm and pitch of 120nm, which is made of Cr with refractive index of n=2.924 (@365nm), acts as the function of mask in exposure. The air gap comes from the vacuum contact between mask and silver layer, which can be viewed as a kind of practical nanolithography technique.

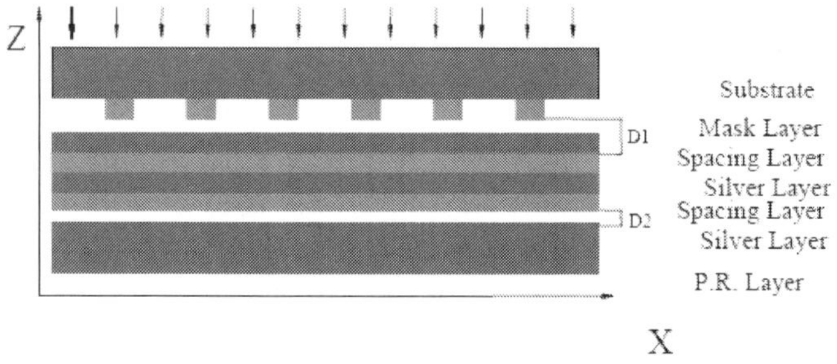

Fig. 10. Sketch of Nanolithography Model.

Materials	Parameters	Thickness
Substrate	Refractive index n=1.6@365nm	100nm
Mask	Refractive index n=2.9@365nm	40nm
Spacing	Refractive index n=1.52@365nm	40nm
Silver	Permittivity $\varepsilon_{Ag} = -2.4 + i0.2488@365nm$	40nm
Photoresist	Refractive index n=1.7@365nm	100nm

Table 1. Physical parameters of the materials in Fig. 10.

In figure 10, the spacing layer with refractive index of n=1.517 (@365nm) acts as the following two functions: 1) to match the surface plasmon polaritons resonating conditions; 2) to protect the surface of silver slab. As for the image recording, we chose the negative photoresist of i-line.

In order to explore the potential valid imaging distance between mask and photoresist, a repeat of spacing layer and silver slab layer was followed after the first silver slab. The sample was exposed in i-line light that shone from the substrate side. For the convenience of description, we defined D1 as the distance between mask and silver slab, and D2 as the distance between silver slab and photoresist.

3.2 When silver slab separated 40nm from mask (i.e. D1=40nm)

In order to save calculating time and PC resources, we computed the distribution of optical near-field intensity by the 2-D FDTD method. We chose cell size of X*Z=2×1(nm), which is much smaller than both the exposure light wavelength and the mask's feature size. The time step, according to Courant condition, should be:

$$TimeStep = 2\sqrt{2}N^{1/3} \tag{5}$$

N is the total cells of computing area. We chose 3500, by which the amplitude of electric field already became steady. The distance of 40nm comes from the spacing layer. At this situation, it is a kind of ideal condition, because mask and spacing layer had a hard contact. The surface plasmons polaritons of two interfaces between silver slab and its surroundings can magnify the evanescent waves that carried the detailed information of object. When D1=40nm, we calculated the following 4 conditions: D2=0nm, 20nm, 40nm, 60nm. Figure 11 show the distribution of electromagnetic (abbreviated to EM afterwards in the paper) field respectively.

It was found that the image of mask can be clear resolved by the method. Figure 11(a) shows better result, however, figure 11(b)-(d) showed worse results due to the exponential decay of the evanescent waves came from the exit side of the interface between silver slab and photoresist. The strong contrast of EM field may come from the edge effect of the evanescent waves. We chose the 10nm cross-section of photoresist layer to compare the imaging result of silver slab. When D1=40nm and D2 varied from 0nm to 60nm, the distribution of optical field in the section was shown in Figure 11. When D2 changed from 0nm to 60nm, the amplitude reduced to about a half under the same condition, but the high contrast of lines still can be clearly observed in figure 12. The amplitudes of lines were relatively uniform when D_2=20nm and 60nm compared with D_2=0nm. If the parameters of photoresist were under better control, the lines of images will be more uniform. However, compare with the condition of D2=0nm, the depth of lines in photoresist will be shallower at the same exposure condition.

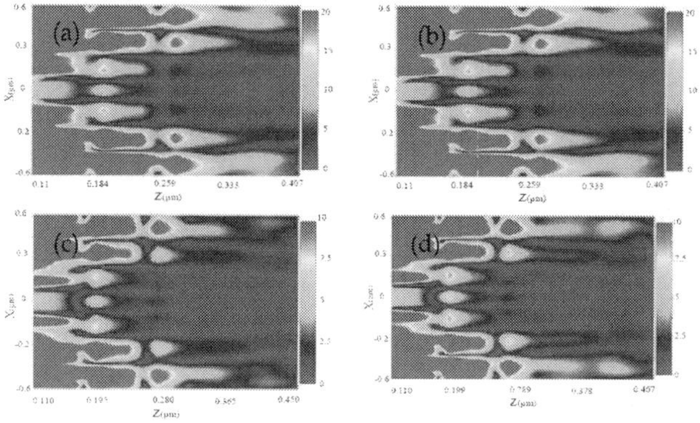

Fig. 11. The distribution of EM field in the model. Photoresist layer lies (a) between Z=0.31μm and Z=0.41μm, (b) between Z=0.33μm and Z=0.43μm, (c) between Z=0.35μm and Z=0.45μm, (d) between Z=0.7μm and Z=0.47μm.

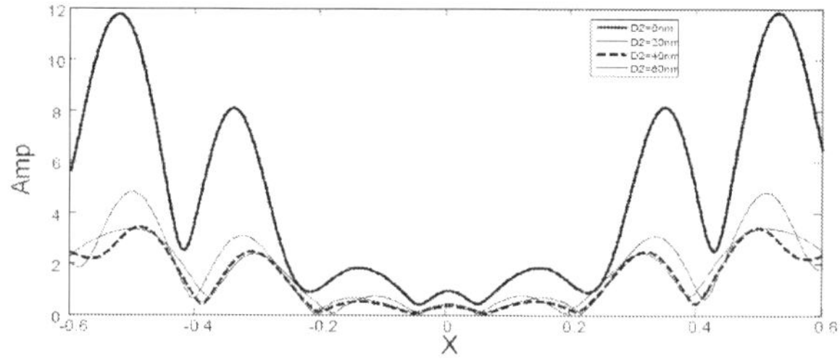

Fig. 12. The distribution of optical field in the 10 nm cross-section of photoresist when D1=40nm.

3.3 When silver slab separated 60nm and 80nm from mask (i.e. D1=60nm, 80nm)

When D1=60nm, 80nm, we calculated 4 conditions respectively for each D1. We still chose the 10nm cross-section of photoresist layer to investigate the optical field. In order to show the clear comparison results, we give the final comparison of the amplitude instead of EM distribution figures for each condition of different D2. Figure 13 and figure 14 showed the result respectively when silver slab separated 60nm and 80nm from mask.

It was found that there came out extra fringes in figure13 and figure 14. This kind of phenomena may be caused by the strong interference effects among evanescent waves. The image of mask still can be resolved in photoresist layer by proper choice of materials and exposure conditions. These results showed a bad conformity between mask and recorded image in photoresist layer. On the other hand, it gave us a hint to realize better resolution of optical lithography by reasonably using the interference effect.

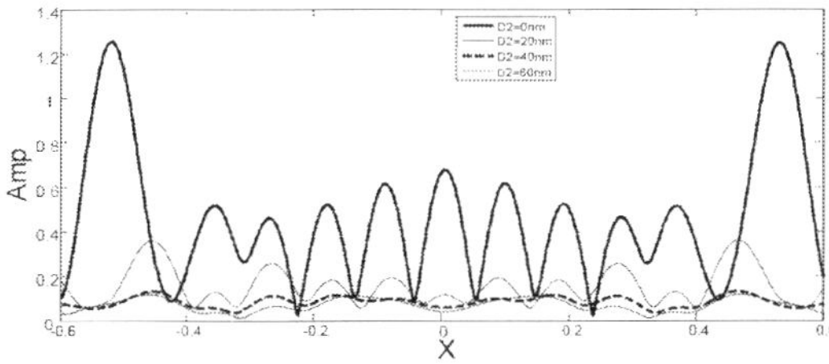

Fig. 13. The final comparison when D_1=60nm while D_2=0, 20, 40 and 60nm.

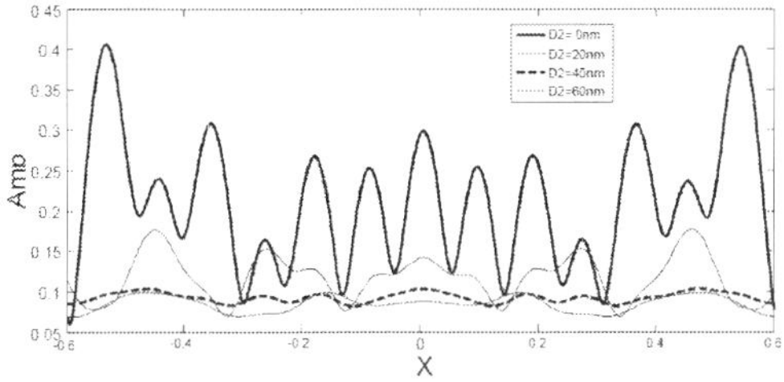

Fig. 14. The final comparison when D_1=80nm while D_2=0, 20, 40 and 60nm.

3.4 When silver slab separated 100nm from mask (i.e. D1=100nm)

The evanescent waves can not propagate to a long distance due to its exponential attenuation. The intensity of evanescent waves decays with a characteristic length Z_0:

$$Z_0 = \frac{n}{k_t\sqrt{\sin^2\theta_1 - n^2}} \qquad (6)$$

where $n = n_2/n_1$ is the relative refractive index of two surrounding media; θ_1 is the incidence angle of light from optically denser media to optically thinner media; where $k_t = 2n_2\pi/\lambda$, λ is the wavelength of incident light. Theoretically, Z_0 can be 100nm by calculation.

In order to explore the potential imaging property of silver slab, we increased D1 as much as possible. Considering the evanescent waves may diminish when D_1=100nm, so we calculated the condition of D2=0nm only, the distribution of electromagnetic filed was shown in figure 15. It was found that the image of mask still can be resolved clearly in photoresist layer with good uniformity of imaged lines. We investigated the distribution of optical field in the 10nm cross-section of photoresist layer. The distribution of optical filed in the transverse section of photoresist was shown in figure 16. The image of mask can be resolved with high contrast. With proper choice of exposure condition and materials, the information of mask can be transferred to photoresist layer, and the image of mask can be reconstructed by the silver slab layer.

In brief, the image of mask can be transferred to the photoresist layer by the enhancement function of surface plasmon polaritons in silver slab. We calculated the 2D distribution of electromagnetic field in our model; the results showed that the image of mask with feature size of 60nm line width can be resolved in photoresist layer when silver slab separated 100nm from mask. By proper design and choice of material, nanolithography with better resolution can be realized by the very function of silver slab, and this technique will be a possible alternative nanolithography technique for the next generation lithography.

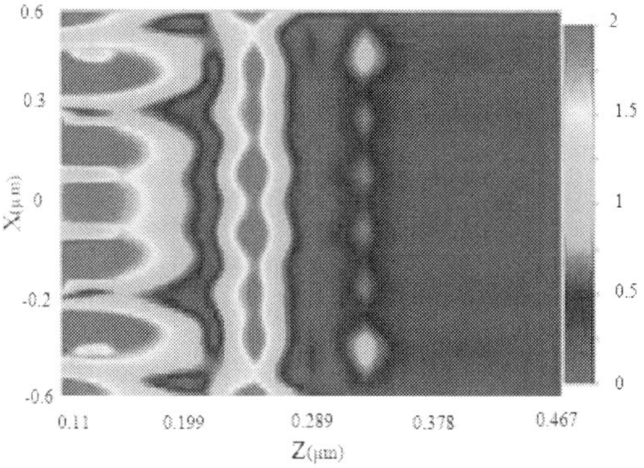

Fig. 15. The distribution of electromagnetic field in the model (Both X and Z is in unit of μm. Photoresist layer lies between Z=0.37μm and Z=0.47μm).

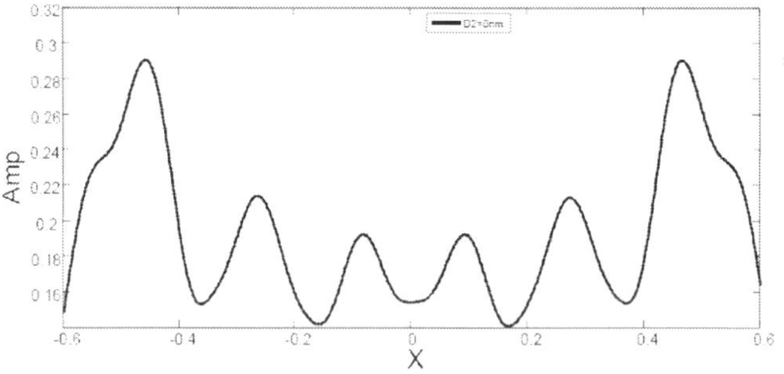

Fig. 16. The distribution of optical field in the 10nm cross-section of photoresist when silver slab separated 100nm from mask.

4. Conclusion

Two types of maskless lithography are discussed in this chapter. The first uses an array of high-numerical-aperture photon sieves as focusing elements in a scanning X-ray maskless nanolithography system. The system operating at wavelength of 0.5~2nm synchrotron light sources radiated, each of a large array of photon sieves focuses incident X-ray into a diffraction-limited on-axis nanoscale spot on the substrate coated photoresist. The X-ray intensity of each spot is modulated by means of a spatial light modulator. Patterns of arbitrary geometry are exposed and written in a dot matrix fashion while the substrate on a stepping stage is precisely driven in two dimensions according to the computer program.

The characteristics of synchrotron radiation light, resolution limits and depth of focus of the lithographic system are discussed. The design and fabrication of photon sieve are illustrated with a low-numerical-aperture amplitude-photon sieve fabricated on a chrome-coated quartz plate by means of laser-beam lithographic process, which minimum size of pinhole was 5.6um. The PSF of photon sieve in terms of side-lobes strength and main lobe width may be controlled by utilizing the apodization window function. The focusing performance of the photon sieve operating at wavelength of 632.8nm was simulated and tested. In combination with the synchrotron light sources, the photon sieve array X-ray maskless nanolithography is a promising tool in the nanometric device and special IC areas for the purpose of high resolution regardless of low yield.

The second focuses on the evanescent near field in nano-filmed noble metals. Sub-diffraction-limited feature size can be resolved by using i-line illumination exposure. Compared with the model of original superlens, we separated the superlens 100nm away from the mask, under the illumination of i-line light, the initial simulation shows that the sub-diffraction-limited feature as small as 60nm line width with 120nm pitch can be clearly resolved without hard contact between mask and nano-filmed noble metal. By proper design of the materials and the parameters of nano-filmed layers, better resolution can be realized.

In brief, a plasmonic structure for imaging and super focusing is a new approach besides the concept of negative refractive index. It is possible to realize imaging resolution beyond diffraction limit with a certain working distance within several wavelengths range. To realize this target, one of technical challenges is that how to transfer the high spatial frequency near-field signals from evanescent wave to propagation wave. The other challenge is that how to amplify the near-field evanescent wave from conventional ~200 nm to be $\sim 1\mu m$ or even several wavelengths in free space.

5. Acknowledgment

This work was supported by the grants from the Guangdong Natural Science Foundation (S2011040000711), the Key Lab of Robotics & Intelligent System, Guangdong Province (2009A060800016), the Guangdong & Chinese Academy of Sciences Cooperation Project (2009B091300160), National Natural Science Foundation of China (60904031; 60776029), Shenzhen Science & Technology Research Funds (2009-203), the Main Direction Program of the Knowledge Innovation of Chinese Academy of Sciences (KGCX2-YW-166), and the Scientific Research Foundation for the Returned Overseas Chinese Scholars of Ministry of Education, China (2008-890).

6. References

Blaikie, R.J., Melville, D.O.S. & Alkaisi, M. M. (2006). Super-resolution near-field lithography using silver lens: a review of recent developments. *Microelectronic Engineering*, Vol.83, No.4-9, pp.723-729, ISSN 0167-9317

Cao, Q., & Jahns, J. (2002). Focusing analysis of the pinhole photon sieve: individual far-field model. *J. Opt. Soc. Am. A*, Vol.19, No.12, pp.2387-2393, ISSN 0740-3232

Cao, Q., & Jahns, J. (2003). Nonparaxial model for the focusing of high-numerical-aperture photon sieves. *J. Opt. Soc. Am. A*, Vol.20, No.6, pp. 1005-1012, ISSN 0740-3232

Cheng, G. & Hu, C. (2011). X-ray Zernike apodized photon sieves for phase-contrast microscopy. *Acta Phys. Sin.* Vol.60, No.8, ISSN 1000-3290

Cheng, G., Hu, C., Xu P. & Xing, T. (2010). Zernike apodized photon sieves for high-resolution phase-contrast x-ray microscopy. *Opt. Lett.,* Vol.35, No.21, pp. 3610-3612, ISSN 0146-9592

Cheng, G., Xing, T., Yang, Y. & Ma, J. (2008). Resolution enhancement of photon sieve based on apodization. *Proc. SPIE,* Vol.6832, pp.83229-83229, ISBN 978-0-8194-7007-2, Beijing, China, Nov. 12-14, 2007

Cheng, G., Xing, T., Lin, W., Zhou, J., Qiu, C., Liao, Z., Yang, Y., Hong, L. & Ma, J. (2007). Photon sieve array x-ray maskless nanolithography. *Proc. SPIE,* Vol.6517, pp.51736-51736, ISBN 978-0-8194-6636-5, San Jose, CA, USA, Feb. 27-Mar. 01, 2007

Cheng, G., Xing, T., Yang, Y. & Ma, J. (2007). Experimental characterization of optical properties of photon sieve. *Proc. SPIE,* Vol.6724, pp.D7240-D7240, ISBN 978-0-8194-6881-9, Chengdu, China, Jul. 08-12, 2007

Cheng, G., Xing, T., Lin, W., Zhou, J., Qiu, C., Liao, Z. & Ma, J. (2006). Design and fabrication of low-numerical-aperture amplitude-photon sieve. *Chinese Journal of Sensors and Actuators,* Vol.19, No.5, pp.2344-2347, ISSN 1004-1699

Fang, N., Lee, H., Sun, C. & Zhang, X. (2005). Sub-diffraction-limited optical imaging with a silver superlens. *Science,* Vol.308, No.5721, ISSN 0036-8075

Kipp, L., Skibowski, M., Johnson, R. L., Berndt, R., Adelung, R., Harm, S. & Seemann, R. (2001) Sharper images by focusing soft X-rays with photon sieves. *Nature,* Vol.414, No.6860, pp. 184-188, ISSN 0028-0836

R. Menon, D. Gil, G. Barbastathis, and Smith, H. (2005). Photon-sieve lithography. *J. Opt. Soc. Am. A,* Vol.22, No.2, pp.342-345, ISSN 1084-7529

Yang, Y., Fu, Y., Yao, H., Hu, S., Zhou, S., Yan, W., Chen, W., Cheng, G. & Li, Z. (2009). Beam Splitter Achieved by Using Metallic Structure with Nanoslits. *J. Comput. Theor. Nanosci.,* Vol.6, No.5, pp.1030-1033, ISSN 1546-1955

Yang, Y., Hu, S., Yao, H., Cheng, G., Zhang, C. & Yan, W. (2007). Nanolithography in the Evanescent Near Field by Using Nano-filmed Noble Metal Layers. *Proc. SPIE,* Vol.6724, pp.A7241-A7241, ISBN 978-0-8194-6881-9, Chengdu, China, Jul. 08-12, 2007

Part 4

EUV Lithography and Resolution Enhancement Techniques

18

Laser-Plasma Extreme Ultraviolet Source Incorporating a Cryogenic Xe Target

Sho Amano
University of Hyogo
Laboratory of Advanced Science and Technology for Industry (LASTI)
Japan

1. Introduction

Optical lithography is a core technique used in the industrial mass production of semiconductor memory chips. To increase the memory size per chip, shorter wavelength light is required for the light source. ArF excimer laser light (193 nm) is used at present and extreme ultraviolet (EUV) light (13.5 nm) is proposed in next-generation optical lithography. There is currently worldwide research and development for lithography using EUV light (Bakshi, 2005). EUV lithography (EUVL) was first demonstrated by Kinoshita et al. in 1984 at NTT, Japan (Kinoshita et al., 1989). He joined our laboratory in 1995 and has since been actively developing EUVL technology using our synchrotron facility NewSUBARU. Today, EUVL is one of the major themes studied at our laboratory.

To use EUVL in industry, however, a small and strong light source instead of a synchrotron is required. Our group began developing laser-produced plasma (LPP) sources for EUVL in the mid-1990s (Amano et al., 1997). LPP radiation from high-density, high-temperature plasma, which is achieved by illuminating a target with high-peak-power laser irradiation, constitutes an attractive, high-brightness point source for producing radiation from EUV light to x-rays.

Light at a wavelength of 13.5 nm with 2% bandwidth is required for the EUV light source, which is limited by the reflectivity of Mo/Si mirrors in a projection lithography system. Xe and Sn are known well as plasma targets with strong emission around 13.5 nm. Xe was mainly studied initially because of the *debris problem*, in which debris emitted from plasma with EUV light damages mirrors near the plasma, quickly degrading their reflectivity. This problem was of particular concern in the case of a metal target such as Sn because the metal would deposit and remain on the mirrors. On the other hand, Xe is an inert gas and does not deposit on mirrors, and thus has been studied as a deposition-free target. Because of this advantage, researchers initially studied Xe. To provide a continuous supply of Xe at the laser focal point, several possible approaches have been investigated: employing a Xe gas puff target (Fiedrowicz et al., 1999), Xe cluster jet (Kubiak et al., 1996), Xe liquid jet (Anderson et al., 2004; Hansson et al., 2004), Xe capillary jet (Inoue et al., 2007), stream of liquid Xe droplets (Soumagne et al., 2005), and solid Xe pellets (Kubiak et al., 1995). Here, there are solid and liquid states, and their cryogenic Xe targets were expected to provide higher laser-to-EUV power conversion efficiency (CE) owing to their higher density compared with the gas state. In addition, a smaller gas load to be evacuated by the exhaust pump system was expected.

We have also studied a cryogenic Xe solid target. In that study, we measured the EUV emission spectrum in detail, and we found and first reported that the emission peak of Xe was at 10.8 nm, not 13 nm (Shimoura et al., 1998). This meant we could only use the tail of the Xe plasma emission spectrum, not its peak, as the radiation at 13.5 nm wavelength with 2% bandwidth. From this, improvements in the CE at 13.5 nm with 2% bandwidth became a most critical issue for the Xe plasma source; such improvements were necessary to reduce the pumped laser power and cost of the whole EUV light source. On the other hand, the emission peak of a Sn target is at 13.5 nm; therefore, Sn intrinsically has a high CE at 13.5 nm with 2% bandwidth. The CE for Sn is thus higher than that for Xe at present, in spite of our efforts to improve the CE for Xe. This resulted in a trend of using Sn rather than Xe in spite of the debris problem. Today, Cymer (Brandt et al., 2010) and Gigaphoton (Mizoguti et al., 2010), the world's leading manufacturers of LPP-EUV sources, are developing sources using Sn targets pumped with CO_2 lasers while making efforts to mitigate the effects of debris.

In the historical background mentioned above, we developed an LPP-EUV source composed of 1) a fast-rotating cryogenic drum system that can continuously supply a solid Xe target and 2) a high-repetition-rate pulse Nd:YAG slab laser. We have developed the source in terms of its engineering and investigated potential improvements in the CE at 13.5 nm with 2% bandwidth. The CE depends on spatial and temporal Xe plasma conditions (e.g., density, temperature, and size). To achieve a high CE, we controlled the condition parameters and attempted to optimize them by changing the pumping laser conditions. We initially focused on parameters at the wavelength of 13.5 nm with 2% bandwidth required for an EUV lithography source, but the original emission from the Xe plasma has a broad spectrum at 5–17 nm. We noted that this broad source would be highly efficient and very useful for many other applications, if not limiting for EUVL. Therefore, we estimated our source in the wavelength of 5–17 nm. Though Xe is a deposition-free target, there may be sputtering due to the plasma debris. We therefore investigated the plasma debris emitted from our LPP source, which consists of fast ions, fast neutrals, and ice fragments. To mitigate the sputtering, we are investigating the use of Ar buffer gas. In this chapter, we report on the status of our LPP-EUV source and discuss its possibilities.

2. Target system – Rotating cryogenic drum

We considered using a cryogenic solid state Xe target and developed a rotating drum system to supply it continuously, as shown in Fig. 1 (Fukugaki et al., 2006). A cylindrical drum is filled with liquid nitrogen, and the copper surface is thereby cooled to the temperature of liquid nitrogen. Xe gas blown onto the surface condenses to form a solid Xe layer. The drum coated with a solid Xe layer rotates around the vertical z-axis and moves up and down along the z-axis during rotation, moving spirally so that a fresh target surface is supplied continuously for every laser shot. A container wall surrounds the drum surface, except for an area around the laser focus point. This maintains a relatively high-density Xe gas in the gap between the container wall and the drum surface so as to achieve a high growth rate of the layer and fast recovery of the laser craters during rotation. The container wall also suppresses Xe gas leakage to the vacuum chamber to less than 5%, and the vacuum pressure inside the chamber is kept at less than 0.5 Pa. The diameter of the drum is 10 cm. Its mechanical rotation and up-down speed are tunable at 0–1200 rpm and 0–10 mm/s in a range of 3 cm respectively.

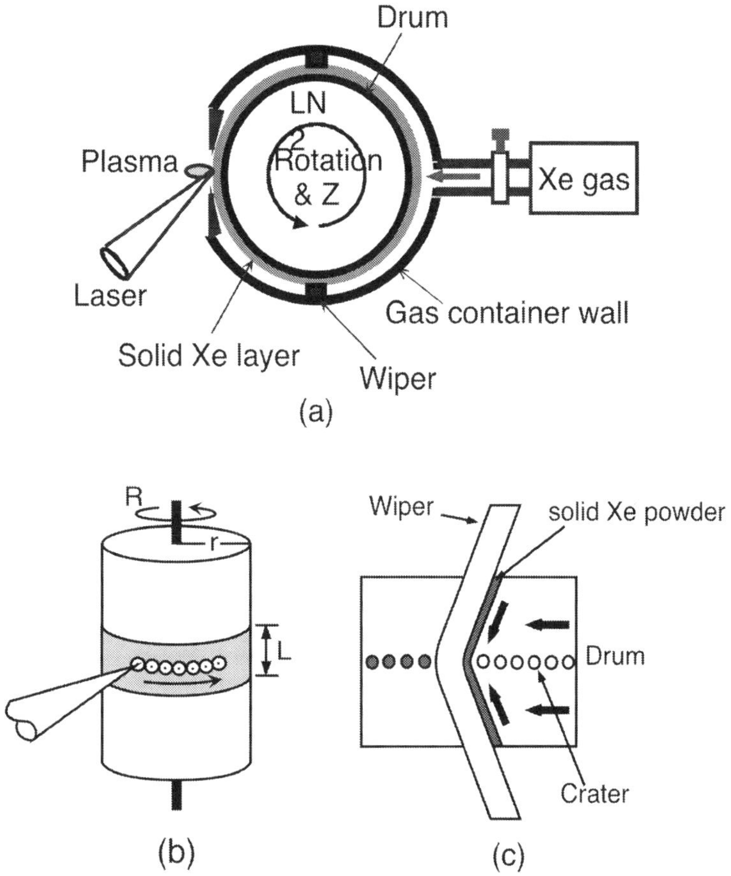

Fig. 1. Illustration of (a) the top view of the rotating cryogenic drum, (b) the side view, and (c) the wiper.

First, we formed a solid Xe layer with thickness of 300–500 μm on the drum surface and measured the size of the laser crater, which depends on the laser pulse energy. The crater diameter was measured directly from a microscope image, and its depth was roughly estimated from the number of shots needed to burn through the known thickness of the layer. A Q-switched 1064 nm Nd:YAG laser was focused on the Xe target surface with a spot diameter of 90 μm. Measured crater diameters D_c and crater depths δ_c are plotted in Fig. 2 for a laser energy range of 0.04–0.7 J. From the results in Fig. 2, a thickness of more than 200 μm was found to be sufficient for a laser shot of 1 J not to damage the drum surface. We then decided the target thickness to be 500 μm.

Two wipers are mounted on the container wall as shown in Fig.1 (a) to adjust the thickness of the solid Xe layer to 500 μm. As shown in Fig. 1 (c), the V-figure wipers also collect the Xe target powder on the craters produced by laser irradiation, thereby increasing the recovery speed. The wipers demonstrated a recovery speed of 150 μm/s up to a rotation speed of 1000 rpm, at a Xe flow rate of 400 mL/min.

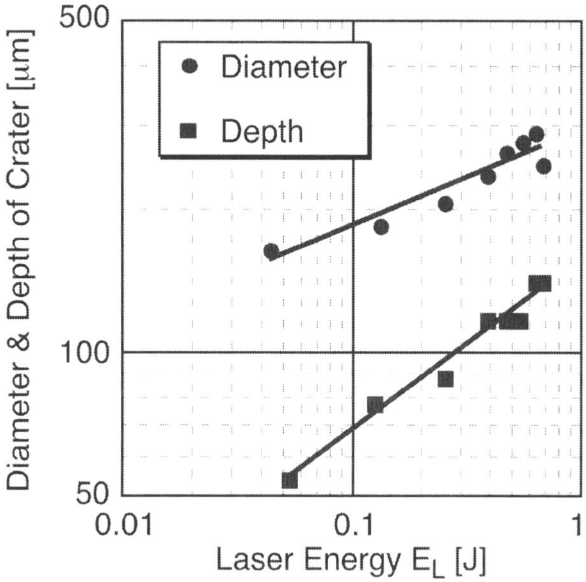

Fig. 2. Measured diameter and depth of a crater as a function of the irradiating laser energy.

Next, operational parameters of the drum are discussed to achieve high-repetition-rate laser pulse irradiation. In Fig. 1(b), R is the rotation speed, r is the radius of the drum, and L is the range of motion (scanning width of the target) along the rotational axis (z-axis). When the laser pulses are irradiated with frequency f, craters form on the target with separation length d between adjacent craters. The recovery time of a crater is T. Under the condition that craters do not overlap, f and T can be written as

$$f = \frac{2\pi r \cdot R}{d} \tag{1}$$

$$T = \frac{2\pi r L}{f \cdot d^2} \tag{2}$$

For example, if we assume laser energy of E_L = 1 J, a formed crater has a diameter of D_c = 300 μm and a depth of δ_c = 160 μm, and d must be at least 300 μm for the craters not to overlap. At r = 5 cm and R = 1000 rpm, we obtain f = 17 kHz from Eq. (1). When f = 10 kHz and L = 3 cm, T is calculated to be 10 s using Eq. (2), and we know that a recovery speed of the crater ($V_c = \delta_c/T$) of 16 μm/s is required. Here, we have already obtained V_c = 150 μm/s via the wiper effect and the required speed has been achieved.

Although flaking of the target layer due to superimposition of shock and/or thermal waves produced by continuous laser pulses was a concern for high-repetition pulse operation, model experiments and calculations show that there is no problem up to 1 J per pulse and 10 kHz (Inoue et al., 2006).
From the above results, we conclude that the rotating drum system we developed can supply the target continuously, achieving the required laser irradiation of 10 kHz and 1 J, and thus realizing a high-average-power EUV light source.

3. Drive laser – Nd:YAG slab laser

High peak power and high focusability (i.e., high beam quality) are required for a driving laser to produce plasma. In addition, high average power is required for high throughput in industrial use such as EUVL. We express such a laser as a *high average and high peak brightness laser,* for which the average brightness and peak brightness are defined as average power/$(\lambda \cdot M^2)^2$ and peak power/$(\lambda \cdot M^2)^2$, respectively; we began studying such lasers in the 1990s (Amano et al, 1997,1999).
We attempted to realize a *high average and high peak brightness laser* using a solid-state Nd:YAG laser (Amano et al., 2001). The thermal-lens effect and thermally induced birefringence in an active medium are serious for such a laser; thus, thermal management of the amplifier head is more critical, and the design of the amplifier system must more efficiently extract energy and more accurately correct the remaining thermally induced wavefront aberrations in the pumping head. To meet these requirements, we developed a phase-conjugated master-oscillator-power-amplifier (PC-MOPA) Nd:YAG laser system consisting of a diode-pumped master oscillator and flash-lamp-pumped angular-multiplexing slab power-amplifier geometry incorporating a stimulated-Brillouin-scattering phase-conjugate mirror (SBS-PCM) and image relays (IR). The system design and a photograph are shown in Fig. 3. This laser demonstrated simultaneous maximum average power of 235 W and maximum peak power of 30 MW with $M^2 = 1.5$. The maximum pulse energy was 0.73 J with pulse duration of 24 ns at a pulse repetition rate of 320 pps. We therefore obtained, simultaneously, both high average brightness of 7×10^9 W/cm^2·sr and high peak brightness of 1×10^{15} W/cm^2·sr.
This peak brightness is enough to produce plasma but the average brightness needs to be higher for EUVL applications. The maximum average power is mainly limited by the thermal load caused by flash-lamp-pumping in amplifiers. The system design rules that we confirmed predicted that average output power at the kilowatt level can be achieved by replacing lamp pumping in the amplifier with laser-diode pumping. Since our work, it seems that there has been no major progress in laser engineering for such *high average and high peak brightness lasers.* Average power of more than 10 kW has been achieved in continuous-wave solid-state lasers using configurations of fibers (ex. IPG Photonics Corp.) or thin discs (ex. TRUMPF GmbH). On the other hand, for the short-pulse lasers mentioned above, the maximum average power remains around 1 kW (Soumagne et al., 2005), which is more than an order of magnitude less than the ~30 kW required for an industrial EUVL source. This is one of the reasons why CO_2 lasers have been preferred over Nd:YAG lasers as the driving laser. To further the industrial use of solid-state lasers, there needs to be a breakthrough to increase the average power.

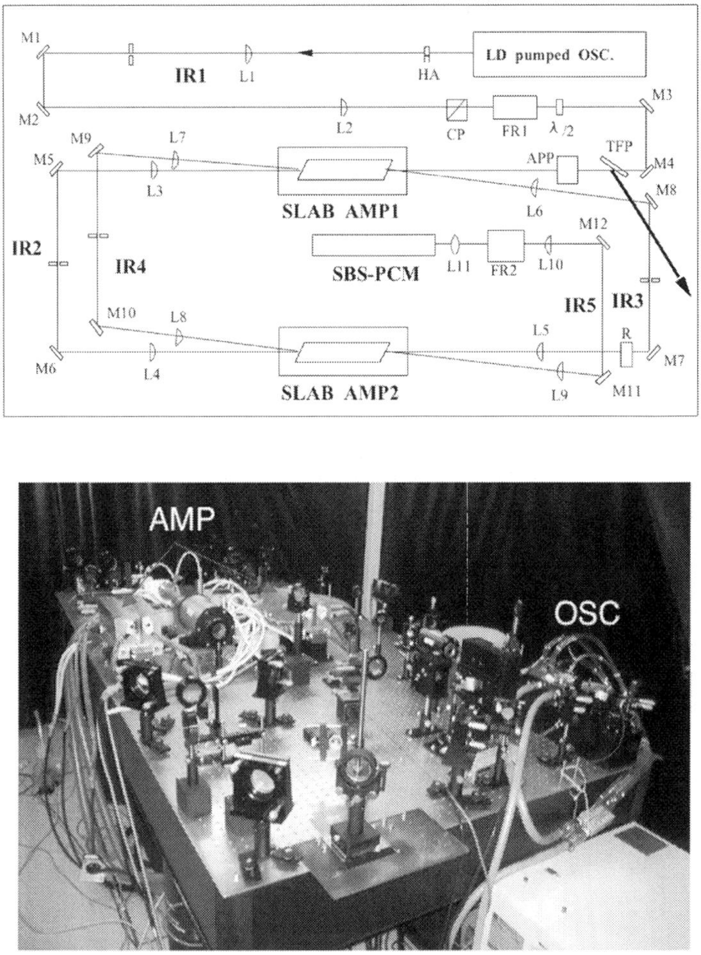

Fig. 3. Experimental setup and photograph of the PC-MOPA laser system.

4. EUV source

Figure 4 is an illustration and a photograph of the LPP-EUV source composed of a rotating cryogenic drum and Nd:YAG slab laser. The drum, detectors, and irradiating samples are installed in a vacuum chamber because EUV light cannot transmit through air. Driving laser pulses passing through the window are focused perpendicularly on the target by the lens so that Xe plasma is produced and EUV radiation is emitted. At a repetition rate of 320 Hz and average power of 110 W, the laser pulses irradiate the Xe solid target on the rotating drum with laser intensity of ~10^{10} W/cm^2. The rotation speed is 130 rpm and the vertical speed 3 mm/s. The Xe target gas is continuously supplied at a flow rate of 400 mL/min. Under these operation conditions, we obtain continuous EUV generation with average power of 1 W at 13.5 nm and 2% bandwidth.

The driving pulse energy was determined to be 0.3 J under the optimal condition that higher CE and lower debris are simultaneously achieved, as detailed below. At present, the maximum achieved CE is 0.9% at 13.5 nm with 2% bandwidth for the optimal condition. Under drum-rotating operation, we found the good characteristics of increased CE and less fast ions compared with the case with the drum at rest. We next detail the EUV and debris characteristics of the EUV source.

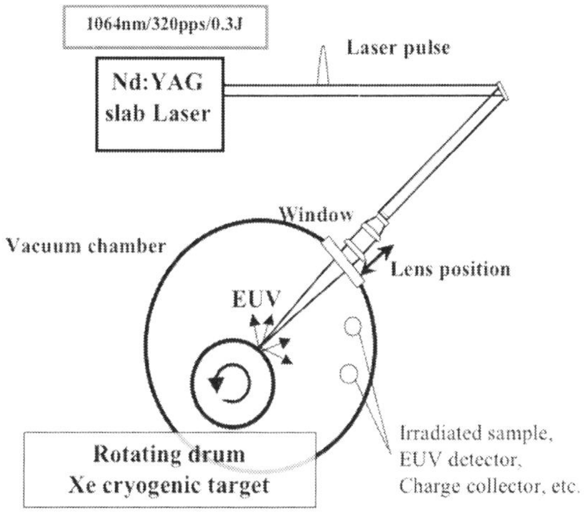

Fig. 4. Experimental setup and photograph of the laser plasma EUV source.

5. Conversion efficiency for EUVL

In this section, we report our studies carried out to improve the CE at 13.5 nm with 2% bandwidth required for the EUVL source (Amano et al., 2008, 2010a). To achieve the highest CE, we attempted to control the plasma parameter by changing the driving laser conditions. We investigated dependences of the CE on the drum rotation speed, laser energy, and laser wavelength. We also carried out double-pulse irradiation experiments to improve the CE.

To obtain data of EUV emission, a conventional Q-switched Nd:YAG rod laser (Spectra-Physics, PRO-230) was used in single-shot operation. By changing the position of the focusing lens to change the laser spot, the laser intensity on the target was adjusted to find the optimum intensity. We note that the lens position (LP) is zero at best focus, negative for in-focus (the laser spot in the target before the focus) and positive for out-of-focus (beyond the focus).

Figure 5(a) shows the CE per solid angle as a function of LP (laser intensity), which was measured by an EUV energy detector calibrated absolutely—*Flying Circus* (SCIENTEC Engineering)—located 45 degrees from the laser incident axis. The laser pulse energy was 0.8 J. We see that the CE was higher under the rotating-drum condition than under the rest condition. Here, the rest condition is as follows. Xe gas flow is stopped (0 mL/min) after the target layer has formed, and the drum rests (0 rpm) during a laser shot and stepwise rotates after every shot so that a fresh target is supplied to the point irradiated by the laser. The rotation condition is as follows. Laser pulses irradiate quasi-continuously the target on the rotating drum (>3 rpm), supplying Xe gas (>40 mL/min) and forming the target layer. The EUV intensity increased immediately with slow rotation (>3 rpm) and appeared to be almost independent of the rotation speed. In Fig. 5(a), we see that the maximum CE per solid angle was for an optimized laser intensity of 1×10^{10} W/cm^2 (LP = -10 mm) during rotation. The EUV angular distribution could be expressed by a fitting curve of $(\cos\theta)^{0.38}$, and taking into account this distribution, we obtained the maximum spatially integrated CE of 0.9% at 13.5 nm with 2% bandwidth. EUV spectra at laser intensity of 1×10^{10} W/cm^2 are shown in Fig. 5(b). It is obvious that the emission of the 13.5 nm band was greater in the case of rotation than it was in the case of rest.

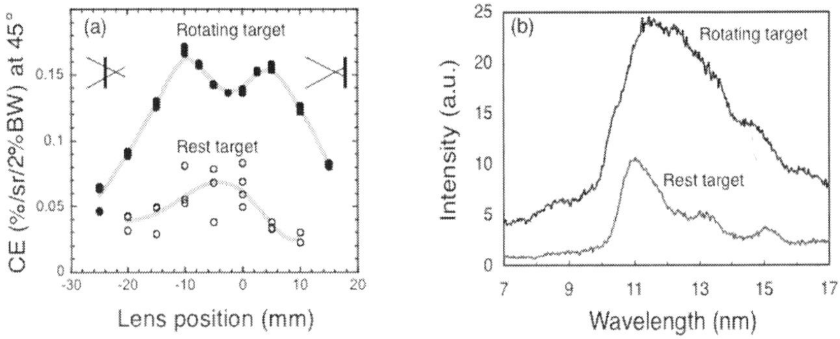

Fig. 5. (a) CE at the wavelength of 13.5 nm with 2% bandwidth as a function of LP under the rotation (130 rpm) and at-rest (0 rpm) conditions. The laser energy was 0.8 J. Insets show the laser beam focusing on the target. (b) Spectra of EUV radiation from the cryogenic Xe drum targets under the rotation (bold line) and at-rest (narrow line) conditions with laser intensity of 1×10^{10} W/cm^2 for LP of -10 mm.

We considered the mechanism for the increase in EUV intensity with rotation of the target. Figure 6 shows photographs of the visible emission from the Xe target observed from a transverse direction. It shows an obvious expansion of the emitting area with longer (optically thicker) plasma in the rotating case compared with the at-rest case. These images indicate the existence of any gas on the target surface. Under the rotation condition, Xe gas is supplied continuously to grow the target layer and the wipers form the layer. However, the wipers are not chilled especially, and the temperature of the target surface might increase owing to contact with the wipers in the rotating case so that the vapor pressure increases. Therefore, the vaporized Xe gas from the target surface was considered as the gas on the target. Although additional Xe gas was added from outside the vacuum chamber, the EUV intensity did not increase and in fact decreased owing to gas absorption. Therefore, it is supposed that Xe gas with adequate pressure localizes only near the target surface. From these results, we conclude that Xe gas on the target surface in the rotating drum produces optically thick plasma that has optimized density and temperature for emitting EUV radiation, and satellite lines of the plasma contribute effectively to increasing the EUV intensity (Sasaki et al., 2004).

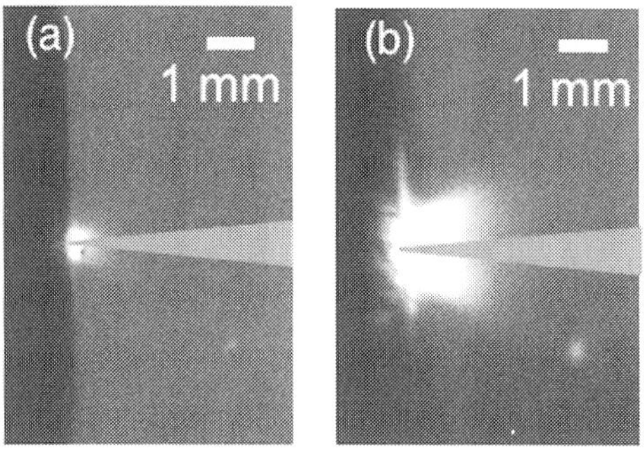

Fig. 6. Images of visible emissions from the plasma on the resting (a) and rotating (b) targets.

Next, the dependence of the laser pulse energy was investigated. We measured the CE as a function of laser energy at different LPs in the rotating drum. For laser energies exceeding 0.3 J, a CE of nearly 0.9% was achieved by tuning the LP with the laser intensity optimized as ~10^{10} W/cm^2. In the energy range, the maximum CE did not depend on the laser energy. At the LP in this experiment, the spot size on the target was larger than 500 μm and plasma energy loss at the edges could be ignored for this large spot. Therefore, the same CE was achieved at the same laser intensity. However, in the lower energy region, the spot size must be small to achieve optimal laser intensity, and edge loss due to three-dimensional expansion in plasma cannot then be ignored and a decrease in the CE was observed. Therefore, it is concluded that laser energy must exceed 0.3 J to achieve a high CE.

The dependence of the laser wavelength was also investigated. Additionally, we carried out 1 ω double-pulse irradiation experiments in which a pre-pulse produces plasma with optimal density and temperature, and after a time delay, a main laser pulse effectively injects emission energy into the expanded plasma to increase the CE. Under the rest condition, there were increases in CE for the shorter laser or the double pulse irradiation (Miyamoto et al., 2005, 2006). In both cases, the long-scale plasmas and their emission spectra were observed to be similar to those under the rotation condition for 1 ω single-pulse irradiation. Therefore, we supposed that in the both cases, the CE was increased by the same mechanism described above. However, when the shorter pulses or the double pulses were emitted under the rotating condition, the CE did not increase but decreased. It is considered that the opacity of the plasma was too great in these experiments and the best condition was not achieved.

In conclusion, the maximum CE was found to be 0.9% at 13.5 nm with 2% bandwidth for the optimal condition.

6. Xe plasma debris

In this section, we report the characteristics of the plasma debris that damages mirrors (Amano et al., 2010b). First, we investigated fast ions, fast neutrals and ice fragments, which constitute the debris.

When we found that EUV radiation was greater for a rotating drum than for a drum at rest, we also found that the number of fast ions decreased simultaneously. Figure 7(a) shows ion signals from a charge collector (CC) with laser pulse energy of 0.5 J and optimal intensity of 10^{10} W/cm^2, for different drum rotation speeds. The ion signal reduces rapidly after the drum starts to rotate (> 4 rpm), after which the signal is almost independent of rotation speed. Ion energy spectra were obtained as shown in Fig. 7(b) using the time-of-flight signals shown in Fig. 7(a). Here, we assume that all ions were doubly charged because we measured the principle charge state of Xe ions to be two with an electrostatic energy analyzer (Inoue et al., 2005). Under the rotation condition, the maximum ion energy decreases to 6 keV and the number of high-energy ions (with energy of a few dozen kilo-electron-volts) also decreases. These are favorable characteristics for the debris problem. The decrease in the ion count under the rotation condition can be explained by a *gas curtain effect* that originates from the Xe gas localized at the target surface. The pressure of this localized Xe gas can be roughly estimated from the peak attenuation (η) in Fig. 7(a); we estimated the product of pressure and thickness to be about 10 Pa·mm.

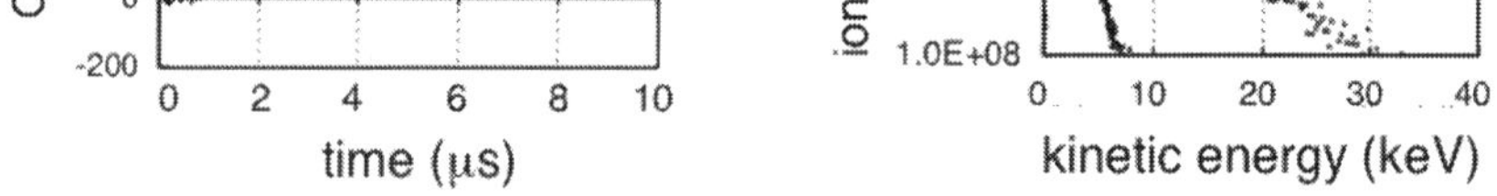

Fig. 7. (a) CC signals of ions and (b) their energy spectra at rotation speeds of 0, 4, 10, 60 and 130 rpm. η in (a) is the loss rate of ions due to the drum rotating. The ion number in (b) was calculated assuming the charge state was two.

Fast neutral particles were measured by the microchannel plate (MCP) detector when the number of fast ions decreased under the rotation condition. The MCP is sensitive to both ions and neutrals, making the use an electric field obligatory to repel ions so that the MCP detects only neutral particles. From the measurement, we found the number of neutrals to be approximately an order of magnitude less than the number of ions.

In the case of solid Xe targets, ice fragments might be produced by shock waves of laser irradiation, whereas this is not the case for gas or liquid targets. In early experiments using a solid Xe pellet, ice fragments were observed and mirror damage due to these fragments was indicated (Kubiak et al., 1995). Since these reports, liquid Xe targets have been preferred over solid Xe targets, with the exception of our group. It is therefore necessary to clarify characteristics of fragment debris from a solid Xe target on a rotating cryogenic drum. After exposing a Si sample to the Xe plasmas pumped by 100 laser pulses, we observed fragment impact damage on its surface using a scanning electron microscope. We observed damage spots on the samples at laser energy of 0.8 J irrespective of whether the drum rotates. Conversely, we did not observe spots at laser energy of 0.3 J. To explain these results, we consider that the fragment speed (kinetic energy) might drop below a damage threshold upon reducing the laser pulse energy because the fragment speed is a function of incident laser energy (Mochizuki et al., 2001). Observing the damage spots, we know that the fragment size was larger than a few microns, and the gas curtain might not be effective for such large fragments. This would explain why the fragment impact damage was independent of the state of drum rotation. From these results, we conclude that fragment impact damage, which occurs especially for the solid Xe target, can be avoided simply by reducing the incident laser pulse energy to less than 0.3 J.

The laser pulse energy was set to 0.3 J to avoid fragment impact damage and the laser repetition rate was 320 pps, giving an average power of 100 W. Next, we investigated damage to a Mo/Si mirror, which was the result of total plasma debris (mainly fast ions) from the laser multi-shots experiments. After 10 min plasma exposure, the sputtered depth was measured to be 50 nm on the surface of a Mo/Si mirror placed 100 mm from the plasma at a 22.5-degree angle to the incident laser beam. Because a typical Mo/Si mirror has 40 layer pairs and the thickness of one pair is approximately 6.6 nm, all layers will be removed within an hour by the sputtering. Although Xe is a deposition-free target, sputtering by debris needs to be mitigated. However, the major plasma debris component is ions, and we believe their mitigation to be simple compared with the case of a metal target such as Sn, using magnetic/electric fields and/or gas. We are now studying debris mitigation by Ar buffer gas. Ar gas was chosen because of its higher stopping power for Xe ions and lower absorption of EUV light, and its easy handling and low cost. After the vacuum chamber was filled with Ar gas, total erosion rates were measured using a gold-coated quartz crystal microbalance sensor placed 77 mm from the plasma at a 45-degree angle, and simultaneously, EUV losses were monitored by an EUV detector placed 200 mm from the plasma at a 22.5-degree angle. Figure 8 shows the erosion rates as a function of Ar gas pressure. The rates were normalized by the erosion N_0 at a pressure of 0 Pa. When the Ar pressure was 8 Pa, we found the erosion rate was 1/18 of that without the gas, but the absorption loss for EUV light was only 8%. The erosion rates (N/N_0) in Fig. 8 can be fitted to an exponential curve:

$$N\left(P_{Ar}\right) = N_0 \cdot \exp\left(-\frac{P_{Ar}}{kT}\sigma l\right) \tag{3}$$

where P_{Ar} is the Ar pressure, k is the Boltzmann constant, T is the gas temperature, σ is the cross section and l is the debris flight length. From this fitting, we obtain $\sigma = 2.0 \times 10^{-20}$ m^2. The Ar buffer gas successfully mitigated the effect of plasma debris with little EUV attenuation. Increasing the Ar pressure, mirror erosion decreases but EUV attenuation increases. Compromising the erosion and EUV attenuation, an optimized pressure is achieved. We should localize the higher density Ar gas to only the debris path so that EUV attenuation is as small as possible. We can design the optimized pressure condition using the σ value obtained and we consider the use of an Ar gas jet. Through this mitigation, we expect that erosion will be reduced by more than two orders of magnitude and the lifetime of the mirror will be extended. We believe the debris problem for Xe plasma will thus be solved.

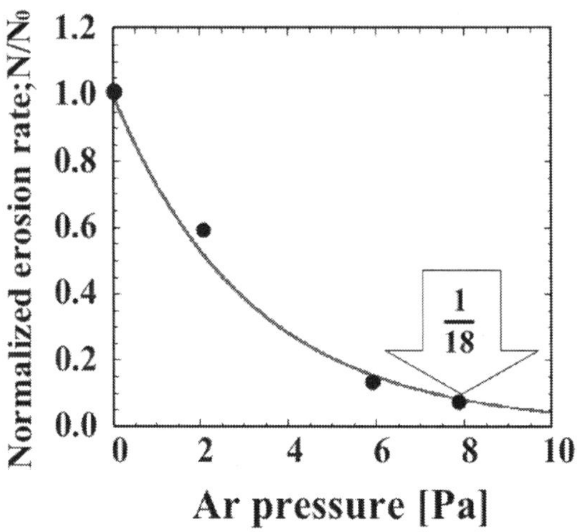

Fig. 8. Normalized erosion rate as a function of Ar pressure. The laser energy was 0.3 J and the rotation speed was 130 rpm.

7. EUV emission at 5-17nm

We began developing the LPP source for EUVL and characterized it at 13.5 nm with 2% bandwidth, but Xe plasma emission has originally a broad continuous spectrum as shown in Fig. 9. If the broad emission is used, our source will be very efficient, not limiting its applications to EUVL. We characterized the source again in the wavelength range of 5–17 nm. Figure 10 shows the CE at 5–17 nm as a function of LP (laser intensity) with laser energy of 0.8 J. The maximum spatially integrated CE at 5–17 nm was 30% for optimal laser intensity of 1×10^{10} W/cm^2. The maximum CE depended on the laser energy and was 21% at 0.3 J. Therefore, high average power of 20 W at 5–17 nm has been achieved for pumping by the slab laser with 100 W (0.3 J at 320 pps). We consider this a powerful and useful source.

Recently, new lithography using La/B_4C mirrors having a reflectivity peak at 6.7 nm was proposed as a next-generation candidate following EUVL using Mo/Si mirrors having a reflectivity peak at 13.5 nm (Benschop, 2009). This means that a light source emitting around 6 nm will be required in a future lithograph for industrial mass production of semiconductors. Because our source emits broadly at 5–17 nm as mentioned above, it can obviously be such a 6 nm light source. We thus next characterized it as a source emitting at 6.7 nm. Here we did not carry out new experiments to optimize the plasma for emitting at 6.7 nm but looked for indications of strong emission at 6.7 nm from the spectrum data already acquired. When making efforts to improve the CE at 13.5 nm, we noticed that emissions around 6 nm became strong at higher laser intensity. When laser energy is 0.8 J and LP = 0 mm (i.e., laser intensity is 4×10^{12} W/cm^2 under the rotation condition), there is a hump around 6 nm as shown in Fig. 9. The spatially integrated CE at 6.7 nm with 0.6% bandwidth is estimated to be 0.1% from this spectrum. Because the bandwidth of 0.6% for the La/B_4C mirror reflectivity is narrower than the 2% for the Mo/Si mirror, the available reflected power is intrinsically small. The CE of 0.1% was not obtained under optimized conditions and higher CE may be achieved in the future. In any event, our source is only one LPP source at present that can generate continuously an emission at 6.7 nm.

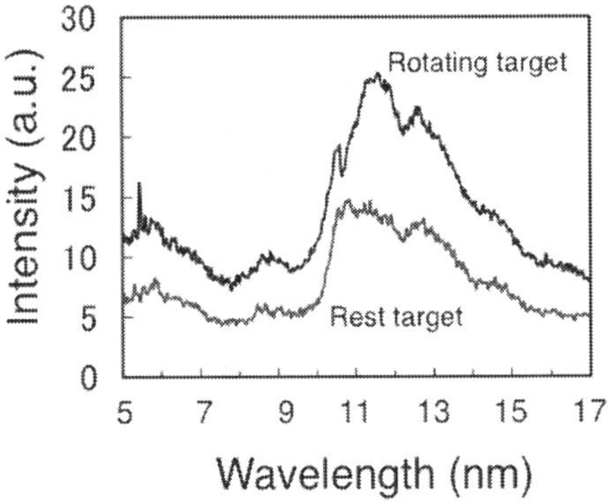

Fig. 9. Spectra of EUV radiation under the rotation (bold line) and at-rest (narrow line) conditions with laser intensity of 4×10^{12}W/cm^2 for best focus (LP = 0 mm). The laser energy was 0.8 J.

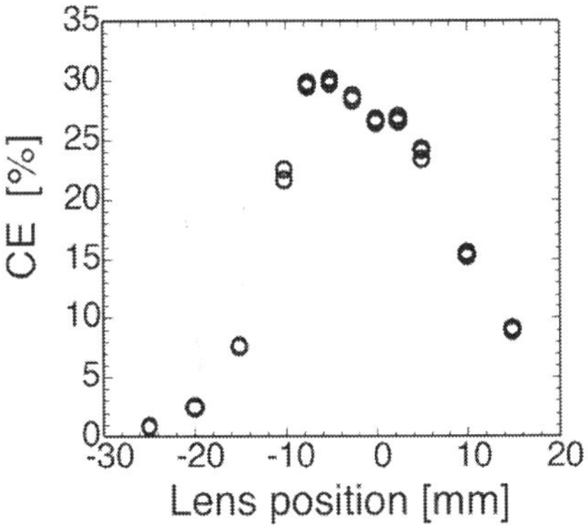

Fig. 10. CE for a wavelength of 5–17 nm as a function of LP under the rotation (130 rpm) condition. The laser energy was 0.8 J.

8. Conclusion

This chapter briefly reviewed our LPP-EUV source. First, we characterized the source at a wavelength of 13.5 nm with 2% bandwidth as an EUVL source and achieved a maximum CE of 0.9%. When the driving laser power is 110 W at 320 pps, the average power of 1 W is obtained at the wavelength and this is thought to be sufficient for the source to be used in various studies. However, the EUV power required for industrial semiconductor products is more than 100 W at present; our power is two orders of magnitude less. To approach the requirements of an industrial EUV source, the remaining tasks are considered. The majority of Xe plasma debris is fast ions, which can be mitigated using gas and/or a magnetic/electric field relatively easily. The drum system can supply the Xe target for laser pulses with energy up to 1 J at 10 kHz. Therefore, a remaining task is powering up the driving laser. A short pulse laser with average power of the order of 10 kW (i.e., *high average and high peak brightness laser*) must be developed and such a breakthrough is much hoped for.

Not limiting the wavelength to 13.5 nm with 2% bandwidth and using the broad emission at 5-17 nm, a maximum CE of 30% is achieved. Pumping with laser power of 100 W, high average power of 20 W is already obtained and the source is useful for applications other than industrial EUVL using Mo/Si mirrors. We are now applying our source to microprocessing and/or material surface modification. Our source also emits around the wavelength of 6 nm considered desirable for the next lithography source. In conclusion, our LPP source is a practicable continuous EUV source having possibilities for various applications.

9. Acknowledgment

Part of this work was performed under the auspices of MEXT (Ministry of Education, Culture, Sports, Science and Technology, Japan) under the contract subject "Leading Project for EUV lithography source development".

10. References

Amano, S., Shimoura, A., Miyamoto, S. & Mochizuki, T. (1997). High-repetition-rate pulse Nd:YAG slab laser for x-ray source by cryogenic target, *1997 OSA Technical Digest Series, Vol.11, Conference Edition, CLEO97,* p.523, Baltimore, USA, May 18-23, 1997

Amano, S., Shimoura, A., Miyamoto, S. & Mochizuki, T. (1999). Development of a high repetition rate Nd:YAG slab laser and soft X-ray generation by X-ray cryogenic target. *Fusion Eng. and Design* 44, pp.423-426

Amano, S. & Mochizuki, T. (2001). High average and high peak brightness slab laser. *IEEE J. Quantum Electron.* 37(2), pp.296-303

Amano, S., Nagano, A.: Inoue, T., Miyamoto, S. & Mochizuki, T. (2008). EUV light sources by laser-produced plasmas using cryogenic Xe and Li targets. *Rev. Laser. Eng.* 36(11), pp.715-720 (in Japanese)

Amano, S., Masuda, K., Shimoura, A., Miyamoto, S. & Mochizuki, T. (2010a). Characterization of a laser-plasma extreme ultraviolet source using a rotating cryogenic Xe target. *Appl. Phys.B* 101, pp.213-219

Amano, S., Inaoka, Y., Hiraishi, H., Miyamoto, S. & Mochizuki, T. (2010b). Laser-plasma debris from a rotating cryogenic-solid-Xe target. *Rev. Sci. Instrum.* 81, pp. 023104-1-023104-6

Anderson, R.J., Buchenauer, D.A., Klebanoff, L., Wood II, O.R. & Edwards, N.V. (2004). The erosion of materials exposed to a laser-pulsed plasma (LPP) extreme ultraviolet (EUV) illumination source, *Proceedings of SPIE, Emerging Lithographic Technologies VIII*, vol.5374, pp.710-719, Santa Clara, USA, February 2004

Bakshi,V.(Ed). (2005) *EUV Sources for Lithography*, SPIE, ISBN:0819458457, Bellingham, WA

Benschop, J. (2009). EUV: past, present and prospects, Keynote I, *2009 International Symposium on Extreme Ultraviolet Lithography*, Prague Czech, October 2009, International Sematech, Available from: <http://www.sematech.org/meetings/archives/litho/index.htm>

Brandt, D.C., Fomenkov, I.V., Partlo, W.N., Myers, D.W., Kwestens, P., Ershov, A.I., Bowering, N.R., Baumgart, P., Bykanov, A.N., Vaschenko, G.O., Khodykin, O.V., Srivastava, S., Hou, R., Dea, S.D., Ahmad, I. & Rajyuguru, C. (2010). LPP EUV source production for HVM, SO-01, *2010 International Symposium on Extreme Ultraviolet Lithography*, Kobe Japan, October 2010, International Sematech, Available from: <http://www.sematech.org/meetings/archives/litho/index.htm>

Fiedorowicz, H., Bartnik, A., Szczurek, M., Daido, H., Sakaya. N., Kmetik, V., Kato, Y., Suzuki, M., Matsumura, M., Yajima, J., Nakayma, T. & Wilhein, T. (1999). Investigation of soft X-ray emission from a gas puff target irradiated with a Nd:YAG laser, *Opt.Comm.*,163(1-3), pp.103-114

Fukugaki, K., Amano, S., Shimoura, A., Inoue, T., Miyamoto, S. & Mochizuki, T. (2006). Rotating cryogenic drum supplying Xe target to generate extreme ultraviolet radiation. *Rev. Sci. Instrum.* 77, pp.063114-1-063114-4

Hansson, B.A.M., Hemberg, O., Hertz, M.H., Berglund, M., Choi, H. J., Jacobsson, B., Janin, E., Mosesson, S., Rymell, L., Thoresen, J. & Wilner, M. (2004). Characterization of a liquid-xenon-jet laser-plasma extreme-ultraviolet source. *Rev.Sci.Instrum.* 75(6), pp.2122-2129

Inoue, T., Kaku, K., Shimoura, A., Nica, P.E., Sekioka, T., Amano, S., Miyamoto, S. & Mochizuki, T. (2005). Studies on laser-produced plasma EUV generation by using fast-supplyunig cryogenic Xe targets, 1-SO-10, *2005 International Symposium on Extreme Ultraviolet Lithography*, San Diego, USA, November 2005, International Sematech, Available from: <http://www.sematech.org/meetings/archives/litho/index.htm>

Inoue, T., Amano, S., Miyamoto, S. & Mochizuki, T. (2006). The stability of a rotating-drum solid-Xe target subjected to high-repetition rate laser irradiation for laser-plasma EUV generation. *Rev. Laser. Eng.* 34(8), pp.570-574 (in Japanese)

Inoue, T., Okino, H., Nica, P.E., Amano, S., Miyamoto, S. & Mochizuki, T. (2007). Xe capillary target for laser-plasma extreme ultraviolet source. *Rev. Sci. Instrum.* 78, pp.105105-1-105105-5

Kinoshita, H., Kurihara, K., Ishii,Y. & Torii, Y. (1989). Soft x-ray reduction lithography using multilayer mirrors. *J. Vac. Sci. Technol.B*, 7(6), pp.1648-1651

Kubiak, G., Krentz. K., Berger, K., Trucano, T., Fisher, P. & Gouge, M. (1995). Cryogenic pellet laser plasma source targets, *OSA Proceedings on Soft X-ray Projection Lithography*, vol.23, pp.248-254, Monterey, USA, September 1994

Kubiak, G., Bernardez, L.J., Krenz, K.D., O'Connell, D.J., Gutowski, R. & Todd, A.M., (1996). Debris-free EUVL sources based on gas jets, *OSA TOPS on Extreme Ultraviolet Lithography*, vol.4, pp.66-71

Miyamoto, S., Shimoura, A., Amano, S., Fukugaki, K., Kinugasa, H., Inoue, T. & Mochizuki, T. (2005). Laser wavelength and spot diameter dependence of extreme ultraviolet conversion efficiency in ω, 2ω, and 3ω Nd:YAG laser-produced plasmas. *Appl. Phys. Lett.* 86(26), pp.261502-1-261502-3

Miyamoto, S., Amano, S.: Inoue, T., Nica, P. E., Shimoura, A., Kaku, K., Sekioka, T. & Mochizuki, T. (2006). EUV source developments on laser-produced plasmas using cryogenic Xe and Lithium new scheme target, *Proceedings of SPIE, Emerging Lithographic Technologies X*, vol.6151, pp.61513S-1-61513S-10, San Jose, USA, February 2006

Mizoguti, H., Abe, T., Watanabe, Y., Ishihara, T., Ohta, T., Hori, T., Kurosu, A., Komori, H., Kakizaki, K., Sumitani, A., Wakabayashi, O., Nakarai, H., Fujimoto, J. & Endo, A. (2010). 1st generation laser-produced plasma 100W source system for HVM EUV lithography, SO-03, *2010 International Symposium on Extreme Ultraviolet Lithography*, Kobe Japan, October 2010, International Sematech, Available from: <http://www.sematech.org/meetings/archives/litho/index.htm>

Mochizuki, T., Shimoura, A., Amano, S. & Miyamoto, S. (2001). Compact high-average-power X-ray source by cryogenic target, *Proceedings of SPIE, Applications of X Rays Generated from Lasers and Other Bright Sources II*, vol.4504, pp.87-96, San Diego, USA, July 2001

Sasaki, A., Nishihara, K., Murakami, M., Koike, F., Kagawa, T., Nishikawa, T., Fujima, K., Kawamura, T. & Furukawa, H. (2004). Effect of the satellite lines and opacity on the extreme ultraviolet emission from high-density Xe plasmas. *Appl. Phys. Lett.* 85(24), pp.5857-5859

Shimoura, A., Amano, S., Miyamoto, S. & Mochizuki, T. (1998). X-ray generation in cryogenic targets irradiated by 1 μm pulse laser. *Appl. Phys. Lett.* 72(2), pp.164-166

Soumagne, G.: Abe, T., Suganuma,T. , Imai, Y., Someya, H., Hoshino, H., Nakano, M., Komori, H., Takabayashi, Y., Ariga, T., Ueno, Y., Wada, Y., Endo, A & Toyoda, K.(2005). Laser-produced-plasma light source for EUV lithography, *Proceedings of SPIE, Emerging Lithographic Technologies IX*, vol.5751, pp.822-828, San Jose, USA, March 2005

19

Irradiation Effects on EUV Nanolithography Collector Mirrors

J.P. Allain
Purdue University
United States of America

1. Introduction

Exposure of collector mirrors facing the hot, dense pinch plasma in plasma-based EUV light sources to debris (fast ions, neutrals, off-band radiation, droplets) remains one of the highest critical issues of source component lifetime and commercial feasibility of nanolithography at 13.5-nm. Typical radiators used at 13.5-nm include Xe, Li and Sn. Fast particles emerging from the pinch region of the lamp are known to induce serious damage to nearby collector mirrors. Candidate collector configurations include either multi-layer mirrors (MLM) or single-layer mirrors (SLM) used at grazing incidence. Due to the strong absorbance of 13.5-nm light only reflective optics rather than refractive optics can work in addition to the need for ultra-high vaccum conditions for its transport.

This chapter presents an overview of particle-induced damage and elucidates the underlying mechanisms that hinder collector mirror performance at 13.5-nm facing high-density pinch plasma. Results include recent work in a state-of-the-art in-situ EUV reflectometry system that measures real time relative EUV reflectivity (15-degree incidence and 13.5-nm) variation during exposure to simulated debris sources such as fast ions, thermal atoms, and UV radiation (Allain et al., 2008, 2010). Intense EUV light and off-band radiation is also known to contribute to mirror damage. For example off-band radiation can couple to the mirror and induce heating affecting the mirror's surface properties. In addition, intense EUV light can partially photoionize background gas used for mitigation in the source device. This can lead to local weakly ionized plasma creating a sheath and accelerating charged gas particles to the mirror surface inducing sputtering. In this overview we will also summarize studies of thermal and energetic particle exposure on collector mirrors as a function of temperature simulating the effects induced by intense off-band and EUV radiation found in EUVL sources. Measurements include variation of EUV reflectivity with mirror damage and in-situ surface chemistry evolution.

In this chapter the details from the EUV radiation source to the collector mirror are linked in the context of mirror damage and performance (as illustrated in Figure 1). The first section summarizes EUV radiation sources and their performance requirements for high-volume manufacturing. The section compares differences between conventional discharge plasma produced (DPP) versus laser plasma produced (LPP) EUV light sources and their possible combinations. The section covers the important subject of high-density transient plasmas and their interaction with material components. The different types of EUV radiators, debris distribution, and mitigation sources are outlined. The second section summarizes the various optical collector mirror geometries used for EUV lithography. A brief discussion on the intrinsic damage mechanisms linked to their geometry is included. The third section summarizes in general irradiation-driven mechanisms as background for the reader and its relation to the "quiescent" plasma collector mirrors are exposed in EUV sources. This includes irradiation-driven nanostructures, sputtering, ion mixing, surface diffusion, and ion-induced surface chemistry. The fourth section briefly discusses EUV radiation-driven plasmas as another source of damage to the mirror. These plasmas are a result of using gases for debris mitigation. The fifth section is a thorough coverage of the key irradiation-driven damage to optical collector mirrors and their performance limitations as illustrated in part by Figure 1.

2. EUV radiation sources

There are numerous sources designed to generate light at the extreme ultraviolet line of 13.5-nm. Historically advanced lithography has considered wavelength ranges from hard X-rays up to 157 nm [Bakshi, 2009]. Radiators of 13.5-nm light rely on high-density plasma generation typically based on discharge-produced configurations with magnetically confined high-density plasmas or laser-produced plasmas. Recently, some sources have combined both techniques (Banine 2011). Generation of high-density plasmas to yield temperatures of the order of 10-50 eV require advanced materials for plasma-facing components in these extreme environments in particular discharge-produced plasma (DPP) configurations. This is due to the need of metallic anode/cathode components operating under high-heat flux conditions. Laser-produced plasmas (LPP) benefits from the fact that no nearby electrodes are necessary to induce the plasma discharge. Further details will be described in section 5.1. One challenge in operating EUV lamps at high power is the collected efficiency of photons at the desired exposure wavelength of 13.5-nm. This particular line has a number of radiators with properties that have consequences on EUV source operation. For example radiators at 13.5-nm include xenon, tin and lithium. The latter two are metals and thus their operation complicated by contamination issues on nearby material components such as electrodes and collector mirrors. Further discussion follows in section 2.2 and 2.3. To contend with the various types of debris that are generated in the plasma-producing volume a variety of novel debris mitigation systems (DMS) have been designed and developed for both DPP and LPP configurations.

2.1 Function and material components

The transient nature of the high-density plasma environment in DPP and LPP systems results in exposure of plasma-facing components to extreme conditions (e.g. high plasma density (~ 10^{19} cm^{-3}) and temperature (~ 20-40 eV). However, in LPP systems since the configuration is mostly limited by the mass of the radiator and the laser energy supplied to it to generate highly ionized plasma with the desired 13.5-nm light. Both configurations rely on efficient radiators of 13.5-nm light, which include: Li, Sn and Xe. In DPP designs a variety of configurations have been used that include: dense plasma focus, capillary Z-pinch, star pinch, theta pinch and hollow cathode among others. For a more formal description of these high-density plasma sources for 13.5-nm light generation the author refers to the recent publications by V. Bakshi in 2006 and 2009 (Bakshi, 2006; Bakshi, 2009).

The in-band and off-band radiation generated in these sources is also a critical limitation in operation of these lamps since on average the off-band radiation is converted into heat on nearby plasma-facing components. There are additional challenges in the design of 13.5-nm light sources that include: high-frequency operation limits driven by the need to extract high EUV power at the intermediate focus (IF) and limited by the available high-throughput power of the plasma device (e.g. laser system or discharge electrode system). Additionally, the scaling of debris with EUV power extraction and the limitation of conversion efficiency (CE) with source plasma size also translate into significant engineering challenges to the design of 13.5-nm lithography source design. Figure 1 illustrates, for the case of the DPP configuration, the primary debris-generating sources that compromise 13.5-nm collector mirrors. The first region depicted on the left is defined here as the "transient plasma region". This is the region described earlier with high-density and high-temperature plasma interacting with the electrode surfaces.

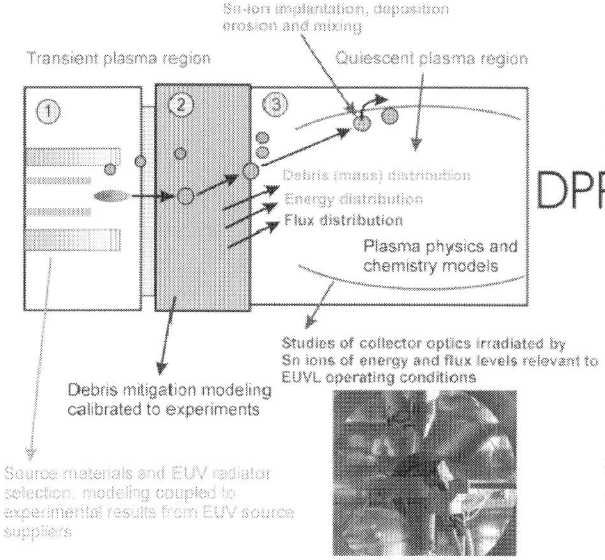

Fig. 1. Illustration of the various components of EUV 13.5-nm radiation source configuration consisting primarily of three major components: 1) plasma radiator section, 2) debris mitigation system and 3) optical collector mirror.

In DPP discharge sources material components that make up the electrode system consist of high-temperature, high-toughness materials. Although DPP source design has traditionally used high-strength materials such as tungsten and molybdenum alloys, the extreme conditions in these systems limit the operational lifetime of the electrode. Significant plasma-induced damage is found in the electrode surfaces, which induce degradation and abrasion over time. Figure 2, for example, shows a scanning electron micrograph of a tungsten electrode exposed to a dense plasma focus high-intensity plasma discharge. The key feature in the SEM image is the existence of plasma-induced damage domains that effectively have induced melting in certain sections of the electrode surface.

The second region depicted in Figure 1 is defined as the debris mitigation zone (DMZ). In this region a variety of debris mitigation strategies can be used to contend with the large debris that exists in operation of the DPP source. For example the use of inert gas to slow-down energetic particles that are generated in the pinch plasma region and/or debris mitigation shields that collect macro-scale particulates when using Sn-based radiators in DPP devices. Radiation-induced mechanisms on the surfaces of the DMZ elements also can lead to ion-induced sputtering of DM shield material that eventually is deposited in the nearby 13.5-nm collector mirror. Therefore care is taken to select sputter-resistant materials for the DM shields used such as refractory metal alloys and certain stainless steels. Design of DM shields also involve computational modeling that can aid in identifying appropriate materials depending on the source operation and generation of a variety of debris types such as clusters, ions, atoms, X-rays, electrons and macroscopic dust particles.

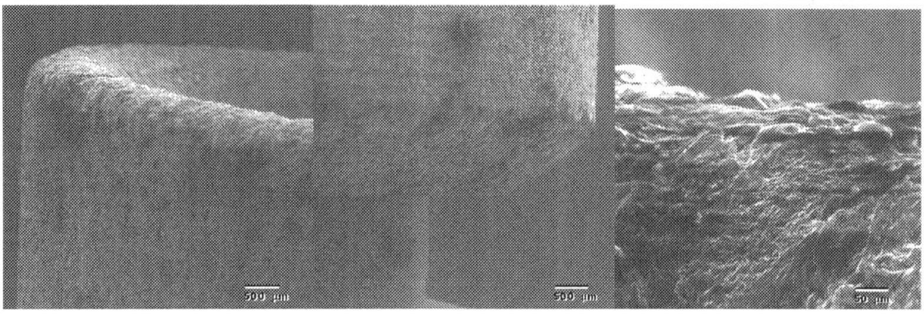

Fig. 2. SEM micrographs of a tungsten electrode exposed to high-intensity plasma during the generation of EUV 13.5-m light.

The third region in Fig. 1 consists of the 13.5-nm light collector mirror. The collector mirror has a configuration to optimally collect as much of the 13.5-nm light as possible. Its function is to deliver EUV power in a specified etendue at the intermediate focus (IF) or the opening of the illuminator. This power is in turn dictated by the specification on EUV exposure of the EUV lithography scanner that must be able to operate with 150-200 wafers per hour (wph) at nominal power for periods of 1-2 years without maintenance (so-called high-volume manufacturing, HVM, conditions). This ultra-stringent requirement is one of the primary challenges to EUV lithography today. Since powers of order 200-300 W at the IF need to be sustained for a year or more, materials at the DPP source and those used for collector mirrors will necessarily require revolutionary advances in materials performance. The third region in Figure 1 also depicts what debris the collector mirror is exposed to during the discharge. A distribution of debris energies (i.e. ions), fluxes and masses will effectively affect the mirror surface performance. The third region is also known as the "condenser or collector optics region".

2.2 Selection of electrode materials in DPP EUV devices

Selection of materials for DPP electrodes depends on the microstructure desired to minimize erosion and maximize thermal conductivity. Figure 3 shows an example of SEM micrographs of materials identified to have promising EUV source electrode properties. The powder composite materials inherited the structural characteristics of the initial powders, determined by the processes of combined restoration of tungsten and nickel oxides (WO_3

and NiO from $NiCO_3$, for instance) and copper molybdate ($MoCuO_4$). Dry hydrogen (the dew point temperature is above 20 ^{0}C) facilitates the formation of the heterogeneous conglomerates in W-Ni-powders, which do not collapse at sintering or saturate the material (Figure 3a), and spheroidizing of molybdenum particles and re-crystallization through the liquid phase in the conditions of sintering the composite consisting of molybdenum and copper (Figure 3b). For comparison, the structure is shown in Figure 3c obtained from tested W-Ni powders. The structure of the materials was studied by means of scanning electron microscopy (SEM) of the secondary electrons. A variety of materials characterization including surface spectroscopy and X-ray based diffraction is used to assess the condition of the materials after processing with sintering-based techniques. The powder composite materials are so-called pseudo alloys, which provide promising high thermal conductivity properties, while displaying sub-unity sputter yields (see Section 4).

Fig. 3. From left to right, (a) the structure of the W-Cu-Ni-LaB6 pseudo alloy (x540), (b) the structure of the Cu-44%Mo - 1%LaB6 pseudo alloy (x2000), and (c) the structure of "irradiated" W-Cu-Ni pseudo alloy produced by class W-Ni powder (x400).

Observations made with secondary mass ion spectrometry (SIMS) on these materials found evidence of hydrogen and beryllium in anode components. Based on these results one can speculate that the hydrogen observed by SIMS after exposing the samples may be caused by that environment, in which the powders are manufactured, sintered, and additionally annealed. In regards to the beryllium observed on the anode surface after exposure to the xenon plasma, one may suppose two possible explanations, each of which requires additional verification. The construction may contain beryllium bronze; or the construction may contain Al_20_3 or BeO based ceramics. Both cases may be the reason for enrichment of the surface samples by these elements during the heating phases.

For systems with the absence of the component interactions, the arc xenon plasma impact to the electrode materials does not cause a noticeable change of durability: for $MoCuLaB_{6:}$ HV = 1600-1690 MPa; and for Cu- Al_2O_3: HV = 660 MPa through the whole height of the anode. In the tungsten and copper based composites, when presence of nickel exists, the mutual dissolution of the elements is increased (W is dissolved in Cu-Ni melt, for instance). At cooling, it may be accompanied by either forming non-equilibrium solid solution, or solidification; which is conformed by the increasing the firmness of the upper part of the anode (3380 MPa compared to 3020 MPa in its lower part). To provide more careful analysis, one should investigate the dependence of electro-conductive composites on heat resistance subject to arc discharges of powerful heat fluxes (up to 10^7 W/m^2). Additional analyses typically conducted include the propagation of cracks, observed on the surface layer of the anode material and deep into the bulk. For that, the precise method of manufacturing is required for further insight on crack development and propagation. These analyses along with erosion material modeling (discussed in Section 4) are mainly used to dictate materials selection for electrode materials in EUV DPP sources.

2.3 EUV radiators, debris generation and debris mitigation systems

One particularly important "coupling" effect between the debris mitigation zone region and the collector optics region is the use of inert mitigation gases (e.g. Ar or He) that in turn are ionized by the expanding radiation field and thus generate low-temperature plasma near the collector mirror surface. This phenomenon is briefly discussed in Section 3. Each candidate radiator (e.g. Li, Sn or Xe or any combination) will result in a variety of irradiation-induced mechanisms at the collector mirror surface. For example, if one optimizes the EUV 13.5-nm light source for Li radiators, the energy, flux and mass distributions will be different compared to Sn. Both of these in turn are also different from the standpoint of contamination given that both are metallic impurities and Xe is an inert gas. The former will lead to deposition of material on the mirror surface. In the case of Xe, thermal deposition would be absent however the energetic Xe implantation on the mirror surface could lead to inert gas damage such as surface blistering and gas bubble production for large doses. Debris mitigation systems would have to be designed according to the radiator used.

3. EUV radiation-driven plasmas

As discussed earlier, Figure 1 shows the general configuration of a DPP system for EUV 13.5-nm light generation. Another "coupling" effect of the DMZ in the source system (e.g. from the electrode materials of the source through the DMZ to the collector mirror) is the fact that the intense EUV and UV radiation generated from the 13.5-nm radiators (e.g. Xe or Sn) can induce a secondary low-temperature plasma at the surface of the collector mirror by ionizing the protective gas used for debris mitigation such as argon or helium [Van der Velden et al, 2006, Van der Velden & Lorenz, 2008]. The characteristic plasma in this region is found to be of low temperature (e.g. 5-10 eV) and moderate densities (e.g. ~ 10^{16} cm^{-3}). The photoionization process can lead to fast electrons that induce a voltage difference the order of 70 V. In addition, due to the sheath region at the plasma-material interface between the plasma and the mirror the ionized gas particles (e.g. Ar^+ or He^+) can be accelerated up to about 50-60 eV. This energy in the case of Ar ions is relatively low and in the so-called sputter threshold regime for bombardment on candidate collector mirror material candidates. In addition, carbon contamination could also be accompanied by this plasma exposure. These candidate materials are typically thin (~20-60 nm) single layers of Ru, Rh or Pd, all of which reflect 13.5-nm light very efficiently. Only few studies have been conducted to elucidate how these low-energy ions may induce changes that can degrade the optical properties of the 13.5-nm collector mirrors. Van der Velden and Allain studied this effect in detail in the *in-situ* experimental facility known as IMPACT to determine the sputter threshold levels at similar energies [Allain et al, 2007]. In the work by van der Velden et al. the threshold sputtering of ruthenium mirror surface films were found to be in close agreement with theoretical models by Sigmund and Bohdansky. The sputter yields varied between 0.01-0.05 atoms/ion for energies about 50-100 eV and models were found to be within 10-15% of these values.

4. Irradiation-driven mechanisms on material surfaces

Before discussion of collector mirror geometry and configuration a brief background on irradiation-driven mechanism on material surfaces is in order. In DPP EUV devices electrodes at the source are exposed to short (10-20 nsec) high-intensity plasmas leading to a variety of erosion mechanisms. Erosion of the electrodes is dictated by the dynamics of the plasma pinch for configurations such as: dense plasma focus, Z-pinch and capillary. The transient discharge deposits 1-2 J/cm^2 per pulse on electrode surfaces. Large heat flux is deposited at corners and edges leading to enhanced erosion. Understanding of how particular materials respond to these conditions is part of rigorous design of DPP electrode systems. Erosion mechanisms can include: physical sputtering, current-induced macroscopic erosion, melt formation, droplet, and particulate ejection [Hassanein et al, 2008]. Erosion at the surface is also governed by the dynamics of how plasma can generate a vapor cloud leading to a self-shielding effect, which results in ultimate protection of the surface bombarded. Determining whether microscopic erosion mechanisms such as: physical sputtering or macroscopic mechanisms such as melt formation and droplet ejection the dominant material loss mechanism remains an open question in DPP electrode design. This is because such mechanisms are inherently dependent on the pinch dynamics and operation of the source. One important consequence of the extreme conditions electrode and collector optics surfaces are exposed is the existence of several irradiation-driven mechanisms that can lead to substantial materials mixing at the plasma-material interface. Bombarment-induced modification of materials can in principle lead to phase transition mechanisms that can substantially change the mechanical properties of the material accelerating degradation.

Conceptually, the phenomenon of bombardment-induced compositional changes is simplest when only athermal processes exist such as: preferential sputtering (PS) and collisional mixing (CM). Preferential sputtering occurs in most multi-component surfaces due to differences in binding energy and kinematic energy transfer to component atoms near the surface. Collisional mixing of elements in multi-component materials is induced by displacement cascades generated in the multi-component surface by bombarding particles/clusters and is described by diffusion-modified models accounting for irradiation damage. Irradiation can accelerate thermodynamic mechanisms such as Gibbsian adsorption or segregation (GA) leading to substantial changes near the surface with spatial scales of the order of the sputter depth (few monolayers). GA occurs due to thermally activated segregation of alloying elements to surfaces and interfaces reducing the free energy of the alloy system. Typically, GA will compete with PS and thus, in the absence of other mechanisms, the surface reaches a steady-state concentration approaching that of the bulk. However when other mechanisms are active, synergistic effects can once again alter the near-surface layer and complex compositions are achieved. These additional mechanisms include: radiation-enhanced diffusion (RED) due to the thermal motion of non-equilibrium point defects produced by bombarding particles near the surface, radiation-induced segregation (RIS), a result of point-defect fluxes, which at sufficiently high temperatures couples defects with a particular alloying element leading to compositional redistribution in irradiated alloys both in the bulk and near-surface regions. Figure 4 shows the temperature regime where these mechanisms are dominant. All of these mechanisms must be taken under account in the design of proposed advanced materials for the electrodes and the collector optics in addition to considering other bombardment-induced conditions (i.e., clusters, HCI, neutrals, redeposited particles, debris, etc...) that can be generated at the 13.5-nm light tool.

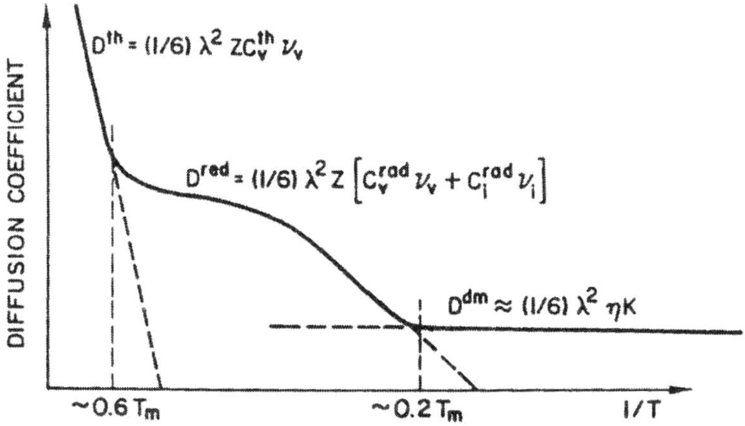

Fig. 4. Schematic plot of the relative importance and temperature dependence of displacement mixing, radiation-enhanced and thermally-activated mechanisms (e.g., Gibbsian segregation).

Modeling of physical sputtering is well known and the field quite mature, see for example work by W. Eckstein (Eckstein 1991) and W. Möller (Möller 1988). For energies above about 100 eV, binary collision approximation (BCA) codes are often used to estimate erosion from various material surfaces. The sputtering yield of 100% Cu from 1 keV Xe^+ bombardment coincides with the experimental result shown for Cu bombardment. Furthermore, example in Figures 5a and 5b, the sputtering from a W-Cu alloy is modeled. The advances in multi-scale and multi-component modeling provided by Monte Carlo damage codes such as TRIM-SP, TRIDYN and ITMC enables scoping studies of candidate materials and their surface response.

An additional mechanism currently missing in plasma-material interaction computational codes is the correlation of surface morphology with surface concentration. Ion-beam sputtering is known to induce morphology evolution on a surface and for multi-component material surfaces plausibly driven by composition-modulated mechanisms [Carter, 2001; Muñoz-Garcia et al., 2009]. Chason et al. have devised both theory and experiments to elucidate on surface patterning due to ion-beam sputtering [Chan & Chason, 2007]. A number of efforts also are attempting to enhance the ability to model ion-irradiation induced morphology and surface chemistry including work by Ghaly and Averback using molecular dynamics and by Heinig et al. using MD coupled KMC (kinetic Monte Carlo) approaches [Ghaly et al, 1999; Heinig et al., 2003]. In spite of these efforts there remains outstanding issues in ion-beam sputtering modification of materials such as the role of mass redistribution that can dominate over surface sputtering mechanisms [Aziz, 2006; Madi et al., 2011]. These developments have important ramifications to the EUV collector mirror operation given the complexity of energetic and thermal particle-surface coupling.

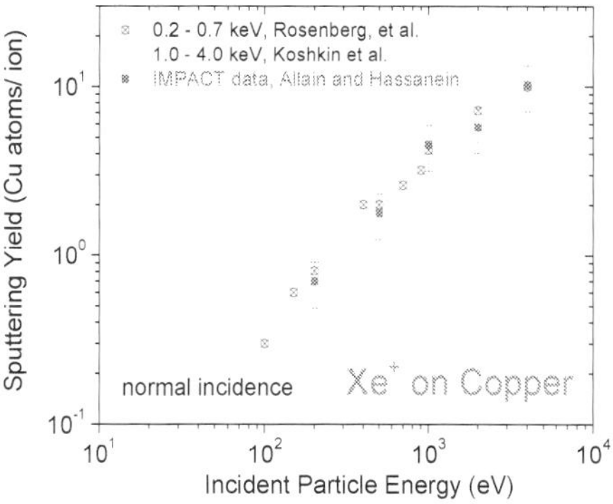

Fig. 5a. Sputtering yield of copper bombarded by singly-charged xenon at normal incidence in the IMPACT (Interaction of Materials with charged Particles And Components Testing) experiment at the Argonne National Laboratory.

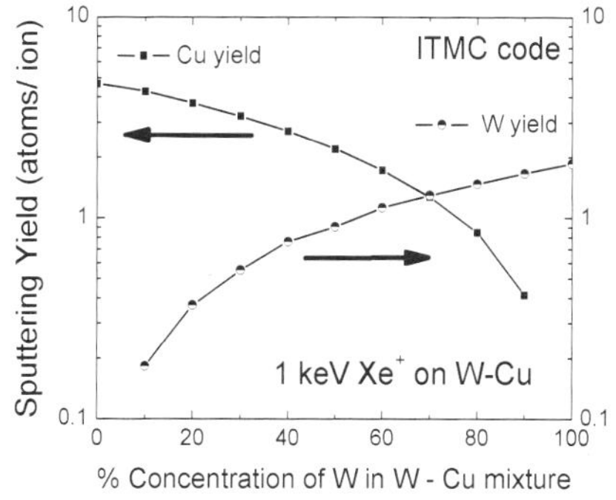

Fig. 5b. The Ion Transport in Materials and Compounds code developed at the Argonne National Laboratory calculates the partial sputtering yield of Cu and W from a W-Cu mixed material bombarded by 1 keV Xe ions at normal incidence. This system is used as a pseudo-alloy with properties able to withstand large heat fluxes in EUV source devices.

5. Collector mirrors for EUV lithography

The nature of the collector mirror damage is largely dictated by the configuration designed to optimize collection of the 13.5-nm light. Due to the refractive index in the X-ray and EUV range being less than unity, total external reflection is possible at angles that are large with respect to the mirror surface plane. If the geometry for collection of the light is such that the mirrors must collect light at more grazing incidence, than the configuration consists of collector mirrors with very thin single-layer coatings of candidate materials such as Ru, Pd or Rh. As discussed earlier the configuration in current EUV source technologies consist of either normal incidence mirrors or grazing incidence mirrors. The latter configuration must use a collection of multiple shell collectors designed to optimize collection of the 13.5-nm light. Media Lario, a lens manufacturer based in Italy, has optimized the multiple shell collector design in recent years.

5.1 Normal incidence mirrors

The normal incidence mirror configuration consists of a multi-layer mirror geometry exposed to 13.5-nm at normal incidence to the mirror surface. Due to the low reflectance fractions at normal incidence 10's of bilayers are stacked on top of each other to improve the reflectivity to the order of 50-60%. The mechanisms of radiation-induced damage depend on the mirror configuration as eluded above. In the case of the multi-layer mirror (MLM) the incident radiation is predominantly at near-normal incidence thus with the highest projected range into the material bulk. Intrinsic in the configuration of MLM collector systems is the inherent energy distribution of energetic particles that emanate from the LPP pinch plasma source. Although it is not a necessary requirement that MLM are used with LPP sources, the limited collection efficiency of grazing incidence mirrors motivate their use. However, in the context of irradiation damage from the nearby plasma MLM systems suffer the greatest losses in optical performance compared to GIM. The reason is two-fold. One the energy distribution from LPP sources tends to be dominant in the keV range of energies typically about 0.5-5-keV. Therefore there is immediate damage and ion-induced mixing at the MLM interfaces critical to the optimum reflectance of these mirrors. The use of Xe or Sn radiators also introduces a second challenge.

5.2 Grazing incidence mirrors

Grazing incidence mirrors are collector mirrors that reflect EUV light at angles that are predominantly inclined along the plane of the mirror surface. Since the collector mirror will have an inherent curvature the incident angle on the surface plane will have a variable incidence angle depending on the sector the light is collected. Furthermore, recent developments in grazing incidence mirror technology (e.g. Media Lario designs) have now optimized grazing incidence mirrors as shells with a hyperbolic, parabolic or elliposoidal geometric curvature that optimizes the light collection. Typically the collector angle is about 5-25 degrees from the surface normal. In the grazing incidence mirror configuration there exists a number of issues in the context of irradiation-induced effects. For example the sputter efficiency of materials increases as the angle of incidence becomes more oblique. Therefore with this configuration there is the concern that the mirror could erode more rapidly. On the other hand, the implanted energetic debris is found closer to the surface, which could in some cases prove to be of benefit. The issue of incidence angle and its impact on both sputtering of the mirror material and the effect on EUV 13.5-nm reflectivity is discussed in later sections. Grazing incidence mirrors also entail only single layer materials in general. This is because the inherent light transport is via reflection and at grazing incidence typically a large fraction (> 60-70%) of the light can be reflected by materials such as: niobium, rhodium, ruthenium and palladium.

6. Irradiation modification of EUV optical properties

During a Sn-based LPP or DPP pinch, metal vapor will expand and reach nearby components including the collector mirror. Sn+ energies ranging from several hundred electron volts up to a few keV can be expected from Sn-based LPP or DPP source configurations and therefore constitute the energy range of interest for EUV collector mirror damage evolution. In the years between 2004 and 2007 Allain et al. conducted a series of pioneering experiments at Argonne National Laboratory. The work included a systematic *in-situ* characterization study in IMPACT of how candidate EUV mirror surfaces evolved under exposure to thermal and energetic Sn. Fig. 6 below depicts the various interactions relevant to the EUV 13.5-nm light source environment with candidate grazing incidence mirror materials: Ru, Pd or Rh. In this section studies on these materials and also candidate multi-layer mirror (MLM) materials are discussed with implications of ion-induced damage.

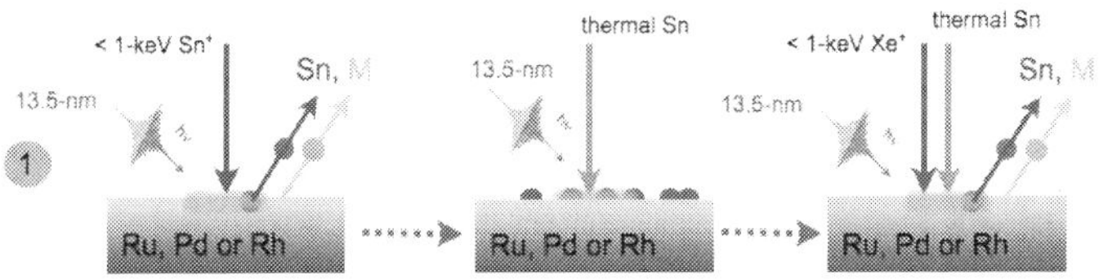

Fig. 6. Schematic of various interactions studied in IMPACT using Sn thermal and energetic particles while *in-situ* characterizing the evolving surface.

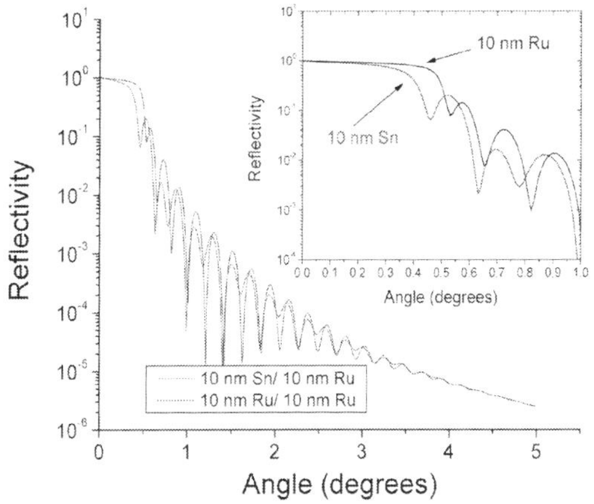

Fig. 7. X-ray reflectivity (8.043 keV X-rays) theoretical response for two different top surfaces: a 10-nm Sn surface on a 10-nm Ru underlayer and a 10-nm Ru surface on a 10-nm Ru underlayer, both with 0.5-nm rms roughness value (CXRO calculations).

The mirror reflectivity response at 13.5-nm light will be sensitive to the thickness of the deposited Sn layer. In addition, the reflectivity response may also be influenced by the structure of the material namely: evaporated porous structure, ion-induced densification phases and possible oxidation effects. All of these can be studied using XRR and in-band EUV reflectivity. When comparing for example a thin Sn layer to a thin Ru layer, theoretically, with enough Sn deposited, the extension of the critical edge will be reduced in the XRR response using CuKα X-rays.

Note the comparison made in Figure 7 showing CuKα (8.043 keV) X-ray reflectivity calculations using CXRO calculations for 10 nm Sn/Ru and 10 nm Ru layers with 0.5 nm rms roughness vs. incident grazing angle [Henke et al, 1993]. In the XRR vs. θ plot, the reflectivity suddenly decreases as θ^{-4} at angles above the critical angle, θ_c, which in this case it is equal to 0.45 degrees and 0.35 degrees for the 10-nm Ru and 10-nm Sn/Ru mirrors, respectively. The presence of the Sn layer effectively reduced the critical edge region and thus its reflectivity performance is reduced. This is because the momentum transfer, Q, is:

$$Q = \left(\frac{4\pi}{\lambda}\right)\sin\theta \tag{1}$$

And the reflectivity response is related by:

$$Q_c^2 = 16\pi\rho_e \propto |R| \tag{2}$$

This reflectivity response can also be assessed for the EUV spectral region (in-band 13.5-nm). Figure 8a shows CXRO calculations of the EUV in-band reflectivity response for same conditions in Figure 7. Note the reduction of the critical edge for the case of Sn deposition with a 10-nm Sn layer on top of a 10 nm Ru SLM.

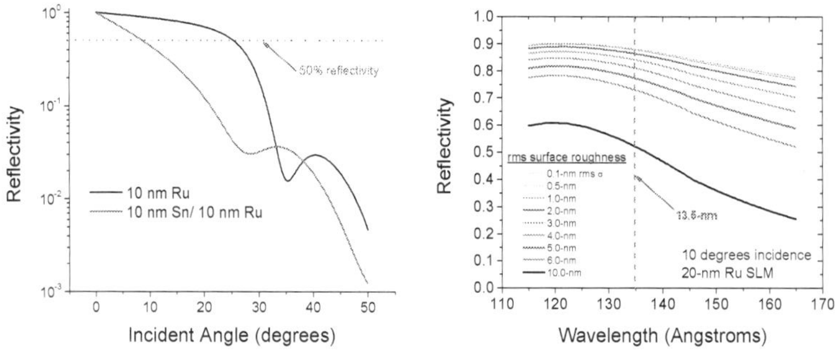

(a) (b)

Fig. 8. (a) In-band EUV (13.5-nm) reflectivity response for a 10-nm Ru mirror and same mirror with a 10-nm Sn cap, and (b) theoretical calculations (CXRO) of in-band EUV reflectivity response versus incident angle at 13.5 nm (92 eV) for Ru and Sn surfaces.

Figure 8b shows the effect that surface roughness (e.g. morphology) can have on the absolute in-band (11-17 nm) EUV reflectivity from a 20-nm mirror Ru film surface. This is a great example of how both multi-component surface concentration (e.g. Sn particles in a Ru mirror surface) can couple with surface morphology evolution during deposition. Both a concentration of Sn and surface roughness can combine to decrease the reflectivity near 13.5-nm. The key question is what is the threshold for damage and can this be mitigated so that in steady-state a tolerable and minimal loss of reflectivity can be managed.

Figure 9 show AES data on a thin Ru-cap MLM before and after exposure to Sn vapor in IMPACT, respectively. In-situ metrology in IMPACT allows us to monitor in real time deposition of Sn on the mirror surface. EUV reflectivity from a MLM is near normal and thus the effect of a thin Sn layer must also be assessed as was done for the grazing incidence mirror data above. Figure 9b shows two major contaminants on the near surface (down to about 50-100 Å), oxygen and nitrogen.

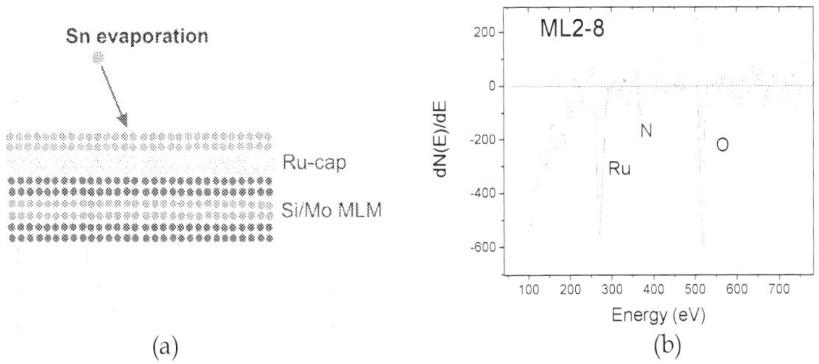

Fig. 9. (a) Schematic of Sn on MLM system. (b) Auger spectra of a thin Ru-cap MLM showing the presence of oxygen on the thin-film Ru cap. This MLM system can reflect up to about 69-72% of EUV light even in the presence of oxygen.

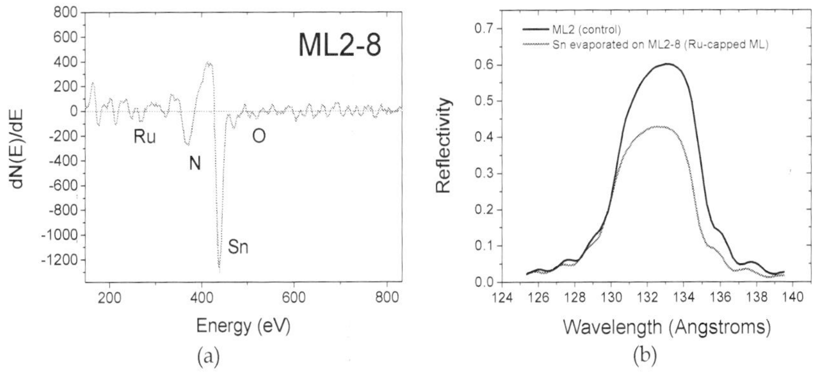

Fig. 10. (a) Auger spectra of a thin Sn layer evaporated on a thin Ru-cap mirror. Note the presence of nitrogen as opposed to oxygen and the strong Sn peak and (b) In-band EUV reflectivity data taken at NIST-SURF facility. Note the noticeable effect on the reflectivity response for the ML2-8 sample.

Oxygen is always found on the surface in the presence of ruthenium due to its high oxygen affinity. When a thin layer of Sn is deposited as shown in Figure 10a, the major contaminant is nitrogen and not oxygen. This is due to tin's high affinity for nitrogen compared to oxygen. Figure 10b shows the effect of an evaporated Sn layer on EUV mirror reflectivity. The EUV in-band reflectivity was measured at the NIST-SURF facility at near-normal incidence. The reduction from about 60% in-band EUV reflectivity to about 40% is consistent with deposition of about a 40-50 Å Sn thin layer. This has been corroborated by calculations on a thick Ru surface layer at near-normal incidence, giving a thickness comparable to about 35 Å.

6.1 Effect of surface roughness on 13.5-nm reflectivity

The effect of the surface evolution (e.g. concentration and morphology) on 13.5-nm reflectivity is a key factor in determining the lifetime of the collector mirror during operation of the high-intensity EUV lamp. A number of *in-situ* characterization studies are conducted to study the evolution of the surface structure, concentration and morphology under relevant EUV light generation conditions. Single-effect studies are presented in this section to illustrate and differentiate effects from the expanding thermal Sn plume and the energetic Sn particles that emanate from the high-density pinch Sn plasma region.

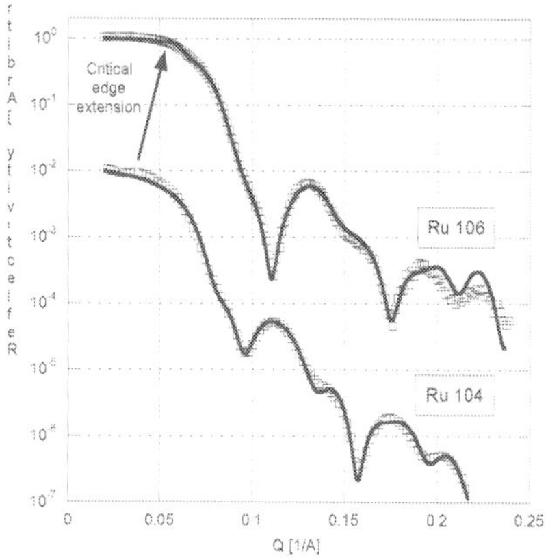

Fig. 11. Fits to Ru 106 and Ru 104. The Ru 104 data set and fit have been shifted downward by a factor of 100 for clarity. The electron density depth profiles from these fits are shown in Figure 12.

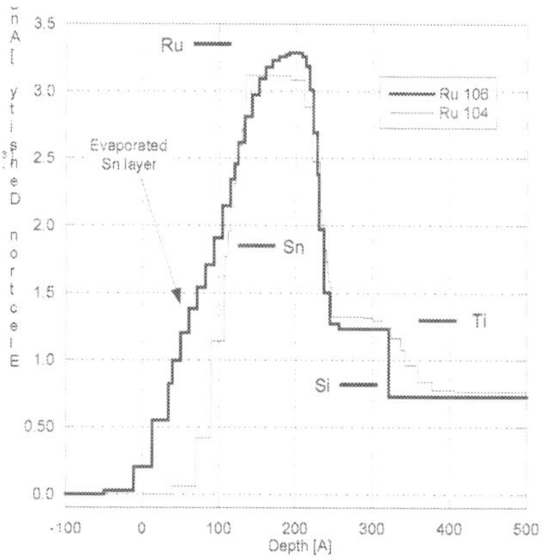

Fig. 12. Electron density profiles for Ru 106 and Ru 104. The presence of a rough Sn layer at the air-film interface of Ru 106 is clear. The bulk density values are shown as horizontal lines.

Fits to Ru 106 (with evaporated Sn layer) and Ru 104 (identical to Ru 106, but without Sn layer) are shown in Figure 11. The electron density depth profiles obtained from these fits are shown in Figure 12. First, the electron density values for Ru 106 and Ru 104 are consistent with the known bulk values. The presence of the Sn layer on Ru 106 is clear. In fact the point at which the profiles for Ru 106 and Ru 104 diverge (near the air-film interface) corresponds to the bulk Sn electron density value. Thus, the Ru 106 data set is consistent with a Sn over-layer approximately 60 Å thick. The air-Sn layer interface is not well defined as determined by the fit of the XRR data. The extension of the critical edge for Ru 106 is evident, an effect due to the Sn layer increasing the total electron inventory of the metal over-layer.

The evaporated Sn layer on this sample is either very rough, has significant internal porosity, or has intermixed with the Ru layer to a large extent. Surface roughness values above 5-nm rms would need to exist to lead to any significant decrease on in-band EUV reflectivity (as shown earlier in Fig. 8b). Significant intermixing is very possible during the low-energy room temperature evaporation. It is possible Sn does not wet Ru adequately and this could lead to a poor surface topography and a rough interface. The blurry Ti-Si interface for the Ru 104 sample probably is not a real effect, but a consequence of an incomplete fit.

The effect of the thin-film Sn layer on in-band (13.5-nm) EUV reflectivity is shown in Figure 13. Measurements were conducted at the NIST-SURF facility. The figure shows two primary cases. One is Ru-104, a virgin 10-nm Ru sample. Both XRR and QCM-DCU (quartz crystal microbalance dual-crystal unit) measurements of this particular batch of Ru SLM measured a Ru film thickness of about 100 Å [Allain et al, 2007]. The EUV in-band 13.5-nm reflectivity data fitted with CXRO calculations is yet a third indication of the Ru thin-film thickness, thus effectively calibrating the QCM-DCU data in *in-situ* characterization. The EUV reflectivity results show that the Ru thin-film thickness is about 90 Å fitting with the CXRO calculations. The sample covered with Sn (Ru-106) is fitted with CXRO calculations using a 2.1-nm Sn surface layer at 20-degrees incidence. This correlates well with estimates from Sn fluences measured in IMPACT giving about a 30-40 Å Sn thin-film layer.

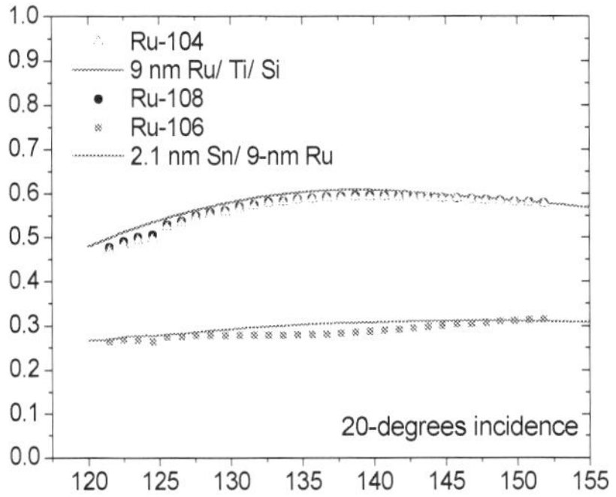

Fig. 13. Two virgin samples, Ru-104 and Ru-108 are shown with their reflectivity response in the EUV in-band 13.5-nm spectral range at 20-degrees with respect to the mirror surface. The reflectivity response of the Sn-covered mirror is also shown.

Fig. 14. SEM image of Rh-313 exposed to similar conditions as sample Ru-106. Therefore Sn coverage is equivalent to about a 2-nm thickness of Sn atoms.

The EUV reflectivity mirror response measured in-situ is correlated to ex-situ surface morphology data using SEM and EDX for electron-based microscopy. Fig. 14 shows SEM data for the case of Rh-313 exposed to 50 nA of Sn evaporation for 15-minutes. The surface morphology is characterized by surface structures that vary in lateral size from 10-100 nm. Observations from BES (backscattering electron spectroscopy) data suggested that the lighter imaged structures correspond to Sn, while darker regions corresponded to Rh. This led to the conclusion that the surface structures are islands of Sn that have coalesced during deposition. The formation of these two-dimensional nanostructures could be associated with diffusion-mediated aggregation of deposited Sn atoms. This is partly due to deposition of tin driving the morphology and structure of the Sn film deposited on the SLM surface far from equilibrium conditions. When one incorporates the kinetic effect of energetic implanted Sn, the net energy available is increased dramatically. This point is further investigated in later sections. The formation and growth of nano-scale tin islands during exposure is a competition between kinetics and thermodynamic equilibrium of deposited Sn atoms on the surface of either of the noble metal used (e.g. Ru or Rh).

The results from a set of thin Ru films exposed to energetic Sn ions are shown in Figure 15. The SLD profiles exhibit the effect of sputter erosion caused by the Sn-ion bombardment. Although the fluence of Ru102 and Ru105 differed by a factor of approximately 20, the profiles are similar. This is probably the effect of greater sputter efficiency for the low fluence Ru 105 case where the ion irradiation angle was 45° instead of normal incidence.

6.2 Effect of fast and thermal particles on MLM reflectivity at 13.5-nm

For MLM systems, Xe^+-bombardment studies in IMPACT demonstrated that the main failure mechanisms were: 1) ion-induced mixing at the interfaces along with significant sputtering of cap material (i.e., Ru) and 2) synergy of energy (1-keV) and high mirror temperature (200° C) leading to mirror reflectivity degradation [Allain et al., 2006]. Therefore, from the point of view of ion-induced damage, MLM systems compared to SLM systems are most susceptible to early failure rates if fast ion and neutral energies are maintained at the 1 keV level or more.

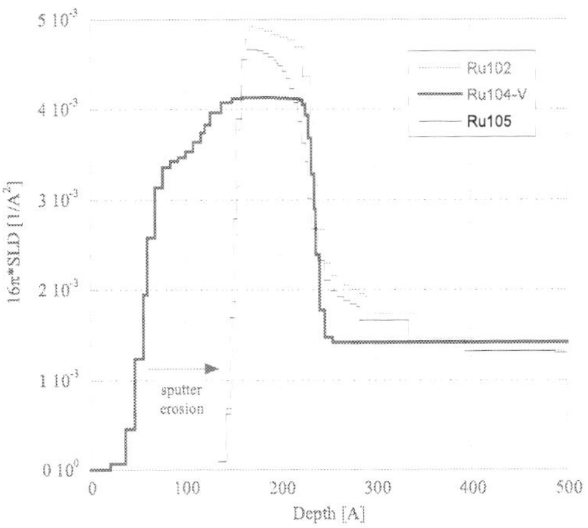

Fig. 15. Electron density depth profiles (ordinate is equal to 16πSLD, where SLD=$r_e\rho_e$, and r_e and ρ_e are the classical electron radius and electron number density, respectively). The overall film thickness for Ru102 and Ru105 has been reduced by sputtering.

Kinematically, Xe and Sn behave similarly, since their mass is very close. However, there is a fundamental difference: unlike Xe (which is inert), Sn can be incorporated into the mirror structure and easily build up on the target. Sn accumulation would be exacerbated if any type of chemical bonding or new phase is formed. The accumulation of Sn is limited during Sn bombardment due to self-sputtering; therefore a steady-state Sn content in the sample is reached. In addition, the overall ion-induced sputtering of the mirror is reduced, since ion-induced sputtering is now shared between the mirror material (i.e., Ru, Rh or Pd) and the previously implanted Sn. Results from Monte Carlo modeling of Sn implantation have shown these trends, and they were later verified by experimental measurements [Allain et al., 2006]. Tests therefore conducted with Xe+ served as an appropriate surrogate for Sn irradiation. Furthermore, since some EUV light sources could in principle use Xe as a 13.5-nm radiatior, these tests were also directly relevant. One particular interesting effect of inert ions such as Xe is that they implant at the near surface and could, if enough vacancy-induced voids are created, lead to Xe bubble accumulation. The work by Allain et al. in fact now has indicated that for a given Xe fluence threshold at 1-keV the stability of small nm-sized bubbles can be created at the near surface of MLM Si/Mo systems. This was indicated by use of XRR tests showing Porod-like scattering of small-angle X-ray scattering experiments.

6.3 Effect of fast and thermal Sn particles on single-layer reflectivity at 13.5-nm

6.3.1 Thermal Sn

Operation of Sn-based EUV lithography DPP sources exposes the collector mirror to two types of Sn contamination: thermal deposition of Sn vapor and bombardment of Sn ions from the expanded plasma. Even with the implementation of debris mitigation mechanisms, some contamination will reach the collector mirror. In the *in-situ* expeirments presented here, both sources of Sn (i.e., energetic and thermal) can be studied on small mirror samples. An electron beam evaporator loaded with Sn supplies the thermal flux. The energetic Sn flux comes from a focused Sn-ion source. Integration of an *in-situ* EUV reflectometer allows monitoring of the reflectivity in real time as the mirror is exposed to Sn.

EUV reflectivity measurements were monitored as the Sn layer was deposited. Results from these Sn exposures are shown in Figure 16. The lower axis corresponds to the Sn fluence and the thickness of the deposited Sn layer (calculated assuming that the film density is equal to the Sn bulk density) in the upper axis. For the case of the Rh sample (Rh-211) the Sn layer thickness is calculated based on fits with the reflectivity code and absolute at-wavelength 13.5-nm data from NIST. For a 15 nA current on an ECN4 evaporator for 2 minutes, sample Rh-213 was used as calibration sample with similar conditions to Rh-211. The sputter rate measured was 0.048 nm/sec or 2.9 nm/min. For Rh-211, the current level used was 5 nA for 34 minutes. This results in a deposition rate of 0.125 nm/min (2.9 divided by a factor of 3 and 7.75) and multiplied by 34 minutes results in a thickness of about 4.25 nm. Ex-situ XRF measurements resulted in an equivalent Sn thickness of 3.14 nm. The result appears consistent between the independent XRF measurement and the known deposition rate measured in the *in-situ* experiments in IMPACT. However, there are two observations with this result when one examines Fig 16 more carefully. One is the fact that the surface atomic fraction never reaches 100% of Sn atoms to Rh for Rh-211. Since LEISS is sensitive only to the first monolayer and the thickness measured is about 4-nm, one would expect LEISS to only scatter from Sn atoms at the surface. The LEISS data shows that instead an equilibrium concentration is reached near 70%. The second issue pertains to the in-situ relative reflectivity measured. For levels of 4-nm Sn deposition one would expect the relative reflectivity loss is of order 40-50% losses. However, the measurements show that losses in reflectivity are only about 20-30%. This is in direct contradiction to theoretical results of a Sn 4-nm layer on Rh. To investigate this further, a different mirror substrate (Pd) is used with similar Sn exposure conditions.

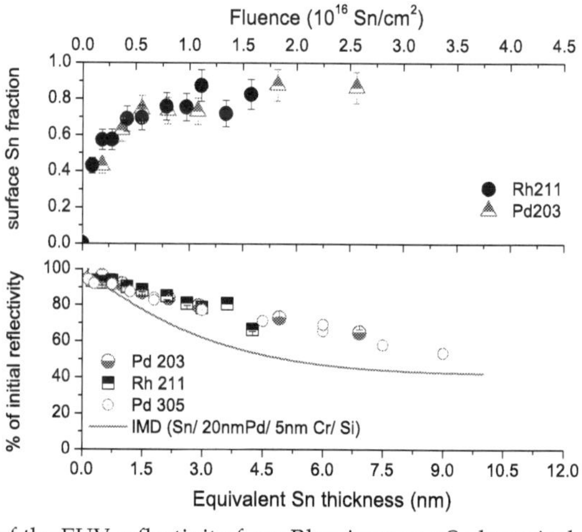

Fig. 16. Evolution of the EUV reflectivity for a Rh mirror as a Sn layer is deposited on the surface compared to deposition on a Pd mirror.

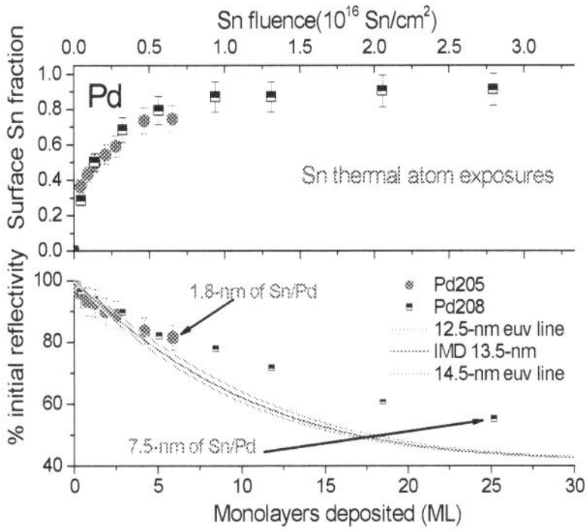

Fig. 17. Evolution of the EUV reflectivity for two Pd mirrors as a Sn layer is deposited on the surface.

Sn thermal deposition on Pd mirrors show similar behavior using about 10 nA of Sn thermal current. This corresponds to a deposition rate of 0.25 nm/min on Pd-203 (2.9 divided by a factor of 1.5 and 7.75) and for a 28 minute exposure a Sn thickness of about 7.5 nm. The XRF measurements resulted in a 6.46 nm equivalent Sn thickness, in reasonable agreement with IMPACT deposition rate measurements. In Figure 16 the relative reflectivity loss is about 35% for Pd-203, much lower than theoretically predicted for deposition of a 7.5-nm Sn layer on Pd.

For the cases of Pd 205 and 208, the deposition rate is 4-5 times less than for Pd-203. This is based on the time of equilibration of the Sn surface atomic fraction measured by LEISS of Pd 205 and 208 compared to Pd 203. Therefore, the deposition rate for Pd 205 and Pd 208 is about 0.0625 nm/min. For Pd-208 and 120 minute exposure the Sn thickness is 7.5 nm and for Pd 205, 28-minute exposure, 1.8 nm. The relative reflectivity losses are 20% and 45% for Pd-208 and Pd-205, respectively as shown in Figure 17. The surface atomic fraction of Pd-208 reaches 85-90% after close to 10^{16} Sn/cm^2 fluence. Before this time, for fluences below 0.6-0.7 x 10^{16} Sn/cm^2 the surface Sn atomic fraction reaches levels of about 70% for Pd-205 and Pd-208 consistent with results for Pd-203. So for exposures below Sn fluences of 10^{16} Sn/cm^2, the relative reflectivity losses are below about 30%. The main difference between Pd-203 and Pd-205, is that for the same exposure time (28 min.), Pd-203 has a "thicker" equivalent Sn layer compared to Pd-205 based on the deposition rate measured. This is an important result in that, although for the fluence exposure one should get "thick" Sn layers, the results from low-energy ion scattering shows otherwise. That is, LEISS is sensitive to the first or second monolayer and the data shows that even in the cases of Pd-203 and Pd-208 about 10-15% of the scattered ions detected, scatter from Pd atoms. Moreover, for lower fluences, scattering from mirror atoms (Pd or Rh) can be as large as 30%. More importantly, the surface Sn fraction seems to reach an equilibrium until the fluence is increased further.

These results imply that Sn is coalescing into nm-scale islands on the substrate surface for Sn exposures below about 10^{16} Sn/cm^2. Surface morphology examination was conducted with scanning electron microscopy (SEM) as a function of the Sn thermal fluence. The results were very important in that it proved that indeed the lower reflectivity loss is attributed to Sn island coalescence.

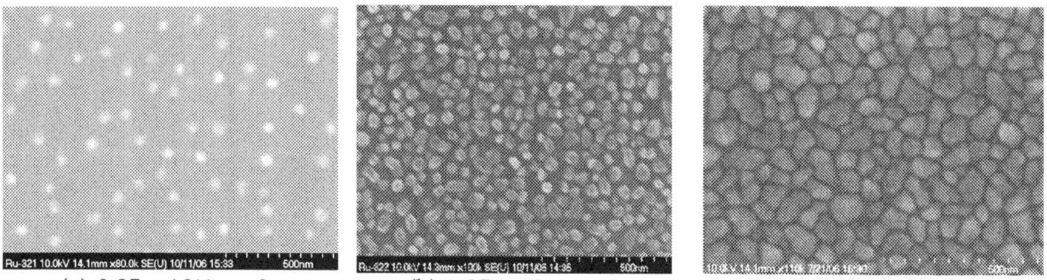

(a) 0.25×10^{16} cm^{-2} (b) 1.25×10^{16} cm^{-2} (c) 3.0×10^{16} cm^{-2}

Fig. 18. SEM micrographs of Sn-deposited Pd thin-film mirrors as a function of the Sn fluence.

6.3.2 Sn ions

EUVL plasma-based Sn sources expose mirrors to both thermal and energetic particles as discussed earlier. In this section we investigate the EUV reflectivity response of grazing incidence mirrors to exposure of Sn ions. The goal of this investigation was to identify failure mechanisms on the performance of Ru mirrors under EUVL source-relevant conditions. Furthermore, these experiments were also designed to elucidate the behavior of energetic Sn particles against results of thermal Sn exposure presented in the previous section. In addition to thermal Sn deposition on the collector mirror in a EUVL source device, the mirror is also subjected to energetic fast-ion and neutral bombardment from expanded plasma that gets through the debris mitigation barrier. The study of this problem is critical to assess the severity of damage induced by fast ion/neutral bombardment on EUV collector mirrors. Ion bombardment induces damage to EUV mirrors with at least three mechanisms: 1) erosion of the mirror material by physical sputtering, 2) modification of surface roughness, and 3) accumulation of implanted material inside the mirror.

These three phenomena have been extensively explored in IMPACT for the case of Xe+ bombardment, both for single-layer and multilayer EUV mirrors (Nieto et al, 2006). For this case, the first mechanism, erosion of the mirror, was determined to be the limiting factor for mirror lifetime. Surface roughness changes induced by ion bombardment in those cases were not large enough to affect the reflectivity in a significant manner. This was consistent with findings of irradiated thin-film surfaces of mirrors fabricated with magnetron sputtering. Typically these films consist of large grain boundary density, and thus surface corrugated structures from ion-beam bombardment are minimized. In regards to accumulation, it was observed that large Xe fluences (>10^{17} Xe^+/cm^2) delivered over a short period of time caused blistering of the mirror most likely due to the formation of bubbles. Xe fuel accumulation in the mirror layer is not regarded as an issue for sources operating with Xe^+ at low EUV power operation. Under high-power HVM (high-volume manufacturing) level operation, with Xe as the EUV radiator, it's unclear how large dose exposures might scale. Suffice to say that if the Xe flux is not controlled and maintained at tolerable levels, significant damage to the grazing incidence mirror is likely, mostly from ion-induced sputtering (Nieto et al, 2006).

Two experiments were performed by exposing two Ru mirrors to 1.3 keV Sn beams with a current of 40 -50 nA. The beams were rastered over a 0.25 - 0.3 cm^2 area, giving a net Sn ion flux of ~ 10^{12} ions cm^{-2} s^{-1}. The mirrors were exposed to this Sn beam for three hours (~104 sec), giving a total fluence of 1016 ions cm^{-2}. Sample ANL-H was manufactured by Philips, and Ru-208 was manufactured by OFM-APS at ANL. Sample ANL-H was bombarded at 60° incidence, while Ru-208 was bombarded at normal incidence. The results of the exposures are presented in Figure 19a and Figure 19b, which show both the Sn surface concentration (upper panels) and relative EUV reflectivity (lower panels).

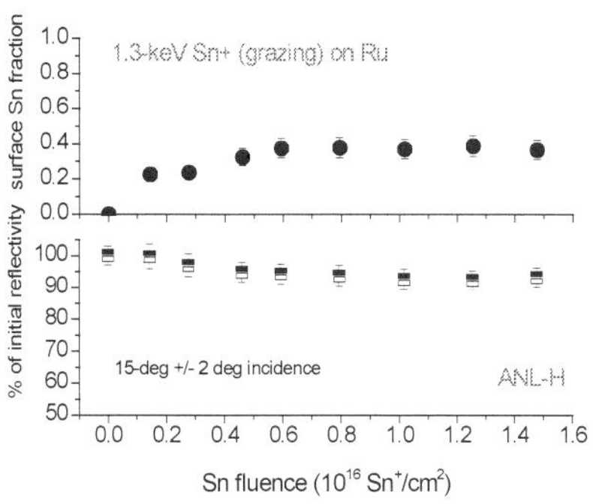

Fig. 19a. Evolution of the surface concentration and the EUV reflectivity of a Ru mirror exposed to a 1.3 keV Sn beam incident at grazing incidence (60°).

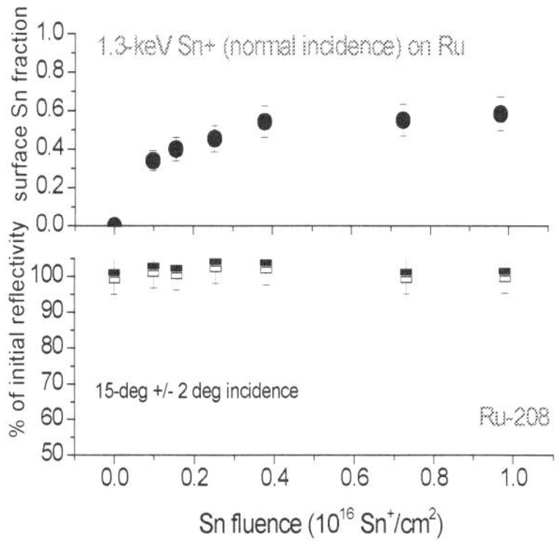

Fig. 19b. Evolution of the surface concentration and the EUV reflectivity of a Ru mirror exposed to a 1.3 keV Sn beam incident at normal incidence (0°).

There are significant differences between the two exposed samples. Regarding the Sn content in the surface, it can be seen that the sample bombarded at grazing incidence (ANL-H) reaches an equilibrium Sn content of 40%, while the sample bombarded at normal incidence has a steady-state Sn surface fraction of 60%. The increase can be explained by an increase of Sn self-sputtering yield. The Sn atomic fraction y_{Sn} on the sample as a function of time is given by:

$$\frac{dy_{Sn}}{dt} = \frac{\Gamma_{sn}}{n_T}\left(1 - y_{Sn} Y_{self\ sp}\right) \tag{3}$$

Equation 3 represents the balance between the implantation and the sputtering flux. The implantation flux is constant, but the sputtered flux is actually a function of the Sn content in the sample, so it gets weighted by the atomic fraction of Sn in the target y_{Sn}. At equilibrium, the time derivative is zero and that condition relates the equilibrium Sn fraction $y_{Sn,eq}$ and the self sputtering yield of Sn, $Y_{self\ sp}$:

$$y_{Sn,eq} = \frac{1}{Y_{self\ sp}} \tag{4}$$

For 60° incidence, the equilibrium fraction is 0.4, which corresponds to a self-sputtering yield of 2.5. For the normal incidence bombardment, the Sn self sputtering yield corresponding to the 0.65 Sn equilibrium atomic fraction is 1.5. These numbers are very close to the ones reported in the literature for Sn self-sputtering. Therefore, this is yet another independent verification of the in-situ EUV reflectivity measurements in IMPACT.

The other interesting observations from [Allain et al, 2007b] and [Allain et al., 2010] relates to the behavior of the EUV reflectivity as Sn is implanted. The effect of implanted Sn is not as drastic as for the case of deposited Sn on the surface, since the change in reflectivity is very small. For the sample irradiated at normal incidence, the reflectivity does not drop at all during the irradiation over a fluence of 10^{16} Sn+/cm^2. For the sample exposed to the beam at 60° incidence, a drop of < 10% in reflectivity is observed. By comparing the fluence scales for figures 18 and 19, it can be seen that the deposited Sn produces a more pronounced drop on reflectivity (15%), a drop at least 3 times larger than the one observed for the samples with implanted Sn. The case for the grazing incidence irradiation produces a larger drop in reflectivity that the normal incidence case, since in the limit of completely grazing incidence (90°), the implantation and thermal deposition cases are basically the same, since there is no penetration into the target.

6.3.3 Sn or Xe ions combined with thermal Sn

To examine the effects of exposure to a more realistic environment in a EUV light tool with both energetic and thermal particles exposing the collector mirror surface, experiments with thermal Sn and energetic Xe^+ were conducted. For these experiments, three samples— Rh 318, Rh 319, and Rh 320 —were each irradiated with a 1 keV ion beam (Xe^+) and exposed to an evaporator (Sn) simultaneously, with a total exposure time of 36 minutes. The target energetic Xe^+ fluences increased by one order of magnitude with each successive sample, beginning at 4.5×10^{15} Xe/cm^2, while target thermal Sn fluences remained constant at 4.5×10^{16} Sn/cm^2. Two control samples, Rh 321 and Rh 323, were used to compare the effects on reflectivity. Rh 321 was exposed to thermal Sn evaporator with a target Sn fluence of 4.5×10^{16} Sn/cm^2 for 36 minutes with no irradiation and Rh 323 was irradiated with an ion beam (Sn^+) at 1.3 keV for 88 mins at a fluence of 1.03×10^{14} Sn^+/cm^2 with no thermal Sn deposition.

Figure 20 shows both relative percent EUV reflectivity and Sn surface fraction versus thermal Sn fluence. A direct correlation between reflectivity loss and surface fraction of Sn is observed. Rh 318 and Rh 319 are fully covered with Sn after 3 minutes of exposure and their relative reflectivity decreased by 41.6% and 48.5%, respectively, after 36 minutes. While reflectivity of Rh 318 and Rh 319 decreased as the experiment progressed, Rh 320 had a local maximum at approximately 2.21×10^{16} Sn cm^{-2} where reflectivity increased to 94.7%. The corresponding Xe^+ fluence, 2.25×10^{16} Xe^+ cm^{-2}, exceeds the final fluences for the other two samples. This suggests that Rh 320 reached a threshold—too high for the other samples—

where the surface changed such that reflectivity could reach a maximum. The control sample, Rh 321, behaved extremely similar to Rh 318 and Rh 319 in both the atomic fraction of Sn and relative reflectivity loss. The relative reflectivity of Rh 323 fluctuated with increasing fluence but was found to only decrease 1.4% at the highest fluence, $5.8x10^{15}$ Sn^{+}/cm^{2}. Figure 5 does not represent the fluence corresponding to the reflectivity of Rh 323 because there was no thermal Sn fluence on the sample. It was plotted purely to show the affects of energetic Sn fluence on reflectivity. The surface fraction of Sn was lowest for this sample when compared to the other Rh samples, which was expected, with the surface fraction of Sn approaching equilibrium at 0.484. This further confirms the direct correlation between reflectivity loss and surface fraction of Sn discussed earlier.

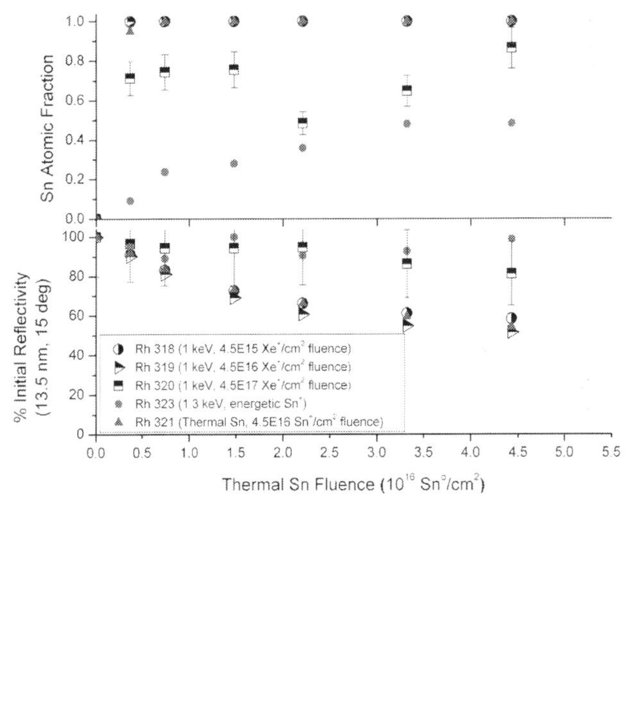

Fig. 20. LEISS data showing the surface Sn fraction versus thermal Sn fluence (top) and 13.5nm EUV reflectivity measurements versus thermal Sn fluence (bottom). Rh 321 had thermal Sn deposition only at a fluence of $4.5E16\ Sn^{0}/cm^{2}$. The fluence of Rh 323 shown, for both cases, is meant for correlation purposes only since there is no thermal Sn fluence on the sample.

	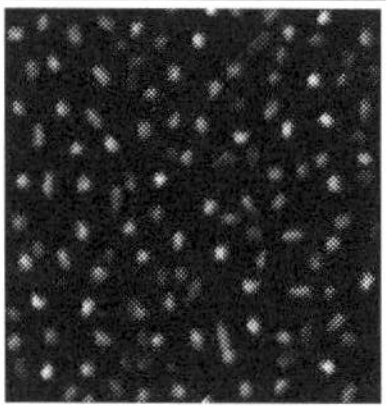 10x10 μm	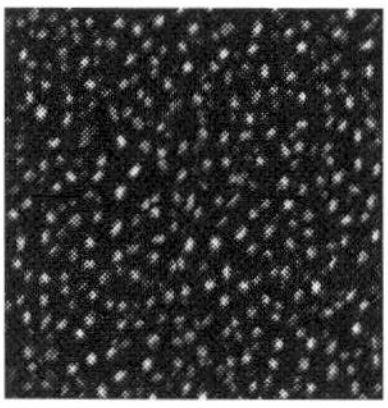 10x10 μm	 10x10 μm
Sample	Rh 318	Rh 319	Rh 320
Area Ra (nm)	18.5	14.8	2.00
Area RMS (nm)	22.4	19.4	3.20
Average height (nm)	60.9	16.7	1.78
Max. height (nm)	137	100	26.4

Table 1. Two-dimensional (10x10 μm) atomic force microscope (AFM) images and roughness values calculated with the AFM computer analysis program.

	 5x5 μm	 5x5 μm
Sample	Rh 321	Rh 323
Area RMS (nm)	14.12	0.11
Max. height (nm)	89.00	2.05
Feature Area (um^2)	22.98	n/a
Feature Coverage (%)	91.9%	n/a

Table 2. Two-dimensional (5x5 μm) atomic force microscope (AFM) images and roughness values calculated with the AFM computer analysis program for Rh thermal Sn only (left) and energetic Sn only (right) samples.

The AFM investigated the morphology of the samples. Table 1 and 2 illustrate the results. As the fluence of the samples is increased, it is found that the height, the roughness and the general size of the features decrease significantly. This is likely due to the increase in sputtering of Sn caused by the higher Xe^+ fluence. Rh 318 has the largest roughness and height values, at 22.4 nm and 137 nm respectively, but has the lowest fluence of the sample set. Rh 320 on the other hand has the smallest values for roughness and height, 3.20 nm and 26.4 nm, but the largest fluence of the set. This shows a direct correlation between the morphology of the samples and the fluence of Xe^+.

There is also a correlation between the morphology and the resulting reflectivity. For these samples, the lower roughness value (and height value) corresponds to the highest reflectivity. This is seen with Rh 320, where the reflectivity drop is 18.7% and the RMS value for the roughness is 3.20 nm, the lowest value for each in the sample set. For Rh 318 and Rh 319, the roughness values are 22.4 nm and 19.4 nm and the height values are 60.9 nm and 16.7 nm with corresponding reflectivity losses of 41.7% and 48.5%, respectively.

The difference in surface morphology for thermal Sn only and energetic Sn only is clearly illustrated in table 2. The roughness for thermal Sn deposition only, Rh 321, was found to be 14.12 nm, almost identical to Rh 319, 14.8 nm, and very close to Rh 318, 18.5 nm. This closeness is surface roughness, as well as similarities in reflectivity loss and Sn atomic fraction, elucidates the correlation between surface morphology and resulting reflectivity. This is further cemented by comparing the reflectivity loss of Rh 323 with its surface morphology. Rh 323 had a surface roughness of only 0.11 nm and maximum feature height of 2.05 nm with its largest drop in reflectivity being only 11%.

7. Conclusion

In conclusion, the success of EUV lithography as a high-volume manufacturing patterning tool remains elusive although great progress has been made in the past half decade. One main challenge is the plasma-facing components (e.g. electrodes, collector mirrors and debris mitigation shields) lifetime that ultimately impact the EUV power available for exposure.

8. Acknowledgements

We acknowledge the Intel Corporation, Dr. Melissa Shell, Dr. Bryan Rice (currently as an Intel at Sematech SUNY Albany), and Dr. Robert Bristol. We thank our strong collaborations with numerous groups including the groups of: Dr. Charlie Tarrio at NIST, Prof. Brent J. Heuser at University of Illinois, Dr. Peter Zink formerly from Philips Research Labs, Dr. Al Macrander of the Optics Laboratory at Argonne National Laboratory, Dr. Sasa Bajt formerly of Lawrence Livermore National Laboratory, and Dr. Vivek Bakshi formerly of Sematech.

9. References

Allain, J. P.; Hassanein, A. et al, "Effect of charged-particle bombardment on collector mirror reflectivity in EUV lithography devices", Proc. SPIE Int. Soc. Opt. Eng. 6151 (2006) 3

Allain, J. P.; Nieto, M.; Hendicks, M.; Harilal, S. S.; Hassanein, A (2007). *Debris-and radiation-induced damage effects on EUV nanolithography source collector mirror optics performance,* SPIE Proceedings, 6586(2007):22, doi: 10.1117/12.723692

Allain, J.P., Nieto, M., Hendricks, M.R., Plotkin, P., Harilal, S.S, & Hassanein, A (2007). IMPACT: A facility to study the interaction of low-energy intense particle beams with dynamic heterogeneous surfaces. Rev. Sci. Instrum. 78, 113105 (2007), ISSN 0034-6748

Allain, J.P.; Nieto-Perez, M.; Hendricks, M.R.; Zink, P.; Metzmacher, C.; & Bergmann, K. (2010). *Energetic Sn+ irradiation effects on ruthenium mirror specular reflectivity at 13.5-nm*. Applied Physics A, 100, 1, pp. 231-237, (July 2010), ISSN 0947-8396

Allain, J.P.; Nieto, M.; & Hassanein, A. (2008). *Specular reflectivity of 13.5-nm light from Sn islands deposited on grazing incidence mirror surfaces. Applied Physics A*, 91, 1, pp. 13-16, (April 2008), ISSN 0947-8396

Aziz, M. J. (2006). *Nanoscale Morphology Control Using Ion Beams*, Proceeding in Ion Beam Science: Solved and Unsolved Problems, Matematisk-fysiske Meddelelser 52, Sigmund, P. (editor), ISBN: 87-7304-330-3

Bakshi, V. Editor (2006). *EUV Sources for Lithography*, SPIE, Bellingham, WA, ISBN 0819458457

Bakshi, V. Editor (2009). *EUV Lithography*, SPIE and John Wiley & Sons, Hoboken, New Jersey, ISBN 978081946949

Banine, V. & Moors R. (2004). *Plasma sources for EUV lithography exposure tools*, J. Phys. D: Appl. Phys. 37 3207

Banine, V.; Koshelev, K.N.; Swinkels G.H.P.M. (2011) *Physical processes in EUV sources for microlithography*, J. Phys. D: Appl. Phys. 44 253001

Benschop, J.; Banine, B.; Lok, S.; & Loopstra, E. (2008). *Extreme ultraviolet lithography: Status and prospects*, J. Vac. Sci. Technol. B 26, 2204, ISSN 1520-8567

Campos, D.; Harilal, S. S.; & Hassanein, A. (2010). *The effect of laser wavelength on emission and particle dynamics of Sn plasma*, J. Appl. Phys. 108, 113305, ISSN 1089-7550

Carter, G., (2001). *The physics and applications of ion beam erosion.* Journal of physics. D, Applied physics 34.3, doi: 10.1088/0022-3727/34/3/201

Chan, W; Chason E (2007). *Making waves: Kinetic processes controlling surface evolution during low energy ion sputtering*, J. Appl. Phys. 101, 121301, DOI:10.1063/1.2749198

Eckstein, W. Comp*uter simulation of ion-solid interactions*, Springer Series in Materials Science, Vol. 10, Springer, Berlin, 1991, ISBN 3-540-190570-0

Fahy, K.; O'Reilly, F.; Scally, E.; & Sheridan, P. (2010). *Robust liquid metal collector mirror for EUV and soft x-ray plasma sources*, Proceedings Vol. 7802 in Advances in X-Ray/EUV Optics and Components V, Goto, S.; Khounsary, A. M.; & Morawe, C., Editors (2010). 78020K, 27 August 2010, Proc. SPIE 7802, 78020K (2010); doi:10.1117/12.860747

Ghaly, M.; Nordlund, N.; Averback, R.S. (1999). *Molecular dynamics investigations of surface damage produced by kiloelectronvolt self-bombardment of solids*, Philosophical Magazine A, vol. 79, Iss. 4, p.795-820, doi: 10.1080/01418619908210332

Harilal, S. S.; O'Shay, B.; Tillack, M. S.; Tao, Y.; Paguio, R.; Nikroo, A.; & Back, C. A. (2006). *Spectral control of emissions from tin doped targets for extreme ultraviolet lithography*, J. Phys. D: Appl. Phys. 39 484

Hassanein, A.; Sizyuk, V.; and Sizyuk, T. (2008). *Multidimensional simulation and optimization of hybrid laser and discharge plasma devices for EUV lithography*, Proc. SPIE 6921, 692113-1-15, DOI:10.1117/12.771218

Heinig, K.H.; Muller, T; Schmidt, B; Strobel, M.; Moller, W. (2003). *Interfaces under ion irradiation: growth and taming of nanostructures*, Appl. Phys. A, vol. 77, iss. 1, pp. 17-25, ISSN: 0947-8396

Henke, B. L.; Gullikson, E. M.; Davis, J.C. (1993). *X-ray interactions: photoabsorption, scattering, transmission, and reflection at E=50-30000 eV, Z=1-92*, Atomic Data and Nuclear Data Tables, Vol. 54, 181-342, Available from: http://henke.lbl.gov/optical_constants

Jurczyk, B. E.; Vargas-Lopez, E.; Neumann, M. N.; & Ruzic, D. N. (2005). *Illinois debris-mitigation EUV applications laboratory*, Microelectronic Engineering, Volume 77, Issue 2, February 2005, Pages 103-109, ISSN 0167-9317

Madi, C. S.; Anzenberg, E.; Ludwig, K.; Aziz M.J. (2011). *Mass Redistribution Causes the Structural Richness of Ion-Irradiated Surfaces*, Physical review letters 106.6, 66101.

Möller, W.; Eckstein, W.; Biersack, J.P. (1988). *Tridyn-binary collision simulation of atomic collisions and dynamic composition changes in solids*, Computer Phys. Comm. 51, No. 3, 355-368.

Muñoz-Garcia, J; Vazquez, L.; Cuerno, R.; Sanchez-Garcia, J.; Castro, M.; Gago, R. (2009). *Self-Organized Surface Nanopatterning by Ion Beam Sputtering*, in Toward Functional Nanomaterials: Lecture Notes in Nanoscale Science and Technology, Whang, Z (Ed.), Springer-Verlag, pp. 323 - 398, ISBN 978-0-387-77716-0

Nieto, M.; Allain, J.P.; Titov, V.; Hendricks, Matthew R.; Hassanein, A.; Rokusek, D.; Chrobak, C.; Tarrio, Charles; Barad, Y.; Grantham, S.; Lucatorto, T.B.; Rice, B. (2006). *Effect of xenon bombardment on ruthenium-coated grazing incidence collector mirror lifetime for extreme ultraviolet lithography*, J. of Appl. Phys., Vol. 100 Issue 5, p053510, ISSN: 00218979

O'Connor, A.; Dunne, P.; Morris, O.; O'Reilly, F.; O'Sullivan, G.; & Sokell, E. (2009). *Investigation of ions emitted from a tin fuelled laser produced plasma source*, J. Phys.: Conf. Ser. 163 012116

Tarrio, C. & Grantham S. (2005). *Synchrotron beamline for extreme-ultraviolet multilayer mirror endurance testing*, Rev. Sci. Instrum. 76, 056101, ISSN 0034-6748

Thompson, K. C.; Antonsen, E. L.; Hendricks, M. R.; Jurczyk, B. E.; Williams, M.; & Ruzic, D. N. (2006). *Experimental test chamber design for optics exposure testing and debris characterization of a xenon discharge produced plasma source for extreme ultraviolet lithography*, Microelectronic Engineering, Vol. 83, Iss. 3, pp. 476-484, (March 2006), ISSN 0167-9317

Van der Velden, M; Lorenz, M. (2008). *Radiation Generated Plasmas: a challenge in modern lithography*, Technische Universeit Eindohoven, Proefshcrift, ISBN: 978-90-386-1258-4

Van der Velden, M. H. L.; Brok, W. J. M.; van der Mullen, J. J. A. M.; & Banine, V. (2006). Kinetic simulation of an extreme ultraviolet radiation driven plasma near a multilayer mirror, J. Appl. Phys. 100 (7) 073303, ISSN 1089-7550

van Herpena, M.M.J.W.; Klundera, D.J.W.; Soera, W.A.; Moorsb, R.; & Banineb, V. (2010). *Sn etching with hydrogen radicals to clean EUV optics*, Chemical Physics Letters, Vol. 484, Iss. 4-6, Pp. 197-199, (January 2010), ISSN 0009-2614

Vargas López, E.; Jurczyk, B. E.; Jaworski, M.A.; Neumann, M. J.; & Ruzic, D. N. (2005). *Origins of debris and mitigation through a secondary RF plasma system for discharge-produced EUV sources*, Microelectronic Engineering, Vol. 77, Iss. 2, pp. 95-102, (February 2005), ISSN 0167-9317

Wagner, C. & Harned, N. (2010). *EUV lithography: Lithography gets extreme*, Nature Photonics 4, 24 - 26 ISSN 1749-4885

Yoshioka, M.; Teramoto, Y.; Zink, P.; Schriever, G.; Niimi, G.; & Corthout, M. (2010). in Proceedings Vol. 7636, Extreme Ultraviolet (EUV) Lithography, 2010, 763610-1

20

High-Index Immersion Lithography

Keita Sakai
Canon Inc.
Japan

1. Introduction

The resolution capability of photolithography is given by Rayleigh's equation.

$$R=k_1 \cdot \lambda / NA\,, \tag{1}$$

where R is the half-pitch resolution of the image, k_1 is a constant that depends on the resist process and exposure method, λ is the exposure wavelength, and NA is the numerical aperture of the projection optic. According to Rayleigh's equation, there are three ways to enhance the resolution. The first is to shorten the exposure wavelength such as extreme ultraviolet lithography (EUVL). The second is to improve the k_1 value, for example, using the double-patterning technique. The third is to increase the numerical aperture (NA) as ArF immersion lithography. It has already realized the NA up to 1.35 and moreover can increase the NA using high-index materials. In this chapter, high-index immersion lithography with the NA over 1.45 is focused.

The NA is actually determined by the acceptance angle of the lens and the refractive index of the medium surrounding the lens and is given by eq. (2).

$$NA=n \cdot \sin\theta\,, \tag{2}$$

where n is the refractive index of the medium surrounding the lens and θ is the acceptance angle of the lens. Therefore, the NA can be enlarged by replacing the air (n=1) with a fluid (n>1) as a medium. In this immersion lithography, the film stack consists of a lens, a fluid layer, and a resist layer. The value of $n \cdot \sin\theta$ is invariant through the film stack because it obeys Snell's Law. Since $\sin\theta$ is smaller than 1, the maximum NA ($=n \cdot \sin\theta$) is limited by the layer with the minimum refractive index. For example, the refractive index of water is 1.44 at 193.4 nm, thus the NA over 1.44 cannot be established as shown in Fig. 1 (a) because of the total reflection. To realize the NA over 1.44, the water must be replaced with a fluid which has a higher refractive index than water (Fig. 1 (b)). In Fig. 1 (b), fused silica has the smallest refractive index in the film stack and it limits the maximum NA. For further increasing the NA, a high-index lens material must be used as a lens material.

As described above, high-index lens materials and high-index immersion fluids are indispensable to realize high-index immersion lithography. One of the candidates of a high-index lens material is lutetium aluminium garnet (LuAG), which has a refractive index of 2.14. Second-generation (G2) and third-generation (G3) fluids are saturated hydrocarbon fluids whose refractive indices are approximately 1.64 and 1.80, respectively.

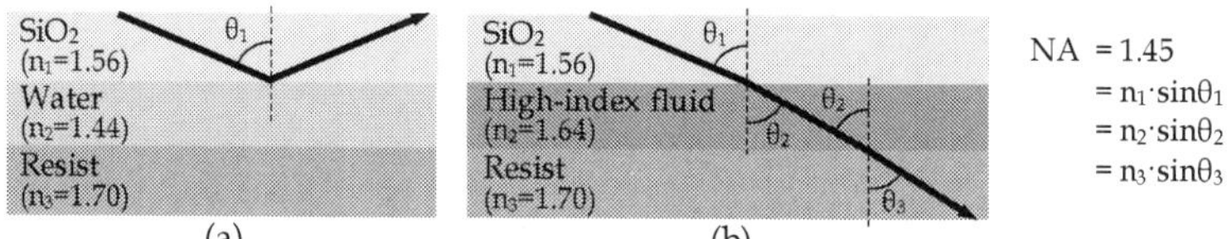

Fig. 1. Light propagation through the film stack at 1.45 NA. (a) The light does not reach the resist layer because of the total reflection at the lens-water boundary. (b) A high-index fluid enables the light to propagate into the resist.

High-index immersion lithography can be classified into three types by combining a lens material and an immersion fluid: fused silica and a G2 fluid (type 1), LuAG and a G2 fluid (type 2), and LuAG and a G3 fluid (type 3). With these types, the maximum NAs are estimated 1.45, 1.55, and 1.70, respectively.

Table 1 shows k_1 values for typical half-pitch and NA. The k_1 is calculated using Rayleigh's equation and needs to be at least 0.25 to resolve the patterns of the half-pitch for the theoretical limit. The resolution of 36 nm is achieved with 1.45 NA optic, which can be realized using only a G2 fluid. Those of 1.55 NA and 1.65 NA can achieve the resolutions of 34 and 32 nm, respectively.

Half-pitch	1.35 NA	1.45 NA	1.55 NA	1.65 NA	1.70 NA
39 nm	0.272	0.292	0.313	0.333	0.343
36 nm	0.251	0.270	0.289	0.307	0.316
34 nm	0.237	0.255	0.272	0.290	0.299
32 nm	0.223	0.240	0.257	0.273	0.281

Table 1. k_1 values for typical half-pitch and NA.

Although an exposure tool of 1.45 NA does not need new materials except for a G2 fluid, customers have little interest in the tool because of its modest gain in resolution. On the other hand, a tool of over 1.65 NA seems attractive for resolution enhancement. However, it would be difficult to realize G3 fluids immediately because no materials meet the requirements. In such a situation, it is a realistic way to develop a tool of 1.55 NA using LuAG and a G2 fluid.

In the next subchapter, projection optics with LuAG are explained. Development of LuAG is a hard work because the specifications of lithography-grade lens materials are extremely stringent. According to the history of LuAG development by Schott Lithotec (Parthier et al., 2008), great progress was achieved in the absorbance but it does not reach the target. The key issue for the optical design with LuAG is a correction of intrinsic birefringence (IBR). An effective method has been developed for IBR correction and it reduces the wave-front aberration to a practical level.

In the third subchapter, an immersion system using a G2 fluid is described. It was demonstrated that fluid absorbance can be kept low enough through an in-line purification unit and an oxygen removal unit. Although lens contamination is an important issue in a G2 fluid system, it was found that contamination can be suppressed by addition of a small amount of water into a G2 fluid. With this water-addition and in-line purification, the necessity for lens cleaning decreases from three times per day to once a week. Fluid confinement is also a challenge for a G2 fluid system because residual fluid is easy to remain on a wafer. Some issues arising with residual fluid, such as fluid darkening due to reentry of oxygen-rich residual fluid, were solved. By accepting residual fluid on a wafer, the scanning speed and the throughput can be raised.

Finally, the remaining challenges to realize high-index immersion lithography are discussed.

2. Projection optics with LuAG

Key parameters for high-index lens materials are refractive index, absorbance, and intrinsic birefringence (IBR). High-index lens materials must have a refractive index to permit NA scaling sufficient to justify the development cost. The absorbance must be sufficiently low to avoid the image degradation by thermal aberrations. The intrinsic birefringence must be minimal to allow a correction to avoid introducing unacceptable aberrations in the final aerial image.

The National Institute of Standards and Technology (NIST) has searched for high-index lens materials that meet the above requirements such as barium lithium fluoride ($BaLiF_3$) and LuAG (Burnett et al., 2006). $BaLiF_3$ developed by Tokuyama is available in various sizes with low absorbance (Nawata et al., 2007). However, the refractive index of $BaLiF_3$ is not high enough to achieve a sufficient NA for the enhancement of the resolution. Only LuAG remains as a candidate for a high-index lens material. LuAG has the intrinsic birefringence over 30 nm/cm and still has a high absorbance caused by impurities. Thus, the status of LuAG should be paid attention to and an IBR correction method should be developed.

2.1 Status of LuAG

Lithography-grade LuAG has been aggressively developed by Schott Lithotec since 2005. The absorbance, which is the biggest issue for LuAG, was largely improved down to 0.035 /cm by purifying the raw material and optimising the crystal growing process as shown in Fig. 2. Since it was found that the intrinsic absorbance of LuAG is 0.00118 /cm (Letz et al., 2010), the absorbance will be less than 0.005 /cm by further reduction of impurities.

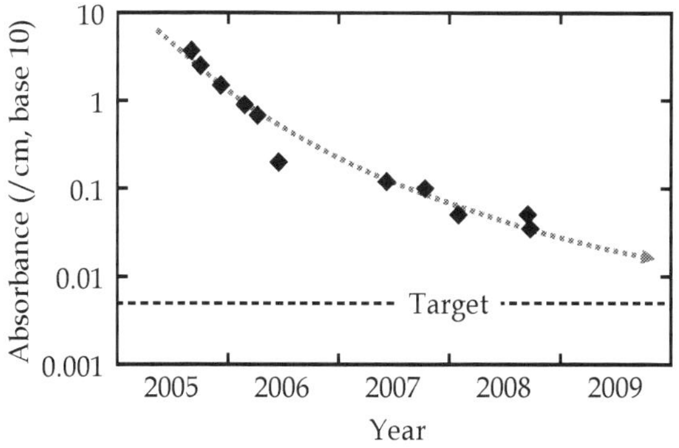

Fig. 2. History of LuAG development by Schott Lithotec. The absorbance of LuAG has been improved down to 0.035 /cm but it has not reached the target (0.005 /cm).

Table 2 shows the target and the status of each requirement. The stress birefringence (SBR) and the homogeneity seem to have gotten closer to the targets. However, these were achieved with a small crystal of yttrium aluminium garnet (YAG). Therefore, it is necessary to confirm the SBR and the homogeneity of large-sized LuAG.
The remaining challenges, such as the absorbance, will be solved if the development is accelerated by strong supports from the industry.

	Target	Status
Absorbance (/cm, base 10)	< 0.005	0.035
Stress birefringence (RMS, nm/cm)	< 0.5	0.73 (40 mmϕ, YAG)
Index homogeneity (-Z36, ppm)	< 0.05	0.03 (45 mmϕ, YAG)

Table 2. Target and status of LuAG.

2.2 Intrinsic birefringence correction

Since the absorbance of LuAG is higher than that of fused silica, it is difficult to use many LuAG lenses in a projection optic. From the viewpoint of optical design, applying LuAG to the final lens is most effective. In the final lens through which light converging at an image plane passes, the range of incident angles to the optical axis is wide. Assuming that the bottom surface of the final lens, which touches an immersion fluid, is flat, the maximum ray angles in LuAG are 46.4 and 52.6° for the NAs of 1.55 and 1.70, respectively.
The IBR distribution depends on the direction of a crystal. The distribution of a (111)-oriented crystal has three-fold rotational symmetry. When the (111)-oriented crystal is used, the angle at the maximum IBR is 35.3° from the (111) axis. On the other hand, a (100)-oriented crystal has four-fold rotational symmetry, and the angle at the maximum IBR is 45° from the (100) axis. In short, the maximum ray angle exceeds both angles of the maximum IBR even in the case of NA 1.55.

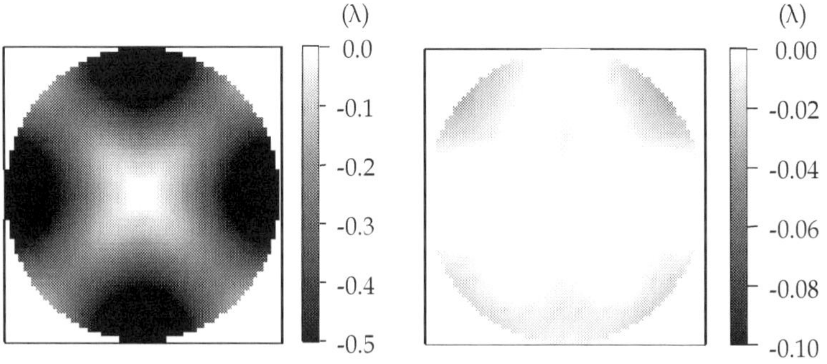

(a) (b)

Fig. 3. Difference between radial and tangential polarized wave front for a projection optic of 1.55 NA; (a) without IBR correction, (b) with an original correction method.

Figure 3 shows wave front aberration maps for a projection optic of 1.55 NA. These show the difference between the radial and the tangential polarized wave fronts. The wave front difference without the correction has a large rms value of 217 mλ and a clear 4θ component. On the other hand, the original correction method reduces the rms difference to 4.2 mλ. This value is within the practical level.
In the case of 1.70 NA, the wave front without the correction has a large rms difference of 680 mλ and the rms difference after the correction is 21.2 mλ. Although this value is still large compared with the case of 1.55 NA, it is close to the practical level. Therefore, it can be concluded that the IBR correction is feasible even at 1.70 NA.

3. Immersion system using a G2 fluid

High-index immersion fluids are categorized as second-generation (G2) or third-generation (G3) fluids. G2 fluids have refractive indices of approximately 1.64 and enable an NA to increase up to 1.55 with LuAG as a final lens material. G3 fluids are targeted to have refractive indices over 1.80 for achieving an NA over 1.65 with LuAG.
DuPont, JSR, and Mitsui Chemicals have reported both G2 and G3 fluids (French et al, 2007, Furukawa et al., 2007, Kagayama et al., 2007). Their G2 fluids are saturated hydrocarbon and have sufficient performance as an immersion fluid. They also developed G3 fluids as some extension of organic G2 fluid materials. This type of G3 fluids has not yet met the requirements of refractive index or absorbance. The other types of G3 fluids, which include nanoparticles, are being developed, but they are still within the research phase (Zimmerman et al., 2008). In the above situation, an immersion system using a G2 fluid has been preferentially developed (Sakai et al., 2008).
To achieve the target cost of $1/layer for a G2 fluid, the recycling of a G2 fluid is required. G2 fluids are easily degraded by laser irradiation and the dissolution of oxygen from the atmosphere. Therefore, a fluid circulation system is desired to keep the fluid absorbance sufficiently low using purification and oxygen removal functions. In addition, lens contamination, bubble, and residual fluid on a wafer are also important issues. An immersion system composed of a fluid circulation system, an immersion nozzle and a cleaning unit should be established to accommodate a G2 fluid.

3.1 Fluid degradation

The absorbance of water does not change largely with dissolved oxygen, and immersion water is not used repeatedly as a G2 fluid. These are the reasons the degradation of water is not a concern as an immersion fluid.
On the other hand, the absorbance of a G2 fluid is easily increased by dissolved oxygen or laser irradiation. Moreover, the dn/dT of a G2 fluid is 5~6 times larger than that of water (French et al., 2006, Santillan et al., 2006, Furukawa et al., 2007). This means that a temperature change causes a larger change in the refractive index and results in larger thermal aberration (Sekine et al., 2007). To suppress a temperature rise with the absorption of exposure light, the fluid absorbance has to remain sufficiently low.

3.1.1 Dissolved oxygen

Although the absorbance of a G2 fluid is lower than that of water (French et al., 2005, Wang et al., 2006), the degradation of a G2 fluid with dissolved oxygen is larger than that of water. The

first reason for this large degradation is the high oxygen solubility. It has been reported that the solubility of oxygen in G2 fluids are 55~70 ppm in air atmosphere (Furukawa et al., 2006). These values are 7~9 times higher than that in water. According to the research at Columbia University (Gejo et al., 2007), dissolved oxygen forms a charge-transfer complex with cycloalkane of a G2 fluid and the complex has strong absorption in the ultraviolet wavelength range. This is the second reason for the large degradation induced by dissolved oxygen.
The induced absorbance against dissolved oxygen in a G2 fluid has been experimentally obtained. Figure 4 shows the induced absorbance of HIL-203 (JSR) compared with water. It was confirmed that the oxygen-rich HIL-203 exhibits strong absorption. Therefore, a removal function of oxygen is necessary to keep the fluid absorbance low in a circulation system.

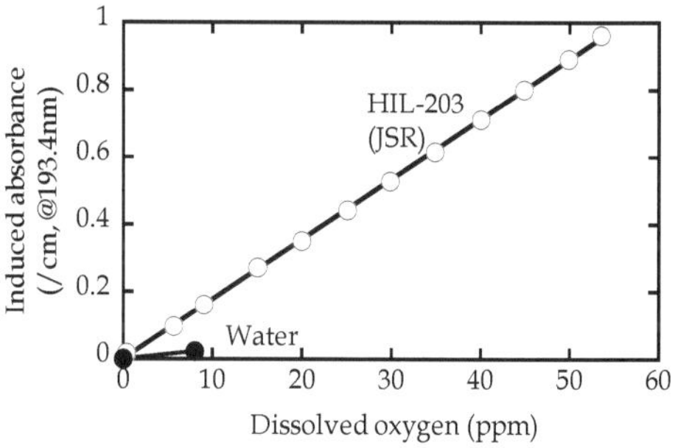

Fig. 4. Experimental results of induced absorbance with dissolved oxygen. Open circles show the induced absorbance of HIL-203 and filled circles show that of water.

An immersion nozzle supplies a G2 fluid under a final lens and sucks the fluid with the neighbouring gas. If the atmosphere around a wafer is air, the oxygen concentration in the recovered fluid increases with the sucked air. Under such a condition, two types of oxygen removal functions were evaluated using an experimental system as shown in Fig. 5. The fluid used in this experiment was HIL-203 (JSR) and the fluid flow rate was 400 mL/min.

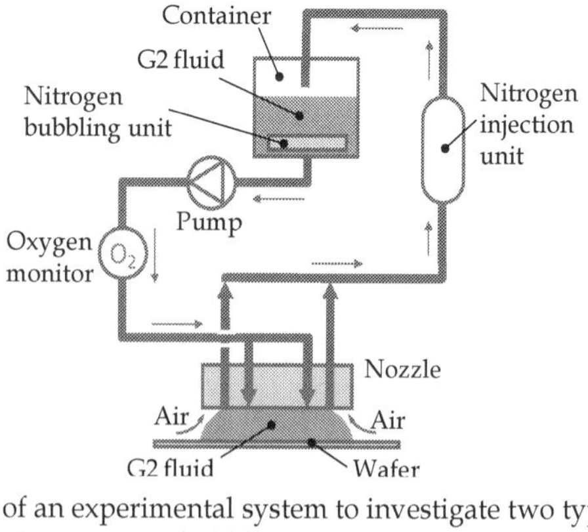

Fig. 5. Schematic view of an experimental system to investigate two types of oxygen removal functions. One is a nitrogen bubbling unit in the fluid container, and the other is a nitrogen injection unit attached to the fluid line. Oxygen concentration in the fluid is measured at the position just before the immersion nozzle.

Figure 6 shows dissolved oxygen using two types of oxygen removal functions. In the case with the nitrogen bubbling unit alone, the oxygen concentration was 4.5 ppm. It corresponds to the absorption of 1.83 %/mm, which can not be permitted. By using the nitrogen bubbling and the nitrogen injection units simultaneously, the oxygen concentration was reduced to 0.2 ppm. Further reduction was achieved increasing the nitrogen flow rate into the injection unit from 5 to 10 L/min. The oxygen concentration of 0.1 ppm is the permissible level because it corresponds to only 0.04 %/mm degradation. The nitrogen injection unit removes the dissolved oxygen from the oxygen-rich fluid just after recovery and prevents the oxygen-rich fluid from going back to the container. That is why the injection unit is efficient for the reduction of dissolved oxygen.

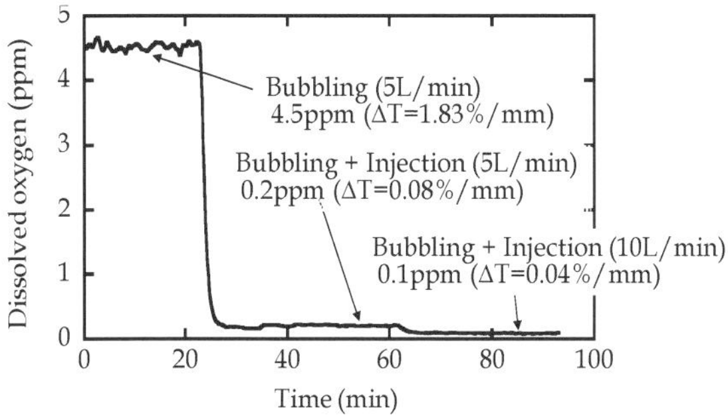

Fig. 6. Experimental results of dissolved oxygen using oxygen removal functions. At first, the dissolved oxygen was not sufficiently low with the nitrogen bubbling unit alone. Then, the large reduction of dissolved oxygen was achieved with the simultaneous use of the nitrogen bubbling and the nitrogen injection units.

3.1.2 Laser irradiation

The absorbance of a G2 fluid increases with the photodecomposed materials of a G2 fluid induced by ArF laser irradiation. To achieve the target cost of $1/layer, it is necessary to remove the photodecomposed materials using an in-line purification unit.

Figure 7 shows the instruments used for a laser irradiation test. ArF excimer laser G41A3 (Gigaphoton), which can irradiate with the rep. rate of up to 3 kHz, is used for irradiation. The irradiation chamber has two beam lines and is purged with nitrogen gas to keep the oxygen concentration less than 1 ppm. The fluid circulation system with the oxygen removal functions can maintain dissolved oxygen at 0.1 ppm and below. The fluid absorbance monitor contains four flow-through cells which can evaluate a number of fluids simultaneously.

The experimental conditions are shown in Table 3. Fluid suppliers have developed in-line purification units for their G2 fluid. Therefore, fluid degradation was evaluated with the combination of a fluid and a purification unit made by the same supplier, such as HIF-A and Unit-A.

The experimental results are shown as the induced absorbance against fluid dose. The fluid dose is defined as an incident dose modified by the volume dilution factor as shown in eq. (3). It has been suggested as the appropriate dose metrics (Liberman et al., 2007).

$$D_{fluid}=I \cdot N \cdot V_{irr}/V_{total}, \quad (3)$$

where D_{fluid} is the fluid dose, I is the fluence per pulse, N is the total pulse count, and V_{irr} and V_{total} are the irradiated volume and the total fluid reservoir volume, respectively.

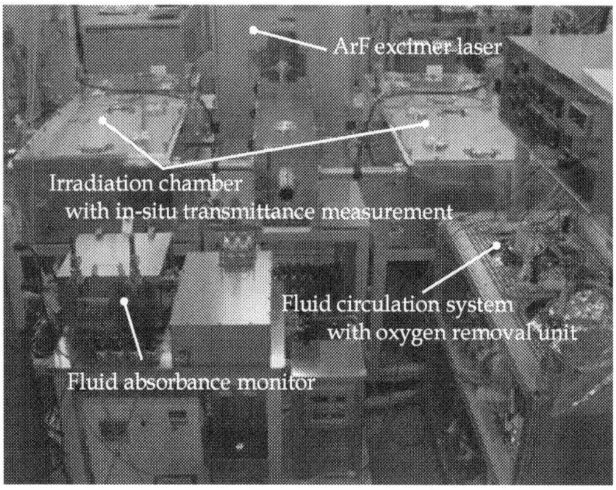

Fig. 7. Instruments for a laser irradiation test. The fluid circulation system consists of a fluid container, a pump, and oxygen removal units.

Immersion fluid / Purification unit	HIF-A / Unit-A HIF-B / Unit-B HIF-C / Unit-C
Fluence	0.5 mJ/(cm^2 · pulse)
Rep. rate	2000 Hz
Irradiation area	8.0 mmϕ
Fluid gap	1.0 mm
Fluid flow rate	100 ~ 350 mL/min

Table 3. Experimental conditions for evaluation of in-line purification units.

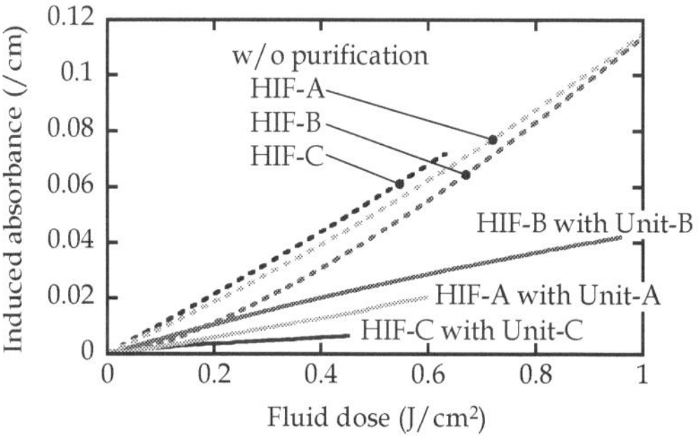

Fig. 8. Induced absorbance of a G2 fluid against the fluid dose. The dashed line shows the induced absorbance without a purification unit. The solid line shows the result with a purification unit. The fluid dose is calculated using eq. (3) and the conditions as shown in Table 3.

Figure 8 shows the induced absorbance of each G2 fluid with and without an in-line purification unit. For example, the degradation rate of HIF-C with Unit-C is one-seventh-fold smaller than that without Unit-C. Using a purification unit, fluid lifetime will be up to 1 week before the fluid absorbance will reach an unacceptable level. Therefore, it can be concluded that recycling of a G2 fluid is feasible using an in-line purification unit.

3.2 Lens contamination

It has been reported that the exposed surface of a final lens is polluted by photodecomposed materials of a G2 fluid (French et al., 2007). This lens contamination diminishes the uniformity of exposure dose on a wafer and would be a source of particles. Thus, the suppression of lens contamination is a serious issue for an immersion system using a G2 fluid.

The suppression of lens contamination was examined by three procedures. The first is to select an appropriate fluid, which has a lower deposition rate. The second is to use an in-line purification, which can remove photodecomposed materials. The third is water-addition into a G2 fluid, which is an original suppression method.

Figure 9 shows a schematic diagram of the experimental setup. The cell absorbance, which consists of the absorbance of windows and a fluid of 1 mm path length in the cell, was measured using two power meters. In addition, the fluid absorbance was measured by a fluid absorbance monitor. Then, the absorbance of windows was calculated by subtracting the fluid absorbance from the cell absorbance as shown in eq. (4).

$$A_{window}=A_{cell}-A_{fluid}, \quad (4)$$

where A_{window} is the absorbance of windows, A_{cell} is the cell absorbance, and A_{fluid} is the fluid absorbance through 1 mm path length.

The laser fluence was 1.0 mJ/(cm^2·pulse), which is almost the same as in an actual exposure tool. The dose estimation on an actual tool is approximately 60 kJ/(cm^2·day). Although the effective cleaning method was proposed (Liberman et al., 2007), the necessity for lens cleaning should be less than once a week for minimizing tool downtime.

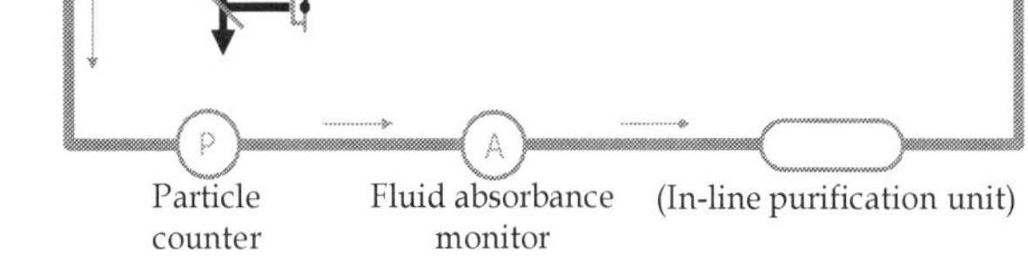

Fig. 9. Schematic diagram of the experimental setup. G2 fluid flows through the 1 mm gap space between the SiO_2 windows of the irradiation cell. The particle counter can detect particles larger than 60 nm in diameter.

3.2.1 Appropriate fluid

Induced absorbance of windows, which correspond to a final lens of a projection optic, was evaluated by using various G2 fluids. Figure 10 shows induced absorbance of windows against exposure dose. HIF-C exhibits a higher deposition rate than HIF-A and D. Therefore, HIF-A or D should be selected from the point of view of lens contamination. However, lens contamination occurs in a few hours on an actual exposure tool, even if using HIF-A or D. It is necessary to introduce some methods for suppressing contamination.

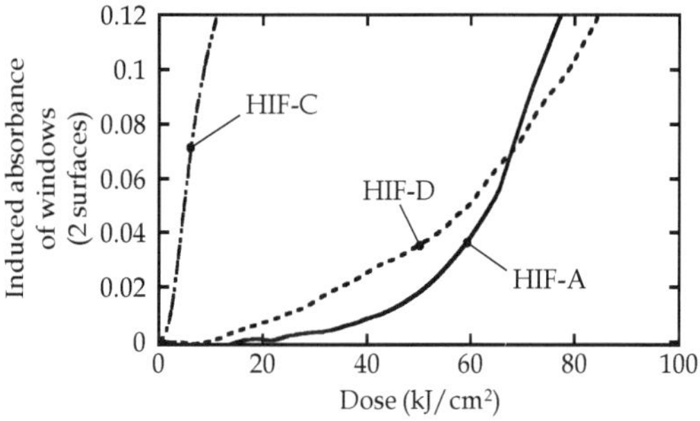

Fig. 10. Induced absorbance of windows using various G2 fluids. The deposition rates of HIF-A (solid line) and HIF-D (dashed line) are lower than that of HIF-C (dashed-dotted line).

3.2.2 In-line purification

It is expected that an in-line purification unit can suppress lens contamination because it can reduce photodecomposed materials. The experiments with several G2 fluids and in-line purification units were carried out using the experimental setup as shown in Fig. 9.

Figure 11 shows experimental results of window absorbance with and without a purification unit. It was confirmed that each purification unit can improve lens contamination. But the performance of purification units are not enough and a final lens needs to be cleaned more than once a day under the practical use conditions. Therefore, the additional method is desired to extend the cleaning interval.

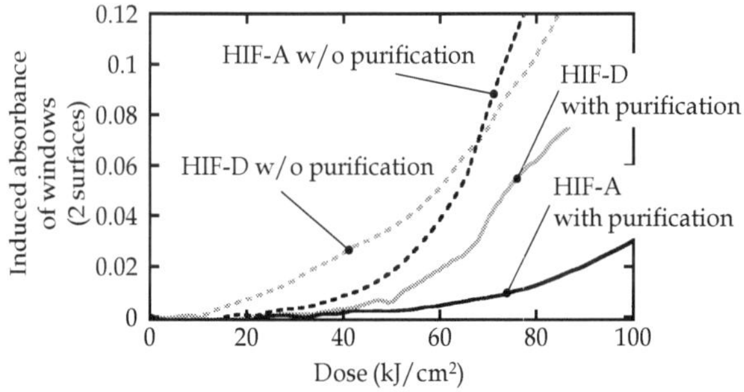

Fig. 11. Induced absorbance of windows with and without a purification unit. The deposition rate with a purification unit (solid line) is lower than that without a purification unit (dashed line).

3.2.3 Water-addition

It was found that lens contamination can be suppressed by addition of a small amount of water into a G2 fluid (Sakai et al., 2008). Since water in a G2 fluid is removed by an oxygen removal unit or an in-line purification unit, a water-addition unit is necessary for a fluid circulation system to keep water in a G2 fluid. Figure 12 shows a schematic diagram of a water-addition unit. Using the water-addition unit, a sufficient amount of water can be solved into a G2 fluid.

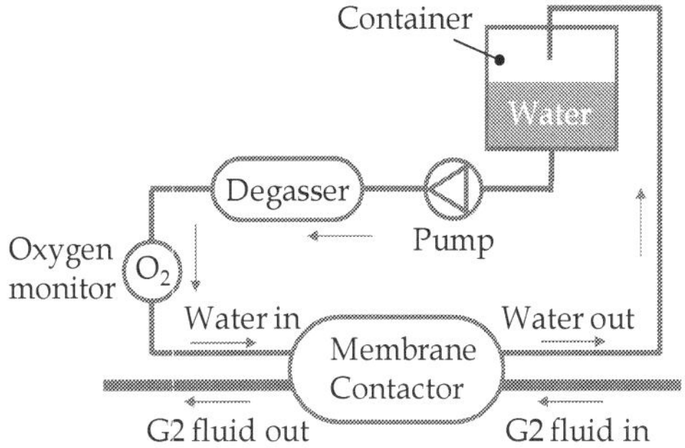

Fig. 12. Schematic diagram of a water-addition unit. The unit consists of a membrane contactor and a circulation unit of degassed water. Degassed water contacts a G2 fluid through the porous membrane with a large area.

Figure 13, 14, and 15 shows the experimental results using HIL-203 (JSR) as a G2 fluid. In the case without water-addition, the absorbance of windows increased gradually (Fig. 13) and many streak contaminations were observed on the window (Fig. 14 (a)). On the other hand, the window was not contaminated in the case with water-addition (Fig. 13, Fig. 14 (b)). A strong impact was also obtained in terms of particle. As shown in Fig. 15, particles increased with lens contamination in the case without water-addition. Some portions of lens contamination would be removed by laser irradiation and flow into the fluid as particles. Using the water-addition unit, there were no particles in the fluid made from lens contamination. It was also confirmed that the water-addition unit does not generate any particles.

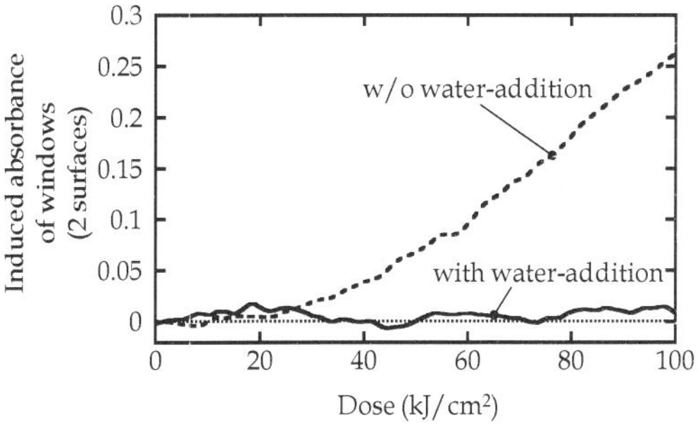

Fig. 13. Induced absorbance of windows against exposure dose. The absorbance of windows increased with the exposure dose in the case without water-addition (dashed line). In contrast, the water-addition unit suppressed contamination (solid line).

As described above, water-addition is the superior method for suppressing lens contamination. In addition, it seems to be a method without any disadvantages. For example, the refractive index of a G2 fluid does not change substantially with water-addition because the solubility of water into a G2 fluid is very low (several tens of ppm).

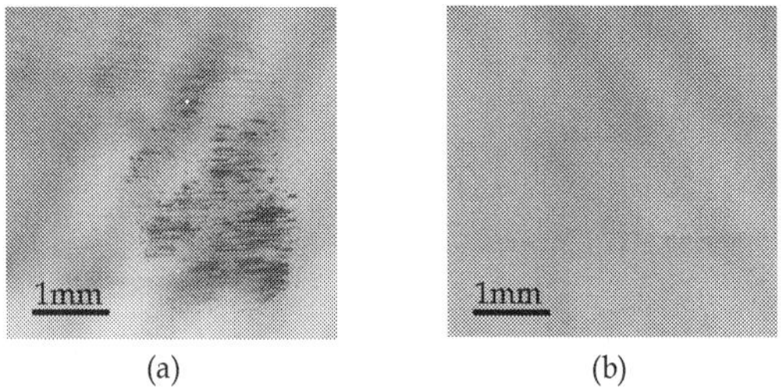

Fig. 14. Micrographs of window surfaces; (a) without water-addition after 100 kJ/cm^2 exposure, (b) with water-addition after 100 kJ/cm^2 exposure.

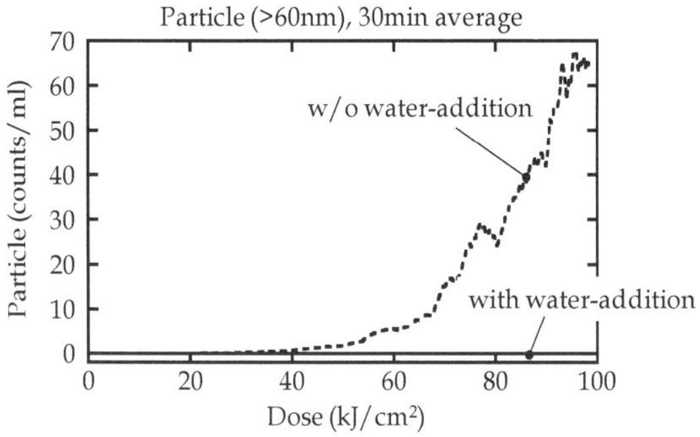

Fig. 15. Particles in a fluid against exposure dose. The number of particles increased with the exposure dose without water-addition (dashed line). In contrast, the water-addition unit suppressed particles induced by lens contamination (solid line).

3.2.4 Long-term test

A long-term test for lens contamination was done with an in-line purification unit and a water-addition unit simultaneously. IF132 (DuPont) and ARP (DuPont) were used as a G2 fluid and an in-line purification unit, respectively.

As shown in Fig. 16, the absorbance of windows started to increase at 20 kJ/cm^2 without any suppression units. The dose estimation on an actual exposure tool is approximately 60 kJ/(cm^2·day). Therefore, lens cleaning must be done three times per day in this case. Even if with ARP, the absorbance started to rise at 60 kJ/cm^2 and a lens needs to be cleaned once a day. The cleaning interval is not allowed in the case with ARP alone. On the other hand, lens contamination was suppressed until 420 kJ/cm^2 using ARP and a water-addition unit simultaneously. The necessity for lens cleaning is once a week and it is practical. Figure 17 shows micrographs of window surfaces after the experiments. The micrographs also exhibit that the simultaneous use of ARP and a water-addition unit has a sufficient performance. It is concluded that lens contamination is not a critical issue anymore.

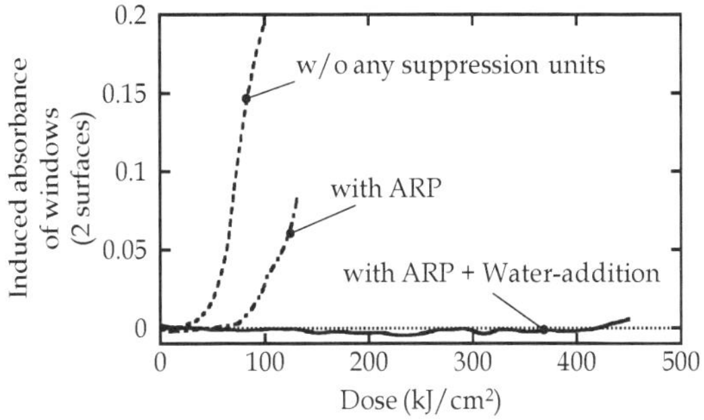

Fig. 16. Induced absorbance of windows against exposure dose. Simultaneous use of ARP and a water-addition unit suppressed lens contamination until 420 kJ/cm^2.

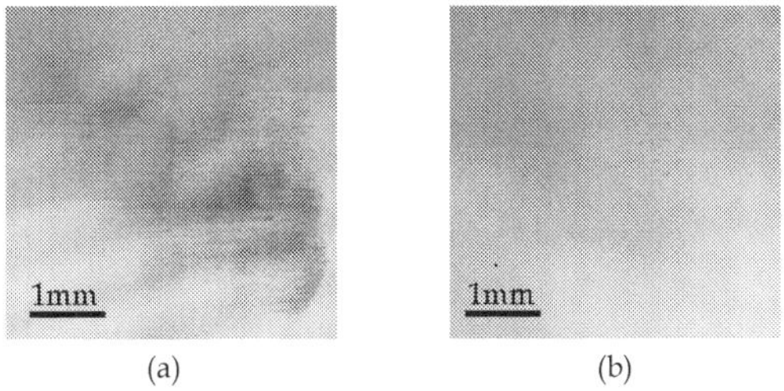

Fig. 17. Micrographs of window surfaces; (a) with ARP after 130 kJ/cm^2 exposure, (b) with ARP and a water-addition unit after 450 kJ/cm^2 exposure.

3.3 Bubble

A bubble is one of the origins of immersion defect. A bubble is entrapped on an air-liquid boundary and is translated with a wafer motion. If the bubble lifetime is sufficiently short, the bubble will be eliminated before reaching an exposure area. Therefore, the reduction of bubble lifetime is important.

The gas in a bubble diffuses from the surface of the bubble into the fluid. This diffusion causes the bubble to shrink with time and to eventually vanish. The bubble lifetime is approximated using eq. (5) (Honda et al., 2004).

$$\tau = \frac{\rho d^2}{8D(C_S - C_\infty)}, \tag{5}$$

where τ is the bubble lifetime, ρ is the density of the gas inside the bubble, d is the diameter of the bubble, D is the diffusion coefficient, C_S is the saturated concentration, and C_∞ is the concentration of the dissolved gas at a position far from the bubble. Figure 18 shows a schematic diagram of the experimental setup for the bubble lifetime measurement. By increasing the pressure of the fluid from 1 to 2 atm, the bubble vanishes at a certain lifetime. It corresponds to the lifetime in the half-degassed fluid. Figure 19 shows the bubble lifetimes in water and two types of G2 fluids. The experimental results in water correlated with the theory fairly well. The bubble lifetime in the G2 fluid was shorter than that in water. Hence, the concept of the bubble elimination method for a water immersion system, i.e., using degassed fluid, is also feasible in an immersion system using a G2 fluid.

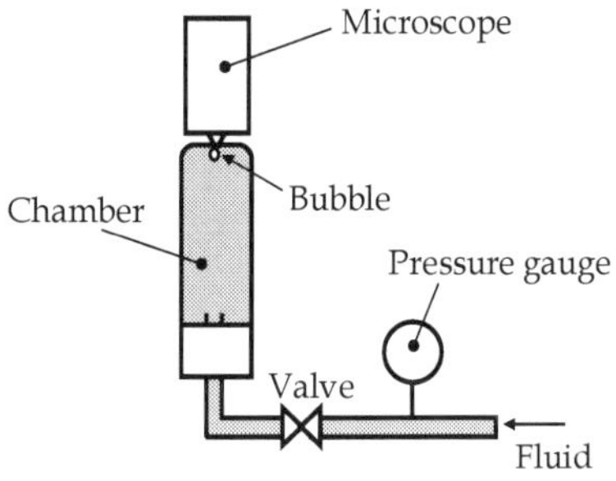

Fig. 18. Schematic diagram of the experiment setup for the bubble lifetime measurement. The size of a bubble is observed using the microscope.

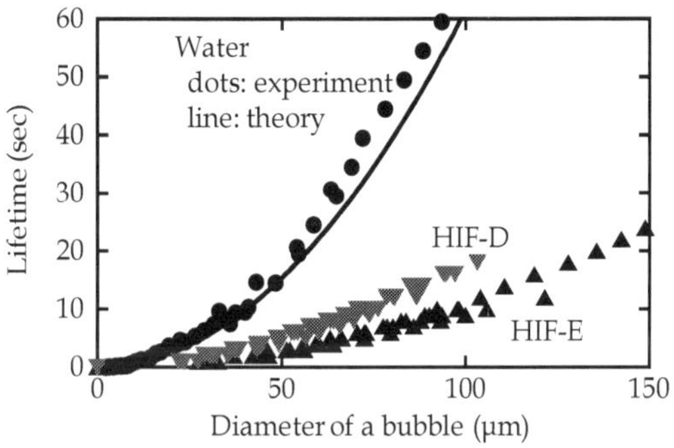

Fig. 19. Bubble lifetimes in various fluids. The lifetimes in G2 fluids (triangles) are shorter than that in water (circles).

3.4 Residual fluid on a wafer

The surface tension and viscosity of water are 72 mN/m and 1 mPa·s, respectively. This extremely high surface tension and low viscosity enable the immersion nozzle to keep water under a final lens (Kubo et al, 2007). On the other hand, it has been reported that a G2 fluid is easy to remain on a wafer (Sewell et al, 2007). After the verification that residual fluid is hard to avoid, some issues arising with residual fluid are discussed. The experimental results for the issues, such as the interaction between a G2 fluid and a resist, metal contamination in a G2 fluid, and fluid darkening due to oxygen-rich residual fluid, are explained.

3.4.1 Scanning test

The surface tension and viscosity of G2 fluids are approximately 30 mN/m and 2~4 mPa·s, respectively. The film pulling velocity (υ_{fp}), which is the minimum scanning velocity to leave fluid behind the nozzle, is proportional to the surface tension (γ) and the inverse of the viscosity (μ) as shown in eq. (6) (Shedd et al., 2006). If a wafer has the same static receding contact angle (SRCA, $\theta_{s,r}$) for water and a G2 fluid, the film pulling velocity for a G2 fluid is approximately one-sixth-fold lower than that for water. Although the higher SRCA can increase the film pulling velocity, the target SRCA over 120° is too high to develop an appropriate topcoat material (Sanders et al., 2008). Therefore, it is difficult to achieve a sufficient scanning speed without residual fluid in an immersion system using a G2 fluid.

$$\upsilon_{fp} = C\frac{\gamma}{\mu}\theta_{s,r}^3, \qquad (6)$$

where C is the empirical constant.

Figure 20 shows the results of the fundamental scanning test using a prototype topcoat for G2 fluids. HIL-203 (JSR) was used as a G2 fluid for the experiment. Even with the scanning speed of 30 mm/sec, the dynamic receding contact angle became 0° and fluid was left behind the nozzle. As a result, it was confirmed that residual fluid should be allowed to realize a sufficient scanning speed. Since the contact time of a G2 fluid and a resist becomes longer in such a system, the interaction between a G2 fluid and a resist should be discussed.

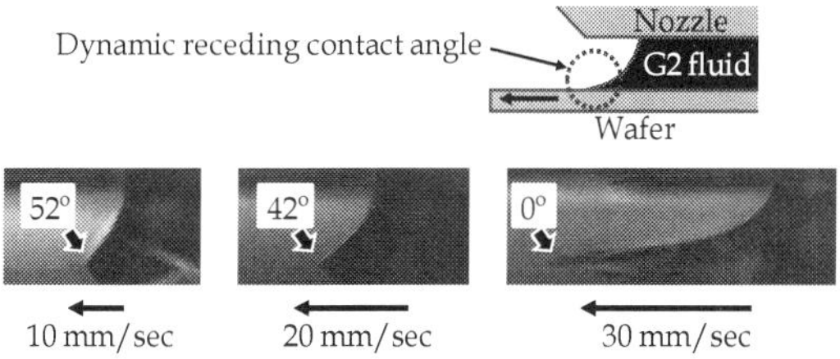

Fig. 20. Side views of the fundamental scanning test. The dynamic receding contact angle on a wafer became 0° with the scanning speed of 30 mm/sec.

3.4.2 Soaking test

To investigate the interaction between a G2 fluid and a resist, the defect inspection of soaked wafers is effective. Figure 21 shows the procedure of a soaking test. Between the exposure and post-exposure bake (PEB), a wafer was soaked in a G2 fluid for 60 seconds and the fluid was removed by spin drying. The defect inspection was carried out using KLA2371 (KLA-Tencor). The defects induced by post-exposure soaking were evaluated with the above procedure.

Figure 22 shows the results of the defect inspection. A large number of defects were found on the resist after soaking in HIF-E. In a microscopic review, there were many stains as shown in Fig. 22. As a result, it was confirmed that it is necessary to use an inert resist for a G2 fluid. On the other hand, it is not apparent that the wafer soaked in HIF-A or HIF-D had more defects than the reference wafer. There were some variations in the number of defects because the experiment was carried out in off-line process. Defect studies should be done in more clean circumstances.

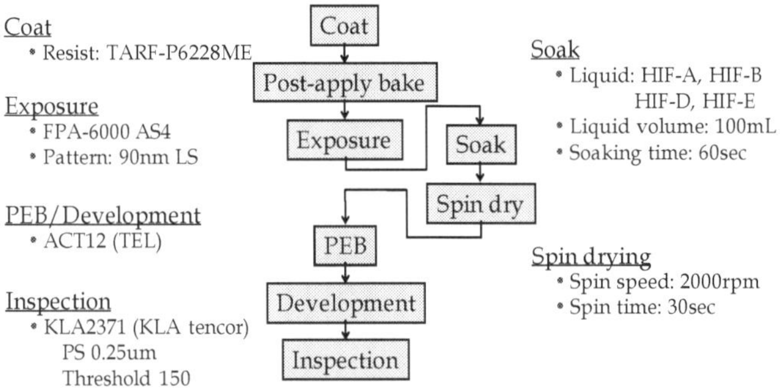

Fig. 21. Experimental procedure of a post-exposure soaking test to investigate the interaction between a G2 fluid and a resist.

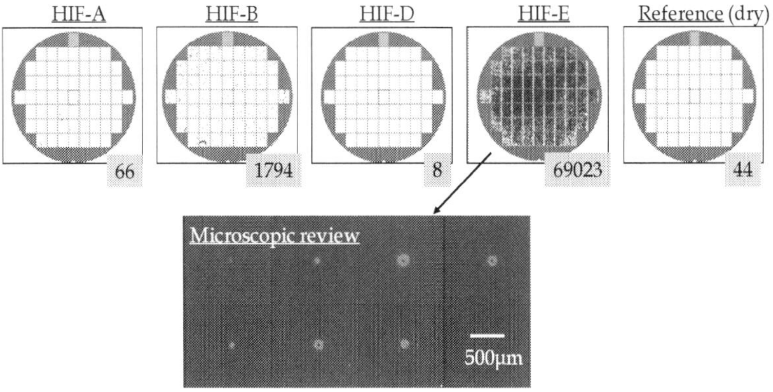

Fig. 22. Results of defect inspection. The number of defects after soaking in HIF-A, B, D, and E are 66, 1794, 8, and 69023, respectively.

3.4.3 Metal contamination

If the fluid contains a lot of metals, a wafer would be polluted by metals in the residual fluid. Thus, the objective of the following experiment is to confirm that G2 fluids are sufficiently clean. The requirement is less than 10^{10} atoms/cm^2 on a silicon wafer.

The conditions for the experiment are shown in Table 4. At first, a silicon wafer was dipped into a fluid for 5 minutes. Then, the wafer was pulled up and was dried by air blowing. Then, the metal concentrations on the wafer were evaluated by vapor phase decomposition inductively coupled plasma mass spectrometry (VPD/ICP-MS). The detection limit of the spectrometry is 0.03×10^{10} atoms/cm^2. The experiment was done twice for each fluid.

Table 5 shows the experimental results. The metal concentrations of the reference mean those on a wafer without dipping into any fluids. In the case of HIF-A, the concentrations of some metals are more than 10^{10} atoms/cm^2 (black-out cells in Table 5). On the other hand, HIF-B and D are sufficiently clean. As a result, the residual fluids of HIF-B and D are acceptable from the viewpoint of metal contamination.

Dipping	Immersion fluid	HIF-A, HIF-B, HIF-D
	Volume of fluid	200 mL
	Dipping time	5 minutes
	Wafer	Bare Si (200 mmϕ)
Analysis	Method	VPD/ICP-MS
	Evaluation area	180 mmϕ
	Detection limit	0.03×10^{10} atoms/cm^2

Table 4. Experimental conditions for metal contamination on a wafer dipped into a G2 fluid.

Fluid		Metal concentration ($\times 10^{10}$atoms/cm^2)													
		Li	Na	Mg	Al	K	Ca	Ti	Cr	Mn	Fe	Ni	Cu	Zn	Pb
Reference	1	<0.03	0.03	<0.03	0.03	<0.03	<0.03	0.43	<0.03	<0.03	<0.03	<0.03	<0.03	<0.03	<0.03
	2	0.06	<0.03	<0.03	0.05	<0.03	<0.03	0.32	<0.03	<0.03	<0.03	<0.03	0.03	<0.03	<0.03
HIF-A	1	0.15	0.30	0.09	1.30	0.28	0.28	0.71	0.29	0.09	1.57	0.87	0.05	0.13	<0.03
	2	0.11	0.92	0.05	0.85	0.97	0.36	0.45	0.13	0.19	0.82	0.40	0.04	0.10	<0.03
HIF-B	1	<0.03	0.45	0.05	0.23	0.13	0.38	0.48	<0.03	<0.03	0.10	<0.03	0.04	0.20	<0.03
	2	<0.03	0.48	0.07	0.29	0.14	0.47	0.58	<0.03	<0.03	0.06	<0.03	0.03	0.40	<0.03
HIF-D	1	<0.03	0.36	0.12	0.57	0.10	0.21	0.51	<0.03	<0.03	0.09	<0.03	0.07	0.06	<0.03
	2	<0.03	0.26	0.09	0.47	0.13	0.17	0.66	<0.03	<0.03	0.10	<0.03	0.04	0.03	<0.03

Table 5. Metal concentrations on a wafer evaluated by VPD/ICP-MS.

3.4.4 Dissolved oxygen

Oxygen in the atmosphere diffuses into the residual fluid on a wafer. When the oxygen-rich residual fluid goes back under the lens, it decreases the transparency of the fluid under the lens. That is the reason why dissolved oxygen is one of the issues arising with a residual fluid. Oxygen concentration in the fluid under the lens was evaluated while a wafer was moving along the sequence as shown in Fig. 23. HIL-203 (JSR) was used as a G2 fluid. A prototype immersion nozzle designed for a G2 fluid was used for the experiment. The processing time for one wafer was 30 seconds and it consisted of a scanning time and a waiting time at the edge of the stage. The fluid under the lens was sucked through the hole opened in the lens and the oxygen concentration in the sucked fluid was measured. The absorbance was calculated by using the relation as shown in Fig. 4.

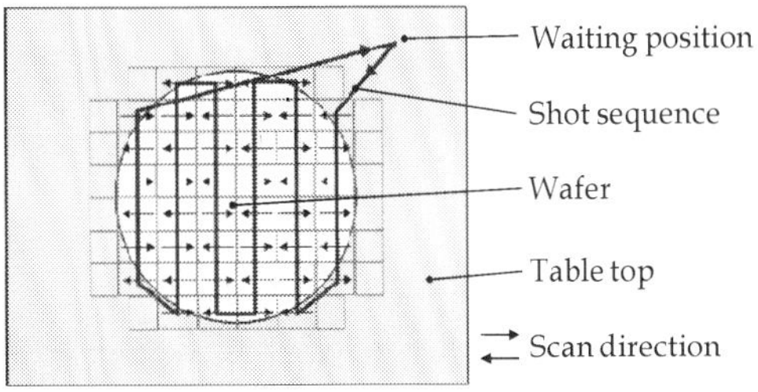

Fig. 23. Sequence of a scanning test on an 8 inches wafer.

Figure 24 shows fluid absorbance induced by dissolved oxygen against the number of scanned wafers. The absorbance rises with the scanning speed but it is low enough even at the scanning speed of 800 mm/sec. This result exhibits that the nozzle can reduce the amount of residual fluid sufficiently and can suppress the fluid darkening. In conclusion, the issue of dissolved oxygen was solved by the immersion nozzle designed for a G2 fluid.

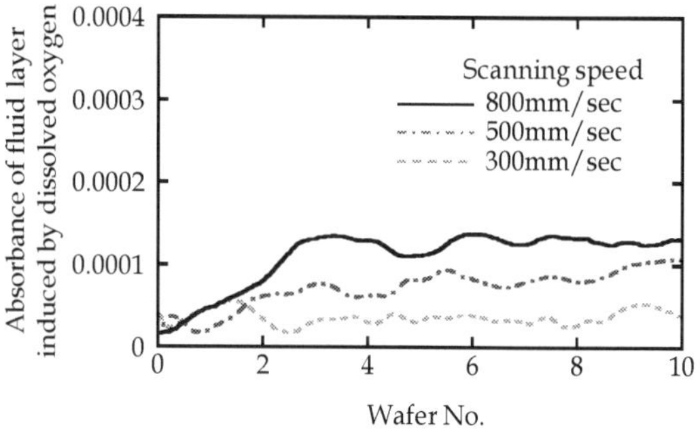

Fig. 24. Fluid absorbance between a lens and a wafer during the scan. The absorbance induced by dissolved oxygen is low enough even at the scanning speed of 800 mm/sec.

4. Remaining challenges

There are some remaining challenges to realize high-index immersion lithography. As described in the second subchapter, the quality of LuAG material does not reach the target specification. Especially, the absorbance of LuAG would be a critical issue. While the intrinsic absorbance of LuAG is lower than the target absorbance, further reduction of impurities is required.

On the other hand, there still remain a few issues to be studied in an immersion system using a G2 fluid. Although fluid absorbance can be kept low enough, dose homogeneity through a fluid layer should be confirmed. Defect study should be done with various resists. It is preferable to examine them with a preproduction tool using a G2 fluid.

For the extendability of high-index immersion lithography, the research activity on new materials such as G3 fluids and high-index resists is needed.

5. Conclusion

It has been discussed the feasibility on a high-index immersion system of 1.55 NA using LuAG and a G2 fluid. Although the IBR correction of LuAG is feasible, the quality of LuAG is not enough and the acceleration of its development is desired. The immersion system using a G2 fluid is being developed without serious issues. It was demonstrated that fluid absorbance can be kept low enough through an in-line purification unit and an oxygen removal unit. Lens contamination can be suppressed by addition of a small amount of water into a G2 fluid. Some issues arising with residual fluid, such as fluid darkening due to reentry of oxygen-rich residual fluid, were solved. By accepting residual fluid on a wafer, the scanning speed and the throughput can be raised.

EUVL is the orthodox candidate of the next generation lithography but still needs real verifications in various items. If EUVL is delayed, high-index immersion lithography will be at the leading edge lithography. It is a common knowledge that strong supports from the industry are indispensable to overcoming the remaining challenges such as LuAG. It is just said, "No market, no tool."

6. Acknowledgment

The author would like to acknowledge DuPont, JSR, and Mitsui Chemicals for supplying G2 fluids and purification units. The author also thanks colleagues for their cooperation.

7. References

Parthier, L.; Wehrhan, G.; Seifert, F.; Ansorg, M.; Aichele, T. & Seitz, C. (2008). High-index lens material LuAG : development status and progress, *presented at SEMATECH Litho Forum*, Bolton Landing, New York, USA, May 12-14, 2008

Parthier, L.; Wehrhan, G.; Seifert, F.; Ansorg, M.; Aichele, T.; Seitz, C. & Letz, M. (2008). Development update of high index lens material LuAG for ArF hyper NA immersion systems, *presented at 5th Int. Symp. on Immersion Lithography*, HI-01, The Hague, Netherlands, September 22-25, 2008

Burnett, J. H.; Kaplan, S. G.; Shirley, E. L.; Horowitz, D.; Josell, D.; Clauss, W.; Grenville, A. & Peski, C. V. (2006). High index materials for 193 nm immersion lithography, *presented at 3rd Int. Symp. on Immersion Lithography*, OO-21, Kyoto, Japan, October 2-5, 2006

Nawata, T.; Inui, Y.; Masada, I.; Nishijima, E.; Mabuchi, T.; Mochizuki, N.; Satoh, H. & Fukuda, T. (2007). High-index fluoride materials for 193 nm immersion lithography, *Proc. SPIE* Vol. 6520, 65201P

Letz, M.; Gottwald, A.; Richter, M.; Liberman, V. & Parthier, L. (2010). Temperature-dependent Urbach tail measurements of lutetium aluminum garnet single crystals, *Phys. Rev. B* 81, 155109

French, R. H.; Liberman, V.; Tran, H. V.; Feldman, J.; Adelman, D. J.; Wheland, R. C.; Qiu, W.; McLain, S. J.; Nagao, O.; Kaku, M.; Mocella, M.; Yang, M. K.; Lemon, M. F.; Brubaker, L.; Shoe, A. L.; Fones, B.; Fischel, B. E.; Krohn, K.; Hardy, D. & Chen, C. Y. (2007). High-index immersion lithography with second-generation immersion fluids to enable numerical aperatures of 1.55 for cost effective 32-nm half pitches, *Proc. SPIE* Vol. 6520, 65201O

Furukawa, T.; Kishida, T.; Miyamatsu, T.; Kawaguchi, K.; Yamada, K.; Tominaga, T.; Slezak, M. & Hieda, K. (2007). High-refractive index materials design for ArF immersion lithography, *Proc. SPIE* Vol. 6519, 65190B

Kagayama, A.; Wachi, H.; Namai, Y. & Fukuda, S. (2007). High-refractive index fluids for second-generation 193-nm immersion lithography, *persented at SPIE Advanced Lithography*, 6519-66, San Jose, California, USA, February 25-March 2, 2007

Zimmerman, P. A.; Byers, J.; Rice, B.; Ober, C. K.; Giannelis, E. P.; Rodriquez, R.; Wang, D.; O'Connor, N.; Lei, X.; Turro, N. J.; Liberman, V.; Palmacci, S.; Rothschild, M.; Lafferty, N. & Smith, B. W. (2008). Development and evaluation of a 193nm immersion generation-three fluid candidates, *Proc. SPIE* Vol. 6923, 69230A

Sakai, K.; Iwasaki, Y.; Mori, S.; Yamada, A.; Ogusu, M.; Yamashita, K.; Nishikawara, T.; Hara, S. & Watanabe, Y. (2008). Feasibility study on immersion system using high-index materials, *Jpn. J. Appl. Phys.* Vol. 47, No. 6, pp. 4853-4861

French, R. H.; Qiu, W.; Yang, M. K.; Wheland, R. C.; Lemon, M. F.; Shoe, A. L.; Adelman, D. J.; Crawford, M. K.; Tran, H. V.; Feldman, J.; McLain, S. J. & Peng, S. (2006). Second generation fluids for 193nm immersion lithography, *Proc. SPIE* Vol. 6154, 615415

Santillan, J.; Otoguro, A.; Itani, T.; Fujii, K.; Kagayama, A.; Nakayama, N.; Tamatani, H. & Fukuda, S. (2006). Novel high refractive index fluids for 193-nm immersion lithography, *Proc. SPIE* Vol. 6154, 61544Q

Sekine, Y.; Kawashima, M.; Sakamoto, E.; Sakai, K.; Yamada, A. & Honda, T. (2007). Feasibility of 37-nm half-pitch with ArF high-index immersion lithography, *Proc. SPIE* Vol. 6520, 65201Q

French, R. H.; Yang, M. K.; Peng, S.; Qiu, W.; Wheland, R. C.; Lemon, M. F.; Crawford, M. K.; Sewell, H.; McCafferty, D. & Markoya, L. (2005). Imaging of 32-nm 1:1 lines and spaces using 193-nm immersion interference lithography with second-generation Immersion fluids to achieve a numerical aperture of 1.5 and a k1 of 0.25, *J. Microlith., Microfab., Microsyst.* Vol. 4, 3, 031103

Wang, Y.; Miyamatsu, T.; Furukawa, T.; Yamada, K.; Tominaga, T.; Makita, Y.; Nakagawa, H.; Nakamura, A.; Shima, M.; Kusumoto, S.; Shimokawa, T. & Hieda, K. (2006). High-refractive-index fluids for the next generation ArF immersion lithography, *Proc. SPIE* Vol. 6153, 61530A

Furukawa, T.; Hieda, K.; Wang, Y.; Miyamatsu, T.; Yamada, K.; Tominaga, T.; Makita, Y.; Nakagawa, H.; Nakamura, A.; Shima, M. & Shimokawa, T. (2006). High refractive index fluid for next generation ArF immersion lithography, *J. Photopolym. Sci. Technol.* Vol. 19, 5, pp. 641-646

Gejo, J. L.; Kunjappu, J. T.; Zhou, J.; Smith, B. W.; Zimmerman, P.; Conley, W. & Turro, N. J. (2007). Outlook for potential third-generation immersion fluids, *Proc. SPIE* Vol. 6519, 651921

Liberman, V.; Rothschild, M.; Palmacci, S. T.; Zimmerman, P. A. & Grenville, A. (2007). Laser durability studies of high-index immersion fluids : fluid degradation and optics contamination effects, *Proc. SPIE* Vol.6520, 652035

Sakai, K.; Mori, S.; Sakamoto, E.; Iwasaki, Y.; Yamashita, K.; Hara, S.; Watanabe, Y. & Suzuki, A. (2008). Progress of high-index immersion exposure system, *presented at 5th Int. Symp. on Immersion Lithography*, HI-06, The Hague, Netherlands, September 22-25, 2008

Honda, T.; Kishikawa, Y.; Tokita, T.; Ohsawa, H.; Kawashima, M.; Ohkubo, A.; Yoshii, M.; Uda, K. & Suzuki, A. (2004). ArF immersion lithography: critical optical issues, *Proc. SPIE* Vol. 5377, pp. 319-328

Kubo, H.; Hata, H.; Sakai, F.; Deguchi, N.; Iwanaga, T. & Ebihara, T. (2007). Immersion exposure tool for the 45-nm HP mass production, *Proc. SPIE* Vol. 6520, 65201X

Sewell, H.; Mulkens, J.; Graeupner, P.; McCafferty, D.; Markoya, L.; Donders, S.; Samarakone, N. & Duesing, R. (2007). Extending immersion lithography with high index materials, *Proc. SPIE* Vol. 6520, 65201M

Shedd, T. A.; Schuetter, S. D.; Nellis, G. F. & Peski, C. V. (2006). Experimental characterization of the receding meniscus under conditions associated with immersion lithography, *Proc. SPIE* Vol. 6154, 61540R

Sanders, D. P.; Sundberg, L. K.; Brock, P. J.; Ito, H.; Truong, H. D.; Allen, R. D.; McIntyre, G. R. & Goldfarb, D. L. (2008). Self-segregating materials for immersion lithography, *Proc. SPIE* Vol. 6923, 692309

21

Double Patterning for Memory ICs

Christoph Ludwig and Steffen Meyer
Q-CELLS SE, Bitterfeld-Wolfen
Germany

1. Introduction

In order to continue technology shrink roadmaps and to provide year by year smaller chips with more functionality, nearly all of the leading edge semiconductor companies have adopted double patterning process technologies in their fabrication lines to bridge the time until next generation EUV lithography reaches production maturity. Double patterning technology can be classified into two major main streams, however its implementation and especially the details of the process integration vary strongly among the semiconductor company and every manufacturer found his own optimum.
Therefore the given schemata will focus on the key principles. We will not stress material combinations, but point out particular challenges, show some detailed analyses, and sketch solutions.

2. Double patterning by litho-etch-litho-etch

The most straightforward approach of doing double patterning is splitting a given pattern into two parts by separating neighbouring patterns (see figure 1a). By doing this the minimum pitch of each split will be enlarged and becomes again printable with standard ARF immersion exposure tools. At the first lithography step, a photo mask containing only the blue part of the pattern will be used. As one can easily see from figure 1b, the pitch of blue only pattern is about two times larger than in the original layout. However, in order to achieve a good optical contrast during imaging as well as to get sufficient (photo resist) process window for the patterning, the pattern will usually be biased on the mask as shown in figure 1c.

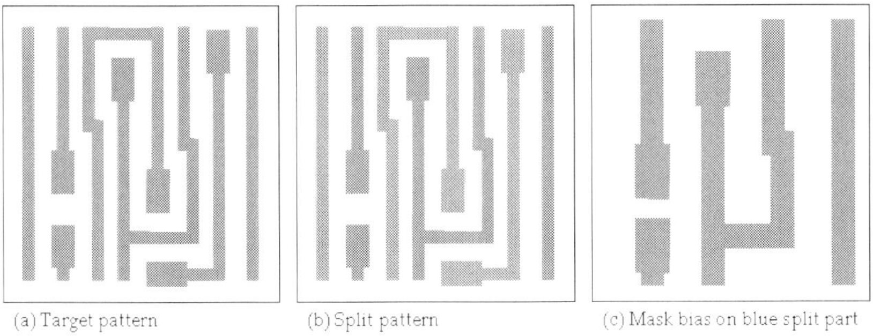

Fig. 1. Example of decomposing an irregular structure for double patterning processing.

Triangular conflicts deserve special attention. They can for instance be solved by using junctions to bridge split part features with each other as shown in figure 2. Another way is to restrict the layout to full splits and to use separate interconnects. Despite the need of such kind of additional rules, LELE can handle complex layout and modern CAD tools do support pitch splitting well.

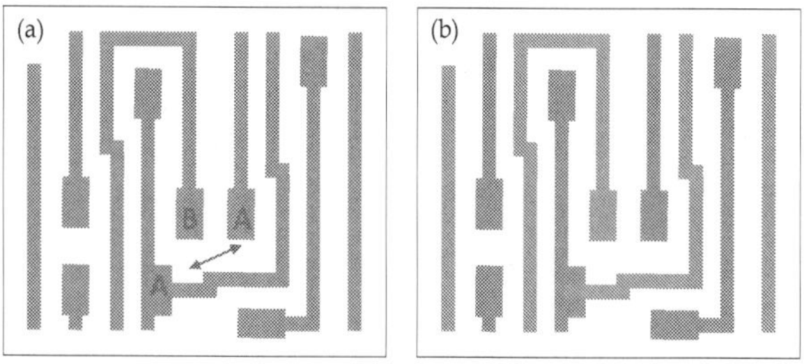

Fig. 2. Topological conflict (a) and a possible decomposition by a junction bridge (b).

To achieve the intended structure dimensions, the applied bias must be removed again. Overexposing the photo resist is commonly used to reduce the bias, but there are limits for process stability reasons. Therefore further bias reduction might be required. Often this will be done in subsequent etch steps, which also have the function of "freezing" the pattern, so that it will not be destroyed when the second part of the pattern will be added. Figure 3 shows a sample sequence.

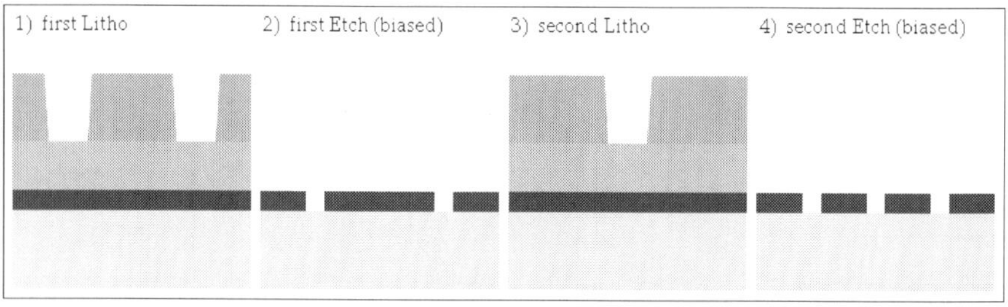

Fig. 3. Process sequence for litho-etch-litho-etch double patterning.

The first pattern freeze is typically done in a thin sacrificial hard mask. Although this thin sacrificial hard mask adds process complexity and thus fabrication cost, there are two good arguments for introducing it. The argument #1 is as follows: It is easier to assemble all pattern elements first within a thin sacrificial layer and later transfer them altogether to the intended layer stack, rather than structuring the full intended layer stack step by step. This is because the intended structures on wafers tend to have large aspect ratios. But once the second lithography gets carried out, the topography of the wafer must be nearly flat. Otherwise resist spin-on gets problematic and one has to struggle with reflective notching and local CD (critical dimension) variation. That means, without using the sacrificial thin hard mask, complex planarization steps would be necessary. However, the thin hard mask topography differences can easily be covered by BARC / bottom layer coatings, which have to be applied for ARF resist systems anyway. The argument #2 is: Hard masks are typically needed for the plasma etch process of the intended layer stack anyway. The limits of the plasma etch selectivities and the aspect ratios require the usage of sophisticated hard mask stacks. Etching a complex stack twice is also more costly than doing the double patterning etch sequences only on the thin hard mask and perform the complex and expensive full stack etch only once. Last for the sake of completeness we want to mention here, that for cost and complexity reasons techniques of freezing the first split pattern directly in photo resist system are showing up in the industry.

Splitting the given pattern into two parts allows complex designs. Therefore this type of double patterning is the preferred choice for logic manufactures. But since two lithography steps must be done to achieve the patterning for one layer, the intra-layer overlay must be controlled with a very high degree of precision. This was challenging in the past, because first and second generation ARF immersion tools could hardly achieve overlay control beyond 6-8nm, which is above the needed budget for 36nm...32nm and 28nm...24nm technologies. Around 2010, lithography tool makers removed this blocking point by introducing a third generation of ARF immersion tools with can meet double patterning overlay control requirements. This double patterning technique is usually referenced with Litho-Etch-Litho-Etch (LELE) but other names such as Brute Force or Pitch Splitting exist as well.

For the extremely regular arrays of memory chips like e.g. NAND flash memories, aggressive double patterning technology nodes have been introduced already in 2008, before exposure tools with improved overlay control became available. Memory chip manufacturers have therefore driven another double patterning main stream, which we will describe in the next paragraph.

3. Self aligned double patterning

The second main stream is often called self aligned double patterning. Some people call this technique pitch fragmentation and you might often hear spacer based double patterning, because for most of this techniques spacer-like approaches are used to double the number of features.

Self aligned double pattering comes with strong layout limitations. However, it is well suited for regular patterns e.g. of memory chips. The design process is more complex than LELE and the pattern on the photo masks look different compared to the final pattern. This creates a first challenge, because layout versus schematic checks require additional CAD tooling to transform the self aligned double patterning mask data into the final layout data. The transformation in the other direction - from final layout to double patterning layout - is even more challenging. Automated tools are rare and manual interaction of the designers is required. This is limiting the application to a manageable number of layout features and is the main reason why self aligned double patterning is mostly only used by memory manufactures.

The big advantage of self aligned double patterning is that only one main feature lithography step has to be done. However, one or two additional assist lithography steps with considerably reduced resolution requirements are used to form the edges of the regular pattern within the memory array as well as the transition from the regular array towards the remaining area of the chip, containing the control function elements of the memory.

The lithography tool overlay requirement is only driven by the inter-layer technology demands and not by the intra-layer alignment as for LELE.

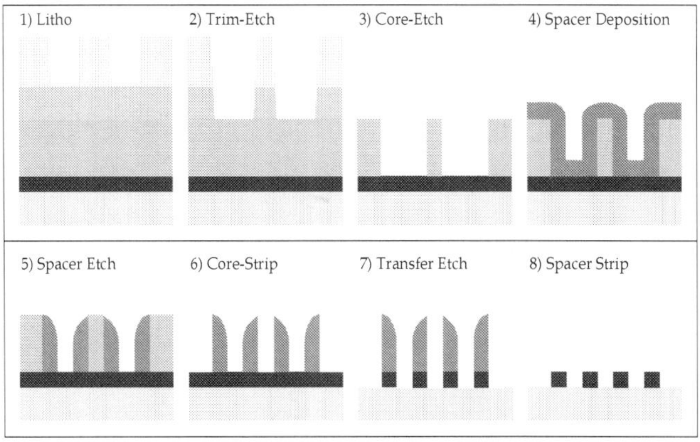

Fig. 4. Sketch of a process sequence for "line by space" double patterning.

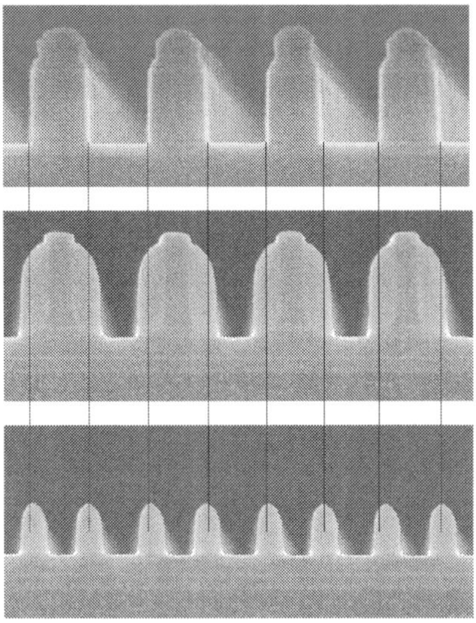

Fig. 5. Example of a process sequence "line by spacer": The SEM cross section at the top reveals the primary line pattern as defined by lithography, after print and dry etch. The cross section in the middle shows the primary lines now covered by (sacrificial) spacers, after spacer deposition and spacer etch. The bottom picture is taken after removal of the primary lines. The sacrificial spacers have become now a regular pattern of twice the spatial frequency compared to the primary pattern, and can be further used as a hard mask for the subsequent patterning of the desired stacks.

With the main lithography step, a so called "primary" or "core" pattern is printed on the wafer. This pattern has a pitch twice as large as the final structures and gets usually biased for process window reasons. Out of this pattern, a tall rectangular-shaped core structure with an aspect ratio of typical 1:3 to 1:5 is formed. This can be done directly in the photo resist system, or - more common - can be transferred by reactive ion etch into a relative thick underlying hard mask. Afterwards a thin conformal liner will be deposited onto the core shape. Now a spacer etch is performed and the core pattern gets stripped out. Finally the spacer structures will be transferred into another under laying thin hard mask and the spacer will be taken off as well. Figure 4 and 5 show such a process sequence. A detailed process description including key process parameters is listed in table 1. For this sequence the final line structure width is defined by the deposited thickness of the sacrificial spacer. Therefore this scheme is called "line by space" (LBS). All created lines will have the width of the sacrificial spacer. There is no degree of freedom for the chip layouter to modify the line width. Only the space between the lines can be varied by layout.

Process Step	Target
Deposition of CVD carbon hard mask	150nm (should to be adjusted to layer requirement)
Anneal carbon	630°C, 60min
Deposition of CVD multi-layer stack (double patterning hard mask)	25nm SiON + 25nm a-Si + 35nm SiON + 150nm a-Si + 20nm TEOS (from bottom to top)
Litho 193nm immersion	72nm line / 72 nm space CD uniformity < 3nm
Trim etch of poly mask	~ 42nm line CD uniformity < 2nm taper within the array: 90 deg
Strip Resist	
Clean Wet	minimum attack of SiON
Etch Wet TEOS	high selectivity, minimum attack of SiON
Deposition CVD HARP liner	36nm thickness thickness uniformity < 0,5nm
Etch Spacer Oxide	
Etch Recess Poly	complete carrier a-Si removal
Etch Plasma SiON	spacer transfer etch, stop on a-Si
Etch Wet Oxide	remove space; minimum attack of SiON
Subtractive Patterning	etch SiON selective to a-Si
Additive Patterning	etch aSi with SiON and resist mask with stop on SiON

Table 1. Process Sequence for Spacer Transfer Scheme (LBS) with process targets for the gate layer in a 36nm NAND flash chip.

Some applications however require lines with well-defined different line widths, but can accept equal spaces. This can be realized by a slight modification of the described processing sequence: filling the pattern after the spacer etch followed by a controlled recess etch and spacer material strip. Figure 6 demonstrates an example for a process flow which is called "line by fill" (LBF).

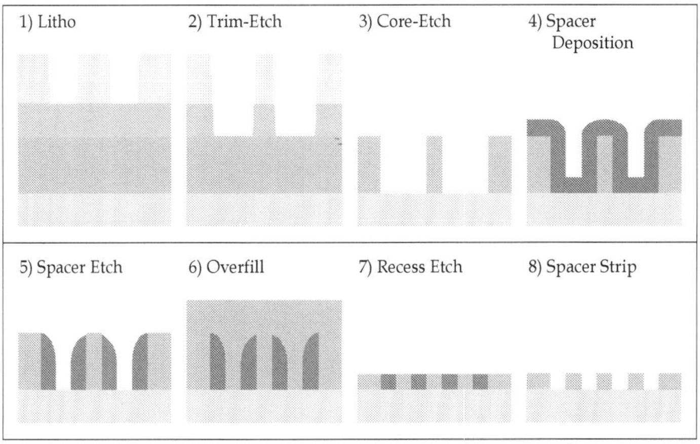

Fig. 6. Sketch of a process scheme for "line by fill" double patterning.

3.1 Line by space (LBS) – Line by fill (LBF)

The decision for either of the two basic self-aligned double patterning schemes LBS or LBF has to be taken for each lithography layer individually. In general, LBS is advantageous if the tolerances in line width are most critical for the electrical functionality of the product, and LBF is the preferred option if the distance between the lines are most critical for the product functionality. Figure 7 illustrates this difference. Therefore, sound knowledge of the fabrication process tolerances and thorough simulation are mandatory.

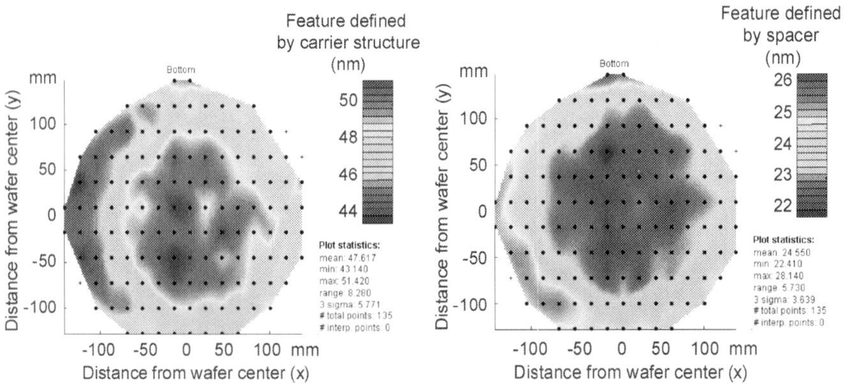

Fig. 7. Critical dimension of the gate hard mask, measured across the full wafer. The left hand graph shows the hard mask spaces, the "space-1", being defined by the primary line, whereas the right hand graph reprents the hard mask "lines", being formed by the sacrificial spacers. The hardmask lines show a much smaller spread in width than the hard mask spaces. Note that the hardmask lines are narrower than the intended gates: this is to compensate for the gate etch bias.

3.1.1 Process tolerances

One of the biggest challenges for self aligned double patterning is the control of the critical dimensions. Compared to the traditional litho, when CD control was more or less managed by lithography, self aligned double patterning relies also on tight process control for etch and deposition processes. For the example sequence sketched in figure 4, the line width is defined by the thickness of the deposited (spacer) liner and the subsequent spacer etch. Extremely low process variations, exact reproducibility, uniformity are mandatory. Therefore, special care has to be dedicated to the search of well controlled deposition processes with reasonable process cost.

The first process choice for a well-controlled deposition process could be LPCVD. However, most of the LPCVD furnace processes require process temperatures above 500°C. Front-end application might allow such temperatures, but in the back-end of line such high temperature would destroy the already manufactured metal lines or contact junctions. Therefore, LPCVD cannot be applied to critical metal layers.

PECVD processes can run at lower temperatures and come with acceptable cost. However, still most of the PECVD deposition processes are not suitable for spacer based double patterning, because the deposition mechanism is transport limited. In other words: the new material deposition reaction is very fast and is limited by how fast and how much new material can be transported out of the plasma to the wafer. The "stream" of new material is limited in time and will be distributed over the wafer surface. As a consequence the thickness of the deposited liner will vary with the surface area. Therefore dense structure areas with larger surface will see thinner liner ticknesses compared to isolated structures. Unfortunately this so-called micro-loading has quite some range and will create dozens of array edge line with different line width. As a result, PECVD processes cannot deliver tight in-chip CD control.

A highly promising process candidate is atomic layer deposition (ALD). They come with extreme good thickness uniformity and can be done at very low temperatures. The first ALD processes hat the drawback of a relatively slow process speed, resulting in high process cost. However, the original ALD with its slow atom-by-atom like deposition has experienced considerable improvements. Nowadays modified ALD processes are available like e.g. catalytic (fast) ALD, spatial ALD, molecular layer depositions or pulsed plasma depositions. Some of these advanced deposition processes can be done at temperature around or even below 100°C. This enables spacer liner depositions directly on photo resist type core materials.

The liner thickness variations are not the only critical parameter. Also for other process steps, the control of critical dimensions is challenging. The CD variation of the "primary" or "core" space results from both litho and core-etch non-uniformity. The line width variation depends on spacer deposition and etch uniformity. The "secondary" or remaining space in principle will suffer from the sum of both variations. In order to keep the "secondary" or remaining space within the allowed technology variation specifications, one has either to run each process with very tight specifications or to implement an advanced feedforward / feedback APC system. Such a system is capable e.g. of tuning the etch-bias as a function of the photo resist CD after the lithography patterning.

The uniformity of the line width is affected by the topography of the environment. There are both short-range effects and long-range effects. Transport processes and reaction kinetics of chemical and plasma reactions are sensitive to the effective surface in the vicinity of a considered line as well as to the macroscopic location on the wafer. In addition to the structure dimension of interest, monitoring structures need to be developed for a fast and reliable monitoring of the mass production.

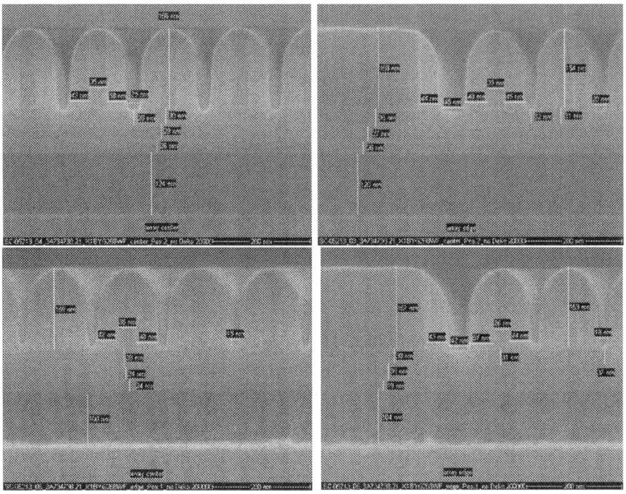

Fig. 8. Micro loading in the double patterning hard mask for the gate level of a 36nm NAND Flash. The hard mask is made for the line by space (LBS) scheme.

We have selected the gate layer of a 36nm flash memory as an example of micro loading. Figure 8 shows measurements taken from cross-sections after the formation of the sacrificial spacer at the sides of the primary lines. The hard mask is not yet structured. An in-line measurement at this process step will reveal the sum of the width of the primary line plus two adjacent spacer thicknesses, equalling roughly the desired line with plus two spaces. An intuitive way of plotting this in-line measurement number is shown in Figure 9.

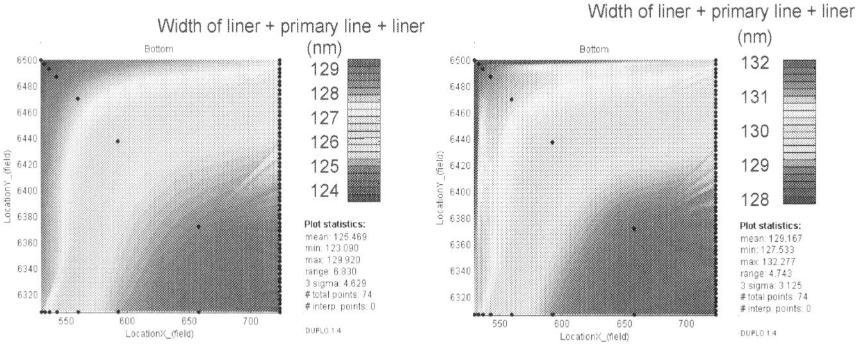

Fig. 9. Plot of the in-line measured line width consisting of the primarily formed line plus two adjacent spacers. The coloured area represents a quarter of the regular gate array. The center of the array is located at the bottom right of the plot, the corner of the array corresponds to the top left corner of the plot. Obviously, the line width increases from the center of the array towards the corners. The right hand graph was taken from an improved spacer deposition process, showing a narrower width distribution.

After removing the primary lines and transforming the sacrificial spacers into the hard mask, one obtains values as given in Table 2.

Wafer Uniformity Center to Edge	3nm
Micro Loading Array to Select Gate Edge	2..3nm
Micro Loading Gate Array Edge	6..10nm
Micro Loading Array to Litho Monitor (nested)	10nm
Micro Loading Array to Litho Monitor (isolated)	17nm

Table 2. Experimental data for uniformity and micro loading and uniformity of the the gate layer of a 36nm NAND flash patterned by LBS.

3.1.2 Gate

In general, the gate length is critical, requiring LBS (line by spacer) for the gate level. Typical electrical parameters of relevance are:

- Narrow distribution of IV characteristics of the transistors, mainly for:
- Large on-currents: short gates are preferable
- Source-Drain leakage: long gates are preferable
- Hot carrier effects: long gates are preferable

In the case of a flash memory, another electrical parameter comes into play: the coupling ratio, which determines mainly the programming and erase characteristics of the flash transistors. The coupling ratio is affected by the capacitance between the floating gate and the active area, i.e. the CD variations of both the (floating) gate and the active area cause a spread in programming and erase performance.

As a result, a carefully controlled balance of the gate length in all kinds of chips is more important than the control of the gate spaces.

3.1.3 Active area / STI

For the active area, the following elements are relevant:

- Transistor on-current: active area width variations translate directly into transistor current variations.
- Coupling ratio of flash transistors: active area width variations translate into the program and erase speed.
- Transistor leakage: The STI (shallow trench isolation) fill turns out to be highly sensitive to the STI trench width. Trap centers at the boundary of transistor channel edge and STI fill can result in undesired leakage currents, both along the channel direction and across the source/drain junction into the substrate.

The balance of the mentioned three arguments led the authors to the decision of LBF (line by fill) for the active area layer.

3.1.4 Bit line contact

NAND-Flash memory arrays require single rows of so called dense bit line contacts. They connect the metal bit lines to the active area, and they are following the critical pitch. The special challenge consists in the lack of any room for overlay tolerances and in the extreme asymmetry of the pattern: the contact row can be considered an isolated structure in one dimension, and in the orthogonal dimension a periodic structure with minimum pitch.

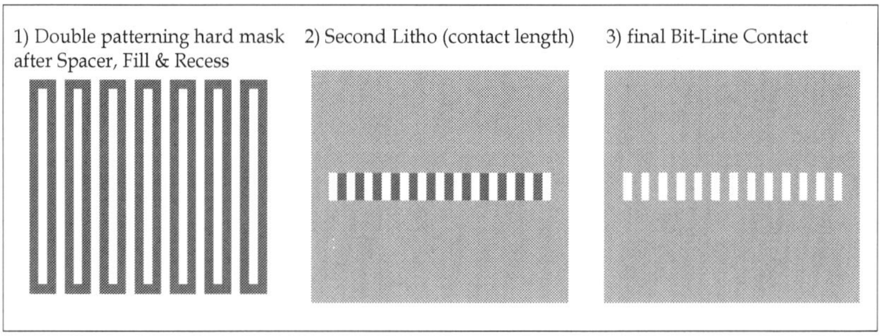

Fig. 10. A two-mask process for the formation of a single row of on-pitch contacts.

Figure 10 shows one example how to realize elongated dense bit line contacts. In a first step a dense array-like hard mask is created by double patterning. The lines and spaces of this array-like hard mask define the bit line contact widths and the contact to contact distances, respectively. The array is much larger than the intended contacts in order to reach a good imaging in the lithography tool. A second lithography layer is used to cut out only a small region of the array-like hard mask. This second litho step prints a long opening in a resist mask across the array-like hard mask. The width of this long opening defines the length of the contacts. The following etch process is to open only the regions which are neither covered by the spacer nor by the photo resist.

3.1.5 Tight pitch metal layer

The predominant issues for the selection of the proper double patterning scheme for metallization are

- dielectric breakthrough,
- variations of capacitive coupling,
- variations of resistance,
- feasibility of landing pads for contacts / vias for the electrical connection to layers beyond or beneath.

Typically, bit lines are supposed to run fast electrical signal pulses with lowest possible loss and lowest possible cross-talk. This requirement suggests that a lower limit of the metal space might be important in order to minimize RC losses. Simulations have been performed for the bit line capacitance in a flash memory for both the LBF (line by fill) and the LBS (line by space) process scheme. The results led to the conclusion that both LBS and LBF show nearly identical variations with respect to the capacitive coupling between adjacent bit lines. The resistivity variations, however, are clearly in favour of the LBS scheme, which has a well controlled metal line width, i.e. the resistance tolerances are much better than for LBF. But in the end, deeper study of process architecture and complexity and cost led to the decision of LBF.

Two-dimensional electric field and capacitance simulations have been performed for LBF (figure 11) and LBS. The calculations include fringing fields for up the third neighbour bit line as well as the capacitance to the top and bottom layers of the chip. Worst and best cases, i.e. largest and lowest capacities have been calculated. For the variations, not only the double patterning induced line and space variations have been considered, but also variations in bit line thickness. The target thickness of the considered example is 60nm for the full metal stack, the tolerance is +/- 10nm.

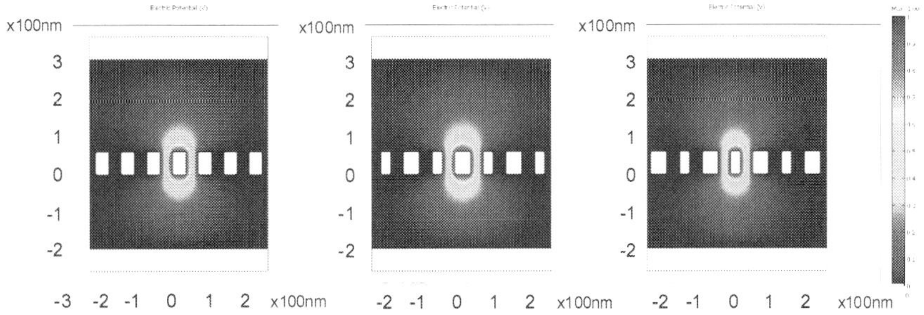

Fig. 11. Electric field simulations for bit line capacitance calculations, depicted for the LBF process.

It is not upfront clear by intuition, which combination of double patterning variations will result in the largest and lowest bit line capacitance, respectively. For example, in the LBS scheme, the smallest bit line capacitance does not occur for the largest primary space. The reason is the linear (inter-)dependence of the four parameters: The sum of two spaces and two lines is fixed to twice the pitch, 144 nm in our example. An illustration of the line / space variations towards lower capacitances is given in figure 12. The lowest possible value for LBS (3 sigma geometrical variations assumed) is 2,01 x 10^{-10} F/m, which is 10% below the value of the target geometry (2,23 x 10^{-10} F/m).

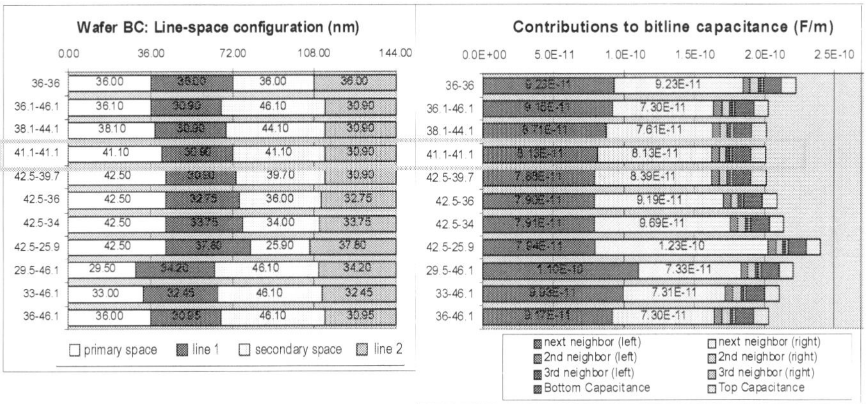

Fig. 12. Bit line capacitance for various extreme geometry parameters in the LBS scheme, for lowest capacitances from process variations ("best case"). A totally symmetric configuration of 41.1 nm for both the primary and the secondary space and a line width of 30.9 nm is found to be the best case, resulting in a total capacitance of 2,01 x 10^{-10} F/m.

A summary of the results is depicted in figure 13 and listed in table 3. The outcome of the considerations is a tiny preference for the LBS scheme.
The feasibility of process architecture finally requires a clear decision between the two double patterning schemes. A reliable connection to the bit line contact and a robust process

margin require some kind of so called landing pad, i.e. a local enlargement (landing pad) of the bit line around the bit line contact. This is only feasibly by LBF, since only in the LBF, the layouter can intentionally modify the width of the bit line. The LBS would ascribe the bit lines a well-defined, fixed width.

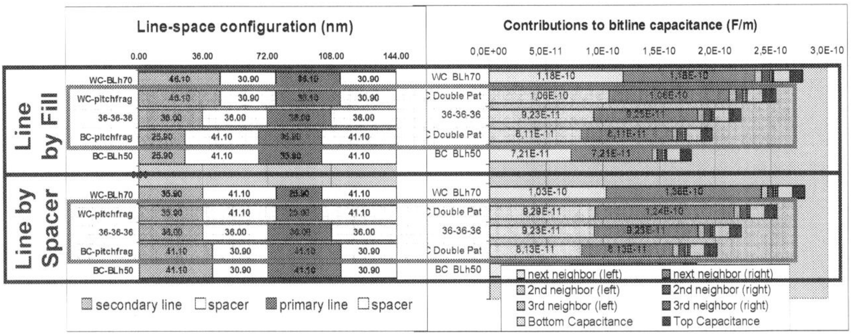

Fig. 13. Bit line capacitance in a 36nm flash M0 layer for the extreme 3 sigma cases of double patterning variations, marked by the red boxes, and in addition bit line height variations of +/- 10nm.

	LBF (10^{-10} F/m)	LBF (%)	LBS (10^{-10} F/m)	LBS (%)
WC incl. BL height (3 σ)	2,61	17	2,61	17
WC (double patt. only) (3 σ)	2,54	14	2,55	14
Target dimension	2,23	0	2,23	0
BC (double patt. only) (3 σ)	1,97	-12	2,01	-10
BC incl. BL height (3 σ)	1,90	-15	1,93	-14

Table 3. Bit line capacitance (numbers absolute and relative to the geometrical target dimensions) for LBF and LBS. All variations represent 3 sigma values. The range between the largest, i.e. worst case (WC) capacitance and the lowest, i.e. best case (BC) capacitance is roughly identical for both double patterning schemes. Besides the pure double patterning effects, the table also considers a calculation of the bit line height variations. Altogether, the two double patterning schemes do not exhibit a significant difference.

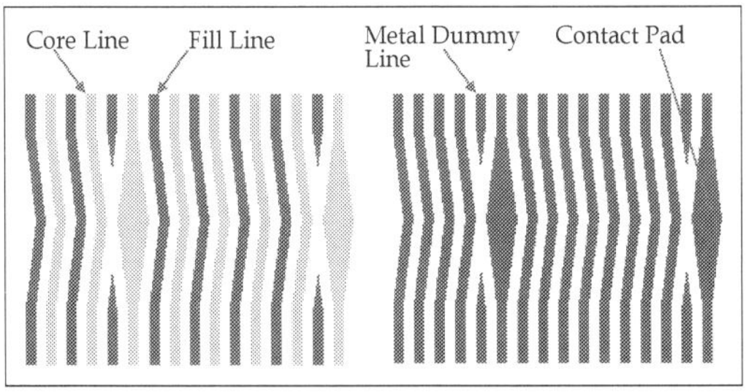

Fig. 14. LBF allows local enlargements of the bit line width.

In an NAND flash, the regular bit line pattern exhibits periodic interruptions which are needed for the source line contacts. This enables the introduction of some dummy extra bit lines. They are used to allow tiny landing pads on top of the bit line contacts. Figure 14 shows an example for bit line landing pads.

3.2 Fan-out

So far, we have focussed our discussion mainly on the regular array. The edge of the array deserves special attention.

Self-aligned double patterning naturally creates lines or spaces which are connected pair-wise. In general, this is not desired. Therefore after the array formation by double patterning, the edge of the array is cut by applying a second uncritical litho & etch step, as illustrated in figure 15. Preferably, this step is applied to the hard mask before etching the full stack.

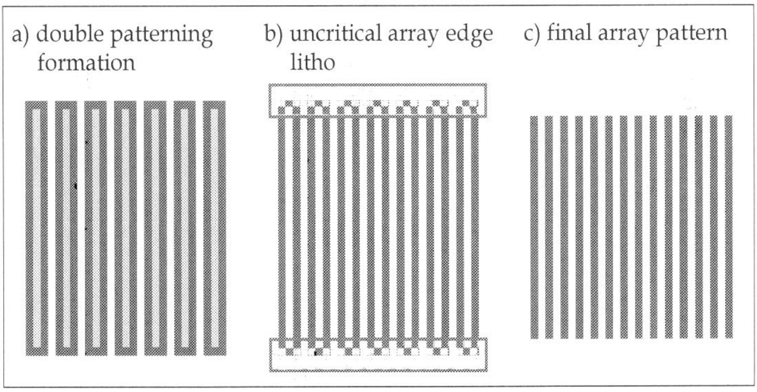

Fig. 15. An uncritical litho & etch step is used to remove the unintended pair-wise connections of the double patterning lines.

The array lines have to be connected electrically to other functional elements on the chip. This requires a so-called fan-out, a topological connection between the double patterning elements and the structures printed by direct lithography.

A possible process sequence of a "Christmas tree" fan-out for the gate layer of a NAND flash is shown in figure 16. The ends of the lines formed by primary lithography are drawn in a shape as depicted in figure 16 (a). The first lithography step prints the primary lines, having twice the array pitch, and forms the fan-out core. The spacer covering the sidewalls of the primary lines follows the core edges and routes each array line into the Christmas tree fan out (figure 16 (b)). The spacer has the intended pitch within the array and is transformed into array lines. They are connected in pairs. Therefore an additional (uncritical) litho & etch cut step must be performed (figure 16 (c)). Finally, landing pads are added for better overall process robustness (figure 16 (d)). The additive structuring requires one more litho & etch, which is usually combined with the periphery patterning. For the additive structuring, the combination of the already formed double patterning hard mask and the add-on photo resist act as an etch mask for the final etch.

A SEM photography of such a "Christmas tree" word line fan-out is given in figure 17.

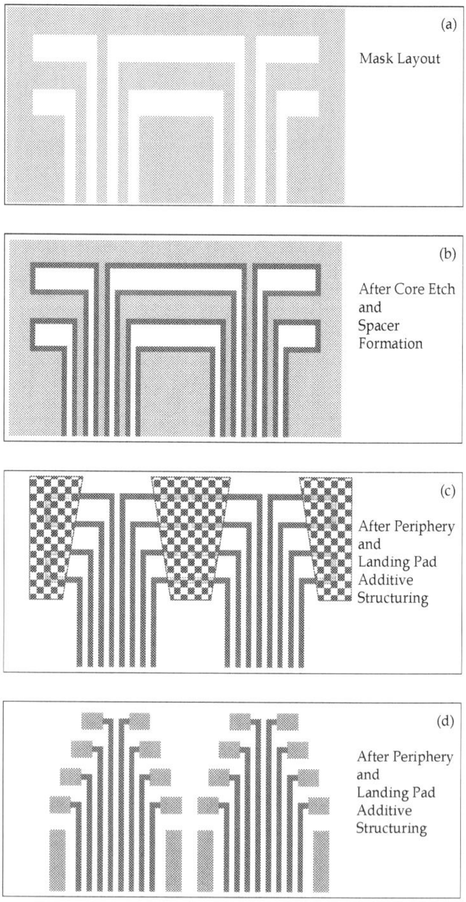

Fig. 16. "Christmas tree" fan-out for the gate layer of a NAND flash.

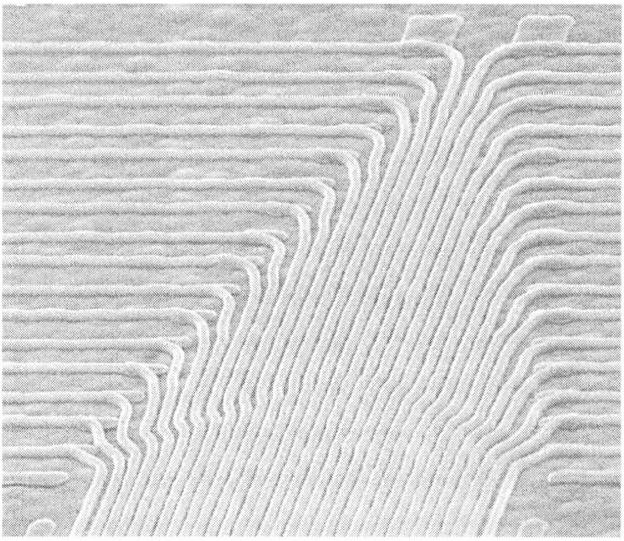

Fig. 17. SEM view of a word line (gate) fan-out for a 36nm half pitch array.

4. Conclusion

Double Patterning has become an important technique for advanced microelectronics devices. Cost-competitive memory chips have structure dimensions which are well below the diffraction limit of any available productive lithography tool. The generation EUV lithography is still on its way towards production maturity.

We have shown a classification of double patterning and have discussed possible implementations. Their specific challenges and advantages have been considered. For a 36nm NAND flash as an example, process flows have been presented.

5. Acknowledgment

This work was financially supported in part by the European Commission in the projects GOSSAMER and PULLNANO, and by the Federal Ministry of Education and Research of the Federal Republic of Germany (project nos. 01M3167 A and 01M3171 A).

The authors would like to express their gratitude to the whole flash development team of the former company Qimonda, especially Tim Höhr for the capacitance simulations.

6. References

Beug, M.F.; Hoehr, T.; Müller, T.; Reichelt, R.; Müller-Meskamp, L.; Geiser, P.; Geppert, T.; Bach, L.; Bewersdorff-Sarlette, U.; Kenny, O.; Olligs, D.; Brandl, S.; Marschner, T. ; Parascandola, S.; Meyer, S.; Riedel, S .; Specht, M.; Manger, D.; Knöfler, R.; Knobloch, K.; Kratzert, P.; Ludwig, C. & Küsters, K.-H. (2008). Pitch Fragmentation Induced Odd/Even Effects in a 36nm Floating Gate NAND Technology,

International Electron Devices Meeting Technical Digest, p.353, San Francisco, CA, USA, December 15-17, 2008, ISSN 8164-2284

Beug, M.F.; Knöfler, R.; Ludwig, C.; Hagenbeck, R.; Müller, T.; Riedel, S.; Höhr, T.; Sachse, J.-U.; Nagel, N.; Mikolajick, T.; Küsters, K.-H. (2008). Charge cross talk in sub-lithographically shrinked 32nm Twin Flash memory cells, *Solid-State Electronics Vol.* 52, pp. 571-576

Ghaida, R.S.; Torres, G.; Gupta, P. (2011). Single-Mask Double-Patterning Lithography for Reduced Cost and Improved Overlay Control, *Semiconductor Manufacturing*, Volume: 24 , Issue:1

Dusa, M.; Arnold, B.; Finders, J.; Meiling, H.; van Ingen Schenau, K.; Chen, A. C. (2008). The lithography technology for the 32nm HP and beyond, *Proceedings of the SPIE*, Volume 7028, pp. 702810-702810-11, DOI: 10.1117/12.796016

Bencher, C.; Chen, Y.; Dai, H.; Montgomery, W.; Huli, L. (2008). 22nm half-pitch patterning by CVD spacer self alignment double patterning, *Proceedings of the SPIE*, Volume 6924, pp. 69244E-69244E-7, DOI: 10.1117/12.772953

Wie,Y.; Brainard, R. L.. (2009). Advanced Processes for 193-nm Immersion Lithography Book, ISBN: 0819475572

Chiou, T.-B.; Socha, R.; Kang, H.-Y.; Chen, A. C; Hsu, S.; Chen, H.; Chen, L.. (2008). Full-chip pitch/pattern splitting for lithography and spacer double patterning technologies, *Proceedings of the SPIE*, Volume 7140, pp. 71401Z, DOI: 10.1117/12.804763

Noelscher, C.; Jauzion-Graverolle, F.; Heller, M.; Markert, M.; Hong, B.-K.; Egger, U.; Temmler, D. (2008). Double patterning down to k1=0.15 with bilayer resist, *Proceedings of the SPIE*, Proceedings of the SPIE, Volume 6924, pp. 69240Q-69240Q-12, DOI: 10.1117/12.772750

Ludwig, C ; Beug, M.F. ; Küsters, K.-H. (2010). Advances in Flash Memories, *Materials Science-Poland*, Vol. 28, No. 1. 105-116

22

Diffraction Based Overlay Metrology for Double Patterning Technologies

Prasad Dasari[1], Jie Li[1], Jiangtao Hu[1], Nigel Smith[1] and Oleg Kritsun[2]
[1]*Nanometrics*
[2]*Globalfoundries*
USA

1. Introduction

193nm optical immersion lithography is approaching its minimum practical single-exposure limit of 80nm pitch [1]. The semiconductor industry has adopted double patterning technology (DPT) as an attractive solution for the low k_1 regime until extreme ultraviolet (EUV) lithography becomes commercialized. DPT also brings additional demands of increased critical dimension uniformity (CDU) and decreased overlay errors. The International Technology Roadmap for Semiconductors (ITRS) [2] target for overlay control at the 32nm DRAM node in single patterned lithography steps is 6nm. The process budget is reduced to 1.1nm for DPT. If 20% of the process error budget is allowed to occur in the metrology tool, as the ITRS states, then the measurement error budget at the 32nm node is 1.2nm for single patterning, and 0.22nm for DPT.

The ITRS defines total measurement uncertainty (TMU) for overlay only in terms of precision, tool-induced shift (TIS) variation and site-to-site tool matching differences. Determining whether a measurement technology is capable of controlling these advanced processes is no longer a case of simple data self-consistency checks on precision, TIS and matching. For example, the error arising from assumptions of a linear change of overlay error with position is significant. This error can be reduced by using very small targets [3] and performing in-device overlay measurements, but the demanding sub-nanometer measurement budget in overlay measurements still remains a considerable challenge.

Recent advances in lithography metrology for advanced patterning have led to the proposal of three different pitch splitting technologies [Fig. 1]. The Litho-Etch-Litho-Etch method (LELE, Fig. 1a) involving two process steps requires very tight overlay control and is both very expensive and slow, making alternative methods attractive. The first alternative process flow is Litho-Freeze-Litho-Etch (LFLE), which reduces the processing cost by replacing the intermediate etch step with a process step in the litho track (Fig. 1b). After exposing the first pattern, the resist is baked in a post-exposure bake (PEB) step and developed. Exposed pattern is coated with material to freeze the resist. The second resist layer is added and the second exposure is done. The freezing material prevents the first resist layer from washing away during the second layer PEB and develop steps. This technique allowed printing 2D logic cells and dense poly lines with two lithography steps, illustrating good resolution and process margin [4].

The next alternative process is Self-Aligned Double Patterning (SADP, Fig. 1c), in which a spacer film is formed on the sidewalls of pre-patterned features. Etching removes all the material of the original pattern, leaving only the spacer material. Since there are two spacers for every line, the line density has now doubled. The spacer approach is unique in that with one lithographic exposure the pitch can be halved indefinitely with a succession of spacer formation and pattern transfer processes. The spacer film deposition process is very uniform and results in extremely good SADP CDU of less than 1nm. The spacer lithography technique has most frequently been applied in patterning fins for FinFETs and metal layers [5].

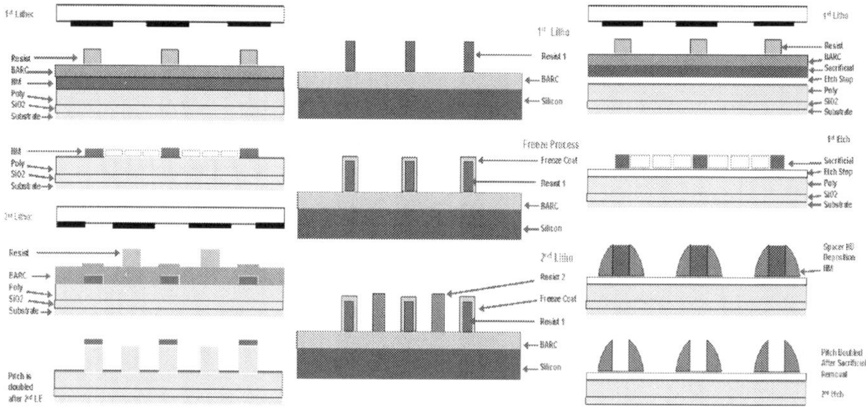

(a) (b) (c)

Fig. 1. Various double patterning schemes: (a) Litho-etch-litho-etch (LELE), (b) Litho-freeze-litho-etch (LFLE), and (c) Self-aligned double patterning (SADP)

These pitch splitting double patterning techniques not only involve more demanding process steps, they also require tighter overlay control than conventional single patterning [2]. Therefore measurement of overlay with much higher certainty is a necessity. As technology transitions toward the 22nm and 16nm nodes using these methods there is serious concern about the capability of the available metrology solutions, both in process development and production control.

High TIS and tool-to-tool matching errors make it difficult to meet the measurement uncertainty requirements using the traditional Image-Based Overlay method (IBO), even though most advanced IBO tools are operating at TMU levels under 1nm. Diffraction-based (scatterometry) overlay (DBO) measurement is an alternative optical measurement technique that has been reported to offer better precision than IBO and near zero TIS [6, 12, 14-15], and is therefore a possible solution to the measurement uncertainty budget. Bischoff *et al.* proposed measuring overlay using the diffraction efficiencies of the first diffracted orders [7]. Chun-Hung Ko used angular scatterometry combined with an experimental library to determine the overlay error on ADI stacks with intermediate poly-silicon lines [8]. H.-T. Huang *et al.* used spectra from reflection symmetry gratings and a rigorous coupled-wave analysis (RCWA) regression approach to calculate the overlay error [9]. W. Yang *et al.* [10] and D. Kandel *et al.* [11] used arrays of specially constructed pads with programmed offsets to determine overlay without the need for model fitting. These DBO methods have the potential to meet the demanding overlay metrology budget for sub-32nm technology nodes. In this chapter, the advantages of DBO for precise and accurate overlay measurement in LELE, LFLE and SADP processes will be shown.

2. Spectroscopic scatterometry

2.1 Experimental setup

Spectroscopic scatterometry is used to measure overlay errors between stacked periodic structures (*e.g.*, gratings). In this technique, broadband linearly polarized light is incident perpendicular to the wafer surface and the zero-order diffracted signal (spectrum) is measured as a function of wavelength. Fig. 2 shows a typical experimental configuration. At normal incidence, different reflectance spectra are obtained for various angles of polarization with respect to that of the periodic structure. Typical data collection involved both TE and TM spectra. A specific advantage of using polarized light is that it provides enhanced sensitivity as both the amplitude and phase differences between the TE and TM spectra can be measured.

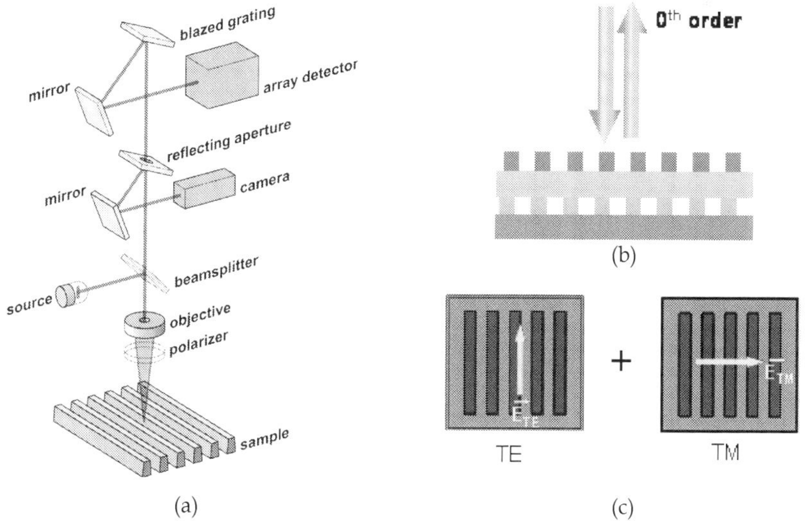

Fig. 2. The figure shows a typical hardware set up for a normal incidence scatterometer (a) spectroscopic reflectometry, (b) normal incidence reflection, and (c) TE TM data acquisition modes.

2.2 Theory

Spectra are obtained from pads, each of which has gratings patterned in both layers between which the overlay error is being measured (Fig. 3). The gratings in each pad are overlaid but by design shifted with respect to each other. Spectra from pads with shifts of equal magnitude but opposite direction are identical due to symmetry:

$$R(+x_0,\lambda) = R(-x_0,\lambda) \tag{1}$$

Here $R(x_0,\lambda)$ is the reflectance spectrum from one pad as a function of wavelength λ and shift $+x_0$. The difference spectra $(\Delta R(\lambda)=R(+x_0,\lambda) - R(-x_0,\lambda))$ from two pads with shifts $+x_0$ and $-x_0$ is zero in the absence of noise in the measuring tool. A small overlay error shifts both upper gratings in the same direction and breaks the symmetry. The resulting differential spectrum is proportional to the direction and magnitude of the overlay error:

$$\Delta R(\lambda) = R(x_0+\varepsilon,\lambda) - R(-x_0+\varepsilon,\lambda) = R(x_0+\varepsilon,\lambda) - R(x_0-\varepsilon,\lambda) \cong 2\varepsilon \left.\frac{\partial R}{\partial x}\right|_{x_0} \tag{2}$$

Here, ε is the overlay error, x_0 is the offset bias, and a Taylor expansion of the reflectance around x_0 has been applied. The overlay error can now be calculated by comparing the measured differential spectrum, ΔR, to a second differential spectrum, $\Delta R'$, acquired from a pair of test pads having a known relative offset. If, for example, a shift of $x_0+\delta$ is designed into a third pad, then within the linear-response range the difference between its spectrum and that from the $+x_0$ pad is:

$$\Delta R'(\lambda) = R(x_0+\delta+\varepsilon,\lambda) - R(x_0+\varepsilon,\lambda) \cong \delta \left.\frac{\partial R}{\partial x}\right|_{x_0} \tag{3}$$

Equation (3) provides the calibration required to calculate the overlay error, ε:

$$\varepsilon = \frac{\delta \Delta R(\lambda)}{2\Delta R'(\lambda)} \tag{4}$$

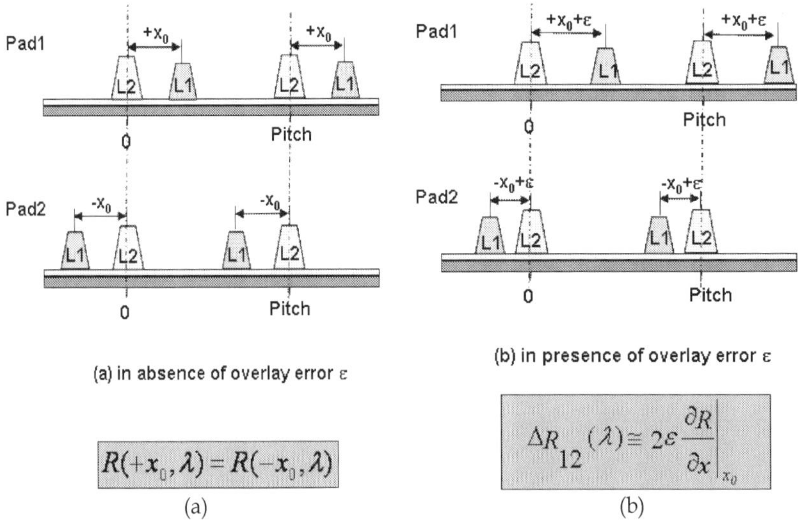

Fig. 3. Illustration of DBO targets design: (a) reflection symmetry with $+x_0$ and $-x_0$ shift (no spectral difference), (b) reflection symmetry broken due to overlay error with $+x_0+\varepsilon$ and $-x_0+\varepsilon$ shift (ΔR_{12} spectral difference between pads 1 and 2).

From equation (4), the ratio $\Delta R(\lambda)/\Delta R'(\lambda)$ must be independent of λ. This arises from our assumption that ε is small and only applies in the linear-response range where this assumption holds.
Equation (4) shows that the overlay can be measured in one direction using a minimum of three pads with suitably defined offsets. In practice four pads are often used (Fig. 4a), with an additional calibration pad with offset $-x_0-\delta$, because the additional data improves precision and provides a check that the overlay error is within the linear response range (Fig. 4b). To measure overlay in two directions two sets of gratings are required. The second set with the gratings rotated by 90° from the first. As overlay is a vector quantity, it is usually measured in both X&Y directions. The nomenclature "2x3 target" and "2x4 target" indicates whether three or four pads are used to measure in each axis.
The nomenclature CD/pitch is also used to indicate the designed CD and pitch of the gratings in each target. For example 65/390 means CD=65nm and pitch=390nm. All targets use the same CD and pitch at both layers.

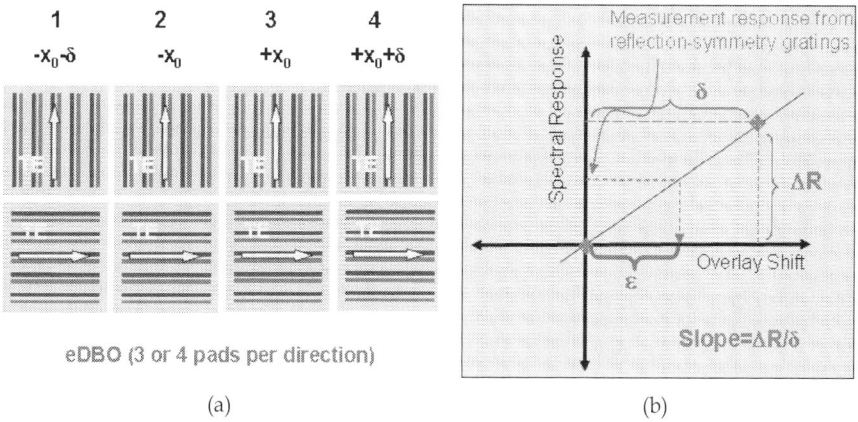

Fig. 4. (a) 2x4 target used for measurement of signal by a normal incidence scatterometer and (b) linear dependency of overlay shift as a function of spectral response.

2.3 Spectral response to overlay

Fig. 5(a) and (b) show the spectral response to overlay in the difference spectra from programmed reflection-symmetry gratings (Equation 2). The black line in the figure is the average $\Delta R'(\lambda)$ from both pairs of calibration pads, scaled by $(2\varepsilon/\delta)$. The scaled signal shows excellent agreement with the measured response, in accordance with equation 4. The maximum response increases from ~15x (Noise) to ~45x (Noise) for a ~3x change in overlay error. As expected, the spectral response also changes in sign with the measurement. The maximum spectral response at any wavelength is about 4x (Noise) per 1 nm overlay error. Measurement uncertainty much better than 0.25nm is possible because the data is summed over all available wavelengths.

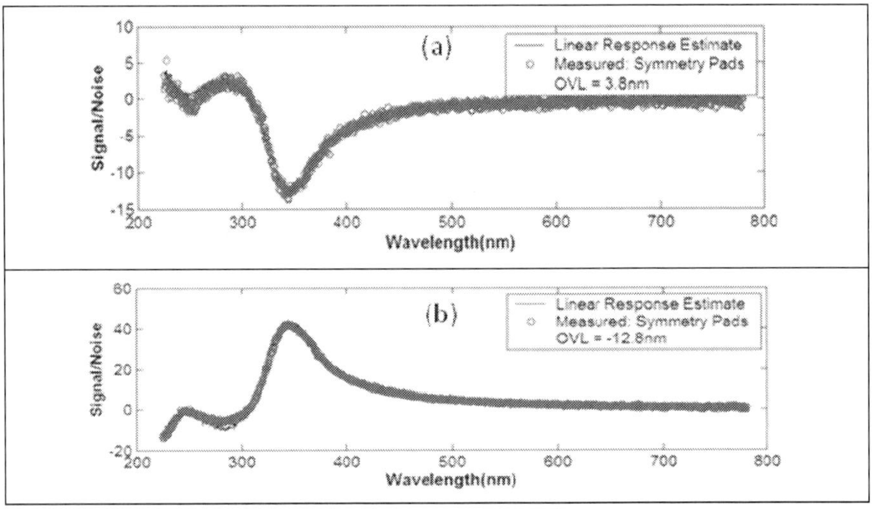

Fig. 5. Spectral response (signal to noise ratio) for corresponding overlay errors (shown in the legend box). The plots (a) and (b) also show response (black line) calculated from the calibration pad spectra.

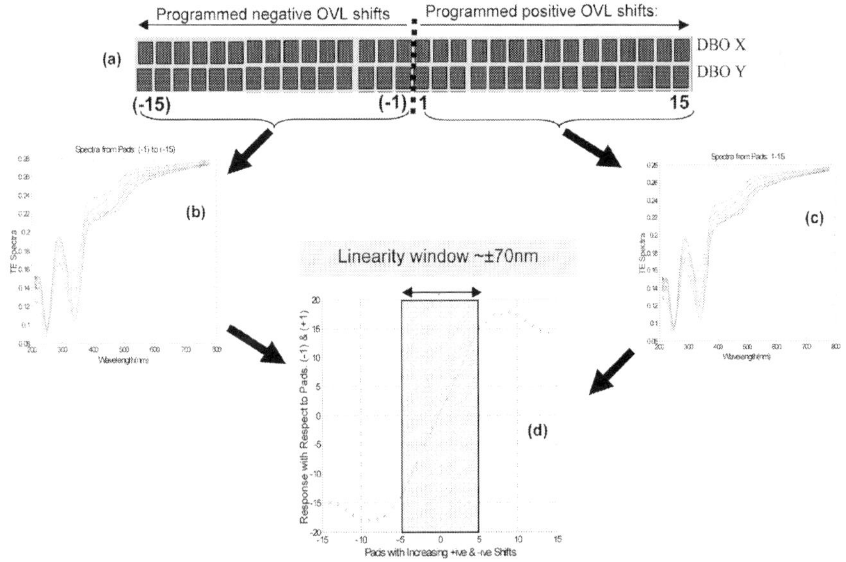

Fig. 6. (a) shows Pads 1-15 with increasing programmed positive overlay shifts and Pads (-1)-(-15) with increasing negative overlay shifts, (b) and (c) show TE spectra from corresponding vertical and horizontal gratings pad sequence respectively, and (d) shows measured spectral response in arbitrary units.

2.4 Range of linearity

The linearity range is tested by printing a sequence of pads with varying overlay shifts. Pads in the right half of the sequence (pads 1-15) have increasing positive overlay shifts (Fig. 6(a)) in 15 nm steps. Pads (-1) to (-15) have increasing negative overlay shifts in 15 nm steps. Fig. 6b and 6c shows raw TE spectra collected from these pads. The difference signal is calculated by subtracting the pad 1 spectrum for pads 2-15, and the pad -1 spectrum for pads -2 to -15. The range of linearity observed (~±70 nm) is significant for this process as shown in Fig. 6(d). A similar linearity range is observed for the horizontal gratings.

3. Litho-etch-litho-etch (LELE)

In the absence of a EUV processing solution below 80nm pitch, DPT using the litho-etch-litho-etch dual step became an attractive solution for the low k_1 regime.

3.1 DPT structures for testing DBO

The first DPT test structure is on a silicon substrate (fig 7a). The structure consists of ~120 nm photoresist lines and ~40 nm nitride lines with silicon over etch. The second DPT stack (Fig. 7b) comes from a gate (bitline)-level patterning step in a NOR flash process.

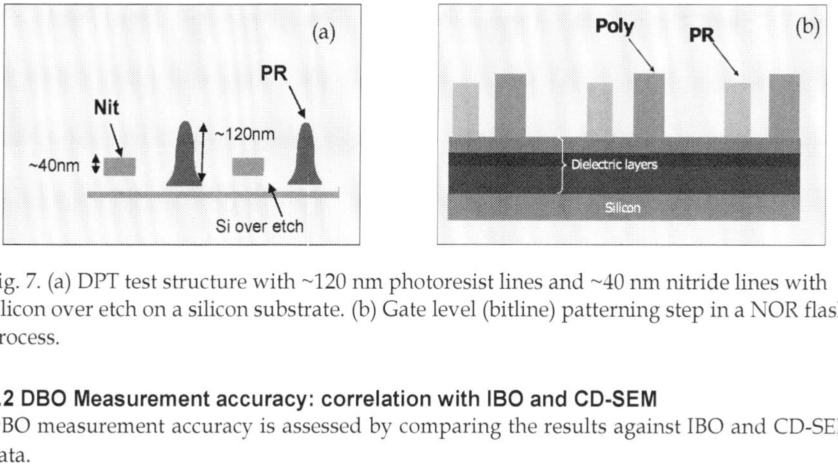

Fig. 7. (a) DPT test structure with ~120 nm photoresist lines and ~40 nm nitride lines with silicon over etch on a silicon substrate. (b) Gate level (bitline) patterning step in a NOR flash process.

3.2 DBO Measurement accuracy: correlation with IBO and CD-SEM

DBO measurement accuracy is assessed by comparing the results against IBO and CD-SEM data.

3.2.1 Correlation with IBO data

Fig. 8(a) shows the comparison between DBO and IBO measurements on all 143 fields on the wafer. There is a very good correlation between DBO and IBO measurements (R^2=0.99) with an offset of ~7 nm. The correlation is good for the subset of measurements less than ±3 nm (Fig. 8b). The inset histogram in Fig. 8a shows the difference in IBO and DBO measurements after removing the ~7 nm constant offset. The distribution is approximately normal with standard deviation of 1.8nm.

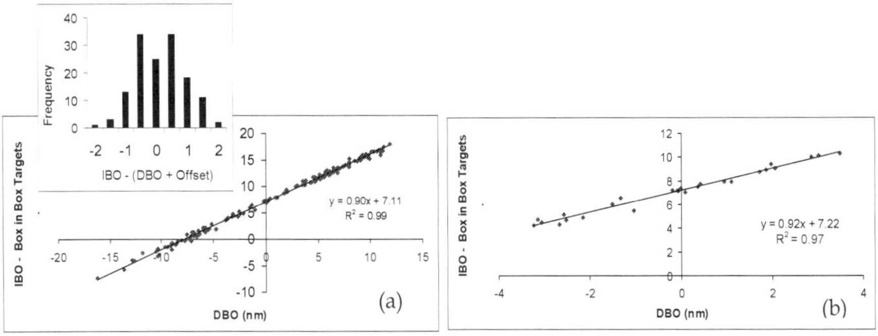

Fig. 8. DBO vs. IBO correlations: (a) shows correlations on 143 targets throughout the wafer, and (b) shows a subset of the same data in a narrow range of ~±3 nm.

3.2.2 Correlation with CD-SEM

Fig. 9a shows a CD-SEM image of the nitride (dark) and photo resist (gray) lines. Fig. 9(b) and 9(c) show the correlation between DBO and CD-SEM measurements. In Fig. 9b, the overlay errors are calculated from the CD-SEM data for the top of the lines, while in Fig. 9c, the bottom CD data is used. The inset histograms in Fig. 9b and 9c show the distributions of the difference between IBO and CD-SEM measurements after subtracting offsets.

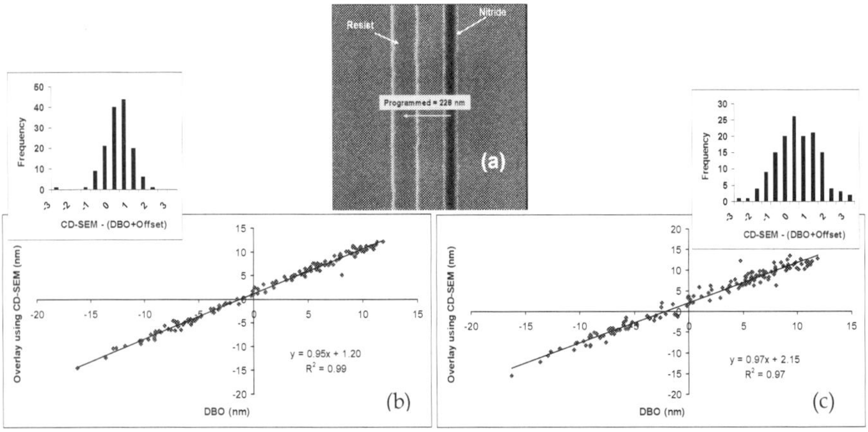

Fig. 9. (a) CD-SEM image of the DPT structure. Dark lines are nitride and the light gray lines are the photo resist, (b) correlations between DBO and top CD-SEM measurements from all 143 fields. The inset histogram shows the histogram of differences between the CD-SEM and DBO results. (c) as (b) but using the bottom CD data.

The slope of the CD-SEM/DBO correlation is 0.95 (for top CD) and 0.97 (bottom CD) compared to 0.9 for the DBO-IBO correlation. The offset between DBO and CD-SEM measurements is in the order of 1 to 2 nm. Concerns of accuracy of IBO measurements have been raised before. Dusa *et al.* reported an offset of ~5 nm between CD-SEM and IBO for a

DPT application [10]. It is possible that this offset might be coming from differences in mask writing errors between the DBO and IBO targets. The CD-SEM results are closer to the DBO values (Fig. 9), which suggests that the accuracy of the DBO technique is better than that of IBO. The observed DBO repeatability is better than CD-SEM. DBO is also non-destructive, as the sample is not subjected to the charging effects that occur in a CD-SEM.

3.3 DBO target types – 4 vs. 3 pads

While scatterometry offers precise and accurate overlay measurements for DPT, the number of reference and sample pads required for such measurements are still a concern. In this section the possibility of reducing the number of pads required without sacrificing the performance is explored using two target types - 2x4 and 2x3 targets (see definitions of target types in Section 2). Fig. 10 shows excellent correlation between overlay measured from these 2 target types. Table 1 shows the root mean square dynamic precision on two 2x3 targets and a 2x4 target. The precision on the 2x3 targets is a factor of ~1.4 (= $\sqrt{2}$) higher than that on the 2x4 target because the number of difference spectra used in the 2x4-target algorithm is twice the number used for the 2x3-target algorithm. The figure shows excellent correlation, suggesting that a 3-pad target is sufficiently accurate for this application. The dynamic (load-unload) and static (no unload) precision is excellent.

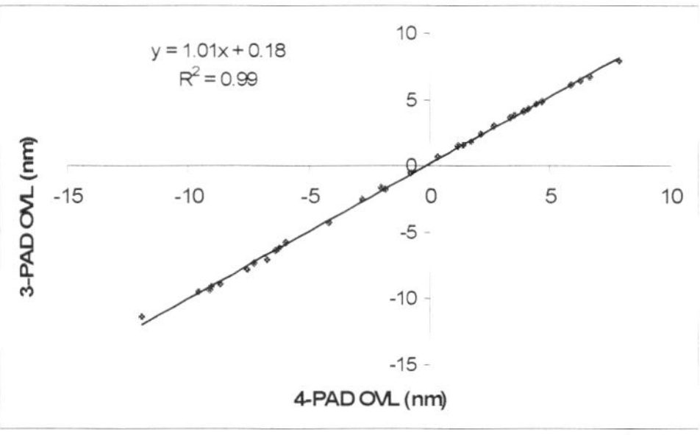

Fig. 10. Correlation between DBO targets with 4 pads/direction and 3 pads/direction.

Target/Measurement Type	RMS Precision 3_{σ} (nm)
2x3 Dynamic, Target-II	0.31
2x3 Dynamic, Target-II	0.32
2x4 Dynamic, Target II	0.20
2x4 Static, Target II	0.11

Table 1. Shows the dynamic precision for the 2x3 target is higher by a factor of ~ $\sqrt{2}$

3.4 DBO performance: precision, TIS, matching, and TMU

DBO capability was further assessed by calculating the measurement uncertainty (TMU) from the measurement data using equation (5). Short-term (dynamic) precision (*DYNP*) is the Root-Mean-Square (RMS) of three times the standard deviation of the 20-cycle measurements at each target, and combines static precision (occurring without any movement within the tool) and the effects of target reacquisition. $TIS3\sigma_T$ is three times the standard deviation of the TIS in the measurements at each target. Where multiple tools are available, site-by-site tool matching (TM) is included. TM is three times the standard deviation of the difference in the average measurement from each tool at the same location. Matching data is not included in the DBO to IBO TMU comparison.

$$TMU = \sqrt{(DYNP)^2 + (TIS3\sigma_T)^2 + (TM)^2} \tag{5}$$

The results for both DPT applications are summarized in Table 2. For the gate level DPT the dynamic precision is less than 0.1 nm and the TMU is 0.26 nm Average TIS is under 0.1nm. DBO matching data is between two tools (Atlas and FLX) with similar reflectometer optical heads. For the DPT structure on a silicon surface precision is 0.2nm. TIS and matching data is not available for this structure.

TECHNOLOGY	Process Step	DYNP (nm)	TIS Avg (nm)	TIS 3σ (nm)	Tool Match	TMU
DBO	DPT Silicon Substrate	0.2				
	DPT Gate Patterning NOR Flash	0.07	-0.04	0.17	0.18	0.18
IBO	DPT Silicon Substrate	0.48	-0.37	0.31		0.57
	DPT Gate Patterning NOR Flash	0.33	-2.05	6.03		6.04

Table 2. Performance summary of DBO on two stacks discussed in Fig. 4.

4. Litho-freeze-litho-etch (LFLE)

While LELE involving two process steps offers an adequate solution for DPT process steps, both are very expensive and slow. The alternative Litho-Freeze-Litho-Etch (LFLE, Fig. 1b) process reduces cost by replacing the intermediate etch step with a process step in the litho track.

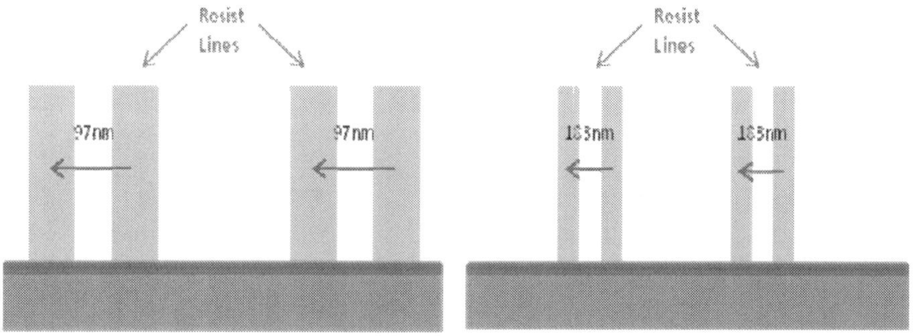

(a) (b)

Fig. 11. DPT structure that has alternative photo resist lines with silicon over etch: (a) 65/390 Line/Pitch ratio, (b) 110/660 Line/Pitch ratio

4.1 DPT Structure

An example LFLE DPT structure consists of ~120 nm photoresist lines with silicon over etch as shown in Fig. 11(a) and (b).

4.2 Prediction based on simulation

The LFLE stack is built on silicon with BARC and two resist lines patterned on top. Simulated spectral response curves (for example from NanoDiffract™, Nanometrics scatterometry software) are used to predict the static precision of overlay measurements in the range of interest for the LFLE DPT stack.

Static precision for the LFLE model can be determined by two different methods. The first method uses analysis of measurement covariance as described by Vagos *et al* [16], and which is referred to as the "Uncertainty and Sensitivity Analysis" method (U&SA). In the second method random noise is introduced into the simulated DBO spectra for all four pads and static precision determined as if the spectra are obtained experimentally. The predicted static precision for the LFLE stack is 0.24nm (3σ) using the U&SA method and 0.30nm by the noise induced method.

4.3 DBO accuracy (freeze process)

4.3.1 Correlation with IBO

The CD-SEM image in Fig. 12a shows the resist lines in one of the four pads of a 110/660 target. To test measurement accuracy, the results from two 2x4 DBO targets (65/390, 110/660) are compared with IBO measurements using nearby Blossom targets. Fig. 12b shows the excellent correlation (R^2 ~0.99) between the DBO and Blossom data.

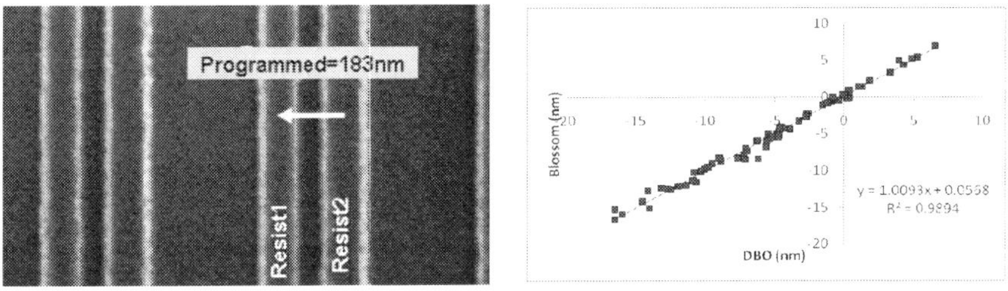

(a) (b)

Fig. 12. Correlation of DBO vs. IBO: (a) 110/660 pad, (b) Correlation of DBO 2x4 targets vs. blossom

4.4 LFLE DBO performance

Good overlay control requires good measurement capability. Table 3 summarizes the precision, TIS, tool matching and measurement uncertainty (TMU) for DBO of LFLE structures. DBO dynamic precision was less than 0.2nm and TMU less than 0.5nm. Average TIS was under 0.1nm.

Table 3 contains site-by-site (SBS) 3σ DBO matching data for this stack from three tools with the same design of reflectometer optical head. Tool matching of 0.14nm or less is achieved without calibration or adjustment of the tools. This is possible because the method of equation (4) is self-calibrating. Absolute spectral matching between tools is not necessary.

TECHNOLOGY	CD:Pitch	Tool	DYNP (nm)		TIS Avg (nm)		TIS 3σ (nm)		Tool Match		TMU	
			X	Y	X	Y	X	Y	X	Y	X	Y
DBO	65:390	Tool1	0.19	0.17	0.06	-0.01	0.42	0.44	0.09	0.14	0.47	0.49
	65:390	Tool2	0.18	0.18	-0.07	0	0.36	0.37	0.09	0.14	0.41	0.44
	110:660	Tool2	0.21	0.18	0.07	0.02	0.28	0.32	0.06	0.07	0.36	0.38
	110:660	Tool3	0.21	0.19	-0.08	-0.08	0.21	0.19	0.06	0.07	0.3	0.28
IBO		Tool4	0.71	0.59	0.18	-0.13	0.56	0.31			0.9	0.67

Table 3. Performance summary of DBO on LFLE stack.

4.5 Prediction vs. observation

The dynamic precision (3σ) of ~0.20nm is slightly better than the predictions made in section 4.2 of 0.24nm (U&SA method) and 0.30nm (noise induced method).

5. Spacer double patterning

While LFLE minimizes the number of process steps needed and thus reduces cost it still requires very tight overlay control. Spacer (SADP) forms lines around pre-patterned features, relaxing the requirement for overlay control and potentially allowing the indefinite pitch halving.

5.1 Spacer 1st layer patterning

In this section we discuss some results from first layer patterning by SADP. Experimental TE and TM spectral data obtained for an SADP stack (Fig. 13a) is fitted to modeled spectra (Fig. 13b). Although the model has not been optimized to improve the fit there is good agreement between the modeled and actual spectrum.

The CD-SEM image in Fig. 13c shows the spacer pattern. The measured bottom CD is around 48nm for layer 1 spacers. Although the spacer structures are expected to be identical for both layer 1 and layer2, the CD may vary depending on the application. The model fitting data (Fig. 13a) is consistent with uniform spacer formation across the wafer. The spacer width at the bottom is 42.5 ±1nm and SWA is 72.3 ±0.2 deg.

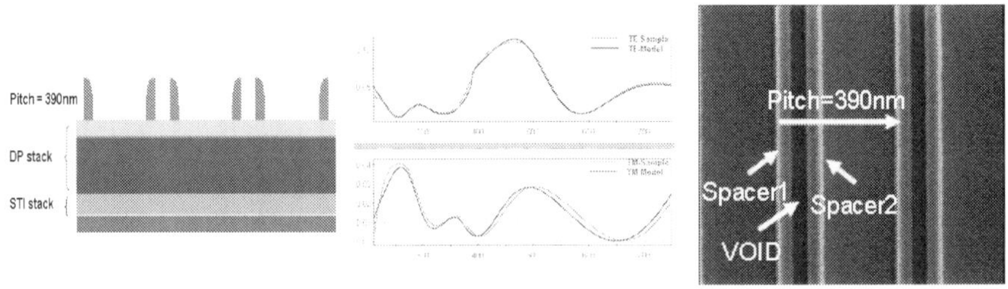

(a) (b) (c)

Fig. 13. (a) STI based spacer Layer 1 stack, (b) Modeled and experimental spectra, and (c) CD-SEM Image of spacer L1.

5.2 Prediction of spacer 2nd layer precision

The second spacer layer patterning characteristics were predicted using simulation. Since the static precision predicted for LFLE (in sections 4.2 and 4.4) was consistent with the experimental data in table 3, predictions for the double layer SADP case should also be valid. The good fit between modeled and experimental TETM spectra in Fig. 13b further supports the validity of this approach.

The final SADP stack after completion of the patterning steps is shown in Fig. 14. For LELE and LFLE, both patterning steps are done on the same layer. In the spacer case, the first patterning step is done on L1 (STI+DPT) and second patterning step is done on L2 (WL + DPT) with a programmed shift. The CD of the two spacers can be different.

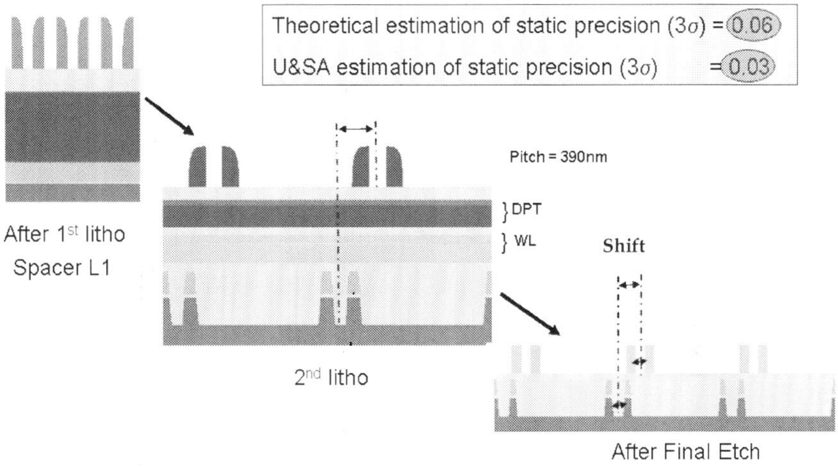

Fig. 14. The spacer stack DPT steps shown for pitch of 390nm and bottom CD ~65nm

The simulated TETM spectra for the spacer stack in Fig. 15(a) shows sensitivity across the spectral region. The TETM spectral response to overlay shift in Fig. 15(b) is linear for overlay around 25% of the grating pitch.

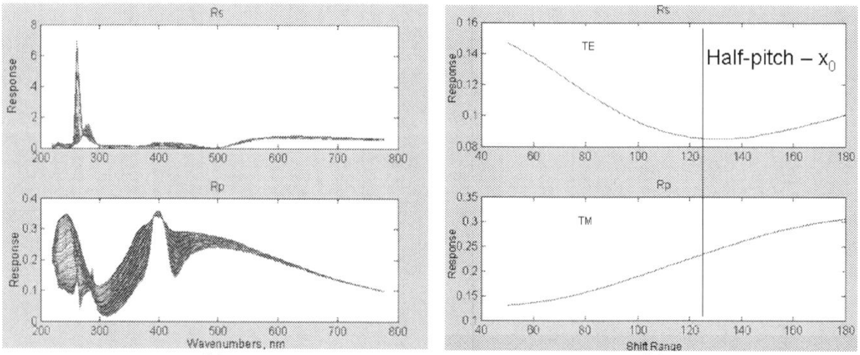

(a) (b)

Fig. 15. (a) RCWA simulated spectra; (b) DBO sensitivity over the shift range

5.3 Spacer DBO prediction vs. expectation

The dynamic precision (3σ) observed for a 2x4 65/390 target in the case of LFLE was better than predicted (sections 4.2-4.4). Assuming the same behavior applies to SADP, precision should be of the order of ~0.05nm.

6. Model-based overlay measurement (mDBO)

The success of scatterometry for CD and profile measurement comes from the ability to model the signal formation process. The signature contains enough information that the measurement can be made by finding those parameters that give the closest fit between modeled and experimental signatures. The same approach can be applied to overlay measurement, reducing the number of measurement pads needed and providing profile data as well as overlay.

6.1 mDBO LELE sample

The first mDBO DPT structure consists of alternate photo resist and nitride lines on a silicon substrate (Fig. 16a). As mentioned in section 3, eDBO measurements are performed on four specially designed pads per direction (Fig. 4) with *D* designed to be around 25-35% of the pitch to ensure maximal overlay sensitivity [6]. The mDBO measurements are performed on two of the pads with shift *+D* and *-D*. For normal incident polarized reflectometry, it is found that the TE spectrum is more sensitive to overlay than TM [6]. To reduce measurement time without compromising sensitivity, only TE spectra are collected and used for data analysis. In mDBO analysis a physical model is first set up using NanoDiffract Software to describe the sample structure. Fig. 16 shows the model of one of the pads with designed shift *+D*. Four parameters, nitride bottom CD (NI_BW), resist bottom CD (PR_BW), resist height (PR_HT) and the distance between the nitride and resist lines (*S*), are floated to optimize the model fit to the measured spectra. When there is an overlay error ε, the distance between the nitride and resist lines, denoted as *S*(+) with + for positive shift, is given by $D+\varepsilon$. The model for the second pad with designed shift *-D* is identical to the first pad except that the shift denoted by *S*(-) is D-ε. In the regression it is assumed that the corresponding thickness and CDs are the same for these two pads due to proximity. *D* is fixed to the designed value of 244nm. Fig. 18(b) displays the experimental spectrum and theoretical calculation at best fit for one of the pads. The agreement is excellent. The shape of the spectrum and fit quality for the second pad is very similar to the first one.

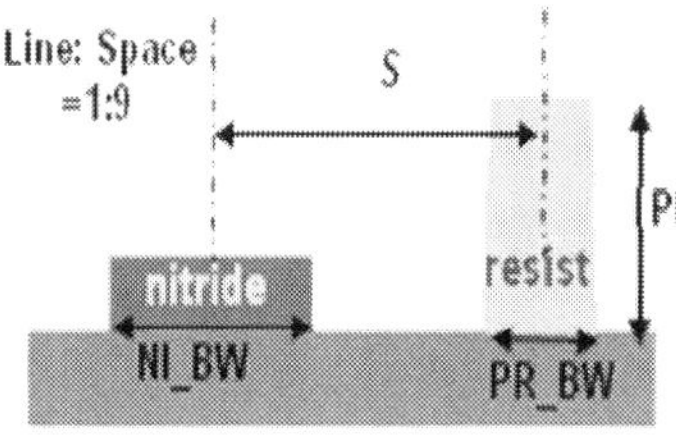

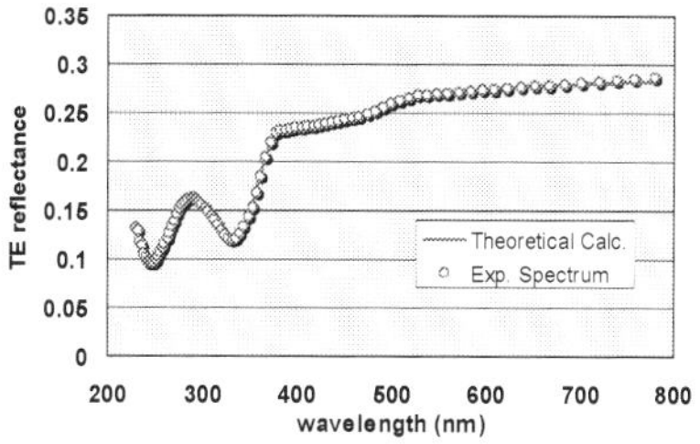

(a) (b)

Fig. 16. (a) DPT structure with alternative photo resist and nitride lines with silicon over etch. Four parameters are floated: nitride bottom CD (NI_BW), resist bottom CD (PR_BW), resist height (PR_HT) and the shift (S) of resist from nitride lines, measured from center to center. (b) Experimental spectrum and theoretical calculation.

To check the stability and performance of the model, uncertainty and sensitivity analysis (U&SA) was performed using NanoDiffract™ software [16]. Fig. 17 shows the signal to noise ratio corresponding to a 2 nm change in the overlay error. Reasonable sensitivity is observed. The parameter correlation matrix and predicted static precision (3σ) are summarized in table 4. No strong correlation is found between overlay and other parameters. The predicted static precision for overlay is 0.16nm (3σ), which compares well with the eDBO result of 0.25nm in section 3.3.

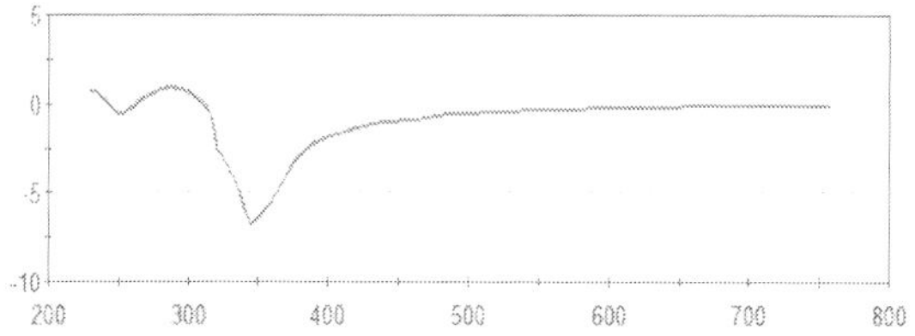

Fig. 17. Overlay signal/noise ratio. The signal corresponds to 2nm change in overlay.

Fig. 18a compares two-pad mDBO measurements with 4-pad eDBO results. Both data sets are from ~140 dies across the wafer. Excellent correlation (R^2 ~0.99) and a slope of 1.00 are achieved. The offset is about 0.1nm. Fig. 18b shows the histogram of the deviation of the data points from the correlation curve shown in Fig. 18a. The distribution follows a normal distribution, indicating the absence of systematic error between these two analysis methods. Standard deviation (3σ) is 1.05nm, which contains measurement uncertainties from both measurement methods.

Parameter Correlation Matrix:					Precision (3σ) (nm)
NI_BW	1				0.091
S (overlay)	-0.19	1			0.16
PR_HT	-0.47	0.54	1		0.047
PR_BW	0.05	-0.68	-0.62	1	0.052

Table 4. Parameter correlation matrix and precision predicted using model shown in Fig. 16 (a).

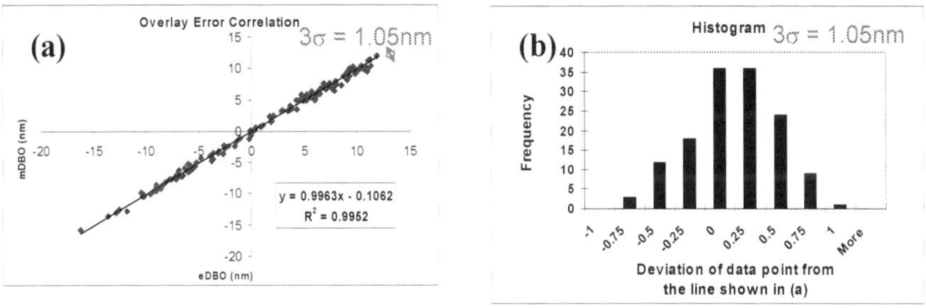

Fig. 18. (a) Correlation of mDBO and eDBO for LELE sample. (b) Histogram of the deviation of the data points form the straight line shown in (a). Data shown here for X and similar overlay error is observed for Y direction

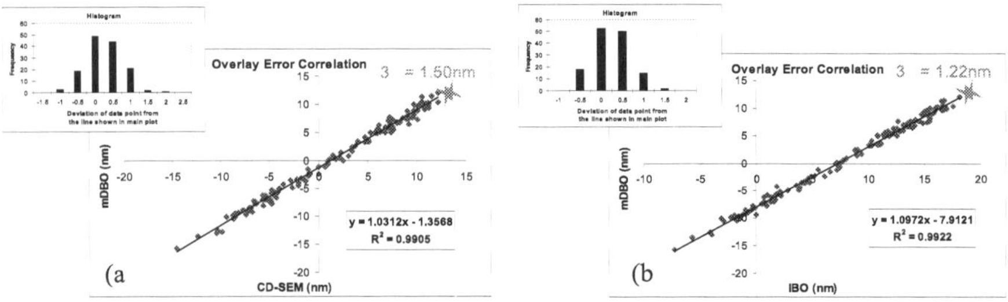

Fig. 19. (a) Correlation of mDBO and CD-SEM for LELE wafer. The inset shows the histogram of the deviation of the data points from the straight line shown in main plot. (b) Correlation of mDBO and IBO. The inset shows the histogram of the deviation of the data points from the straight line shown in main plot.

To further evaluate the accuracy of scatterometry measurement, mDBO results are compared with other metrology techniques, i.e., The CD-SEM data is from the DBO targets. Image based overlay (IBO) measurements are made on standard box-in-box targets nearby. Correlations of mDBO to these two techniques are shown in Fig. 19. A good correlation (R^2=0.99) and a slope of 1.03 are observed between eDBO and CD-SEM. The offset between eDBO and CD-SEM measurements is ~1.3nm. A good correlation (R^2=0.99) is also observed between eDBO and IBO. However, there is an offset of ~7.9 nm between the two methods.

The source of the offset is not clear. The deviation of the data points from the linear correlation curve is 1.50nm 3σ between mDBO and CD-SEM, and 1.22nm 3σ between mDBO and IBO.

6.2 mDBO LFLE sample

Targets composed of only one pad are desirable because they further reduce total target size. 2D gratings that are sensitive to overlay errors in both X and Y directions may be used [13, 17]. One example is shown in Fig. 20. A 2D lattice (similar to an IBO box-in-box target) is formed with a period on the order of hundreds of nanometers, chosen to maximize diffraction efficiency and overlay sensitivity. For IBO targets , the scale of the boxes is on the order of microns to a few tens of microns. The size of the IBO target is limited by optical resolution.

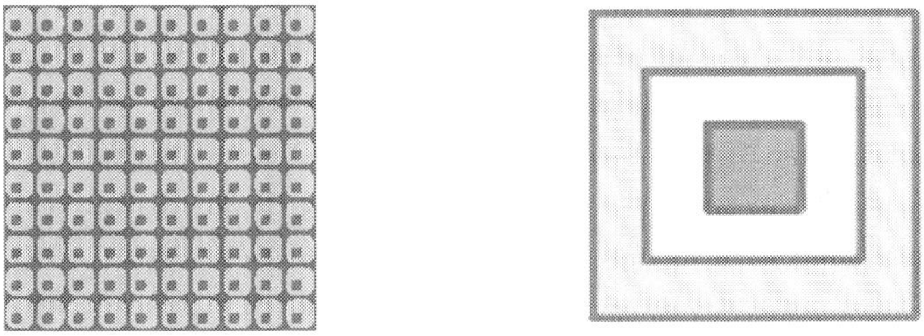

(a) (b)

Fig. 20. (a) mDBO 2D grating target; (b) IBO box-in-box target

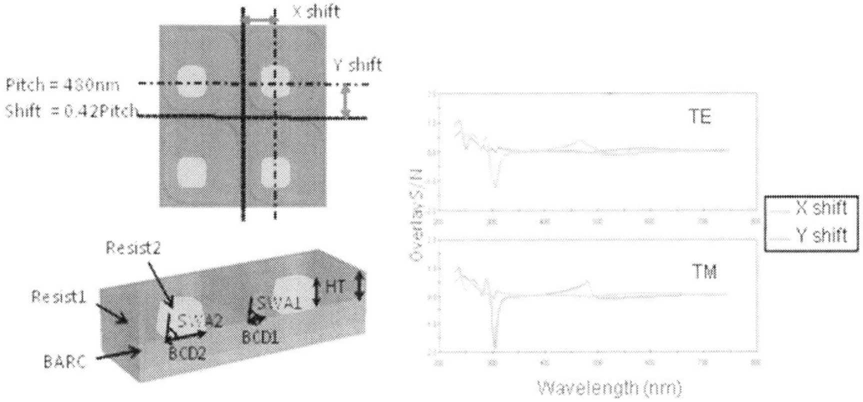

(a) (b)

Fig. 21. (a) 2D DBO targets for the LFLE sample. Seven parameters are floated: resist1 BCD, SWA and HT, resist2 BCD, SWA and HT (coupled to resist1 HT), X shift and Y shift defined from the center of the grids (resist1) to the center of the squares (resist2). (b) Overlay S/N spectrum. The signal corresponds to 0.5nm change in X and Y overlay.

The overlay errors are extracted from the DBO target using the modeling approach. The sample structure and modeling details are shown in Fig. 21a. The pitch is 480nm and the nominal values of X shift and Y shift are 42% of the pitch. The design is symmetric in X and Y so that resist side wall angles (SWA) and bottom critical dimensions (BCD) can be coupled between X and Y directions. Seven parameters are floated: resist1 BCD, SWA and height (HT), resist2 BCD, SWA and HT (coupled to resist1 HT), X shift and Y shift defined from the center of the grids (resist1) to the center of the squares (resist2). Sensitivity analysis shows that TE spectra are more sensitive to X overlay while TM spectra are more sensitive to Y overlay (Fig. 21b). Therefore, both TE and TM spectra are used in the measurement.

The experimental spectra and RCWA fits are shown in Fig. 22. In fig. 21b, the overlay S/N corresponds to 0.5nm change in overlay. Sensitivity to overlay of the 2D targets (Fig. 11), is about half of that of 1D targets for the most sensitive wavelength, if both are normalized to 1nm. This is reasonably understood considering the reduction in the target size.

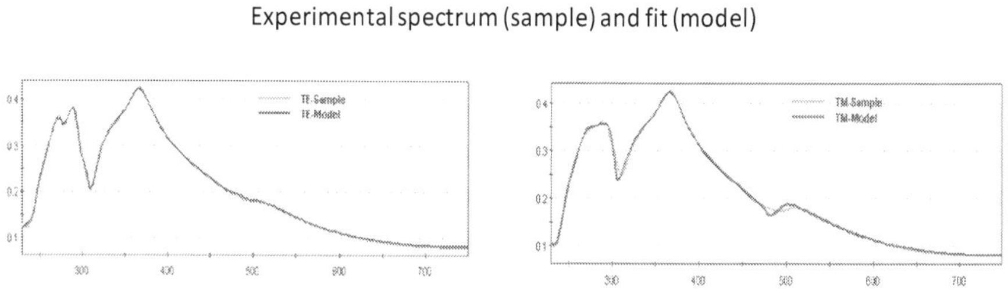

Fig. 22. Experimental TE and TM spectra and theoretical fits for the structure in Fig. 21a.

Parameters	Resist1 SWA	Resist Ht	Resist1 BCD	Y Shift	Resist2 SWA	X Shift	Resist2 BCD
Resist1 SWA	-1						
Resist Ht	-0.79	-1					
Resist1 BCD	0.95	-0.85	-1				
Y Shift	-0.03	-0.34	-0.04	-1			
Resist2 SWA	-0.62	0.28	-0.6	0.19	-1		
X Shift	-0.03	-0.37	-0.05	0.43	0.18	-1	
Resist 2 BCD	-0.5	0.001	-0.64	0.29	0.83	0.32	-1

Table 5. Parameter correlation matrix for the model shown in Fig. 21(a).

The parameter correlation matrix is shown in table 5. There are no strong correlations between overlay and the other parameters. The accuracy is first verified by measuring a series of five targets with designed shifts increasing by 2nm between two neighboring targets. The correlation of the measurement and the programmed overlay is displayed in Fig. 23a. R^2 is 0.996 and slope is 0.996. An offset of -11.36nm is observed. It comes from the local registration error due to scanner alignment errors. This can be corrected by adding an overlay offset between the layers during exposure. IBO measurements on a Blossom target next to the DBO targets show a registration error of -11.18nm, which agrees with the offset within 0.2nm. The DBO accuracy is further verified by measuring 49 fields across the wafer and correlating to Blossom measurements. The correlation plot is shown in Fig. 23b, with R^2 of ~ 0.98 and slope of ~0.97.

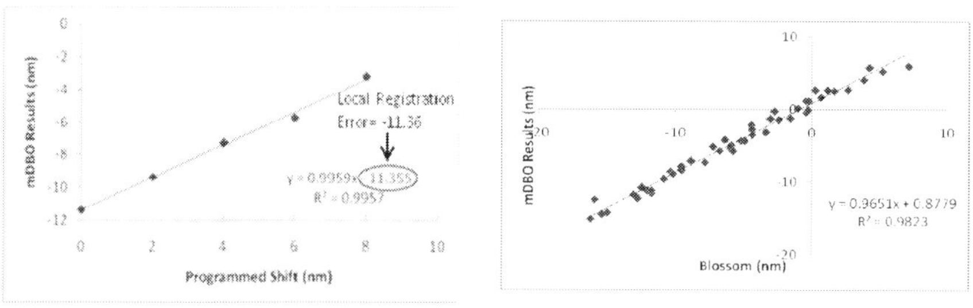

(a) (b)

Fig. 23. (a) Correlation of mDBO 2D target measurements with programmed shifts.
(b) Correlation of mDBO 2D targets with IBO blossom measurements.

6.3 mDBO LFLE performance

The dynamic repeatability (DYN 3σ), TIS mean, and TIS 3σ are reported in table 5. The dynamic repeatability is measured from 15 load/unload cycles on multiple fields (9 fields for all 1D targets, and 15 fields for the 2D targets). DYN 3σ is reported as the average of the 3σ-precisions from the measurement sites. TIS is defined as in eq. (6),

$$TIS = \frac{OVL_0 + OVL_{180}}{2} \tag{6}$$

where OVL_0 is the overlay measurement result at 0° loading angle, and OVL_{180} is the measurement result at 180° loading angle. The reported TIS mean is measured over 71 sites for 1D target and 49 sites for 2D targets across the wafer. For all DBO targets, TIS mean is

nearly zero. TIS 3σ reported is from multiple fields (9 fields for all 1D targets, and 15 fields for the 2D targets); with OVL_0 and OVL_{180} results averaged over 15 load/unload cycles respectively. By removing the contribution from dynamic variations for each loading angle, TIS 3σ is very small (on the order of 0.01nm).
All three types of standard 1D DBO targets have shown excellent performance with TMU <0.1nm and the 2D 1x1 target has a TMU ~0.2nm (not including tool matching). It is worth mentioning that the mDBO 2x2 target has better TMU, which is a good balance between measurement performance, target size, and measurement time. Similar performance is also observed for Y.

Technology	DYNP (nm)	TIS Avg (nm)	TIS 3_σ (nm)	TMU*
Targets	X	X	X	X
eDBO 1D 2x4 target	0.088	-0.006	0.029	0.092
mDBO 1D 2x2 target	0.050	-0.006	0.028	0.058
mDBO 1D 2x1 target	0.085	0.005	0.042	0.095
mDBO 2D 1x1 target	0.172	0.057	0.120	0.209

Table 6. Performance summary of eDBO and mDBO targets (*TMU does not include tool matching)

7. Conclusion

The multi-pad empirical diffraction-based overlay (eDBO) technique is capable of controlling the overlay in double patterning optical lithography processes (DPT). The usable range of LELE DPT eDBO is ±70nm. eDBO results agree well with traditional image-based overlay (IBO) results and with overlay calculated from CD-SEM data. While good correlation and linearity between DBO and IBO was observed, a significant systematic offset can occur that appears to originate in the IBO data. Reduction in the number of pads from 2x4 to 2x3 results in only a small deterioration in precision.
DBO measurements have near-zero TIS. TMU (including tool-to-tool matching) is less than 0.5nm for both the LELE and freeze process. The overlay errors determined by eDBO (4 pad measurement) agree well with modeled DBO (2 pad measurement) data. This ability to model the signal formation process allows model-generated spectra to be used to predict measurement precision with good success.
Simultaneous model-based measurement of overlay in X&Y (2D mDBO), is possible with good results and also allows reduction in the overall target size. mDBO requires knowledge of the film stack, material optical properties and target layout and consequently more effort in creating recipes than eDBO but provides significant value in reducing measurement time

and target size. In addition to overlay data, mDBO provides CD measurements and profile data for the target, which is not possible with other methods. The multi-pad DBO approach is a good method of overlay process control, especially if combined with in-chip measurements using an alternative technique [3].

8. Acknowledgement

The authors thank C. Saravanan and Nagesh Avadhany (*Nanometrics*), for their contribution to this article.

9. References

[1] C. Ludwig and S. Meyer, "Double Patterning for Memory ICs", "Lithography / Book 2/Chapter IX", ISBN 978-953-307-1356-4.
[2] ITRS, "http://www.itrs.net/Links/2009Summer/PresentationsPDF/Litho_7-2009_SF-V2.pdf"
[3] Yi-Sha Ku, Chi-Hong Tung and N. P. Smith, "In-chip overlay measurement by existing bright-field imaging optical tools," Proc. SPIE 5752, 43 (2005)
[4] P. Dasari et al., "Diffraction Based Overlay Metrology for Double Patterning Technologies," Proc SPIE 7272, 41 (2009)
[5] R. Kim et. al., "22nm half-pitch patterning by CVD spacer self alignment double patterning (SADP)," Proc SPIE 7973, 22 (2011).
[6] C. Saravanan et al., "Evaluating Diffraction Based Overlay Metrology for Double Patterning Technologies," Proc SPIE 6922, 10 (2008)
[7] J. Bishoff, R. Brunner, J. Bauer and U. Haak, "Light diffraction based overlay measurement," Proc SPIE 4344, 222 (2001)
[8] Chun-Hung Ko et al., "Comparisons of overlay measurement using conventional bright-field microscope and angular scatterometer," Proc. SPIE 5752, 987 (2005)
[9] H.-T. Huang et al., "Scatterometry-based Overlay Metrology," Proc SPIE 5038, 126 (2003)
[10] W. Yang, et al., "Novel Diffraction-based Spectroscopic Method for Overlay Metrology," Proc SPIE 5038, 200 (2003)
[11] Daniel Kandel et al., "Differential Signal Scatterometry Overlay Metrology: An Accuracy Investigation," Proc SPIE 6616, 0H1 (2007)
[12] Jie Li et al., "Advancements of Diffraction-Based Overlay Metrology for Double Patterning," Proc SPIE 7971, 70 (2011)
[13] M Dusa et al., "Application of optical CD for characterization of 70mm dense lines," Proc SPIE 5752, 30 (2005)
[14] N.P. Smith et al., "Overlay metrology at the crossroads," Proc SPIE 6922, 2 (2008)
[15] Jie Li et al., "Simultaneous Overlay and CD Measurement for Double Patterning: Scatterometry and RCWA Approach," Proc SPIE 7272, 4 (2009)
[16] P Vagos et al., "Uncertainty and sensitivity analysis and its application in OCD measurements", Proc SPIE 7272, 65 (2009)

[17] B. Schultz, US Patent 7099010 "Two-dimensional structure for determining an overlay accuracy by means of scatterometry"

Part 5

Other Lithographic Technologies: Scanning Probe, Nanosphere, Inkjet Printing, etc.

23

Nanolithography Study Using Scanning Probe Microscope

S. Sadegh Hassani and H. R. Aghabozorg
Research Institute of Petroleum Industry
Iran

1. Introduction

Recently, the interest to design useful nanostructures in science and technology has rapidly increased and these technologies will be superior for the fabrication of nanostructures (Iwanaga & Darling, 2005; Martin et al., 2005; Sadegh Hassani et al., 2010). The patterning of material in this scale is one of the great importances for future lithography in order to attain higher integration density for semiconductor devices (Sadegh Hassani & Sobat, 2011).
Conventional lithography techniques, i.e., those divided to optical and electron beam lithography are either cost-intensive or unsuitable to handle the large variety of organic and biological systems available in nanotechnology. Hence, the various driving forces have been considered for development of nanofabrication techniques (Geissler & Xia, 2004; Quate, 1997; Sadegh Hassani & Sobat, 2011).
Applying of these techniques has started approximately since 1990 and it has given rise to the establishment of different nanolithography methods, which one of the most important method is scanning probe based lithography. An interesting way of performing nanometer pattern is direct scratching of a sample surface mechanically by a probe. The controlled patterning of nanometer scale features with the scanning probe microscope (SPM) is known as scanning probe lithography (SPL) (Irmer et al., 1998). Many reports have been presented about various lithographic methods by this technique (Garcia, 2004; Garcia, 2006). SPL would also be ideal for evaluation of mechanical characteristic of surfaces.
Scanning probe microscopy, such as scanning tunneling microscopy (STM) and atomic force microscopy (AFM), has become a standard technique for obtaining topographical images of surface with atomic resolution (Hyon et al., 1999). In addition, it may be used to study friction force, surface adhesion and modifying a sample surface (Sundararajan & Bhushan, 2000; Burnham et al., 1991; Aime et al., 1994; Sadegh Hassani & Ebrahimpoor Ziaie, 2006; Ebrahimpoor Ziaie et al., 2008). Manipulating surfaces, creating atomic assembly, fabricating chemical patterns and characterizing various mechanical properties of materials in nanometer regime are enabled by this technique (Hyon et al., 1999; Sadegh Hassani & Sobat, 2011; Bouchiat & Esteve, 1996).
Nanolithography with AFM is also a tool to fabricate nanometer-scale structures with at least one lateral dimension between the size of an individual atom and approximately 100 nm on silicon or other surfaces (Wilder & Quate, 1998). This technique is used during the fabrication of leading edge semiconductor integrated circuits (Sugimura & Nakagiri, 1997)

or nanoelectromechanical systems (NEMS). This method is not restricted to conductive materials (Fonseca Filho et al., 2004). The advantages of this technique are high resolution and alignment accuracy, which could not be achieved by conventional lithographic techniques (Sheehan & Whitman, 2002; Martin et al., 2005). Moreover, the AFM nanolithography technique takes advantages of the ability to move a probe over the sample in a controllable way (Samori, 2005). Nanolithography using AFM can be done in various modes (Jones et al., 2006): chemical and molecular patterning (DPN), mechanical patterning by scratching or nanoindentation, local heating, voltage bias application and manipulation of nanostructures. Most popular AFM lithographic techniques are resist film lithography (Li et al., 1997) and lithography by oxidation (Sheglov et al., 2005; Sugimura et al., 1993; Sadegh Hassani et al., 2008a; Dubois & Bubbendroff, 1999; Avouris et al., 1998; Snow et al., 1999; Lemeshko et al., 2005; Avouris et al., 1997). The atomic force microscope has also become an increasing popular tool for manipulating thin films of many different types of materials. Lithography techniques can be carried out on the film of polymers such as polymethylmethaacrylate (PMMA), chloromethyl phenyltrichlorosilan (CMPTS), polyethylene (PE) and others (Lee et al., 1997; Sadegh Hassani et al., 2008b, Yoshimura et al., 1993; Chen et al., 1999; Huang et al., 2001). This capability can potentially be extended to evaluate nano-scale material response to indentation and would be ideal for evaluation of mechanical characteristic of surfaces (Burnham & Colton, 1989; Hues et al., 1994; Sadegh Hassani et al., 2008b).

To apply force optimally for making nano scratches, we require to understand the underlying behavior and parameter that control it, a tip which is optimized for applying force under the experimental conditions and scanning techniques which allows one to use these tips and retain desired properties (Yasin et al., 2005; Sadegh Hassani et al., 2008a; 2010). Some factors such as resolution, accuracy of alignment and reproducibility are important in this way. By reducing wear of AFM tip and controlling variables such as applying force, scan speed and environment, it can be systematically calibrated the size of features that is written by AFM tip. So, the reproducibility of issues can be controlled.

In this chapter, it is focused on the use of lithography process to build the desired nanostructures and nanolithography on surface of different substrates by AFM. Creating the scratches on various surfaces by silicon nitride and diamond tips using contact mode is discussed. For scratching, the mechanical action of the tip as a sharp pointed tool in order to produce fine scratches is used (Notargiacomo et al., 1999; Sadegh Hassani et al., 2010). The direct scratching is possible with high precision but low quality results are obtained due to probe wear during lithographic process.

Silicon nitride cantilever tip with average spring constant is used to investigate soft surfaces including poly methyl methacrylate (PMMA) (LG-IH 830) thin film coated on the silicon and glass substrates (Sadegh Hassani et al., 2008a). A diamond cantilever tip with high spring constant is used for hard surfaces including highly-oriented pyrolytic graphite (HOPG) and polyethylene substrate (Sadegh Hassani et al., 2010). Since its hardness is much more than silicon nitride, the direct formation of nanoscratches could easily be achieved.

Effects of applied normal force, time of applying pressure, speed and number of scratching cycles on the geometry and depth of scratches are studied. This study shows that there is a critical tip force to remove material from various surfaces (Sadegh Hassani et al., 2008a; 2008b; 2010).

2. Force lithography

An interesting way of performing nanometer pattern is force lithography which based on direct mechanical impact produced by a sharp probe on the sample surface (Lyuksyutov et al., 2003; Park et al., 2000; Sadegh Hassani et al., 2008a). The probe tip pressure on the surface is sufficient to cause plastic deformation of the substrate surface. This type of modification has been used in nanoelectronics, nanotechnology, material science, etc. It enables the fabrication of electronic components with active areas of nanometer scale, super dense information recording and study of mechanical properties of material.

In force lithography no bias voltage is required to produce nanostructures. The nanostructure formation normally occurs as a result of AFM tip motion above the polymer surface with set point magnitude constraining the tip to come closer to the surface (Lyuksyutov et al., 2004; Sadegh Hassani et al., 2008a; 2008b; 2010). In order to apply sufficient normal load to reach plastic deformation of surface, a three-side pyramidal single crystalline diamond tip or another tip with high spring constant is used and pressed against a desired surface (Santinacci et al., 2003; Sadegh Hassani et al., 2010). Much higher forces are achieved by accordingly increasing the applied voltage to piezo-scanner. By scanning the sample in the X or Y direction at various conditions (such as different scanning velocity and number of cycles) grooves are created. However, the protrusions along the edges are formed, which indicates clearly stress deformation during the scratching process (Santinacci et al., 2005).

It is shown that by applying a little force (several μN), removing an amount of material from a metal or polymer film is possible (Bruckl et al., 1997).

Use of cantilevers with high spring constant could apply the desired amount of force without large bending. When tip move toward the substrate or reverse direction, up or down bending of cantilever occurs, respectively. Since an angle of about 10° is typically set between the cantilever and the substrate (see Fig.1. b), this bending influence the tip-substrate interaction, so the geometry and size of scratches are affected in this way. However, increase of applied force cause cumulating of material along or at the end of the grooves. This deformity is occurred because of cantilever bending at the start point of moving tip through the surface. In this way cantilever reach the desired force to create scratch. (Notargiacomo et al., 1999; Sadegh Hassani & Sobat, 2011).

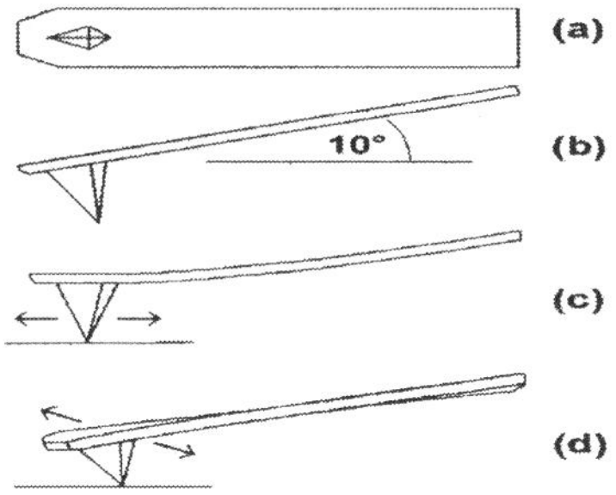

Fig. 1. Typical silicon cantilever with pyramidal tip: (a) upper view; (b) lateral view showing the 10° angle formed with the substrate surface; (c) cantilever bending and (d) torsion (Notargiacomo et al., 1999).

Resolution will be major challenge in lithographic fabrication and the limiting factor for resolution is the tip quality. Sharp silicon tips deliver brilliant and reproducible results. To even further achieve the fine lithographic structure, electron beam deposited tips (EBD tips) can be additionally sharpened in oxygen plasma (Wendel et al., 1995).

Wearing of probe led to low-quality results and reduced the repeatability of produced scratches. Indeed, by using the same tip at another experiment, the sample surface could experience two completely different values of pressure, because the amount of produced pressure depends on the shape of tip (Hu et al., 1998). To decrease wearing of probe, a soft resist polymer film (usually PMMA film) is coated on the surface. On the other hand, the roughness of surface is very important to take high quality scratches. Observations show that surface roughness is strongly influenced by its thickness as while; the surface roughness increases with the increase of the thickness. So, to perform the lithography process, the smoothest surface has to be chosen (Yasin et al., 2005; Fonseca et al., 2004).

Studies show that in the case of AFM, the possibility of directly machining a surface has been explored in two ways, i.e. by either using a static approach in which the microscope is operated in conventional contact mode (Magno & Bennett, 1997; Sumomogi et al., 1995) or using a dynamic approach in which the microscope is operated in the tapping mode (Heyde et al., 2001; Davis et al., 2003). Usually the lithography developed using both static and dynamic approaches are employed to pattern a resist layer, subsequently the patterned layer is used as an etch mark. Both techniques are giving lithography resolution of the order of tens of nanometer (Wendel et al., 1994; Quate, 1997).

An advantage of the vibration in the tapping mode is that very small lateral forces stress the tips, resulting in very slow tip degradation (Wendel et al., 1996).

3. Force curve

In lithographic experiments, it is often critical to know the tip pressure on the desired sample. To estimate the pressure corresponding to a specified level of the probe impact, the force created by the probe has to be determined. It can be calculated from force spectroscopy data.

The normal force between tip and sample is estimated from cantilever deflection (nA) curve plotted against Z-displacement of the cantilever and converting this curve to Force-Distance curve (Vanlandingham, 1997; Yeung et al., 2004; Carallini et al., 2003; Santinacci et al., 2005; Sadegh Hassani et al., 2008a; Argento & French, 1996). To take the force curve, to avoid punching surface, it is essential that the sample has a rigid surface such as silicon or polycrystalline substrate. By performing spectroscopy in a point, force curve is obtained.

The conversion factor for converting nA to nm was obtained from the slope of the linear portion of the deflection-distance curve. There was also one conversion needed for the X-axis values. The change in piezo height, which has been used for the distance between the tip and the sample, was corrected for the deflection of the cantilever by subtracting the cantilever deflection from the piezo height.

On the other hand, there are two measurements required to convert photo detector signal into a quantitative value of force. The first stage is to calibrate the distance that the cantilever actually deflects for a certain measured changes in photo detector voltage. This value depends on type of cantilever and the optical path of the AFM detection laser. When every cantilever is mounted in the instrument, this value will be slightly different. Once the deflection of cantilever is known as a distance Z, the spring constant k, is needed to convert this value into a force F, using Hook's law (F = k × ΔZ) (Heimberg & Zandbergen, 2004; Ebrahimpoor Ziaie et al., 2005; Carpick & Salmern, 1997; Sadegh Hassani & Ebrahimpoor Ziaie, 2006).

4. Nanolithography on various substrates

In some reported experiments, a commercial scanning probe microscope (Solver P47H, NT-MDT Company), operated in AFM contact and noncontact modes, equipped with (NSG11) and (DCP20) cantilevers were used to perform the lithography of desired surfaces (Sadegh Hassani et al., 2008a; 2008b; 2010). The NSG11 cantilever made of silicon nitride had a rectangular shape, and its length, width and thickness were 100 ± 15 µm, 35 ± 3 µm and 1.7–2.3 µm, respectively. Its normal bending constant measured by supplier was 11.5 nN/nm. Another cantilever which was used in this process was DCP20 Cantilever made of diamond with the length, width and thickness of 90±5 µm, 60±3 µm and 1.7-2.3 µm, respectively. The normal bending constant measured by supplier was 48 nN/nm.

These two types of cantilever were selected to reach deformation of different types of surfaces and also for obtaining good images of scratches. These experiments were designed to fabricate scratches on the various surfaces with the different rigidity.

The lithography process was executed with the use of lithography menu supported by the microscope software. The AFM tip was brought into contact with the sample surface using the smallest force possible to minimize any undesired surface modification. An image of surface was prepared in order to choose a suitable surface free of defects for lithography; then the nanolithography process was executed under various specific and controlled conditions to analyze the effect of lithography important factors on the shape of scratches.

For studying force effect, the force was increased by applying a higher voltage to the piezo-scanner in order to reach the cantilever deflection (ΔZ) corresponding to the force (F) range where plastic deformation of polymeric surface occurred. (Santinacci et al., 2003; Notargiacomo et al., 1999; Sadegh Hassani et al., 2008 b). Scratches were made in Y direction on various substrates in different conditions (Sadegh Hassani et al., 2008a; 2008b; 2010), so in this way the influences of applied normal force, scanning velocity, time of applying pressure and number of scratching cycles were investigated. Finally surface was scanned by atomic force microscope in non-contact mode to observe and evaluate the shape and depths of scratches. If the contact mode had been chosen to image the scratches, the surface of substrates would have probably been damaged.

4.1 Nanolithography on PMMA thin films

Sadegh Hassani et al. (2008a; 2008b; 2010) reported lithography performance on PMMA thin films. In this regard, soft thin films of PMMA polymer on the silicon and glass substrates were prepared. For making PMMA (LG-IH830) thin film on silicon and glass substrates, these substrates were washed and sonicated in acetone/ethanol (50-50 % vol.) for 15 minutes at room temperature. Then a very small amount of diluted PMMA/$CHCl_3$ solution was coated over the silicon and glass surfaces using spin coater with 6000 rpm for 30 seconds. The coated substrates were dried in an oven at 130 °C for 30 minutes. The thickness of these coated layers was ~150 nm, measured by atomic force microscope.

In order to choose suitable area for nanolithography process, the topography images and the roughness of the surfaces of PMMA thin films were investigated with the AFM (Sadegh Hassani et al., 2008a). It was reported (Notargiacomo et al., 1999) that "a high value of the surface roughness could produce unwanted features and inhomogeneous results during patterning". At the first step, it was necessary to evaluate the substrate surface after cleaning. In Figure 2 the evolution of the topography images and profiles of the silicon surface after cleaning and PMMA thin films are presented. It is seen that the roughness of PMMA thin film is low and its surface profile is appropriate for lithography. An accurate study was performed on the samples in order to find the optimum patterning conditions for the PMMA film.

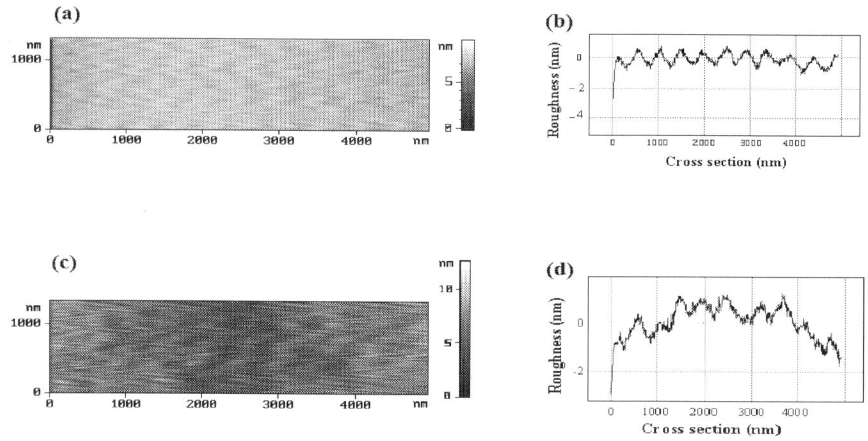

Fig. 2. The evolution of the topography images and profiles of (a & b) silicon surface after cleaning and (c & d) PMMA thin film (Sadegh Hassani et al., 2008a).

For PMMA coated on silicon substrate, scratches were performed using the NSG 11 tip. The main factor in pattern formation was the magnitude of the force applied to the sample. The influence of the applied normal force on the scratches created on the PMMA film coated the silicon substrate had been investigated. In Figure 3 (a-e) some of surface profiles of nanoscratches are presented, which are formed with the constant scanning velocity of 140 nm/s, number of scratching cycle of 10 within 25 ms at various forces (2350, 2700, 3050, 3400, and 3510 nN).

These profiles indicate that the increase of applied normal force, leads to the deeper scratches. The scratches are V-shape; however protrusions are visible along some of the scratches indicating the presence of permanent deformation. It was found that the optimum value for applied normal force was about 3050 nN. The scratch made by this force is shown in Figure 4. In Figure 5, the scratch depths are plotted as a function of the applied normal force. As expected, the scratch size increases with increasing the force load. The depth varies from 4 to 32 nm by increasing force load from 1300 to 3510 nN. However, Notargiacomo et al. claimed (1999) that as the applied force increases, curved cuts ("tails") become visible at the ends of the lines. It has to be mentioned that due to the convolution effect of the tip and substrate topography, the scratch depth may appear smaller by AFM imaging than their actual size (Sadegh Hassani et al., 2008a).

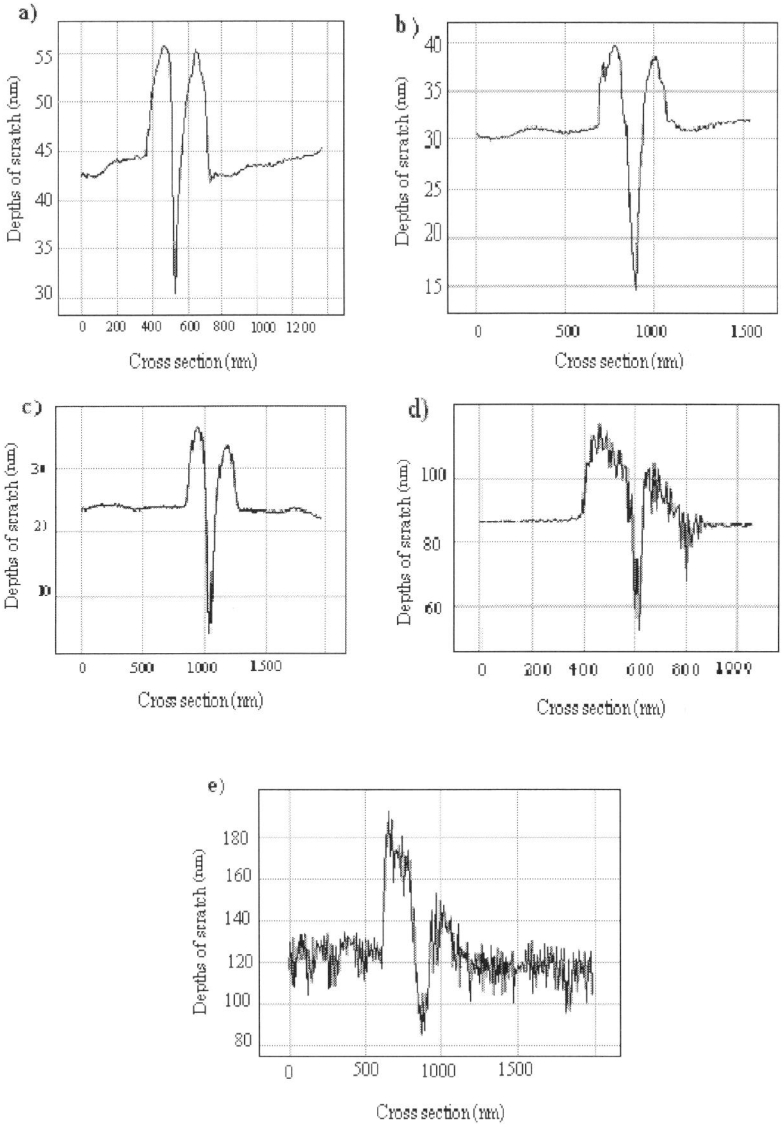

Fig. 3. (a-e) Surface profiles for scratched PMMA film (Sadegh Hassani et al., 2008a) (N= 10 cycles, T=25 ms, V=140 nm/s and F is equal to (a) 2350 nN, (b) 2700 nN, (c) 3050 nN, (d) 3400 nN and (e) 3510 nN).

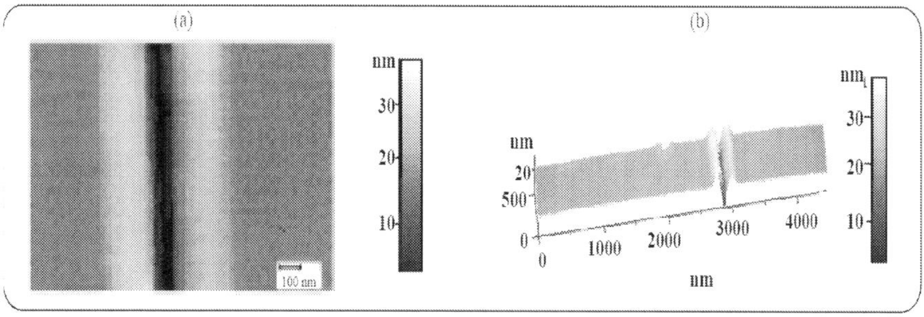

Fig. 4. Topography images of the scratch performed on the PMMA / silicon (Sadegh Hassani et al., 2008a; 2008b; 2010) at N= 10 cycles, T=25 ms, V=140 nm/s and F= 3050 nN. (a) Two dimensional image and (b) Three dimensional image

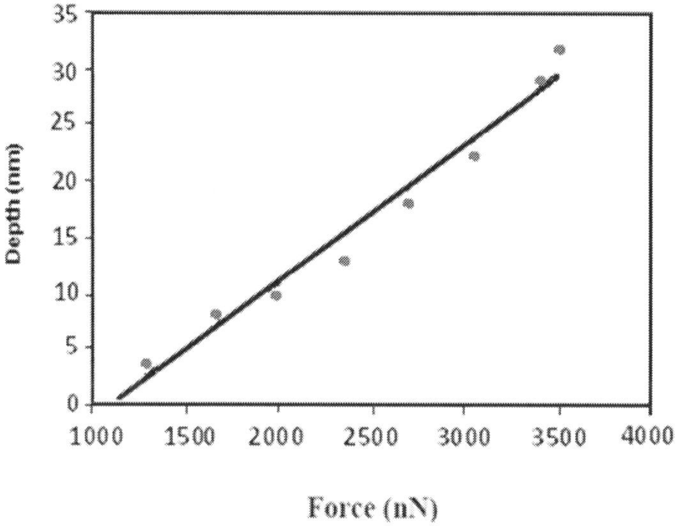

Fig. 5. Dependence of the scratch depth, created on PMMA/silicon, to the applied normal force (Sadegh Hassani et al., 2008a; 2008b; 2010). (The time of applying pressure, number of scratching cycle and scanning velocity are 25 ms, 10 and 140 nm/s, respectively.)

For PMMA coated glass substrate, scratches were performed with exerting various normal forces using NSG 11 tip. In Figure 6, the groove depths are plotted as a function of the applied normal force for PMMA on glass substrate. The most uniform scratches were achieved by applying 3,000 nN force load, while scanning velocity, number of scratching cycle and time of applying pressure were 1400 Å/s, 10 and 25 ms, respectively. Topography image of this scratch is shown in Figure 7. The uniformity of scratches on PMMA coated on silicon and glass is comparable. However, the depth of scratch on the PMMA/glass at the same conditions is more than that of on the PMMA/silicon.

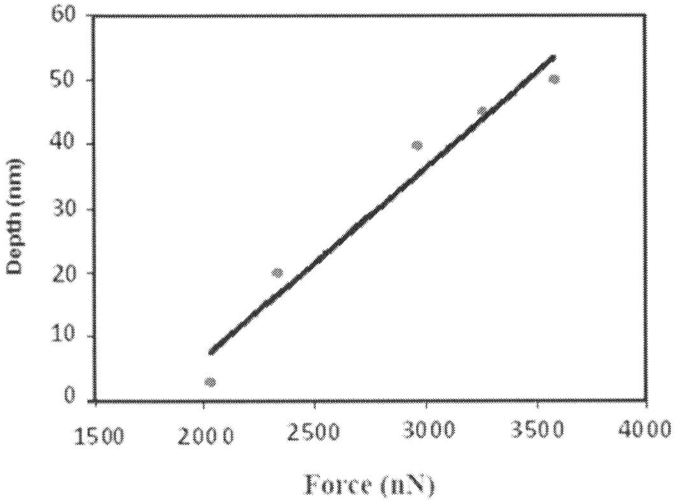

Fig. 6. Dependence of scratches depth, created on PMMA/glass (Sadegh Hassani et al., 2010), with the applied normal force while scanning velocity, number of scratching cycle and time of applying pressure were 140 nm/s, 10 and 25 ms, respectively.

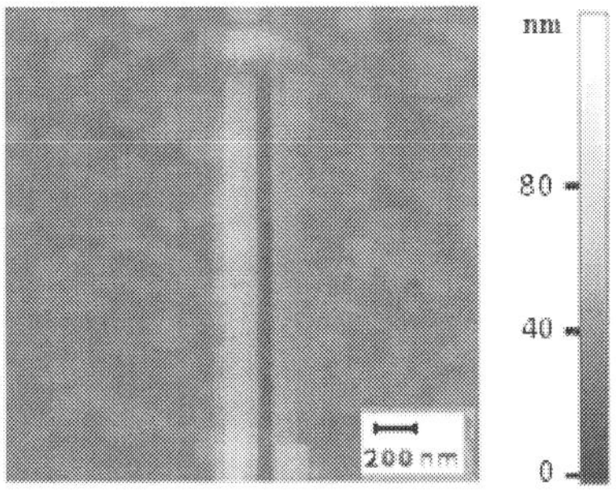

Fig. 7. Topography image of the scratch created on PMMA/glass (Sadegh Hassani et al., 2010), while the applied normal force, scanning velocity, number of scratching cycle and time of applying pressure were 3000 nN, 140 nm/s, 10 and 25 ms, respectively.

The influence of the number of scratching cycles was also investigated by scratching experiments. In Figure 8 dependence of the scratch depth to the number of cycles (N = 1, 5, 10, 15, 20, 25, and 30) in a constant applied normal force of 2350nN, scanning velocity of 140 nm/s in 25 ms is presented. This figure shows that the depth varies from 4 to 30 nm by increasing the number of cycles. As expected, the depths of scratches increase with N linearly. This linear relationship between depth and number of cycles confirms layer-by-layer removal mechanism (Sadegh Hassani et al., 2008a). This result is in agreement with that of obtained by Santinacci and coworkers (2003) for performing nanolithography on p-Si (100) substrate.

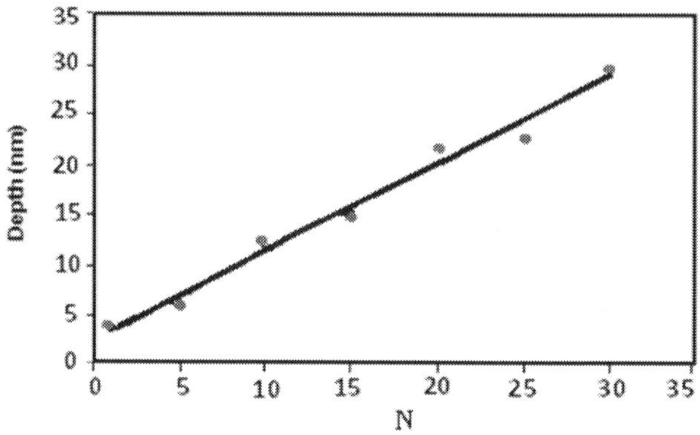

Fig. 8. Dependence of scratch depth created on PMMA/Si to the number of cycles (Sadegh Hassani et al., 2008a; 2008b) (The applied normal force, scanning velocity and time of applying pressure are 2350 nN, 140 nm/s, and 25 ms, respectively.)

The influence of the scanning velocity on the lithography pattern, taken at a normal force of 2350 nN is presented in Figure 9. As it is shown in this figure, the depth varies from 24 to 8 nm by decreasing scanning velocity from 140 to 540 nm/s. It is observed that the increase of the scanning velocity induces a decrease in the scratch depth. Thus, slower scans seem to generate higher pressure and as a result deeper scratch pattern are obtained. However, it could not be determined whether the depth decreases linearly or exponentially with the increase of the scanning velocity (see Fig. 9). To analyze the time effect, nanoindentations are performed on the PMMA surface. The indentation depth created on PMMA/Si depth as a function of time of applying pressure using constant force is presented in Figure 10. It can be seen that by increasing the time of applying pressure, indentation depth is increased. In the other words, the plastic deformation on the PMMA film is time dependent. In this case, to accumulate the tip-induced stress, dilation changes such as defects created or absorbed near the vicinity of the deformed region on the surface occur. This effect leads to an additional plastic deformation of the film.

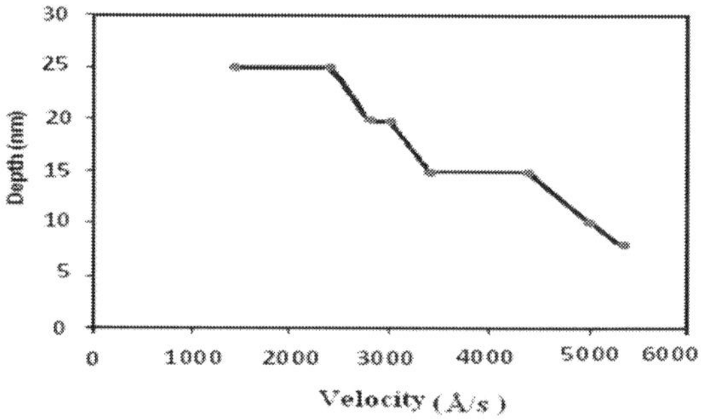

Fig. 9. The influence of the scanning velocity on the depth of the lithographed pattern created on PMMA/Si (Sadegh Hassani et al., 2008a) (The applied normal force, time of applying pressure and number of cycle are 3125 nN, 25 ms and 10, respectively.)

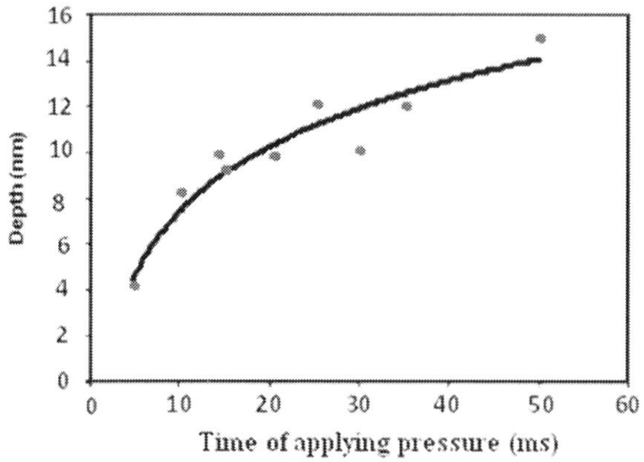

Fig. 10. The indentation depth created on PMMA/Si as a function of the time of applying pressure (Sadegh Hassani et al., 2008a; 2008b) (The applied normal force, scanning velocity and number of cycle are 2350 nN, 140 nm/s and 10, respectively.)

4.2 Nanolithography on polyethylene substrate

This experiment was performed on the polyethylene (PE) substrate (Sadegh Hassani et al., 2008b; 2010). Polyethylene surface was cleaned by washing and sonicating in acetone-ethanol (50-50%Vol.) for 15 minutes at room temperature. This substrate was more inflexible than PMMA thin layer, so performing any modification over PE needed more rigid cantilever tip. The results verified this comment. Scratches were just made by maximum amount of force load, which was equal to 4 μN for NSG11 cantilever tip that was the threshold of force for modifying the PE substrate. The investigation of force effect on the PE substrate was continued by DCP20 cantilever with diamond tip. The force load created by this tip was sufficient to make modification on the PE substrate because of higher spring constant.

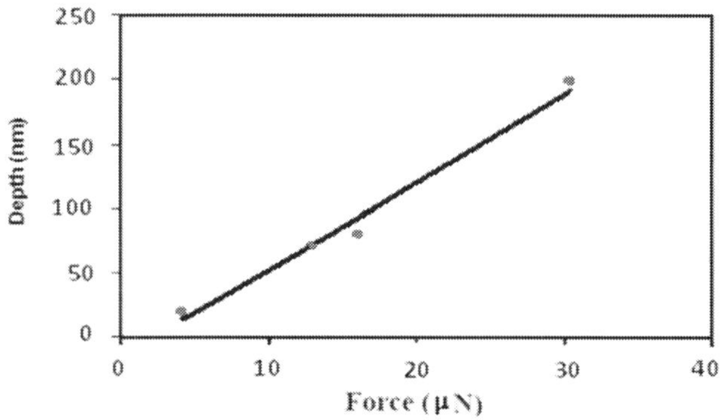

Fig. 11. Dependence of scratches depth with the applied normal force on the PE substrate (Sadegh Hassani et al., 2008b; 2010), while scanning velocity, number of scratching cycle and time of applying pressure are 140 nm/s, 20 and 50 ms, respectively.

A topography image of nanoscratche made on the PE showed that the best quality of scratch was obtained by applying of a 4μN force load. The obtained results showed that the uniformity of the scratches reduced by increasing the force load. Accumulation the vicinity of scratches was occurred, because increasing the applied force induced additional plastic deformation. Figure 11 shows the linear increase of scratches depth on the PE substrate as a function of applied normal force. Meanwhile, increase of applied force caused cumulating of material at the start and end point of the grooves. This deformity was occurred because of cantilever bending at the start point of moving tip through the surface (Notargiacomo et al., 1999; Sadegh Hassani et al., 2088b).

In Figures 12 and 13, the indentation depths are plotted as a function of the time of applying pressure and number of cycles for PE substrate, respectively. Figure 12 shows that the dependence of depth to time is not quite linear. This result was also reported by Santinacci et al. (2005). Figure 13 shows the linear increase of scratches depth with number of cycles, as expected (Santinacci et al., 2003; Santinacci et al., 2005; Sundararajan & Bhushan, 1998).

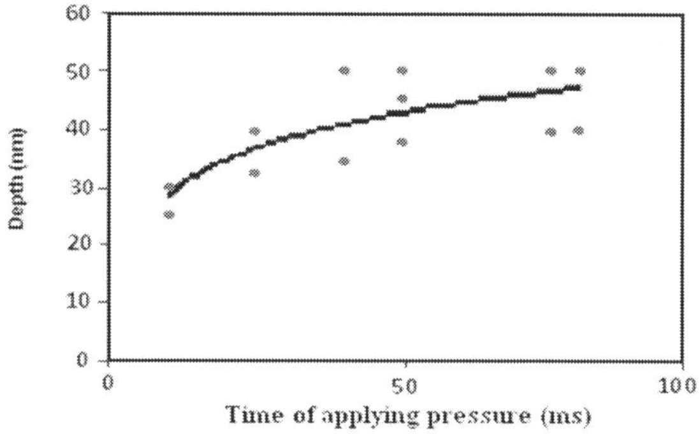

Fig. 12. Indentation depth dependence to time of applying pressure for PE substrate (Sadegh Hassani et al., 2008b), while applied normal force, scanning velocity and number of scratching cycle are 4μN, 140 nm/s, 10, respectively.

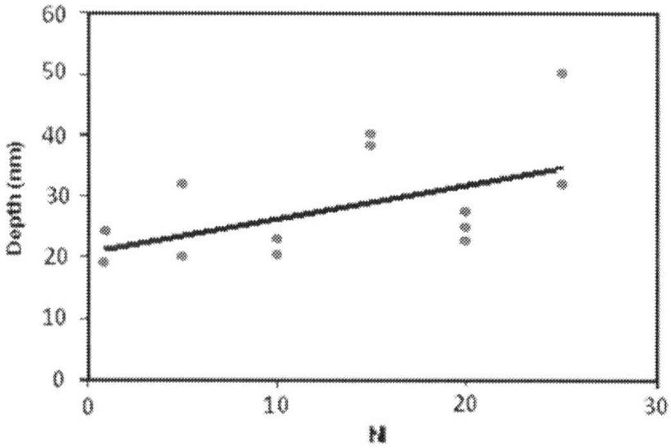

Fig. 13. Indentation depth dependence to number of scratching cycle for PE substrate (Sadegh Hassani et al., 2008b), while applied normal force, scanning velocity and time of applying pressure are 4μN, 140 nm/s, 25 ms, respectively.

According to Figures 12 and 13, repeatability of results for PE substrate is less than PMMA layer. It refers to roughness and flexibility of PMMA thin film. In the other words, making scratch over very uniform and flexible PMMA thin layer is much more successful than PE substrate.

4.3 Nanolithography on HOPG

This experiment was performed on the HOPG substrate as a completely hard surface and by exerting applied normal force ranging from 5.5 to 50.5 μN (Sadegh Hassani et al., 2010). Moreover, in comparison with PMMA thin films and polyethylene substrate, the scanning velocity for performing nanolithography on HOPG surface had to be increased which would led to wearing tip very fast. However, the time of applying pressure on HOPG was much less than those applied on PMMA thin films and polyethylene substrate. Hence, after some trial experiments, 10000 nm / s was chosen for scanning velocity on HOPG substrate at 1 ms. HOPG surface was cleaned using double-sided tape and removing one layer of it. The results of this experiment are obtained by DCP20 cantilever and are presented in Figure 14. Because HOPG surface was very uniform with very low roughness, shape of made scratches were completely V-form. Topography image and cross section of one of the scratches performed on the HOPG is shown in Figure 15. Meanwhile, increase of applied normal force caused cumulating of material at the start and end point of the grooves. This deformity occurred because of cantilever bending at the start point of moving tip through the surface (Wendel et al., 1996).

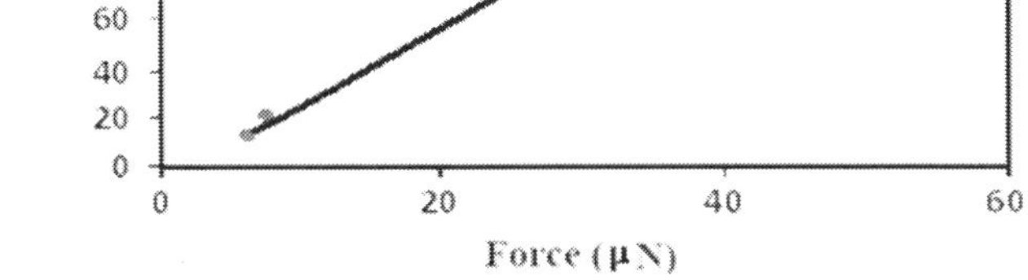

Fig. 14. Dependence of scratches depth to the applied normal force executed by DCP20 cantilever on the HOPG (Sadegh Hassani et al., 2010) substrate, while scanning velocity, number of scratching cycle and time of applying pressure are 10000 nm / s, 1 and 1 ms, respectively.

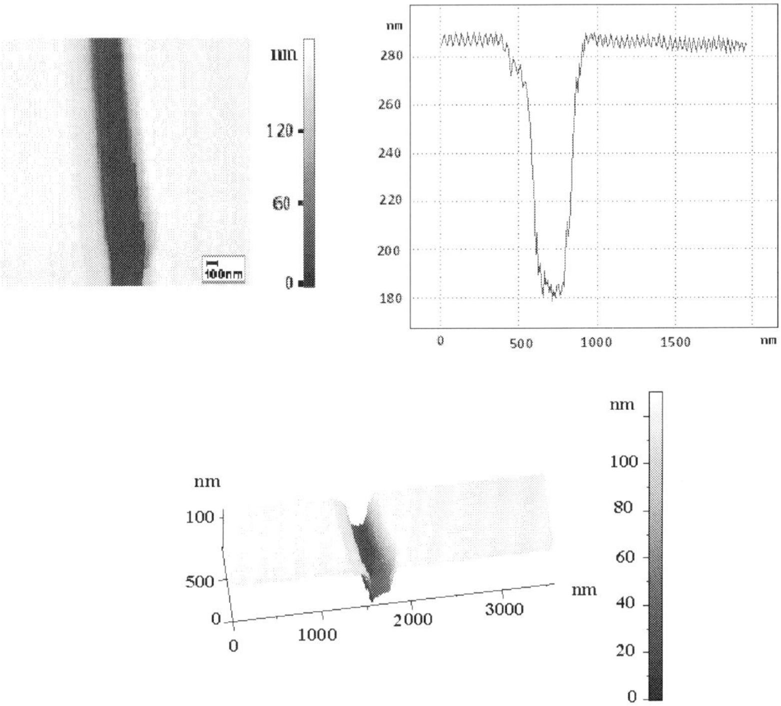

Fig. 15. Two and three-dimensional topography images and cross-section of one of the scratches performed on the HOPG (Sadegh Hassani et al., 2010) , while scanning velocity, number of scratching cycle, time of applying pressure and applied normal force are 10000 nm / s, 1, 1 ms and 50.4 μN, respectively.

5. Conclusion

This chapter is focused on the study of the Scanning Probe Lithography in a controlled way on various substrates. The load force produced by silicon nitride (NSG11) tip with average spring constant is sufficient to deform and make scratch on the PMMA thin film. The analysis of the roughness of the surface shows that the concept using a thin insulting film of PMMA on silicon and glass surfaces as a scratching mask can be successfully performed for nanopatterning. Drawing patterns are being controlled by the necessary parameters such as normal force, scanning velocity, time of applying pressure and number of scratching cycle. It is shown that the depth of the lithography mark increases linearly with the increase of the applied normal force. The uniformity of scratches on the PMMA coated on silicon and glass is comparable.

The load force applied by NSG11 tip is not sufficient to exert scratch on the hard surface and is disabled to perform any changes, so diamond tip with much higher spring constant is required. It must be mentioned that the minimum necessary force to modify the PE surface is about 4 μN that can be achieved by NSG11 tip and with maximum force load. Therefore, for investigation of the effect of higher forces, DCP20 tip is used for PE substrate. The experimental results show that depth of the lithography pattern increased with the increase of the applied normal force with a linear trend for all of the applied substrates. The increase of applied normal force caused accumulating of material at the start and end point of the grooves. This deformity occurred because of cantilever bending at the start point of the tip moving through the surface.

It is presented that the increase of the scanning velocity induces a decrease in the scratch depth. Thus, slower scans seem to generate higher pressure and as a result, deeper scratch pattern are obtained. However, it could not be determined whether the depth decreases linearly or exponentially with the increase of the scanning velocity.

The increase of the lithography depth with the loading time suggests that the plastic deformation on PMMA layer is time dependent. However, the results show that the dependence of depth to the loading time is not quite linear. By the tip-induced stress, dilation changes such as defects created or absorbed near the vicinity of the deformed region on the surface might occur, which would lead to an additional plastic deformation of the film.

It is shown that the depth of the lithography mark increases with the increase of the number of scratching cycle. The depths of scratches increase linearly with the number of scratching cycle.

Finally, due to the convolution effect of the tip and substrate topography, the scratch depth may appear smaller by AFM imaging than their actual size.

6. References

Aime, J. P.; Elkaakour, Z.; Odin, C.; Bouhacina, T.; Michel, D.; Curely, J. & Dautant, A. (1994). Comments on the use of the force mode in atomic force microscopy for polymer films. *Journal of Applied Physics*, 76, 754-762.

Argento, C. & French, R.H. (1996). Parametric tip model and force-distance relation for Hamaker constant determination from aromic force microscopy. *J. Appl. Phys.*,80 (11) 6081-6090.

Avouris, Ph.; Martel, R.; Hertel, T. & Sandstrom, R. (1998). AFM tip induced and current local oxidation of silicon and metals. *Appl. Phys. A* 66, S659-S667.

Avouris, Ph.; Hertel, T. & Martel, R. (1998). Atomic force microscope tip- induced local oxidation of silicon: kinetics, mechanism and nanofabrication. *Appl. Phis. Lett.* 71(2) 285-287.

Bouchiat, V. & Esteve, D. (1996). Lift-off lithography using an atomic force microscope, *Appl. Phys. Lett* ,69(20) 3098-3100.

Bruckl, H.; Rank, R.; Vinzelberg, H.; Monch, I.; Kretz, L. & Reiss, G. (1997). Observation of coulomb-blockade effects in AFM-machined tunnel junctions. *Surf. Interface Anal.* 25, 611–613.

Burnham, N. A. & Colton, R.J. (1989). Measuring the nanomechanical properties and surface forces of materials using an atomic force microscope. *Journal of Vacuum Science Technology*, A7, Issue 4, 2906-2913.

Burnham, N. A.; Colton, R. J. & Pollock, H. M.(1991). Interpretation issues in force microscopy. *Journal of Vacuum Science & Technology A: Vacuum, Surfaces, and Films*, Vol.9, Issue 4, 2548-2556.

Carallini, M.; Biscarini, F.; Leon. S.; Zerbetto, F.; Bottari, G. & Leigh, D.A. (2003). Information Storage using Supramolecular Surface Patterns. *Science* 299, 531.

Carpick, R. W. & Salmern, M. (1997). Scratching the surface: Fundamental investigations of tribology with atomic force. *Chem. Rev.* 97, 1163-1194.

Chen, J.; Read, M. A.; Raweltt, A. M. & Tour, J. M. (1999). Large on- off ratios negative differential resistance in a molecular electronic device. *Science,* 286, 1550-1552.

Davis, Z. J.; Abadal, G.; Hansen, O.; Borise, X.; Barniol, N.; Perez-Murano, F. & Boisen, A. (2003). AFM lithography of aluminum for fabrication of nanomechanical systems. *Ultramicroscopy* 97, 467-472.

Dubois, E. & Bubbendroff, J. L. (1999).Nanometer sacle lithography on silicon, titanium and PMMA resist using scanning probe microscopy. *Solid-state Electronics,* 43, 1085-1089.

Dienwiebel, M.; Verhoeven, G. S.; Pradeep, N.; Frenken, J.W.; Heimberg, J. A. & Zandbergen, H. W.,(2004). Superlubricity of graphite. *Phys. Rev. Lett.,* 92(12), (126101-1)-(126101-4).

Ebrahimpoor Ziaie, E.; Rachtchian, D. & Sadegh Hassani, S. (2008). Atomic force microscopy as a tool for comparing lubrication behavior of lubricants. *material science: An Indian journal,* 4(2), 111-115.

Ebrahimpoor Ziaie, E.; rachtchian, D. & Sadegh Hassani, S. (2005). Atomic Force Microscopy as a tool for comparing lubrication behavior of lubricants. *IV international conference on tribochemistry,* 3-5 october, Poland.

Fonseca Filho, H. D.; Mauricio, M. H. P.; Ponciano, C.R. & Prioli, R. (2004). Metal layer mask patterning by force microscopy lithography. *Material Science and Engineering* B., 112, 194.

Garcia, R.; Martinez R.V. & Martinez, J. (2006). Nano-chemistry and scanning probe nanolithographies. *Chem. Soc. Rev,* 35, 29-38.

Geissler, M. & Xia, Y. (2004). Patterning: Principles and Some New Development. *Adv. Mater.* 16, 1249-1269.

Heimberg, J. A. & Zandbergen, H. W. (2004). Superlubricity of graphite. *Physical Review Letters* 92 (12), 126101 (1) -126101.

Heyde, M.; Rademann, K.; Cappella, B.; Guess, M.; Strum, H.; Spanegenberg, T. & Niehus, H. (2001). Dynamic plowing nanolithography on polymethylmethacrylate using an atomic force microscope. *Rev. Sci. Instrum.* 72, 136-141.

Hu, S.; Hamidi; A.; Altmeyer; S.; Koster, T.; Spangenberg, B. & Kurz, H. (1998). Fabrication of silicon and metal nanowires and dots using mechanical atomic force lithography. *J. Vac. Sci. Technol.* B16, 2822-2824.

Huang, Y.; Duan, X.; Wie, Q. & Lieber, C. M. (2001). Directed assembly of one-dimensional nanostructures into functional networks. *Science,* 291, 630-633.

Hues, S. M.; Draper, C. F. & Colton, R.J., (1994). Measurement of Nanomechanical Properties of Metals Using the Atomic-Force Microscope. *J. of Vacuum Science Technology,* B12, 2211-2214.

Hyon, C. K.; Choi, S. C.; Hwang, S.W.; Ahn, D.; Kim, Y. & Kim, E. K. (1999). Direct nanometer scale patterning by the cantilever oscillation of an atomic force microscope. *Appl. Phys. Lett.,* Vol. 75, No. 2, 292–294.

Irmer, B.; Kehrle, M.; Lorenze, H. & Kothaus, J. P.,(1998). Nanolithography by non-contact AFM-induced local oxidation: fabrication of tunnelling barriers suitable for single-electron devices, *Semicond. Sci. Technol.,* 13, A79-A82 .

Iwanaga, Sh. & Darling, R.B. (2005). Stable and erasable patterning of vanadium pentoxide thin films by atomic force microscope nanolithography. *Applied Physics Letters*, Vol. 86, 133113–133113-2.

Jones, A. G.; Balocco, C.; King, R. & Song, A. M. (2006). Highly tunable, high-throughput nanolithography based on strained regionregular conducting polymer films. *Appl. Phys. Lett.* 89, 013119 (1-3).

Lee, H. T.; Oh, J. S.; Park, S. J.; Park, K. H.; Ha, J. S.; Yoo, H. J. & Koo, J. Y. (1997). Nanometer-scale lithography on H-passivated Si(100) with an atomic force microscope in air. *Journal of Vacuum Science Technology*, A15(3), 1451.

Lemeshko S.; Saunin S. & Shevyakov V. (2005). Atomic Force Microscope As a Tool for Nanometer Scale Surface Surface Patterning. *Nanotechnology*, 2, 719-721.

Li, S. F. Y.; Ng, H. T.; Zhang, P. C.; Ho, P. K. H.; Zhou, L.; Bao, G. W. & Chan, S. L. H. (1997). Submicrometer lithography of a silicon substrate by machining of photoresist using atomic force microscopy followed by wet chemical etching. *Nanolithography*, 8, 76-81.

Lyuksyutov, S. F.; Paramonov, P.B.; Dolog, I. & Ralich, R.M. (2003). Peculiarities of an anomalous electronic current during atomic force microscopy assisted nanolithography on n-type silicon. *Nanotechnology*, 14, 716-721 .

Lyuksyutov, S. F.; Paramonov, P. B.; Sharipov, R. A. & Sigalov, G. (2004). Induced nanoscale deformation in polymers using atomic force microscopy. *Physical Review B*, 70, 174110-174110(1-8).

Magno, R. & Bennett, B. R. (1997). Nanostructure patterns written in III -V semiconductors by an atomic force microscope. *Appl. Phys. Lett.* 70, 1855-1857.

Martin, C.; Rius, G.; Borrise, X. & Perez-Murano, F. (2005). Nanolithography on thin layers of PMMA using atomic force microscopy. *Nanotechnology*, Vol. 16, pp.1016–1022.

Notargiacomo, A.; Foglietti, V.; Cianci, E.; Capellini, G.; Adami, M.; Faraci, P.; Evangelisti, F. & Nicolini, C. (1999) Atomic Force Microscopy Lithography Study towards the Development of Nanodevices. *Nanotechnology*, 10, 458-463.

Park, J. & Lee, H.,(2004). Effect of surface functional groups on nanostructure fabrication using AFM lithography. *Material Science and Engineering* C. 24, No.1, 311-314.

Quate, C. F. (1997). Scanning probe as a lithography tool for nanostructures, *surface science*. 386, 259-264.

Sadegh Hassani, S. & Ebrahimpoor Ziaie, E. (2006). Application of Atomic Force Microscopy for the study of friction properties of surfaces. *material science: An Indian journal*, 2(4-5), 134-141.

Sadegh Hassani, S. & Sobat, Z. (2011). Studying of various nanolithography methods by using Scanning Probe Microscope. *Int .J. Nano .Dim* 1(3) , Winter 159-175.

Sadegh Hassani, S.; Sobat, Z. & Aghabozorg, H. R. (2010). Force nanolithography on various surfaces by atomic force microscope. *Int. J. Nanomanufacturing*, Vol. 5, Nos. 3/4, 217-224.

Sadegh Hassani, S.; Sobat, Z. & Aghabozorg, H. R. (2008a). Nanometer-Scale Patterning on PMMA Resist by Force Microscopy Lithography. *Iran. J. Chem. Chem. Eng.* Vol. 27, No. 4, 29-34.

Sadegh Hassani, S.; Sobat, Z. & Aghabozorg, H. R. (2008b). Scanning probe lithography as a tool for studying of various surfaces. *Nano Science and Nano Technology: An Indian journal*, Volume 2 Issue (2-3),94-98.

Samori, D. (2005). Exploring supramolecular interactions and architectures by Scanning Force Microscopies. *Chem. Soc. Rev.*, 34, 551-561.

Santinacci, L., Djenizian, T., Hildebrand, H., Ecoffey, S., Mokdad, H., Campanella, T. & Schmuki, P.,(2003). Selective Pd electrochemical deposition onto AFM-scratched silicon surfaces. Electro chimica Acta, 48, 3123-3130.

Santinacci, L.; Zhang, Y. & Schmuki, P. (2005). AFM scratching and metal deposition through insulating layers on silicon. *Surface Science*, 597(1-3), 11-19.

Sheehan, P. E. & Whitman, L. (2002). Thiol Diffusion and the Role of Humidity in Dip Pen Nanolithography. *J. Physical Review Letters*, 88, (156104-1)-(156104-4) .

Sheglov, D. V.; Latyshev, A. V. & Aseev, A. L. (2005). The deepness enhancing of an AFM-tip induced surface nanomodification. *Applied Surface Science*, 243(1-4), 138-142 .

Snow, E. S.; Campbell P.M. & Perkin F. K. (1999). High speed patterning of a metal silicide using scanned probe lithography. *Appl. Phys. Lett.* 75(101476-1478..

Sumomogi, T.; Endo, T.; Kuwahara, K. & Kaneko. R. (1995). Nanoscale layer removal of metal surfaces by scanning probe microscope scratching. *J. Vac. Sci. Technol.* B13, 1257-1260.

Sugimura, H.; Uchida, T.; Kitamura, N. & Masuhara, H.,(1993). Tip- induced anodization of titanium surfaces by scanning tunneling microscopy: a humidity effect of nanolithography. *Appl. Phys. Lett.*, 63, 1288.

Sugimura, H. & Nakagiri, N. (1997). AFM lithography in constant current mode, *Nanotechnology*, 8 A15-A18.

Sundararajan S. & Bhushan B. (1998). Micro/Nanotribological Studies of Polysilicon and SiC Films for MEMS Applications, Wear 217, 251-261.

Sundararajan, S. & Bhushan, B. (2000). Topography-Induced Contributions to Friction Forces Measured Using an Atomic Force/Friction Force Microscope. *Journal of Applied Physics*, 88(8), 4825.

Vanlandingham, M. R. (1997). The effect of instrumental uncertainties on AFM indentation measurements. *Microscopy Today* 97 (10) 12-15.

Wendel, M., Kuhn, S., Lorenz, H., Kotthaus, J. P. & Holland, M. (1994). Nanolithography with an Atomic Force Microscope for Integrated Fabrication of Quantum Electronic Devices. *Appl. Phys. Lett.* 65, 1775-1777.

Wendel, M.; Irmer, N.; Cortes, J.; Kaiser, R.; Lorenz, H.; Kotthaus, J. P. & Lorke, A. (1996), Nanolithography with an Atomic Force Microscope. *Superlattices and Microstructures* 20 (3), 349-356.

Wendel, M.; Lorenz, H. & Kotthaus. J. P. (1995). Sharpened electron beam deposited tips for high resolution atomic force microscope lithography and imagine. *Appl. Phys. Lett.* 67, 3732-3734.

Wilder, K. & Quate, C. F. (1998). Noncontact nanolithography using the atomic force microscope *Appl. Phys. Lett.*, 73 (17) 2527-2529.

Yasin, Sh.; Khalid, M. N.; Hasko, D.G. & Sarfraz, S. (2005). Correlation of surface roughness with edge roughness in PMMA resist. *Microelectronic Engineering*, 78-79, 484-489 .

Yeung, K. L. N. & Yao, J.(2004). Scanning probe microscopy in catalysis. *Nanosci. Nanotechol.*, 4, 1-44.

Yoshimura, T.; Shiraishi H.; Yamamoto J. & Okazaki Sh. (1993). Nano edge roughness in polymer resist patterns. *Appl. Phys. Lett.* 63,(6), 764-766.

24

Scanning Probe Lithography on Organic Monolayers

SunHyung Lee[1], Takahiro Ishizaki[2], Katsuya Teshima[1], Nagahiro Saito[3] and Osamu Takai[3]
[1]*Faculty of Engineering, Shinshu University,*
[2]*National Institute of Advanced Industrial Science and Technology (AIST),*
[3]*EcoTopia Science Institute, Nagoya University,*
Japan

1. Introduction

Lithographic technologies for the surface modification of inorganic and organic surfaces have been developed for various devices such as sensing, data memory, single molecule electronics and biological systems. Nano and micropatterning of organic monolayers have attracted attentions for applications to biological systems in which proteins or DNA are fixed. Photolithography, microcontact printing, and electron beam lithography have usually been used as patterning techniques for organic monolayers (Hayashi et al., 2002; Hong et al., 2003; Saito et al., 2003; Hahn et al., 2004; Kidoaki & Matsuda, 1999). Although the electron-beam lithography can fabricate very small patterns, it requires an ultra-high vacuum system (Harnett et al., 2001). The resolution of photolithography is limited by the light wavelength. Moreover, these methods are based on destructive lithography, i.e., they cause damages to the organic materials.

In particular, nano-lithographic technologies have evolved in order to satisfy persistent demands for miniaturization and high-density integration of semiconductor electric circuits. Scanning probe microscopy (SPM) has been a key tool in achieving this goal. SPM can be used not only as the means that observe surface structure at sub-molecule level by a probe but also as the means that control the atomic and molecular arrangement on a substrate. As a local nano-fabrication means, the lithography technique in nanoscale range by using SPM is called to the scanning probe lithography (SPL) (Kaholek et al., 2004; Blackledge et al., 2000; Tello et al., 2002; Liu et al., 2002). In particular, a variety of lithographic techniques using SPM probe can fabricate nano-scale patterns on an organic monolayer, such as nanoshaving, nanografting, anodization SPL, dip-pen nanolithography (DPN), and electrochemical SPL. The lithography technique is used to break the material surface by using various energy sources. SPM can also be used to break the organic monolayer. For instance, nanoshaving involves mechanical scratching by physical pressure of the probe, and anodization lithography involves anodic oxidation of the substrate surface by an applied bias voltage (above 9 V) between the probe and the substrate (Jang et al., 2002; Kaholek et al., 2004; Sugimura, & Nakagiri, 1995). In the case of the anodization lithography, the oxide layer can be fabricated by anodic oxidation. As another SPL technique, the nanografting procedure combines the fabrication of a nanopattern and the binary SAM using atomic force microscopy (AFM) (Amro et al., 2000; Xu et al., 1999; Liu et al., 2000). As the other process, a probe is scanned on the matrix SAM at a high force. The matrix SAM is removed and simultaneously replaced by another molecule in the scanning area. Fig. 1 shows the nanoshaving, nanografting, and anodization lithography techniques.

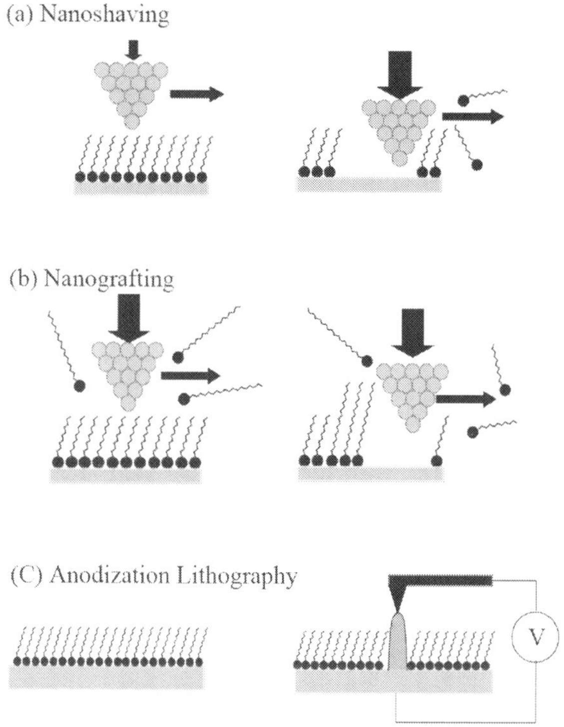

Fig. 1. Various lithographic methods using SPM probe on the organic monolayer. :
(a) nanoshaving, (b) nanografting and (c) anodization lithography

In addition, dip-pen nanolithography (DPN) is a nanopatterning technique with a probe which delivers molecules to a surface via a water meniscus in the ambient atmosphere (Pena et al., 2003; Schwartz,2002; Maynor. et al., 2004). This direct-writing technique offers high-resolution patterning capabilities for a number of moleculars and biomolecular materials (ink) on a variety of substrates (paper). Nanografting and DPN are recently developed as "soft" lithographic methods, which do not cause any damage to the organic monolayers. Electrochemical SPL is also one of the soft lithographic methods. This is performed through the water column condensed between the tip of the SPM and the substrate surface. This water column can be used as a minute electrochemical cell. When a bias voltage is applied, a redox reaction occurs on the substrate surface. In the case of SAM, the functional group is converted by this redox reaction and a nanopattern is formed on the SAM surface. Fig. 2 shows the DPN and electrochemical SPL techniques.

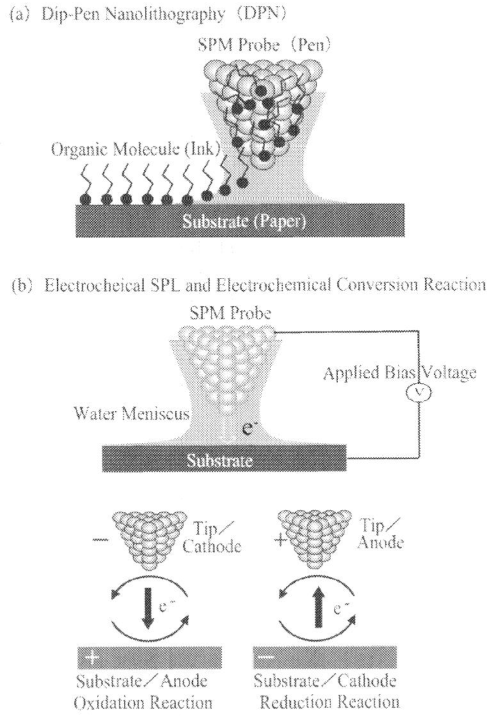

Fig. 2. Schematic illustration of dip-pen nanolithography (DPN) and electrochemical SPL.

In this chapter, we introduce anodization SPL and electrochemical SPL for fabricating the nanostcuture and controlling the function group on the various organic monolayers, such as organosilane self-assembled monolayers (SAMs) and the monolayer covalently attached to silicon through Si-C bond . First, anodization SPL techniques are expected to fabricate nanostructures on surfaces of electronic device. Their technique can remove an organic monolayer and fabricate an oxide structure by applying highly bias voltage between probe and substrate. When a high positive bias voltage is applied to the probe, an oxidation reaction proceeds and an oxide structure forms on the surface. The electrochemical reactions on SPM probe and substrate are shown in Eqs., respectively (Lee et al., 2009).

$$4H_2O + 4e^- \rightarrow 2H_2 + 4OH^- \quad (1)$$

$$M + 2H_2O \rightarrow MO_2 + 4H^+ + 4e^- \ (M : \text{substrate}) \quad (2)$$

In these reactions, electric characteristics of coated materials on the probes are important to formation of an oxide structure. The oxide structure is easily fabricated on the substrate by anodization SPL since the silicon oxide insulator layer is removed. However, the oxide structure is difficult to control and is not attained to single nanometer level though many researchers have reported on anodization SPL. We introduced fabricating technique of a three-dimensional nanostructure (nanoline structure) of silicon oxide on the hydrogen-

terminated Si substrates by anodization SPL, and the effects of coated materials of SPM probe on the sizes of oxidized structures (Lee et al., 2009).
Second, we introduce a novel soft lithography based on electrochemistry through scanning probe electrochemistry for controlling the functional group on the organic monolayers (Lee et al., 2007; Lee & Ishizaki et al., 2007; Saito et al., 2005; Sugimura et al., 2004). Scanning probe electrochemistry in which materials surfaces locally oxidized or reduced by a tip of SPM is a promising technique for constructing nanostructures consisting of organic molecules. This electrochemical SPL is performed through water column which is condensed between the tip of SPM and the substrate surface. This water column can be used as a microscopic electrochemical cell. In the case of electrochemical conversion using a nanoprobe, the electrochemical reactions which proceed at the probe-sample junction are governed by the applied bias voltage and its polarity. When the substrate is polarized positively, anodic reactions, that is, oxidation reactions, proceed on its surface. On the contrary, when the substrate is polarized negatively, cathodic reactions, that is, reductive reactions, proceed. The method is expected as a key technology for future molecular nanodevices. In addition, organosilane self-assembled monolayers (SAMs) have been applied to a resist material for SPL. Here we report the chemical conversion of an organic molecular monolayer in a reversible manner using SPM. The chemical state of the monolayer from its oxidized state to reduced state or vice versa is controlled. First, an amino-terminated SAM was chemically converted into an oxidized SAM by SPL at positive-bias voltages. Moreover, this oxidized SAM was then reconverted into an amino-terminated SAM by SPL at negative bias voltages. We examine the chemical changes undergone in the scanned area from the viewpoint of surface-potential reversibility. Additionally, we introduce a electrochemical SPL to fabricate -COOH groups on an organic monolayer directly attached to silicon, which was synthesized from 1,7-octadien (OD) and hydrogen-terminated silicon. The -COOH groups on the OD-monolayer were also synthesized by the properties of surface produced by SPL.

2. Anodization scanning probe lithography on the organic monolayer on the hydrogen-terminated Si substrate

In this chapter, we introduced fabricating technique of a three-dimensional nanostructure of organic monolayer on the hydrogen-terminated Si substrates by anodization SPL. First, we fabricated the organic monolayer on the hiydrogen-termicated Si(111) wafers with an electrical resistance of 10.0-20.0 Ω-cm. Si substrates were sonicated in acetone and ethanol for 10 min, and then, cleaned by an ultraviolet (UV) light/ozone cleaning method. The substrates were exposed to vacuum UV (VUV) light (172 nm) from an excimer lamp for 30 min under atmospheric pressure and room temperature. Subsequently, the substrates were cleaned in piranha solution (H_2SO_4:H_2O_2 = 3:1) at 100 °C for 10 min and rinsed in ultrapure water. They were then etched for 15 min by immersing in 40 % aqueous ammonium fluoride solution (NH_4F). As a result, the native silicon oxide layer was removed from the substrates, and hydrogen-terminated surfaces were formed. 1-decane monolayer was prepared by a liquid phase method. The hydrogen-terminated substrates were immersed in the solution of 1-decane molecules at 150 °C for 3 h . After the immersion, the organic monolayer coated substrates were cleaned in toluene, acetone, and ethanol, and rinsed in ultrapure water.
Firstly, the 1-decane monolayer was fabricated on the hydrogen-terminated Si substrates through a liquid phase method at 150 °C. Fig. 3(a) shows the chemical structure of 1-decane

molecule. The static water contact angle and film thickness of the monolayer was about 107° and 1.12 nm, respectively. A smooth surface coated with densely packed CH_3 groups (alkyl monolayer) shows a water contact angle of approximately 110°. In addition, its thickness was considered reasonable since the chain length of 1-decane molecule was estimated to be 1.32 nm. It is indicated that the 1-decane molecules formed an organic monolayer on the silicon substrate. Fig. 3(b) shows the XPS Si2p spectrum of the Si surface coated with the 1-decane monolayer. As clearly shown in Fig. 3(b), the peak of native silicon oxide was not observed at all. These results indicate that the 1-decane molecules reacted directly to the hydrogen-terminated silicon surfaces and formed the organic monolayer there. Fig. 3(c) and 3(d) show the AFM topographic and friction images of the Si surfaces after immersing in 1-decene solution. The flat terraces with steps for silicon one-atom and the changes of high friction on steps were, respectively, observed on the topographic and friction images. In the topographic image [Fig. 3(c)], the intervals of the steps and the terraces were 180 nm and 3.2 ± 0.3 Å, respectively. In addition, the difference of friction force between the flat terraces and steps was about 10 mV. These results indicated to the surface profiles of hydrogen-terminated silicon. It was therefore found that the 1-decane molecules were vertically and densely assembled, and their monolayer was formed through doubly-boned terminated group attached to the hydrogen-terminated silicon surface.

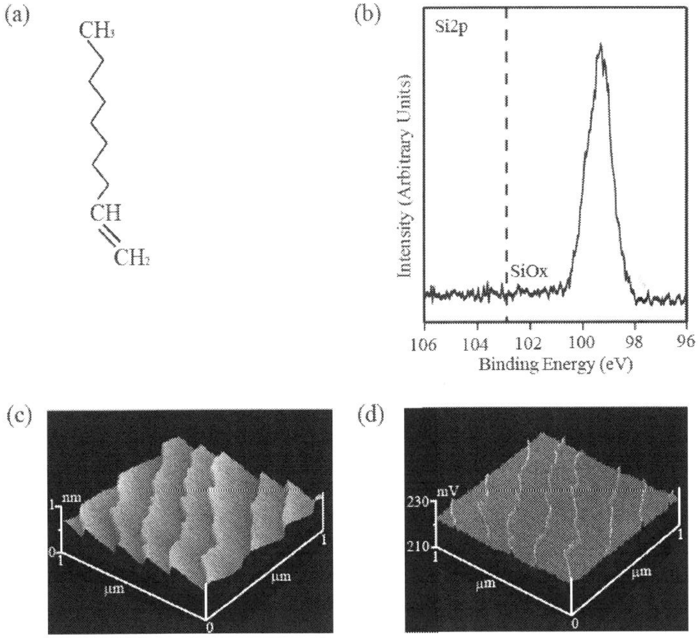

Fig. 3. (a) The chemical structure of 1-decane (b) XPS Si2p spectra (c) topographic image (d) friction image of sample surfaces after prepared 1-decane monolayer. [SH. Lee, N. Saito, O. Takai, Highly reproducible technique for three-dimensional nanostructure fabrication via anodization scanning probe lithography, Appl. Surf. Sci., 255, 7302-7306 (2009). Copyright@ELSEVIER (2009)]

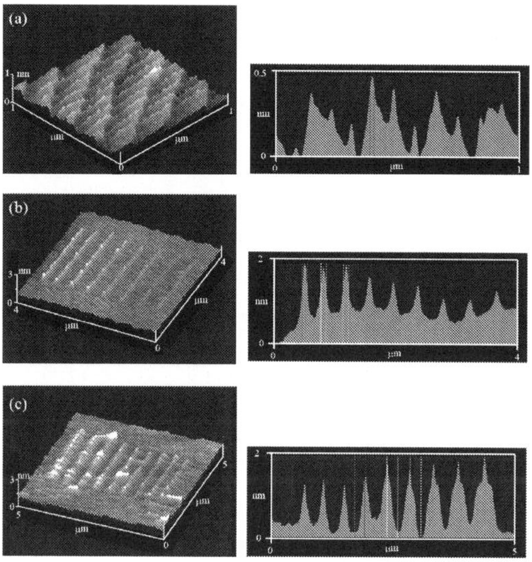

Fig. 4. The topographic images of SiOx line structure fabricated on the 1-decane monolayer and profiles for each topographic image by using various tip: (a) diamond coated Si tip, (b) Si tip and (c) Au coated Si tip. [SH. Lee, N. Saito, O. Takai, Highly reproducible technique for three-dimensional nanostructure fabrication via anodization scanning probe lithography, Appl. Surf. Sci., 255, 7302-7306 (2009). Copyright@ELSEVIER (2009)]

Next, we investigated that the effect of scanning rate on nanoline width of silicon oxide fabricated by anodization SPL. An anodization SPL was carried out on the 1-decane monolayer in air (humidity in ranging from 30 to 40 %) at the applied bias voltage of 9 V. In the anodization SPL, 1-decane monolayer was removed and the SiOx nanoline structures were formed by scanning the probes. Fig. 4(a)-4(c) show the topographic images and profiles of SiOx nanoline structure fabricated by the anodization SPL. In these experiments, various probes were used to fabricate oxide nanostructures [Fig. 4(a); diamond-coated Si probe, Fig. 4(b); Si probe (i.e., uncoated Si probe), Fig. 4(c); Au-coated Si probe]. Additionally, we investigated the effect of coated materials on the formation of the oxide nanostructure. The ranging of scanning rates for SPL is 0.1 to 5 μm/s. In these topographic images, the nanoline structure and flat terraces with steps were obviously observed under all scanning conditions. However, the widths of nanoline structures were changed with the scanning rates. Fig. 5 shows the variation in the width of nanoline structures with scanning rates and surface compositions of probes. When the diamond coated probe is used for anodization SPL, the nanoline widths were found to remain constant under all scanning conditions, and the scanning rates had no effect on the line width. The width was approximately 15 nm, which is one of the finest nanostructures in the field of SPL technique. The highly reproducible structure was fabricated by anodization SPL using the diamond-coated Si probe, and this technique is thought to be able to apply various industrial fields. On the other hand, when the Si probe (uncoated-Si) and Au-coated Si probe were used for anodization SPL, the line widths were markedly changed with the scanning rates. In the case of Au-coated Si probe, as the scanning rate increased from 0.1 to 5 μm/s, the width of line structures drastically decreased from 375 to 125 nm. For Si probe (uncoated Si probe), the line width gradually decreased with increasing scanning rate, reaching about 100 nm at 5 μm/s. The width variations were Au-coated Si probe > Si probe (uncoated) > diamond-coated Si probe. These variations of the line width could be explained the band-gap energy of the coated materials. The band-gap energies of Au, Si and diamond are respectively about 0, 1.2 and 5.47 eV. That is to say, their conductive property is thought to be greatly dependent on the line width. These results indicate that high reproducibility of oxide nanoline structures is attainable by means of anodization SPL using the diamond-coated probe or a probe which has relatively low conductivity.

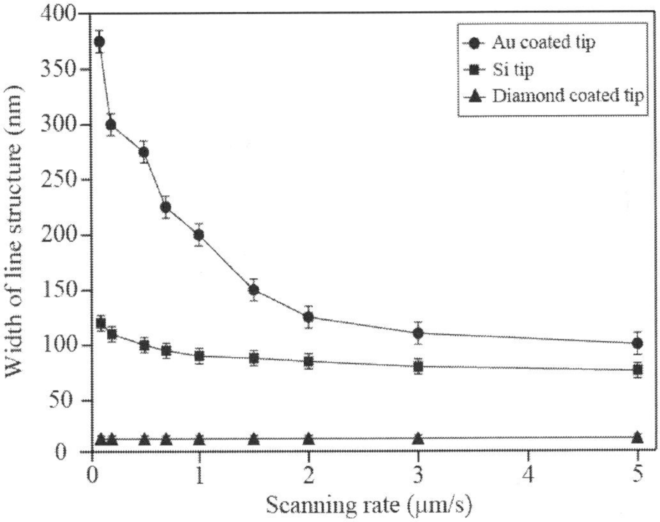

Fig. 5. The change of the line width with increasing scanning rate [SH. Lee, N. Saito, O. Takai, Highly reproducible technique for three-dimensional nanostructure fabrication via anodization scanning probe lithography, Appl. Surf. Sci., 255, 7302-7306 (2009). Copyright@ELSEVIER (2009)]

Finally, the effect of applied bias voltage on nanoline width of silicon oxide was investigated by use of various SPM probes. Fig. 6(a)-(c) show the topographic images of SiOx nanoline structures fabricated by diamond-coated Si, Si (uncoated) and Au-coated Si probes, respectively. The applied bias voltage and scanning rate were, respectively, fixed at 7 V and 1 μm/s. The obtained widths were 15, 60 and 100 nm in Fig. 6(a)-(c), respectively. The line width fabricated at the bias voltage of 7 V decreased compared to that at 9 V when Au-coated Si and uncoated Si probes were used. For the diamond-coated probe, the nanoline structure maintained the width of 15nm, that is, the applied bias voltages had no effect on the nanoline width. These results are also attributed to the band-gap energy of the coated materials. In particular, three-dimensional oxide nanostructures fabricated by the diamond-coated probe showed highly reproducibility even though various scanning rates and applied bias voltages are used in aondization SPL. Therefore, this technique is expectable to be applied to fabricate a wide variety of nanodevices, that have three-dimensional structures, in various industrial fields.

(a)

(b)

(c)

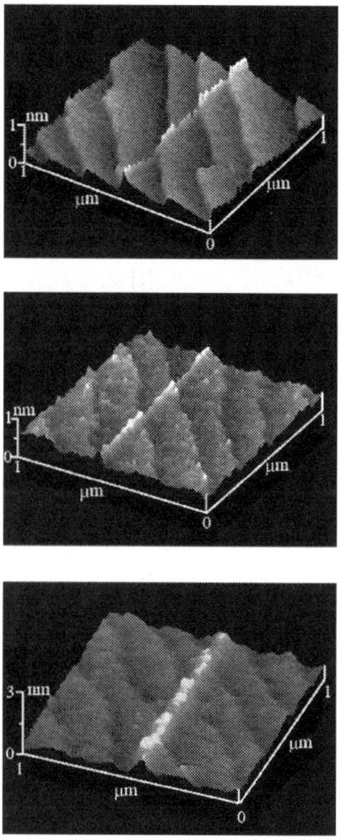

Fig. 6. Topographic images of line structure fabricated by (a) diamond coated Si tip, (b) Si tip and (c) Au coated Si tip, respectively. The applied bias voltage and scanning rate are 7 V and 1µm/s, respectively. [SH. Lee, N. Saito, O. Takai, Highly reproducible technique for three-dimensional nanostructure fabrication via anodization scanning probe lithography, Appl. Surf. Sci., 255, 7302-7306 (2009). Copyright@ELSEVIER (2009)]

3. Electrochemical scanning probe lithography on organic monolayer

A future tool of nanolithography may be nanopattern drawing based on SPM. Several reports describe SPL in which a sample surface is chemically converted or mechanically deformed at the probe-sample junction. The characteristic feature of all these methods is that they allow permanent nanopatterns to be drawn on the substrate. Scanning probe electrochemistry in which materials surfaces locally oxidized or reduced by a tip of SPM is a promising technique for constructing nanostructures consisting of organic molecules. This electrochemical SPL is performed through water column which is condensed between the tip of SPM and the substrate surface. This water column can be used as a microscopic electrochemical cell. When the bias voltage is applied, redox reaction proceeds on the substrate surface. The method is expected as a key technology for future molecular nanodevices. In addition, organosilane SAMs have been applied to a resist material for SPL.

4. Nano-probe electrochemistry on amino-terminated self-assembled monolayers toward nano memory

In this chapter, we investigated that an amino-terminated SAM was electrochemically converted into an oxidized SAM by SPL at positive bias voltages. Moreover, this oxidized SAM was then reconverted into an amino-terminated SAM by SPL at negative bias voltages. The chemical conversions of amino groups were confirmed by Kelvin probe force microscopy (KPFM), atomic force microscopy (AFM) and the site-selective adsorption of carboxyl-modified fluorescent spheres. We examined the chemical changes undergone in the scanned area from the viewpoint of surface-potential reversibility. Fig. 7 schematically illustrates of the experimental procedure. An amino-terminated SAM was prepared by chemical vapor deposition (CVD) from *p*-aminophenyltrimethoxysilane (APhS) on n-type silicon (111) wafers with electrical resistivity of 4-6 Ω/cm. First, the silicon substrate was cleaned in acetone, ethanol, and deionized water, in that order. After cleaning, the silicon substrate was irradiated by 172 nm vacuum ultraviolet light in air for 20 min. This removed organic contaminants and introduced silanol groups onto the substrate surface. Next, each cleaned silicon substrate was placed together with a glass vessel filled with APhS liquid in a Teflon container. The Teflon container was sealed and placed in an oven with the temperature kept at 100 °C. The reaction time was 1 h. The heated APhS liquid vaporized and hydrolyzed. The hydrolyzed APhS reacted with the silanol groups on the silicon substrate resulting in the fabrication of an amino-terminated SAM.

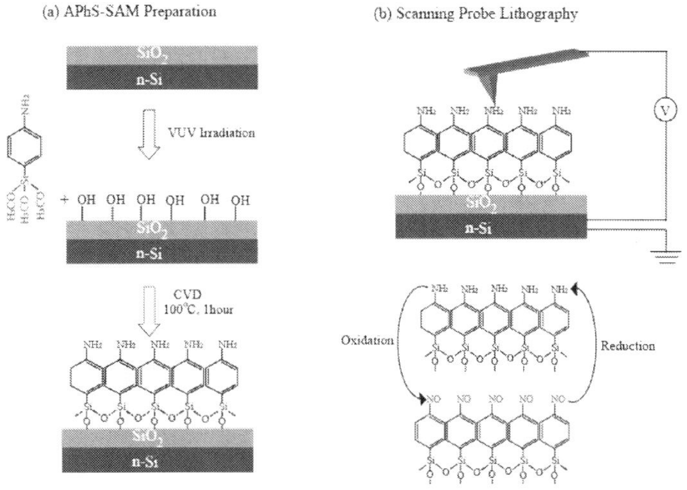

Fig. 7. Preparation and electrochemical scanning probe lithography of APhS-SAM [N. Saito, SH. Lee, T. Ishizaki, J. Hieda, H. Sugimura, O. Takai, Surface potential reversibility of an amino-terminated self-assembled monolayer based on nanoprobe chemistry, J. Phys. Chem. B, 109, 11602-11605 (2005) . Copyright@American Chemical Society (2005)]

We fabricated the APhS-SAM through the CVD method. The formation of APhS SAM was confirmed by the ellipsometry, water contact angle and X-ray photoelectron spectroscopy (XPS) measurement. In our SPL system, electrons were transferred between a gold-coated probe and the silicon substrate through the APhS SAM and an adsorbed water layer. The adsorbed water played the same role as water in a typical, macroscopic electrochemical cell system. The electrochemical conversion was conducted with a gold-coated silicon nanoprobe with a force constant of 1.8 N/m and a resonance frequency of 23.2 kHz. The probe was scanned in air at a relative humidity of 35% to 40%.

Firstly, the formation of APhS SAM was confirmed by the ellipsometry, water contact angle and XPS measurement. The static water contact angle of the sample surface after preparation was about 60°. Its film thickness obtained by ellipsometry was ca. 0.58 nm. This thickness was considered reasonable since the chain length of the APhS molecule was estimated to be 0.6 nm by semiempirical molecular orbital calculation using the AM1 Hamiltonian. The XPS N1s spectrum of the sample seen in Fig. 8 shows that the N1s binding energy is 399.6 eV. This binding energy agrees approximately with a previously reported value (400.0 eV). These results indicated the formation of APhS SAM on the SiO_2 substrate.

To show the effect of adsorbed water, the nano-probe was scanned at the pressure of 10^{-6} Pa. Fig. 9 shows topographic and surface-potential images scanned at 2 and 5 V in both air and vacuum. On the basis of these images, it is obvious that the chemical reaction does not proceed in vacuum. Thus, adsorbed water is considered necessary for the chemical conversion.

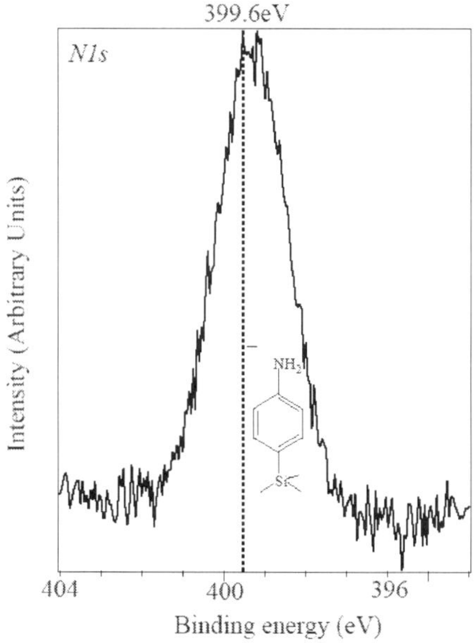

Fig. 8. XPS *N1s* spectrum of the Si surface covered with APhS-SAM.

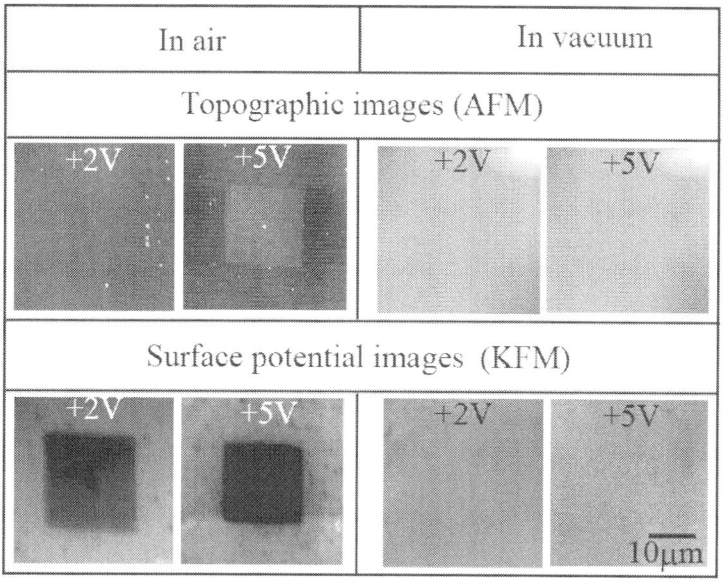

Fig. 9. Topographic and surface-potential images scanned at +2 and +5 V in air and vacuum. [N. Saito, SH. Lee, T. Ishizaki, J. Hieda, H. Sugimura, O. Takai, Surface potential reversibility of an amino-terminated self-assembled monolayer based on nanoprobe chemistry, J. Phys. Chem. B, 109, 11602-11605 (2005) . Copyright@American Chemical Society (2005)]

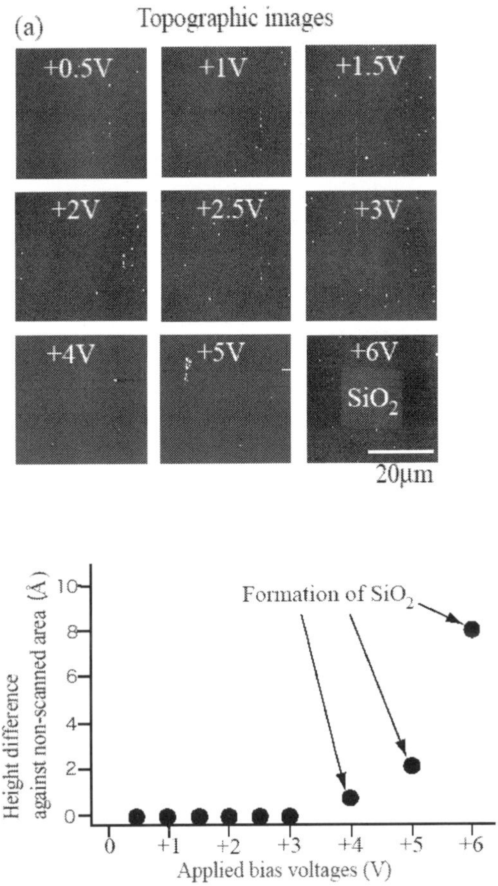

Fig. 10. (a) AFM topographic images and (b) the height difference against the non-scanned area after scanning at positive bias voltages.

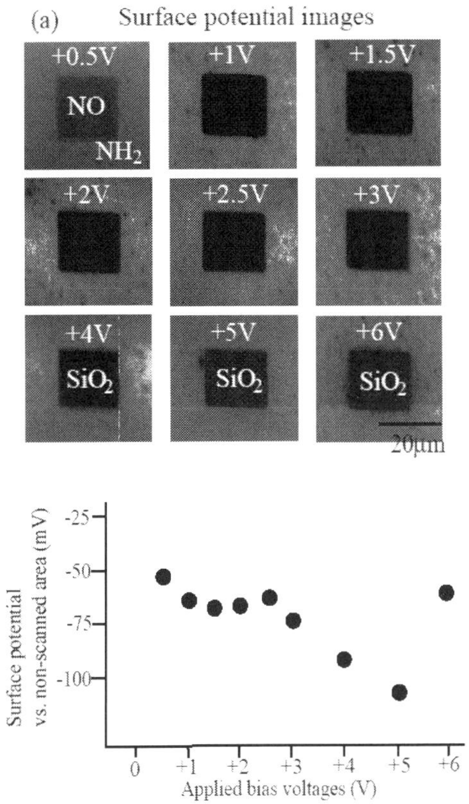

Fig. 11. (a) Surface potential images obtained by KFM and (b) the change in surface potential on the area scanned at positive bias voltages.

The nanoprobe was scanned across the sample surface over an area of 20μm X 20μm at bias voltages of 0.5-6 V. Fig. 10 shows both AFM topographic images and the height difference against the nonlithographic area after scanning. A slight protuberance can be observed in the topographic images of samples scanned at the bias voltages of 4 to 6 V. This protuberance became much higher at the bias voltages of 4 to 6V. These protuberances are due to the production of SiO_2, which resulted from the decomposition of the as-deposited APhS SAM and the oxidation of silicon. This demonstrates that there is no possibility of chemical reversibility at these bias voltages since the SAM has been damaged. Specifically, there is no possibility of surface-potential reversibility. On the other hand, no change can be observed in the topographic images of samples scanned at the bias voltages of 0.5-3 V. At these voltages, it is possible that the framework of the SAM molecules remained intact. Fig. 11 shows both surface-potential images obtained by KFM and the change in surface potential on the scanned region. The surface potential shifted negatively, which can be roughly explained by the apparent dipole moment of the SAM. The apparent dipole moment of the untreated APhS SAM is in the direction from the sample surface to the substrate. This can explain the negative shifts of the surface potential in the scanned area. In addition, amino surfaces were converted into nitroso surface at the bias voltages of 1 V to 3 V because surface potential contrast was nearly constant. In surface potential image, the surface potential reversed with the applied bias voltage.

To confirm surface-potential reversibility, a nanoprobe scanning series was performed as follows. At first, (a) a 60 µm × 60 µm square region was oxidized, and (b) a 20 µm × 20 µm square region in the 60 µm × 60 µm square region was reduced. The scan rates were 0.5 and 1.0 Hz, respectively. Fig. 12 shows illustrates of these (a) experimental processes and (b) surface potential image of scanned area. In surface potential image, the surface potential reversed with the applied bias voltage. Fig. 13 shows (a) schematic illustration of selective adsorption of carboxyl- modified fluorescent spheres after AFM lithography and (b) dark field image acquired by optical microscope after immersion of the sample in Fig. 12 in a pH 4 solution containing carboxyl-modified fluorescent spheres. The -COOH and -NH_2 groups in the pH 4 solution were converted into -COO^- and -NH_3^+ ion groups. Thus, the selective adsorption of fluorescent spheres onto the NH_2 regions proceeded due to attractive electrostatic interaction. In the pH 4 solution, amino-modified fluorescent spheres were repulsed in regions with -NO terminated groups. Therefore, the bright and dark areas correspond to the NH_2 and NO surface, respectively. These indicated that the nitroso terminated surfaces were reconverted into amino terminated surfaces with a negative bias voltage.

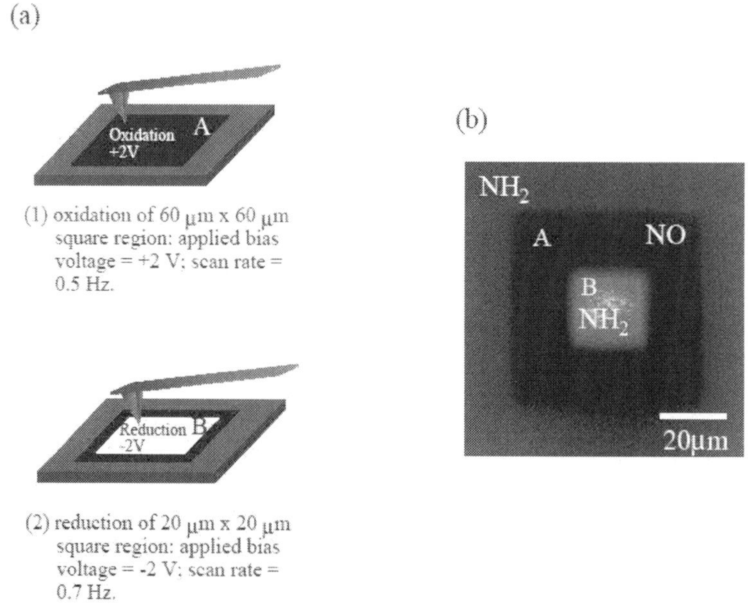

Fig. 12. (a) Schematic illustrations for a series of the lithography and (b) the obtained surface potential image.

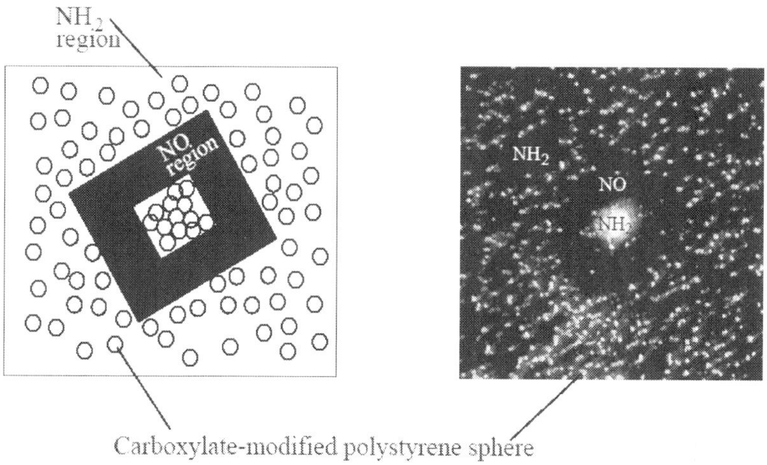

Fig. 13. (a) Schematic illustration of selective adsorption of carboxylate-modified polystyrene spheres after AFM lithography and (b) dark field image acquired by optical microscope after immersion in a pH 4 solution containing carboxylate-modified polystyrene spheres, followed by successive scanning probe lithography.

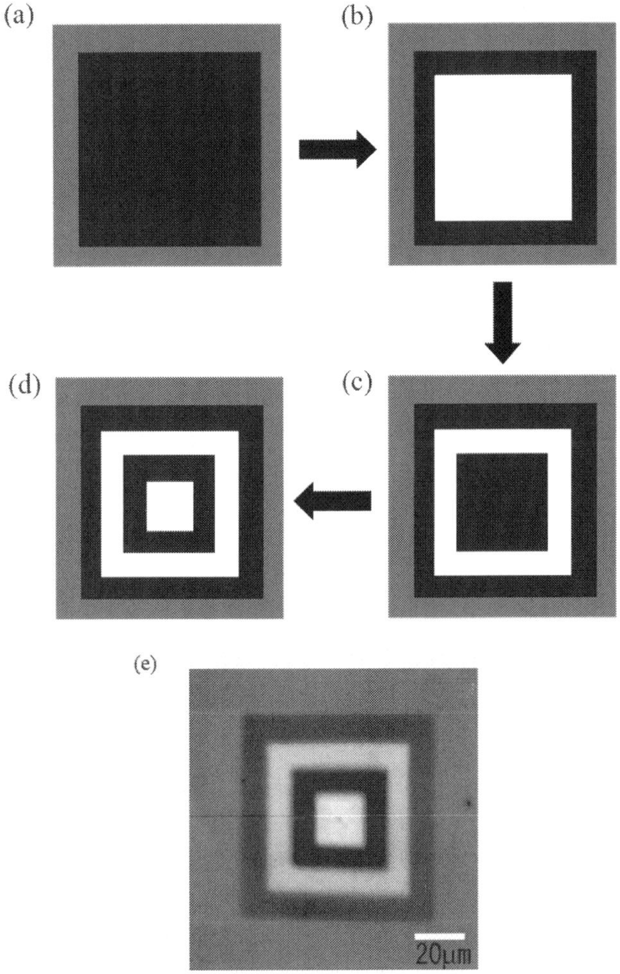

Fig. 14. Illustrations of reversibility processes ((a)-(d)) and the surface potential image of multu-scanned area (e).

In addition, we investigated the multi-reversible conversion of the APhS surface. Firstly, (a) a 80 μm × 80 μm square region was oxidized, and (b) a 60 μm × 60 μm square region in the 80 μm × 80 μm square region was reduced. Moreover, (c) the 40 μm × 40 μm square region in the 60 μm × 60 μm square region was oxidized, and then (d) the 20 μm × 20 μm square region the 40 μm × 40 μm square region was reduced. The scan rates were 0.5, 0.7, 1.0, and 2.0 Hz, respectively. Fig. 14 (a) and (b) show illustrations of these experimental processes and the surface potential image of the scanned area, respectively. In the surface potential image, the surface potential reversed with the applied bias voltage. These indicated the multi-reversible conversion of amino terminated surfaces. Thus, we can control the surface-potential reversibility on the amino-terminated SAM by controlling the applied bias voltage. This reaction formula is as follows .

$$-NH_2 + H_2O \leftrightarrow -NO + 4H^+ + 4e^- \quad (3)$$

Finally, by making use of this phenomenon of surface-potential reversibility, we demonstrated surface-potential memory. First, a square region 10 μm × 10 μm was oxidized at the bias voltage of 2 V. Next, dotted areas in the oxidized region were selectively reduced at the bias voltage of -2 V. Finally, the 10 μm × 10 μm square region was again oxidized at the bias voltage of 2 V. Fig. 15 shows the surface potential changes corresponding to (a) "writing" and (b) "erasing" with the experimental process. In Fig. 15 (a), sixteen bright areas corresponding to the chemically converted region can be observed. These sixteen areas disappeared after "erasing," as shown in Fig. 15 (b). Although this "surface potential memory" has not yet been highly integrated, it has the potential to perform as ultra-integrated memory.

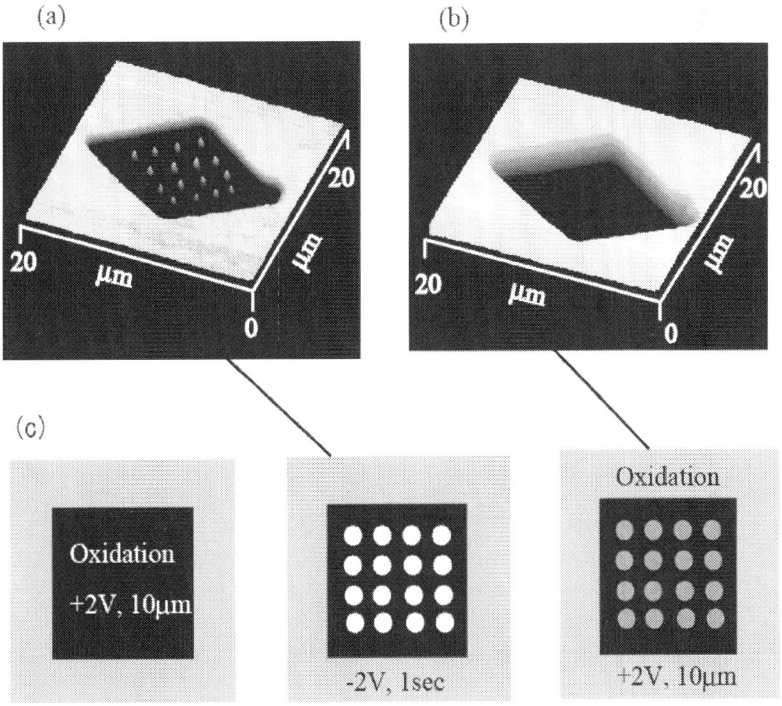

Fig. 15. Surface-potential memory: (a) "writing" state, (b) "erasing" state (c) and the experimental process.

4.1 Electrochemical lithography of 1,7-octadiene monolayers covalently linked to hydrogen-terminated silicon using scanning probe microsoopy

The organic monolayer covalently attached to silicon through Si-C bond has been expected to have better chemical resistivity compared to organosilane monolayers. The Si-C interface provides a good electronic property for molecular devices constructed on the silicon. In particular, the construction of hybrid organic- molecule/silicon devices is a promising approach for the future molecular devices. To realize such devices, it is vital to establish the fabrication technology for microstructure of the organic monolayer. The electrochemical SPL is performed through the water column condensed between the tip of SPM and the substrate surface. This water column can be used as a minute electrochemical cell. When the bias voltage is applied, a redox reaction proceeds on the substrate surface. Through reversible chemical SPL, we successfully controlled this redox reaction so that an NH_2-terminated organosilane monolayer surface was converted into an NO-terminated surface. However, this organosilane monolayer suffered from electrical defects due to the presence of SiOx. A 1,7-octadiene (OD) monolayer was directly formed on a hydrogen-terminated silicon surface by radical reaction. In this section, we report the electrochemical conversion of the vinyl-terminated groups on an OD monolayer induced by a nanoprobe.

Si (111) with electrical resistivity of 10.0-20.0 Ω/cm was used as for the substrates. Fig. 16 shows schematic illustrations of the experimental process. The substrates were cleaned in piranha solution (H_2SO_4 : H_2O_2 = 3 : 1) at 100 °C for 10 min and rinsed in ultrapure water. They were then etched for 15 min by immersing in 40 % aqueous ammonium fluoride solution (NH_4F). The surface was hydrogen-terminated and silicon oxide was removed by this immersion. The OD monolayer was prepared by the liquid phase method. The substrates were immersed in OD solution at 120 °C for 1h. After immersion, the samples were cleaned in toluene, acetone, ethanol and rinsed in ultrapure water.

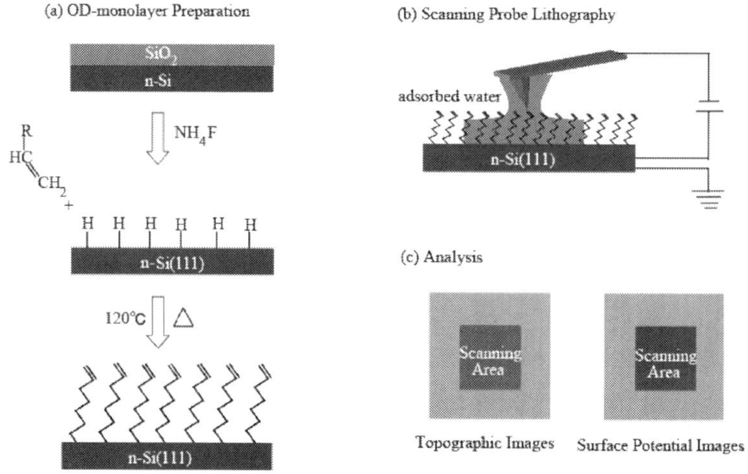

Fig. 16. Schematic illustrations of the experimental process: (a) preparation of the OD-monolayer, (b) electrochemical scanning probe lithography and (c) chemical conversion analysis. [SH. Lee, T. Ishizaki, N. Saito, O. Takai, Electrochemical Soft Lithography of an 1,7-octadiene Monolayer Covalently Linked to Hydrogen-Terminated Silicon using Scanning Probe Microscope, Surf. Sci., 601, 4206-4211 (2007). Copyright@ELSEVIER (2007)]

Fig. 17 (a) shows the XPS Si *2p* spectra of silicon substrate surfaces before and after the immersion in 40 % aqueous ammonium fluoride solutions (NH_4F). In the spectrum of the sample surface before the immersion, the peak at 104 eV attributed to SiO_2 was observed, while no appreciable peak related to the oxide was observed in the spectrum after the immersion. This indicates that the native oxide layer on the silicon substrate is completely removed after the immersion. Fig. 17 (b) show the AFM topographic image of the silicon substrate surface. The AFM image has flat terraces with the steps for silicon one-atom. The distance and the height difference between steps were evaluated to be 180 nm and 0.32±0.03 nm, respectively, as shown in Fig. 17 (b). These results reveal that the silicon surfaces are terminated with hydrogen. In order to deposit the OD monolayers, the substrate was immersed in the OD solution heated at 120 °C for 1 h. After the immersion, the water contact angle of the OD monolayers became saturated at approximately 80° at the reaction time of 1 hour. Furthermore, the film thickness of 1.2 nm corresponded approximately to the distance from Si to $-CH_2$ end groups in the precursor, as determined by ellipsometry. Fig. 17 (c) shows the topographic image of the OD monolayer surface. The distance and the height difference between steps, as shown in Fig. 17 (c), were 120 nm and 0.28±0.03 nm, respectively. These values are well in agreement with that of hydrogen-terminated silicon surface. The OD monolayers were stably attached to the hydrogen-terminated Si surfaces, since the parallel monoatomic steps were observed on the OD molecule. This means that the OD molecule was deposited at a monolayer on the substrate. These surfaces are very stable and can be stored for several weeks without any change in the topographic properties.

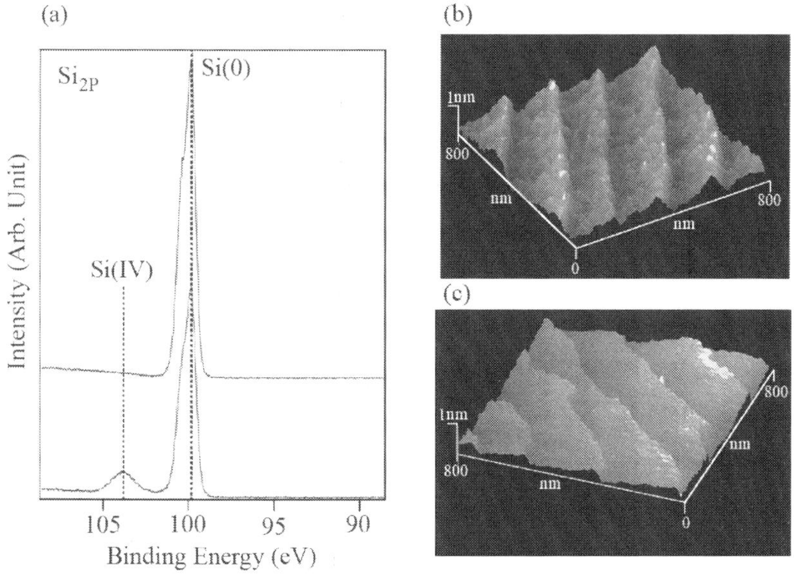

Fig. 17. (a) Si2p XPS spectra of the Si(111) surface after and before etching in 40% NH4, (b)AFM image (800nm×800nm)of a H-terminated Si(111) surface, (c) AFM image (800nm×800nm) of a OD-monolayer surface [SH. Lee, T. Ishizaki, N. Saito, O. Takai, Local Generation of Carboxyl Groups on an Organic Monolayer through Chemical conversion using Scanning Probe Anodization: Mater. Sci. Eng. C, 27, 1241-1246 (2007) Copyright@ELSEVIER (2007)]

The OD monolayers were selectively oxidized in air through the photomask by vacuum ultraviolet (VUV) light irradiation for 10 min. The areas irradiated were converted into -COOH groups due to photochemical oxidation, thus dividing the small surface into distinct $-CH_2$ and -COOH end groups regions. Fig. 18 (a) shows the XPS *C1s* spectra of OD monolayer surfaces before and after the irradiation. Its peak of 289.6 eV is assigned to the carboxyl group. Fig. 18 (b) shows the surface potential image (KFM) of the OD monolayer. The dark and bright regions correspond to the CPD images of low and high surface potential, respectively. In this figure, the surface potential for the irradiated OD monolayer surfaces was 20 mV lower than that of the unirradiated surfaces. The change of the surface potential indicates that $-CH_2$ end groups on the OD monolayer were chemically converted into -COOH end groups. The end groups of the OD monolayers were confirmed by the selective adsorption of amino-modified fluorescence spheres in a pH 4 solution. The -COOH and $-NH_2$ groups in the pH 4 solution were converted into -COO- and $-NH_3^+$ ion groups, so that the selective adsorption of fluorescence spheres on to the substrate proceeded due to their attractive interaction to the surface. Under this pH condition, the regions of $-CH_2$ end groups on the surface were not negatively charged and the amino-modified polystyrene fluorescence spheres did not adsorb onto it. Fig. 18 (c) shows an image acquired by dark field microscopy of the micropatterned CH_2 / COOH sample after immersion. The lighter areas between the dark rectangular regions correspond to the COOH terminated regions. This dark-field image indicates that amino-modified polystyrene fluorescence spheres selectively adsorbed on the -COOH end group regions since scattered light due to surface roughness can be observed. From these results, we determined that an $-CH_2$ end group had been successfully converted into the COOH terminated surface through chemical lithography, i.e., photolithography.

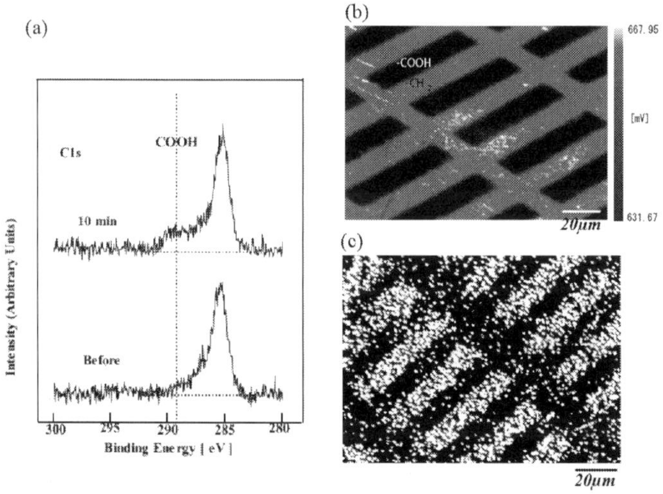

Fig. 18. (a) C1s XPS spectra for VUV irradiation of 0 and 10min, (b) KFM image (150μm×150μm) of OD-monolayer irradiated for 10min, (c) optical microscope image of the irradiated OD-monolayer which adsorbed amino terminated particles.[SH. Lee, T. Ishizaki, N. Saito, O. Takai, Local Generation of Carboxyl Groups on an Organic Monolayer through Chemical conversion using Scanning Probe Anodization: Mater. Sci. Eng. C, 27, 1241-1246 (2007). Copyright@ELSEVIER (2007)]

A gold-coated probe was scanned in air on OD monolayers at the bias voltage of 1 V and the scanning rate of 2 Hz. The areas scanned were a square, 20 μm on a side. Fig. 19 (a) and (b) shows the topographic and the surface potential images, respectively. The surface potential for the area scanned was 20 mV more negative than that of the area non-scanned and no change was observed in the topographic images. The value of surface potential difference between the scanned and non-scanned areas by SPL is well in agreement with that by photo oxidation, i.e., VUV light. These results indicate that the scanned OD monolayer surfaces are completely oxidized and chemically converted into COOH terminated surfaces without decomposition of the OD monolayer and siloxane networks. In order to verify that the chemical conversion with SPL was electrochemically proceeded on the areas scanned through the water column, the SPL was carried out in vacuum. The conditions were as follows: the bias voltage : 1 V, the pressure : 10^{-4} Pa, the scanned rate : 2 Hz, and the scanned areas : a square of 20 μm on a side. Fig. 20 shows the topographic and the surface potential images on the scanned area. In both images, no change was observed. This means that the chemical conversion on the scanned areas does not proceed under such a condition. Therefore, we can conclude that the chemical conversions with SPL are based on electrochemical reactions through the water column.

Electrochemical SPL was performed on the OD monolayer at the scanning rate of 2 Hz and at bias voltages of -3 V to 3 V. Fig. 21 shows representative surface potential images, topographic images, the changes in surface potential, and the height difference against the non- lithographic regions. In topographic images, no change in height difference was observed at any of the bias voltages. This indicates that the scanning caused no change in microscopic morphology under these conditions. On the other hand, the surface potential changed remarkably. The dark and bright regions in the surface potential images correspond to low and high surface potential regions, respectively. When the bias voltage was positively applied, oxidation proceeded on the substrate. At the positive bias voltage, the scanned regions were oxidized and showed lower surface potential than the unscanned areas. The surface potential contrast was nearly constant at approximately -18 mV at bias voltages of 1 V to 2 V. However, the surface potential contrast at bias voltages of 2 V to 3 V gradually decreased in proportion to voltage evolution.

On the other hand, the surface potential contrast at negative bias voltages gradually increased in proportion to voltage evolution. The surface potential contrast at bias voltages of 0 V to -1 V was negative, indicating that some oxidation proceeded under these conditions. This was due to the difference in contact potential between the Au-coated probe and the Si substrate. The surface potential constant was positive at bias voltages from -1.5 V to -3 V. This change in surface potential indicates that reduction reactions occurred on the substrate surface due to the applied negative bias voltage. Considering these AFM and KPFM results, the electrochemical conversion of the vinyl-terminated groups is believed to have been governed by the applied bias voltage.

Using XPS, we investigated the conversion of vinyl terminated groups at each bias voltage. Fig. 22 (a) and (b) show XPS Si*2P* and C*1s* spectra for sample surfaces after probe-scanning at bias voltages of 1 V and 3 V. The C*1s* spectrum in Fig. 22 (b) shows an additional peak from -COOH groups at 288.5 eV at the bias voltage of 1 V, but not at 3 V. In addition, a silicon oxide peak at 103.4 eV can be seen in Fig. 22 (a) at the bias voltage of 3 V, but not at 1 V. The intensity of the alkyl chain peak in Fig. 22 (b) at the bias voltage of 3 V was observed to be lower than that at 1 V. These XPS results indicate that the vinyl functional groups were oxidized and converted into carboxyl groups at the bias voltage of 1 V. The surface potential of the carboxyl surface was lower than that of the vinyl-terminated surface because the carboxyl groups had more negative dipole moment. This agrees with the KFM results. In addition, OD molecules on the sample surface were decomposed and silicon oxide was formed at the bias voltage of 3 V. However, the fact that there was no change of AFM morphology at the bias voltage of 3 V was probably due to the formation of "depthless" silicon oxide, that is, the partial decomposition of OD molecules, and to the effect of absorbed water on the sample surface.

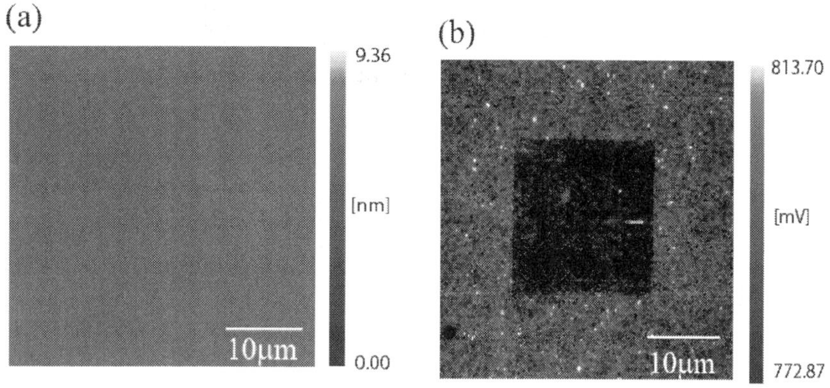

Fig. 19. (a) Topographic images and (b) surface potential image after the scanning at the rate of 2Hz. [SH. Lee, T. Ishizaki, N. Saito, O. Takai, Local Generation of Carboxyl Groups on an Organic Monolayer through Chemical conversion using Scanning Probe Anodization: Mater. Sci. Eng. C, 27, 1241-1246 (2007). Copyright@ELSEVIER (2007)]

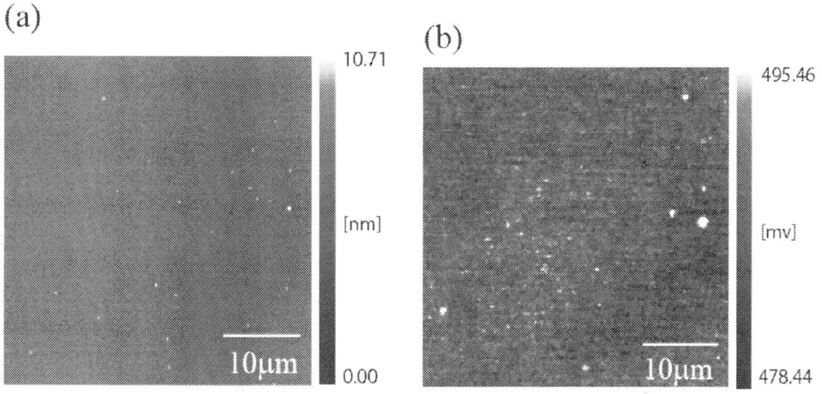

Fig. 20. (a) Topographic image and (b) surface potential image after the scanning in vacuum.[SH. Lee, T. Ishizaki, N. Saito, O. Takai, Local Generation of Carboxyl Groups on an Organic Monolayer through Chemical conversion using Scanning Probe Anodization: Mater. Sci. Eng. C, 27, 1241-1246 (2007). Copyright@ELSEVIER (2007)]

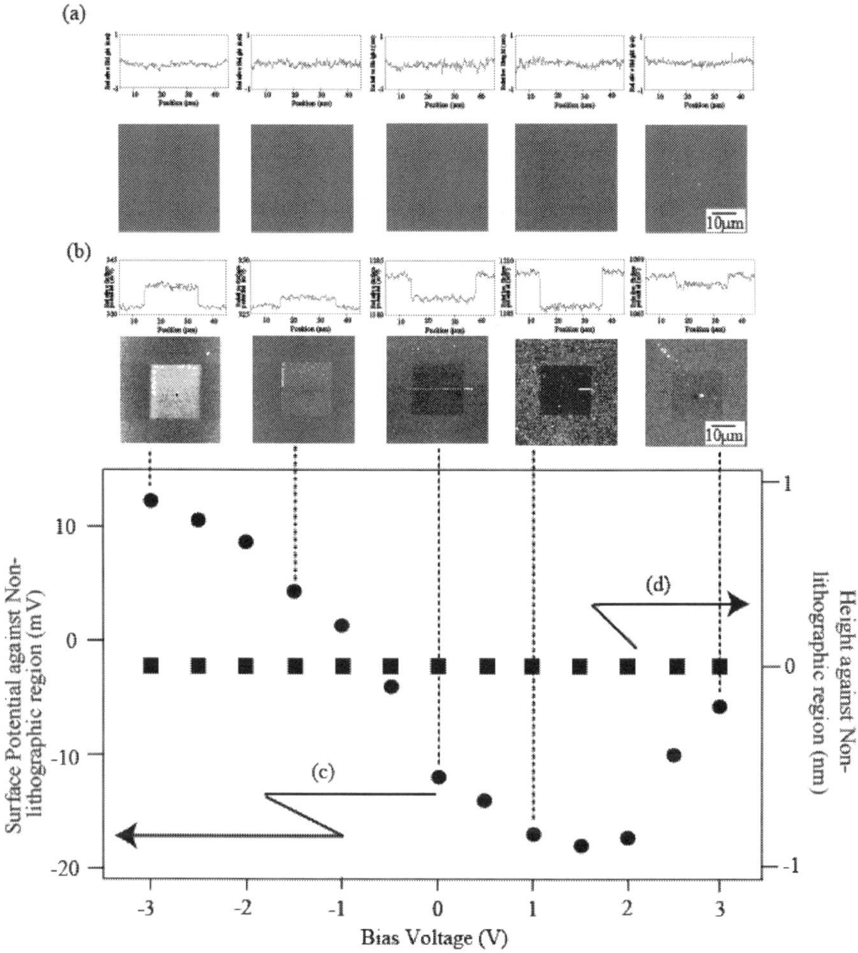

Fig. 21. (a) Representative topographic images, (b) representative surface potential images, (c) change of surface potential against non-lithographic regions; and (d) height difference against non-lithographic regions. Electrochemical SPL was performed at bias voltages of -3V to 3V. [SH. Lee, T. Ishizaki, N. Saito, O. Takai, Electrochemical Soft Lithography of an 1,7-octadiene Monolayer Covalently Linked to Hydrogen-Terminated Silicon using Scanning Probe Microscope, Surf. Sci., 601, 4206-4211 (2007). Copyright@ELSEVIER (2007)]

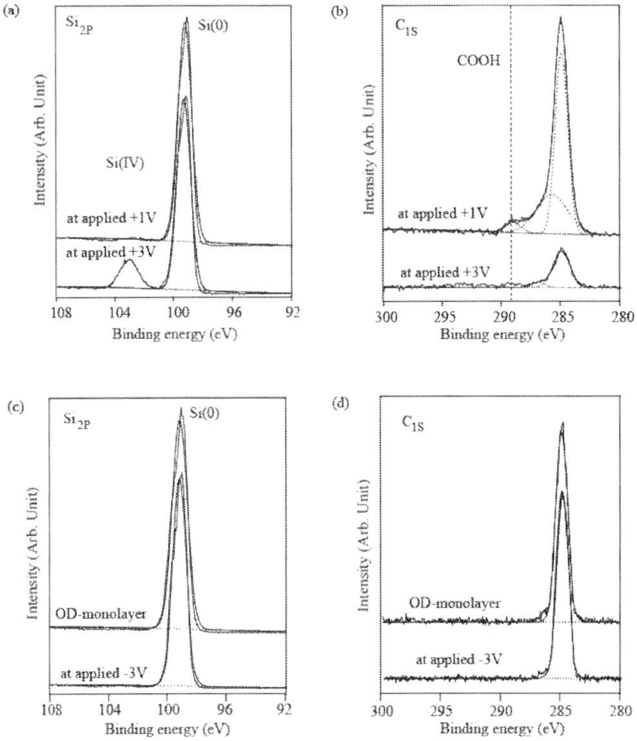

Fig. 22. (a) XPS Si2*P* spectra and (b) XPS C*1s* spectra of sample surfaces after probe-scanning at bias voltages of 1V and 3V. (c) XPS Si2*P* spectra and (d) XPS C*1s* spectra of an OD-monolayer surface and a sample surface after probe-scanning at the bias voltage of -3V. [SH. Lee, T. Ishizaki, N. Saito, O. Takai, Electrochemical Soft Lithography of an 1,7-octadiene Monolayer Covalently Linked to Hydrogen-Terminated Silicon using Scanning Probe Microscope, Surf. Sci., 601, 4206-4211 (2007). Copyright@ELSEVIER (2007)]

Fig. 22 (c) and (d) show XPS Si2*P* and C*1s* spectra of an unscanned OD-monolayer surface and a sample surface scanned at the bias voltage of -3 V. In the C*1s* XPS spectrum in Fig. 22 (d), the intensity of the alkyl chain peak for the surface scanned at the bias voltage of -3 V was the same as that for the OD monolayer. The silicon oxide peaks in the Si2*p* spectra were not observed in all cases [Fig. 22 (a) and (c)]. These XPS results indicate that OD molecules were not decomposed at negative bias voltage. We believe that the vinyl functional groups were reduced and converted into cyclobutane rings. No peak for these cyclobutane rings was observed in the C*1s* spectra since such a peak is generally weak. However, in view of the AFM and KFM results, we consider that cyclobutane rings form in the same manner as they are known to in photochemical and thermal reactions.

Fig. 23 shows a schematic illustration of the mechanism of electrochemical SPL on the OD monolayer. In this Section, two factors were considered: the alkyl radical reaction from frictional heat due to the probe scanning, and the redox reaction on the sample surface caused by polarization due to the applied bias voltage. First, alkyl radicals were formed by frictional heat. Next, the redox reaction occurred on the sample surface in the radical atmosphere. With positive bias voltages, the oxidation reaction easily occurred on the sample surface due to polarization in adsorbed water. Thus, the conversions on the sample surface were governed by the oxidation reaction. The vinyl-terminated groups of the OD monolayer were converted into carboxyl groups at positive bias voltage. However, the reduction reaction on the sample surface rarely occurred at negative bias voltages because the dissolved oxygen was preferentially reduced in adsorbed water. Thus, in this case, the surface reaction was governed by alkyl radicals. The formation of cyclobutane rings was considered to have occurred due to alkyl radical combinations at the negative bias voltage.

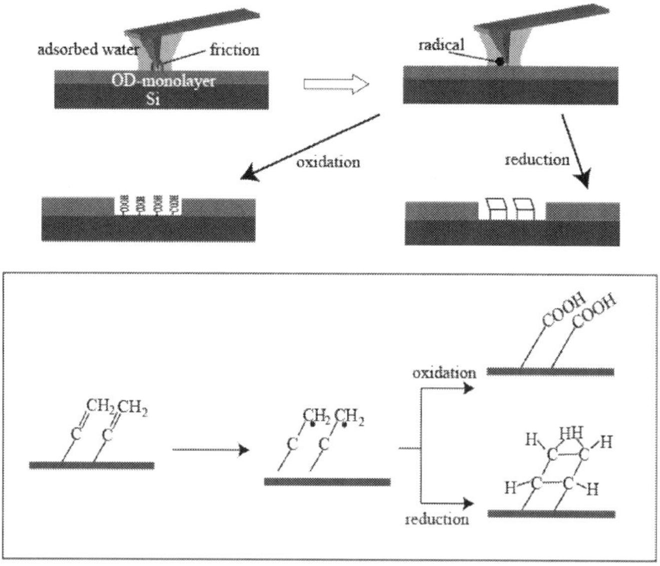

Fig. 23. Schematic illustration of the redox reaction induced by electrochemical SPL [SH. Lee, T. Ishizaki, N. Saito, O. Takai, Electrochemical Soft Lithography of an 1,7-octadiene Monolayer Covalently Linked to Hydrogen-Terminated Silicon using Scanning Probe Microscope, Surf. Sci., 601, 4206-4211 (2007). Copyright@ELSEVIER (2007)].

In support of the SPM and XPS results, the oxidized groups of the OD monolayer were confirmed by the selective adsorption of amino-modified fluorescent spheres. The -COOH and $-NH_2$ groups in the pH 4 solution were converted into $-COO^-$and $-NH_3^+$ ion groups. Thus, the selective adsorption of fluorescent spheres onto the COOH regions proceeded due to attractive electrostatic interaction. In the pH 4 solution, regions with vinyl terminated groups were not negatively charged, and the amino-modified polystyrene fluorescent spheres were repulsed. Fig. 24 (a) shows the mechanism of this selective adsorption of the amino-modified fluorescence spheres. Fig. 24 (b) shows a dark field image of samples scanned at the bias voltage of 1 V after immersion in the solution of amino-modified fluorescent spheres. The bright areas correspond to the areas scanned at the bias voltage of 1 V which site-selectively adsorbed the fluorescent spheres. This confirms that vinyl terminated groups of the OD monolayer were converted into COOH terminated groups by scanning at the applied bias voltage of 1 V.

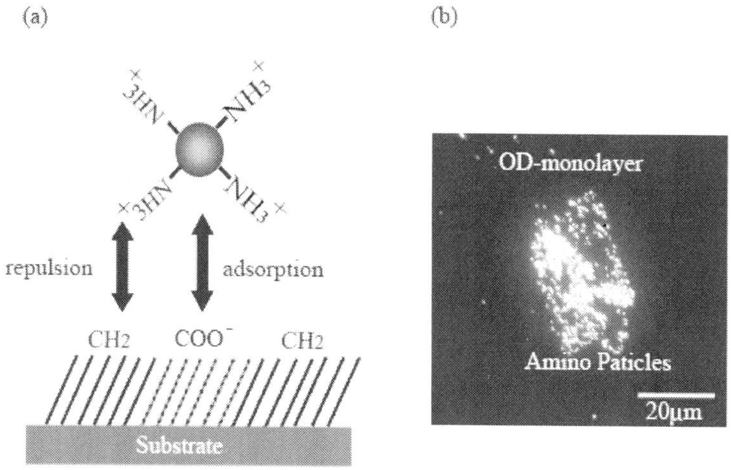

Fig. 24. (a) Selective adsorption of amino-modified fluorescent spheres on a surface scanned at the bias voltage of 1V; (b) dark field image of a patterned surface after immersion in a pH 4 solution containing amino-modified fluorescent spheres [SH. Lee, T. Ishizaki, N. Saito, O. Takai, Electrochemical Soft Lithography of an 1,7-octadiene Monolayer Covalently Linked to Hydrogen-Terminated Silicon using Scanning Probe Microscope, Surf. Sci., 601, 4206-4211 (2007). Copyright@ELSEVIER (2007)].

5. Conclusion

In this chapter, we introduced the chemical conversion of functional groups on the organic monolayer by electrochemical SPL. The three-dimensional nanostructures of silicon oxide were successfully fabricated by decomposing the 1-decane monolayer and subsequent oxidizing the hydrogen-terminated Si surfaces via anodization SPL. The size and reproducibility of oxide nanoline structures were greatly dependent on the sorts of probes for anodization SPL. In the case of Au-coated Si and uncoated Si probes, the obtained nanoline structures were changed with the scanning rates and the applied bias voltages. On the other hand, the nanotexture fabrication using the diamond-coated probe showed one of the finest structures (15 nm nanoline) and highly reproducibility even though any fabrication conditions such as scanning rate and applied bias voltage are used in the anodization SPL.The amino surface on SAM was oxidized and converted into a nitroso surface at bias voltages of 0.5 to 3 V. The functional groups on APhS SAM were reversibly converted by controlling the applied bias voltage. It was also demonstrated that the surface-potential memory was based on surface potential reversibility. In addition, the vinyl-terminated groups of the OD monolayer were site-selectively oxidized and chemically converted into carboxyl groups at bias voltages of 1 to 2 V. OD molecules on the sample surface were decomposed and silicon oxide was formed at bias voltages greater than 3 V. On the other hand, CH_2-terminal groups were converted into cyclobutane rings at bias voltages of less than −1.5 V. Recently, the research into applications of SAMs has progressed rapidly because of their ability to modify surfaces functionally and provide hydrophobicity, hydrophilicity, or biocompatibility. However, the reproducibility of the formation of SAMs is difficult. In this chapter, the formation mechanism of SAMs has been studied for high reproducibility. The control of surface properties by fabrication of micro/nanosized domains composed of SAMs is expected to be applied in the field of biomaterials. In addition, electrochemical SPL is also expected to be applied to various devices.

6. References

Hayashi, K.; Saito, N.; Sugimura, H.; Takai, O. & Nagagiro, N. (2002). Regulation of the Surface Potential of Silicon Substrates in Micrometer Scale with Organosilane Self-Assembled Monolayers. *Langmuir,* Vol.8, No.20 (September 2002), pp. 7469-7472, ISSN 0743-7463.

Hong, L.; Sugimura, H.; Furukawa, T. & Takai, O. (2003). Photoreactivity of Alkylsilane Self-Assembled Monolayers on Silicon Surfaces and Its Application to Preparing Micropatterned Ternary Monolayers. *Langmuir,* Vol.19, No.6 (February 2003), pp. 1966-1069, ISSN 0743-7463.

Saito, N.; Wu, Y.; Hayashi, K.; Sugimura, H. & Takai,O. (2003). Principle in Imaging Contrast in Scanning Electron Microscopy for Binary Microstructures Composed of Organosilane Self-Assembled Monolayers. *The Journal of Physical Chemistry B,* Vol.107, No.3 (December 2002), pp. 664–667, ISSN 1089-5647.

Hahn, J. & Webber, S. E. (2004). Graphoepitaxial Deposition of Cationic Polymer Micelles on Patterned SiO_2 Surfaces. *Langmuir,* Vol.20, No.4 (January 2004), pp. 1489-1494, ISSN 0743-7463.

Kidoaki, S. & Matsuda, T. (1999). Adhesion Forces of the Blood Plasma Proteins on Self-Assembled Monolayer Surfaces of Alkanethiolates with Different Functional Groups Measured by an Atomic Force Microscope. *Langmuir,* Vol.15, No.22 (September 1999), pp. 7639-7646, ISSN 0743-7463.

Harnett, C. K.; Satyalakshmi, K. M. & Craighead, H. G. (2001). Bioactive Templates Fabricated by Low-Energy Electron Beam Lithography of Self-Assembled Monolayers. *Langmuir,* Vol.17, No.1 (December 2000), pp. 178-182, ISSN 0743-7463.

Kaholek, M.; Lee, W. K.; LaMattina, B.; Caster, K. C. & Zauscher, S. (2004). Fabrication of Stimulus-Responsive Nanopatterned Polymer Brushes by Scanning-Probe Lithography. *Nano Letters,* Vol.4, No.2 (January 2004), pp. 373–376, ISSN 1530-6984.

Blackledge, C.; Egebretson, D. A. & McDonald, J. D. (2000). Nanoscale Site-Selective Catalysis of Surface Assemblies by Palladium-Coated Atomic Force Microscopy Tips: Chemical Lithography without Electrical Current. *Langmuir,* Vol.16, No.22 (September 2000), pp. 8317–8323, ISSN 0743-7463.

Tello, M.; García, F. & García, R. (2002). Linewidth Determination in Local Oxidation Nanolithography of Silicon Surfaces. *Journal of Applied Physics,* Vol.92, No.7 (September 2002), pp. 4075-4080, ISSN 1089-7550.

Liu, S.; Maoz, R.; Schmid, G. & Sagiv, J. (2002). Template Guided Self-Assembly of [Au_{55}] Clusters on Nanolithographically Defined Monolayer Patterns. *Nano Letters,* Vol.2, No.10 (August 2002), pp. 1055-1060, ISSN 1530-6984.

Jang, C. H.; Stevens, B. D.; Carlier, P. R.; Calter, M. A. & Ducker, W. A. (2002). Immobilized Enzymes as Catalytically-Active Tools for Nanofabrication. *Jounal of the American Chemical Society,* Vol.124, No.41 (September 2002), pp. 12114-12115, ISSN 0002-7863.

Kaholek, M.; Lee, W. K.; Ahn, S. J.; Ma, H.; Caster, K. C.; LaMattina, B. & Zauscher, S. (2004). Stimulus-Responsive Poly(N-isopropylacrylamide) Brushes and Nanopatterns Prepared by Surface-Initiated Polymerization. *Chemical of Materials,* Vol.16, No.19 (August 2004), pp. 3688-3696, ISSN 1520-5002.

Sugimura, H. & Nakagiri, N. (1995). Degradation of a Trimethylsilyl Monolayer on Silicon Substrates Induced by Scanning Probe Anodization. *Langmuir,* Vol.11, No.10 (October 1995), pp. 3623-3625, ISSN 0743-7463.

Amro, N. A.; Xu, S. & Liu, G. Y. (2000). Patterning Surfaces Using Tip-Directed Displacement and Self-Assembly. *Langmuir,* Vol.16, No.7 (March 2000), pp. 3006-3009, ISSN 0743-7463.

Xu, S.; Miller, S.; Laibinis, P. E. & Liu, G. Y. (1999). Fabrication of Nanometer Scale Patterns within Self-Assembled Monolayers by Nanografting. *Langmuir,* Vol.15, No.21 (August 1999), pp. 7244-7251, ISSN 0743-7463.

Liu, G. Y.; Xu, S. & Qian, Y. (2000). Nanofabrication of Self-Assembled Monolayers Using Scanning Probe Lithography. *Accounts of Chemical Research,* Vol.33, No.7 (March 2000), pp. 457-466, ISSN 1520-4898.

Pena, D. J.; Raphael, M. P. & Byers, J. M. (2003). "Dip-Pen" Nanolithography in Registry with Photolithography for Biosensor Development. *Langmuir,* Vol.19, No.21 (September 2003), pp. 9028-9032, ISSN 0743-7463.

Schwartz, P. V. (2002). Surface Functionalization and Stabilization of Mesoporous Silica Spheres by Silanization and Their Adsorption Characteristics. *Langmuir,* Vol.18, No.10 (April 2002), pp. 4014-4019, ISSN 0743-7463.

Maynor, B. W.; Li, J.; Lu, C. & Liu, J. (2004). Site-Specific Fabrication of Nanoscale Heterostructures: Local Chemical Modification of GaN Nanowires Using Electrochemical Dip-Pen Nanolithography. *Jounal of the American Chemical Society,* Vol.126, No.20 (May 2004), pp. 6409-6413, ISSN 0002-7863.

Lee, SH.; Saito, N. & Takai, O. (2009). Highly reproducible technique for three-dimensional nanostructure fabrication via anodization scanning probe lithography. *Applied Surface Science,* Vol.255, No.16 (May 2009), PP. 7302-7306, ISSN 0169-4332.

Lee, SH.; Ishizaki, T.; Saito, N. & Takai, O. (2007). Electrochemical soft lithography of an 1,7-octadiene monolayer covalently linked to hydrogen-terminated silicon using scanning probe Microscopy. *Surface Science,* Vol.601, No.18 (September 2007), pp. 4206-4211, ISSN 0039-6028.

Lee, SH.; Ishizaki, T.; Saito, N. & Takai, O. (2007). Local generation of carboxyl groups on an organic monolayer through chemical conversion using scanning probe anodization.

Materials Science and Engineering C, Vol.27, No.5-8 (September 2007), pp. 1241-1246, ISSN 0928-4931.

Saito, N.; Lee, SH.; Ishizaki, T.; Hieda, J.; Sugimura, H. & Takai, O. (2005). Surface-Potential Reversibility of an Amino-Terminated Self-Assembled Monolayer Based on Nanoprobe Chemistry. *The Journal of Physical Chemistry B,* Vol.109, No.23 (May 2005), pp. 11602-11605, ISSN 1089-5647.

Sugimura, H.; Saito, N.; Lee, SH. & Takai, O. (2004). Reversible nanochemical conversion. *Journal of Vacuum Science & Technology B*, Vol.22, No.6 (November 2004), pp. L44-46, ISSN 1520-8567.

25

Controlled Fabrication of Noble Metal Nanomaterials via Nanosphere Lithography and Their Optical Properties

Yujun Song
Key Laboratory for Aerospace Materials and Performance (Ministry of Education), School of Materials Science and Engineering, Beihang University, Beijing, China

1. Introduction

Since ancient time, noble metal has been used to make ornaments, jewelry, high-value tableware, utensils, currency coins and medicines due to its brilliant metallic luster, stability in air and water and anti-bacteria and anti-fungi properties (Jain, Huang et al. 2007) (Erhardt 2003; Daniel and Astruc 2004; Brayner 2008; Maneerung, Tokura et al. 2008). In fact, noble metal is also valuable due to its unique physicochemical properties, the highest electrical and thermal conductivity, the lowest contact resistance, and the highest optical reflectivity (particularly in ultra-violet region) of all metals(Edwards and Petersen 1936 ; Hammond 2000). Its d-electron configuration endows them with active chemical properties, for example, 3 variable oxidation states for silver, the most common of which is the +1 state, as in $AgNO_3$, the +2 state as in silver(II) fluoride AgF_2, and the +3 state as in compounds such as potassium tetrafluoroargentate $K[AgF_4]$, and suitability as catalysts by losing one or two more 4d electrons (Dhar, Cao et al. 2007). Silver and gold have the stable face-centered cubic (fcc) crystal structures but readily absorbs free neutrons due to its massive nucleus, which make them good absorbers for nucleus raidation. These unique features have enabled them to be applied to diverse applications such as those mentioned above, medical and dental applications, photography, electronics, nuclear reactors, catalysts, clothing and foods (http://en.wikipedia.org/wiki/Silver).

The intrinsic features of noble metal also endow their nanoscale species with attractive physicochemical properties due to the size and shape effects, including unique optical properties (e.g. Localized Surface Plasmon Resonance: LSPR; Surface Enhanced Raman Scattering: SERS), catalytic/electric properties and bio-functions (Percival, Bowler et al. 2005; Jain, Huang et al. 2007; Schwartzberg and Zhang 2008; Zhou, Qian et al. 2008; Vo-Dinh, Wang et al. 2009). Although ancient people used some features of Ag or Au nanocolloids (e.g. optical property) in fabrication of ceramic glazes for lustrous or iridescent effect in ancient Persia, they did not realize that these effects were due to nanoscale effects from size, shape and surface morphology dependent physicochemical properties of silver materials (Erhardt 2003; Brayner 2008). As materials science has progressed down to nanoscale, the unique properties of nanoscaled noble metal materials are only now being recognized and realized intentionally. These properties have shown vast applications in microelectronics, photonic devices, optoelectric coupling, catalytic processes, biomedical engineering and medicines. As understanding that the intrinsic properties (e.g. optical, catalytic) of nanomaterials on the size, shape, surface spatial morphology and arrangement (Ahmadi, Wang et al. 1996; Jensen, Duval Malinsky et al. 2000; Mock, Barbic et al. 2002; Haynes, Mcfarland et al. 2003; Noguez 2007; Song 2009) has increased, fabrication of silver or gold nanoparticles (NPs) and their arrays with controlled three-dimensional (3D) morphologies, interspacing and orientation has become a very significant research stream in recent years. A variety of fabrication techniques, such as thermal decomposition, metal salt reduction, photo reduction/conversion, template assisted growth and deposition, γ-ray-irridation, as well as microfluidic processes have been developed. As a result, significant progresses have been achieved in the tailoring of the 3-dimension (3D) morphologies (size, shape and surface morphology), crystal structures and spatial arrangement of noble metal nanomaterials as desired.

Template assisted (TA) lithography (LIGA) has developed to a powerful physical technique that enables the production of surface morphology confined NPs and NPs arrays with controlled shapes, sizes and interparticle spacing (Jensen, Duval Malinsky et al. 2000; Lee, Morrill et al. 2006; Zhang, Whitney et al. 2006; Lombardi, Cavallotti et al. 2007; Zhu, Li et al. 2008; Song 2009). Lots of templates have been developed for these purposes, such as porous polymers(Lombardi, Cavallotti et al. 2007), porous Al_2O_3 foils(Chong, Zheng et al. 2006; Lee, Morrill et al. 2006; Xu, Meng et al. 2009), or nanosphere arrays (polymers or ceramics)(Zhang, Whitney et al. 2006; Song 2009; Song and Elsayed-Ali 2010; Song , Zhang et al. 2011), resulting in varieties of template-assisted lithography, correspondingly as porous polymers LIGA (PP-LIGA), porous anodic Al_2O_3 LIGA (PAA-LIGA), or nanosphere-LIGA (NSL). Among them, the most popular and well-developed method may be NSL. In this chapter, recent progresses in NSL, for controlled producing noble metal nanomaterials will be summarized. The first discussion involves in this technique for size, shape and surface morphology controlled fabrication of noble metal nanoparticles (NPs) and nanoarrays. Then four distinct progresses in the development of NSL techniques: (1) Fabrication of hierarchically ordered nanowire arrays on substrates by combination of NSL and Porous anodic alumina (PAA); (2) Identification of single nanoparticles and nano-arrays by combination of NSL and multi-hierarchy arrayed micro windows; (3) Fabrication of biosensing system based on the combination of the noble metal nanoparticles and nanoarrays fabricated by NSL and microfluidic techniques; (4) Synthesis of solution-phased nanoparticles by the transfer of the surface confined NPs fabricated by NSL into solutions, will be discussed. In (2) and (3), the related 3D morphologies and arrangement dependent optical properties, and comparison between the numerical and experimental results, revealing their intrinsic quantum mechanism, such as LSPR will be analyzed. These researches are fundamental requirements for the discovery of novel properties and applications of noble metal NPs, as well as for paving the theory development. Finally, issues and perspectives in the controlled fabrication of noble metal nanomaterials by NSL, and investigation of their 3D morphologies and arrangement dependent optical properties for future potential applications will be highlighted and discussed in closing.

2. Size and shape controlled fabrication of nanomateials via NSL

Nanospheres have been used to form uniformly arranged layers as templates to produce perfect triangle nanoprisms on substrates (Haynes and van Duyne 2001; Song and Elsayed-

Ali 2010). The routine procedure for the production of triangular shaped nanoprisms, based on the nanosphere LIGA, is described in Figure 1 (a: cross-section view; b: top view) (Hulteen, Treichel et al. 1999; Haynes and van Duyne 2001; Song 2009). The hexagonal arranged nanosphere mono layer is first formed on the substrate by a coating process (e.g. dip-coating, rotating-coating or spinning-coating) (Step 1: a). The interstitials among any three adjacent nanospheres will form triangle shaped voids (Step 1: b) as templates. The desired noble metal (e.g. Ag) will then be deposited on the triangle shaped interstitials among the nanospheres to form triangle shaped Ag NPs (Step 2: a and b). After nanospheres are released by sonication or other methods, surface-confined triangular Ag nanoprisms can be obtained (Step 3: a). By this nanosphere LIGA process, uniform hexagonal-arrayed triangle nanoprisms can be fabricated on a variety of substrates (e.g. glass, mica, silica wafer, PMMA, etc.). Step 3-b is an Atomic Force Microscope (AFM) image of Ag triangular nanoprisms fabricated by our group using a self-assembled monolayer of 300 nm polystyrene nanospheres as the template (Song 2009).

The initial critical step in NSL is the formation of a uniform large scale nanosphere template. Both drop-coating or spin-coating can produce uniform templates on a glass, silica wafer or mica substrate. The uniformity of the nanosphere template produced by drop coating depends on the nanosphere type and concentration, the hydrophilic properties of the substrate, the environmental humidity and temperature, and the drying speed. A monolayer colloidal polystyrene nanosphere mask can be prepared by drop-coating of ~3.0-4.0 μL, 3-10 times diluted nanosphere solution (conc. 4.0 wt.%) onto the glass support and leaving them to dry overnight. A detailed procedure to fabricate the nanosphere mask using drop-coating is as follows. The glass substrates are cleaned by sonication with a mixture of sulfuric acid and hydrogen peroxide (3:1 = conc. H_2SO_4: 30% H_2O_2, Volume ratio) at 80 °C for 30 min and washed using sufficient nanopure water. Then, the glass substrates are sonicated in a mixture of ammonia and hydrogen peroxide (5:1:1 = H_2O: NH_4OH (37%): 30% H_2O_2, volume ratio) to increase the hydrophilic property on the surface of the glass substrates. Finally, the glass substrates are washed using sufficient nanopure water again and stored in the nanopure water for future use. When drop coating is to be performed, the glass substrate is picked up from the nanopure water from one of its edges. The remaining water droplets on the glass substrate are removed by touching the opposite edge on filter paper. The substrate is then left flat in a clean Petri-dish with a tilt angle of ~3-5°. A 15 μL of PS nanosphere solution is added on the surface of the glass substrate using a droplet. The water spreads over the whole glass substrate to form a semi-ellipsoidal shaped water spot. The Petri-dish is left for enough time to allow the water to evaporate. During evaporation, the temperature is kept at 18±3 °C and the humidity is kept ~50±5%. In our group, a near-uniform monolayer nanosphere template can be prepared on almost the whole glass substrate (18 mm diameter). Figure 2 shows one typical area of a near-uniform monolayer template over scale ~20 μm. From the magnified image, a selected area shown in the inset, no lattice defects can be observed. Using this template, uniform Ag nanoprisms can be fabricated by vapor deposition process. One typical area fabricated by my group is shown in Step 3-b in Figure 1, where these nanoprisms have very uniform edge length of 67 ± 4 nm (STDEV% of 6%) and thickness of 20.0±1.0 nm (STDEV% of 5%) (Song 2009).

Recent progress in nanosphere lithography (NSL) has shown that it provides a good template for other shape (besides triangle) controlled fabrication of surface confined NPs by a combination of deposition angle tilting, multi-step deposition and different post treatment methods(Haynes and van Duyne 2001; Song and Elsayed-Ali 2010). A new class of NSL structures has been fabricated by varying the deposition angle, θ_{dep}, between the nanosphere mask and the beam of material being deposited, which is hereafter referred to as angle-resolved NSL (AR NSL)(Haynes and van Duyne 2001). The size and shape of the three-fold interstices of the nanosphere mask change relative to the deposition source as a function of

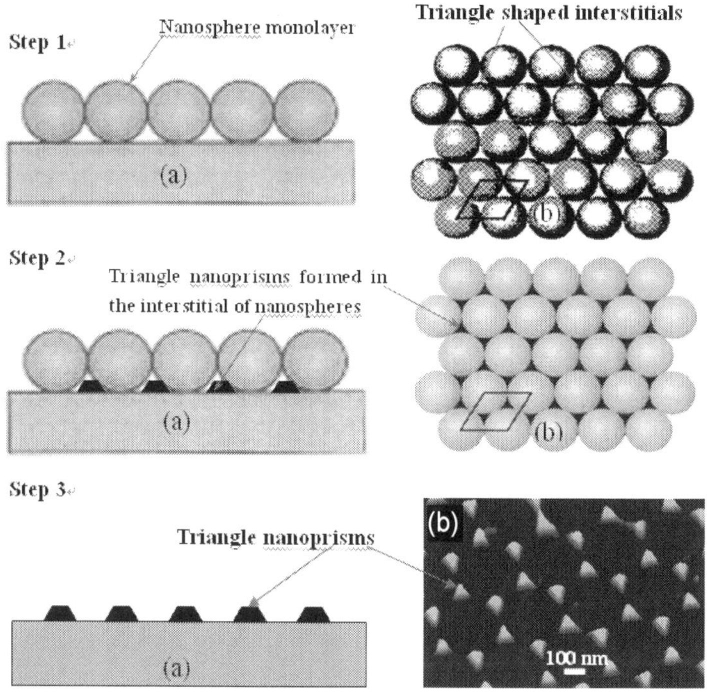

Fig. 1. The NSL process for triangular NPs fabrication. Step 1a: The hexagonal arranged nanosphere mono layer is first formed on the substrate by coating process; Step 1b: The interstitials among any three adjacent nanospheres will form triangle shaped voids as templates; Step 2a-b: the Ag metal will be deposited on the triangle shaped interstitials among the nanospheres to form triangle shaped Ag NPs; Step 3a: The nanospheres will be released by sonication or other methods, leaving the triangle shaped Ag nanoprisms on the substrates, by this nanosphere LIGA process, the hexagonal arrayed uniform triangle nanoprisms can be fabricated on variety of substrates (e.g. glass, mica, silica wafer, PMMA, etc.); Step 3b: The AFM image for Ag triangle nanoprisms fabricated by monolayer template from 290 nm polystyrene nanospheres in my group, these nanoprisms have very uniform edge length of 67 ± 4 nm (STDEV% of 6%) and thickness of 20.0 ± 1.0 nm (STDEV% of 5%). (a): cross-section view; (b) top 3D view. (Adapted in part from Song, Y. China Patent, CN200910085973.9; Haynes, C. L.; van Duyne, R. P., *J. Phys. Chem. B* 2001 105, 5599, Figure 2, Copyright (2001) American Chemical Society; and Hulteen, J. C.; et al., *J. Phys. Chem. B* **1999** 103, 3854, Figure 1, Copyright (1999) American Chemical Society.)

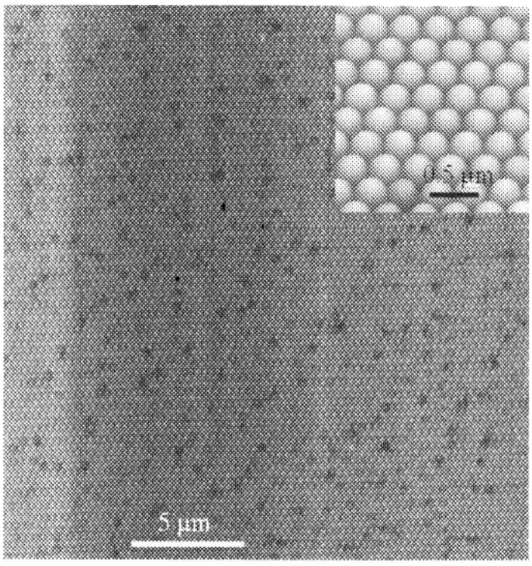

Fig. 2. Nanosphere templates based on 290 nm spherical polystyrene nanospheres for Ag nanoparticle fabrication (Reprinted from Song et al., Appl. Surf. Sci. 2010 256, (20), 5961, Figure 1. Copyright (2010) Elsevier.).

θ_{dep}, and accordingly, the deposited nanoparticles' shape and size are controlled directly by θ_{dep} and the diameter of nanosphere. Figure 3 schematically describes the effect of angle-resolved deposition on the interstices of a NSL mask from the top view. (Figure 3). As a convention, θ_{dep} = 0° represents a substrate mounted normal to the evaporation beam (Figure 3A), and all variations of θ_{dep} are made by mounting the substrates on machined aluminum blocks. It is clear from this illustration that an increase in θ_{dep} causes the projections of the interstices onto the substrate to decrease and shift (Figure 3B and 3C). At high values of θ_{dep} (e.g. 45°, Figure 3C), the projections of the interstices close, completely blocking the substrate to line of sight deposition.

One very important consequence of AR NSL, beyond the increased flexibility in nanostructure architecture, lies in the decrease in nanoparticle size. Before AR NSL, the only way to fabricate nanoparticles in the 1-20 nm size range with NSL required self-assembly of nanospheres with diameters on the order of 5-100 nm. Not only synthesis of uniform nanospheres at this range is usually difficult, but self-assembly of such small nanospheres into well ordered 2D arrays is extremely challenging because of problems with greater polydispersity and the surface roughness of substrates. However, with AR NSL, increasing θ_{dep} from 0° to 20° will halve the in-plane dimension of nanosphere templates, leading to the success in small nanoparticle preparation by NSL. In addition, nano-overlapped, nano-gapped and nano-chained structures can be addressed by multi-step AR NSL, which is fulfilled by depositing materials through a nanosphere mask mounted at different θ_{dep} below the overlap threshold value of θ_{dep} several times. Van Duyne et al have used two step AR NSL to fabricate over-lapped and gapped Ag nanostructures through a nanosphere mask with D = 542 nm onto mica substrates by a first deposition at θ_{dep} = 0° and a second deposition at an increased θ_{dep}. The importance in the fabrication of over-lapped nanoparticles theoretically exists in the enhanced optical properties due to their increased aspect ratio (in-plane width/out-of-plane height) nanoparticles(Kreibig and Vollmer 1995). Nano-overlapped structures can give an significantly increased sensitivity of optical response since they allow predictable aspect ratio to increase up to double of the original value (Haynes and van Duyne 2001). One of the interests for gapped nanostructures may exist in the investigation of the distance dependent LSPR coupling amomng gapped nanostructures.

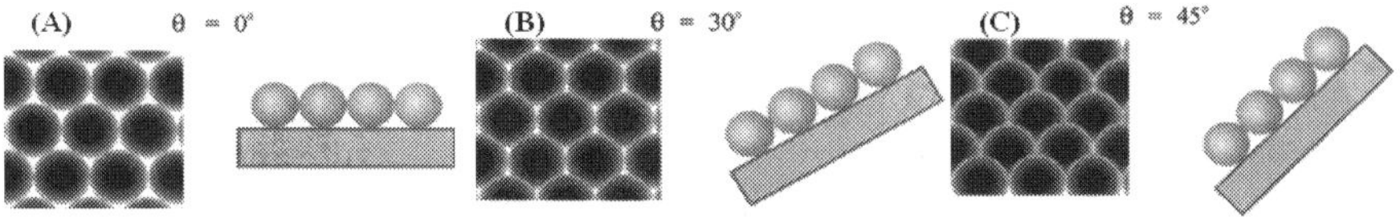

Fig. 3. Scheme of the angle resolved deposition process. (A) Samples viewed at 0°. The interstices in the nanosphere mask are equally spaced and of equal size. (B) Sample viewed at 30°. The interstices in the nanosphere mask follow a pattern including two different interparticle spacing values, and the interstitial area is smaller. (C) Sample viewed at 45°. The interstices are now closed to line of sight deposition. (Adapted from Haynes, C. L.; van Duyne, R. P., *J. Phys. Chem. B* 2001 105, 5599, Figure 5, Copyright (2001) Amercian Chemical Society.)

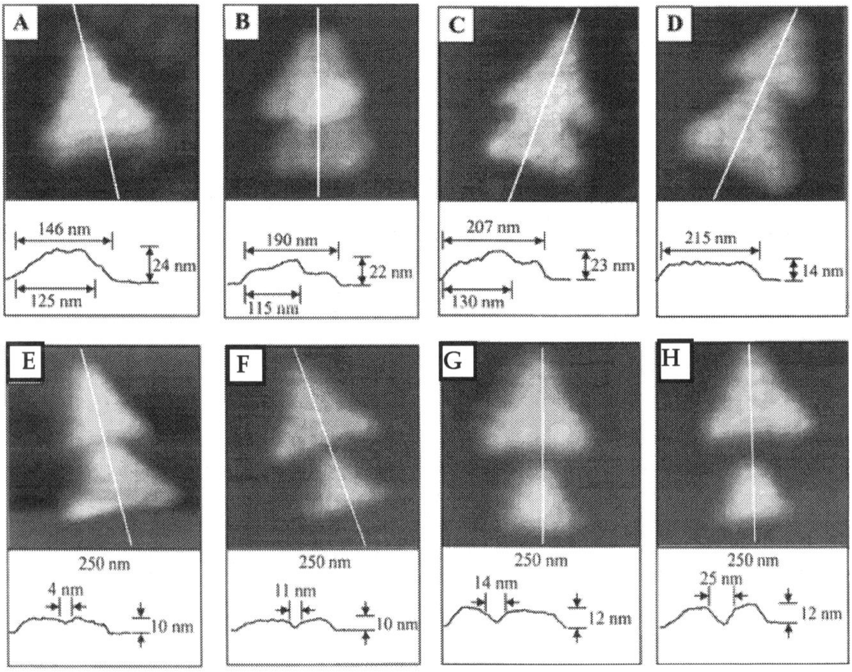

Fig. 4. Contact mode AFM images of typical Ag nano-overlapped and nano-gapped structures fabricated on mica substrates using the single layer nanosphere template. (A) 300 nm - 300 nm image, D = 542 nm, dm = 20 nm, θ_{dep} = 0° and 6°. (B) 250 nm - 250 nm image, D = 542 nm, dm = 20 nm, θ_{dep} = 0° and 10°. (C) 300 nm - 300 nm image, D = 542 nm, dm = 20 nm, θ_{dep} = 0° and 15°. (D) 250 nm-250 nm image, D = 542 nm, dm = 20 nm, θ_{dep} = 0° and 20°. (E) 250 nm - 250 nm image, D = 542 nm, dm = 20 nm, θ_{dep} = 0° and 22°. (F) 250 nm - 250 nm image, D = 542 nm, dm = 20 nm, θ_{dep} = 0° and 23°. (G) 250 nm - 250 nm image, D = 542 nm, dm = 20 nm, θ_{dep} = 0° and 24°. (H) 250 nm - 250 nm image, D = 542 nm, dm = 20 nm, θ_{dep} = 0° and 26°. (Adapted from Haynes, C. L.; van Duyne, R. P., N *J. Phys. Chem. B* 2001 105, 5599, Figure 6 and Figure 7, Copyright (2001) Amercian Chemical Society.)

The overlap percent of nanoparticles can be adjusted by θ_{dep} at a certain mass deposition thickness (e.g. 20 nm) and nanosphere diameter (e.g. 542 nm). As shown in Figure 4A-D, the overlap percent decreases with the increase of θ_{dep} from 0° to 20°. The θ_{dep} at 20° is the threshold deposition angle since neither overlap nor gap is visible by AFM investigation at this point (Figure 4D). When the second deposition angle is more than the threshold angle (e.g. 20 ° based on the fabrication condition using nanosphere mask with D = 542 nm and mass thickness d_m = 20 nm(Haynes and van Duyne 2001)), nano-gapped structures can be formed. With the same experimental parameters defined above, the gap between nanoparticles increases as θ_{dep} is increased from 22° to higher values up to thecritical θ_{dep} value at which the interstitial projections are closed to line-of-sight deposition. Figure 4E-H shows the AFM images of the typical nano-gapped Ag structures with different gap distances when the second deposition angles change from 22°, to 23°, to 24° and to 26°. Another of the applications proposed by van Duyne is to use the nanogap architecture to measure the electrical conductivity of a single molecule or nanoparticle (Haynes and van Duyne 2001). If one side of the nanogap is insulated from a conductive substrate while the other side of the nanogap is in contact with a conductive substrate, the conductance of the junction should be measurable with a scanning tunneling microscopy probe.

Clearly, like the two depositions at different values of θ_{dep}, three or more depositions will further extend the range of nanoparticle architectures accessible by AR NSL. An endless number of nanostructures are possible when one combines the ability to vary θ_{dep} and to perform multiple material depositions. As an example, the nanochain motif with three-connected-nanoparticle chains can be fabricated by three consecutive depositions. The first deposition is done at θ_{dep} = -15°, whereas the second and third depositions will be further done at θ_{dep} = 0° (tilted forward) and θ_{dep} = 15° (tilted backward). An AFM image of the nanochain structure is shown in Figure 5 and gives a typical domain where the sample tilt axis is aligned with the triangular base of the nanoparticles. Possible applications of the nanochain architecture include sub-100 nm near-field optical waveguides, chemical and biological sensors based on the LSPR of these high aspect ratio nanoparticles, and the fabrication of nanowires.

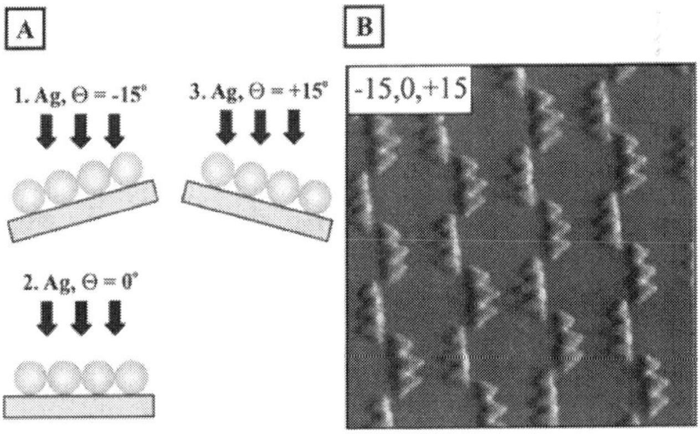

Fig. 5. (A) Schematic fabrication process and (B) contact mode AFM image for three deposition nanochain structure on mica. 1.6 μm - 1.6μm area, D = 542 nm, dm = 10 nm, θ_{dep} = +15°, 0°, and -15°. (Reprinted from Haynes, C. L.; van Duyne, R. P., *J. Phys. Chem. B* 2001 105, 5599, Figure 9, Copyright (2001) Amercian Chemical Society.)

We recently developed a modified NSL process to fabricate Ag NPs with controlled shapes on substrates. The modification in NSL is performed by thermally annealing the triangular nanoprisms, and sonication to remove weak tips, followed by removing debris and small broken parts around the NPs on the substrates(Song and Elsayed-Ali 2010). The detailed process is shown in the following: (1) Releasing the nanospheres by immersing the cover slip into a 5% HCl solution for 30 minutes, then immersing the glass substrates into CH_2Cl_2 for 30 s, then sonication for ~20-60 s; (2) The fabricated Ag nanoprisms on the glass substrates are annealed at 100-300 °C for 2-5 hours; (3) Then Ag nanoprisms are cleaned by immersing the glass cover slip into 5% HNO_3 for 10-20 s to remove any surface contamination and dissolve debris around the NPs, and then washed by large amount of nanopure water. Comparing the AFM images in Step 3-b of Figure 1 showing the NPs without above post treatment, tip-rounded triangle nanoprisms, square-shaped and trapezoidal Ag NPs (Figure 6) can be obtained via one or two of the above treatment. We observe that thermal annealing results in much more uniform NP surfaces without the thin, weak tips and edges (Figure 6(a-1)). From the magnified AFM plane image in the inset of Fig. 6(a-1) and the 3D image in Figure 6(a-2), the NPs still show triangular prism shape with

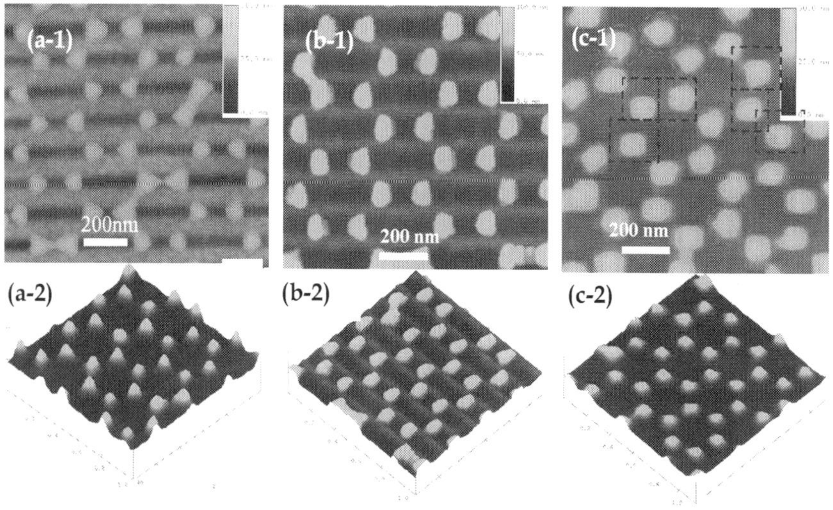

Fig. 6. Surface-confined Ag NPs with controlled shapes fabricated by the modified NSL process. (a-1) AFM image of triangular prism Ag NPs with rounded tips after thermal annealing at 200 °C for 4 hours, cleaning by 5% nitric acid, and washing by nanopure water. (a-2) The 3D image of the triangular prism Ag NPs with rounded tips. (b-1) Flat trapezoidal Ag NPs after sonication to remove one tip, thermal annealing, cleaning by 5% nitric acid, and washing by nanopure water. (b-2) The 3D image of the trapezoidal Ag NPs with one snipped tip. (c-1) The quadrilateral or pentagon shaped Ag NPs after sonication intensively to remove two tips, thermal annealing, cleaning by 5% nitric acid and washing by nanopure water. Dashed circles: pentagonal Ag NPs with one sharp tip left; dashed squares: quadrilateral Ag NPs. (c-2) is the 3D image of the quadrilateral and pentagon shaped Ag NPs. (Song Y.; Elsayed-Ali H. E., Appl. Surf. Sci. 2010 256, (20), 5961, Figure 3, Copyright (2010) Elsevier.)

rounded edges and little surface defects. Alternatively, if we sonicate the NPs produced by NSL for ~30-45 s to remove a weak tip, anneal them at 200 °C for 1- 4 hours, then wash them with 5% nitric acid, trapezoidal shaped NPs with rounded edges are formed, as shown in Figure 6(b-1) and 6(b-2). If the sonication time is increased to more than 2 min, the NPs lose their two sharp tips and form quadrilateral or pentagon shaped NPs. After thermal annealing for 1-4 hours and washing with 5% nitric acid, their edges and corners become rounded, as shown in Fig. 6(c-1), which show quadrilateral NPs (in dashed squares) or pentagon (in dashed circles). The 3D AFM image, Figure 6(c-2), shows that these NPs have rounded edges and corners. Clearly, even after thermal annealing, they are still showing prism shapes with increased thickness from their edges to centers according to their 3D AFM images.

The work described above demonstrates that NSL, broadly defined to include AR NSL and some modified post treatment after deposition of the desired materials, is manifestly capable of creating far more than arrays of nanotriangles, nanodots as previously supposed. The progresses in NSL endow much potential in the size and shape controlled fabrication of nanoparticles and nanoarrays, which gives NSL a bright future since the ability of NSL to synthesize monodisperse, size- and shape- tunable nanoparticles can be exploited to precisely investigate the size- and shape- dependent physiochemical properties of nano-optics and nanoarrays.

3. Investigation of optical properties of specific noble metal nanoparticles and nanoarrays by the combination of NSL and multi-hierarchy arrayed micro windows

The physicochemical properties of nanomaterials significantly depend on their three-dimensional (3D) morphologies (sizes, shapes and surface topography), their surrounding media, and their spatial arrangement. Systematically and precisely correlating these parameters with the related physicochemical properties of specific single nanoparticles (NPs) or nanoarrays is a fundamental requirement for the discovery of their novel properties and applications, as well as for advancing the fundamental and practical knowledge required for the design and fabrication of new materials(Song , Zhang et al. 2011). The lack of effective means of fabricating recognizable 3D morphologies controlled NPs and nanoarrays and correlating their structure parameters with their physicochemical properties as observed by different characterization techniques represents an obstacle for studying the 3D morphology-dependent properties of individual NPs and nanoarrays(Song, Zhang et al. 2011). Most current studies investigate the physicochemical properties of the NP ensemble, but not of a single NP(Jin, Cao et al. 2001; Kelly, Coronado et al. 2003; Haes, Zou et al. 2004; Song , Zhang et al. 2011). The ensemble of NPs is typically heterogeneous, because the morphologies of individual NPs prepared by routine chemical synthesis or physical vapor fabrication methods are rarely identical at the nanometre or sub-nanometre scale (Song , Zhang et al. 2011). Effective methods for 3D morphology controlled fabrication of nanomaterials, and to correlate their 3D morphology of single NPs or nanoarrays with their physicochemical properties are also essential to address fundamental and practical questions related to the single NPs (Song , Zhang et al. 2011).

An important research area in nanoscale plasmonic optics is single NP identification and characterization of their 3D morphologies and space-orientation dependent physicochemical properties(Yang, Matsubara et al. 2007; Song 2009; Song , Zhang et al. 2011). Recently, much attention has been given to the localized surface plasmon resonance (LSPR) of metal NPs because of their promising applications in plasmonic circuits, optoelectronic transducers, optical bioprobes, and surface plasmon resonance interference lithography(Shen, Friend et al. 2000; Prasad 2004; Ozbay 2006; Song 2009; Song , Henry et al. 2009; Song , Jin et al. 2010; Song, Sun et al. 2010; Song , Zhang et al. 2011). Since the plasmonic properties of metal NPs intrinsically rely on their size, shape, surface topography, crystal structure, inter-particle spacing and the dielectric environment around them, methods to correlate their plasmonic properties with the above structural and environmental parameters have become one of the most rapidly developing research directions (Song , Zhang et al. 2011).

In the precise investigation of the relationship between the LSPR properties and their 3D morphologies of specific nanoparticles and nanoarrays, two kinds of methods have been developed recently, or the *in situ* method and the spatial-localization method (Song , Zhang et al. 2011). The *in situ* method combines at least two different instruments together to conduct the structure and property characterization simultaneously: one can be used to characterize the 3D morphology (e.g. AFM or STEM) of NPs and the others will be used to chatacteize the LSPR-related optical properties of the same NPs (e.g. Dark-field microscope and spectroscopy). The spatial-localization method requires using markers to recoganize the same single nanoparticle in different instruments. We have also developed one spatial-localization method to precisely investigate the 3D morphologies dependent LSPR properties of specific NPs and nanoarrays by the combination of NSL and traditional UV-LIGA, where Ag NPs and nanoarrays can be fabricated by NSL in the pre-formed multi-hierarchy arrayed transparent micro-windows on the substrates (e.g., glass cover slip) by the UV-LIGA(Song 2009; Song , Zhang et al. 2011). This technique permits easy characterization of the 3D morphologies of single NPs by AFM or SEM and their LSPR spectra using dark-field optical microscopy and spectroscopy (DFOMS). It is also possible to investigate the local morphology dependence of the LSPR spectra of the single NPs and nanoarrays. In this method, multi-hierarchy arrayed micro windows are first fabricated on a glass cover slip using the standard photolithography, whose details are shown in reference 27. Fig. 7A and Fig. 7B show one example of the designed multi-hierarchy arrayed micro windows (3 tiers) and the typical final micro-windows (Fig. 7C) pattern after printing. The multi-hierarchy arrayed micro-windows on the glass cover slip are used to identify the location and orientation of single NPs, whose tiers can be determined by the observed field at desired resolution. For example, in the first tier of the multi-hierarchy arrayed micro windows (Fig. 7A), each local area can be discerned by marking its X and Y number, such as the shaded area X1-Y2. Then, in the second tier of the multi-hierarchy arrayed micro windows (Fig. 7B), the scale can be reduced by M or N times and each local area can also be marked by x and y number. If this area is the sub-tier in the shaded area of the first tier, it can be labeled as X1-Y2-x3-y3. In a similar way, step-by-step, we can reach the last tier with several transparent micro windows available (Fig. 7C), in which the desired nanoparticle can be made by different fabrication methods (e.g., electron beam lithography or nanosphere lithography). Nanoparticles less than 10 nm of different shapes synthesized by a wet-chemical process can be immobilized by a routine diluted deposition process. Consequently, the same nanoparticle in each window can be identified by comparing the images taken by the optical microscope with those characterized by the AFM. Finally, in each window, the same nanoparticle can be characterized by different techniques (e.g., DFOMS and AFM) allowing correlation of its 3D morphology with its optical response(Song 2009; Song , Zhang et al. 2011).

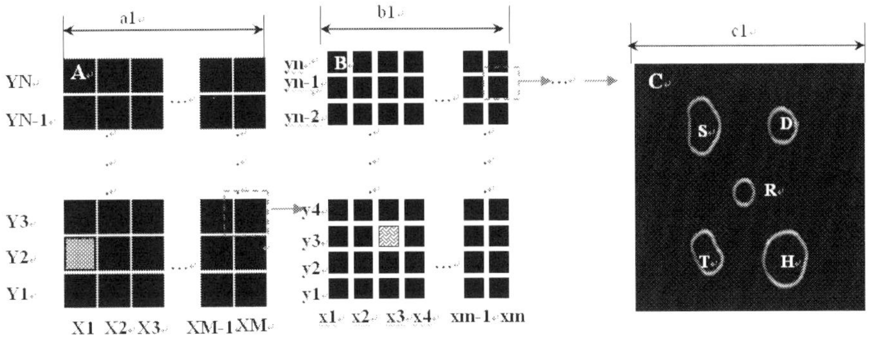

Fig. 7. The multi-hierarchy arrayed micro windows on the substrate (e.g., glass cover slip). (a) The first tier of the multi-hierarchy arrayed micro window, each local area can be discerned by marking its X and Y number, such as the red-dashed square area of X1–Y2. (b) The second tier of the multi-hierarchy arrayed micro windows, whose scale can be reduced by M or N times, whose local area can also be marked by x and y numbers. If this net area is the sub-tier n the red area of the first tier, it can be labeled as X1–Y2–x3–y3. Step-by-step, the last tier with several unique-shaped transparent windows can be reached. The open windows can be made with different shapes. (c) The nanoparticles can be fabricated on the micro-pattern by various methods (e.g., nanosphere lithography). In each window, the same nanoparticle can be identified by comparing the images taken by optical microscopy, AFM, or other microscopy methods. Finally, the structural parameters (size, shape, orientation, interparticle spacing, and thickness) can be correlated with their optical responses (Reprinted from Song Y.; et al., Nanoscale 2011, 3, 31-44, Figure 7, copyright (2011) from the Royal Society of Chemistry.)

A typical example to identify NPs and nanoarrays using both AFM and DFOMS is illustrated in Figure 8. Triangular Ag NPs and hexagon-arranged nanoarrays fabricated on the surface of glass cover slips within the nearly circle-shaped micro window can be identified and characterized using AFM (Figure 8A, 8B is the 3D AFM image of the dash-squared area in 8A) and DFOMS equipped with a color camera (Figure 8 C) and charge-coupled device (CCD) camera (Figure 8D). The CCD camera offers higher spatial resolution than the color camera, while the color camera provides the real colors of individual Ag NPs that are generated by LSPR. The center of each individual NP in the optical images recorded by the CCD is located with a single-pixel resolution (each pixel can be 125 nm or 67 nm depending on the CCD resolution and equipment setup) by determining the address of the pixel with the highest intensity of the NP. The positions of individual NPs of interest (e.g. the circled one) within the micro window in the optical images (Figure 8C and D) are then determined with a spatial resolution limited by the optical diffraction limit (~200 nm) and an orientation angle resolution of about 1.0 degree. This approach allows us to correlate AFM images of individual NPs (as the one circled in each image) with the same NP shown in its corresponding optical image and to investigate its 3D morphological-dependent LSPR properties. Clearly, these triangle nanoparticles in this window almost show the same scattering color (Figure 8C) and intensity contrast (Figure 8D). By comparing their scattering color images (Figure 8C) with their AFM images (Figure 8A and B) of these nanoparticles, it is once again showing that NSL is powerful method in the fabrication of uniform triangular nanoparticles and nanoarrays.

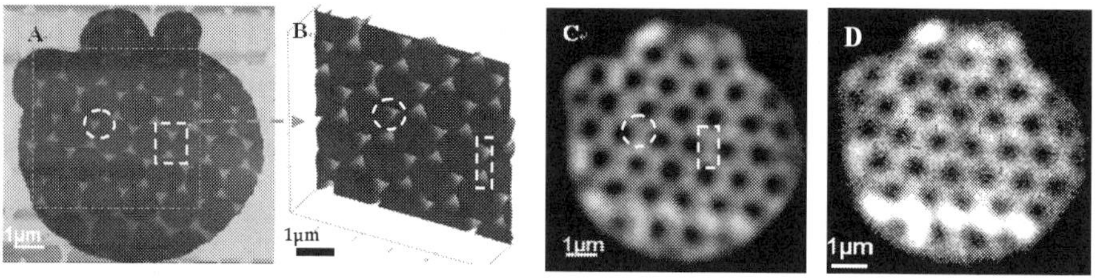

Fig. 8. One example for the identification of the specific nanoparticles and nanoarrays in different instruments via multi-hierarchy arrayed micro windows based on Ag triangle nanoparticles and nanoarrays fabricated by NSL in one nearly circle-shaped window. (A) The plan view of the hexagon-arrayed triangle Ag NPs in one circle micro-window scanned by atomic force microscope (AFM); (B) the 3D view of the hexagon-arrayed triangle Ag NPs marked in the large pink dash-square in (A); (C) the real scattering color of these hexagon-arrayed triangle Ag NPs observed under a dark-field microscope; (D) the CCD images of the scattering light of these hexagon-arrayed triangle Ag NPs recorded by a CCD camera equipped in the dark-field microscope. The dashed circles in each image refer to the same specific particle and the dashed squares in each image refer to the same specific nanoparticle pair. (Adapted from reference Y. Song, China Patent, Appl. No. CN200910085973.9).

We have used it to investigate size- and shape-dependent LSPR spectra of single Ag NPs by the analysis of the experimental results with the theoretical calculation (i.e. DDA simulation)(Song , Zhang et al. 2011). Figure 9 gives the AFM images of one specific triangle-shaped Ag NPs characterized by multi-hierarchy arrayed micro windows. The AFM image of the triangular silver NP shows that it has the edge length of 375-420 nm (Figure 9A) and the out-of-plane height of about 16.1 nm (Figure 9B). This NP shows multi LSPR scattering colors (Figure 9C), as further evidenced by its multi-mode LSPR peaks at 562.3 nm, 659.9 nm and 759.6 nm (Figure 9D-b). The peak wavelengths, peak ratios, and line widths (FWHM) at 562.3 nm and 659.9 nm from experiment are in good agreement with DDA simulation for its LSPR scattering (Figure9D-c), as have been summarized together with other shaped nanoparticles fabricated by the modified NSL in reference 27. In general, the DDA simulation shows best agreement with the experimental spectra for NPs, hence their shapes can be accurately modeled. However, it can also be seen that for wavelengths longer than 650 nm for the investigated NPs, the experimental result has a lower intensity than the simulation(Song , Zhang et al. 2011). By analysis the instrument errors and the wavelength dependent CCD quantum efficiency, these deviations are deduced by the precision in the shape construction during DDA simulations. From these results, it was also found that when the shapes and 3D morphologies of the NPs became more complicated, the deviation between the DDA simulation and the experimental result increased (Song , Zhang et al. 2011). This is due to the geometrical deviation between the real NPs and the regular species used in the calculations. If these two instrumental factors and the geometrical deviation of NPs are considered, the corrected experimental results will match with the DDA simulation very well. This result also confirms that our experimental method (DFOMS), based on the far field detection, preserves the ability to detect the near-field LSPR signal.

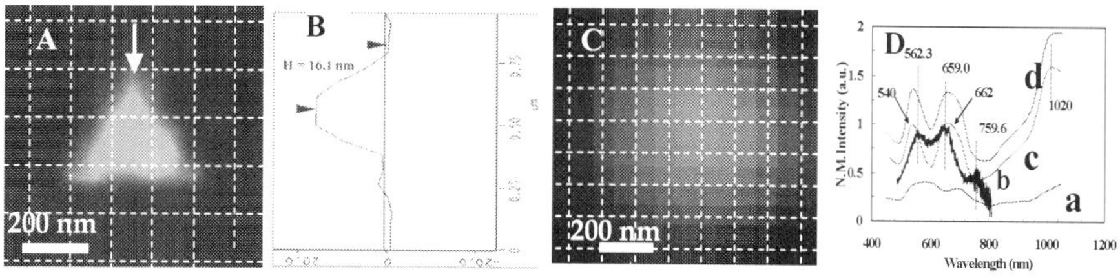

Fig. 9. (A): The AFM image for one single triangular shaped Ag nanoparticle with edge length of 375-420 nm; (B) the height mapping of the triangle shape Ag nanoparticle along the direction of the arrow in Fig. 9(A), showing the out-of-plane height of this nanoparticle of 16.1 nm; (C) the real scattering color image of this triangular shaped Ag nanoparticle; (D-a): the LSPR absorption spectrum of this nanoparticle by DDA; (D-b) the LSPR scattering by experiment; (D-c) the LSPR scattering by DDA; (D-d): the LSPR extinction by DDA. In order to identify the location and the orientation of these positions around the NPs, the AFM image and color image were netted by dashed lines with each square unit of 125 nm ×125 nm after their distances and orientations were corrected. (Adapted from references: Y. Song, China Patent, Appl. No. CN200910085973.9; Song Y.; et al., Nanoscale 2011, 3, 31-44, Figure 13, copyright (2011) from the Royal Society of Chemistry. Adapted with permission.).

This combined method based on the NSL and the multi-hierarchy arrayed micro windows also allows us to investigate the 3D morphology dependent tip-tip LSPR coupling of triangular nanoparticle pairs. The zoom-in AFM image for the detailed 3D morphology of one typical Ag nanoprism pair is shown in Figure 10A. The nanoprisms have almost the same edge size ~ 375 nm and maximum out-of-plane height ~ 17.1 nm shown in Figure 10B by the typical height map along the arrowed tip-tip direction in Figure 10A. The real scattered color for the nanoprism pair, taken from dark-field microscopy, is shown in Figure 10C. Both of the nanoprisms in the pair give red color with different brightness, which might be due to variation in their surface roughness, slightly difference in the underlying surrounding dielectrics, and the focusing distance during image recording. The middle area between the two nanoprisms clearly shows more reddish color than the optical centers of the two nanoprisms. The LSPR spectrum for the middle area of the two optical centers (representing the tip-tip-coupling) is recorded in Figure 10D using their CCD image (not shown here) for the location identification, together with that obtained by the discrete dipole approximation (DDA) calculation of the nanoprism pair. According to its 3D morphology of the nanoparticle pair, the two nanoprisms can be treated as regular triangular nanoprisms with the bottom edge length of 375 nm, the top edge length of 125 nm and out-of-plane height of 17.1 nm for conducting the DDA calculation of the nanoprism pair. The recorded LSPR spectrum (Figure 10D: a) at the middle optical center of the two nanoprisms shows three distinct peaks, one strongest peak at 605 nm, one shoulder at 536 nm, and one secondary strong peak at 754 nm. By comparing the experimental result for the tip-tip coupling of the nanoprism pair with the DDA calculation (Figure 10D: b), it can be deduced that the peak at 605 nm represents the in-plane quadrupole resonances originated from the two source nanoprisms and the peak at 536 nm is from the out-of-plane quadrapole resonances of the two source nanoprisms. Although the DDA simulation does not show one distinct peak at 754 nm, our experiment result suggest one strong peak at this wavelength, which is probably from the strong tip-tip coupling. In order to reveal whether the peak at 754 nm is mainly from the tip-tip coupling or not, the LSPR spectra from the optical centers of the source nanoprisms, are recorded (not shown here), showing one strong peak at the same position. Generally, one can see that the peak positions and shape resonances for the two nanoprisms are almost the same, suggesting that the nanosphere lithography process is very powerful in the fabrication of the nanoprisms with almost identical 3D morphologies and surroundings. Both of the two triangle nanoprisms do not give the peak at 754 nm as strong as the pair, confirming that the additional peak at 754 nm indeed is from the tip-tip coupling. However, previous investigations did not show additional strong peak due to tip-tip coupling (Su, Wei et al. 2003; Zhao, Kelly et al. 2003). The reason for this significant coupling between the nanopair may be caused by the unique size of our particles that is just lying in the range of half wavelength of visible light, which can cause a strong long-range electrodynamic interaction among light and the collective electrons on the particle surfaces.

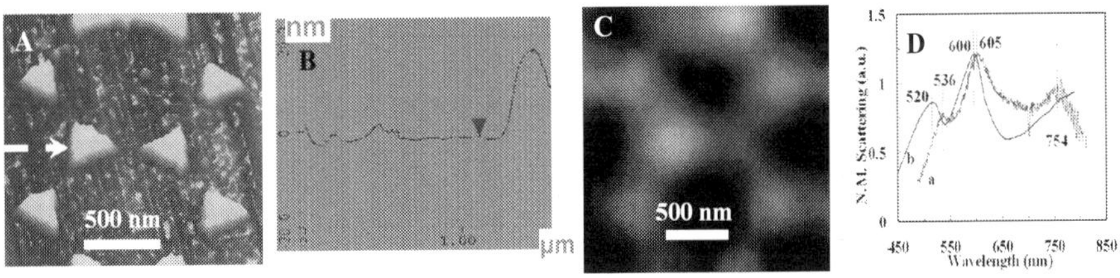

Fig. 10. Tip-tip coupling of one typical pair of triangular Ag nanoprisms in the arrays with interspacing of 103 nm is characterized using (A) AFM; (B) the scan of out-of-plane height of the two nanoprisms along the arrow direction in (A); (C) the color image taken from dark-field microscopy; (D: a) the LSPR scattering spectrum at the central locations (3 pixels) of the source nanoprisms and (D:b) the LSPR scattering spectrum of the pair calculated by DDA.

In additon, our experimental observations show that nanoprism coupling does not affect the quadrupole mode in LSPR significantly, resulting in little shifts in the highest peak at 598-605 nm (the in-plane quadrupole mode). However, one additional peak (i.e. 754 nm) as compared with the in-plane quadrupole mode can be observed. This peak resulted from LSPR coupling is in good agreement with the prediction by the semianalytical model by Schatz et al. (Zhao, Kelly et al. 2003) In the present study, the edge lengths of the triangular nanoprisms are more than $\lambda/2\pi$ (64-128 nm), which is more than the critical scale in the semianalytical model in the DDA.(Zhao, Kelly et al. 2003) Therefore, the long-range electrodynamic interaction, not electrostatic effects, will be dominant in the LSPR of the two nanoparsms. The center-to-center interspacing of the two nanoprisms is ~ 532 nm, more than the critical interspacing. As a consequence, the coupling will be mainly dertermined by the long-range radiative dipolar interactions (or radiative damping effects),(Zhao, Kelly et al. 2003) and phase retardance effects,(Su, Wei et al. 2003) resulting in one new peak with wavelength more than the highest peak for the two nanoprisms.

Based on this combined method, we have investigated the distance dependent tip-tip coupling between triangular Ag nanoprism pairs with dimensions at the range of half wavelength of visible light and distance ranging from 100 nm to 400 nm. It has been found that the coupling peak wavelength increases and the coupling intensity decreases with the increased tip-tip distance, and finally the coupling disappears (no coupling peak) when the tip-tip distance is more than about 400 nm due to the coupling intensity becomes extremely low.

Generally, the combination of NSL and the multi-hierarchy arrayed micro windows fabricated by the routine UV-LIGA shows a powerful ability not only in the identification of nanoparticles and nanoarrays but also in the precise investigation of the fundamental theory related to the 3D morphology dependent LSPR and LSPR coupling. In our study, the detector is far-field while the DDA calculation is based on the near-field. Thereby, the results indicate that the near-field LSPR of single NPs and the coupling signals of nanoarrays can be detected by the far-field detector if the 3D morphologies of NPs or nanoarrays can be precisely accounted for in the DDA model.

4. Microfluidic biosensing system based on NSL and microfluidic reactor fabrication

Recently, Song has developed a high-throughput single Ag NPs biosensing device by coupling a variety of functionalized Ag NPs fabricated by NSL into a series of microfluidic channels (Song 2009). The designed microfluidic biosensing system based on Ag single nanoparticles and nanoparticle arrays is illustrated in Figure 11. Samples were fabricated by the combination of NSL and the traditional UV-LIGA process for the microfluidic reactor fabrication (Song 2009; Song 2010; Song and Elsayed-Ali 2010). In this biosensing system, the corresponding microfluidic channels are fabricated on the designed patterns where series of single Ag NPs or arrays (Figure 11: a) have been fabricated by careful alignment. The glass cover is then connected with glass optical fiber binding on the top of the microfluidic channels after careful alignment with the desired single Ag NPs or nanoarrays. In order to alleviate the non-specific absorption in the microfluidic channels, the channels are modified by polyvinylacholol (PVA) or polyethylene glycol (PEG) solution. After that, the single Ag NPs will be surface modified by a mixture of at least two thiol compounds with one having carboxyl group or amine group as the conjugating compound (e.g. 11-mercaptoundecanoic acid: MUA), and another thiol compound without carboxyl group or amine group as spacer (e.g. 6-mercapto-1-hexanol: 6-MCH, 1-octanethiol: 1-OT). The modification reaction is shown in equation (1). The Ag NPs can then be functionalized with biomolecules, as reporter (e.g. IgG), by a conventional 1-ethyl-3-(3- dimethylaminopropyl)-carbodiimide (EDC) coupling process to form the f, as shown in equation (2) and (3) for the functionalization of Ag NPs (Figure 11: a and b).

The number per Ag NPs can be controlled by the ratio of the conjugation compounds and spacers, which can be used to calculate the number of the responding biomolecules (e.g. Protein A) that can bind with the reporters, which can be directly sensed by the LSPR peak shift. As shown in Figure 11, the solution having a specific concentration of the corresponding detected biomolecules can be delivered into the microfluidic channels (Figure 11: g). The channel widths are designed from several hundreds micro meter to ten micrometers that will play a role like a dark-field condenser for incident white light. The scattering color changes and the LSPR spectrum variations of Ag NPs (a) caused by the binding of the detected biomolecules on the reporters, (b) will be collected in the opened windows, (d) and transported into the detector and analyzer, (f) by the glass fiber, (e) after signal magnification.

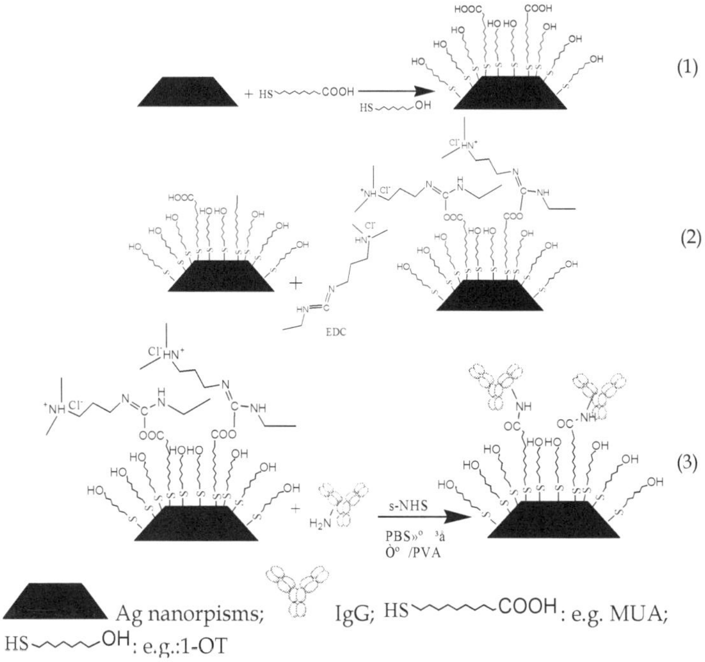

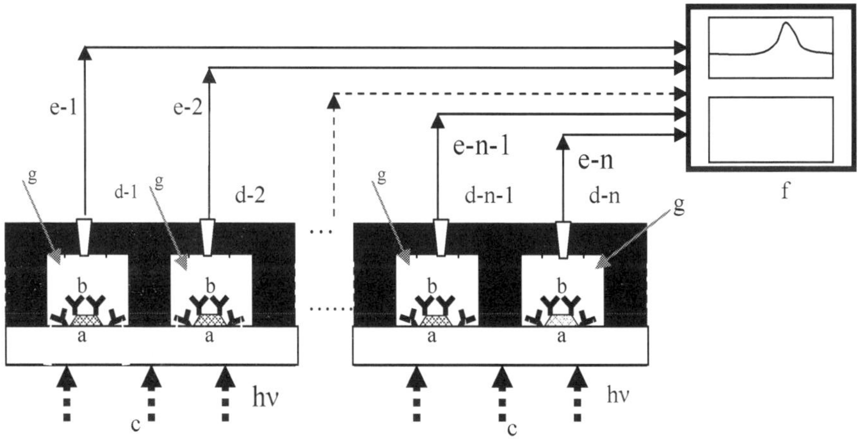

Fig. 11. a: nanoparticles (e.g. triangular Ag NPs); b: biomolecule (e.g. antibody); c: induced light (e.g. white light from tungsten lamp); d-1 - d-n) : Series of detecting windows connecting with optical microfibers in different microchannels ; e-1 – e-n : glass microfibers for signal transport ; f : Optical spectroscopy (e.g. micro optical fiber spectroscopy-S2000, Spectropro-150, surface enhanced Raman spectroscopy; g: the sealed microchannels). (Adapted from reference Y. Song, China Patent, Appl. No. CN200910085973.9)

Song has investigated the efficiency of this kind of biosensing system. Using the color change and LSPR spectra shifts based on the binding of one model biomolecule pairs (anti-body: IgG is firstly functionalized on the Ag surfaces by EDC process, then the antigen: protein-A buffer solution is pumped into the microfluidic channels) as model, it can be seen that the spectra shift and the color changes from the binding of the model biomolecule pairs depend on the concentration of biomolecules and the running time. Up to now, the detection resolution of this kind of biosensors based on the scattering from single Ag NPs has reached 2 nm peak shift per 1 nM concentration change and the resolution for one single NP biosensor is calculated as 10-20 biomolecules per Ag NP(Song 2009). This result suggests another persepctive application by the combination of NSL, microfluidics and bio-functionalization process.

5. Fabrication of hierarchically ordered nanowire arrays on substrates by combination of NSL and Porous anodic alumina (PAA)

In some applications of nanomaterials, the NPs need to be arranged in some particular patterns, architectures or motifs with controlled interspacing, or conjugated with some other kinds of materials (e.g. polymers) (Chong, Zheng et al. 2006; Song , Zhang et al. 2011). The controlled arrangement and immobilization of Ag NPs on substrates will be very crucial to enable some fascinating and delicate applications, particularly in electronic circuit based electro-optical devices and long term functional composites for biological applications. Many methods have been explored for this purpose. Among them, template-assisted LIGA or structure controlled artificial fabrication methods (e.g. E-beam LIGA, NSL, PAA-LIGA) may be the most convenient techniques(Song , Zhang et al. 2011). In the NSL development, the suitability and powerful ability in the architecture and interspacing controlled fabrication of NPs and nanoarrays can be expanded extremely if the NSL can be combined with other template-assisted LIGA methods. Here we just show one example to fabricate hierarchically ordered nanowire arrays on substrates by the combination of NSL and porous anodic alumina (PAA) LIGA(Chong, Zheng et al. 2006).

Like NSL, porous anodic alumina (PAA) templates have attracted intense attention in nanodevice-oriented fabrication in recent years(Xu, Meng et al. 2009). As a well-developed template, PAA offers amazing simplicity and convenience for nanofabrication due to the capabilities of forming high-density, well-aligned, and hexagonally packed sub-100-nm pores, the ability to control the 3D pore structures by simply varying the anodization conditions, and the ease of selectively removing the template after fabrication(Chong, Zheng et al. 2006). As shown in Figure 12, one typical PAA-LIGA process includes(Lombardi, Cavallotti et al. 2007): (a) formation of a 300 nm thick PAA film on Al by a two step anodization process in 0.3M oxalic acid; (b) dissolution of unoxidized Al; (c) barrier layer etching in 5wt% phosphoric acid; (d) transfer of the PAA mask onto Au-coated Si followed by a thermal treatment to improve the adhesion of the films to the substrate; (e) Ag electrodeposition through the PAA pores; (f) PAA mask removal. By carefully controling the sizes and interpore spacing of the nanoholes, very uniform Ag nanorods with controlled interspacing can be fabricated by electroplating. The typical Ag nanorods prepared by this template assisted electrodeposition process can give a much uniform size and interparticle spacing distribution, with a standard size deviation less than 5% and a spacing deviation less than 7%.

Progresses in PAA-LIGA have shown its abilities not only in the synthesis of the traditional high aspect ratio nanomaterials, such as nanorods (Pan, Zeng et al. 2000; Lombardi, Cavallotti et al. 2007; Xu, Meng et al. 2009)and nanowires (Hong, Bae et al. 2001; Xu, Zhang et al. 2005), but also some unique nanostructures, such as porous metallic nanorods by galvanic exchange reaction (Mohl, Kumar et al. 2010), Y-junction nanowires and multiply branched nanomaterials due to its flexibility in the pore structure control according to the principle $1/\sqrt{n} \times V_s$, where V_s is the anodizing voltage for stem pores and n is the number of branched pores from that stem(Meng, Jung et al. 2005; Xu, Meng et al. 2009).

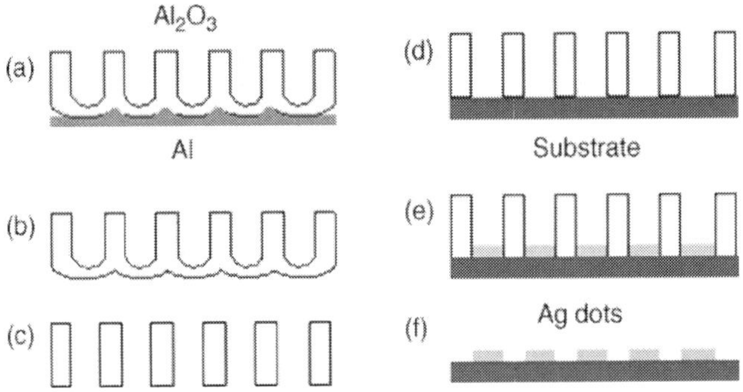

Fig. 12. Fabrication scheme for Ag nanoparticle arrays: (a) formation of a PAA film on Al by a two step anodization process in 0.3Moxalic acid; (b) dissolution of unoxidized Al; (c) barrier layer etching in 5 wt.% phosphoric acid; (d) transfer of the PAA mask onto Au-coated Si followed by a thermal treatment to improve the adhesion of the films to the substrate; (e) Ag electrodeposition through the PAA pores; (f) PAA mask removal. (Lombardi I.; et al., Sensors and Actuators B: chemical 2007, 125, 353-356, Figure 1. Copyright (2007) Elsevier.)

The combination of NSL and PAA-LIGA has been further developed to create hierarchically ordered nanowire arrays, as schematically shown in Figure 13A(Chong, Zheng et al. 2006). A monolayer of self-assembled polystyrene nano or microspheres as masks is first used to deposit periodic porous gold films on silicon substrates (Figure 13A: i-ii). Next, PAA films are fabricated on top of the porous gold film/substrate (Figure 13A: iii). Nanowires are then selectively electrodeposited into the pores of the alumina using the porous gold film as a working electrode (Figure 13A: iv-vi). In detail, a drop of polystyrene sphere suspension (e.g. 1 μm in diameter, 10 wt % aqueous dispersion) is spin-coated onto a pretreated substrate (e.g. Si, glass or mica) to form close-packed microsphere monolayers. The size of the nano or microspheres can be tuned using O_2 reactive ion etching (RIE) with a suitable O_2 flow (e.g. 20 SCCM, SCCM denotes cubic centimeter per minute at STP) at a certain pressure (e.g. 15 mTorr) and a power density (e.g. 110 W) for 6–10 min. After RIE, isolated nano or microsphere monolayers with tunable spacing will be formed. Then, about 5 nm Ti (as an adhesion layer) and 40 nm gold films in turn deposit onto the substrate using the RIE reduced nano or microspheres as masks. After removal of the mask by sonicating in a solvent (e.g. toluene) for 3 min, a porous gold film will be formed on the substrate. After that, an aluminum film with a thickness of ~500 nm deposits onto the porous gold film. Prior to anodization, the aluminum film is subjected to an imprinting step, in which a free-standing PAA with a thickness of ~10 μm fabricated by the process in Figure 12 is used as a mold for imprinting. Finally, the imprinted aluminum film is anodized in 0.3*M* oxalic acid at 2 °C and the barrier layer at the bottom is removed in 5 wt % H_3PO_4 for 60 min. The gold nanowires can be deposited at −1.0 V versus standard calomel electrode (SCE) from a commercial bath (Orotemp 24, Technic) for varying amounts of time. Alumina templates can be removed in 1M KOH for some time (e.g. 10 min) to obtain hierarchically patterned free-standing nanowires or nanorods.

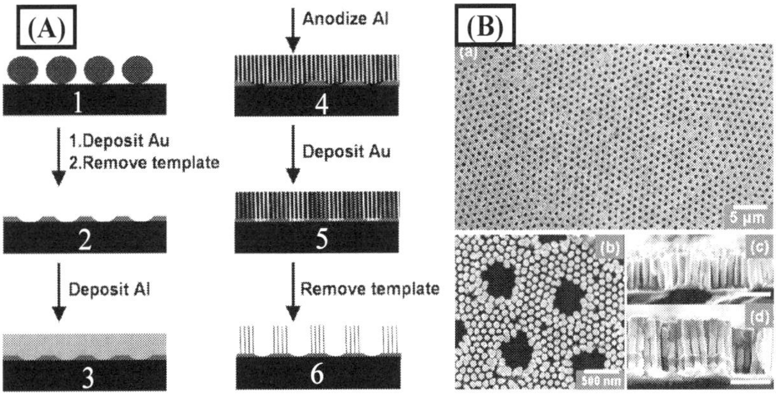

Fig. 13. (A: 1-6) Schematic of method to create hierarchical nanowire arrays on substrates. (B) SEM images of hierarchical nanowire arrays on substrate:(a) Top view of the nanowire arrays with hexagonally organized microvoids over large areas; (b) High magnification SEM image of (a); (c) Side view of the cleaved sample from (a), the concaves caused by voids are clearly apparent; (d) Side view of Au/Ni/Au/Ni segmented nanowire arrays. The side view shows clear contrast; brighter segment is gold portion. Scale bars in (c) and (d): 500 nm. (Chong, M. A. S.; et al., Appl. Phys. Lett. 2006, 89, 233104. Figure 1 and Figure 3. Copyright (2006) from American Institute of Physics. Adapted with permission.)

Figure 13B (a) shows the top view of one typical nanowire arrays with hexagonally organized microvoids (e.g. 500 nm in diameter) over large areas (Chong, Zheng et al. 2006). One of its high magnification images as shown in Figure 13B(b) clearly demonstrates the arranged patterns by individual gold nanowires. The nanowires with uniform size, replicated from the PAA, are hexagonally packed at the nanoscale. The hierarchical nanowire arrays standing on the substrate are further checked from a cleaved sample by side view SEM images (Figure 13B(c), suggesting the well aligned nanowires and the concave features from the nanoscale voids. In addition to fabricating pure gold nanowire arrays, this combinational template is also suitable for selectively electrodepositing other materials for functional device applications. In particular, the vertical structure/composition along the length of the nanowire is also tunable. For example, multilayer Au/Ni/Au/Ni nanowire arrays with in-plane hierarchical structure can be fabricated by the alternating deposition twice of nickel from a Watt's bath (300 g/L $NiSO_4 \cdot 6H_2O$, 45 g/L each H_3BO_3 and $NiCl_2 \cdot 6H_2O$) and of gold at −1.0 V versus SCE electrode. Figure 13B(d) shows the cross section of Au (bottom, brighter segment) /Ni (darker segment)/Au/Ni (top) nanowire arrays.

6. Solution phased nanomaterials by the releasing of nanoparticles fabricated by NSL

Recent progress in nanosphere lithography (NSL) has shown that this method provides a good template for shape-controlled fabrication of surface confined NPs(Zhang, Whitney et al. 2006; Song , Zhang et al. 2011), which also allows for flexible functionalization of these NPs on the clean surface (as cartooned in Figure 14A) using the routine functionalization process from equation 1 to 3. After the surface functionalization of surface confined nanoparticles fabricated by NSL, they can be dislodged into solution phase (as schemed in Figure 14B). This dislodging process provides a useful alternative to synthesis uniform solution phase NPs besides the wet chemical process. Van Duyne *et al.* have developed this process and used it to fabricate solution phase NPs in ethanol(Amanda, Zhao et al. 2005). However, their results indicated that most of the NPs in the solution have nonuniform surface morphologies with truncated tips in addition to the presence of debris and some of the NPs attached together on the glass substrate surface causing the agglomeration of the released NPs. In addition, aqueous phase NPs are expected to be more biocompatible than those in ethanol. Therefore, technology development to obtain aqueous-stable nanocolloids via surface modification and releasing of the surface-confined NPs fabricated by NSL into water solution are much desired.

Our group recently developed a modified NSL process to fabricate Ag NPs with controlled shapes on glass substrates and with the ability to release them into the aqueous solution without any obvious agglomeration(Song and Elsayed-Ali 2010). Three modifications of the standard procedure of nanosphere lithography were made in order to obtain stable NPs with different shapes. The modification to the process were the following: (1) Releasing the nanospheres by immersing the cover slip into a 5% HCl solution for 30 minutes, then immersing the glass substrates into CH_2Cl_2 for 30 s, then sonication for ~20-60 s; (2) The fabricated Ag nanoprisms on the glass substrates were annealed at 100-300 °C for 2-5 hours then cleaned by immersing the glass cover slip into 5% HNO_3 for 10-20 s to remove any surface contamination and dissolve debris around the NPs, and then washed by large amount of nanopure water; (3) The glass substrates were immersed into 5-10wt.% HF and HCl acid mixture (HF: HCl = 1:1) for 30-60 s or 10% NaOH solution for 60-120 s to etch part of the glass substrate under the Ag NPs, and then the substrates were washed with sufficient amounts of nanopure water. Finally, the glass substrates with the Ag NPs were dried by inert gas flow and kept in desiccators. The surfaces of the Ag NPs can be modified by chemicals containing thiol groups (such as 1-OT, MUA, 6-MCH and Tiopronin (TP)) forming strong sulfur-silver covalent bonds. We used 1-OT and MUA as functional reagents. The functional solution was prepared by dissolving 0.049 g 1-OT and 0.073 g MUA into 100 mL pure ethanol in a 100 mL volume certificated flask to form 2mM 5:1 1-OT /11-MUA solution. The Ag NPs were once again cleaned using 5 % nitric acid and then immersed into the 2 mM 5:1 1-OT /11-MUA solution and left overnight. The releasing aqueous solution contains 5 V% of 2mM 5:1 1-OT /11-MUA in nanopure water. The glass substrates with surface modified Ag NPs were removed from the functional solution and immersed into the releasing solution. The NPs was then sonicated for 30-120 s to remove them from the substrates into the releasing solution. For a 2-4 mL releasing solution, 4-8 glass substrates were used in order to reach a NP concentration suitable for optical property measurements.

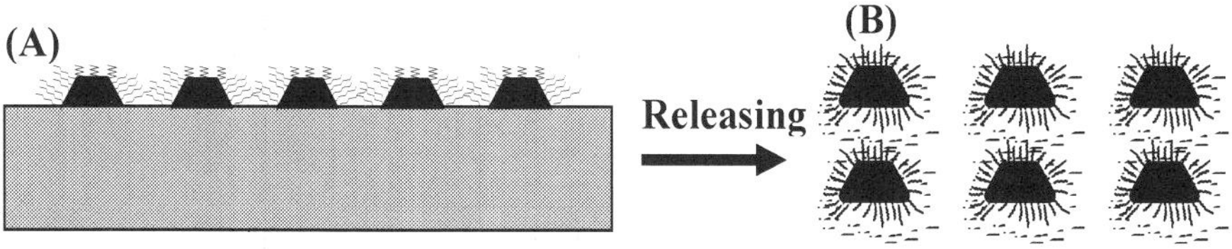

Fig. 14. (A) the surface-confined triangular Ag NPs are functionalized by chemicals with thiol groups and (B) can be further released into water or other solvents forming solution-phased nanocolloids. (Adapted from reference Y. Song, China Patent, Appl. No. CN200910085973.9)

Comparing with those Ag NPs via the traditional dislodging method(Amanda, Zhao et al. 2005; Song and Elsayed-Ali 2010), the shape integrity of the heat-treated NPs after releasing them into water can be retained perfectly. Figure 15(a) shows TEM image of the Ag NPs after thermal annealing without pre-sonication. Most of those Ag NPs show triangular shapes with rounded tips (doted circles in Fig. 15(a)) and some with snipped tips (dashed circles in Fig. 15(a)). The inset is a magnified image of these NPs, clearly showing a triangular shape with rounded tips. The histogram for these Ag NPs (Fig. 15(b)) gives a mean size of 39.6 ± 4.9 nm with much narrower size distribution of STDEV % = 12.4 % than those obtained from surface-confined Ag NPs without any post annealing (STDEV % = 41.7 %)(Song and Elsayed-Ali 2010). Fig. 15(c) is a TEM image for Ag NPs that were thermally annealed after removing two tips by sonication, whose histogram gives a mean size of about 33.9 ± 6.8 nm (Fig. 15(d)), less than that for those triangular shaped NPs with rounded tips after post-annealing. Most of these NPs show quadrilateral shapes (dashed circles) or pentagon shapes as shown more clearly in the inset of Fig. 15(c). These NPs have a similar shape as those observed by AFM images in Figs. 6(c-1) and (c-2). From the TEM images in Fig. 15(c), some of the NPs give less contrast in their central parts (NPs labeled by dashed circles). We believe that the lighter centers in these NPs are from a thinner center resulting from adhesion of the center of these NPs to the glass substrate during annealing. AFM observation of the glass substrate after removal of the NPs show debris forming hexagonal shaped arrangements. This observation is consistent with adhesion of the central part of the triangular nanoprisms to the substrate.

Variations in the shape, surface modification and surrounding environment of these NPs give significant shifts in their UV-vis absorption spectra for the surface confined NPs before and after tip rounding, after surface modification, and after dislodging into water, as shown in Figure 16. The absorption spectrum for the surface-confined Ag NPs fabricated by NSL without any tip rounding and surface modification has two distinct peaks at 476 nm and 672 nm (Figure 16a). The absorption peak at 476 nm is primarily from the higher-order mode surface plasmon resonance (e.g., quadrupole) of the NPs, and the peak at 672 nm is mainly from the dipole resonance of the NPs. We note that the higher-order resonance peak has almost the same intensity as that for the dipole resonance for all types of NPs, although the higher order modes are expected to be much weaker than the dipole resonance. Since the substrate is continuously covered by a hexagonally arranged array of Ag NPs with tip-tip distance less than 100 nm, we postulate that the particle-particle coupling will contribute to the LSPR spectrum. This particle-particle interaction effect could be responsible for the observed spectrum. When the tips in the Ag triangular nanoprisms are rounded, the tip-tip

LSPR coupling effects are alleviated, as indicated by the disappearance of the peak at 672 and the red-shift of the peak at 476 nm to 504 nm representing the higher-order surface plasmon resonance mode (Figure 16b). The absorption spectrum for the surface modified Ag NPs (Fig. 16c) shows a slight blue shift at the peak of 476 nm (to 470 nm) and a significant blue shift at 672 nm (to 626 nm) with reduced intensities. This spectrum was expected to give a red shift due to the increased dielectric constant from the adsorbed thiol compounds(Amanda, Zhao et al. 2005). We attribute this blue shift to shape variation (e.g., increased height, smooth surface topography) during surface modification by immersion that was similar to solvent annealing which results in blue-shift of LSPR since any solvent annealing has not been done on our NPs(Jensen, Duval Malinsky et al. 2000; Malinsky, Kelly et al. 2001). These variations have been observed by the slightly reduced NP size and rounded shapes observed in the TEM image of Figure 15 when compared with the AFM image of Figure 1 and Figure 6. In addition, when the Ag NPs are covered by thiol groups, the surface free electron density may be reduced, leading to weaker surface plasmon resonance in single NPs and surface plasmon resonance coupling among nanoparticle arrays(Kelly, Coronado et al. 2003). This will result in a blue shift of the LSPR peak and a reduced LSPR intensity.

The UV-vis absorption spectrum of the Ag NPs after release in water, shown in Fig. 16d, was compared to other surface confined NPs. The aqueous Ag NPs give a main peak at 532 nm and a very weak peak at 352 nm. The main peak at 532 nm appears to be from LSPR by the triangular nanoprisms with rounded tips and is blue shifted from that obtained for NPs with a LSPR peak at 605 nm fabricated by the routine NSL and released from the surface. This is attributed to the reduced size and rounded tips. The peak at 352 nm in Fig. 16d becomes much weaker and narrower than that for the aqueous Ag NPs released from the surface confined Ag NPs as fabricated by routine NSL, obviously due to the shape variation of NPs and almost no small spherical shaped debris observed in the aqueous Ag NPs released from the surface-confined Ag NPs fabricated by the modified NSL (Figure 16a). By comparing the TEM images for the two kinds of Ag NPs, it can be deduced that the peak at 352 nm in Fig. 16d is mainly from the out-of-plane quadrupole resonance of Ag nanoprisms with rounded tips according to the previous investigation(Jin, Cao et al. 2001; Amanda, Zhao et al. 2005; Zhang, Li et al. 2005). The peak intensity ratio between the main peak at 532 nm and the weak peak at 352 nm for these NPs is ~11.5:1 (after subtracting the background), which is much higher than that for the NPs obtained by the routine NSL and releasing process (1:3.6)(Song and Elsayed-Ali 2010). Clearly, the number of the small debris caused by the sonication is greatly reduced using the modified NSL and releasing process. The modified NSL process favors the formation of uniform Ag NPs with rounded tips with significant reduction in Ag debris, as shown in Fig. 15. In addition, 1-OT and 11-MUA can be substituted by the combination of 1-BT and TP, or MCH and MUA if more water-soluble NPs are desired.

Clearly, Ag NPs with controlled shapes and reduced defect density can be fabricated by a modified NSL process. Upon dislodging these NPs into a solution, they retain their shapes significantly better than NPs produced by routine NSL. Thus, aqueous phase Ag NPs with relatively uniform size and shape distribution can be fabricated. The UV-vis absorption spectra for surface confined NPs show two distinct absorption peaks (Figure 16a), comparing with those with rounded tips (Figure 16b). After surface modification, the central wavelengths of the two absorption peaks blue shifted and showed reduced intensities. The aqueous phased Ag NPs produced by the modified NSL method show a main peak and another peak with very low intensity attributed mainly to small debris produced during the dislodging process. The noticeable reduction in the intensity of the short wavelength peak for the modified NSL method compared to the routine method is due to the significant reduction in Ag debris. TEM images show that the uniformity of Ag NPs can be improved significantly by the modified NSL and releasing processes.

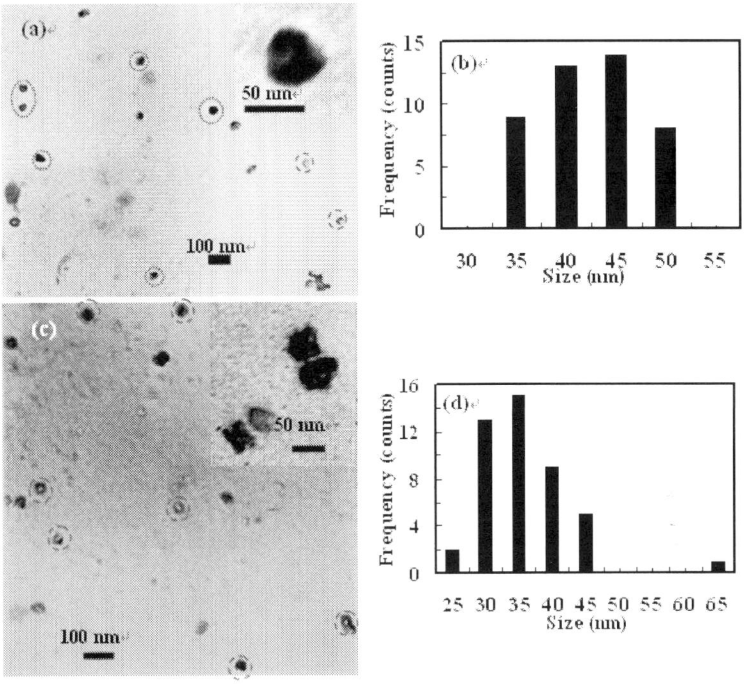

Fig. 15. TEM images of the aqueous phase Ag NPs after surface modification by thiol compounds and dislodging from the glass substrate. (a) Triangle Ag NPs with rounded tips. Dashed circles: Ag NPs with rounded tips; dotted circle: Ag NPs with slightly rounded tips. (b) Histogram of triangular shaped Ag NPs with rounded tips based on 45 NPs giving a mean size of 39.6±4.9 nm. (c) Quadrilateral and pentagon shaped Ag NPs. Dashed circles: some typical Ag NPs with quadrilateral shapes. (d) Histogram of quadrilateral and pentagon shaped Ag NPs based on 45 NPs giving a mean size of 33.9±6.8 nm. (Reprinted from Song et al., Appl. Surf. Sci. 2010 256, (20), 5961, Figure 4. Copyright (2010) Elsevier.)

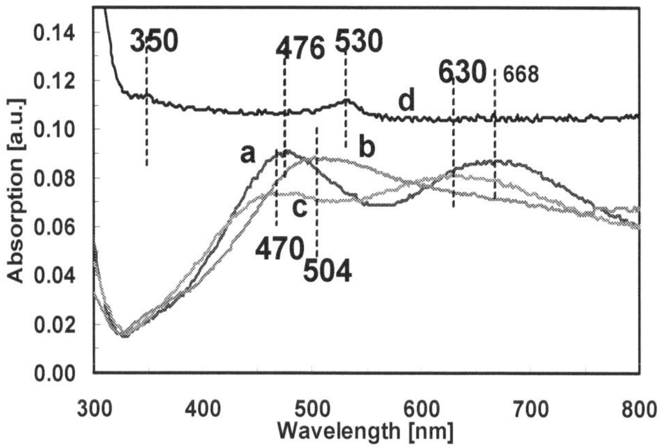

Fig. 16. UV-vis optical absorption of Ag NPs. (a) Surface confined Ag NPs before tip rounding and surface modification. (b) Surface confined Ag NPs after tip rounding (c) Surface confined Ag NPs with rounded tips after surface modification with thiol. (d) Aqueous phase Ag NPs after releasing the surface confined NPs into water. (Adapted from Song et al., Appl. Surf. Sci. 2010 256, (20), 5961, Figure 5. Copyright (2010) Elsevier.)

7. Perspective for the NSL in the controlled fabrication of nanomaterials

The great progress in controlled synthesis/fabrication of noble metal NPs by NSL, and the increase in the experimental and theoretical achievements in control of their size, shape, surface morphology and 3-dimensional space orientation dependent physicochemical properties and functions suggest expanding application in many fields because of the potential for essential breakthroughs by researchers and engineers for more advanced applications of NSL. Particularly, the developed multi-step angle resolved NSL and the modified NSL incorporated with suitable post-treatments have enabled us to obtain uniform surface-confined overlapped and nano-gapped nanostructures, the tip-rounded triangular nanoprisms, the square-shaped and the trapezoidal nanoprisms, besides the common triangular nanoprisms. Besides the marvelous progresses in the surface-confined nanostructures fabrication, a modified NSL process has also been developed to dislodge these uniform nanomaterials into the desired solvents (e.g. water, ethanol) without any obvious agglomeration as in the solution-phased nanocolloids synthesis.

Progresses in the incorporation of NSL with other LIGA techniques have shown that the suitability and ability in the architecture and interspacing controlled fabrication of NPs and nanoarrays. Their applications can thus be expanded extremely. When the multi-hierarchy arrayed micro windows fabricated by the traditional UV-LIGA process is joined in NSL, one powerful method for single nanoparticle identification will be born, resulting in the possibility of the precise investigation of the 3D morphology dependent LSPR of nanoparticles and LSPR coupling in nanoarrays. By collaboration with UV-LIGA microfabrication, uniform noble metal nanoparticles or nanoarrays can be fabricated into the targeted micro channels, leading to a much sensitive optical biosensing system after their surfaces are modified by the traditional functionalization process. By combination of PAA-LIGA and NSL, the possibility for building hierarchically ordered multi-segment nanowires or nanorods will be realized conveniently.

Summarizing from the recent progresses and discussion on NSL presented in this chapter, four main researches thrust that includes several active and challenging topics may form the primary research focuses and directions in this particular field. One is the fabrication technique development for the formation of monolayer of nanospheres with uniform area as large as several centimeter squares, which founds the basis of NSL. Another is the convenient and practical process in the releasing of these surface confined nanomaterials into solvent with perfectly retained 3D morphologies, which is still challenging but a desired alternative to obtain the uniform nanomateirals besides the well-developed wet chemical process. The third is the advanced incorporation of NSL with other fabrication techniques besides LIGA processes for the building more complex 3D hierarchically ordered nanostructures, which will definitely make a breakthrough in the nanoscale device and assemble development. The fourth may be the fabrication of tunable hetero-structure-composition nanocomposites, such as sandwich discs or multi-layer nanostructures, which will produce hetero-nanostructures with multi-functions (e.g. magntic, optical, electronic, etc). Consequently, outcomes of these challenging researches will result in the discovery of many exciting and versatile techniques for nanomaterials fabrication, and theoretical breakthrough in their novel physicochemical properties and for advanced applications.

8. Acknowledgement

The author appreciates the support from the basic research Vision Funds (YWF-11-03-Q-002) and, Chinese Scholarship Council (File No. 2010307428) NSFC (Grant No. 50971010).

9. References

Ahmadi, T. S.; Wang, Z. L.; et al. (1996). "Shape-Controlled Synthesis of Colloidal Platinum Nanoparticles." Science 272: 1924-1927.

Amanda, H. J.; Zhao, J.; et al. (2005). "Solution-Phase, Triangular Ag Nanotriangles Fabricated by Nanosphere Lithography." J. Phys. Chem. B 109(22): 11158-62.

Brayner, R. (2008). "The toxiccological impact of nanoparticles." Nano Today 3(1-2): 48-55.

Chong, M. A. S.; Zheng,Y. B.; et al. (2006). "Combinational template-assisted fabrication of hierarchically ordered nanowires arrays on substrates for device applications." Appl. Phys. Lett. 89: 233104-1-3.

Daniel, M.-C. and Astruc, D. (2004). "Gold Nanoparticles: Assembly, Supramolecular Chemistry, Quantum-Size-Related Properties, and Applications toward Biology, Catalysis, and Nanotechnology." Chem. Rev. 104: 293-346.

Dhar, P.; Cao, Y.; et al. (2007). "Autonomously Moving Local Nanoprobes in Heterogeneous Magnetic Fields." J. Phys. Chem. C 111: 3607-3613.

Edwards, H. W. and Petersen, R. P. (1936). "Reflectivity of evaporated silver films." Phys. Rev. 9: 871.

Erhardt, D. (2003). Materials conservation: Not-so-new technology Nature Materials 2: 509-510.

Haes, A. J.; Zou, S.; et al. (2004). "A Nanoscale Optical Biosensor: The Long Range Distance Dependence of the Localized Surface Plasmon Resonance of Noble Metal Nanoparticles." J. Phys. Chem. B 108: 109-116.

Hammond, C. R. (2000). The Elements, in Handbook of Chemistry and Physics 81th edition, CRC press.

Haynes, C. L.; Mcfarland, A. D.; et al. (2003). "Nanoparticle Optics: the Importance of Radiative Dipole Coupling in Two-Dimensional Nanoaprticle Arrays." J. Phys. Chem. B 107: 7337-7342.

Haynes, C. L. and van Duyne, R. P. (2001). "Nanosphere Lithography: A Versatile Nanfabrication Tool for Size Dependent Nanoparticle Optics." J. Phys. Chem. B 105: 5599.

Hong, B. H.; Bae, S. C. et al. (2001). "Ultrathin single-crystalline silver nanowire arrays formed in an ambient solution phase." Science 294: 348-351.

http://en.wikipedia.org/wiki/Silver.

Hulteen, J. C.; Treichel,D. A.; et al. (1999). "Nanosphere Lithography: Size-tunable silver nanosparticle and surface cluster arrays." J. Phys. Chem. B 103: 3854-3863.

Jain, P. K.; Huang, X.; et al. (2007). "Review of some interesting surface plasmon resonance-enhanced properties of noble metals nanoparticles and their applicaitons to biosystems." Plasmonics 2: 107-118.

Jensen, T. R.; Duval Malinsky, M.; et al. (2000). "Nanosphere Lithography: tunable localized surface plasmon resonance spectra of silver nanoparticles." J. Phys. Chem. B 104: 10549-10556.

Jin, R.; Cao,Y. W.; et al. (2001). "Photoinduced conversion of silver nanospheres to nanoprisms." Science 294: 1901-1904.

Kelly, K. L.; Coronado, E.; et al. (2003). "The optical proeprties of metal nanoparticles: the influence of size, shape and dielectric environment." J. Phys. Chem. B 107(3): 668-677.

Kreibig, U. and Vollmer, M. (1995). Optical Properties of Metal Clusters. Berlin, Springer.

Lee, S. J.; Morrill, A. R.; et al. (2006). "Hot spots in silver nanowire bundles for surface-enhanced Raman spectroscopy." J. Am. Chem. Soc. 128: 2200-2201.

Lombardi, I.; Cavallotti, P. L.; et al. (2007). "Template assisted deposition of Agnanoparticle arrays for surface-enhanced Raman scattering applications." Sensors and Actuators B: Chemical 125: 353-356.

Malinsky, M. D.; Kelly, K. L.; et al. (2001). "Chain length Dependence and sensing capabilities of the localized surface plasmon resonance of silver nanoparticles chemically modified with alkanethiol self-assembled monolayers." J. Am. Chem. Soc. 123: 1471-1482.

Maneerung, T.; Tokura, S.; et al. (2008). "Impregnation of silver nanoparticles into bacterial cellulose for antimicrobial wound dressing." Carbohydrate Polymers 72: 43-51.

Meng, G.; Jung,Y. J.; et al. (2005). "Controlled fabrication of hierarchically branched nanopores, nanotubes and nanowires." PNAS 102(20): 7074-7078.

Mock, J. J.; Barbic, M.; et al. (2002). "Shape effects in plasmon resonance of individual colloidal silver nanoparticles." J. Chem. Phys. 116(15): 6755-6759.

Mohl, M.; Kumar, A.; et al. (2010). "Synthesis of catalytic porous metallic nanorods by galvanic exchange reaction." J. Phys. Chem. C 114: 389-393.

Noguez, C. (2007). "Surface Plasmons on Metal Nanoparticles: the Influence of Shape and Physical Environment." J. Phys. Chem. C 111(10): 3806-3819.

Ozbay, E. (2006). "Plasmonics:Merging Photonics and Electronics at Nanoscale Dimensions." Science 311: 189-193.

Pan, S. L.; Zeng, D. D.; et al. (2000). "Preparation of ordered array of nanoscopic gold rods by template method and its optical properties." Appl. Phys. A 70: 637-640.

Percival, S. L.; Bowler, P. G.; et al. (2005). "Bacterial resistance to silver in wound care." Journal of Hospital Infection 60: 1-7.

Prasad, P. N. (2004). Nanophotonics. Hoboken, New Jersey, John Wiley & Sons, Inc.

Schwartzberg, A. M. and Zhang J. Z. (2008). "Novel Optical Properties and Energing Applications of Metal Nanostructures." J. Phy. Chem. C 112(28): 10324-10337.

Shen, Y.; Friend, C. S.; et al. (2000). "Nanophotonics: Interaction, Materials and application." J. Phys. Chem. B 104: 7577-7587.

Song, Y. (2009). Fabrication of high throughput biosensors based on single Nanoparticles and Nanoparticle arrays, China Patent, Appl. No. CN 200910085973.9.

Song, Y. (2010). "Fabrication of Multi-level 3-Dimension Microstructures by Phase Inversion Process." Nano-Micro Letters 2(2): 95-100.

Song, Y. and Elsayed-Ali, H. E. (2010). "Aqueous Phase Ag Nanoparticles with Controlled Shapes Fabricated by a Modified Nanosphere Lithography and their Optical Properties." Appl. Surf. Sci. 256(20): 5961-5967.

Song , Y.; Henry, L. L.; et al. (2009). "Stable Cobalt Amorphous Nanoparticles Formed by an In-situ Rapid Cooling Microfluidic Process " Langmuir 25 (17): 10209-10217.

Song , Y.; Jin, P.; et al. (2010). "Microfluidic Synthesis of Fe Nanoparticles." Mater. Lett. 64: 1789-1792.

Song, Y.; Sun, S.; et al. (2010). "Synthesis of Worm and Chain-like Nanoparticles by a Microfluidic Reactor Process." J. Nanopart. Res. 12: 2689-2697.

Song , Y.; Zhang, Z.; et al. (2011). "Identification of Single Nanoparticles." Nanoscale 3: 31-44.

Su, K.-H.; Wei, Q.-H.; et al. (2003). "Interparticle Coupling Effects on Plasmon Resonances of Nanogold Particles." Nano lett. 3(8): 1087-1090.

Vo-Dinh, T.; Wang, H.-N.; et al. (2009). "Plasmonic nanoprobes for SERS biosensing and bioimaging." J. Biophoton.: 1-14.

Xu, C.-l.; Zhang, L.; et al. (2005). "Well-dispersed gold nanowire suspension for assembly applicaiton." Appl. Surf. Sci. 252: 1182-1186.

Xu, Q., Meng, G.; et al. (2009). "A Generic Approach to Desired Metallic Nanowires Inside Native Porous Alumina Template via Redox Reaction." Chem. Mater. 21: 2397-2402.

Yang, Y.; Matsubara, S.; et al. (2007). "Solvothermal synthesis of multiple shapes of silver nanoparticles and their SERS properties." J. Phys. Chem. C 111: 9095-9104.

Zhang, J.; Li, X. et al. (2005). "Surface enhanced raman scattering effects of silver colloids with different shapes." J. Phys. Chem. B 109: 12544-12548.

Zhang, X.; Whitney, A. V.; et al. (2006). "Advances in Contemporary Nanosphere Lithographic Techniques." J. Nanosci. Nanotech. 6: 1920-1934.

Zhao, L. L.; Kelly, K. L.; et al. (2003). "The extinction spectra of Ag nanoparticle arrays: influence of array structure on plasmon resonance wavelength and widths." J. Phys. Chem. B 107: 7343-7350.

Zhou, Q.; Qian, G.; et al. (2008). "Two-dimentional assembly of silver nanoparticles for catalytic reduction of 4-nitroaniline." Thin Solid Films 516: 953-956.

Zhu, S.; Li, F.; et al. (2008). "A localized surface plasmon resonance nanosensor based on rhombic Ag nanoparticle array." Sensors and Actuators B: Chemical 134: 193-198.

26

A Feasible Routine for Large-Scale Nanopatterning via Nanosphere Lithography

Zhenyang Zhong, Tong Zhou, Yiwei Sun and Jie Lin
Fudan University/Department of Physics Shanghai, China

1. Introduction

Due to the unique electronic, optical, catalytic and biological properties, well ordered nanostructures have attracted enormous interest. They have potential applications in photonic crystal devices (Yablonovitch, 1987), large-density magnetic recording devices (Chou et al., 1994), novel electronic devices (Schmidt & Eberl, 2001), synthesis of DNA electrophoresis mediate (Volkmuth & Austin, 1992) , nanocontainers (Chen et al., 2008), surface-plasmon resonance biosensors (Brolo et al., 2004), antireflective coatings for solar cells (Yae et al., 2005), and etc. Such broad applications of nanostructures were intimately associated with their unique properties, which are sensitively dependent on their size and/or shape. It is well-established that magnetic (Shi et al., 1996 ; Zhu et al.,2004), optical (Aizpurua et al., 2003; Larsson et al. 2007), electrocatalytic (Bratlie et al., 2007 ; Narayanan & El-Sayed, 2004), optoelectronic (Chovin et al., 2004), data storage (Ma, 2008), thermodynamic (Volokitin et al., 1996; Wang et al., 1998) and electrical transport (Andres et al., 1996 ; Bezryadin et al., 1997) properties of the nanostructures are affected by the shape and the size, as well as the interfeature spacing.

In general, there are two approaches to realize ordered nanostructures with desired size, shape and arrangement. One is the "bottom up" approach on pre-patterned substrates (Zhong et al, 2007; Zhong et al., 2008). The other is the "top-down" approach (Ito & Okazaki, 2000). Both of these two approaches are always based on lithographic technology. In the first approach, lithographic techniques were employed to fabricate various patterned substrates, on which ordered nanostructures can then be realized by subsequent growth of desired materials. The main reason for this approach is to suppress defects in the nanostructures. In the second approach, ordered nanostructures can be directly fabricated by lithographic techniques. Several standard lithographic techniques are frequently exploited to fabricate desired surface nanostructures, including holographic lithography, electron-beam lithography (EBL) and ion-beam lithography (IBL) (Arshak et al., 2004 ; Ebbesen et al., 1998; Ito & Okazaki, 2000). Recently, a new extreme ultraviolet (EUV) lithography was developed, which is a potential candidate for achieving critical dimensions below 100 nm (Service, 2001). In addition, there are some other lithographic techniques applied in the fabrication of nanostructures (Haynes & Van Duyne, 2001). However, fabrication of nanostructures in a regular arrangement over large areas is still a major challenge in modern nanotechnologies. There is substantial interest in developing new technologies to facilitate pattern fabrication.

2. Nanosphere lithography (NSL)

Enormous efforts have been devoted to investigate alternative nanolithography approaches. One of promising methods is nanosphere lithography (NSL) (Fuhrmann et al., 2005; Haynes & Van Duyne, 2001; Hulteen & Van Duyne, 1995; Kosiorek et al., 2004; Sinitskii et al., 2007), which is a highly accessible, low cost, parallel fabrication process capable of producing nanostructured surfaces over large areas and with high resolution. In NSL, self-assembled nanospheres can be served directly as ordered nanostructures (Park et al., 1998) or a mask for the subsequent fabrication of nanostructures, which can be realized by deposition of desired materials, or by etching on desired substrates. A rich varieties of ordered nanostructures have been achieved by NSL, such as triangular structures (Winzer et al., 1996), metallic rings (Boneberg et al., 1997), nanopillars (Weeks et al., 2004), and multilayer with modified topography (Albrecht et al., 2005), nanodots (Chen et al., 2009; Weekes et al., 2007), 3D nanostructure (Zhang et al., 2007), discs (Hanarp et al., 2003) and nanoscale crescents (Gwinner et al., 2009; Retsch et al., 2009; Vogel et al, 2011). The shape, the size and the arrangement of ordered nanostructures can be readily controlled in combination of NSL and the subsequent deposition of desired materials (Haynes & Van Duyne, 2001; Zhang et al., 2007; Vogel et al, 2011).

2.1 Main features of NSL

To obtain desired nanostructures by NSL, one monolayer of self-assembled nanospheres is always obtained first and served as a mask for the subsequent fabrication of nanostructures. The material of the nanosphere can be nanoscale polystyrene (PS), SiO_2, or polydimethylsiloxane (PDMS) (Choi et al, 2009), and etc. The shape of the resulting pattern is most often spherical. Using PDMS, the shape and the feature size of the pattern can be modulated by changing the stretching axis and ratio of the PDMS replica. The nonspherical shaped patterns, such as rectangular or elongated hexagonal shaped patterns, can then be obtained (Choi et al, 2009). An additional noteworthy feature of the PDMS is that different pattern can be produced from a single PDMS replica mold (Choi et al, 2009). In addition, binary nanospheres composed of two different-size colloidal particles can be self-assembled both in hexagonal lattices via a two-step process (Kim et al., 2009), forming binary colloid crystals (BCCs). Such BCCs may have potential applications in the fabrication of photonic crystal structures, theoretical models of phase transition, and templates of inverse structure (Kim et al., 2009). The arrangement of self-assembled nanospheres can be close-packed or non-close-packed (Vogel et al., 2011). In general, the self-assembled nanospheres are arranged in a hexagonal lattice. Using more sophisticated processes, squarely ordered array of nanospheres can also be realized, which is speculated to be metastable structures between more stable hexagonal structures (Sun et al., 2009).

In the simplest NSL, a monolayer of close-packed nanospheres in a hexagonal lattice is first obtained on the substrate, which can be served as a mask for the subsequent deposition of desired materials. For generally vertical deposition, the three-fold interstices allow deposited material to reach the substrate, giving rise to an array of triangularly shaped nanoparticles with P6mm symmetry (Haynes & Van Duyne, 2001). The perpendicular bisector of the triangular nanoparticles, a, and the interparticle spacing, d_{ip}, are proportional to the nanosphere diameter, D, which can be simply calculated by,

$$a = \frac{3}{2}(\sqrt{3} - 1 - \frac{1}{\sqrt{3}})D, \quad d_{ip} = \frac{1}{\sqrt{3}}D$$

If double layer of nanospheres are employed as the mask, both a and d_{ip} will be changed (Haynes & Van Duyne, 2001). In addition, circular shaped interstice particles are frequently obtained in the case of small nanospheres mainly because some materials are not perpendicularly deposited in the interstices, and the general hot materials can diffuse in some region. More interestingly, for angle resolved deposition, some particularly-shaped nanostructures can be realized, such as nanochain structures (Haynes & Van Duyne, 2001) and nanocrescents (Vogel et al., 2011). If the monolayer of nanospheres is non-close-packed, more complex nanostructures can be obtained by changing the incidence angle of the material vapor beam and the azimuth angle of the vapor beam with respect to the normal direction of the nanospheres mask (Zhang et al., 2007).

The critical step of NSL is to form monolayer of ordered nanospheres on desired substrates. Several methods have been developed to form regularly arranged nanospheres on substrates, including transferal coating (Weekes et al., 2007), vertically dipping coating (Choi et al., 2009), spin coating (Hulteen & Van Duyne, 1995), drop coating (Hulteen et al., 1999), and thermoelectrically cooled angle coating (Micheletto et al., 1995). All of these formation methods are based on the ability of the nanospheres to freely diffuse to seek their lowest energy configuration. The diffusion processes and the interaction among nanospheres can be influenced by chemically modifying the nanosphere surface with a negatively charged functional group such as carboxylate or sulfate. Such a modification of the surface features of nanospheres can be easily realized for polystyrene (PS) spheres (Weekes et al., 2007). The self-assembled monolayer of nanosphere masks always include a variety of defects that arise as a result of nanosphere polydispersity, site randomness, point defects (vacancies), line defects (slip dislocations), and polycrystalline domains (Haynes & Van Duyne, 2001). These defects are always remained in the finally obtained nanostructures, which will degrade the properties of the ordered nanostructures. Therefore, it is important to try to get rid of those defects in the monolayer of self-assembled nanospheres.

2.2 NSL based on transferal coating

It was found that the transferal coating is much easier in operation to obtain ordered nanospheres in large areas than the other methods. The domain size of ordered PS spheres can be up to 1 cm^2 (Weekes et al., 2007). The key step of the transferal coating is to self-assemble highly ordered monolayer of PS spheres at the interface between water and air (or oil). In general, the suspensions (1-10 wt%) of PS spheres in de-ionized (DI) water are diluted in a 1:1 ratio in some spreading agent, such as ethanol or methanol. Drops of the diluted suspension of PS sphere are then introduced into water surface via a titled glass from a pipet. On contact with the water, the PS spheres immediately form a momolayer and start to assemble. The inherent mechanism for the ordering of PS spheres at a liquid interface has been studied by several groups (Aubry & Singh, 2008; Boneva et al., 2009; Larsen & Grier, 1997; Nikolaides et al., 2002; Pieranski, 1980; Trau et al., 1996; Yeh et al., 1997). A reasonable model has been provided to account for the ordering of PS spheres at the interface (Nikolaides et al., 2002). It was suggested that the ordering arrangement of PS spheres resulted from the balance between an electrostatic repulsion and an additional capillary attraction among PS spheres. The former is originated from the negative charges on PS spheres (Weekes et al, 2007). The latter is due to the deformation of liquid meniscus by electrostatic stresses at interface. Both of these two forces were associated with an electric dipolar field, which resulted from an asymmetric charge distribution on particles at the interface due to mismatch in dielectric constant of adjacent fluids. Such a creation of the attractive capillary force is crucial because spheres with diameters of less than 5 μm generally do not have sufficient weight to deform the liquid meniscus by means of gravity (Kralchevsky & Nagayama, 2000).

Considering the importance of the electrical field in self-assembling PS spheres at interface, it is natural to find ways to change the electric field to control the arrangement of PS spheres. One way is to apply an external electric field during the self-assembly of PS spheres at the interface (Aubry & Singh, 2008; Boneva et al., 2009; Nikolaides et al., 2002; Trau et al., 1996). This external electrical field will change the charge distribution on the surface of PS Spheres, leading to the change of the electric dipole field around PS spheres. It may also exert an additional electric force on the negatively charged PS spheres, which will affect the deformation of the surface. In a word, the external electric field will affect both the attraction force and the repulsion force mentioned above. It has been found that the inter-particle distance can be remarkably changed by the electric field (Nikolaides et al., 2002). As a result, the ordering of self-assembled PS spheres can be improved considerably, as shown in Fig. 1. In addition, the effect of an external electric field on the arrangement of PS spheres is more pronounced for smaller spheres.

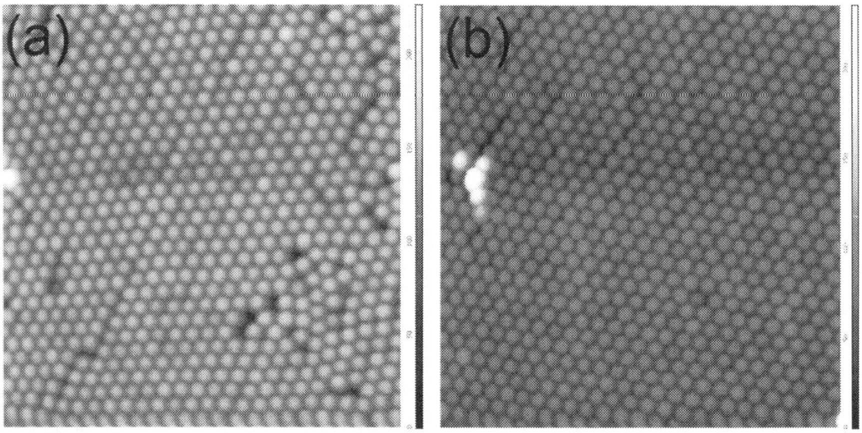

Fig. 1. Self-assembled monolayer of PS spheres (5 x 5 μm^2) with an external electric field of, (a) 0 v/m, (b) 5 x 10^4 v/m. The diameter of PS sphere is 200 nm. The electric field is upward and perpendicular to the interface of water and air. The external electric field can efficiently reduce defects in the regular arrangement of PS spheres, remarkably improving the ordering of the self-assembled monolayer of PS spheres.

Another way to modify the balance between the attraction force and the repulsion force among PS spheres is to change the surface chemistry of PS spheres or the electrostatic environment of the water-air interface (Sirotkin et al., 2010). By adding some electrolyte, e.g. acid (H_2SO_4) and NaCl, into the water, effective surface charge density of PS spheres and /or effect of electric screen of PS spheres can be changed, which give rise to the change of the interaction among PS spheres. It has been found that the ordering of self-assembled PS spheres at interface can be considerably improved by adding suitable acid (H_2SO_4) in water (Sirotkin et al., 2010), as shown in Fig. 2. Given that the charges on PS spheres are related to the diameter of PS spheres, the suitable amount of acid or other electrolyte is dependent on the size of PS spheres. In addition, the temperature of the water to some degree also affect the self-assembly of the PS spheres at the interface. It was found that ordering of PS was improved on the water of ~ 4 °C. Such an improvement may be related to the increase of water surface tension and the suppression of the Brownian motion of the PS spheres and dust clusters in the water.

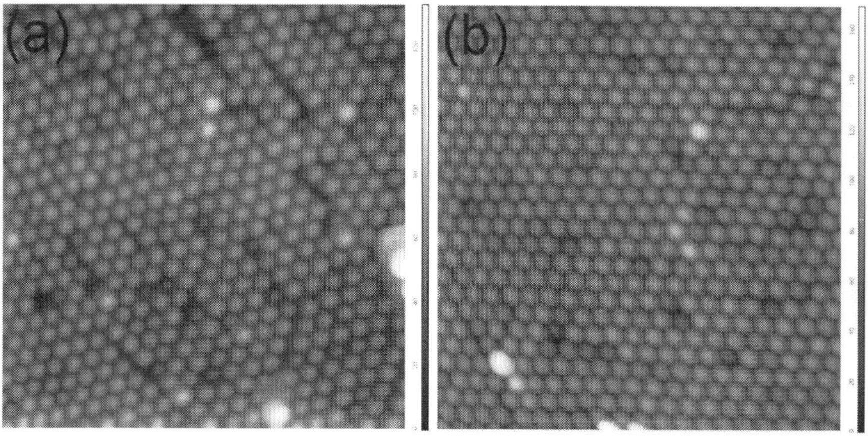

Fig. 2. Self-assembled monolayer of PS spheres (5 x 5 μm2) at the interface of air and the solution of de-ionized water and H_2SO_4 with PH value of, (a) ~7, (b) ~ 5.3. The diameter of PS sphere is 240 nm. The H_2SO_4 can provide some additional ions around PS spheres, which can effectively change the interaction among PS spheres. Under certain PH value of the solution, considerable improvement of ordering of PS spheres can be made.

The self-assembled monolayer of PS spheres can be finally transferred to varieties of smooth substrates underneath the water by draining off the water. This process can be affected by some charges on the substrates. In addition, some cracks may appear once the monolayer was disturbed by movements of the water during draining. The PS spheres nearby the cracks slightly displaced from the ideal sites of a hexagonal lattice. In this case, the long-range ordering of the subsequent structure is degraded. The monolayer of ordered PS spheres on substrates can then serve as a mask for the subsequent fabrication of ordered nanostructures by deposition of varieties of materials or by etching.

2.3 Periodic pit-pattern obtained by NSL and chemical etching

Periodic pit-pattern can be obtained in combination of NSL and selective chemical etching (Chen et al., 2009). The processes mainly involves three steps: (i) self-assembling monolayer of PS spheres on hydrogenated Si surface; (ii) forming a novel net-like Au-Oxide mask via Au catalyzed oxidation; (iii) resulting in periodic pits by selective chemical etching of Si in KOH solution.

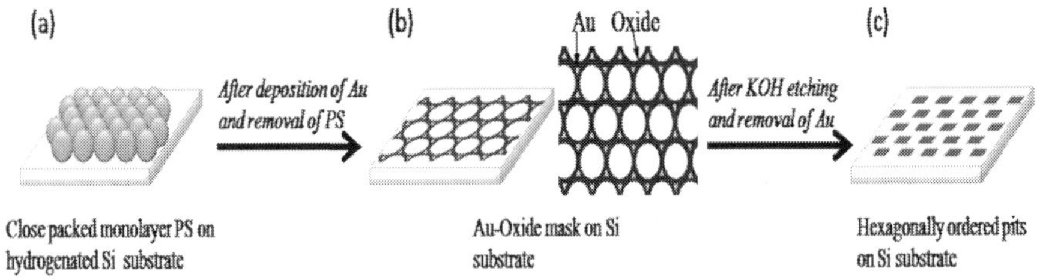

Fig. 3. Schematic illustration for the fabrication of ordered pit-pattern using nanosphere lithography.

The experimental processes are schematically illustrated in Figure 3. It starts with self-assembling monolayer of PS spheres onto hydrogenated Si (001) substrates. The PS spheres with different diameters of 100nm, 200nm, 500nm, 600nm or 1.6 μm were used. All PS sphere suspensions were purchased from Duke Scientific Corporation. The Si (001) substrates were chemically cleaned and hydrogenated by a subsequent HF dip. The close-packed monolayer of PS spheres was self-assembled on the surface of DI water as mentioned above, which was then transferred onto Si (001) substrates immersed in DI water by draining away the DI water. Fig. 4 shows a large-area highly ordered PS spheres on a hydrogenated Si substrate.

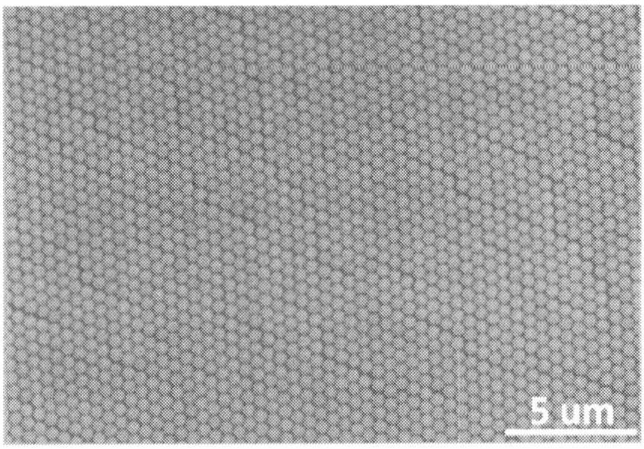

Fig. 4. Representative SEM image of self-assembled monolayer of PS spheres with a diameter of 500 nm on a hydrogenated Si (001) substrate.

The ordered PS spheres in a hexagonal lattice on Si substrate then serves as a mask for thermal evaporation of Au. After deposition of about 1 nm Au onto the PS spheres covered substrates, six Au particles around each PS spheres on Si surface were obtained. Because Au is directly deposited onto the Si surface without SiO_2, Si adjacent to Au particles then electrochemically oxidizes, or anodizes upon exposure to air (Robinson et al., 2007). As a result, an Au-oxide mask was naturally formed, as illustrated in Fig. 3(b). To avoid oxidation of the Si surface underlying PS spheres, the samples were immediately rinsed in tetrahydrofuran to remove PS spheres and then etched in 20% KOH solution at room temperature. Since Au-oxide mask protects the underlying Si from KOH etching, well ordered and uniformed 2D pit-pattern was then formed.

The periodicity and the size of the pits were determined by the diameter of PS spheres. The period of the patterned pits is in fact equal to the diameter of PS spheres. This indicates that the period of ordered pits can be readily changed by using PS with corresponding diameter, as demonstrated in Fig 5. In addition, the lateral size of inverted-pyramid-like pits was essentially decreased linearly with the diameter of PS, as shown in Fig.6. This is mainly because the area underlying the PS spheres for etching is nearly proportional to the projection of PS spheres on Si substrates. Moreover, according to the fitting line in Fig. 6, the lateral size of pits will decrease to be zero when the diameter of spheres equals to ~40nm. This result can be explained by the fact that the Au-catalyzed SiO_2 partially fill up the region underlying the PS. It means that the minimum period of patterned pits can be down to 40 nm by present method. On the other hand, any dispersion of the diameter of PS will degrade the ordering of the pits near this limit. Thus, before approaching the limit of the nanosphere lithography, more uniformed PS is required so that homogenous and ordered small pits can be obtained.

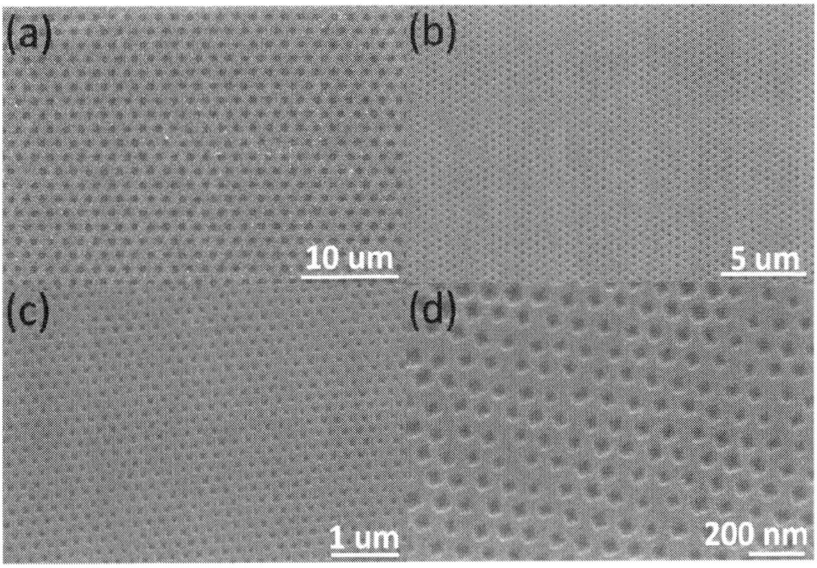

Fig. 5. Representative SEM images of pit-patterns with a periodicity of: (a) 1.6 um (after 1 minute etching), (b) 500 nm (after 3 minutes etching), (c) 200 nm (after 3 minutes etching), (d) 100 nm (after 1 minute etching) .

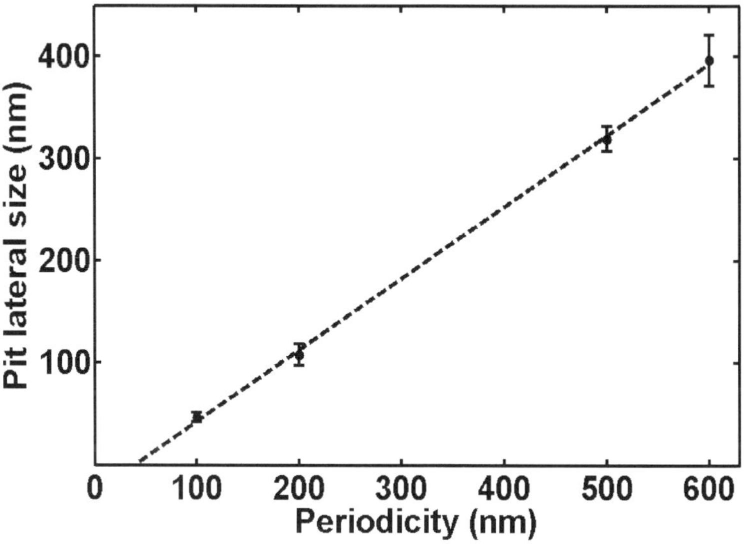

Fig. 6. Mean lateral size vs periodicities of pits. The broken curve is a fitting line.

Moreover, the shape of pits can be tuned by controlling KOH etching time. Depending on the etching time, three types of pits can be obtained. At the beginning of etching, shallow pits with rounded open-mouth (type I pits) can be obtained, as shown in Fig. 7 (a). For an intermediate etching time, inverted truncated-pyramid-like pits with {111} facets and flat (001) bottom (type II pits) are obtained, as shown in Fig. 7 (b). After sufficiently long etching time, inverted pyramid-like pits with {111} facets (type III pits) are obtained, as shown in Fig. 7 (c) and (d). Such a shape evolution with etching time is related to the anisotropic etching rate of Si by KOH solution (up to 100:1 for the etching of Si along <100> and <111> direction at room temperature). As a result, {111} facets will finally appear in the pits. The depth of type III pits approximates to be $w / 2 \times tg(54.7^{\circ})$, where w is the lateral size of pits with sidewalls of {111} facets having slope angle of 54.7°. In addition, the etching time corresponding to each type of pits depends on the periodicity of the patterned pits. As described above, a larger periodicity of patterned pits is accompanied with a larger area region for etching， which gives rise to a larger pits. It takes time to form the sidewalls of pits completely with {111} facets from their appearance. The larger pits will have larger area sidewalls, as results in longer etching time corresponding to different types of pits. Therefore, the pits with a periodicity of 1.6 μm in Fig. 5 (a) are of type I. The pits with a periodicity of 500 nm in Fig. 5 (b) are actually of type II. The pits with a periodicity of 200 nm and 100 nm in Fig. 5 (c) and (d) are of type III. It has been found that some materials growth on Si substrates was orientation-dependent (Zhang et al., 2008). Such patterned Si substrates with the coexistence of spatially ordered (001) surface and {111} facets may provide potential templates to form ordered unique nanostructures of orientation dependence.

Well ordered GeSi nano-islands were obtained by deposition of Ge on such pit-patterned Si (001) substrates using molecular beam epitaxy (Chen et al., 2009), as shown in Fig. 8(c). Such preferential formation of GeSi nano-islands within each pit is energetically favorable under the assistance of growth kinetics (Zhong et al, 2007; Zhong et al., 2008). In comparison, GeSi nano-islands on a flat substrate under identical growth conditions are random, as shown in Fig 8(d). With decreasing the periodicity of the pit-pattern by using small PS spheres, higher density of smaller GeSi nano-islands in a hexagonal lattice are expected, which can facilitate the investigation of size-dependent quantum confinement effect of nano-islands.

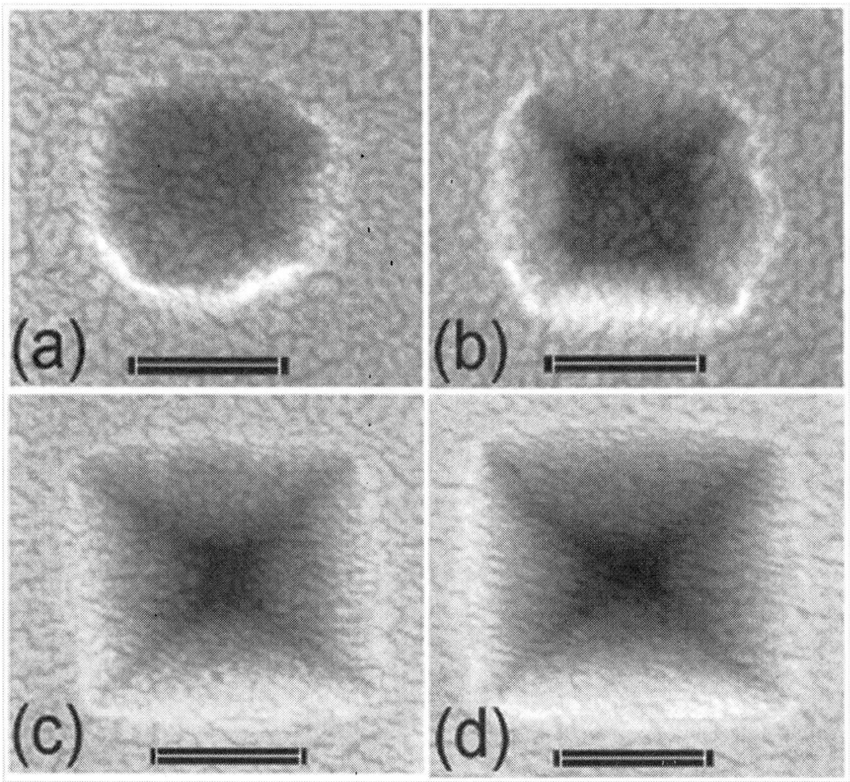

Fig. 7. SEM image of a pit (the diameter of PS spheres used is 600 nm) obtained after etching of (a) 2 minutes, (b) 3 minutes, (c) 5 minutes, (d) 7 minutes. The scale bar is 200 nm.

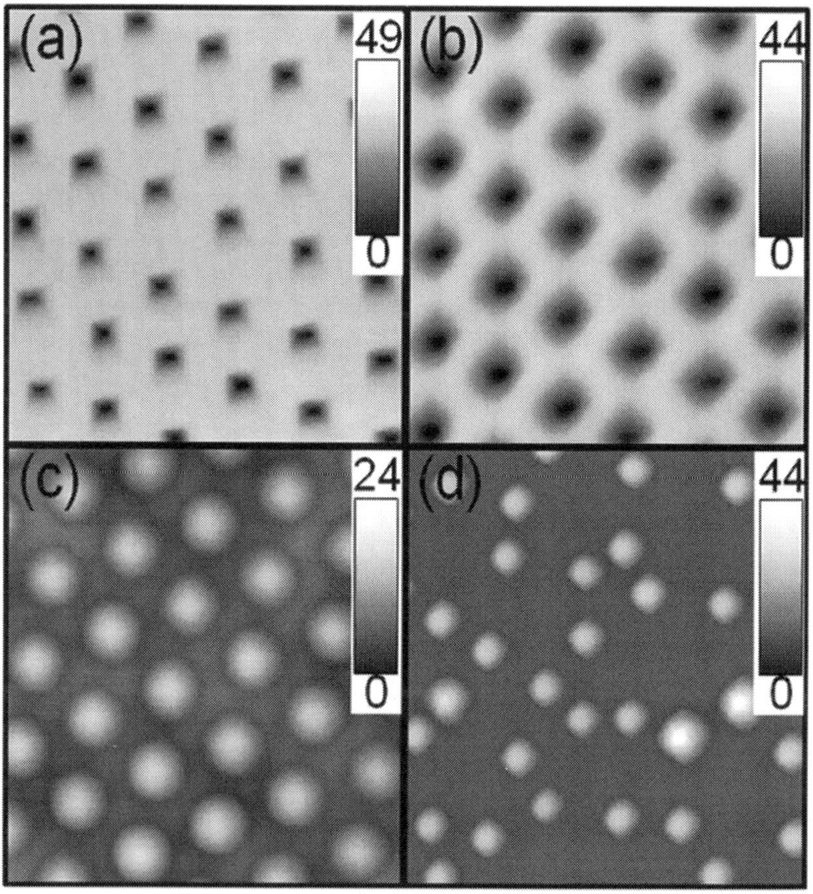

Fig. 8. AFM images (1×1 um^2) of (a) pit-pattern with a periodicity of 200 nm, (b) pit-pattern after Si buffer layer growth, (c) ordered GeSi nano-islands after deposition of 10 monolayers Ge by molecular beam epitaxy on a pit-patterned Si (001) substrate, (d) Randomly distributed GeSi nano-islands grown under the same conditions on a flat Si (001) substrate. The unit of height bar is nm.

3. Conclusions

A scalable approach to fabricate periodic nanopatterning in a large-scale area with controllable periodicity using nanospheres, so called NSL, has been developed. The NSL generally started with self-assembling monolayer of PS spheres on the substrates, which can be obtained by various methods. One potential routine to obtain ordered monolayer of PS spheres is via self-assembling PS spheres at the interface between water and air. Such a regular arrangement of monolayer PS spheres in a hexagonal lattice resulted from the balance between an electrostatic repulsion among adjacent spheres and a capillary attraction due to the deformation of liquid meniscus by electrostatic stresses. An external electrical field perpendicular to the water surface, which affected the interaction between PS spheres, could efficiently improve the ordering of PS spheres, particularly of small PS spheres. The interplay among PS spheres can also be affected by changing the surface chemistry of PS spheres or the electrostatic environment of the water-air interface, which can be readily realized by mixing some electrolytes, such as solution of H_2SO_4 or NaCl. In addition, it was found that the ordering of PS spheres was improved on the water of ~ 4 °C mainly due to the increase of water surface tension and the suppression of the Brownian motion of the PS spheres and dust clusters in the water. This ordered monolayer PS spheres could be transferred onto the substrate placed previously inside water by draining off the water. This method facilitates large-area highly ordered monolayer of PS spheres on substrates，which can act as a mask or a template for subsequent lithography to obtain ordered nano-wires or nano-pits, or for subsequent growth of desired nanostructures. Two-dimensionally ordered nanopattern with a periodicity equal to the diameter of PS spheres in the range of several micrometers to less than 100 nm could be readily obtained. The geometrical profiles of the nanopattern could be modulated by controlling the etching conditions. NSL has been exploited in fabricating ordered nano-wires and nano-dots. This technique was characterized by its low-cost，high throughput, and easy manipulation for producing large-scale periodic patterns. More interestingly，NSL can be applied to obtain nanostructures of various materials on many kinds of substrates, which will facilitate the production of varieties of ordered nanostructures.

4. Acknowledgements

This work was supported by the special funds for Major State Basic Research Project No. 2011CB925600 of China.

5. References

Aizpurua, J.; Hanarp, P.; Sutherland, D. S.;K€all, M.; Bryant, G. W.; Garcia, D.; & Abajo, F. J. (2003) Optical Properties of Gold Nanorings. *Phys. Rev. Lett.* 90, 057401 (1-4)

Albrecht, M.; Hu, G. H.; Guhr, I. L.; Ulbrich, T. C.; Boneberg, J.; Leiderer, P.; & Schatz, G. (2005) Magnetic multilayers on nanospheres. *Nat. Mater.* 4, 203-206

Andres, R. P.; Bielefeld, J. D.; Henderson, J. I.; Janes, D. B.; Kolagunta, V. R.; Kubiak, C. P.; Mahoney, W. J.; & Osifchin, R. G. (1996) Self-Assembly of a Two-Dimensional Superlattice of Molecularly Linked Metal Clusters. *Science* 273, 1690-1693

Arshak, K.; Mihov, M.; Arshak, A.; McDonagh, D.; & Sutton, D. (2004) Novel dry-developed focused ion beam lithography scheme for nanostructure applications. *Microelectron. Eng.* 73/74, 144-151

Aubry, N.; & Singh, P. (2008) Physics underlying controlled self-assmbly of micro-and nanoparticles at a two-fluid interface using an electric field. *Phys. Rev. E* 77, 56302 (1-11)

Bezryadin, A.; Dekker, C.; & Schmid, G. (1997) Electrostatic trapping of single conducting nanoparticles between nanoelectrodes. *Appl. Phys. Lett.* 71, 1273-1275

Boneberg, J.; Burmeister, F.; Schafle, C.; Leiderer, P.; Reim, D.; Fery, A.; & Herminghaus, S. (1997). The Formation of Nano-Dot and Nano-Ring Structures in Colloidal Monolayer Lithography. *Langmuir* 13, 7080-7084

Boneva, M. P.; Danov, K. D.; Christov, N. C.; & Kralchevsky, P. A. (2009). Attraction between Particles at a Liquid Interface Due to the Interplay of Gravity- and Electric-Field-Induced Interfacial Deformations. *Langmuir* 25, 9129-9139

Bratlie, K. M.; Lee, H.; Komvopoulos, K.; Yang, P.; & Somorjai, G. A. (2007). Platinum Nanoparticle Shape Effects on Benzene Hydrogenation Selectivity. *Nano Lett.* 7, 3097-3101

Brolo, A. G.; Gordon, R.; Leathem, B.; & Kavanagh, K. L. (2004). Surface Plasmon Sensor Based on the Enhanced Light Transmission through Arrays of Nanoholes in Gold Films. *Langmuir* 20, 4813-4815

Chen, P.; Fan, Y.; & Zhong, Z. (2009) The fabrication and application of patterned Si(001) substrates with ordered pits via nanosphere lithography. *Nanotechnology* 20, 095303 (1-5)

Chen, X.; Wei, X.; & Jiang, K. (2008) Large-scale fabrication of ordered metallic hybrid nanostructures. *Opt. Express* 16, 11888-11893

Choi, H. K.; Im, S. H.; & Park, O (2009) Shape and Feature Size Control of Colloidal Crystal-Based Patterns Using Stretched Polydimethylsiloxane Replica Molds. *Langmuir* 25, 12011-12014

Chou, S. Y.; Wei, M. S.; Krauss, P. R.; & Fischer, P. B. (1994) Single-domain magnetic pillar array of 35 nm diameter and 65 Gbits/in.2 density for ultrahigh density quantum magnetic storage. *J. Appl. Phys.*, 76, 6673-6675

Chovin, A; Garrigue, P.; Manek-H€onninger, I.; & Sojic, N. (2004) Fabrication, Characterization, and Far-Field Optical Properties of an Ordered Array of Nanoapertures. *Nano Lett.* 4, 1965-1968

Ebbesen, T. W.; Lezec, H. J.; Ghaemi, H. F.; Thio, T.; & Wolff, P. A. (1998) Extraordinary optical transmission through sub-wavelength hole arrays. *Nature* 391, 667

Fuhrmann, B.; Leipner, H. S.; Hoche, H. R.; Schubert, L.; Werner, P.; Gösele, U. (2005) Ordered Arrays of Silicon Nanowires Produced by Nanosphere Lithography and Molecular Beam Epitaxy. *Nano Lett.* 5, 2524-2527

Gwinner, M. C.; Koroknay, E.; Fu, L. W.; Patoka, P.; Kandulski, W.; Giersig, M.; & Giessen, H. (2009) Periodic Large-Area Metallic Split-Ring Resonator Metamaterial Fabrication Based on Shadow Nanosphere Lithography. *Small* 5, 400-406

Hanarp, P.; Kall, M.; & Sutherland, D. S. (2003) Optical Properties of Short Range Ordered Arrays of Nanometer Gold Disks Prepared by Colloidal Lithography. *J. Phys. Chem. B* 107, 5768-5772

Haynes, C. L.; & Van Duyne, R. P. (2001) Nanosphere Lithography: A Versatile Nanofabrication Tool for Studies of Size-Dependent Nanoparticle Optics. *J. Phys. Chem. B* , 105, 5599-5611

Hulteen, J. C.; & Van Duyne, R. P. (1995) Nanosphere lithography: A materials general fabrication process for periodic particle array surfaces. *J. Vac. Sci. Technol. A* 13, 1553-1558

Hulteen, J. C.; Treichel, D. A.; Smith, M. T.; Duval, M. L.; Jensen,T. R.; & Van Duyne, R. P. (1999) Nanosphere Lithography: Size-Tunable Silver Nanoparticle and Surface Cluster Arrays. *J. Phys. Chem. B* 103, 3854-3863

Ito, T; & Okazaki, S. (2000) Pushing the limits of lithography. *Nature* 406, 1027-1031

Kim, J. J.;Li, Y.; Lee, E. J.; § Cho, S. O. (2009) Fabrication of Size-Controllable Hexagonal Non-Close-Packed Colloidal Crystals and Binary Colloidal Crystals by Pyrolysis Combined with Plasma−Electron Coirradiation of Polystyrene Colloidal Monolayer. *Langmuir* 27, 2334-2339

Kosiorek, A.; Kandulski, W.; Chudzinski, P.; Kempa, K.; & Giersig, M. (2004) Shadow Nanosphere Lithography: Simulation and Experiment. *Nano Lett* 4 1359-1363

Kralchevsky, P. A.; & Nagayama, K. (2000) Capillary interactions between particles bound to interfaces, liquid films and biomembranes. *AdV. Colloid Interface Sci.* 85, 145-192

Larsen, A. E.; & Grier, D. G. (1997) Like-charge attractions in metastable colloidal crystallites *Nature* 385, 230-233

Larsson, E. M.; Alegret, J.; K€all, M.; & Sutherland, D. S. (2007) Sensing characteristics of NIR localized surface plasmon resonances in gold nanorings for application as ultrasensitive biosensors. *Nano Lett.* 7, 1256-1263

Ma, X. (2008) Memory properties of a Ge nanoring MOS device fabricated by pulsed laser deposition. *Nanotechnology* 19, 275706 (1-4)

Micheletto, R.; Fukuda, H.; & Ohtsu, M. (1995) A simple method for the production of a two-dimensional, ordered array of small latex particles. *Langmuir*, 11, 3333-3336

Narayanan, R.; & El-Sayed, M. A. (2004) Shape-Dependent Catalytic Activity of Platinum Nanoparticles in Colloidal Solution. *Nano Lett.* 4, 1343-1348

Nikolaides, M. G.; Bausch, A. R.; Hsu, M. F.; Dinsmore, A. D.; Brenner, M. P.; & Weitz, D. A. (2002) Electric-field-induced capillary attraction between like-charged particles at liquid interfaces. *Nature* 420, 299-301

Park, S. H.; Qin, D.; & Xia, Y. (1998) Crystallization of Mesoscale Particles over Large Areas. *Adv. Mater.* 10, 1028-1032

Pieranski, P. (1980) Two-Dimensional Interfacial Colloidal Crystals. *Phys. ReV. Lett.* 45, 569-572

Retsch, M.; Tamm, M.; Bocchio, N.; Horn, N.; Forch, R.; Jonas, U.; Kreiter, M. (2009) Parallel Preparation of Densely Packed Arrays of 150-nm Gold-Nanocrescent Resonators in Three Dimensions. *Small* 5, 2105-2110

Robinson, J. T.; Evans, P. G.; Liddle, J. A.; & Dubon, O. D. (2007) Chemical Nanomachining of Silicon by Gold-Catalyzed Oxidation. *Nano Lett.* 7, 2009-2013

Schmidt, O. G.; & Eberl, K. (2001) Self-assembled Ge/Si dots for faster field-effect transistors. *IEEE Trans. Elec. Devices* 48, 1175-1179

Service, R. F. (2001) Optical Lithography Goes to Extremes--And Beyond. *Science* 293, 785-786

Shi, J.; Gider, S.; Babcock, K.; & Awschalom, D. D (1996) Magnetic Clusters in Molecular Beams, Metals, and Semiconductors. *Science* 271, 937-941

Sinitskii, A.; Neumeier, S.; Nelles, J.; Fischler, M.; & Simon, U. (2007) Ordered arrays of silicon pillars with controlled height and aspect ratio. *Nanotechnology* 18, 305307 (1-6)

Sirotkin, E.; Apweiler, J. D.; & Ogrin, F. Y. (2010) Macroscopic Ordering of Polystyrene Carboxylate-Modified Nanospheres Self-Assembled at the Water−Air Interface *Langmuir* 26, 10677-10683

Sun, C.; Min, W.; Linn, N. C.; Jiang, P.; Jiang, B. (2009) Large-scale assembly of periodic nanostructures with metastable square lattices. *J. Vac. Sci. Technol. B* 27, 1043-1047

Trau, M.; Saville, D. A.; & Aksay, I. A. (1996) Field-Induced Layering of Colloidal Crystals. *Science* 272, 706-709

Vogel, N; Fischer, J.; Mohammadi, R.; Retsch, M; Butt, H.; Landfester, K.; Weiss, C. K.; & Kreiter, M. (2011) Plasmon Hybridization in Stacked Double Crescents Arrays Fabricated by Colloidal Lithography. *Nano. Lett.* 11, 446-454

Volkmuth, W. D.; & Austin, R. H. (1992) DNA electrophoresis in microlithographic arrays *Nature* 358, 600-602

Volokitin, Y.; Sinzig, J.; de Jongh, L. J.; Schmid, G.; Vargaftik, M. N.; & Moiseev, I. I. (1996) Quantum-size effects in the thermodynamic properties of metallic nanoparticles. *Nature* 384, 621-623

Wang, Z. L.; Petroski, J. M.; Green, T. C.; & El-Sayed, M. A. (1998) Shape Transformation and Surface Melting of Cubic and Tetrahedral Platinum Nanocrystals. *J. Phys. Chem. B* 102, 6145-6151

Weekes, S. M.; Ogrin, F. Y.; & Murray, W. A. (2004) Fabrication of Large-Area Ferromagnetic Arrays Using Etched Nanosphere Lithography. *Langmuir* 20, 11208-11212

Weekes, S. M.; Ogrin, F. Y.; Murray, W. A.; & Keatley, P.S. (2007) Macroscopic Arrays of Magnetic Nanostructures from Self-Assembled Nanosphere Templates. *Langmuir 23, 1057-1060*

Winzer, M.; Kleiber, M.; Dix, N.; & Wiesendanger, R. (1996) Rapid communication Fabrication of nano-dot-and nano-ring-arrays by nanosphere lithography. *Appl. Phys. A* 63, 617-619

Yablonovitch, E. (1987) Inhibited Spontaneous Emission in Solid-State Physics and Electronics. *Phys. Rev. Lett.* 58, 2059-2062

Yae, S.; Tannka, H.; Kobayashi, T.; Fukumuro, N.; & Matsuda, H. (2005) Porous silicon formation by HF chemical etching for antireflection of solar cells. *Phys. Status Solidi c* 2, 3476-3480

Yeh, S.; Seul, M.; & Shraiman, B. I. (1997) Assembly of ordered colloidal aggregrates by electric-field-induced fluid flow. *Nature* 386, 57-59

Zhang, Z.; Wong, L. M.; Ong, H. G.; Wang, X. J.; Wang, J. L.; Wang, S. J.; Chen, H.; & Wu, T. (2008) Self-Assembled Shape- and Orientation-Controlled Synthesis of Nanoscale Cu3Si Triangles, Squares, and Wires. *Nano Lett.* 8, 3205-3210

Zhang,G.; Wang, D.; & Mohwald, H. (2007) Fabrication of Multiplex Quasi-Three-Dimensional Grids of One-Dimensional Nanostructures via Stepwise Colloidal Lithography. *Nano. Lett.* 7, 3410-3413

Zhong, Z.; Chen, P.; Jiang, Z.; & Bauer, G. (2008) Temperature dependence of ordered GeSi island growth on patterned Si (001) substrates. *Appl. Phys. Lett.* 93, 043106 (1-3)

Zhong, Z.; Schwinger, W.; Schäffler, F.; Bauer, G.; Vastola, G.; Montalenti, F.; & Miglio, L. (2007). Delayed Plastic Relaxation on Patterned Si Substrates: Coherent SiGe Pyramids with Dominant {111} Facets. *Phys. Rev. Lett.* 98, 176102 (1-4)

Zhu, F. Q.; Fan, D.; Zhu, X.; Zhu, J. G.; Cammarata, R. C.; Chien, C. L. (2004) Ultrahigh-Density Arrays of Ferromagnetic Nanorings on Macroscopic Areas. *Adv. Mater.* 16, 2155-2159

27

Electrohydrodynamic Inkjet – Micro Pattern Fabrication for Printed Electronics Applications

Kyung-Hyun Choi, Khalid Rahman, Nauman Malik Muhammad, Arshad Khan, Ki-Rin Kwon, Yang-Hoi Doh and Hyung-Chan Kim
Jeju National University
Republic of Korea

1. Introduction

In electronic industry the manufacturing of conductive patterning is necessary and ineluctable. Traditionally, lithography is widely used for fabrication of the conductive patterns. However, lithographic processes require the complicated equipments, are time consuming and the area throughput is limited. In order to reduce the material usage, process time and large area fabrication, different fabrication technique is required. Non-lithographic-direct fabrication method (Pique & Chrisey, 2001) such as inkjet (Gans et al., 2004) and roll-to-roll (Gamota et al., 2004) printing (also known as printed electronics) are predominant examples for reasonable resolution and high throughput as compared to lithography techniques. This direct fabrication technology can be further classified into two different technologies depending on the fabrication method as contact (gravure, offset or flexographic etc) and non-contact (inkjet) method. Non-contact inkjet printing method has moved beyond graphic printing as a versatile manufacturing method for functional and structural materials.

Commercially available inkjet printer can be divided into two modes based on the ejection of the fluid: Continuous, where jet emerges from the nozzle which breaks in stream of droplets or Drop-on-Demand, the droplet ejects from the nozzle orifice as required (Lee, 2002). Inkjet printing offers the advantages of low cost, large area throughput and high speed processing. The most prominent examples of inkjet printing includes the direct patterning of, printed circuit board, conductive tracks for antenna of radio frequency identification tags (RFID) (Yang et al., 2007), Photovoltaic (Jung et al., 2010), thin film transistors (Arias et al., 2004), micro arrays of the DNA (Goldmann & Gonzalez, 2000), biosensors, etc. In case of continuous inkjet printing, the deflector directs the stream of droplets into a waste collector or onto substrate, for start and stop of the printing. This wastage of the ink issue has been addressed by the introduction drop-on-demand inkjet printing (thermal and piezoelectric). In drop-on-demand, thermal or vibration pulse are used to eject the liquid droplet from the nozzle to the substrate. However, the current printing technologies have constrained due to limitation of the ink viscosity, clogging of small size nozzles, generation of pattern smaller than the nozzle size and limitation of material to be deposited (Le, 1998). In order address these limitations, many researchers are focusing on electrohydrodynamic inkjet printing (continuous and drop-on-demand) (Park et al., 2007). Electrohydrodynamic jet printing uses electric field energy to eject the liquid from the nozzle instead of thermal or acoustic energy (Hartman, 1998). Based on the applied electric field energy, the electrohydrodynamic jetting can be used for continuous patterning, drop-on-demand printing and thin film deposition (electrospray). Electrohydrodynamic drop-on-demand, jetting or atomization has numerous applications in inkjet printing technology (Wang, 2009), thin film deposition (Jaworek, 2007), bio-application (Park 2008), mass spectrometry (Griss, 2002), etc.

2. Electrohydrodynamic jetting

In electrohydrodynamic printing the liquid is pulled out the nozzle rather than the pushing out as in case of conventional inkjet systems. When the liquid is supplied to nozzle without applying the electric field, a hemispherical meniscus is formed the nozzle due to the surface tension at the interface between the liquid and air. When the electric field is applied between the liquid and the ground plate (located under the substrate), the ions with same polarity move and accumulate at surface of the meniscus. Due to ions accumulation, the Maxwell electrical stresses are induced by the Coulombic repulsion between ions. The surface of the liquid meniscus is mainly subjected to surface tension σ_s, hydrostatic pressure σ_h and electrostatic pressure σ_e. If the liquid is considered to be a pure conductor, then the electric field will be perpendicular to the liquid surface and no tangential stress component will be acting on the liquid surface. The liquid bulk will be neutral and the free charges will confined in a very thin layer. This situation can be summarized in the following equations.

$$\sigma_h + \sigma_e + \sigma_s = 0 \tag{1}$$

Since the liquid is not a perfect conductor, the resultant electric stress on the liquid meniscus has two components, i.e. normal and tangential as shown in figure1. This repulsion force (electrostatic force) when exceeds the certain limit deforms the hemispherical meniscus to a cone. This phenomenon is known as the cone-jet transition, which refers to the shape of meniscus (Poon, 2002).

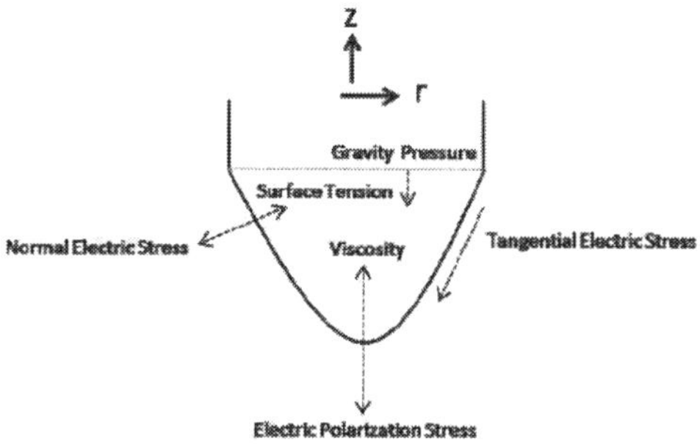

Fig. 1. Stresses due to different forces on the liquid meniscus (Kim et al. 2011)

For specific configuration and constant flow rate, there are different modes of electrohydrodynamic jetting as a function of applied voltage (Cloupeau & Foch, 1994). It should be noted that not for all liquids each mode can occur, because of the properties of the liquid. The different modes of the electrohydrodynamic jetting are discussed as follows:

2.1 Dripping, micro-dripping, spindle and intermittent cone-jet mode

In dripping mode, when the liquid is pumped in to the nozzle or capillary without applying the electric field, the droplets disintegrate from the orifice, the size of the droplets are larger than the size of the nozzle orifice. As the electric field is increased, the frequency of the droplet generation is also increased and size of the droplet decreases. At relative low flow rate, the droplet disintegrate in much smaller size as compare to the inner diameter of nozzle, this mode is known as micro-dripping mode, the frequency of the droplet increases with increase in applied electric field and size decreases. Depending on the liquid properties, increasing further electric potential, the spindle mode observed. In spindle mode, the jet extended from the meniscus and breaks up into larger droplet and satellites droplets are also observed. Further increasing in applied voltage, with relative high flow-rate intermittent cone-jet mode occurs, causing the pulsating cone-jet modes due to the high space charges reduce the electric field on the liquid jet and causing relaxation of the cone-jet into hemispherical meniscus. The pulsation in the intermittent cone-jet mode increases with increase in the applied voltage.

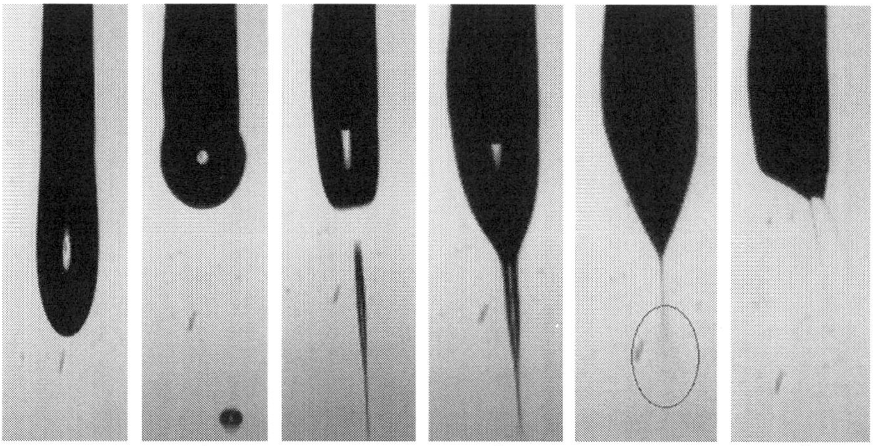

(a) (b) (c) (d) (e) (f)

Fig. 2. Modes of electrohydrodynamic jetting captured through high speed camera (a) dripping, (b) micro-dripping, (c) spindle mode, (d) pulsating cone jet, (e) stable cone jet, and (f) multi-jet mode

2.2 Cone-jet mode

Further increase in voltage, the meniscus deforms into cone and thin stable jet emerges from the apex of the jet. This mode is known as cone-jet mode. In cone-jet mode, the intact jet used to fabricate the patterns on the surface of the substrates. The main advantage of cone-jet as compared to conventional method of ejection of the liquid is its large ratio between diameter of the nozzle and the jet. The typical jet diameters are about two orders of magnitude smaller than that of nozzle; this enables patterning at very fine resolution. However, the cone-jet also has shortcoming, it is very difficult to stabilize and control the trajectory of thin jet under an electric-field, and jet disintegrate into droplets and cause spray due to electrostatic repulsive forces between themselves.

2.3 Multi-jet mode

If the applied voltage is further increased, the cone becomes smaller and smaller. With increasing the applied voltage, second jet emerges from the cone. With further increase in applied voltage, more and more jet emerges from the cone, this mode is called multi-jet mode.

2.4 Parameters affecting the cone-jet

The parameters that influence the formation and transition of stable cone-jet mode can be divided in two groups. Operating parameters i.e. flow-rate, electric field and nozzle diameter, and liquid properties i.e. electric conductivity, viscosity and surface tension (Poon, 2002).

2.4.1 Flow-rate

The flow-rate has a significant effect on the jet diameter and stability of the jet in cone-jet transition. It is the main parameters to control the jet diameter for the patterning process. The flow-rate also affects the applied potential requirement for development of cone-jet (operating envelop) and the resulting the shape of the cone-jet. In electrohydrodynamic jetting, flow-rate also affect the stability of the jet, at low flow-rate the jet is stable, whereas the high flow-rate in cone-jet region destabilize the jet resulting in shorter jet length. This is due to amount of charge carrying at high flow-rate which destabilizes the jet. The typical effects of the flow-rate on stable cone-jet mode are shown in figure 3.

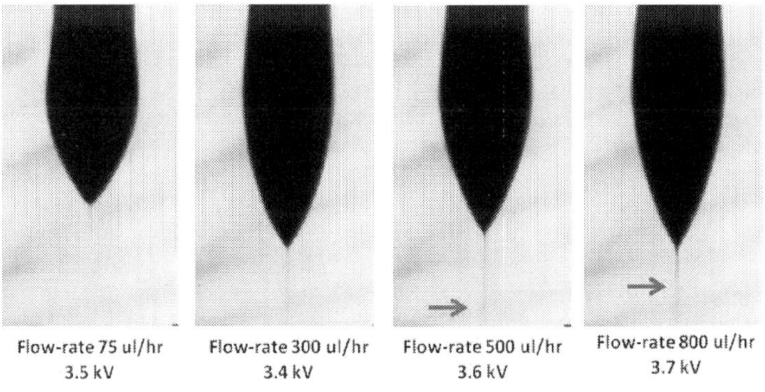

Fig. 3. Effect of flow rate on the cone-jet transition, as shown the jet diameter increases with increase in flow-rate. The red arrow indicates the jet break-up point.

2.4.2 Electric field (applied voltage)

Electric field affects the morphology of the cone, by increasing the electric field strength in steady cone-jet mode, the cone-jet recedes towards the nozzle. However, there is less effect on the jet diameter by increasing the applied voltage. At relatively low flow-rate and high electric field, the jet disintegrated into mist of small droplets also known as electrospray atomization, this behavior is used for the thin film deposition of functional material. The typical shape of the come at different applied voltage is shown in figure 4.

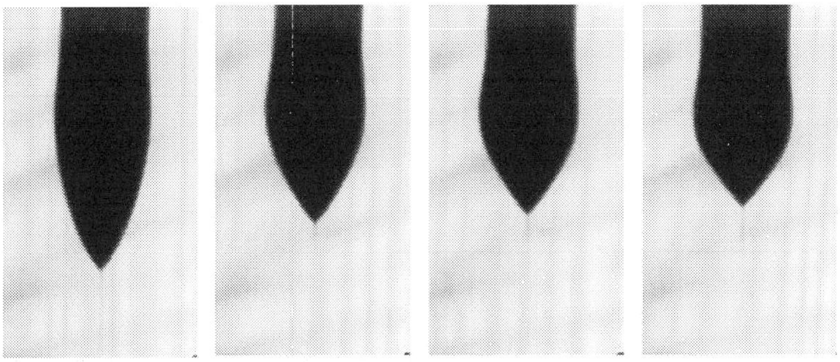

3.4 kV 3.6 kV 3.8 kV 4.0 kV

Fig. 4. Effect of the applied voltage on shape of the cone-jet at constant flow rate (200µl/hr)

2.4.3 Nozzle diameter

Nozzle diameter has significant influence on the operating envelope of the stable cone-jet region. Smaller nozzle diameters extend the lower and higher value of flow-rate limit. The voltage requirement for the cone-jet also decreases with decrease in nozzle diameter for any given liquid.

2.4.4 Conductivity

The liquid conductivity affects both the shape of the liquid cone and stability of the jet, due to the amount of electric charge on the liquid surface and jet produces is also very unstable because of high radial electric field. Highly conductive liquid deforms into sharp cone-jet shape. Liquid with very low conductivity do not deform into cone-jet by applying electric field, only dripping mode is observed. Liquids with intermediate conductivity range produce steady cone-jet.

2.4.5 Viscosity

The role viscosity is in the stabilization of the jet and diameter of the jet produced. In high viscous liquid, the jet is stable for larger portion of the length but also produces the thicker diameter. This is due to charge mobility, which is reduced significantly in high viscosity liquid, and causes decrease in conductivity.

2.4.6 Surface tension

The formation of the jet occurs when the electrical forces overcomes the surface tension on the apex of the meniscus. The required applied voltage will be increased with increase in surface tension of the liquid.

2.5 Operating envelop

In order to perform the patterning of any liquid containing nanoparticles, the operating parameters of flow rate corresponding to applied voltage for stable cone-jet has to be determined. Starting with high flow rate formation of stable cone-jet is determined by applying different voltages. Then for each flow-rate, the range of applied voltage is investigated at which the stable cone-jet is observed. This creates an operating envelop of certain liquid in electric field and voltage domain. Based on the operating envelop, behavior of the jetting is observed and parameters for patterning are determined. Operating envelop along with different modes of electrohydrodynamic jetting of the liquid containing Copper nanoparticles is shown in figure 5.

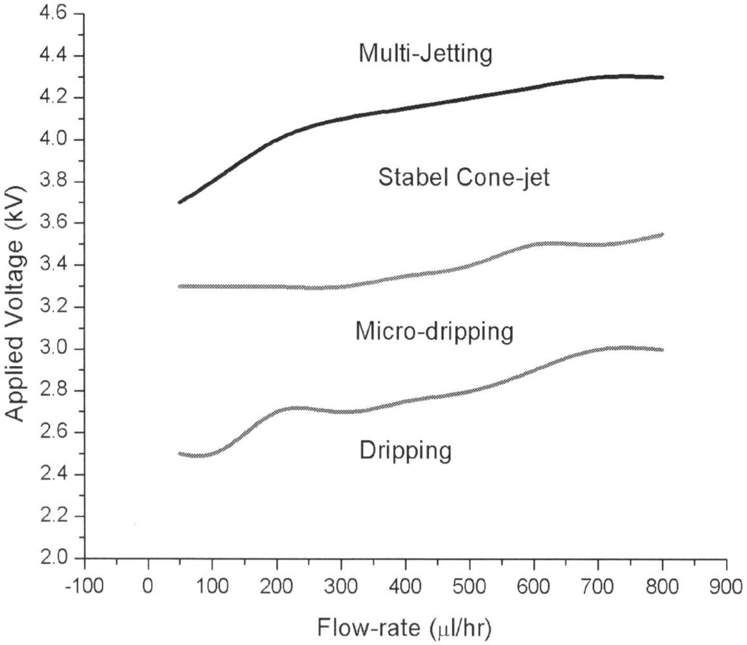

Fig. 5. Operating envelop (stable con-jet region) along with different electrohydrodynamic jetting mode of ink containing copper nanoparticles, using metallic capillary of inner diameter 410μm, external diameter 720μm and capillary to ground distance is 2mm.

3. Patterning setup

For direct patterning through electrohydrodynamic printing, lab developed system was used. The equipment used for patterning is consistent of high voltage power supply, function generator, 5 channel voltage distributor, syringe-pump for ink supply, X-Y stage with motor controller, substrate holder, high-speed camera and nozzle holder with Z-axis controller, which are connected to National Instruments PXI-1042Q hardware system. Which is controlled through lab made software based on LabView. Positive potential is applied to the nozzle head while ground is applied to the conductive plate. The substrate is place on the top of metallic ground plate. The schematic of experiment setup along with the actual system is shown in Figure 6.

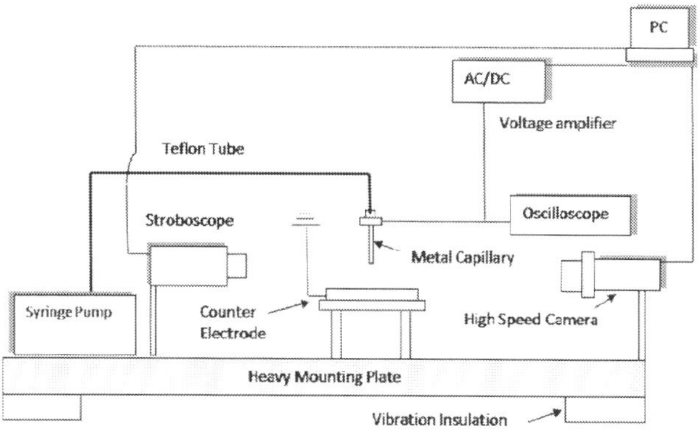

(a)

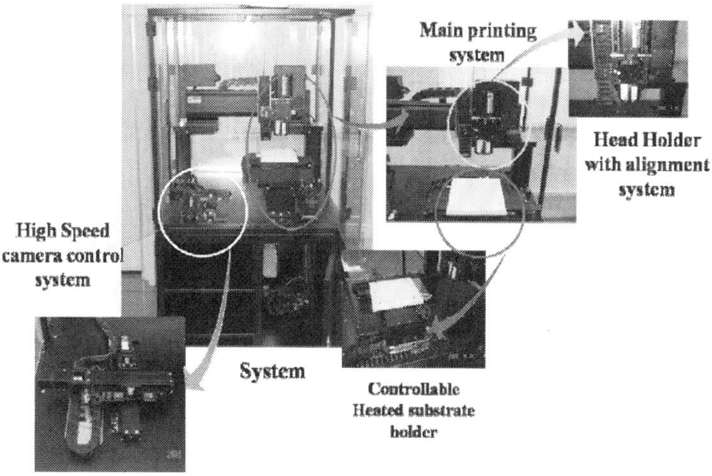

(b)

Fig. 6. (a) Schematic of lab developed system and (b) Photograph of the electrohydrodynamic inkjet system used for patterning

4. Electrohydrodynamic printing

Electrohydrodynamic printing can be divided in two different categories, depending on the ejection of liquid from the nozzle, continuous and drop-on-demand. The experiment setup for both the modes is same.

4.1 Continuous electrohydrodynamic printing

For continuous printing, ink containing copper nanoparticles is used. Initial patterning is performed by using metallic capillary of internal diameter of 210μm and outer diameter of

410μm. The distance between capillary and ground is kept 1.5mm. Patterning is performed on the glass substrate of 0.5mm thickness placed on the top of metallic ground plate by applying the lower limit of the applied voltage (ranging from 3.1kV to 3.6kV) and corresponding applied flow-rate, which is investigated for stable cone-jet operating envelop, the speed of substrate is kept at 25mm/sec for all the experiment. In electrohydrodynamic printing the jet diameter is more dependent on the flow-rate, in a cone-jet region the diameter of the jet increases with increase in flow-rate. The applied voltage has minor effect on the diameter of the jet in cone-jet mode. After patterning, the samples are placed in dry oven for sintering at 80°C for 30min. After sintering the pattern width is measured with the help of digital microscope. The pattern width with respect to applied flow-rate is shown in graph at figure 7. As shown in the graph the pattern width increases with increase in the applied flow-rate. However, pattern at high flow-rate is irregular as compared to pattern at low flow-rate, because the length of the jet decreases with increase in flow-rate at cause destabilization to the jet in cone-jet mode. The minimum pattern width of 116μm is achieved after the sintering.

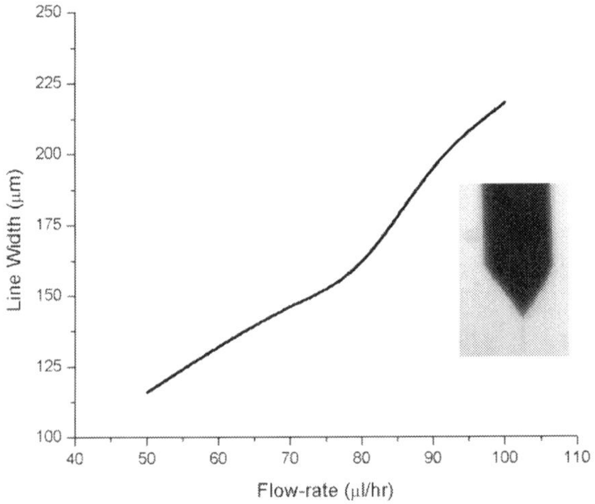

Fig. 7. Pattern width with respect to flow-rate on glass substrate

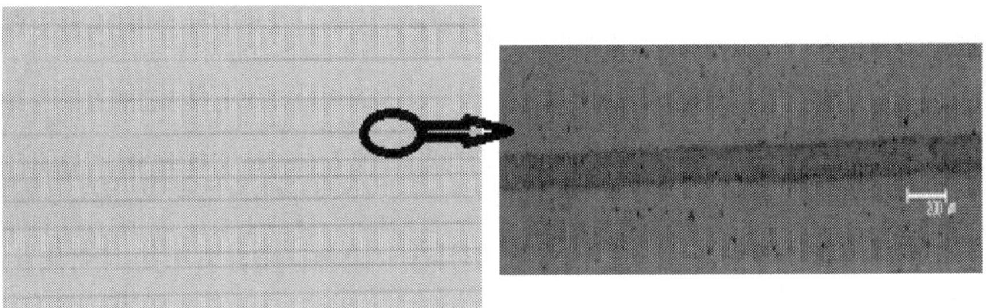

Fig. 8. Camera image and microscopic image of the copper pattern on glass substrate

Since the substrate i.e. glass has a hydrophilic surface; as a result the deposited jets were able to spread out so that the width of the lines became larger than the original size of the generated jets. Figure 8 shows the high zoom static camera and optical microscope image of continuous copper tracks on glass substrate without any defects such as bulges or coffee-ring effects.

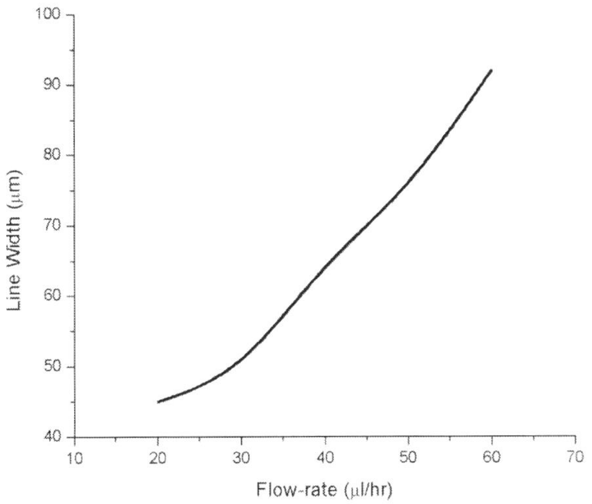

Fig. 9. Pattern width with respect to flow-rate on glass substrate using 60μm glass capillary

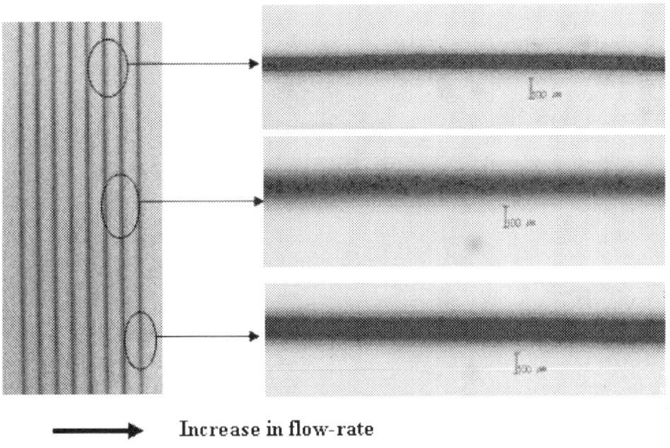

Fig. 10. Pattern images of the copper nanoparticles on glass substrate through 60μm internal diameter glass capillary nozzle

For application in printed electronics, higher resolution pattern size is required. In order to reduce the pattern size, the pattern is performed through tapered glass capillary of 60μm inner diameter and 80μm is used. The glass capillary tubes of 1.5mm outer diameter and

0.75mm inner diameter (BF 150, Sutter Instrument) was pulled and micro-nozzles is formed, with sharp tip by using a micropipette puller (P-97, Sutter Instrument). The voltage is applied by inserting the copper wire of 500µm in the glass capillary and connected to high voltage power supply. Before performing the patterning, the operating envelop for stable cone-jet is investigated. The gap between nozzle and glass substrate is also reduced to 500µm, because of small size of jet produces and is difficult to control the smaller diameter jet at larger length. The applied voltage range for developing the stable cone-jet is also reduces (ranging from 1.8kV to 2.5kV) with respect to applied flow-rate (ranging from 20µl/hr to 80µl/hr). The pattern width with respect to flow-rate by applying minimum value of required voltage, after sintering is shown in figure 9, the minimum pattern width achieved is approximately 42µm.

Figure 10 shows the image of the pattern on the glass substrate along with the microscopic images of the pattern after sintering. The images show that the pattern width of copper nanoparticles increases with increase in flow-rate.

4.2 Drop-on-demand electrohydrodynamic printing

In continuous electrohydrodynamic printing mode, the stabilization of the micron size jet is very difficult (Hohman, et al., 2001), and also problems in the placement of the jet at start point and end point of the pattern. In order to address these issues, the electrohydrodynamic drop-on-demand printing is the alternative technology. In electrohydrodynamic drop-on-demand mode, the printing is performed through time dependent generation of cone-jet by applying the pulsed voltage to the liquid in capillary or nozzle. When the pulse voltage is applied to the capillary, the meniscus of the liquid deforms into cone-jet and generating a thin jet, as the voltage is switched-off the jet breaks up and generates small droplet, and this phenomena is pulsating by generating single droplet at each pulse. The size of droplet and the frequency of the droplet depend on the amount of pulsed voltage applied, frequency of the pulse, diameter of the nozzle, viscosity of the liquid and conductivity of the liquid (Stachewicz et al., 2009). In previous researches, for drop-on-demand, the researchers have applied simple square wave pulse voltage or by superimposing AC on applied DC voltage, in both the cases the pulse is square either with zero or bias-voltage (Li et al., 2009; Kim et al., 2008). This square pulse induces unnecessary vibrations in meniscus causing problems in the placement of the droplet on the substrate. In order to avoid this vibration in meniscus, multi-step voltage is suggested for electrohydrodynamic drop-on-demand printing (Rahman et al., 2011; Kim et al. 2011). Multi-step voltage is applied by super-imposing two square waves with same frequency but different duty cycle on each other. The multi-step voltage is consist of bias-voltage "V_a" for initialization of the meniscus; intermediate-voltage "V_b" for deformation of the meniscus into cone shape and ejection-voltage "V_c" for steady droplet generations. The applied voltage is in the form of two step functions, with "V_a" consists of 25% of the pulse, "V_b" consists 50% of the pulse and "V_c" consists 25% of the pulse.

The ejection behavior at 75µl/hr at 200Hz frequency by applying square voltage (V_a=2kV and V_b=3kV) with 50% of Duty Cycle and multi-step voltage (V_a=2kV, V_b=2.5kV and V_c=3kV) is shown in figure 11 through pictures taken with high speed camera. The main benefit of the multi-step pulse voltage as compared to the square pulse voltage is time of application of bias voltage to ejection voltage. As shown in figure 11, in square voltage there is sudden change in applied voltage to ejection voltage, which causes disturbance in meniscus and instability in ejection phenomena. By multi-step pulse voltage there is intermediate voltage which ramp the effect of applied voltage, which avoids the sudden application of high voltage to the meniscus and also induced less vibration to the meniscus hence stabilization of the ejection process. The other advantage of the multi-step pulse voltage is the related to high voltage switching hardware, in square the switching time is less as compared to multi-step pulse voltage due to intermediate voltage.

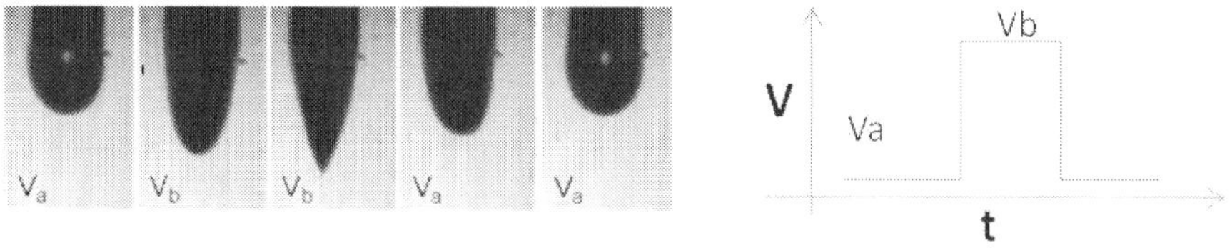

(a)

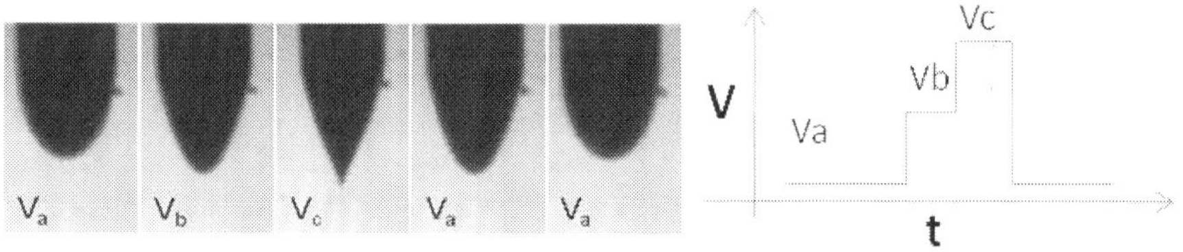

(b)

Fig. 11. The behavior ejection from 210μm metallic capillary, nozzle to ground distance 1.5mm at flow-rate 75μl/hr and frequency 200Hz (a) Square wave V_a=2kV and V_b=3kV with 50% Duty Cycle and (b) Multi-step voltage V_a=2kV, V_b=2.5kV and V_c=3kV (Rahman et al. 2011)

For drop-on-demand patterning on the glass substrate, initial experiments were performed to find lower and upper values for the applied DC voltage at which stable cone-jet formed and corresponding flow-rate. The average lower value of DC voltage at which cone-jet formed was 3.2kV and average upper value is 4.1kV. The dimension of the pattern was measured through digital microscope after sintering the Ag nanoparticles ink for 1hour at 250°C. Following can be described as input parameters for drop-on-demand printing:

1. "V_a" bias-voltage must be closed to the value of lower limit of stable cone-jet region.
2. "V_b" intermediate-voltage must be at middle range of stable cone-jet region.
3. "V_c" ejection-voltage must be closed to the value of upper limit of stable cone-jet region.
4. Determine the optimal value of "V_a", "V_b" and "V_c", at low flow-rate and frequency at which the phenomena is stable.
5. After investigating the optimal value of "V_a", "V_b" and "V_c", optimal frequency and flow-rate should be obtained.
6. Determine the optimal substrate speed based on the drop-on-demand frequency and droplet spacing on the substrate.

Figure 12 shows the static high zoom camera and microscopic images of printed dots and lines on the glass substrate by using EHD drop-on-demand ink jet printing technique of ink containing copper nanoparticles, using metallic capillary of internal diameter 410μm and external diameter 720μm. Patterning has been carried out at a constant flow rate and at pulse frequencies of 10Hz, 25Hz, 50Hz, and 100Hz with a constant linear motor speed (substrate speed) of 25mm/s. Effect of biased and pulse voltages on droplet size has been analyzed by varying the biased and step voltages. figure 12(a) shows the deposited droplets which are generated at applied pulse of 50Hz frequency with 50% duty cycle and at voltages V_a, V_b and V_c of 2.5kV, 4.5kV (first pulse of 2kV) and 5.5kV (second pulse of 1.5kV) respectively. The average diameter size of droplets is 780μm. Comparatively smaller droplets have been generated by increasing the magnitude of biased voltage and decreasing the magnitude of step voltages. The deposited droplets in figure 12(b) are generated at the same frequency and duty cycle as that of figure 12(a) i.e. 25Hz and 50% respectively but at low pulse voltages i.e. V_a, V_b and V_c; 3.5kV, 4.5kV (first pulse of 1kV) and 5.0kV (second pulse of 0.5kV) respectively. The average diameter size of deposited droplets is 780μm. The reason of this relatively smaller drop generation is that with low biased voltage and high pulsed voltages, energy gain per unit area of the liquid and the tangential electric stress at the liquid meniscus increases more quickly than the normal electric stress. As a result, greater pulsed voltages (V_b and V_c) are more likely to produce a temporary jet rather than a drop-on-demand mode which generates relatively large sized droplets. Similarly figure 12(c) shows the patterned droplets which are generated at applied pulse of 10 Hz frequency with 75% duty cycle and at voltages V_a, V_b and V_c 3.5kV, 4.5kV(first pulse of 1kV) and 5.0kV(second pulse of 0.5kV) respectively. The deposited droplets have an elliptical shape rather than round due to high duty cycle of the pulse voltage. High duty cycle (75%) increases the application time of the triggering pulse which results in a temporary jet rather than a droplet. Since the substrate speed is constant and relatively high than the speed of ejection of temporary jet which does not allowing the temporary jet to accumulate to a large round shape drop on the substrate. As a result, temporary generated jet forms an oval shape drop after deposition on the substrate. Using this high duty cycle (75%) of pulsed voltage, conductive lines are patterned at applied pulse of 100 Hz frequency and at voltages V_a, V_b and V_c of 3.5kV, 4.5kV (first pulse of 1kV) and 5kV (second pulse of 0.5kV) respectively. The printed lines shown in figure 12(d) have an average size of 780μm. It can be concluded from the printed results shown in figure 12(d) that EHD drop-on-demand can also be used for printing of conductive lines for metallization in printed circuit boards and backplanes of printable transistors if the substrate speed and frequency of droplet generation get synchronized.

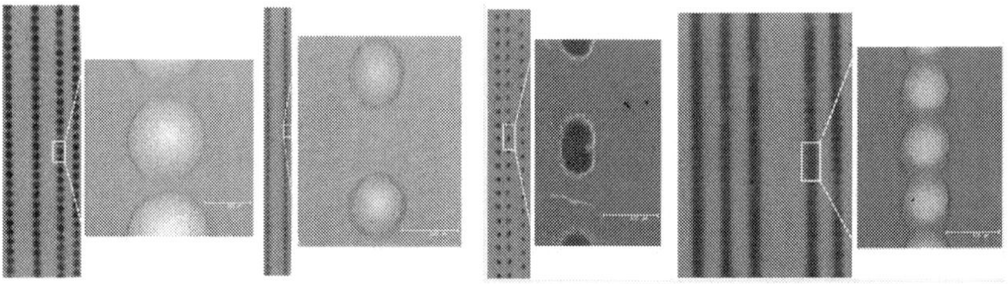

(a) (b) (c) (d)

Fig. 12. Camera and microscopic images of deposited droplets and line patterning resulted by drop-on-demand ejection at constant flow rate and constant substrate speed: (a) 50Hz and 50% duty cycle (b) 25Hz and 50% duty cycle (c) 10Hz and 75% duty cycle (d) the line pattern by drop-on-demand at 100Hz and 75% duty cycle (Kim et al. 2011)

The drop-on-demand experiment is also performed by using silver nanoparticles ink. The numbers of experiments are performed by changing the value of V_a, V_b and V_c. The frequency of the multi-step voltage is at 50Hz and with flow-rate 75μl/hr. The droplet diameter on the substrate at different multi-step voltage is shown in table 1. As shown in table 1, in Case-1 the droplet diameter is larger due to spray because at high voltage the droplet caries more charge. As in Case-5 and 6 the droplet diameter is smaller due to more stable phenomena. The microscopic image of the printed droplet on glass substrate after sintering for Case-1 and Case-6 is shown in figure 13.

Case	V_a (kV)	V_b (kV)	V_c (kV)	Approx. Droplet Diameter (μm)
1	0	2	4.1	200
2	1	2	4.1	189
3	2	3	4	160
4	3	3.5	4	126
5	3	3.2	3.7	106
6	3.2	3.5	3.8	82

Table 1. Droplet Diameter at different applied multi-step voltage

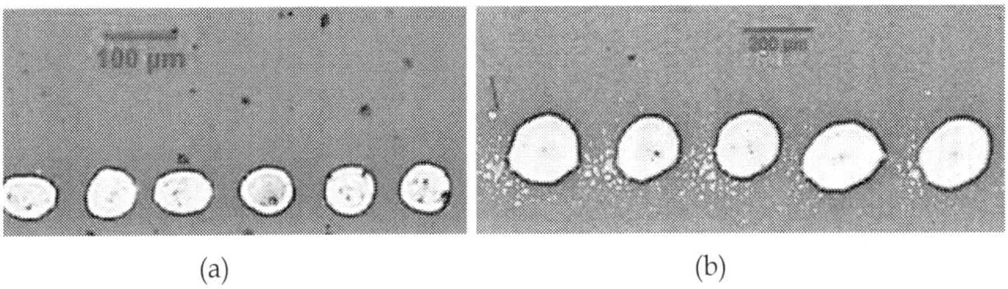

Fig. 13. (a) case-1 and droplet diameter approximately 200μm and (b) case-6 and droplet diameter approximately 82μm (Rahman et al. 2011)

In order to analyze the effect of frequency on the drop-on-demand pattering, the experiments are performed by changing the applied frequency. The applied multi-step voltage is kept as Case-6 (V_a=3.2kV, V_b= 3.5kV and V_c=3.8kV) and flow-rate 75μl/hr. The measured droplet diameter on the glass substrate against the frequency is shown in graph at figure 14. The droplet diameter is decreased from 120μm to 40μm as the frequency increases from 10Hz to 350Hz. The maximum applied frequency at which the drop-on-demand phenomena observed is 350Hz for the silver ink by using 210μm inner diameter capillary. The reason is due to short pulse times at high frequency, the voltage required to generate the jet is applied in shorter time i.e. jetting time decreases due to which droplet diameter decreases at higher frequencies. Figure 15 shows the microscopic images of the droplets after sintering by applying 10Hz and 350Hz frequency. The result also indicates the size of the droplets is much smaller than the nozzle size which is the main advantage of electrohydrodynamic drop-on-demand technique.

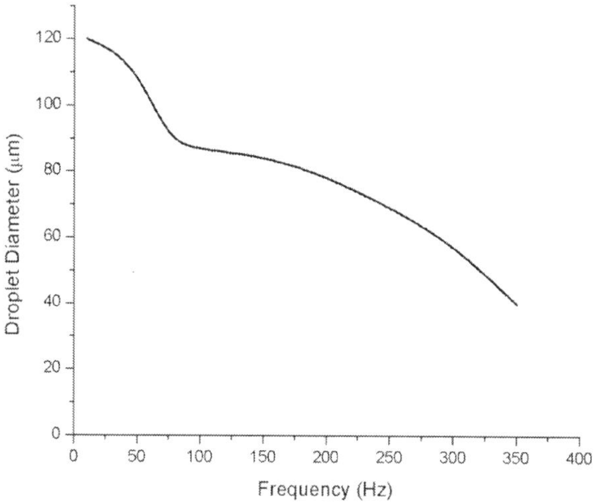

Fig. 14. Droplet diameter vs. applied frequency (Rahman et al. 2011)

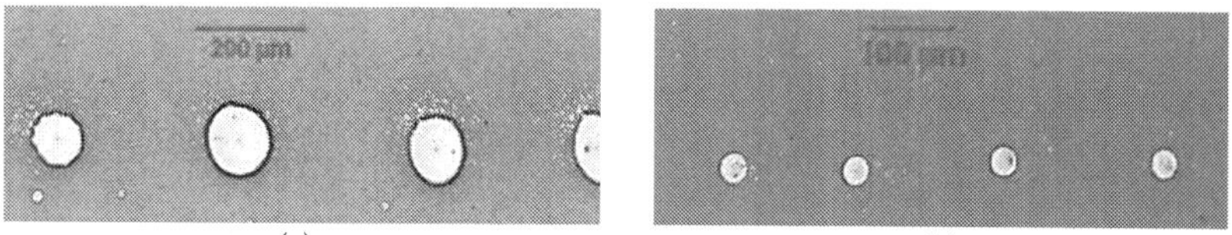

(a) (b)

Fig. 15. Microscopic images of the sintered droplet by applying V_a=3.2kV, V_b=3.5kV, V_c=3.8kV and flow-rate 75μl/hr (a) applied frequency 10Hz and diameter approximately 120μm (b) applied frequency 350Hz and droplet diameter approximately 40μm (Rahman et al. 2011)

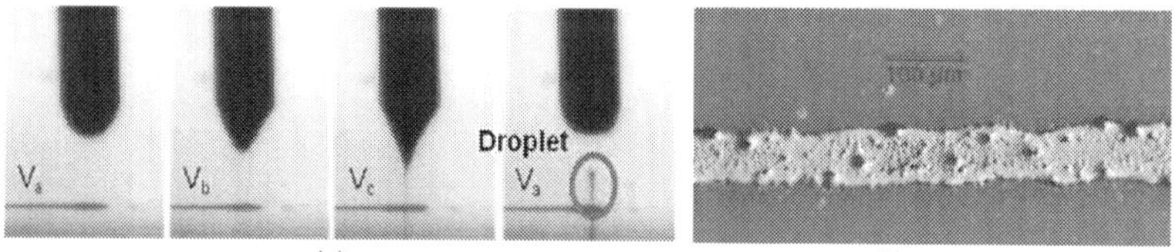

(a) (b)

Fig. 16. (a) Sequential images of the drop-on-demand pattern on the glass substrate with respect to applied multi-step pulsed voltage and (b) microscopic image of the line pattern after sintering process (Rahman et al. 2011)

The pattern line is formed after synchronizing the substrate speed with the drop-on-demand frequency by analyzing the droplet spacing on the substrate. The sequential images of the drop-on-demand line patterning with respect to applied multi-step pulse voltage at V_a=3.2kV, V_b=3.5kV, V_c=3.8kV, flow-rate 75μl/hr and frequency 100Hz along with the

pattern line are shown in figure 16. The pattern size shown is approximately 95μm, the line pattern size is greater than the droplet size (87μm), which is due to the over lapping of the droplets to create the line pattern.
Figure 17 shows the XRD spectrum of the printed line pattern on the glass substrate. The XRD spectrum peaks shows the existence of the silver only which confirms the deposited material was consist of silver nanoparticles.

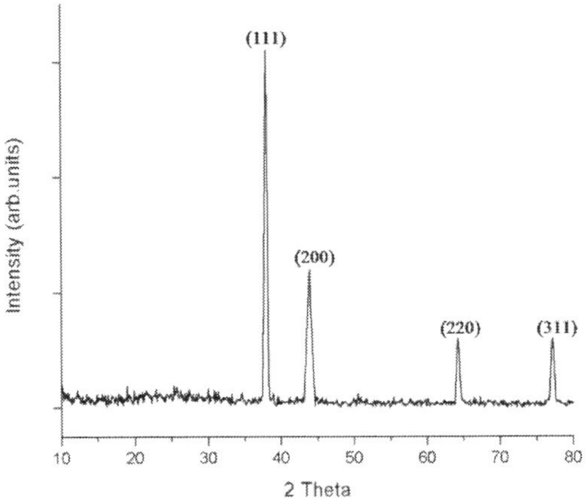

Fig. 17. XRD spectrum of the line pattern (Rahman et al. 2011)

The resistance of the pattern line is measured by 4-point probe method by measuring the voltage drop ΔV across 2mm long segment of the line pattern by applying the different current intensity of 10μA, 20μA, 50μA, 75μA and 100μA. The pattern line showed the linear ohmic behavior with resistance of 0.39Ω. I-V curve obtain through 4-point measurement is shown in figure 18.

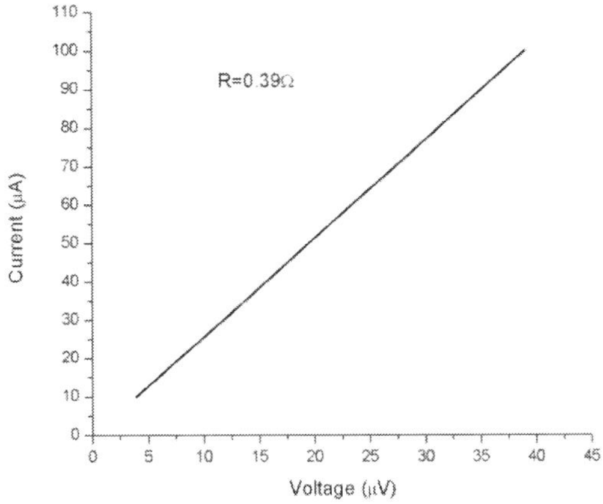

Fig. 18. I-V curve of the pattern line (Rahman et al. 2011)

4.3 Multi-nozzle electrohydrodynamic printing

The main draw-back of single nozzle electrohydrodynamic printing is the limitation of low throughput. In order to address this drawback and attain high production efficiency electrohydrodynamic inkjet printing process for industrial production of printed displays, PCBs, printed TFTs and printed solar cells, many researcher are working on a multi-nozzle electrohydrodynamic inkjet printing process(Lee et al., 2008). However, due to the interaction (cross-talk) between the electrically charged neighboring jets, it is difficult to precisely control and reproducible multi-nozzle EHD inkjet printing process. To overcome the limitation of low throughput of EHD inkjet printing process, a multi-nozzle EHD inkjet printing head consisted of three nozzles is fabricated and successfully tested by printing simultaneously conductive lines of silver nanoparticles ink onto glass substrate (Arshad et al 2011). Multi-nozzle electrohydrodynamic inkjet printing is consisted of three parts i.e. PDMS (Polydimethylsiloxane) holder, glass capillaries and copper electrodes. PDMS holder is manufactured through molding technique with the channels for capillaries, ink-supply and electrodes. The schematic of the mold and multi-nozzle head is shown in figure 19(a) and (b). The tapered glass capillaries of 100μm internal diameter and 120μm are then inserted in the outlet channels of the PDMS part. Finally, three copper electrodes having outer diameter of 500μm are inserted from the top of the PDMS holder. The nozzle to nozzle distance is kept at 3mm. The complete schematic of the multi-nozzle EHD inkjet printing head is shown in Figure 19(c).

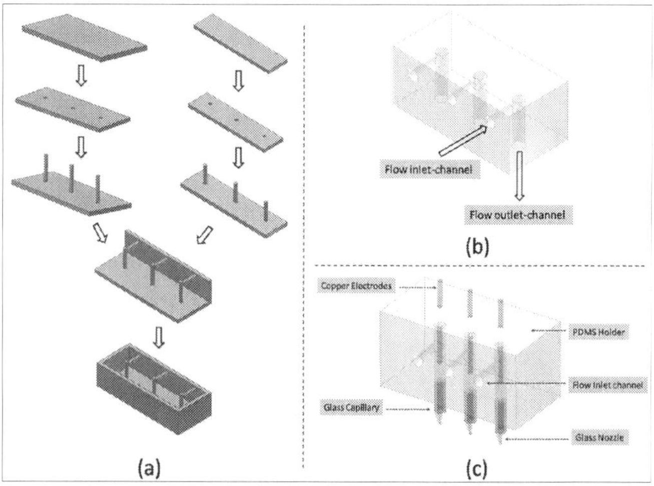

Fig. 19. Schematic illustration of the multi-nozzle EHD inkjet printing head fabrication process: (a) Simplified fabrication steps of mold preparation (b) Resulted PDMS holder having L-shaped channels for ink supply and clasping of glass capillaries (c) Complete multi-nozzle EHD inkjet printing head (Khan et al 2011)

Figure 20 shows images of stable meniscus formed at the tip of individual nozzles. Cone-jets are formed at the tip of each nozzle i.e. nozzle 1, nozzle 2 and nozzle 3 at an applied DC voltage and flow rate of 3.5kV and 20μl/h respectively.

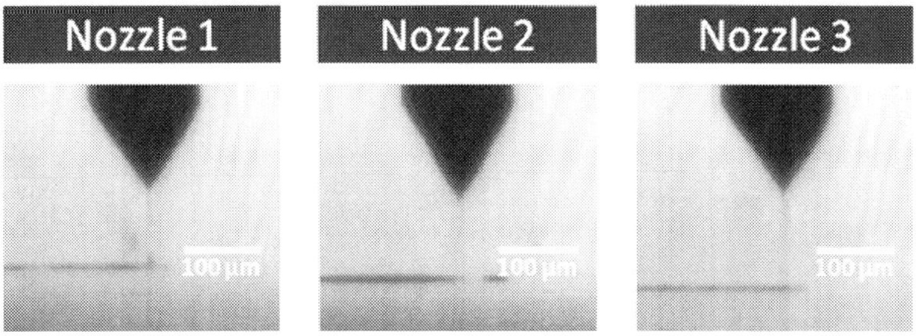

Fig. 20. Photographs of axisymmetric cone-jet established at the nozzles tips of multi-nozzle electrohydrodynamic inkjet printing head (Khan et al 2011)

Printing is performed by applying a DC voltage of 3.5kV and flow rate of 20µl/h to each nozzle. The nozzles to substrate distance is set as 300µm while substrate speed is kept constant at 10mm/sec. Figure 21 shows the high zoom static camera and optical microscope image of continuous silver tracks simultaneously printed by three nozzles on glass substrate without any defects such as bulges or coffee-ring effects. The average line width of the printed lines is 140µm.

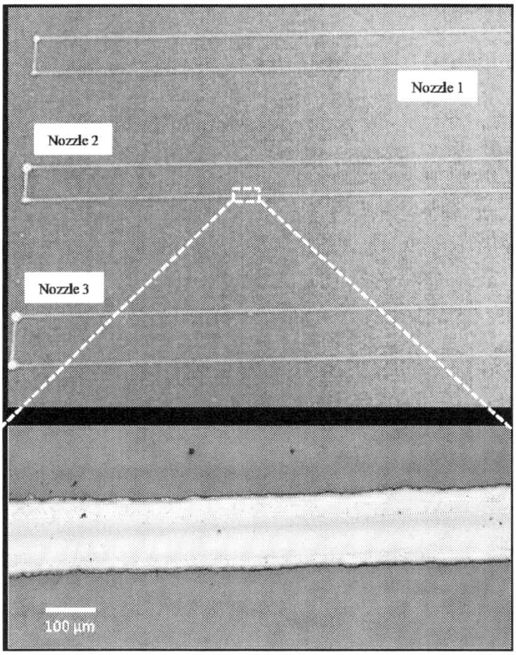

Fig. 21. Camera and optical microscope image of continuous silver patterns printed on glass substrate by multi-nozzle printing at an applied voltage of 3.5 kV and flow rate of 20µlh (Khan et al 2011)

Moreover, the SEM images of a typical printed line as shown in figure 22 also illustrates that nanoparticles are three-dimensionally interconnected with each other, which favorably affect the electrical conductivity.

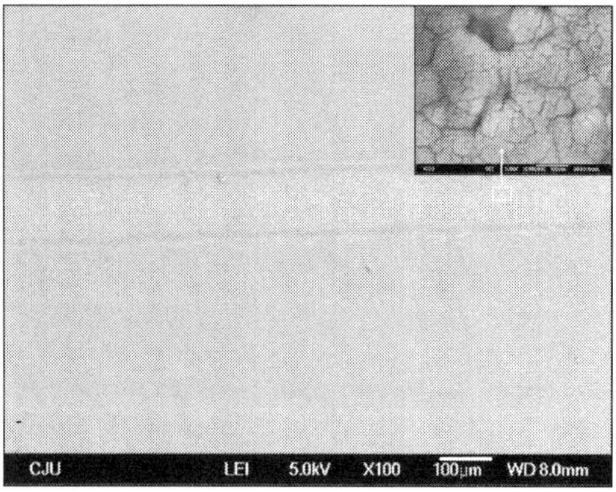

Fig. 22. SEM image of continuous silver pattern deposited by multi-nozzle electrohydrodynamic inkjet printing (Khan et al 2011)

4.4 Thin film deposition

The main benefit of the electrohydrodynamic atomization in cone-jet mode also known as electrospray deposition can be used for thin film deposition for functional materials on different substrates. The same experiment setup for printing is applicable for the thin film deposition through electrospray. In electrospray mode, the nozzle to substrate distance is larger than the patterning setup; this allows the jet to disintegrate into small monodisperse droplets of equal size due to the repulsive forces in the droplets carrying the charges. These mondisperse droplets landed on the substrate and generate the uniform thin layer of that material. The thickness and area of the layer depends on the flow-rate, distance between nozzle and substrate, and also the time of the spray.

For thin film deposition of CIS (Copper-Indium di-Selenide) through electrospray deposition is performed by using 430μm internal diameter metallic nozzle (Muhammad et al., 2010). Initially the operating envelop for stable cone-jet is investigated for the ink containing CIS nanoparticles. The operating envelop for CIS ink along with different electrohydrodynamic modes is shown in figure 23. For spray purpose higher value of applied voltage is selected. Different experiments are performed by changing the flow rate and also the distance between nozzle and substrate distance. The quality and thickness of deposition of thin layers using ESD depends upon three main factors i.e. distance from nozzle to substrate, spraying time or substrate speed and the flow rate. Therefore as less is the distance between the nozzle and the substrate, higher will be the layer quality and also the layer thickness will be on the higher side and vice versa. Similarly as much is the spraying time, or as less is the substrate speed in this case, the layer quality will be on the higher side, however concerns will be on the thickness of the layer which can negatively affect the device efficiency. Figures 24 present the layer morphology obtained by SEM at a flow rate of 150μl/h and varying nozzle to substrate distances with substrate speed of

0.5mm/s. As clear from the figures, the best layer quality is achieved at a nozzle to substrate distances of 6 mm and 8 mm the dense layers is produced and particles are completely intact and no pores or islands are visible as in the case of nozzle to substrate distances of 13mm the layer is porous and contain voids. These pores will case defect in the functionality of the deposited layer.

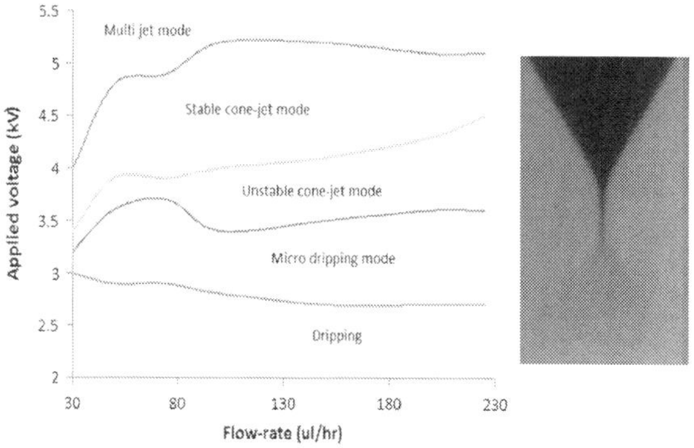

Fig. 23. Operating envelope for the CIS ink, showing different modes (Muhammad et al., 2010)

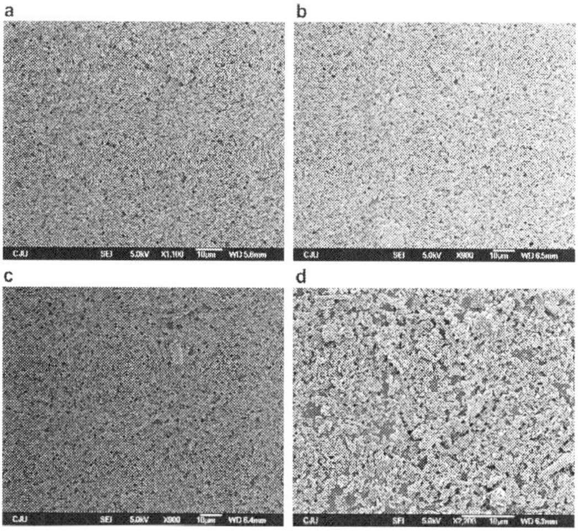

Fig. 24. FE-SEM micrographs of the deposited layer at stand-off distance of a. 6 mm, b. 8 mm, c. 10 mm, d. 13 mm, at substrate speed of 0.5 mm/s and flow rate of 150 ml/h (Muhammad et al., 2010)

5. Conclusions

Electrohydrodynamic inkjet printing is relatively new but very power tool and process for the direct patterning of the functional materials on substrate. Electrohydrodynamic inkjet printing can be used in continuous mode as well as drop-on-demand mode. The main benefit of the electrohydrodynamic inkjet printing is generation of pattern smaller than the nozzle size, because the printing is performed by pulling the liquid rather than pushing of the liquid, which is limitation in other inkjet technology such as thermal and piezoelectric printing. The advantage of electrohydrodynamic printing over conventional printing is high resolution can be obtained. The other advantage of this direct patterning method is the flexibility in the process in terms of material to be used as well as patterns can be made on different kinds of substrate. In continuous mode the patterning is performed by achieving through thing continuous jet in the stable cone-jet mode, however the thin is difficult to stabilize due to electric field. The drop-on-demand mode can be used as alternative to continuous electrohydrodynamic printing. In drop-on-demand the stable cone-jet is achieved for shorter period of time by applying the pulse voltage. Single nozzle electrohydrodynamic printing has low throughput, in order to increase the efficiency multi-nozzle printing can be performed as reported. The other advantage of electrohydrodynamic printing is thin film deposition in electrospray mode. Thin films can be deposited on the surface of different substrate without changing the experimental setup. This combination of both the technologies (patterning and thin film deposition) can help in fabrication of the electronic devices such as TFT, OLED or Solar Cells etc. through single technology.

6. Challenges and future trends

There is lot of research is being performed in the field of the electrohydrodynamic printing by patterning the functional materials even in submicron level (Schirmer et al. 2010). However the functional materials meeting these requirements are a challenge, which can be used for the patterning purpose. The inks containing these functional materials impact the morphology, adhesion, chemical and environmental stability, these factors affect the performance. In addition to this the material used in direct patterning technologies has drawbacks in functional performance such as ion mobility, on-off voltage, threshold voltage and off current. Many researchers are working in the field of chemical and material technologies to overcome these limitations, by introducing the different materials that can be used for fabrication of devices through direct fabrication technology. Inorganic material such as carbon nanotube (single-wall and multi-wall) and metal nanoparticles (silver, gold and copper) based inks and pastes are commercially available for the direct patterning applications. Recently there researches are also working on the organic materials that can be used as the insulation material for electronic devices and also conductive organic material such as PEDOT, which can used for the fabrication of the conductive tracks. Recently composite polymers have received interests in the field of direct patterning material due to improvement in mechanical, thermal, optical and conductive properties. But the major drawback is still the performance of devices fabricated through the direct patterning technology as compared to conventional electronic.

In electrohydrodynamic inkjet printing, for large area manufacturing the design of multi-nozzle printing head is another bottle neck due to interfering of electric field between nozzles. The miniaturization of the nozzles dimensions cause problems in obtaining the symmetric and stable cone-jet for printing. Other limitation is the fabrication of such small size nozzles. To optimize the performance of small nozzles, the physics and chemistry of the nozzle has to optimize to ensure the printability of different functional materials, with damaging the nozzles. When using the micron size nozzle, the issue of the clogging arises due to accumulation of the nanoparticles on the nozzle opening and cause blocking, which affects the performance of the printing. For micron size nozzle special inks are required from which printing can be performed.

In general, electrohydrodynamic inkjet printing is powerful tools for direct patterning of the functional materials and can be used for high resolution printing. Developing this technology will allow exploring the potential application of electrohydrodynamic printing in high resolution fabrication of electronic devices and appear to be promising direction for the future.

7. Acknowledgement

This work was supported by the National Research Foundation of Korea (NRF) grant funded by the Korea government (MEST) (No.2010-0026163).

8. References

Pique, Alberto. & Chrisey, D. B. (2001). *Direct-Write Technologies for Rapid Prototyping: Application to Sensors, Electronics, and Passivation Coatings*, ISBN 978-0121742317, San Diego, USA

Gamota D., Brazis. P., Kalyansundaram, K. & Zhang, J. (2004). *Printed Organic and Molecular Electronics*. ISBN 1-4020-7707-6, Massachusetts, USA

Gans, B. J., Duineveld, P. C. & Schubert, U. S. (2004). *Inkjet printing of polymers: State of the art and future development*. Advanced Materials, 16, 203–213.

Lee, E., R. (2003). *Microdrop Generation*. ISBN 0-8493-1559-X, Florida USA

Yang, L., Rida, A., Vyas, R. & Tentzeris,M. M. (2007). *RFID Tag and RF Structures on a Paper substrate Using Inkjet-Printing Technology*. IEEE Transactions on Microwave Theory and techniques, vol. 55, no. 12

Jung, J., Kim, D., Lim, J., Lee, C. & Yoon, S. C. (2010). *Highly Efficient Inkjet-Printed Organic Photovoltaic Cells*. Japanese Journal of Applied Physics, 49, 5, 05EB03-05EB03-5

Arias, A. C., Ready, S. E., Lujan, R., Wong, W. S., Paul, K. E., Salleo, A., Chabinyc, M. L., Apte, R. & Street R. A. (2004). *All jet-Printed Polymer Thin-Film Transistor Active-Matrix Backplanes*. Applied Physics Letters, 85, 15, 3304-3306.

Goldmann, T., Gonzalez, J. S. (2000), DNA-Printing: *Utilization of a Standard Inkjet Printer for the Transfer of Nucleic Acids to Solid Supports*. Journal of Biochemical and Biophysical Methods, 42, 105-110.

Le, H. P. (1998). *Progress and Trends in Ink-jet Printing Technology*. The Journal of Imaging Science and Technology, 42, 1, 49-62

Park, J.U., Hardy, M., Kang, S.U. , Barton, K., Adair, K., Mukhopadhyay, D.K., Lee, C.Y., Strano, M. S., Alleyne A. G., Georgiadis, J. G., Ferreira, P. M. & Rogers, J. A. (2007). *High-Resolution Electrohydrodynamic Jet Printing*. Nature Materials, 6, 782-789

Hartman, R. P. A. (1998). *Electrohydrodynamic Atomization in the Cone-jet Mode from Physical Modeling to Powder Production*. PhD Thesis, TU Delft

Wang, K., Paine, M. D. & Stark, J. P. W. (2009). *Fully Voltage-Controlled Electrohydrodynamic Jet Printing of Conductive Silver Tracks with a Sub-100μm Linewidth*. Journal of Applied Physics, 106, 024907

Jaworek, A., (2007). *Micro- and Nanoparticle Production by Electrospraying*. Powder Technology, 176, 18–35

Park, J.U., Lee, H. L., Paik, U., Lu, L., & Rogers, J. A. (2008). *Nanoscale Patterns of Oligonucleotides Formed by Electrohydrodynamic Jet Printing with Applications in Biosensing and Nanomaterials Assembly*. Nano Letters, 8 (12), 4210-4216

Griss, P., Melin, J., Sjodahl, J., Roeraade, J., Stemme, G., (2002). *Development of Micromachine Hollow Tips for Protein Analysis Based on Nanoelectrospray Ionization Mass Spectrometry*. Journal Micromechanics and Microengineering, 12, 682–687

Poon, H. F. (2202). *Electrohydrodynamic Printing*. PhD thesis, Princeton University

Cloupeau, M. & Foch B. P., (1994). *Electrohydrodynamic Spraying Functioning Modes: A Critical Review*. Journal of Aerosol Science, 24 (6), 1021-1036

Hohman, M. M., Shin, M. Rutledge, G. & Brenner, M. P. (2001). *Electrospinning and electrically forced jets. I. Stability theory*. Physics of Fluid, 13 (8), 2201-2220

Chen, C. H., Kelder, E. M., Van-Der-Put J.J.M. & Schoonman, J. (1996). *Morphology Control of Thin $LiCoO_2$ Films Fabricated Using the Electrostatic Spray Deposition (ESD) technique*. Journal of Materials Chemistry, 6, 765-771

Stachewicz U., Yurteri C.U., Marijnissen J.C.M., & Dijksman, J.F., (2009). *Stability Regime of Pulse Frequency for Single Event Electrospraying*. Applied Physics Letters 95, 224105

Li, J.L., & Zhang, P., (2009). *Formation and Droplet Size of EHD Drippings Induced by Superimposing an Electric Pulse to Background Voltage*. Journal of Electrostatic, 67, 562-567

Kim, J. H., Oh, H. C., & Kim, S. S. (2008). *Electrohydrodynamic Drop-on-Demand Patterning in Pulsed Cone-Jet Mode at Various Frequencies*. Journal of Aerosol Science, 39, 819-825

Rahman, K., Khan, A., Nam, N. M., Choi, K. H. & Kim D. S. (2011). *Study of Drop-on-Demand Printing Through Multi-step Pulse Voltage*. International Journal of Precision Engineering and Manufacturing, 12 (4)

Kim D. S., Khan, A., Rahman,, K., Khan, S., Kim, H. C. & Choi, K. H. (2010). *Drop-on-Demand Direct Printing of Colloidal Copper Nanoparticles by Electrohydrodynamic Atomization*. Materials and Manufacturing Processes (doi: 10.1080/10426914.2011.551956)

Lee J.S., Kim S.Y., Kim Y.J., Park J., Kim Y., Hawng. J., & Kim Y.J. (2008). *Design and Evaluation of a Silicon Based Multi-Nozzle for Addressable Jetting Using a Controlled Flow Rate in Electrohydrodynamic Jet Printing*. Applied Physics Letter, 93, 243114

Khan A., Rahman K., Hyun M.T., Kim D.S., & Choi K.H. (2011). *Multi-Nozzle Electrohydrodynamic Inkjet Printing of Silver Colloidal Solution for the Fabrication of Electrically Functional Microstructure*. Applied Physics A (doi:10.1007/s00339-011-6386-0)

Muhammad N. M., Sundharam, S., Dang H. W., Lee, A., Ryud B. H. & Choi K. H. (2010). *CIS Layer deposition through electrospray process for solar cell fabrication*. Current Applied Physics, 11 (1), S68-S75

Niklas C. Schirmer, N. C., Kullmann, C., Schmid, M. S., Burg, B. R., Schwamb, T., & Poulikakos D. (2010). *On Ejecting Colloids Against Capillarity from Submicrometer Openings: On-Demand Dielectrophoretic Nanoprinting*. Advanced Materials, 22, 4701-4705

Lithography-Free Nanostructure Fabrication Techniques Utilizing Thin-Film Edges

Hideo Kaiju[1,2], Kenji Kondo[1] and Akira Ishibashi[1]
[1]*Research Institute for Electronic Science, Hokkaido University*
[2]*PRESTO, Japan Science and Technology Agency*
Japan

1. Introduction

Fabricating nanoscale patterns with sub-10 nm feature size has been an important research target for potential applications in next-generation memories, microprocessors, logic circuits and other novel functional devices. Typically, according to the International Technology Roadmap for Semiconductor (ITRS) from 2009, an 8.9 nm node device is targeted for the year 2024. To achieve this milestone, liquid immersion lithography and extreme ultraviolet (EUV) lithography can be expected to be among the most commonly used techniques for the fabrication of nanopatterns. With liquid immersion lithography using a wavelength of 193 nm and a high numerical aperture (NA), it has been demonstrated that 32 nm features can be patterned (Finders et al., 2008; Sewell et al., 2009). EUV lithography using a short wavelength of 13.5 nm and 0.3-NA exposure tool has also enabled the printing of 22 nm half-pitch lines (Naulleau et al., 2009).

On the other hand, attractive patterning techniques, such as a superlattice nanowire pattern transfer (SNAP) method (Melosh et al., 2003; Green et al., 2007), a mold-to-mold cross imprint (MTMCI) process (Kwon et al., 2005) and a surface sol-gel process combined with photolithography (Fujikawa et al., 2006), are currently proposed and pursued actively. The SNAP method, which is based on translating thin film growth thickness control into planar wire arrays, has enabled the production of molecular memories consisting of 16 nm wide titanium/silicon nanowires. The MTMCI process using silicon nanowires formed by spacer lithography, in which nanoscale line features are defined by the residual part of a conformal film on the edges of a support structure with the linewidth controlled by the film thickness, has been used to produce a large array of 30 nm wide silicon nanopillars. With the surface sol-gel process combined with photolithography, where the linewidth is determined by the thickness of coating silica layer on the resist pattern, the size reduction and the large area of sub-20 nm silica walls have been achieved.

Recently, we have proposed a double nano-baumkuchen (DNB) structure, in which two thin slices of alternating metal/insulator nano-baumkuchen are attached so that the metal/insulator stripes cross each other, as part of a lithography-free nanostructure fabrication technology (Ishibashi, 2003 & 2004; Kaiju et al., 2008; Kondo et al., 2008). The schematic illustration of the fabrication procedure is shown in Fig. 1. First, the metal/insulator spiral heterostructure is fabricated using a vacuum evaporator including a film-rolled-up system. Then, two thin slices of the metal/insulator nano-baumkuchen are cut out from the metal/insulator spiral heterostructure. Finally, the two thin slices are attached together face to face so that each stripe crosses in a highly clean environment (Ishibashi et al., 2005; Kaiju et al., 2005; Rahaman et al., 2008). Utilizing this DNB structure, we can expect to realize high density memory devices, the crossing point of which can be scaled down to ultimate feature sizes of a few nanometers thanks to their atomic-scale resolution of the film thickness determined by the rate of metal deposition, ranging from 0.01 to 1 nm/s. This DNB structure also gives a huge potential impact and importance of uniting bottom-up structures with top-down systems (Ishibashi, 2003). One element of the DNB structure is called a quantum cross (QC) device, which consists of two metal thin films (nanoribbons) having the edge-to-edge configuration as shown in Fig. 1 (Ishibashi, 2004; Kondo et al., 2006; Kaiju et al., 2008). In this QC device, the area of the crossed section is determined by the film thickness, in other words 1-20 nm thick films can produce 1×1-20×20 nm^2 nanoscale junctions. Since the vacuum evaporation has good spatial resolution of one atomic layer thickness, the junction size of QC devices could ultimately be as small as a few ångströms square ($10^{-2}nm^2$). This method offers a way to overcome the feature size limit of conventional optical lithography. When molecular-based self-assembled monolayers (SAMs), such as rotaxanes (Chen et al., 2003), catenanes (Balzani et al., 2000) and pseudorotaxanes (Pease et al., 2001), are sandwiched between the two thin metal films, QC devices can serve as novel non-volatile memory devices and switching devices. Moreover, when magnetic materials, such as Fe, Co and Ni, are used for the two thin metal films, QC devices can work as nanoscale spin injectors and tunneling magnetoresistance (TMR) devices. Among these devices, the resistance of the electrodes (thin metal films) can be reduced down to ~kΩ since the width of films can be easily controlled to the one as long as ~mm. This makes it possible to realize a highly sensitive detection for a junction resistance and to apply QC devices to high-frequency devices.

In this chapter, we present structural and electrical properties of Ni/polyethylene naphthalate (PEN) films used as electrodes in QC devices (Kaiju et al., 2009 & 2010) and current-voltage (*I-V*) characteristics for three types of QC devices. The three types of QC devices are as follows: (i) Ni/Ni QC devices (17 nm linewidth, 17×17 nm^2 junction area), in which two Ni thin films are directly contacted with their edges crossed (Kondo et al., 2009; Kaiju et al., 2010), (ii) Ni/NiO/Ni QC devices (24 nm linewidth, 24×24 nm^2 junction area), in which NiO thin insulators are sandwiched between two Ni thin-film edges (Kaiju et al., 2010) and (iii) Ni/poly-3-hexylthiophene (P3HT): 6, 6-phenyl C61-butyric acid methyl ester (PCBM)/Ni QC devices (16 nm linewidth, 16×16 nm^2 junction area), in which P3HT:PCBM organic molecules are sandwiched between two Ni thin-film edges (Kaiju et al., 2010; Kondo et al., 2010). In our study, we have successfully fabricated various types of QC devices with nano-linewidth and nano-junctions, which have been obtained without the use of electron-beam or optical lithography. Our method will open up new opportunities for the creation of nanoscale patterns and can also be expected as novel technique beyond conventional lithography. Furthermore, we present the calculation results of the electronic transport in Ni/organic-molecule/Ni QC devices and discussed their possibility for switching devices. According to our calculation, a high switching ratio in excess of 100000:1 can be obtained under weak coupling condition. These results indicate that QC devices fabricated using thin-film edges can be expected to have potential application in next-generation switching devices with ultrahigh on-off ratios.

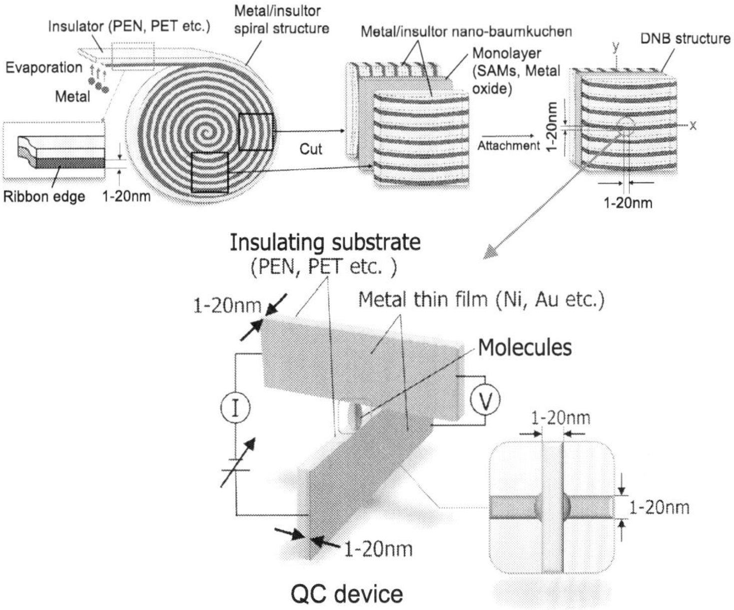

Fig. 1. Fabrication procedure of DNB structures and a schematic illustration of QC devices.

2. Fabrication and evaluation method of QC devices

2.1 Fabrication method of QC devices

The fabrication method of QC devices is shown in Fig. 2. First, Ni thin films were thermally evaporated on PEN organic substrates (2 mm width, 10 mm length and 25 μm thickness) in a high vacuum chamber at a base pressure of ~10^{-8} torr. PEN organic substrates of TEONEX Q65 were supplied by Teijin DuPont Japan and cut down from 5 to 2 mm width using a slitter in a clean environment. A boron nitride crucible N-1, made by DENKA, and a tungsten filament, made by CRAFT, were used for the thermal evaporation of Ni thin films. A heat-block stainless plate with a hole was inserted between the Ni vapor source and the PEN substrate. The length of the crucible and the aperture size in the stainless plate were designed using a geometrical simulation to evaporate uniform Ni films in-plane to PEN substrates. The temperature near PEN substrates was less than 62 °C, which was lower than the glass transition temperature T_g of 120 °C for PEN substrates. The pressure during the evaporation was 10^{-5} torr and the growth rate was 1.0-1.5 nm/min at an evaporation power of 280-350 W.

Then, fabricated Ni/PEN films were sandwiched between two polymethyl methacrylate (PMMA) resins using epoxy. The volume of the PMMA resin was 6 × 3 × 3 mm^3. The edge of PMMA/Ni/PEN/PMMA structure was polished by chemical mechanical polishing (CMP) methods using alumina (Al_2O_3) slurries with particle diameters of 0.1, 0.3 and 1.0 μm. The polishing pressure was 6.5 psi and the platen rotation speed was 75 rpm. Finally, two sets of polished PMMA/Ni/PEN/PMMA structures were prepared and attached together with their edges crossed in a highly clean environment of ISO class minus 1. The attachment pressure was 0.54 MPa and no glue was used. This is a basic process in fabricating QC devices. When

Ni/Ni QC devices are fabricated, two Ni thin films are directly contacted with their edges crossed, as shown in Fig. 2(a). For the fabrication of Ni/NiO/Ni QC devices, after two sets of polished PMMA/Ni/PEN/PMMA structures were prepared, the Ni edge in one of the two samples was oxidized by O_2 plasma at a power of 5 W (= 5 mA and 1 kV) and then the oxidized and the unoxidized Ni edges were attached together with their edges crossed, as shown in Fig. 2(b). Also, for the fabrication of Ni/P3HT:PCBM/Ni QC devices, after two sets of polished PMMA/Ni/PEN/PMMA structures were prepared, P3HT:PCBM organic molecule blend was sandwiched between two sets of PMMA/Ni/PEN/PMMA structures whose edges were crossed, as shown in Fig. 2(c). P3HT and PCBM organic molecules were separately dissolved in monochlorobenzene, then blended together with a weight ratio of 1:1 to form a 20 mg/ml solution. This P3HT:PCBM solution was dripped on one of the polished PMMA/Ni/PEN/PMMA structure, then sandwiched by the other of the polished PMMA/Ni/PEN/PMMA structure. These three types of QC devices are shown in Fig. 2.

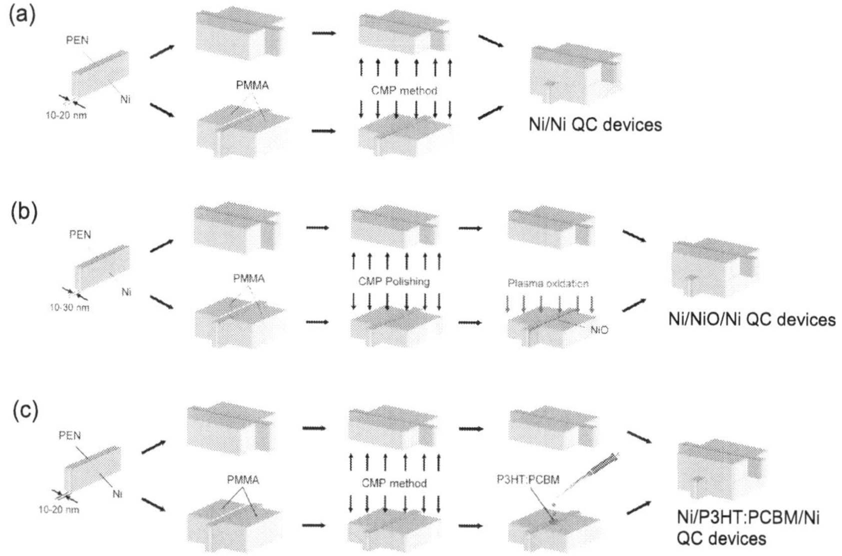

Fig. 2. Fabrication process of (a) Ni/Ni QC devices, (b) Ni/NiO/Ni QC devices and (c) Ni/P3HT:PCBM/Ni QC devices.

2.2 Evaluation methods of Ni/PEN films and QC devices

The Ni thickness was measured by a mechanical method using the stylus surface profiler DEKTAK and an optical method using the diode pumped solid state (DPSS) green laser at a wavelength of 532 nm and the photodiode detector. The surface morphologies of Ni/PEN samples were analyzed by atomic force microscope (AFM) Nanoscope IIIa. The microstructures as well as the Ni/PEN interfacial structures were examined using a JEOL JEM-3000F transmission electron microscope (TEM) operating at 300 kV. The cross-sectional TEM samples were prepared by a combination of mechanical polishing and Ar ion thinning. To reduce the beam-heating effects during ion thinning, the sample stage was cooled to -160

°C by liquid nitrogen conduction cooling. The *I-V* characteristics of QC devices were measured by a four-probe method at room temperature.

3. Ni/PEN films used as electrodes in QC devices

3.1 Cross-sectional TEM images of Ni/PEN films

Figs. 3(a) and (b) show the cross-sectional TEM images for Ni (20 nm)/PEN films. It can be seen that there is no diffusion of Ni into the PEN layer, resulting in clear and smooth formation of the Ni/PEN interface. Here, it should be noted that some researchers have reported that metal atoms diffuse into organic layers in the process of the metal evaporation onto organic layers (Tarlov, 1992; Hirose et al., 1996; Ito et al., 1999; Dürr et al., 2002). For example, the metastable atom electron spectroscopy (MAES) spectra of Au on the p-sexiphenyl (6P)/Au system shows that the features of 6P remain even though Au was deposited to about 20 nm thickness (Ito et al., 1999). This indicates that Au atoms or clusters penetrate into the 6P films. The soft x-ray photoemission spectroscopy (SXPS) investigation of the interface between evaporated indium and perylenetetracarboxylic dianhydride (PTCDA) also demonstrates that the interfacial region is very wide, ranging from 7 to 60 nm, and this means that the metal atoms of indium diffuse into PTCDA organic layers (Hirose et al., 1996). Moreover, according to studies on the interaction between evaporated Ag and octadecanethiol (ODT) on Au films using XPS, Ag deposited at 300 K migrates through the ODT layer and resides at the ODT/Au interface (Tarlov, 1992). As compared with above results, such a metal diffusion into organic layers does not occur in Ni/PEN interface. This indicates that Ni thin films on PEN organic substrates are suitable for metal/organic films used in QC devices. It can also be confirmed that the surface of Ni films is smooth, and this smoothness is in good agreement with the results of the AFM observation, where the surface roughness R_a is 1.1 nm. Fig. 3(c) shows the electron diffraction (ED) pattern for the same specimen. Ni thin films on PEN films have been shown to be face-centered-cubic structures, which are equal to those in bulk Ni structures. The ED pattern also shows that Ni thin films have polycrystalline structures, which can be recognized from the cross-sectional TEM image of Fig. 3(a). Thus, Ni /PEN films are suitable for QC devices from the viewpoint of the Ni/PEN interfacial and internal structures.

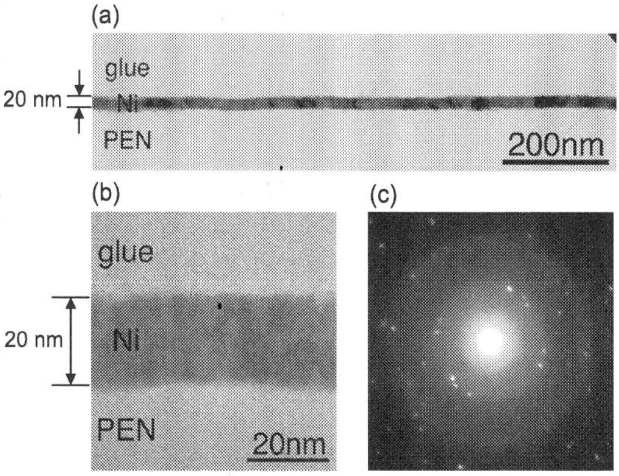

Fig. 3. (a) Cross-sectional TEM image, (b) high-resolution cross-sectional TEM image and (c) ED pattern for Ni (20 nm)/PEN films.

3.2 AFM surface morphology of Ni/PEN films

Fig. 4 shows the three-dimensional (3-D) surface images obtained from AFM observation for (a) PEN, (b) Ni (16 nm) /PEN and (c) Au (14 nm)/PEN. From the 3-D images, which are 500×500 nm^2 in area, mound-like surfaces are observed in Ni (16 nm)/PEN and Au (14 nm)/PEN, and are classified by the surface roughness R_a. Here, the surface roughness R_a is defined by

$$R_a = \frac{1}{L_x L_y} \int_0^{L_x} \int_0^{L_y} |h(x,y)|\, dxdy, \qquad (1)$$

where $h\,(x,y)$ is the height profile as a function of x and y and $L_{x(y)}$ is the lateral scanning size in the x (y) direction. R_a of PEN is 1.3 nm, which is smaller than that of widely-used organic films, such as polyethylene terephthalate (PET) and polyimide. R_a of Ni (16 nm)/PEN is also as small as 1.22 nm. In contrast, R_a of Au (14 nm)/PEN is as large as 2.53 nm. Fig. 5 shows the surface roughness as a function of the metal film thickness for Ni/PEN and Au/PEN. R_a increases up to 3.8 nm for a film thickness of 21 nm for Au/PEN. In comparison, R_a decreases slightly down to 1.1 nm with increasing the film thickness for Ni/PEN. Here, we consider the growth mode of Ni/PEN, and discuss their feasibility in QC devices from the viewpoint of the surface roughness. Fig. 6 shows the scaling properties of R_a for PEN and Ni/PEN. The inset represents the scaling properties of the root mean square (RMS) surface roughness R_q. R_q obeys a scaling law, $R_q = w(L) \propto L^{\alpha}$, where $w(L)$ is the interface width corresponding to the standard deviation of the surface height, L is the system size and α is the growth scaling exponent. The growth scaling exponent for roughening, $w(L) \propto L^{\alpha}$, has been widely used to characterize the growth of a solid from a vapor, such as the epitaxial growth of Fe/Si (111) (Chevrier et al., 1991), growth of evaporated Ag/quartz (Palasantzas et al., 1994) and molecular beam epitaxial growth of $CuCl/CaF_2$ (111) (Tong et al., 1994), as described by the Kardar-Parisi-Zhang (KPZ) equation (Kardar et al., 1986). As for PEN and Ni/PEN, α shows the almost constant value of 0.67-0.68, as seen from the similar roughness slope in any sample. This indicates that the surface morphology of Ni/PEN exhibits almost the same behaviour as that of PEN and these results are consistent with the 3-D AFM observation in Fig. 4. We have also found that the surface is described as self-affine due to $\alpha \neq \beta$, where β is the dynamical exponent in a scaling law, $R_q = w(L) \propto t_h^{\beta}$. Here, t_h is a growth thickness. As one can see from Fig. 5, β is the negative value since the surface roughness slightly decreases with increasing the thickness for Ni/PEN. This results in $\alpha \neq \beta$, which shows the self-affine growth and it can also be seen in sputtered copper films (Ohkawa et al., 2002) and evaporated silver films on silicon substrates (Thompson et al., 2004). The growth process itself of Ni thin films on PEN organic substrates is of great interest and is rich in physics, so detailed work including the dynamic physical mechanism, such as the random deposition and ballistic deposition, will be reported elsewhere. Here, we consider their feasibility in QC devices from the viewpoint of the surface roughness. Since the junction area in QC devices is determined by the film thickness, we need to clarify the surface roughness in the same scanning scale as the thickness size. As shown in Fig. 6, R_a's of Ni (16 nm)/PEN and PEN are 0.34 nm and 0.44 nm, respectively, which correspond to 2-3 atomic layers, in the scanning scale of 16 nm. This result suggests that the number of molecules sandwiched between two metal thin films in QC devices can be strictly determined in a high resolution of 2-3 atoms by controlling the thickness of Ni thin films and it leads to a high product yield of memory devices and switching devices due to the reduction of the fluctuation in a junction resistance. These experimental results indicate that Ni/PEN films are suitable for QC devices from the viewpoint of the surface, interface and internal structures.

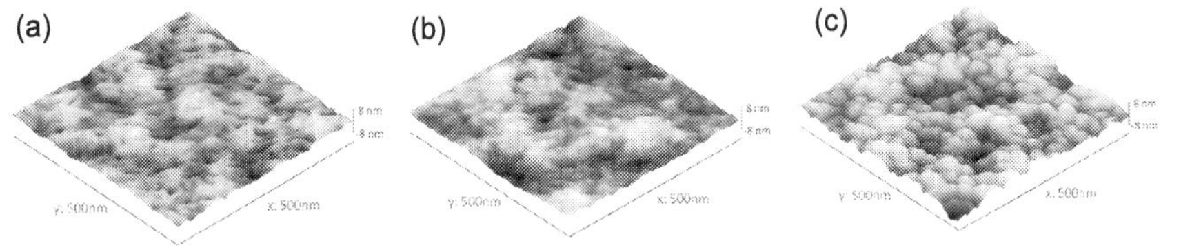

Fig. 4. 3-D surface images obtained from AFM observation for (a) PEN, (b) Ni (16 nm)/PEN and (c) Au (14 nm)/PEN.

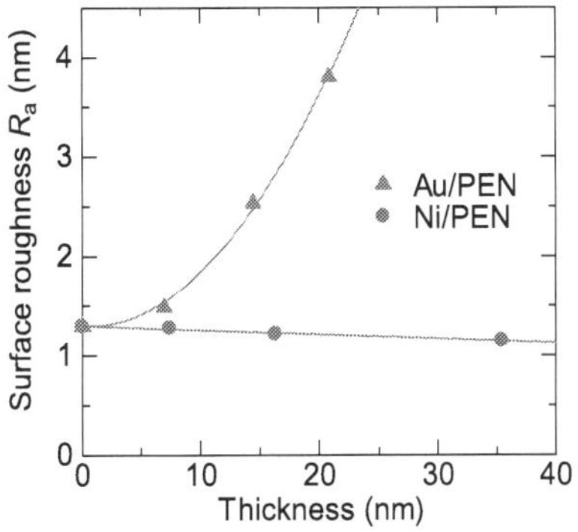

Fig. 5. Surface roughness as a function of the metal film thickness for Ni/PEN and Au/PEN.

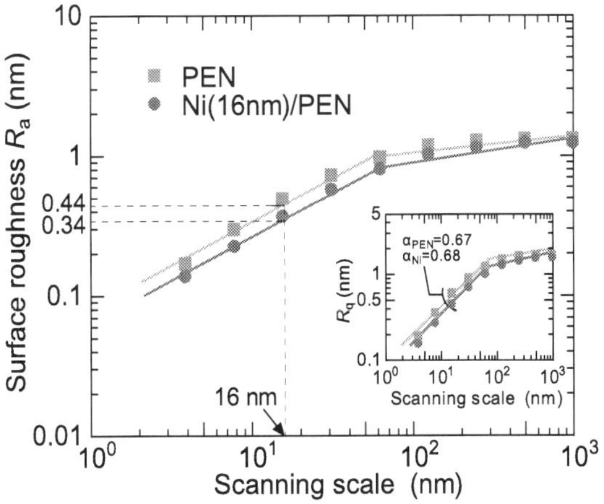

Fig. 6. Scaling properties of the surface roughness R_a for PEN and Ni/PEN. The inset represents the scaling properties of the RMS surface roughness R_q.

3.3 Electric resistivity of Ni/PEN films

Fig. 7 shows the Ni thickness dependence of the electric resistivity for Ni thin films on PEN substrates. The electric resistivity ρ_{Ni} increases with decreasing the Ni thickness d. In order to explain this experimental result quantitatively, we have calculated the electric resistivity using Mayadas-Shatzkes model (Mayadas et al., 1970). According to Mayadas-Shatzkes model, the electric resistivity ρ_{Ni} is expressed by

$$\rho_{Ni} / \rho_0 = \left[1 - \frac{3}{2}\alpha + 3\alpha^2 - 3\alpha^3 \ln(1 + \frac{1}{\alpha}) \right]^{-1}, \quad (2)$$

$$\alpha = \frac{\lambda}{D} \frac{R_g}{1 - R_g}, \quad (3)$$

where λ is the electron mean free path, D is the average grain diameter, R_g is the reflection coefficient for electrons striking the grain boundary and ρ_0 is the electric resistivity for bulk Ni. The electron mean free path λ is 11 nm for bulk Ni. The average grain diameter D is 3 nm, which has been obtained from the high-resolution TEM image and the ED pattern. The reflection coefficient R_g is 0.71-0.95, which is the extrapolation value obtained from R_g in Ni thin films with the thickness of 31-115 nm (Nacereddine et al., 2007). From Fig. 7, the experimental result shows good agreement with the calculation result quantitatively. This means that the main contribution to the electric resistivity comes from the electron scattering at grain boundaries in Ni thin films on PEN substrates. Here, we discuss the use of Ni thin films on PEN substrates for electrodes of QC devices. As can be seen from Fig. 7, the electric resistivity of Ni thin films on PEN substrates is 1-2 orders larger than that of bulk Ni. This large resistivity could produce high-resistance electrodes in QC devices. However, as stated in the introduction of this chapter, the electrode resistance can be reduced down since the film width can be controlled to the one as long as ~mm. Fig. 8(a) shows the Ni electrode resistance as a function of the linewidth l, which corresponds to the Ni thickness d, in QC devices. The schematic illustration of QC devices is shown in Fig. 8(b). In Fig. 8(a), Ni electrode resistances in the conventional cross-bar structures are also shown. The black solid line, dashed line and dotted line represent Ni electrode resistances estimated in conventional cross-bar structures with aspect ratios of 1:1, 3:1 and 5:1, respectively, where the resistivity is assumed to be the reported value in Ni thin films on glass substrates (Vries, 1987). The schematic illustration of conventional cross-bar structures is shown in Fig. 8(c). From Fig. 8(a), Ni electrode resistances in the conventional cross-bar structures are larger than 0.74, 1.2 and 3.7 MΩ in aspect ratios of 5:1, 3:1 and 1:1, respectively, for a linewidth of less than 20 nm. In contrast, QC devices show electrode resistances as small as 0.3-2 kΩ for a linewidth of 10-20 nm. This resistance reduction in Ni electrodes makes it possible to detect the resistance of sandwiched materials between two edges of Ni thin films very strictly and precisely and also this result indicates that QC devices have potential application in high-frequency devices. Thus, Ni/PEN films are suited for electrodes in QC devices from the viewpoint of the electrical properties as well as the surface, interface and internal structures.

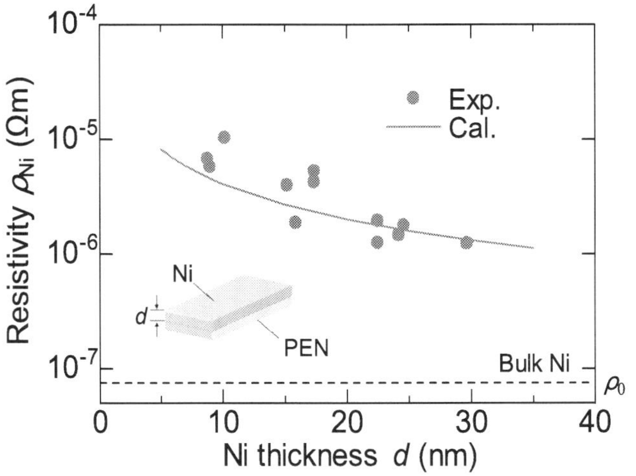

Fig. 7. Ni thickness dependence of the electric resistivity for Ni thin films on PEN substrates.

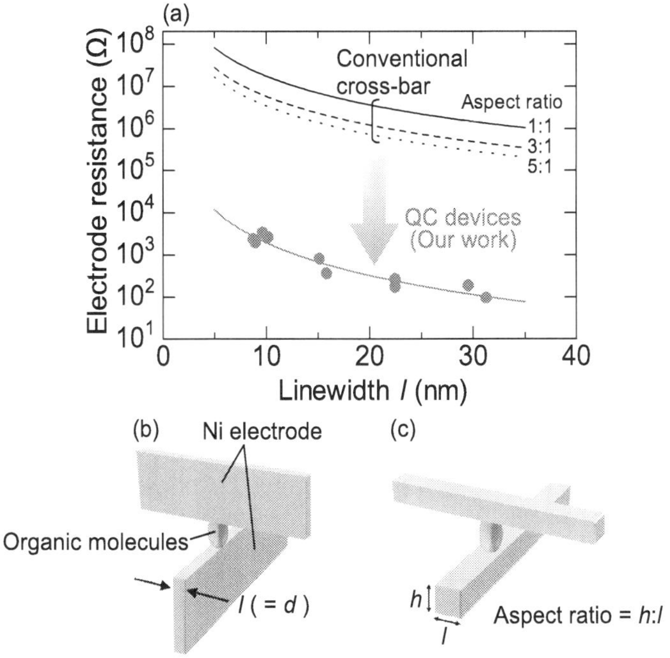

Fig. 8. (a) Ni electrode resistance as a function of the linewidth l, which corresponds to the Ni thickness d, in QC devices. Schematic illustration of (b) QC devices and (c) conventional cross-bar structures.

4. Electrical properties of QC devices

4.1 I-V characteristics of Ni/Ni QC devices

Fig. 9 shows the *I-V* characteristics for Ni/Ni QC devices fabricated using Ni/PEN films discussed above. The inset shows the experimental setup for the four-probe method. Since the Ni thickness is 17 nm, the cross-sectional area between two Ni thin films is 17×17 nm^2. As seen from Fig. 9, the ohmic *I-V* characteristics have been obtained for both positive and negative bias at room temperature. Fig. 10 shows the aging properties for the voltage with constant currents of 0.1, 0.2, 0.3, 0.5, 0.7 and 1.0 μA, respectively. The voltage is stable in any current and the standard deviation ΔV of the voltage is 22-25 mV, which corresponds to the signal-to-noise (SN) ratio of 34-52 dB, where the SN ratio is defined by 20 $\log V/\Delta V$. Here, it should be noted that the fabrication of nanojunctions using the film edges had been challenged by other researchers before (Nawate, et al., 2004). According to their attempts, Co and Ni thin films were evaporated onto glass substrates using vacuum evaporation and then they were cleaved and stuck to each other with their edges crossing. Although the current flowed across the junction, there remained a few problems: that the edge angle had to be inclined at a 15-25° and the film thickness had to be larger than 50 nm. Furthermore, the current was slightly changed as time passed although the current flowed. In contrast, such problems have not occurred in our experiments, and we have obtained stable ohmic characteristics, where there has been no change with time, for the 17 nm size junction. These experimental results indicate that our method using thin-film edges can be expected to work as a new nanostructure fabrication technology beyond conventional lithography.

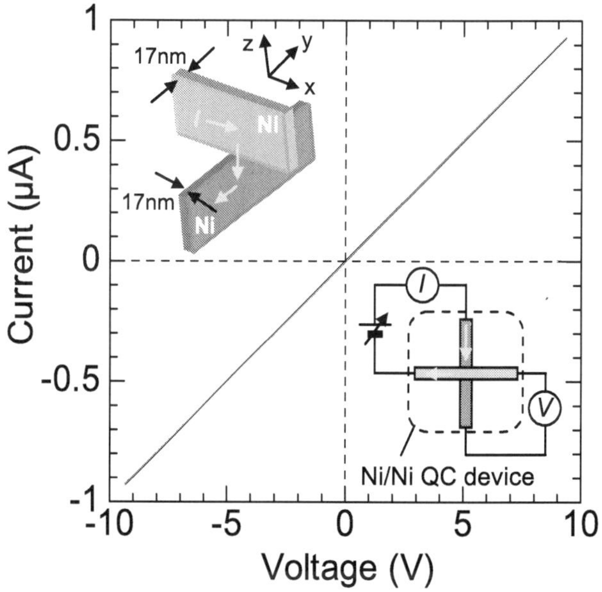

Fig. 9. *I-V* characteristics for Ni/Ni QC devices with a junction area of 17×17 nm^2. The inset shows the experimental setup for the four-probe method.

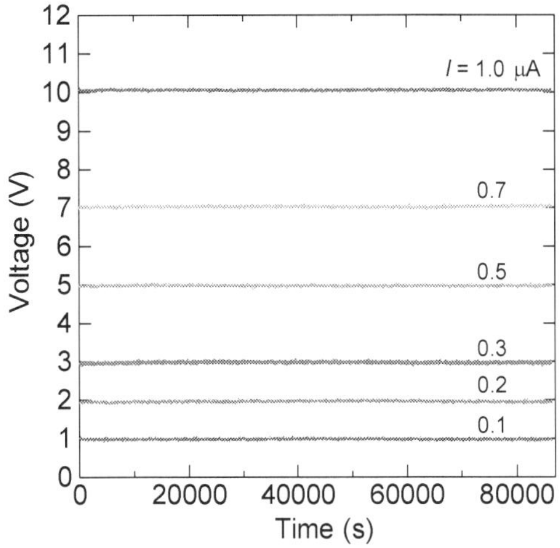

Fig. 10. Aging properties for Ni/Ni QC devices with a junction area of 17×17 nm^2.

4.2 J-V characteristics of Ni/NiO/Ni QC devices

In this section, we present current density-voltage (*J*-*V*) characteristics in Ni/NiO/Ni QC devices, which consist of NiO tunnel barriers sandwiched between two Ni thin films whose edges are crossed. First, we introduce the background of sub-micrometer scale tunnel junctions and the motivation for fabricating QC devices with tunnel barriers. Then, we show the derivation of a formula for *J*-*V* characteristics of QC devices with tunnel barriers and finally demonstrate experimental results of their *J*-*V* characteristics.

4.2.1 Background of sub-micrometer scale tunnel junctions

Sub-micrometer scale tunnel junctions have attracted much interest due to their potential application in magnetic random access memories (MRAMs), fast detectors of terahertz (THz) and IR radiation and superconducting quantum interference devices (SQUIDs). Magnetic tunnel junctions (MTJs), which consist of two ferromagnetic metals separated by thin insulators, can be expected as ultrahigh-density MRAM devices because of the giant TMR effect at room temperature (Miyazaki et al., 1995; Moodera et al., 1995; Yuasa et al., 2004; Parkin et al., 2004). CoFeB/MgO/CoFeB MTJs with a small junction area of 0.02 μm^2 exhibit a large TMR ratio of 98 %, where a clear current-induced magnetization switching (CIMS) with a low switching current density of 3.6 MA/cm^2 have been observed (Hayakawa et al., 2008). Antenna-coupled tunnel junction devices, which consist of metal/insulator/metal tunnel junctions coupled to a thin-film metal antenna, can also be expected as fast detectors of THz and IR radiation (Sanchez et al., 1978; Kale, 1985; Hobbs et al., 2005). Ni/NiO/Ni tunnel junctions with a junction area of 0.16 μm^2 coupled to thin-film metal antennas can serve as IR detectors and frequency mixers in the 10 μm band (Wilke et al., 1994; Fumeaux et al., 1996). Moreover, much effort has been devoted to the development of sub-micrometer scale SQUIDs, which are very promising devices with a high magnetic flux sensitivity (Rugar et al., 2004; Troeman et al., 2007; Huber et al., 2008; Kirtley et al., 2009). Aluminum SQUIDs with an effective area of 0.034 μm^2 display a high flux sensitivity of $1.8\times10^{-6}\ \Phi_0/\mathrm{Hz}^{1/2}$, where $\Phi_0=h/2e$ is the flux quantum, and operates in fields as high as 0.6 T (Finkler et al., 2010). Thus, sub-micrometer scale tunnel junctions can find a wide range application in various functional electronic devices.

As shown in the previous section, our nanostructure fabrication method using thin-film edges offers a way to overcome the feature size limit of conventional lithography. This method can be extended to fabricating nanoscale tunnel junctions. The realization of nanoscale tunnel junctions leads to the enhancement of the performance in MRAMs, fast detectors of THz and IR radiation and SQUIDs. From this motivation, it is of great importance to fabricate QC devices with tunnel barriers and very meaningful to investigate their *J-V* characteristics.

4.2.2 Derivation of a formula for J-V characteristics of QC devices with tunnel barriers

We derive a formula for *J-V* characteristics in QC devices with tunnel barriers, which consist of thin insulating barriers sandwiched between two thin metal films whose edges are crossed, shown in Fig. 11. The thicknesses of the top and bottom metal films are l_z and l_y, respectively. The barrier height and the barrier thickness are ϕ and d, respectively. The number of electrons per unit area and unit time from the top to bottom metal is given by

$$\begin{aligned} N_S^T &= \int_0^\infty v_x n(v_x)\,|T(v_x)|^2\,dv_x \\ &= \frac{1}{2\pi}\int_0^\infty \frac{1}{\hbar}\frac{\partial E}{\partial k_x} n(k_x)\,|T(k_x)|^2\,dk_x\,, \end{aligned} \tag{4}$$

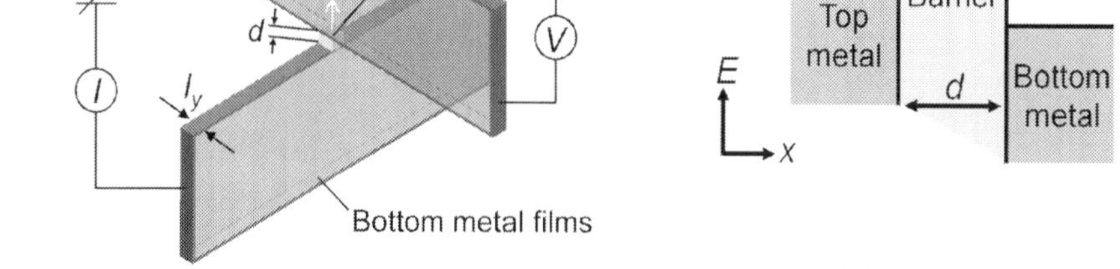

Fig. 11. (a) QC devices with tunnel barriers and (b) the potential profile of the barrier, which is used for the calculation of the transmission probability.

where v_x is the x-direction velocity, $n(v_x)$ is the number of electrons per unit volume and unit velocity, k_x is the x-direction wave number, $n(k_x)$ is the electron density, $\hbar$ is Planck's constant divided by 2π and E is the total energy of electrons, which is expressed by

$$\begin{aligned} E &= E_x + E_y + E_z \\ &= \frac{\hbar^2}{2m}\{k_x{}^2 + k_y{}^2 + (\frac{n\pi}{l_z})^2\}, \end{aligned} \tag{5}$$

where E_x, E_y and E_z are the x, y and z components of the total energy of electrons, respectively, m is the electron mass, k_y is the y-direction wave number and n is the positive integer. $T(k_x)$ is the transmission probability of electrons through the barrier. $T(k_x)$ is calculated by the well-known Wentzel-Kramers-Brillouin (WKB) approximation. The potential profile is illustrated in Fig. 11(b). Here, the electron density is given by

$$n(k_x) = \frac{1}{2\pi^2}\int_0^\infty \int_0^\infty f(E)dk_y dk_z \,, \tag{6}$$

where $f\,(E)$ is the Fermi-Dirac distribution function. Substituting Eqs. (5) and (6) in Eq. (4) yields

$$N_S^T = \frac{1}{4\pi^3\hbar}\int_0^\infty |T(k_x)|^2 \,[\int_0^\infty \int_0^\infty f(E)dk_y dk_z]dE_x. \tag{7}$$

Since the density of states in the k_y direction for the top metal is equal to the solution of one-dimensional density of states, the relation between k_y and E_y is given by

$$dk_y = \frac{m}{\hbar\sqrt{2mE_y}}dE_y. \tag{8}$$

Substituting Eqs. (5) and (8) in Eq. (7) gives

$$N_S^T = \frac{1}{4\pi^3\hbar}\int_0^\infty |T(k_x)|^2[\sum_{n=1}^{\infty}\int_0^\infty f(E)\frac{m}{\hbar\sqrt{2mE_y}}\theta(E_y - \frac{\hbar^2}{2m}(\frac{n\pi}{l_z})^2)dE_y]dE_x, \tag{9}$$

where θ(x) is the unit step function. The number of electrons per unit area and unit time from the bottom to top metal is also derived in a similar manner. When the positive bias V is applied to the bottom metal, the Fermi-Dirac distribution function is written by $f\,(E+eV)$. Thus, the number of electrons from the bottom to top metal is

$$N_S^B = \frac{1}{4\pi^3\hbar}\int_0^\infty |T(k_x)|^2[\sum_{n=1}^{\infty}\int_0^\infty f(E+eV)\frac{m}{\hbar\sqrt{2mE_z}}\theta(E_z - \frac{\hbar^2}{2m}(\frac{n\pi}{l_y})^2)dE_z]dE_x. \tag{10}$$

Consequently, we obtain for the net current density J through the barrier

$$\begin{aligned} J &= e(N_S^T - N_S^B) \\ &= \frac{e}{4\pi^3\hbar}\int_0^\infty |T(k_x)|^2[\sum_{n=1}^{\infty}\int_0^\infty f(E)\frac{m}{\hbar\sqrt{2mE_y}}\theta(E_y - \frac{\hbar^2}{2m}(\frac{n\pi}{l_z})^2)dE_y \\ &\quad - \sum_{n=1}^{\infty}\int_0^\infty f(E+eV)\frac{m}{\hbar\sqrt{2mE_z}}\theta(E_z - \frac{\hbar^2}{2m}(\frac{n\pi}{l_y})^2)dE_z]\,dE_x. \end{aligned} \tag{11}$$

We can calculate *J-V* characteristics as a function of the metal thickness, the barrier height and the barrier thickness using Eq. (11).

4.2.3 J-V characteristics of Ni/NiO/Ni QC devices

Fig. 12 shows the *J-V* characteristics for Ni/NiO/Ni QC devices at room temperature. Since the Ni thickness is 24 nm, the junction area is 24×24 nm^2. In Fig. 12(a), solid circles are experimental data and dashed, solid and dotted lines are calculation results for barrier heights of 0.6, 0.8 and 1.0 eV, respectively, with a constant barrier thickness of 0.63 nm. In Fig. 12(b), solid circles are experimental data and dashed, solid and dotted lines are calculation results for barrier thicknesses of 0.42, 0.63 and 0.84 nm, respectively, with a barrier height of 0.8 eV. As seen from Fig. 12, experimental results are in good agreement with calculation results for a barrier height of 0.8 eV and a barrier thickness of 0.63 nm. Since the crystal structure of NiO is a NaCl-type structure with a lattice constant a = 0.42 nm, 2 monolayer of NiO is formed as shown in the inset of Fig. 12. Here, we discuss the barrier height of NiO insulating layers. The bulk NiO is a charge-transfer insulator, which has been approved by X-ray absorption (Kuiper et al., 1989) and X-ray photoemission and bremsstrahlung isochromat spectroscopy (Sawatzky et al., 1984; Elp et al., 1992). According

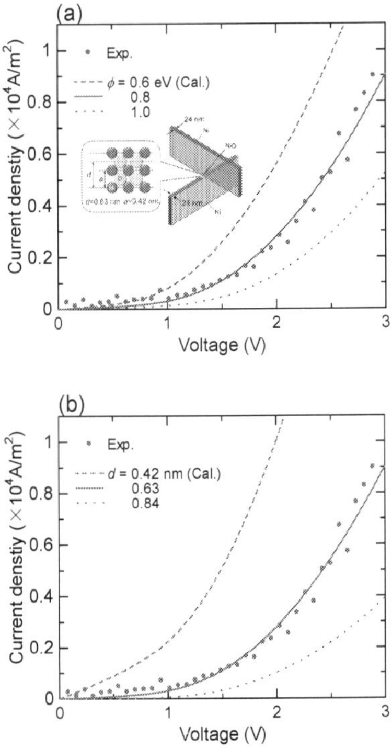

Fig. 12. (a) Barrier height dependence and (b) barrier thickness dependence of *J-V* characteristics for Ni/NiO/Ni QC devices with a junction area of 24×24 nm^2 at room temperature.

to these experiments, the energy gap is 4.0-4.3 eV, which agrees with recent calculation results performed using the full-potential linearized augmented plane wave (FLAPW) method within the LSDA(GGA)+U (LSDA, local-spin-density approximation; GGA, generalized gradient approximation; U, on-site Coulomb interation) (Cai et al., 2009). Since the Fermi energy of NiO is located in the level of 1.0 eV higher than the top of the valence band (Sawatzky et al., 1984), the barrier height in metal/NiO/metal tunnel junctions is estimated to be 3.0-3.3 eV. However, several researchers have reported that the barrier height in metal/NiO/metal tunnel junctions tends to become smaller than the estimated value of 3.0-3.3 eV (Doudin et al., 1997; Ono et al., 1997; Hobbs et al., 2005). For example, the barrier height of Ni/NiO/Ni tunnel junctions with a junction area of 0.3 μm^2 is as low as 0.18-0.22 eV with a barrier thickness of 2.5 nm (Hobbs et al., 2005). Electrodeposited Ni/NiO/Co tunnel junctions with a cross section of 0.01 μm^2 exhibit a low barrier height of 0.2-0.4 eV with a barrier thickness of 2-3 nm (Doudin et al., 1997). Thus, the barrier height of NiO thin films is reported to be smaller than that of NiO bulk materials. In our experiments, the barrier height in Ni/NiO/Ni QC devices is also estimated to be as low as 0.8 eV. This means that the derived formula is useful for the evaluation of the barrier height and the barrier thickness in QC devices with tunnel barriers. This fact also indicates that high-quality NiO thin films were formed and Ni/NiO/Ni QC devices worked well as nanoscale tunnel devices. Therefore, our method to fabricate nanoscale junctions utilizing thin-film edges can be expected to work as a new technique for the creation of nanoscale tunnel junctions.

4.3 I-V characteristics of Ni/P3HT:PCBM/Ni QC devices

In this section, we present *I-V* characteristics in Ni/P3HT:PCBM/Ni QC devices, which consist of P3HT:PCBM organic molecules sandwiched between two Ni thin films whose edges are crossed. First, we introduce the background of molecular electronic devices and the motivation for fabricating QC devices with organic molecules. Then, we show the theoretical calculation for *I-V* characteristics of QC devices with organic molecules and compare them with experimental results. Finally, we discuss their possibility for switching devices with high on-off ratios.

4.3.1 Background of molecular electronic devices

Molecular electronics have stimulated considerable interest as a technology that offers the prospect of scaling down device dimensions to a few nanometers and that also promotes a practical introduction to high-density memory applications (Chen et al., 1999; Reed et al., 2001; Chen et al., 2003; Lau et al., 2004; Wu et al., 2005; Mendes et al., 2005; Green et al., 2007). Especially, in the ITRS 2009 edition, molecular memory devices have been expected as candidates for beyond-CMOS devices since they offer the possibility of nanometer-scale components. For examples, Au/2′-amino-4-ethynylphenyl-4′-ethynylphenyl-5′-nitro-1-benzenethiolate/Au molecular devices with a junction area of 30×30-50×50 nm^2 have been fabricated using nanopore methods, and have exhibited a negative differential resistance and a large on-off peak-to-valley ratio (Chen et al., 1999). Molecular devices, which comprise a monolayer of bistable [2]rotaxanes sandwiched between two Ti/Pt metal electrodes with a linewidth of 40 nm, have also been fabricated using nanoimprint lithography, and have shown bistable *I-V* characteristics with high on-off ratios and reversible switching properties (Chen et al., 2003). Moreover, cross-bar molecular devices consisting of a monolayer of bistable [2]rotaxanes sandwiched between top Ti and bottom Si electrodes (16 nm wide at 33 nm pitch) have been fabricated using SNAP methods, and have acted as 160 kbit memories patterned at a density of 100 Gbit/cm[2] (Green et al., 2007). Thus, molecular electronic devices can be expected as next-generation high-density memories.

As demonstrated in the previous section, our method using thin-film edges provides the fabrication of nanoscale tunnel junctions. This method can be extended to fabricating nanoscale molecular devices. The realization of nanoscale molecular devices leads to the application in next-generation high-density memories. From this motivation, it is of great significance to fabricate QC devices with organic molecules and very interesting to investigate their J-V characteristics.

4.3.2 Theoretical calculation and experimental results of I-V characteristics in Ni/P3HT: PCBM/Ni QC devices

We calculate the I-V characteristics of QC devices with the molecule within the framework of Anderson model (Kondo et al., 2008 & 2009; Kondo, 2010). QC devices with the molecule consist of the molecule sandwiched between two thin metal films whose edges are crossed. The molecule is assumed to have two energy levels. In this calculation model, the Anderson Hamiltonian is given by

$$H = H_{\text{electrodes}} + H_{\text{molecules}} + H_{\text{t}}. \quad (12)$$

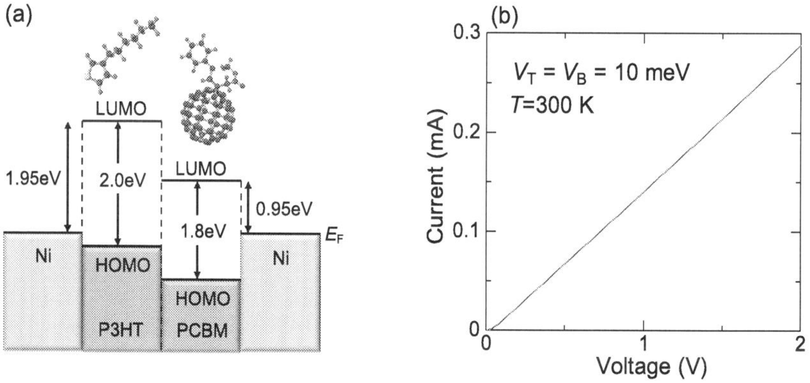

Fig. 13. (a) Energy diagram used in the calculation of I-V characteristics for QC devices and (b) the calculated I-V characteristics for QC devices under the strong coupling limit.

Here, $H_{\text{electrodes}}$ is the Hamiltonian of both of the metal electrodes, $H_{\text{molecules}}$ is the Hamiltonian of the molecule sandwiched between the metal electrodes and H_t is the transfer Hamiltonian between the sandwiched molecule and each electrode. $H_{\text{electrodes}}$, $H_{\text{molecules}}$ and H_t can be expressed by

$$\begin{aligned} H_{\text{electrodes}} &= \sum_{\alpha=T,B}\sum_{k,\sigma} \varepsilon_{k\sigma} c^{+}_{\alpha,k\sigma} c_{\alpha,k\sigma}, \\ H_{\text{molecules}} &= \sum_{i,\sigma} \varepsilon_0(i)\, a^{+}_{i,\sigma} a_{i,\sigma}, \\ H_{\text{t}} &= \sum_{\alpha=T,B}\sum_{k,\sigma}\sum_{i,\sigma} (V_\alpha c^{+}_{\alpha,k\sigma} a_{i,\sigma} + h.c.). \end{aligned} \tag{13}$$

Here, $\varepsilon_{k\sigma}$ is the free electron energy of $\hbar^2 k^2 / 2m$, where m is the free electron mass, $\hbar$ is Planck's constant divided by 2π and k is a two-dimensional wavenumber. $c^{+}_{\alpha,k\sigma}$ and $c_{\alpha,k\sigma}$ are creation and annihilation operators for electrons of wavenumber k and spin index σ in α electrode. α indicates the top or bottom metal electrode. $\varepsilon_0(i)$ represents the ith energy level of eigenstates of the molecule. $a^{+}_{i,\sigma}$ and $a_{i,\sigma}$ are creation and annihilation operators for electrons of spin index σ in the ith energy level of the molecule. V_α is the transfer matrix between the molecule and the α electrode. In this calculation, we assume the molecule has two energy levels of $\varepsilon_0(1)$=0.95 eV and $\varepsilon_0(2)$=1.95 eV, which are estimated from Fermi levels E_F of the metal electrode, imagining that QC devices consist of the Ni electrodes and the sandwiched molecule of a P3HT:PCBM organic molecule, as shown in Fig. 13(a) (Eastman, 1970; Thompson et al., 2008). E_F is assumed to be 9.071 eV for Ni electrodes (Wang et al., 1977). Considering H_t as a perturbation, we can derive a formula for the I-V characteristics from the top to the bottom electrode using the many-body perturbation technique. As a result, the current flowing from the top to the bottom electrode can be expressed by

$$I = \frac{2e^2}{h}\int_{E_F}^{E_F+eV} d\varepsilon \sum_i \left(\frac{4\Gamma_T(\varepsilon)\Gamma_B(\varepsilon)}{(\varepsilon-\varepsilon_0(i))^2 + (\Gamma_T(\varepsilon)+\Gamma_B(\varepsilon))^2} \right) [f(\varepsilon - eV - E_F) - f(\varepsilon - E_F)], \tag{14}$$

where e is the elementary charge and $f(\varepsilon)$ is the Fermi-Dirac distribution function. $\Gamma_{T(B)}(\varepsilon)$ is the coupling strength between the top (bottom) Ni electrode and the P3HT:PCBM organic molecule, which is given by

$$\Gamma_{T(B)}(\varepsilon) = \pi D_{T(B)}(\varepsilon)\,|V_{T(B)}|^2, \tag{15}$$

where $D_{T(B)}(\varepsilon)$ is a density of states of electrons for the top (bottom) Ni electrode and $V_{T(B)}$ is the coupling constant between the top (bottom) Ni electrode and the P3HT:PCBM organic molecule. Fig. 13(b) shows the calculated I-V characteristics for Ni/P3HT:PCBM/Ni QC devices under the strong coupling limit. $V_{T(B)}$ is assumed to be 10.0 meV, corresponding to $\Gamma_{T(B)}$ of 3927 meV. We have obtained the ohmic I-V characteristics with a resistance of 6.7 kΩ. Fig. 14 shows the experimental results of I-V characteristics for Ni/P3HT:PCBM/Ni QC devices at room temperature. The inset represents the experimental setup. The Ni thickness is 16 nm. Therefore, the junction area is 16×16 nm^2. We have obtained ohmic characteristics with a junction resistance of 32 Ω. Here, we compare this experimental value with the calculation result. The calculation result has demonstrated that the resistance is 6.7 kΩ, where the junction area is 1×1 nm^2, which is expected as a size of one P3HT:PCBM organic molecule. The number of the conductance channel is four, taking into consideration the spin degeneracy. On the other hand, in experiments, the junction area of P3HT:PCBM organic molecules is 16×16 nm^2, which corresponds to 1024 (=4×16×16) conductance channels.

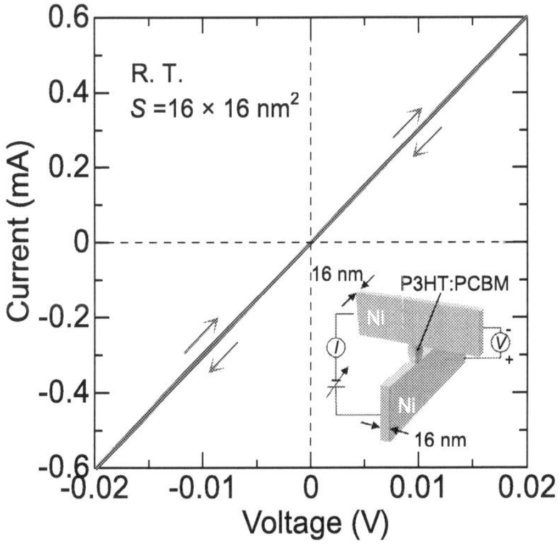

Fig. 14. Experimental results of *I-V* characteristics for Ni/P3HT:PCBM/Ni QC devices with a junction area of 16×16 nm^2 at room temperature. The inset represents the experimental setup.

Therefore, the junction resistance in a size of 16×16 nm^2 is calculated to be 26 Ω (=6.7kΩ/16/16), which is in good agreement with the experimental value of 32 Ω. This result indicates that electrons in nanoscale junctions can transport through the molecules in the ballistic regime without any scattering. This also demonstrates that our method to fabricate nanoscale junctions utilizing thin-film edges can be a useful tool for the creation of nanoscale molecular devices.

4.3.3 Possibility of Ni/P3HT:PCBM/Ni QC devices for switching devices

Finally, we have discussed the possibility of Ni/P3HT:PCBM/Ni QC devices for switching devices with high on-off ratios. Fig. 15 shows the calculated *I-V* characteristics for Ni/P3HT:PCBM/Ni QC devices under the weak coupling condition. $V_{T(B)}$ is assumed to be 0.2 meV, corresponding to $\Gamma_{T(B)}$ of 1.57 meV. From Fig. 15, the calculated result shows the sharp steps at the positions of the energy level of the P3HT:PCBM organic molecule. The off-state current I_0 is 3.8 pA at the voltage V_0 of 0.03 V, and the on-state current I_1 is 0.57 μA at the voltage V_1 of 1.03 V. As we estimate the switching on-off ratio, the I_1/I_0 ratio is found to be in excess of 100000:1. Here, it should be noted that it is essentially important that the junction area is as small as nanometer scale in order to obtain such a high on-off ratio. When the junction area is as large as micrometer scale, the number of molecules sandwiched between the electrodes is large, so the energy level can be broadened. In contrast, when the junction area is as small as nanometer scale, the number of molecule is small, so the energy level can be discrete. This discrete energy level leads to the sharp steps in the *I-V* characteristics, which can produce such a high switching ratio. Thus, Ni/P3HT:PCBM/Ni QC devices utilizing thin-film edges can be expected to have potential application in switching devices with high on-off ratios.

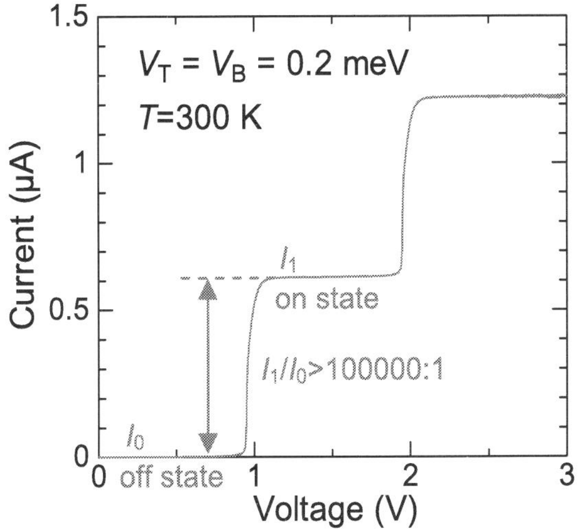

Fig. 15. Calculated *I-V* characteristics for Ni/P3HT:PCBM/Ni QC devices under the weak coupling condition.

5. Conclusions

In this chapter, we have introduced structural and electrical properties of Ni/PEN films used as electrodes in QC devices and *I-V* characteristics for three types of QC devices. The three types of QC devices are as follows: (i) Ni/Ni QC devices (17 nm linewidth, 17×17 nm^2 junction area), in which two Ni thin films are directly contacted with their edges crossed, (ii) Ni/NiO/Ni QC devices (24 nm linewidth, 24×24 nm^2 junction area), in which NiO thin insulators are sandwiched between two Ni thin-film edges and (iii) Ni/P3HT:PCBM/Ni QC devices (16 nm linewidth, 16×16 nm^2 junction area), in which P3HT:PCBM organic molecules are sandwiched between two Ni thin-film edges. In our study, we have successfully fabricated various types of QC devices with nano-linewidth and nano-junctions, which have been obtained without the use of electron-beam or optical lithography. Our method will open up new opportunities for the creation of nanoscale patterns and can also be expected as novel technique beyond conventional lithography. Furthermore, we have presented the calculation results of the electronic transport in Ni/organic-molecule/Ni QC devices and discussed their possibility for switching devices. According to our calculation, a high switching ratio in excess of 100000:1 can be obtained under weak coupling condition. These results indicate that QC devices fabricated using thin-film edges can be expected to have potential application in next-generation switching devices with ultrahigh on-off ratios.

6. Acknowledgment

This research has been partially supported by the Management Expenses Grants for National Universities Corporations and a Grant-in-Aid for Young Scientists from The Ministry of Education, Culture, Sports, Science and Technology (MEXT), Precursory Research for Embryonic Science and Technology program from Japan Science and Technology Agency (JST) and a Grant-in-Aid for Scientific Research from Japan Society for the Promotion of Science (JSPS). The authors would like to express their sincere appreciation to Dr. M. Hirasaka of Teijin Ltd. and Research Manager K. Kubo of Teijin DuPont Films Japan Ltd. for supplying PEN organic films, Prof. Y. Hirotsu, Assoc. Prof. M. Ishimaru and Assist. Prof. A. Hirata at Osaka University for investigating cross-sectional TEM images and ED patterns, Y. Inoue of Meiwafosis Co., Ltd. for examining the fabrication of NiO insulators and Prof. M. Yamamoto, Assist. Prof. K. Matsuda, A. Ono, N. Basheer, N. Kawaguchi, S. White, H. Sato and M. Takei at Hokkaido University for their cooperation and helpful discussions.

7. References

Balzani, V., Credi, A., Mattersteig, G., Matthews, O. A., Raymo, F. M., Stoddart, J. F., Venturi, M.; White, A. J. P. & Williams, D. J. J. (2000). *Org. Chem.* 65: 1924.

Cai, T., Han, H., Yu, Y., Gao, T., Du, J. & Hao, L. (2009). *Physica B* 404: 89.

Chen, J., Reed, M. A., Rawlett, A. M. & Tour, J. M. (1999). *Science* 286: 1550.

Chen, Y., Ohlberg, D. A. A., Li, X., Stewart, D. R., Williams, R. S., Jeppesen, J. O., Nielsen, K. A., Stoddard, J. F., Olynick, D. L. & Anderson, R. (2003). *Appl. Phys. Lett.* 82: 1610.

Chevrier, J., Lethanh, V., Buys, R. & Derrien, J. (1991). *Europhys. Lett.* 16: 737.

Doudin, B., Redmond, G., Gilbert, S. E. & Ansermet, J. -Ph. (1997). *Phys. Rev. Lett.* 79: 933.

Dürr, A. C., Schreiber, F., Kelsch, M., Carstanjen, H. D. & Dosch, H. (2002). *Adv. Mater.* 14: 961.

Eastman, D. E. (1970). *Phys. Rev. B* 2: 1.

Elp, J., Eskes, H., Kuiper, P. & Sawatzky, G. A. (1992). *Phys. Rev. B* 45: 1612.

Finders, J., Dusa, M., Vleeming, B., Megens, H., Hepp, B., Maenhoudt, M., Cheng, S. & Vandeweyer, T. (2008). *Proc. SPIE* 6924: 692408.

Finkler, A., Segev, Y., Myasoedov, Y., Rappaport, M. L., Ne'eman, L., Vasyukov, D., Zeldov, E., Huber, M. E., Martin, J. & Yacoby, A. (2010). *Nano Lett.* 10: 1046.

Fujikawa, S., Takaki, R. & Kunitake, T. (2006). *Langmuir* 22: 9057.

Fumeaux, C., Herrmann, W., Rothuizen, H., Natale, P. D. & Kneubuehl, F. K. (1996). *Appl. Phys. B* 63: 135.

Green, J. E., Choi, J. W., Boukai, A., Bunimovich, Y., Johnston-Halperin, E., Delonno, E., Luo, Y., Sheriff, B. A., Xu, K., Shin, Y. S., Tseng, H.-R., Stoddart, J. F. & Heath, J. R. (2007). *Nature* 445: 414.

Hayakawa, J., Ikeda, S., Miura, K., Yarnanouchi, M., Lee, Y. M., Sasaki, R., Ichimura, M., Ito, K., Kawahara, T., Takemura, R., Meguro, T., Matsukura, F., Takahashi, H., Matsuoka, H. & Ohno, H. (2008). *IEEE Trans. Mag.* 44: 1962.

Hirose, Y., Kahn, A., Aristov, V. & Soukiassian, P. (1996). *Appl. Phys. Lett.* 68: 217.

Hobbs, P. C. D., Laibowitz, R. B. & Libsch, F. R. (2005). *Appl. Opt.* 44: 6813.

Huber, M. E., Koshnick, N. C., Bluhm, H., Archuleta, L. J., Azua, T., Bjornsson, P. G., Gardner, B. W., Halloran, S. T., Lucero, E. A. & Moler, K. A. (2008). *Rev. Sci. Instrum.* 79: 053704.

Ishibashi, A. (2003). Jpn. Pat. 3974551.

Ishibashi, A. (2004). *Proceedings of International Symposium on Nano Science and Technology*, pp. 44-45, Tainan, Taiwan, November 20-21, 2004.

Ishibashi, A., Kaiju, H., Yamagata, Y. & Kawaguchi, N. (2005). *Electron. Lett.* 41: 735.

Ito, E., Oji, H., Furuta, M., Ishii, H., Oichi, K., Ouchi, Y. & Seki, K. (1999). *Synthetic Metals* 101: 654.

Kaiju, H., Kawaguchi, N. & Ishibashi, A. (2005). *Rev. Sci. Instrum.* 76: 085111.

Kaiju, H., Ono, A., Kawaguchi, N. & Ishibashi, A. (2008). *Jpn. J. Appl. Phys.* 47: 244.

Kaiju, H., Ono, A., Kawaguchi, N. & Ishibashi, A. (2008). *J. Appl. Phys.* 103: 07B523.

Kaiju, H., Ono, A., Kawaguchi, N., Kondo, K., Ishibashi, A., Won, J. H., Hirata, A., Ishimaru, M. & Hirotsu, Y. (2009). *Appl. Surf. Sci.* 255: 3706.

Kaiju, H., Kondo, K., Ono, A., Kawaguchi, N., Won, J. H., Hirata, A., Ishimaru, M., Hirotsu, Y. & Ishibashi, A. (2010). *Nanotechnology* 21: 015301.

Kaiju, H., Kondo, K., Basheer, N., Kawaguchi, N., White, S., Hirata, A., Ishimaru, M., Hirotsu, Y. & Ishibashi, A. (2010). *Mater. Res. Soc. Symp. Proc.* 1252: J0208.

Kaiju, H., Basheer, N., Kondo, K. & Ishibashi, A. (2010). *IEEE Trans. Magn.* 46: 1356.

Kaiju, H., Kondo, K. & Ishibashi, A. (2010). *Jpn. J. Appl. Phys.* 49: 105203.

Kale, B. M. (1985). *Opt. Eng.* 24: 267.

Kardar, M., Parisi, G. & Zhang, Y. -C. (1986). *Phys. Rev. Lett.* 56: 889.

Kirtley, J. R. (2009). *Supercond. Sci. Technol.* 22: 064008.

Kondo, K. & Ishibashi, A. (2006). *Jpn. J. Appl. Phys.* 45: 9137.

Kondo, K., Kaiju, H. & Ishibashi, A. (2008). *Mater. Res. Soc. Symp. Proc.* 1067: B0301.

Kondo, K., Kaiju, H. & Ishibashi, A. (2009). *J. Appl. Phys.* 105: 07D522.

Kondo, K., Kaiju, H. & Ishibashi, A. (2010). *Mater. Res. Soc. Symp. Proc.* 1198: E0701.

Kondo, K. (2010). *J. Appl. Phys.* 107: 09C709.

Kuiper, P., Kruizinga, G., Ghijsen, J., Sawatzky, G. A. & Verweij, H. (1989). *Phys. Rev. Lett.* 62: 221.

Kwon, S., Yan, X., Contreras, A. M., Liddle, J. A., Somorjai, G. A. & Bokor, J. (2005). *Nano Lett.* 5: 2557.

Lau, C. N., Stewart, D. R., Williams, R. S. & Bockrath, M. (2004). *Nano Lett.* 4: 569.

Mayadas, A. F. & Shatzkes, M. (1970). *Phys. Rev. B* 1: 1382.

Melosh, N. A., Boukai, A., Diana, F., Gerardot, B., Badolato, A., Petroff, P. M. & Heath, J. R. (2003). *Science* 300: 112.

Mendes, P. M., Flood, A. H. & Stoddart, J. F. (2005). *Appl. Phys. A* 80: 1197.

Miyazaki, T. & Tezuka, N. (1995). *J. Magn. Magn. Mater.* 139: L231.

Moodera, J. S., Kinder, L. R., Wong, T. M. & Meservey, R. (1995). *Phys. Rev. Lett.* 74: 3273.

Nacereddine, C., Layadi, A., Guittoum, A., Cherif, S. -M., Chauveau, T., Billet, D., Youssef, J. B., Bourzami, A. & Bourahli, M. -H. (2007). *Mater. Sci. Eng. B* 136: 197.

Naulleau, P. P., Anderson, C. N., Chiu, J., Denham, P., George, S., Goldberg, K. A., Goldstein, M., Hoef, B., Hudyma, R., Jones, G., Koh, C., Fontaine, B. L., Ma, A., Montgomery, W., Niakoula, D., Park, J.-o., Wallow, T. & Wurm, S. (2009). *Microelectron. Eng.* 86: 448.

Nawate, M., Shinohara, K., Honda, S. & Tanaka, H. (2004). *Trans. Mater. Res. Soc. Jpn.* 29: 1599.

Ohkawa, K., Nakano, T. & Baba, S. (2002). *J. Vac. Soc. Jpn.* 45: 134.

Ono, K., Shimada, H. & Ootuka, Y. (1997). *J. Phys. Soc. Jpn.* 66: 1261.

Palasantzas, G. & Krim, J. (1994). *Phys. Rev. Lett.* 73: 3564.

Parkin, S. S., Kaiser, C., Panchula, A., Rice, P. M., Hughes, B., Samant, M. & Yang, S.-H. (2004). *Nat. Mater.* 3: 862.

Pease, A. R., Jeppesen, J. O., Stoddart, J. F., Luo, Y., Collier, C. P. & Heath, J. R. (2001). *Acc. Chem. Res.* 36: 433.

Rahaman, M. D., Kaiju, H., Kawaguchi, N. & Ishibashi, A. (2008). *Jpn. J. Appl. Phys.* 47: 5712.

Reed, M. A., Chen, J., Rawlett, A. M., Price, D. W. & Tour, J. M. (2001). *Appl. Phys. Lett.* 78: 3735.

Rugar, D., Budakian, R., Mamin, H. J. & Chui, B. W. (2004). *Nature* 430: 329.

Sanchez, A., Davis, C. F., Liu, K. C. & Javan, A. (1978). *J. Appl. Phys.* 49: 5270.

Sawatzky, G. A. & Allen, J. W. (1984). *Phys. Rev. Lett.* 53: 2339.

Sewell, H., Chen, A., Finders, J. & Dusa, M. (2009). *Jpn. J. Appl. Phys.* 48: 06FA01.

Tarlov, M. J. (1992). *Langmuir* 8: 80.

Thompson, B. C. & Fréchet, J. M. (2008). *J. Angew. Chem. Int. Ed.* 47: 58.

Thompson, C., Palasantzas, G., Feng, Y. P., Sinha, S. K. & Krim, J. (1994). *Phys. Rev. B* 49: 4902.

Tong, W. M., Williams, R. S., Yanase, A., Segawa, Y. & Anderson, M. S. (1994). *Phys. Rev. Lett.* 72: 3374.

Troeman, A. G. P., Derking, H., Borger, B., Pleikies, J., Veldhuis, D. & Hilgenkamp. H. (2007). *Nano Lett.* 7: 2152.

Vries, J. W. C. (1987). *Thin Solid Films* 150: 209.

Wang, C. S. & Callaway, J. (1977). *Phys. Rev. B* 15: 298.

Wilke, I., Oppliger, Y., Herrmann, W. & Kneubuehl, F. K. (1994). *Appl. Phys. A* 58: 329.

Wu, W., Jung, G.-Y., Olynick, D. L., Straznicky, J., Li, Z., Li, X., Ohlberg, D. A. A., Chen, Y., Wang, S. -Y., Liddle, J. A., Tong, W. M. & Williams, R. S. (2005). *Appl. Phys. A* 80: 1173.

Yuasa, S., Nagahama, T., Fukushima, A., Suzuki, Y. & Ando, K. (2004). *Nat. Mater.* 3: 868.

Extremely Wetting Pattern by Photocatalytic Lithography and Its Application

Yuekun Lai[1,2], Changjian Lin[1*] and Zhong Chen[2*]
[1]*State Key Laboratory for Physical Chemistry of Solid Surfaces and College of Chemistry and Chemical Engineering, Xiamen University*
[2]*School of Materials Science and Engineering, Nanyang Technological University*
[1]*China*
[2]*Singapore*

1. Introduction

Wettability is an important property governed by not only chemical composition, but also geometrical structure as well (Ichimura et al., 2000; Lai et al., 2009a; Wang et al., 1997). Super-wetting and antiwetting interfaces, such as superhydrophilic and superhydrophobic surfaces with special liquid-solid adhesion have recently attracted worldwide attention. Superhydrophilicity and superhydrophobicity are defined based on the conventional water contact angle experiment. If the contact angle is smaller than 5°, the surface is said to be superhydrophilic. Superhydrophobic refers to surface with contact angle greater than 150°. Such two extremely cases have attracted much interest due to their importance in both theoretical research and practical application (Lafuma & Quéré, 2003; Liu et al., 2010; Gao & Jiang, 2004).

In recent years, patterned thin films have received considerable attentions due to their interesting properties for a range of applications, such as optoelectronic devices, magnetic storage media, gas sensors, and fluidic systems. Compared to the conventional thin film technology, such as physical vapor deposition (Li et al., 2006; Zhang & Kalyanaraman, 2004), chemical vapor deposition (Jeon et al., 1996; Slocik et al., 2006) and sputtering (Rusponi et al., 1999), solution-based deposition method is becoming popular for the fabrication of patterning films due to the low temperature process under ambient environment, less energy and time consumption, and easier control of the experimental parameters (Lai et al., 2010a; Liu et al., 2007; Yoshimura & Gallage, 2008). Although traditional photolithographic technique is excellent for preparing sub-micrometer or even only sub-100-nanometer pattern (Cui & Veres, 2007; Li et al., 2009), it is a complex multi-step process (wafer cleaning; barrier layer formation; photoresist coating; soft-baking; mask alignment; exposure and development; and hard-baking) and needs to remove part of the film and all the photoresist used. Direct and selective assembly of nanostructured materials from precursors paves a new avenue for the fabrication of electronic optical microdevices.

Wetting micropatterns with different physical or chemical properties, without the need for ultra-precise positioning, have frequently been acted as templates for fabricating various functional materials in a large scale. The great difference in contact angle of the two extreme cases provides a potentially powerful and economical platform to directly and precisely construct patterned nanostructures in aqueous solution. In general, wetting micropatterns with low contact angle contrast (≤120°) on smooth substrates can be formed by photolithography (Falconnet et al., 2004; Kobayashi et al., 2011), microcontact lithography (Csucs et al., 2003; Kumar et al., 1992), colloidal patterning (Michel et al., 2002; Bhawalkar et al., 2010), electron beam lithography (Wang & Lieberman, 2003; Zhang et al., 2007), nanoimprint lithography (Jiao et al., 2005; Zhang et al., 2006), dip-pen nanolithography (Huang et al., 2010a; Lee et al., 2006; Xu & Liu, 1997), and so on. Among these methods, photocatalytic lithography employing semiconductors to photocatalytic decompose of organic monolayer is one of the most practical techniques because it able to accurately transfer an entire photomask pattern to a target substrate at a single exposure time under environmental condition (Bearinger et al., 2009; Lee & Sung, 2004; Nakata et al., 2010; Tatsuma et al., 2002; Wang et al., 2011). Moreover, it can greatly reduce the photoresist waste. The resolution of the patterning is greatly dependant on the mask alignment and light source exposure. Under optimal condition, a resolution of micrometer- or submicrometer-scale pattern of alkylsiloxane self-assembled monolayers can be achieved with UV light projection irradiation. Once patterned on the surface, organic monolayer has been applied in various ways to restrict corrosion or induce nanostructures growth. Firstly, patterned layer itself may serve as etching mask to protect the substrate to generate pattern with certain thickness/aspect ratio. Secondly, patterned organic layer may be employed as barrier to inhibit the liquid phase deposition of nanostructures to generate functional composite pattern with diverse shape and density. So far, only a few reports have been available on the fabrication and application of superhydrophilic–superhydrophobic patterning by photocatalytic lithography under ambient conditions (Lai et al., 2008a; Nishimoto et al., 2009; Zhang et al., 2007).

Upon UV irradiation, the electron-hole pairs in semiconductor TiO_2 can be generated and migrated to its surface, where the hole reacts with OH- or adsorbed water to produce highly reactive hydroxyl radicals (Zhao et al., 1998). These hydroxyl radicals can further oxidize and decompose most organic compounds. Recently, we found that the pollutant solution can be rapidly decomposed on a nanotube array TiO_2 film with UV irradiation (Lai et al., 2006, 2010b; Zhuang et al., 2007). Considering its effectiveness for the photocatalytic decomposition of organic compounds, the photocatalysis of such TiO_2 nanotube film can be a promising way to decompose the low energy hydrophobic fluoroalkyl chains. So it is possible to achieve a conversion from superhydrophobicity to superhydrophilicity due to the amplification effect of the rough aligned nanotube structure. By using a patterned photomask to control the site-selective decomposition by UV light, that is photocatalytic lithography, superhydrophilic cells can be accurately transferred to a target substrate at a single exposure time under environmental condition. Therefore, these two types of extreme wettability coexist on the surface directly to make up of superhydrophilic-superhydrophobic pattern.

In this chapter, we firstly discuss the wettability on TiO_2 nanostructure film by electrochemical anodization. Secondly, we demonstrate using a novel synthetic process to prepare wetting pattern with a high contrast (superhydrophilic–superhydrophobic) on TiO_2 nanotube structured film by a combination of SAM technique and photocatalytic lithography. The resultant micropattern has been characterized with scanning electron microscopy, optical microscopy, electron probe microanalyzer and X-ray photoelectron spectroscopy. Finally, we focus on the technological details and potential future application of wetting template to induce and direct the assembly of functional nanostructure to form uniform micropatterns. For example, the patterning, biomedical and sensing application of the wetting template and functional composite nanostructure pattern (TiO_2, ZnO, OCP and CdS) (Lai et al., 2009b, 2010a-d).

2. Wettability on TiO_2 nanostructures by electrochemical anodization

Wettability of solid surfaces is a very important property of solid surface. Surfaces with extreme wetting properties, e.g. superhydrophilic and superhydrophobic, can be prepared by introducing certain rough structures on the originally "common" hydrophilic and hydrophobic surfaces. Various ways of preparing TiO_2 semiconductor films on the different solid substrates have been developed, including sol-gel technique (Shen et al., 2005), sputtering (Takeda et al., 2001), chemical vapor deposition (Rausch & Burte, 1993), liquid phase deposition (Katagiri et al., 2007), hydrothermal (Yun et al., 2008), and electrochemical anodizing. Among them, the electrochemical anodizing is verified to be a convenient technique for fabricating nanostructured TiO_2 films on titanium substrates (Lai et al., 2004, 2008b, 2009c; Gong et al., 2003). Moreover, the conductive titanium support substrate can be an advantage for fabricating functional material composites through electrochemical depositions to further improve their photoelectrochemical activities.

Figure 1a shows a typical FESEM image of the titanium substrate before electrochemical anodization. The surface of the substrate was relatively smooth, with features of parallel polished ridges and grooves at the micron scale (Lai et al., 2010a). Figure 1b shows the top view SEM image of the typical TiO_2 nanotube array film by anodizing under 20 V for 20 min. After anodization, shallow cavities as large as several micrometers in diameter were present on the surface of the sample. This is probably due to the anisotropic oxidation of the underlying Ti grains (Crawford & Chawla, 2009; Yasuda et al., 2007). From the high magnification image (Fig. 1c), it can be seen that vertically aligned TiO_2 nanotubes with inner diameter of approximately 80 nm covered the entire surface including the shallow polygonal micropits. The side view image shows that the self-assembled layers of the TiO_2 nanotubes were open at the top and closed at the bottom with thickness about 350 nm (inset of Fig. 1c). Water droplet can quickly spread and wet the as-grown vertically aligned TiO_2 nanostructure film due to capillary effect caused by the rough porous structure, indicating such TiO_2 structure film by electrochemical anodizing is superhydrophilic. A more hydrophobic behaviour, on the other hand, was obtained after coating the TiO_2 film with fluoroalkyl silane. The inset of Fig. 1b shows the intrinsic contact angle (CA) on the as-prepared vertically aligned TiO_2 nanotube surface and its corresponding 1*H*,1*H*,2*H*,2*H*-perfluorooctyltriethoxysilane (PTES, Degussa Co., Ltd.) modified surface is nearly 0° (superhydrophilic) and 156° (superhydrophobic), respectively. However, the CA for the "flat" TiO_2 surface and its corresponding PTES modified sample is about 46° (hydrophilic) and 115° (hydrophobic), respectively. From these results, we know the top surface of the vertically aligned nanotubes has an amplification effect to make hydrophilic and hydrophobic surfaces become superhydrophilic and superhydrophobic, respectively. After UV irradiation for 30 min, the water CA on the TiO_2 nanotube film and "flat" TiO_2 film decreased to 0° and 26°, as a consequence of the photocatalytic activity of TiO_2 films (Balaur et al., 2005; Lai et al., 2010a). Moreover, the sample showed hydrophobic character once again when it was treated with PTES. Therefore the surface can be reversibly switched between superhydrophobic and superhydrophilic by alternating SAM and UV photocatalysis on the rough TiO_2 nanotube arrays (shown in Fig. 1d). Compared with the large wettability contrast on this type of rough surface (larger than 150°), the wettability of a "flat" TiO_2 film can only be reversibly changed within the small range between 26° and 115°.

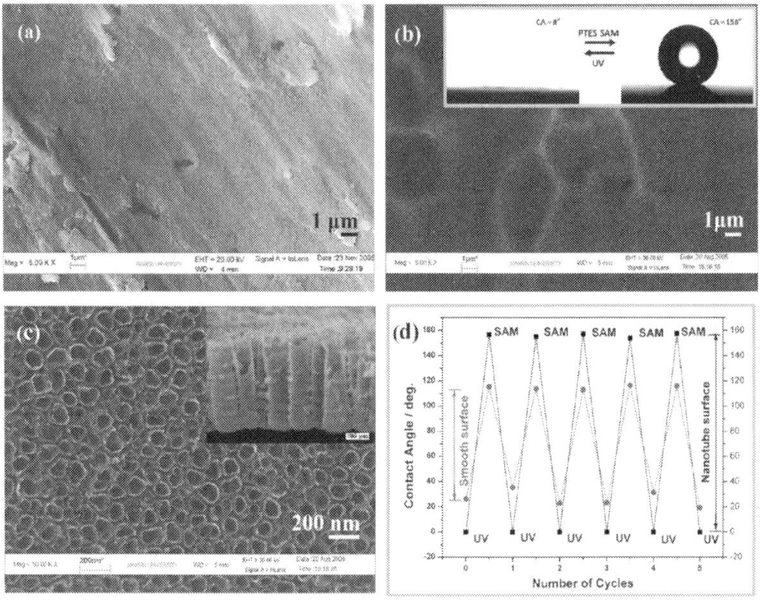

Fig. 1. SEM images of the mechanically polished and cleaned titanium substrate (a), low magnification of a nanotube structured TiO_2 film (b), and a higher magnification of the TiO_2 nanotube array film (c). Reversible surface wettability on a "flat" TiO_2 film and rough nanotube TiO_2 film by alternating SAM and UV photocatalysis (d). The inset of (b) shows the shape of a water drop on the PTES-modified and UV-irradiated TiO_2 nanotube array film. The inset of (c) shows the side view of a TiO_2 nanotube array film.

Recently, we designed three types of superhydrophobic nanostructure models consisting of a nanopore array (NPA), a nanotube array (NTA), and a nanovesuvianite structure (NVS) apply a facile electrochemical process (Figure 2) (Lai et al., 2009a). Based on basic principles of roughness-enhanced hydrophobicity and capillary-induced adhesion, these different porous structures were expected to create interfaces with decreasing adhesive forces. The surface adhesive forces could be effectively tuned by solid-liquid contact ways at the nanoscale and air-pocket ratio in open and sealed systems. The magnitude of the adhesive force of a droplet for a superhydrophobic surface descends in the order "area-contact" > "line-contact" > "point-contact". A continuous three-phase (solid-air-liquid) contact line (TCL) generates serious CA hysteresis and surface adhesion, while a discrete TCL is energetically advantageous to drive a droplet off a superhydrophobic surface, showing lower surface adhesion. Therefore, the water droplet behavior on these superhydrophobic surfaces could be greatly changed from pinning to sliding by adjusting the solid-liquid contact way.

Capillary adhesive force plays a dominant role in imparting adhesive behavior on NPA and NTA nanostructures will sealed cells, while the open NVS nanostructure, which had extremely low adhesion capacity for water, acted solely by van der Waals attraction between water and PTES molecules. A possible explanation is as follows. As the droplet gradually retracted from the sample surface, the meniscus on each nanotube nozzle would be changed from concave to convex (Figure 3a). This could result in an increased volume of air sealed in each nanotube by the liquid/air interface. According to Boyle's law (West, 1999), there is an inverse relationship between the volume (V) and pressure (P) for an ideal gas under the conditions of constant temperature and quality. Therefore, this expansion of air would result in the formation of a negative pressure (ΔP). In this case, the volume of air sealed in the nanotubes was varied by their depths, so longer tubes would be expected to have lower air-expansion ratios (ΔV/V), thus lesser negative pressures. For a fixed nanotube diameter, a longer nanotube would therefore require a smaller pulling-off force, and the total surface adhesive force would be smaller.

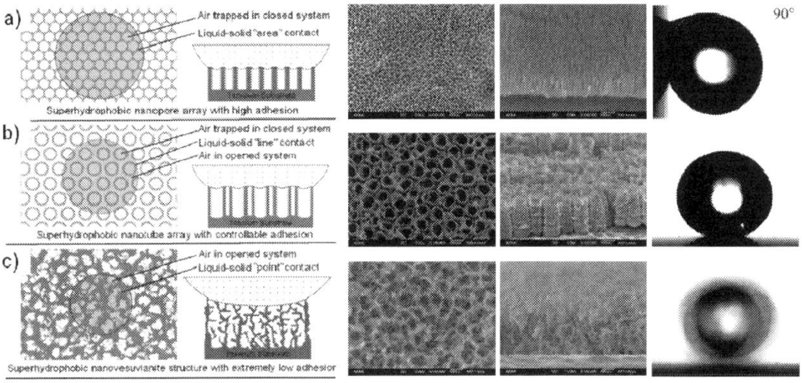

Fig. 2. Schematic models, SEM images and corresponding water behavior on three types of superhydrophobic porous-nanostructure with water adhesive forces ranging from high to low. (a) Superhydrophobic NPA with high adhesion. (b) Superhydrophobic NTA with controllable adhesion. (c) Superhydrophobic NVS with extremely low adhesion.

Figure 3b shows the curves of the water CAs and adhesive forces with respect to the diameter of individual nanotube. When the nanotube diameter decreases, the force drastically increased, while the CA slightly decreased. When the diameter was tuned from 78 nm down to 38 nm, the surface adhesive force of the superhydrophobic NTA film increased 2.06 times, while the decline in magnitude of the water CA was not more than 2%, showing that the negative pressure caused by the volume change of air sealed in the nanotubes could effectively tune the surface adhesive force. The NTA structures in this study had variable length, with values of (0.35 ± 0.04) μm, (0.76 ± 0.05) μm, (1.18 ± 0.07) μm, while their diameter (~80 nm) was fixed. Figure 3c shows the curves for water CAs and adhesive forces obtained with PTES-modified NTA-nanostructure surfaces differing in nanotube length. With lengths extending from 0.35 μm to 0.76 μm and 1.18 μm, the CA change was very small, not more than 2%, which could be due to the minor variation in nanotube diameter. However, the adhesive force linearly decreased from 21.5 μN down to 16.7 μN and 12.2 μN for the above increases in length, respectively. It was evident that the water adhesive force of the superhydrophobic NTA-nanostructure surfaces could be tuned by varying the diameters and also lengths of the nanotubes. These findings are valuable to deepen insight into the roles of nanostructures in tailoring surface water-repellent and adhesive properties for exploring new applications.

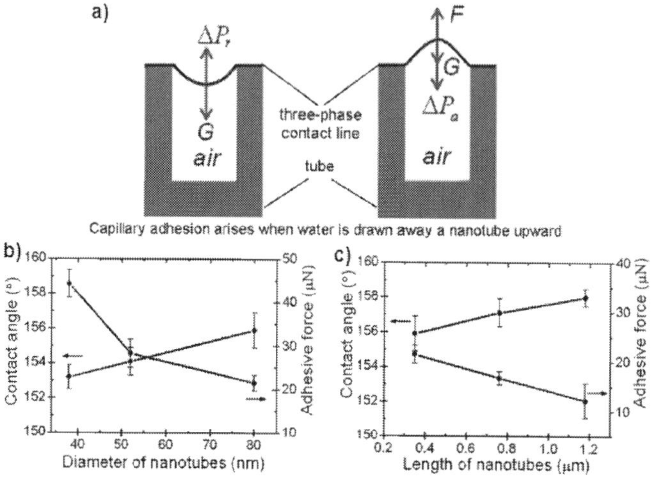

Fig. 3. (a) Capillary adhesion arises when a water droplet sitting on the tube nozzle is gradually drawn upwards because the convex air/liquid interface produces an inward pressure ΔP. (b,c) The curves of water contact angles and adhesive force on the superhydrophobic NTA nanostructures with respect to the diameter and length of nanotubes.

3. Wetting pattern by photocatalytic lithography

A novel approach for constructing superhydrophilic-superhydrophobic micropattern on the nanotube structured TiO_2 films has developed by using photocatalytic lithography (Figure 4a) (Lai et al., 2008a). At the first step, the as-prepared amorphous TiO_2 nanotubes by electrochemical anodizing of titanium sheets were calcinated at 450°C to form anatase phase, then treated with a methanolic solution of hydrolyzed 1 wt% PTES for 1 h and subsequently heated at 140°C for 1 h, and at the second step, the superhydrophobic film is selectively exposed to UV light through a copper grid (photomask) to photocatalytically cleave the fluoroalkyl chain. It is noteworthy, from the characterization of chemical composition before and after UV irradiation by X-ray photoelectron spectroscopy, that the intensities of the F1*s* and F*KLL* are decreased greatly and those of the Ti2*p* and O1*s* are increased after exposing the PTES modified surface to UV light for 20 min (Fig. 4b). From the inset high-resolution spectra (Fig. 4c), the peaks of $-CF_2$ (at 291.8 eV) and $-CF_3$ (at 294.1 eV) are obviously vanished after UV light irradiation, while the strength of silicon peaks in the XPS spectra remains unchanged but shifts from 102.8 to 103.3 eV, suggesting that Si-O-Si networks have already formed due to UV irradiation. According to these results, we believe that the hydrophobic fluoroalkyl chains have been completely decomposed and removed by the photocatalytic reactions at TiO_2 nanotube films. Similarly, a serial of fluoroalkyl silane monolayer pattern (e.g. heptadecafluorodecyltrimethoxysilane, octadecyltriethoxysilane, and methyltriethoxysilane) can be successfully fabricated in our case. Although various monolayer patterns can be prepared with a resolution about micro-scale or submicro-scale under optimal condition, we will focus on the application of the PTES micro-pattern with a general TEM copper grid as a photomask by photocatalytic lighography.

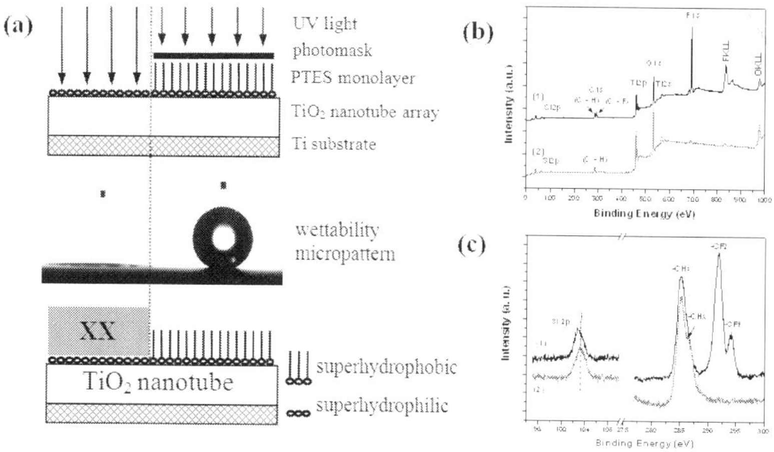

Fig. 4. Schematic outline of the procedures to fabricate nanostructured patterning film by electrochemical deposition based on superhydrophilic-superhydrophobic micropattern (a). Survey-scan X-ray photoelectron spectra of the PTES modified nanotube TiO_2 films before (1) and after (2) 20 min UV irradiation (b). The high-resolution spectra of Si2*p* and C1*s* regions (c).

Figure 5 shows the optical micrograph of the as-obtained superhydrophilic-superhydrophobic pattern by focusing on the droplet within the superhydrophilic regions. A uniform pattern is formed due to the site-selective wetting by water droplets within the superhydrophilic regions (Fig. 5a). A light dot array (inset of Fig. 5a) is seen when focusing on the top of the droplets, indicating that the confined droplet has a hemispherical dome. To further verify the resulting micropatterns with an extreme wettability contrast, fluorescein sodium was used as a probe to label the surface of the films. Figure 5b shows the fluorescent micrograph of the resultant superhydrophilic–superhydrophobic micropatterns on the TiO_2 nanotube array films. As shown, geometrically identical square superhydrophilic regions and dark superhydrophobic regions transferred well from the photomask to form a well-defined pattern. The UV-irradiated regions become superhydrophilic owing to the photocatalytic cleavage of the PTES molecule and the enhanced roughness of the nanotube structure, while the non-irradiated parts remain superhydrophobic without any change. Because the difference in the water CA between the irradiated and non-irradiated regions is larger than 150°, the liquid containing the fluorescent probe selectively appears only on the uniform superhydrophilic grids and not on the neighboring superhydrophobic regions. Therefore, a clear, well-defined fluorescent pattern in line with the dimensions of the Cu grid can be obtained. These results indicate that the micropatterned template composed of superhydrophilic and superhydrophobic regions was fabricated successfully.

The UV irradiation times had a great effect on the quality of the resulting pattern. For example, it cannot exhibit a sufficient wettability contrast between the irradiated and non-irradiated regions to form a well-defined pattern within 5 min. This is attributed to the hydrophobic fluoroalkyl chain in the PTES molecule that was not efficiently cleaved under a short-time UV irradiation. However, with a long-time UV irradiation (i.e., 60 min), the adjacent PTES molecule covered by the Cu grid can be remotely oxidized by a TiO_2 nanotube photocatalyst or the diffusion, scattering, and diffraction of the incident light (Haick & Paz, 2001; Kubo et al., 2004). Therefore, to obtain a higher pattern resolution, the optimized UV irradiated time in our case was controlled in the range of 10-30 min.

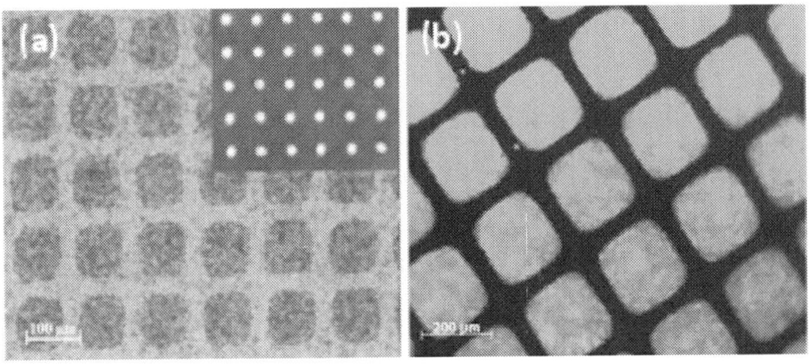

Fig. 5. (a) Optical micrograph of the as-obtained superhydrophilic-superhydrophobic pattern by focusing on the water droplet within the superhydrophilic regions. (b) Fluorescence microscopy image of the fluorescein probes on the as-prepared superhydrophilic-superhydrophobic micropattern.

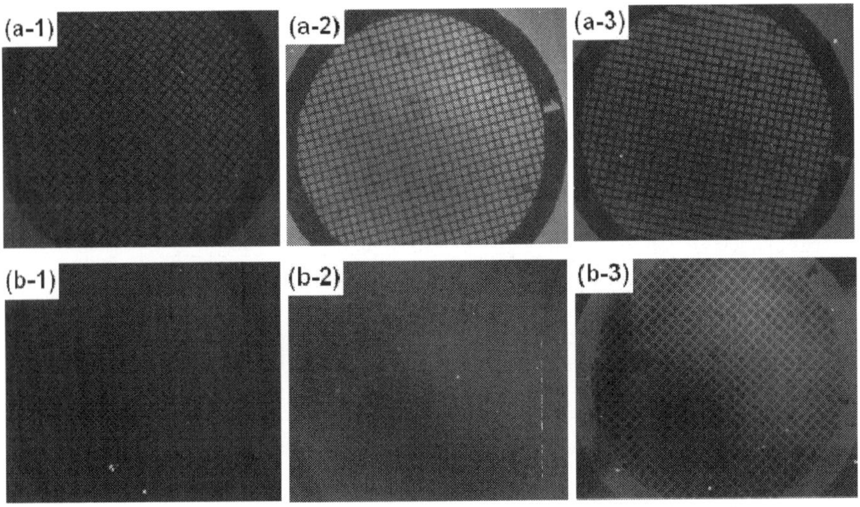

Fig. 6. Fluorescence microscope images generated by blue light excitation on the different micropatterned templates with the adsorption of FITC-BSA, (a-1, a-2, a-3) the PTES template with a pH value of 7.5, 4.5 and 2.5, respectively; (b-1, b-2, b-3) the PTES-APTS template with a pH value of 7.5, 4.5 and 2.5, respectively.

Based on the molecular self-assembly and photocatalytic lithography techniques, micropatterned templates of PTES or PTES-APTS(3-aminopropyltriethoxysilanes) with different wettabilities were fabricated on the titania film. The adsorption behavior of bovine serum albumin (BSA) on the above two templates was investigated using fluorescent labeling (FITC) in buffer solution with different pH values. The results showed that, for the PTES template with great wettability differences, BSA would preferentially adsorb on the superhydrophilic regions. For the APTS-PTES template with smaller differences in wettability, competitive adsorption phenomenon on the super-hydrophobic regions was found due to the hydrophobic interaction force between the albumin and the surface. As the pH value decreased to 2.5, the phenomenon of competitive adsorption was prominent with the albumin adsorbed in the super-hydrophobic areas. The adsorption feature of the albumin may be closely related to the wettability and surface energy of the materials. This technique has promising applications in bio-compatible coatings where drugs could be encapsulated in specific areas of the coating using simple microfabrication methods.

4. Application of wetting pattern

Uniform self-assembly of functional inorganic nanomaterials is a fundamental challenge. Nature adopts a superior approach in biomineralization, where "matrix" macromolecules induce nucleation of inorganic crystals at specific locations with controlled size and morphology, and sometimes even with defined growth orientation. We apply the biomimetic principles derived from liquid phase processes to the assembly of nanoscale functional materials into microscale systems. We carefully control surface wettability to promote etching or heterogeneous nucleation at designated superhydrophilic regions while completely suppress these processes elsewhere (superhydrophobic regions), therefore enable the controlled top-down or bottom-up assembly of inorganic nanomaterials directly from solution. Following this principle, arrays of crystalline TiO_2 nanotube, ZnO nanorods, CdS semiconductor materials and octacalcium phosphate (OCP) biomaterials were nucleated and assembled directly from solution onto Ti substrates at the desired precise locations and then fabricated into arrays of photodetector or matrix devices for large-area microelectronic applications. This strategy of micropatterned nanocomposites will be helpful to develop various micropatterned functional nanostructured materials.

4.1 Template for preparing functional pattern

Figure 7 shows an optical microscopy image of the TiO_2 nanotube micropattern produced using a grid micropattern with different wet etching times (Lai et al., 2009b). A patterning with a clear outline was formed in a short time for 30 s (Fig. 7a). With an increase in the wet etching time (Fig. 7b and c), identical micropatterns with higher aspect ratios can be fabricated. When the etching was prolonged to 240 s (Fig. 7d), the size of the grids increased slightly, indicating that the PTES-SAM layer at the edge of the superhydrophobic lines is more easily etched as compared to the inner superhydrophobic area. This is due to the loose and disordered SAMs resulting from the scattered UV light photocatalytic degradation and the transfer of the active hydroxyl radicals at the edge of the grids. Moreover, the isotropic etching of the Ti substrate underneath the boundary leads to the collapse of the upper nanotube array structures. Therefore, the pattern can be obtained with a clear boundary in a short time after the wet etching in the aqueous solution, and the depth of the etching can be controlled simply by adjusting the etching time.

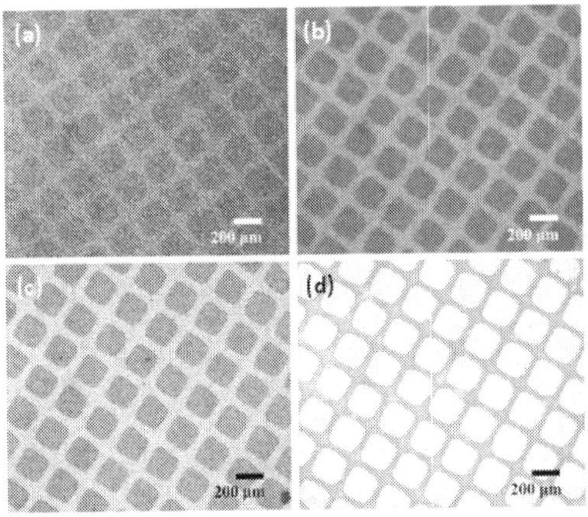

Fig. 7. Optical micrograph images of the time-resolved evolution process of the resulting vertical aligned TiO_2 nanotube micropattern on a superhydrophilic-superhydrophobic template by wet chemical etching in 0.1 wt% HF solution: (a) 30, (b) 60, (c) 120, and (d) 240 s.

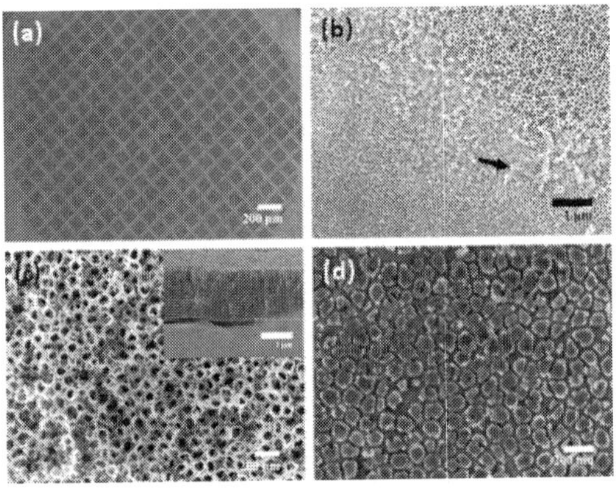

Fig. 8. (a) The SEM micrograph of the ordered vertical aligned TiO_2 nanotube array pattern by vapor-condense etching for 10 min. Magnified images of the corresponding (b) superhydrophilic-superhydrophobic boundary, (c) superhydrophobic area, and (d) superhydrophilic area. The inset figure shows the corresponding cross-sectional image.

The micropattern with a higher aspect-ratio can also be fabricated by a developed vapor etching technique. As water evaporated, the vapor containing HF dewetted the superhydrophobic lines while condensing on the superhydrophilic grids to etch the vertical aligned TiO_2 nanotube layer. Figure 8a and b shows the as-prepared TiO_2 micropattern with a clear boundary by water vapor etching for 10 min. This resulted from the selective condensing of the vapor from the water solution containing 5 wt% HF in the superhydrophilic regions. Some collapsed residue (indicated by the black arrow) covers the boundary due to the rapid etching of the bottom nanotubes and the resultant bubbles;

however, they can be easily removed by sonication. The vertical aligned TiO_2 nanotube film with a length of about 1.53 μm shows no obvious change within the superhydrophobic regions (Fig. 8c), while the TiO_2 nanotubes in the superhydrophilic regions are etched completely for 20 min (Fig. 8d), that is to say, a TiO_2 micropattern with a high aspect ratio of ~20 can be obtained. This technique is particularly attractive in generating large-area functional nanostructure patterns in a high throughput fashion with a high aspect-ratio.

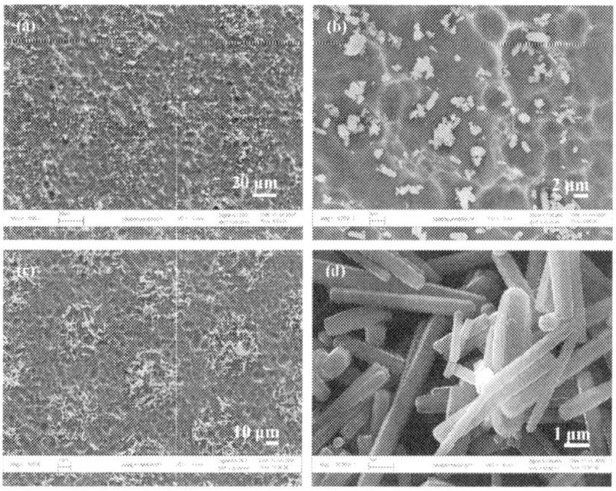

Fig. 9. SEM images of the patterned ZnO nanostructured micropatterns by liquid phase deposition at 90°C for different times: (a, b) 30 min; (c, d) 90 min.

Figure 9 shows representative top-view FESEM images of the ZnO/TiO_2 micropatterns by liquid-phase deposition after different times. After growth for 30 min (Fig. 9a,b), the nucleation and growth of ZnO crystals with various morphologies and sizes (nanoparticles and nanorods) are sparsely dispersed within the predefined superhydrophilic regions. In the superhydrophobic regions, the nanotube structure is retained with its original morphology due to the indirect contact with the solution and the effective protection by the PTES monolayer. Upon further increase in the deposition time to 90 min (Fig. 9c,d), it was observed that ZnO nanorods were the predominant structural features. The average diameter of the grown ZnO nanorods increases greatly to about 400-800 nm in diameter and 3-5 μm in length, which may be due to lower nucleation rate and higher growing space for ZnO nanorods. The superhydrophilic microdots are almost covered with randomly lying ZnO nanorods. Moreover, there are different growth orientations and some connections into adjacent nanorods. It is of interest to note that a two dimensional (2D) pattern with smaller density of randomly packed ZnO nanorods, instead of the 3D pattern consisted of well-aligned vertical nanorods, which were confined and grown within the superhydrophilic regions.

For practical application of thin film devices, the position and orientation-control of ZnO nanorods are very important because they directly relate to their physical and chemical performances (Koumoto et al., 2008; Masuda et al., 2006). In order to precisely control the spatial orientation of the ZnO nanostructures, we developed a new technique which is able to make the ZnO nanorods grow along the vertically aligned titania nanotubes rather by disordered deposition on the titania nanotube array surface. In this technique, resistance discrepancy was adopted to make the entrance of the tubes more conductive than the bottom of the tubes, to induce the epitaxial growth with spatial organization of uniform

ZnO nanorods along the direction of the nanotubes. The quasi-perpendicular ZnO nanorods nucleate and grow uniformly and selectively throughout the superhydrophilic regions of the TiO_2 nanotube surface by electric field assisted deposition at 90°C for 3 min, while no nanorods are observed in the superhydrophobic regions (Fig. 10a,b) (Lai et al., 2010a). The EDS spectra also reveal that the presence of Zn, Ti and O elements on the superhydrophilic regions, while the elemental components in the superhydrophobic areas are only Ti and O. The inset of Fig. 10b shows the hexagonal end facet of a vertically aligned ZnO nanorod with a diameter about 100-150 nm growing on top of the TiO_2 nanotube array surface. Therefore, the density, size and orientation of ZnO nanorods are very sensitive to the presence of electric fields. A 3D AFM profile image (Fig. 10c) shows that the microscopic structure of the ZnO crystal deposition consisted in dense column arrays, which are induced and directed by the wettability template. The thickness of vertical ZnO nanorod film is in the range of 800-900 nm. Furthermore, the three dimensional confocal microscopy image (Fig. 10d) also shows that the growth of the ZnO nanorod pattern is identical with the superhydrophilic/superhydrophobic template.

On the basis of the versatile superhydrophilic-superhydrophobic template, we can successfully control the growth of ZnO nanostructures in the superhydrophilic regions under mild reaction conditions and in the absence of seed and noble metal catalyst. In the superhydrophobic regions, the growth is suppressed. This special template can be utilized to generate different nanostructured ZnO patterns with clearly defined edges. Hence, it is expected that this novel micropatterned technique based on the superhydrophilic-superhydrophobic template will become a powerful tool for fabricating various types of micropatterned nanomaterials and devices.

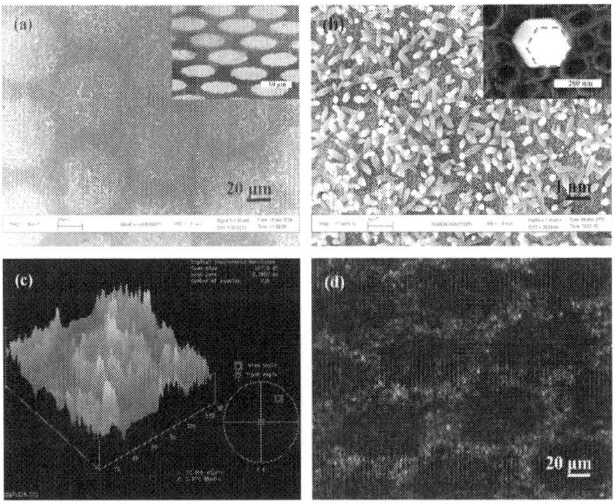

Fig. 10. (a,b) Typical SEM images of the vertically aligned ZnO nanorods selectively grown on superhydrophilic patterning regions by the developed electric field assisted deposition technique at 90°C for 3 min. The inset in (a) shows the side view SEM image of the corresponding ZnO nanorod micropattern. The inset in (b) shows the higher magnified SEM image of a ZnO nanorod with hexagonal end facet. (c) 3D AFM image of the ZnO/TiO_2 micropattern. (d) Confocal microscopy image of the perpendicular ZnO nanorod array.

4.2 Biomedical arrays

The typical SEM image of TiO_2 nanotube array surface before and after the deposition of OCP film by electrochemical technique for 5 min is shown in Fig. 11a,b. The superhydrophobic-superhydrophilic micropatterned TiO_2 was used as a micro-template to selectively deposit nano-OCP crystals on the superhydrophilic regions by an electrochemical deposition to form a special micropatterned nano-OCP. The deposition electrolyte was consisted of 0.042 mol/L $Ca(NO_3)_2$ and 0.025 mol/L $NH_4H_2PO_4$. The pH value was adjusted to approximately 4.2 with 0.05 mol/L NaOH solution. The precipitation was carried out galvanostatically at a cathodic current of 0.5 mA cm^{-2} under 67.5°C for a certain time (Wang et al., 2008). It can see that quasi-perpendicular ribbon-like crystals of several hundred nanometers in width are uniformly grown on the TiO_2 nanotube array surface. Fig. 11c shows a typical fluorescence microscope of the superhydrophilic-superhydrophobic micropattern on TiO_2 nanotube surface. As can be seen, the green dot patterns are clearly imaged through the fluorescence contrast between the UV-irradiated superhydrophilic and photomasked superhydrophobic regions. The photoirradiated dot exhibiting a uniformly stronger fluorescence against the surrounding dark background is due to the highly affinity to solution resulting in the absorption of fluorescent probes into the irradiated nanotube array films. Therefore, a clear well-defined fluorescence pattern based on the superhydrophobic-superhydrophilic pattern is obtained. Fig. 11d displays the identical patterning of OCP biomaterials deposited on the superhydrophilic-superhydrophobic patterns on TiO_2 nanotube array surface. It is obvious that the size of the white OCP dots is equal to that of the superhydrophilic area on template, indicating the deposited regions were only located within the superhydrophilic dots where photocatalytic degradation of PTES-SAMs was performed.

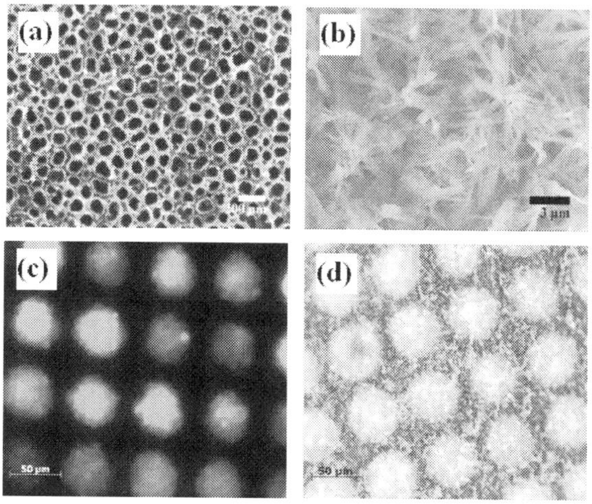

Fig. 11. SEM images of the (a) TiO_2 nanotube array film fabricated by electrochemical anodization; (b) OCP nanostructure layer on TiO_2 nanotube array film by electrochemical deposition for 5 min. Optical fluorescence pattern of the superhydrophilic-superhydrophobic template (c) and patterned OCP thin films selectively deposited in pre-defined superhydrophilic regions by electrochemical deposition for 5 min (d).

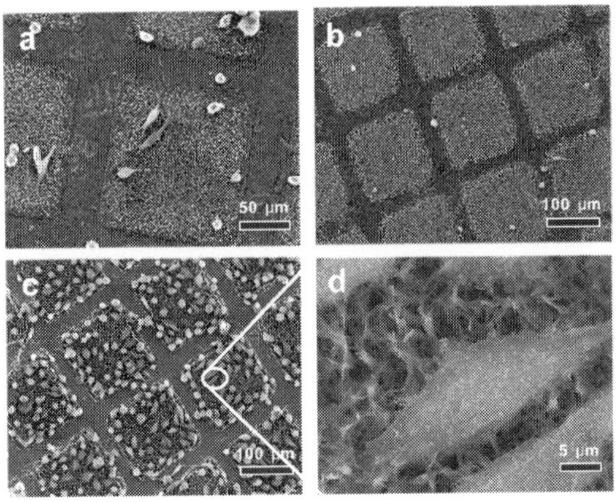

Fig. 12. SEM micrographs of MG-63 cells cultured on the patterned OCP coatings with different deposition time for 6 h, (a) 1 min; (b) 3 min; (c) 5 min; (d) higher magnification.

The *in-vitro* MG-63 cell tests were used to study the biological performance of the as-obtained OCP micropatterns (Huang et al., 2010b). The results showed that MG-63 cells were found preferentially attached on the superhydrophilic regions with OCP thin films, while the superhydrophobic regions with the PTES monolayers can effectively prevented the adhesion of cells on the surface, indicating that the cells had the selective adhesion action on the tiny units of OCP films. Moreover, the cells adhered on the OCP film deposited for a longer period (5 min) are more active to spread on the OCP nanobelt covering surface. It is promising for developing a new cell chip for high throughput evaluation of the cell behaviors.

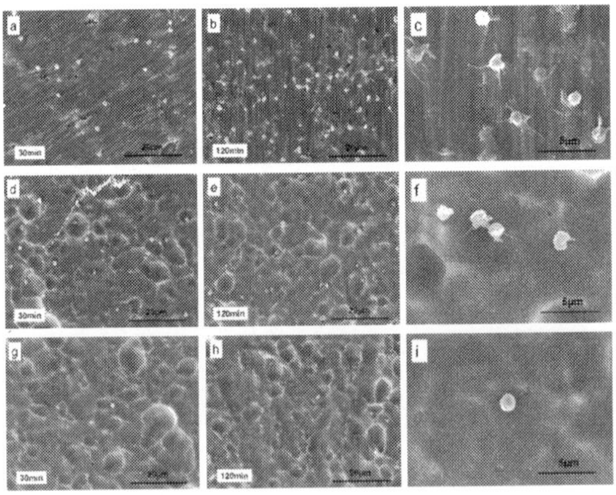

Fig. 13. SEM images of adhered platelets on various kinds of surfaces at 37◦C for different periods. (a-c) mechanically polished and cleaned Ti substrate; (d-f) superhydrophilic surface; (g-i) superhydrophobic surface; (a,d,g) 30 min; (b,e,h) 120 min; (c,f,i) magnified images of the corresponding images of (b,e,h).

The *in vitro* experimental indicated that the superhydrophobic nanotube TiO_2 layers exhibit a remarkable resistance to platelets attachment (Fig. 13) (Yang et al., 2010). It is indicated that the superhydrophobic nanotube TiO_2 layers exhibit a remarkable resistance to platelets attachment. As shown in Fig. 13a,d,g, abundant platelets adhere to both the plain Ti surface and the superhydrophilic TiO_2 nanotube layers after 30 min incubation. Comparatively, after 120 min incubation, a large number of platelets adhered and spread out on both the plain Ti surface (77 ± 7.4 per 5000 μm^2, Fig. 13b,c) and the superhydrophilic surface which was obtained by exposing the TiO_2 nanotubes under a UV irradiation (22 ± 1.5 per 5000 μm^2, Fig. 13e,f), only very few of platelets (1 ± 0.8 per 5000 μm^2) was found to adhere on the superhydrophobic TiO_2 nanotube layers (Fig. 13h). Moreover, even though some platelets were occasionally seen attached on the superhydrophobic surface, they looked smooth without any growth of pseudopods (Fig. 13i), implying that the platelets adhered on the superhydrophobic TiO_2 nanotube surface remain inactive and hardly grow and spread out for a long period. The quantities and morphologies of adhered platelets and their corresponding interactions on the different samples are illustrated in Figure 14. Therefore, the construction of superhydrophobic surface on biomedical implants could pave a way to improve the blood compatibility of the biomedical devices and implants.

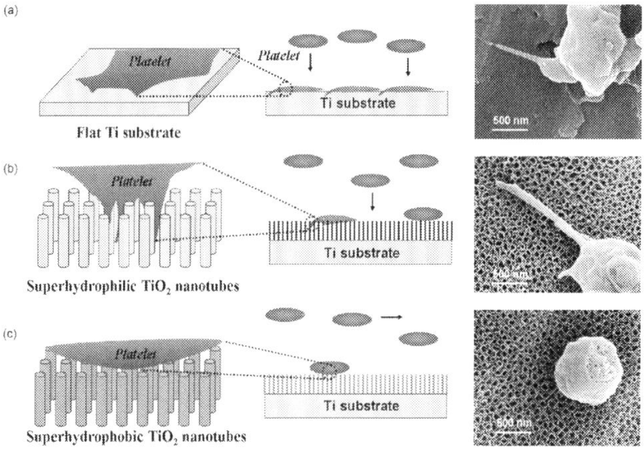

Fig. 14. Schematic illustration of the quantity and morphology of platelet and corresponding interactions on the three kinds of surfaces. (a) Plain Ti substrate; (b) Superhydrophilic TiO_2 nanotubes; and (c) Superhydrophobic TiO_2 nanotubes.

4.3 Sensing devices

Based on photocatalytic lithography, we demonstrate a facile, rapid and practical approach to fabricate Ag nanoparticle (NP) patterns on TiO_2 films by means of pulse-current electrodeposition technique (Huang et al., 2011). The size and density of as-deposited Ag NPs can be controlled by changing deposition charge density. Moreover, the resultant patterned Ag NP films exhibited particle size-as well as density-dependent UV-vis absorption and SERS enhancement effect. It was found that the patterned Ag NP films produced under the deposition charge density of 2.0 C cm^{-2} exhibited the intense UV-vis and Raman peaks. Two dimensional surface enhanced Raman scattering (SERS) mapping of

Rhodamine 6G (R6G) on the patterned Ag NP films demonstrated a high throughput localized molecular adsorption and micropatterned SERS effect.

Furthermore, the elemental distributions of the as-prepared Ag NPs arrays were also observed by electron probe microanalyzer, which are shown in Fig. 15a-c. Figure 15a shows the Ag element distribution map. As shown in the map, the green dot patterns are clearly images obtained through element concentration contrast between the UV-irradiated superhydrophilic and photomasked superhydrophobic regions. The green dots exhibiting a uniform Ag concentration against the surrounding black regions indicate that Ag NPs are uniformly deposited and confined to the superhydrophilic regions. Figure 15b,c shows the element distribution maps of Ti and O, respectively, which are also in line with the dimensions of the photomask. The blue superhydrophilic regions (dot patterns) show lower Ti and O concentrations due to the preferential deposition of Ag NPs in the superhydrophilic areas, while the yellow and red superhydrophobic regions have higher Ti and O concentration. The corresponding line-scan signal intensity profiles of Ag, O, and Ti elements across the dot pattern (red line direction). The intense Ag signals in the superhydrophilic regions as well as the Ti and O signals in the superhydrophobic regions suggest that Ag NPs are deposited only in the dot areas and that the other regions are the exposed TiO_2 nanotube films. The consistent intensity of the Ag signals indicates that Ag NPs are uniformly deposited on the superhydrophilic regions.

In addition, a two-dimensional point-by-point SERS mapping of the patterned Ag NP film whose deposited charge density is 2.0 C cm^{-2} was obtained using R6G as the probe molecule. Figure 15d,e show the optical image and the corresponding SERS mapping image of the patterned Ag NP film. The mapping area was approximately 140 × 100 μm^2 and the data acquisition time was 1 s. A signal to baseline from 594.0 to 623.4 cm^{-1} was chosen for the acquisition of the SERS mapping. The bright and dark areas respectively represent higher and lower intensity of the SERS signal. It is clear that the geometrically identical gray superhydrophilic areas (circle) with a strong SERS activity and the dark superhydrophobic areas without any SERS activity form a high-resolution SERS intensity distribution map. As can be seen from the SERS mapping results, most SERS peak area is in a very narrow intensity window as shown by the contrast in color codes. Furthermore, the SERS peak area is uniform over the superhydrophilic region with several high intensity spots represented by white color codes. On comparing the SERS mapping with the SEM detection, it is reasonable to conclude that the homogeneous SERS signal in the circle areas reflects the uniform dispersion of Ag NPs on the superhydrophilic areas. The high-resolution SERS intensity distribution and micropatterned SERS effect of the Ag NP film might make it potentially useful in high-throughput molecule detection and bio-recognition.

To gain insight into the dependence of SERS enhancement on the size and density of Ag NPs, the SERS spectra of R6G absorbed on the different patterned Ag NP films were detected, which are shown in Fig. 15f. Because of the small particle size and low density, which are not the optimum size and distribution for SERS, the signal enhancement is rather weak below the charge density of 0.5 C cm^{-2}. The enhancement behavior of the substrate, however, is obviously improved under the charge density of 1.0 C cm^{-2}. In particular, the patterned Ag NP film prepared under a charge density of 2.0 C cm^{-2} exhibits the highest intensity, which is attributed to large size and high density of Ag NPs, as shown by the SEM results. On increasing the charge density to 2.5 C cm^{-2}, the signal becomes weaker. The size-correlated enhancement may be explained by the EM mechanism (Zeng et al., 2008). The intensity of the SERS signal might also be controlled by the NPs density, which changes the interparticle spacing as well as the hot spots among the NPs (Felidj et al., 1999; Lu et al., 2005). The average surface enhancement factor for pyridine on the Ag NP film with a charge density of 2.0 C cm^{-2} was calculated to be 1.3×10^5.

Fig. 16a shows the typical SEM micrographs of the CdS nanosphere micropatterns after 3 min deposition on the superhydrophobic-superhydrophilic template of TiO_2 nanotube films (Lai et al., 2010d). The bright rectangular areas corresponded to the deposition of CdS nanosphere crystals on the superhydrophilic regions. The boundary between the CdS pattern and the surrounding superhydrophobic regions is clearly visible at a higher magnification (inset). The dispersed CdS nanosphere crystals grew on the top of TiO_2 nanotube arrays within the rectangular superhydrophilic region (Fig. 16c). Most of the crystals were less than 90 nm in diameter due to the confinement by the inner diameter of nanotube, though a few larger spheres (~120 nm) were seen across neighboring tube openings. While on the superhydrophobic areas (Fig. 16d), there was almost no CdS crystal. The high growth selectivity was also confirmed by the EDS analysis, revealing that CdS spheres easily nucleate and grow on the hydroxyl groups (-OH) terminated regions (Fig. 16e), but not on the $-CF_3$ terminated areas (Fig. 16f). Since the difference of the water contact angle between the superhydrophilic and superhydrophobic regions is larger than 150°, electrolyte solution is preferentially presented on the uniform superhydrophilic dots. No water droplets go to the neighboring superhydrophobic regions. Although a few CdS particles resulted from homogeneous precipitation attached onto superhydrophobic surface due to van der Waals interactions and gravity, they can be easily removed by ultrasonication. Therefore, a clear and well-defined CdS pattern in line with the dimensions of the superhydrophobic-superhydrophilic template has been obtained.

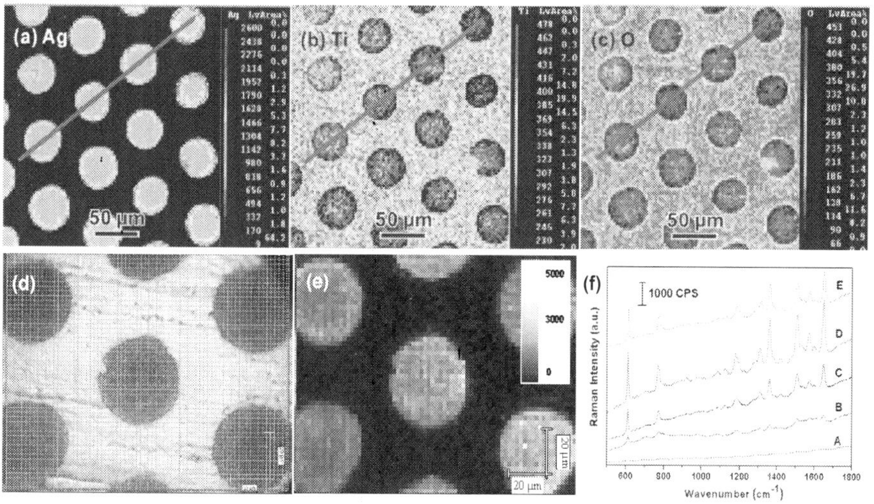

Fig. 15. Typical element distribution maps of Ag (a),Ti (b), O (c), and optical microscopy(d) and the corresponding R6G SERS mapping (e) of the patterned Ag NP films with an area of 140 × 100 μm^2 using the peak area at 614 cm^{-1} as the reference. (f) Raman spectra of the patterned Ag NP films from different charge density: curve A, bare subsrate; curve B, 0.5 C/cm^2; C, 1.0 C/cm^2; D, 2.0 C/cm^2; and E, 2.5 C/cm^2.

Fig. 16b shows the photocurrent spectra of the couple CdS/TiO_2 nanotube array electrode prepared under different electrodeposition times. It is apparent that the pure TiO_2 nanotube array samples have a photo-response wavelength lower than 400 nm due to its band-gap of 3.2 eV (curve a). The decoration of CdS nanospheres with a smaller energy band-gap (2.4 eV) can significantly extend the photo-response range from 380 nm to about 500 nm.Moreover, the CdS modified TiO_2 nanotube array electrodes can also greatly increase the photocurrent response under UV light, especially for the samples obtained under 2 min electrodeposition (curve b), which thus would be the optimal deposition time. This is attributed to the uniform dispersed CdS nanospheres with suitable size decorated onto the TiO_2 nanotubes. This allows for more efficient electron transfer and lower electron-hole recombination rate which leads to enhanced light harvesting at the directly grown CdS/TiO_2 heterojunctions. With the increase of time (curve c and d), more CdS particles with bigger size started to randomly distribute on top of TiO_2 nanotube arrays. Such composite nanostructures would weaken the light absorption of the uniform CdS/TiO_2 heterojunction underlayer, which has resulted in a lower photocurrent in both UV and visible light region.

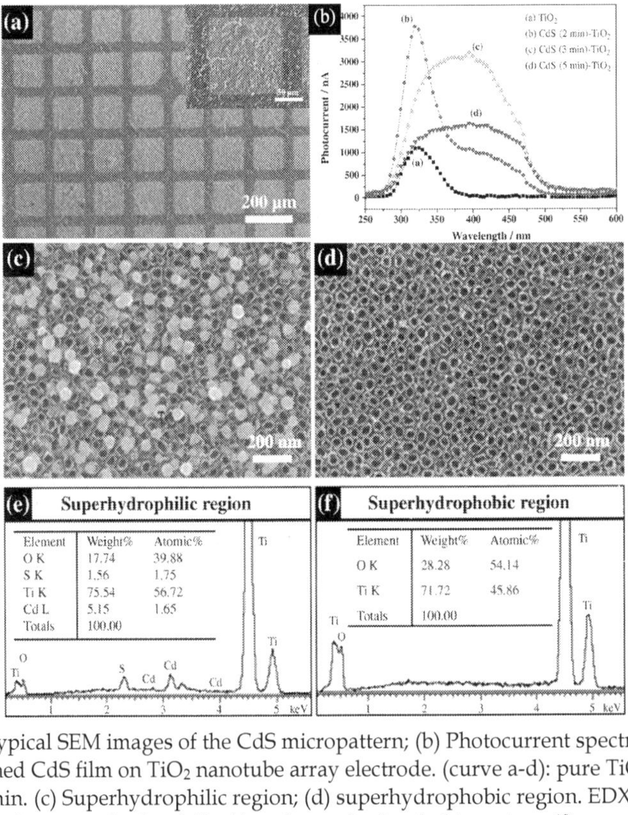

Fig. 16. (a) Typical SEM images of the CdS micropattern; (b) Photocurrent spectra of micropatterned CdS film on TiO_2 nanotube array electrode. (curve a-d): pure TiO_2; 2 min, 3 min; and 5 min. (c) Superhydrophilic region; (d) superhydrophobic region. EDX spectrum of the corresponding superhydrophilic (e) and superhydrophobic regions (f).

5. Summary and outlook

Extremely wetting micropattern (superhydrophilic/superhydrophobic) on TiO_2 nanostructure surface by using SAM technique and photocatalytic lithography has been studied intensely as it provides a cost effective template to construct well defined functional composited pattern. Numerous potential applications have also been proposed and investigated in biomedical, sensors and micro-nano devices. We believe that the photocatatlytic lithography patterning technique presented in this chapter should be general to create micro-scale wetting pattern on other semiconductor substrates and these developments will open the door for more widespread application of the wetting pattern in practical fields.

6. Acknowledgments

The authors thank the National Natural Science Foundation of China (grants 51072170, 21021002), the National Basic Research Program of China (grant 2007CB935603) and the National High Technology Research and Development Program of China (grant 2009AA03Z327), and the Environment and Water Industry Programme Office (EWI) under the National Research Foundation of Singapore (grant MEWR651/06/160) for their financial supports.

7. References

Balaur, E., Macak, J.M., Taveira, L. & Schmuki, P. (2005). Tailoring the wettability of TiO_2 nanotube layers. *Electrochem. Commun.*, Vol. 7, No. 10, 1066-1070, 1388-2481

Bearinger, J.P., Stone, G, Hiddessen, A.L., Dugan, L.C., Wu, L.G., Hailey, P., Conway, J.W., Kuenzler, T., Feller, L., Cerritelli, S. & Hubbell, J.A. (2009). Phototocatalytic lithography of poly(propylene sulfide) block copolymers: Toward high-throughput nanolithography for biomolecular arraying applications. *Langmuir*, Vol. 25, No. 2, 1238-1244, 0743-7463

Bhawalkar, S.P., Qian, J., Heiber, M.C. & Jia, L. (2010). Development of a colloidal lithography method for patterning nonplanar surfaces. *Langmuir*, Vol. 26, No. 22, 16662-16666, 0743-7463

Crawford, G.A. & Chawla, N. (2009). Porous hierarchical TiO_2 nanostructures: Processing and microstructure relationships. *Acta Mater.*, Vol. 57, No. 3, 854-867, 1359-6454

Csucs, G., Kunzler, T., Feldman, K., Robin, F. & Spencer, N.D. (2003). Microcontact printing of macromolecules with submicrometer resolution by means of polyolefin stamps. *Langmuir*, Vol. 19, No. 15, 6104-6109, 0743-7463

Cui, B. & Veres, T. (2007). Fabrication of metal nanoring array by nanoimprint lithography (NIL) and reactive ion etching. *Microelectron. Eng.*, Vol. 84, No. 5-8, 1544-1547, 0167-9317

Falconnet, D., Koenig, A., Assi, T. & Textor M. (2004). A combined photolithographic and molecular-assembly approach to produce functional micropatterns for applications in the biosciences. *Adv. Funct. Mater.*, Vol. 14, No. 8, 749-756, 1616-301X

Felidj, N., Aubard, J. & Levi, G. (1999). Discrete dipole approximation for ultraviolet-visible extinction spectra simulation of silver and gold colloids. *J. Chem. Phys.*, Vol. 111, No. 3, 1195-1208, 0021-9606

Gao, X.F. & Jiang, L. (2004). Water-repellent legs of water striders. *Nature,* Vol. 432, No. 7013, 36-36, 0028-0836

Gong, D.W., Grimes, C.A. & Varghese O.K. (2001). Titanium oxide nanotube arrays prepared by anodic oxidation. *J. Mater. Res.,* Vol. 16, No. 12, 3331-3334, 0884-2914

Haick, H. & Paz, Y. (2001). Remote photocatalytic activity as probed by measuring the degradation of self-assembled monolayers anchored near microdomains of titanium dioxide. *J. Phys. Chem. B,* Vol. 105, No. 15, 3045-3051, 1089-5647

Huang, L., Braunschweig, A.B., Shim, W., Qin, L.D., Lim, J.K., Hurst, S.J., Huo, F.W., Xue, C., Jong, J.W. & Mirkin, C.A. (2010). Matrix-assisted dip-pen nanolithography and polymer pen lithography. *Small,* Vol. 6, No. 10, 1077-1081, 1613-6810

Huang, Y.X., Lai, Y.K., Lin, L.X., Sun, L. & Lin, C.J. (2010). Electrochemical construction and biological performance of micropatterned CaP films. *Acta Phys. -Chim. Sin.,* Vol. 26, No. 8, 2057-2060, 1000-6818

Huang, Y.X., Sun, L., Xie, K.P., Lai, Y.K., Liu, B.J., Ren, B. & Lin, C.J. (2011). A SERS study of Ag nanoparticles electrodeposited on patterned TiO_2 nanotube films, *J. Raman Spectrosc.,* DOI: 10.1002/jrs.2830, 0377-0486, Vol. 42, No. 5, 986-991

Ichimura, K., Oh, S.K. & Nakagawa, M. (2000). Light-driven motion of liquids on a photoresponsive surface. *Science,* Vol. 288, No. 5471, 1624-1626, 0036-8075

Jeon, N.L., Clem, P.G., Payne, D.A. & Nuzzo, R.G. (1996). A monolayer-based lift-off process for patterning chemical vapor deposition copper thin films. *Langmuir,* Vol. 12, No. 22, 5350-5355, 0743-7463

Jiao, L.Y., Gao, H.J., Zhang, G.M., Xie, G.Y., Zhou, X., Zhang, Y.Y., Zhang, Y.Y., Gao, B., Luo, G., Wu, Z.Y., Zhu, T., Zhang, J., Liu, Z.F., Mu, S.C., Yang, H.F. & Gu, C.Z. (2005). Fabrication of metallic nanostructures by negative nanoimprint lithography. *Nanotechnology,* Vol. 16, No. 12, 2779-2784, 0957-4484

Katagiri, K., Ohno, K., Masuda, Y. & Koumoto, K. (2007). Growth behavior of TiO_2 particles via the liquid phase deposition process. *J. Ceram. Soc. Jpn.,* Vol. 115, No. 1348, 831-834, 1882-0743

Kobayashi, T., Shimizu, K., Kaizuma, Y. & Konishi, S. (2011). Formation of superhydrophobic/superhydrophilic patterns by combination of nanostructure-imprinted perfluoropolymer and nanostructured silicon oxide for biological droplet generation. *Appl. Phys. Lett.,* Vol. 98, No. 12, 123706, 0003-6951

Koumoto, K., Saito, N., Gao, Y.F., Masuda, Y. & Zhu, P.X. (2008). Nano/micro patterning of inorganic thin films. *Bull. Chem. Soc. Jpn.,* Vol. 81, No. 11, 1337-1376, 0009-2673

Kubo, W., Tatsuma, T., Fujishima, A. & Kobayashi, H. (2004). Mechanisms and resolution of photocatalytic lithography. *J. Phys. Chem. B,* Vol. 108, No. 9, 3005-3009, 1520-6106

Kumar, A., Biebuyck, H.A., Abbott, N.L. & Whitesides, G.M. (1992). The use of self-assembled monolayers and a selective etch to generate patterned gold features. *J. Am. Chem. Soc.,* Vol. 114, No. 23, 9188-9189, 0002-7863

Lafuma, A. & Quéré, D. (2003). Superhydrophobic states. *Nat. Mater.,* Vol. 2, No. 7, 457-460, 1476-1122

Li, L.Y., Yang, Y., Yang, G.L., Chen, X.M., Hsiao, B.S., Chu, B., Spanier, J.E. & Li, C.Y. (2006). Patterning polyethylene oligomers on carbon nanotubes using physical vapor deposition. *Nano Lett.,* Vol. 6, No. 5, 1007-1012, 1530-6984

Lai, Y.K., Sun, L., Zuo, J. & Lin, C.J. (2004). Electrochemical fabrication and formation mechanism of TiO_2 nanotube array on metallic titanium surface. *Acta Phys. -Chim. Sin.*, Vol. 20, No. 9, 1063-1066, 1000-6818

Lai, Y.K., Sun, L., Chen, Y.C., Zhuang, H.F., Lin, C.J. & Chin, J.W. (2006). Effects of the structure of TiO_2 nanotube array on Ti substrate on its photocatalytic activity. *J. Electrochem. Soc.*, Vol. 153, No. 7, D123-127, 0013-4651

Lai, Y.K., Lin, C.J., Wang, H., Huang, J.Y., Zhuang, H.F. & Sun, L. (2008). Superhydrophilic-superhydrophobic micropattern on TiO_2 nanotube films by photocatalytic lithography. *Electrochem. Commun.*, Vol. 10, No. 3, 387-391, 1388-2481

Lai, Y.K., Lin, C.J., Huang, J.Y., Zhuang, H.F., Sun, L. & Nguyen, T. (2008). Markedly controllable adhesion of superhydrophobic spongelike nanostructure TiO_2 films. *Langmuir*, Vol. 24, No. 8, 3867-3873, 0743-7463

Lai, Y.K., Gao, X.F., Zhuang, H.F., Huang, J.Y., Lin, C.J. & Jiang, L. (2009). Designing superhydrophobic porous nanostructures with tunable water adhesion. *Adv. Mater.*, Vol. 21, No. 37, 3799-3803, 0935-9648

Lai, Y.K., Huang, J.Y., Gong, J.J., Huang, Y.X., Wang, C.L., Chen, Z. & Lin, C.J. (2009). Superhydrophilic-superhydrophobic Template: A simple approach to micro- and nanostructure patterning of TiO_2 films. *J. Electrochem. Soc.*, Vol. 156, No. 11, D480-484, 0013-4651

Lai, Y.K., Zhuang, H.F., Sun, L., Chen, Z. & Lin, C.J. (2009). Self-organized TiO_2 nanotubes in mixed organic-inorganic electrolytes and their photoelectrochemical performance. *Electrochim. Acta*, Vol. 54, No. 26, 6536-6542, 0013-4686

Lai, Y.K., Lin, Z.Q., Huang, J.Y., Sun, L., Chen, Z. & Lin, C.J. (2010). Controllable construction of ZnO/TiO_2 patterning nanostructures by superhydrophilic/superhydrophobic templates. *New J. Chem.*, Vol. 34, No. 1, 44-51, 1144-0546

Lai, Y.K., Huang, J.Y., Zhang, H.F., Subramaniam, V.P., Tang, Y.X., Gong, D.G., Sundar, L., Sun, L., Chen, Z. & Lin, C.J. (2010). Nitrogen-doped TiO_2 nanotube array films with enhanced photocatalytic activity under various light sources. *J. Hazard. Mater.*, Vol. 184, No. 1-3, 855-863, 0304-3894

Lai, Y.K., Huang, Y.X., Huang, J.Y., Wang, H., Chen, Z. & Lin, C.J. (2010). Selective formation of ordered arrays of octacalcium phosphate ribbons on TiO_2 nanotube surface by template-assisted electrodeposition. *Colliods Surf. B*, Vol. 76, No. 1, 117-122, 0927-7765

Lai, Y.K., Lin, Z.Q., Chen, Z., Huang, J.Y. & Lin, C.J. (2010). Fabrication of patterned CdS/TiO_2 heterojunction by wettability template-assisted electrodeposition. *Mater. Lett.*, Vol. 64, No. 11, 1309-1312, 0167-577X

Lee, J.P. & Sung, M.M. (2004). A new patterning method using photocatalytic lithography and selective atomic layer deposition. *J. Am. Chem. Soc.*, Vol. 126, No. 1, 28-29, 0002-7863

Lee, S.W., Oh, B.K., Sanedrin, R.G., Salaita, K., Fujigaya, T. & Mirkin, C.A. (2006). Biologically active protein nanoarrays generated using parallel dip-pen nanolithography. *Adv. Mater.*, Vol. 18, No. 9, 1133-1136, 0935-9648

Li, Z.W., Gu, Y.N., Wang, L., Ge, H.X., Wu, W., Xia, Q.F., Yuan, C.S., Chen, Y., Cui, B. & Williams, R.S. (2009). Hybrid nanoimprint-soft lithography with sub-15 nm resolution. *Nano Lett.*, Vol. 9, No. 6, 2306-2310, 1530-6984

Liu, S.H., Wang, W.C.M., Mannsfeld, S.C.B., Locklin, J., Erk, P., Gomez, M., Richter, F. & Bao, Z.N. (2007). Solution-assisted assembly of organic semiconducting single crystals on surfaces with patterned wettability. *Langmuir*, Vol. 23, No. 14, 7428-7432, 0743-7463

Liu, M.J., Zheng, Y.M., Zhai, J. & Jiang, L. (2010). Bioinspired super-antiwetting interfaces with special liquid-solid adhesion. *Accounts Chem. Res.*, Vol. 43, No. 3, 368-377, 0001-4842

Liu, X.J., Ye, Q.A., Yu, B., Liang, Y.M., Liu, W.M. & Zhou, F. (2010). Switching water droplet adhesion using responsive polymer brushes. *Langmuir*, Vol. 26, No. 14, 12377-12382, 0743-7463

Lu, Y., Liu, G.L. & Lee, L.P. (2005). High-density silver nanoparticle film with temperature-controllable interparticle spacing for a tunable surface enhanced Raman scattering substrate. *Nano Lett.*, Vol. 5, No. 1, 5-9, 1530-6984

Masuda, Y., Kinoshita, N., Sato, F. & Koumoto, K. (2006). Site-selective deposition and morphology control of UV- and visible-light-emitting ZnO crystals. *Cryst. Growth Des.*, Vol. 6, No. 1, 75-78, 1528-7483

Michel, R., Reviakine, I., Sutherland, D., Fokas, C., Csucs, G., Danuser, G., Spencer, N.D. & Textor, M. (2002). A novel approach to produce biologically relevant chemical patterns at the nanometer scale: Selective molecular assembly patterning combined with colloidal lithography. *Langmuir*, Vol. 18, No. 22, 8580-8586, 0743-7463

Nakata, K., Nishimoto, S., Yuda, Y., Ochiai, T., Murakami, T. & Fujishima, A. (2010). Rewritable superhydrophilic-superhydrophobic patterns on a sintered titanium dioxide substrate. *Langmuir*, Vol. 26, No. 14, 11628-11630, 0743-7463

Nishimoto, S., Sekine, H., Zhang, X.T., Liu, Z.Y., Nakata, K., Murakami, T., Koide, Y. & Fujishima, A. (2009). Assembly of self-assembled monolayer-coated Al_2O_3 on TiO_2 thin films for the fabrication of renewable superhydrophobic-superhydrophilic structures. *Langmuir*, Vol. 25, No. 13, 7226-7228, 0743-7463

Rausch, N. & Burte, E.P. (1993). Thin TiO_2 films prepared by low-pressure chemical vapor-deposition. *J. Electrochem. Soc.*, Vol. 140, No. 1, 145-149, 0013-4651

Rusponi, S., Costantini, G., de Mongeot, F.B., Boragno, C. & Valbusa, U. (1999). Patterning a surface on the nanometric scale by ion sputtering. *Appl. Phys. Lett.*, Vol. 75, No. 21, 3318-3320, 0003-6951

Shen, G.X., Chen, Y.C. & Lin, C.J. (2005). Corrosion protection of 316 L stainless steel by a TiO_2 nanoparticle coating prepared by sol-gel method. *Thin Solid Films*, Vol. 489, No. 1-2, 130-136, 0040-6090

Slocik, J.M., Beckel, E.R., Jiang, H., Enlow, J.O., Zabinski, J.S., Bunning, T.J. & Naik, R.R. (2006). Site-speciric patterning of biomolecules and quantum dots on functionalized surfaces generated by plasma-enhanced chemical vapor deposition. *Adv. Mater.*, Vol. 18, No. 16, 2095-2100, 0935-9648

Takeda, S., Suzuki, S., Odaka, H. & Hosono, H. (2001). Photocatalytic TiO_2 thin film deposited onto glass by DC magnetron sputtering. *Thin Solid Films*, Vol. 392, No. 2, 338-344, 0040-6090

Tatsuma, T., Kubo, W. & Fujishima, A. (2002). Patterning of solid surfaces by photocatalytic lithography based on the remote oxidation effect of TiO_2. *Langmuir*, Vol. 18, No. 25, 9632-9634, 0743-7463

Wang, H., Lin, C.J., Hu, R., Zhang, F. & Lin, L.W. (2008). A novel nano-micro structured octacalcium phosphate/protein composite coating on titanium by using an electrochemically induced deposition. *J. Biomed. Mater. Res. A*, Vol. 87A, No. 3, 698-705, 1549-3296

Wang, H., Duan, J.C. & Cheng, Q. (2011). Photocatalytically patterned TiO_2 arrays for on-plate selective enrichment of phosphopeptides and direct MALDI MS analysis. *Anal. Chem.*, Vol. 83, No. 5, 1624-1631, 0003-2700

Wang, R., Hashimoto, K., Fujishima, A., Chikuni, M., Kojima, E., Kitamura, A., Shimohigoshi, M. & Watanabe, T. (1997). Light-induced amphiphilic surfaces. *Nature*, Vol. 388, No. 6641, 431-432, 0028-0836

Wang, Y.L. & Lieberman, M. (2003). Growth of ultrasmooth octadecyltrichlorosilane self-assembled monolayers on SiO_2. *Langmuir*, Vol. 19, No. 4, 1159-1167, 0743-7463

West, J.B. (1999). The original presentation of Boyle's law. *J. Appl. Physiol.*, Vol. 87, No. 4, 1543-1545, 8750-7587

Xu, S. & Liu, G.Y. (1997). Nanometer-scale fabrication by simultaneous nanoshaving and molecular self-assembly. *Langmuir*, Vol. 13, No. 2, 127-129, 0743-7463

Yang, Y., Lai, Y.K., Zhang, Q.Q., Wu, K., Zhang, L.H., Lin, C.J. & Tang, P.F. (2010). A novel electrochemical strategy for improving blood compatibility of titanium-based biomaterials. *Colloids Surf. B*, Vol. 79, No. 1, 309-313, 0927-7765

Yasuda, K., Macak, J.M., Berger, S., Ghicov, A. & Schmuki, P. (2007). Mechanistic aspects of the self-organization process for oxide nanotube formation on valve metals. *J. Electrochem. Soc.*, Vol. 154, No. 9, C472-478, 0013-4651

Yoshimura, M. & Gallage, R. (2008). Direct patterning of nanostructured ceramics from solution - differences from conventional printing and lithographic methods. *J. Solid State Electrochem.*, Vol. 12, No. 7-8, 775-782, 1432-8488

Yun, H., Lin, C.J., Li, J. Wang, J.R. & Chen, H.B. Low-temperature hydrothermal formation of a net-like structured TiO_2 film and its performance of photogenerated cathode protection. *Appl. Surf. Sci.*, Vol. 255, No. 5, 2113-2117, 0169-4332

Zeng, J.B., Jia, H.Y., An, J., Han, X.X., Xu, W.Q., Zhao, B. & Ozaki, Y. (2008). Preparation and SERS study of triangular silver nanoparticle self-assembled films. *J. Raman Spectrosc.*, Vol. 39, No. 11, 1673-1678, 0377-0486

Zhang, C. & Kalyanaraman, R. (2004). In situ lateral patterning of thin films of various materials deposited by physical vapor deposition. *J. Mater. Res.*, Vol. 19, No. 2, 595-599, 0884-2914

Zhang, G.M., Zhang, J., Xie, G.Y., Liu, Z.F. & Shao H.B. (2006). Cicada wings: A stamp from nature for nanoimprint lithography. *Small*, Vol. 2, No. 12, 1440-1443, 1613-6810

Zhang, G.J., Tanii, T., Kanari, Y. & Ohdomari, I. (2007). Production of nanopatterns by a combination of electron beam lithography and a self-assembled monolayer for an antibody nanoarray. *J. Nanosci. Nanotechnol.*, Vol. 7, No. 2, 410-417, 1533-4880

Zhang, X.T., Jin, M., Liu, Z.Y., Tryk, D., Nishimoto, S., Murakami, T. & Fujishima, A. (2007). Superhydrophobic TiO_2 surfaces: Preparation, photocatalytic wettability conversion, and superhydrophobic-superhydrophilic patterning. *J. Phys. Chem. C*, Vol. 111, No. 39, 14521-14529, 1932-7447

Zhao, J.C., Wu, T.X., Wu, K.Q., Oikawa, K., Hidaka, H. & Serpone, N. (1998). Photoassisted degradation of dye pollutants. 3. Degradation of the cationic dye rhodamine B in aqueous anionic surfactant/TiO_2 dispersions under visible light irradiation: Evidence for the need of substrate adsorption on TiO_2 particles. *Environ. Sci. Technol.*, Vol. 32, No. 16, 2394-2400, 0013-936X

Zhuang, H.F., Lin, C.J., Lai, Y.K., Sun, L. & Li, J. (2007). Some critical structure factors of titanium oxide manotube array in its photocatalytic activity. *Environ. Sci. Technol.*, Vol. 41, No. 13, 4735-4740, 0013-936X